FRONTIERS IN TRANSITION METAL-CONTAINING POLYMERS

BICENTENNIAL

1807 WILEY 2007

THE WILEY BICENTENNIAL–KNOWLEDGE FOR GENERATIONS

Each generation has its unique needs and aspirations. When Charles Wiley first opened his small printing shop in lower Manhattan in 1807, it was a generation of boundless potential searching for an identity. And we were there, helping to define a new American literary tradition. Over half a century later, in the midst of the Second Industrial Revolution, it was a generation focused on building the future. Once again, we were there, supplying the critical scientific, technical, and engineering knowledge that helped frame the world. Throughout the 20th Century, and into the new millennium, nations began to reach out beyond their own borders and a new international community was born. Wiley was there, expanding its operations around the world to enable a global exchange of ideas, opinions, and know-how.

For 200 years, Wiley has been an integral part of each generation's journey, enabling the flow of information and understanding necessary to meet their needs and fulfill their aspirations. Today, bold new technologies are changing the way we live and learn. Wiley will be there, providing you the must-have knowledge you need to imagine new worlds, new possibilities, and new opportunities.

Generations come and go, but you can always count on Wiley to provide you the knowledge you need, when and where you need it!

WILLIAM J. PESCE
PRESIDENT AND CHIEF EXECUTIVE OFFICER

PETER BOOTH WILEY
CHAIRMAN OF THE BOARD

FRONTIERS IN TRANSITION METAL-CONTAINING POLYMERS

Edited by

Alaa S. Abd-El-Aziz
University of British Columbia Okanagan, Kelowna,
British Columbia, Canada

Ian Manners
School of Chemistry, University of Bristol, Bristol, United Kingdom

BICENTENNIAL 1807 WILEY 2007 BICENTENNIAL

WILEY-INTERSCIENCE
A John Wiley & Sons, Inc., Publication

Published by John Wiley & Sons, Inc., Hoboken, New Jersey
Published simultaneously in Canada

Library of Congress Cataloging-in-Publication Data

Frontiers in transition metal-containing polymers / edited by Alaa S. Abd-El-Aziz and Ian Manners.
p. cm.
Includes index.
ISBN-13: 978-0-471-73015-6 (cloth)
ISBN-10: 0-471-73015-7 (cloth)
1. Transition metals. 2. Organometallic polymers. 3. Polymeric composites.
I. Abd-El-Aziz, Alaa S. II. Manners, Ian.
QD411.8.T73F76 2007
620.1′92—dc22

2006023013

Printed in the United States of America

10 9 8 7 6 5 4 3 2 1

PREFACE

Many of the interesting properties and functions of solid-state and biological materials can be attributed to the presence of metallic elements. Examples include magnetic materials used in data storage, superconductors, electrochromic materials, and catalysts including metalloenzymes. Following the rapid emergence of synthetic organic polymers in the first part of the twentieth century, the exciting possibilities for the development of new, easily processed materials via the incorporation of metal atoms into synthetic polymer chains intrigued scientists. Although the first metal-containing polymer, poly(vinylferrocene), was reported as early as 1955, the synthetic difficulties associated with the creation of macromolecular chains possessing metal atoms as key structural components were significant deterrents to rapid progress in the field until the last decade. Characterization during the early stages was also a key problem, and this led to the assignment of many structures without the type of convincing evidence by techniques such as NMR that would be deemed absolutely necessary today.

Many factors have contributed to the rapid and substantial recent emergence of the metallopolymer area as one of the most exciting in the whole advanced materials field. First, since the 1990's, many of the obstacles to the preparation of high molecular weight and soluble materials have been overcome through creative new synthetic approaches. Second, the recent ready access to soluble and stable well-characterized high molecular weight metallopolymers in usable quantities has allowed detailed studies of their properties and applications by polymer materials scientists and their wide variety of state-of-the-art characterization tools. This has been critical to the emergence of metallopolymers as *functional* materials. And third, the recent discovery of controlled or "living" polymerization routes to metallopolymers, which allow molecular weights to be controlled, narrow molecular weight distributions to be obtained, and block copolymers to be prepared, have provided a substantial impetus to the area. Block copolymers, for example, self-assemble in thin films or block selective solvents to form nanostructured materials with phase-separated nanodomains and micellar aggregates, respectively. Self-assembly of metal-containing block copolymers thereby provides access to metal-rich nanostructures with a variety of potential applications in nanoscience, such as in nanoparticle catalysis and nanopatterning.

The book aims to survey recent research at the frontiers of the subject and to achieve this we have been fortunate to be able to assemble an exceptional group of contributors.

Chapter 1, written by Pittmann and Carraher, two of the early pioneers of the metallopolymer field, surveys the developments in the area in the 1960s and 1970s when metallopolymers were almost unknown. Chapter 2, by Abd-El-Aziz and Shipman, surveys the recent progress in the synthetic metallopolymer field. Chapter 3, by Rider and Manners, covers the recently emerging field of block copolymers with transition metal atoms in the main chain of at least one of the blocks. Synthetic routes to metal-containing block copolymers and applications in the field of nanoscience are described. In Chapter 4, MacLachlan covers the exciting area of π-conjugated metallopolymers and demonstrates useful properties of π-conjugated organic macromolecules that can be supplemented further by the presence of metal centers. In Chapter 5, Chan and Cheng cover the important field of metal coordination polymers and their applications in nanofabrication and include within the survey block copolymers with metals coordinated to side group substituents. The important field of rigid rod polymers containing metal atoms in the main chain—polymetallynes—is reviewed by Wong and Ho in Chapter 6. These materials have attracted much recent attention with substantial synthetic expansion and their investigation as, for example, liquid crystals, photoconductors, and photoluminescent materials (triplet emitters). Chapter 7, by Tyler, covers the intriguing field of polymers containing metal–metal bonds, which are of interest as photodegradable materials and have many other potential uses. Chapter 8, by Harvey, covers the expanding field of metallopolymers with diphosphine and diisocyanide linkers, which are of interest as a result of their photoluminescent properties and have potential in many other areas, such as liquid crystals and ceramic precursors. The redox properties of metallopolymers are covered by Nishihara in Chapter 9, which describes materials which exhibit conductivity and photoconductivity, ferromagnetic spin coupling, and electro-, photo-, and thermochromism as a consequence of the mixed valence states and unpaired electrons on the metal centers. An overview of the field of metallodendrimers and their broad applications as catalysts, luminescent materials, and sensors is provided by Hwang and Newkome in Chapter 10. Redox-active metallodendritic iron complexes and their use in design, catalysis, and molecular recognition is the special focus of Chapter 11, by Astruc. The emerging field that involves combining metallopolymers with biopolymers to prepare new functional bioconjugates is represented in Chapter 12, by Mahmoud and Kraatz, and in Chapter 13, by Shionoya. Chapter 12 describes conjugates of peptides and metallopolymers with a particular emphasis on ferrocene systems (which are currently by far the most well-developed), and Chapter 13 focuses on metal arrays based on nucleic acid and peptides.

Clearly, metal-containing polymers represent an important emerging field. We look forward to further rapid progress in the next decade.

Alaa S. Abd-El-Aziz
Ian Manners

CONTENTS

CONTRIBUTORS

Alaa S. Abd-El-Aziz, Department of Chemistry, University of British Columbia Okanagan, Kelowna, British Columbia, Canada

Didier Astruc, Laboratoire de Chimie Organique et Organométallique, Université Bordeaux I, Talence, France

Charles E. Carraher, Jr., Department of Chemistry, Florida Atlantic University, Boca Raton, Florida

Wai Kin Chan, Department of Chemistry, The University of Hong Kong, Hong Kong, China

Kai Wing Cheng, Department of Chemistry, The University of Hong Kong, Hong Kong, China

Pierre D. Harvey, Department of Chemistry, University of Sherbrooke, Sherbrooke, Québec, Canada

Cheuk-Lam Ho, Department of Chemistry, Centre for Advanced Luminescence Materials, Hong Kong Baptist University, Hong Kong, China

Seek-Ho Hwang, Department of Polymer Science, The University of Akron, Akron, Ohio

Heinz-Bernhard Kraatz, Department of Chemistry, University of Saskatchewan, Saskatoon, Saskatchewan, Canada

Mark MacLachlan, Department of Chemistry, University of British Columbia, Vancouver, British Columbia, Canada

Khaled A. Mahmoud, Department of Chemistry, University of Saskatchewan, Saskatoon, Saskatchewan, Canada

Ian Manners, School of Chemistry, University of Bristol, Bristol, United Kingdom

George R. Newkome, Department of Polymer Science, The University of Akron, Akron, Ohio

Hiroshi Nishihara, Department of Chemistry, School of Science, University of Tokyo, Tokyo, Japan

Charles U. Pittman, Jr., Department of Chemistry, Mississippi State University, Mississippi State, Mississippi

David A. Rider, Department of Chemistry, University of Toronto, Toronto, Ontario, Canada

Mitsuhiko Shionoya, Department of Chemistry, Graduate School of Science, University of Tokyo, Tokyo, Japan

Patrick O. Shipman, Department of Chemistry, University of British Columbia Okanagan, Kelowna, British Columbia, Canada

David R. Tyler, Department of Chemistry, University of Oregon, Eugene, Oregon

Wai-Yeung Wong, Department of Chemistry, Centre for Advanced Luminescence Materials, Hong Kong Baptist University, Hong Kong, China

CHAPTER 1

Organometallic Polymers: The Early Days

CHARLES U. PITTMAN, Jr.

Mississippi State University, Mississippi State, Mississippi

CHARLES E. CARRAHER, Jr.

Florida Atlantic University, Boca Raton, Florida

I. INTRODUCTION

Historically, polymers only contained the elements C, H, N, O, S, Cl, Br, and occasionally P. The pioneering studies of Rochow in the 1940s on polysiloxanes,[1,2] followed by the rapid development of polysiloxanes by industry in the 1960s, expanded that list of elements to Si. Subsequently, marine antifouling coatings incorporated tin into polymeric systems, through the $—OSnBu_3$ function for the most part. However, the vast majority of elements had not found their way into polymers. Since more than 40 metals exist and many have several available oxidation states, sporadic efforts were made in the 1940s, 1950s, and 1960s to introduce metals into repeating units of polymers through the use of coordination chemistry.[3,4,5] The U.S. Air Force sponsored a large effort to find thermally stable polymers during World War II and after that centered on metal chelation reactions, but useful materials did not emerge. Low molecular weight, intractable, and uncharacterizable materials were the rule. Insolubility was a problem. Some superb scientists, including Carl Marvel[6] and John Bailar[7] were prominent in this difficult effort. Since the mid-1950s, metal coordination polymers have become a large field and recent reviews are now available.[8,9,10]

Organometallic polymers, in contrast to coordination polymers, have metal-to-carbon bonds. They are the topic of this chapter. Three seminal events provided the foundation for the field of organometallic polymers to develop. The landmark discovery of ferrocene by Kealy and Pauson in 1951 marked the first organometallic compound.[11] This was quickly followed by the full elucidation of its structure and an understanding of its reactivity by Wilkinson, Rosenblum, Whiting, and Woodward.[12,13] This history was celebrated in a 2001 feature article in *Chemical and Engineering News*.[14] Finally, the first polymerization of an organometallic compound was reported by Arimoto and Haven at Dupont Co. in 1955.[15,16] Vinylferrocene **1**

Frontiers in Transition Metal-Containing Polymers, edited by Alaa S. Abd-El-Aziz and Ian Manners.

was synthesized from its acetyl derivative and subjected to homopolymerization in the presence of radical initiators.[15,16] Details of the characterization never appeared and this report seemed to attract little early attention. A special edition of the *Journal of Inorganic and Organometallic Polymers and Materials* appeared in March 2005 to celebrate the 50th anniversary of this milestone.[17] It was only in the 1960s that organometallic polymers seemed to come alive. Early contributors included Neuse and Rosenberg in the United States[18] and Korshak and Sosin in Russia,[19] the latter signaling active Soviet interest in extending the scope of organometallic chemistry into polymers. In the United States much of the early activity came from the groups of Carraher and Pittman. A review of ferrocene condensation polymers in 1967 covered the work of others.[20]

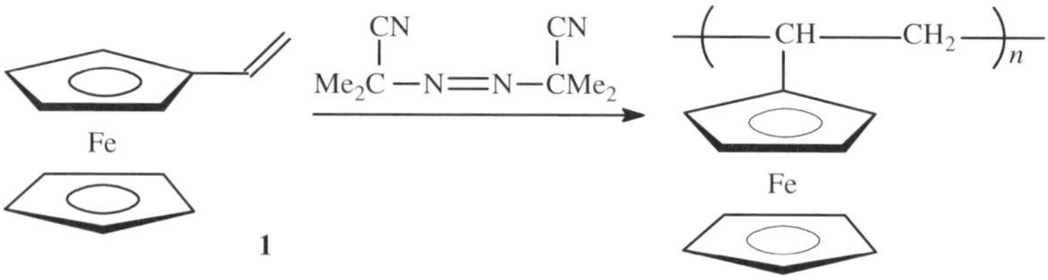

1

The early work in Pittman's group predominantly employed vinyl addition polymerizations while that of Carraher featured condensation polymerizations. Thus, this chapter has been divided accordingly. However, both groups had a similar vision, that of increasing the numbers of elements that could usefully appear in polymers. As he began his lectures, Carraher often jokingly stated that their efforts were all about increasing the periodic table for polymer chemists. Pittman's laboratory featured a banner with a hypothetical dream structure of a single polymer containing every transition metal. These musings simply emphasized to students, and anyone else who would listen, the opportunities that lay unexplored in the mid-to-late 1960s.

In the late 1960s and early 1970s great progress was achieved in both labs due to the application of instrumentation that, by today's standards, were primitive. At the University of Alabama, Pittman conducted kinetic studies using dilatometry,[21,22] accurate but tedious. Soluble polymers were examined by membrane osmometry and vapor-pressure osmometry to obtain absolute values and Q factors for calibrating gel permeation chromatography.[23] Gel permeation chromatography (GPC) was employed with 16 foot-long GPC column banks.[23] Next, intrinsic viscosity studies were employed when the Universal calibration method was developed by Benoit.[24] Light scattering instrumentation was primitive and difficult to use since many polymers had molecular weights that were not above 50,000 to 60,000. Nuclear magnetic resonance (NMR) and infrared (IR) spectroscopy assisted the work. The use of Mössbauer spectroscopy for the analysis of iron-containing polymers and particularly mixed valence state polymers with ferrocene and ferricinium was employed by Pittman, Good, et al.[25,26] Carraher's group brought the use of high-resolution mass spectrometry (HR-EI-MS)[27] and matrix-assisted laser desorption ionization to organometallic polymers.[28] These and other methods led to publications of reasonably well-characterized organometallic polymers and foreshadowed the huge variety of sophisticated instrumental techniques being regularly employed today to characterize a staggering array of metal-containing polymer structures.

Pittman's introduction to organometallic chemistry occurred while he was a young officer on active duty at the U.S. Army's Solid Rocket Propulsion Laboratory in Redstone Arsenal, Alabama, in 1966. *n*-Butylferrocene was being used as a burning-rate catalyst for $NaClO_4$/Al/organic binder-based solid propellants. Fast-burning propellants were needed for fast acceleration, especially for portions of the Nike Zeus missile defense system. However, *n*-butylferrocene, which was the iron-containing catalyst of choice for this purpose, frequently phase separated from the binder when rockets were stored for long periods in the heat. Thus, Pittman began to synthesize ferrocene-containing polymers to prevent phase separation and to elucidate the mechanism by which iron catalyzed propellant combustion.[27,28] Thus, a variety of successful and unsuccessful polymerizations (polycondensations and addition polymerizations) were performed,[29,30] many of which could only appear in the security classified (ICRPG) literature. Without any polymer training (or misgivings), Pittman plunged into this field and continued efforts in this direction upon his arrival at the University of Alabama in 1967. His group set out to (1) synthesize new monomers, (2) develop well-characterized organometallic polymers, and (3) to uncover the basic physical–organic chemistry controlling the vinyl addition polymerization of example organometallic monomers.

Since the Arimoto/Haven report of vinylferrocene polymerization was not detailed, this monomer was made and both its homopolymerization and its copolymerization were studied with a variety of organic comonomers such as styrene, methylacrylate, maleic anhydride, acrylonitrile, methyl methacrylate, *N*-vinylpyrolidone, vinyl acetate, and so on.[31–38] The polymers were as well characterized as possible, and copolymer compositions were obtained versus feed mole ratios.

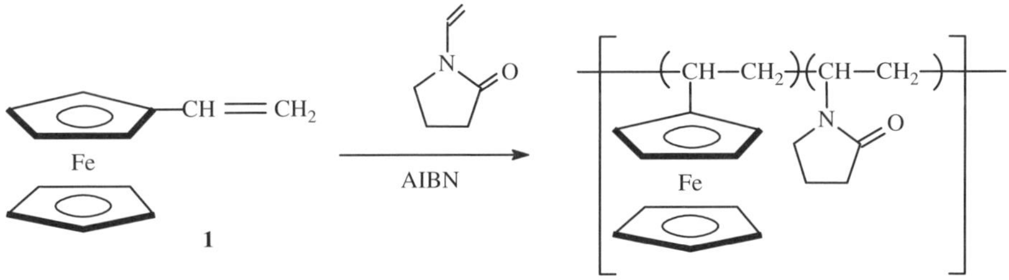

Studies of the homopolymerization kinetics that we carried out were curious and appeared to be greater than half-order in initiator and the polymerizations were sluggish under radical initiated conditions. The reason for this was cleared up by the excellent and precise homopolymerization kinetic studies of George and Hayes, who clearly demonstrated the rate was essentially first order in both initiator and monomer.[39,40,41] What could cause such a rate law? Could the iron center (formally Fe(II)) contribute? Was a redox reaction involved that would be essentially impossible for organic monomers like styrene? The observed rate law: $r = k[M]^{1.1}[I]^{1.1}$ stands in sharp contrast to the normal half-order in initiator concentration found in most vinyl addition polymerizations. Apparently, first-order chain termination had occurred rather than classic bimolecular termination. Indeed, the iron atom was playing a key role.

Internal electron transfer from iron to the radical center generated an anionic chain end, which was quickly quenched, bringing an end to chain growth (see Scheme 1.2). Thus, more initiator had to be consumed to begin a new chain. The lesson was clear: While vinylferrocene was styrene-like, the presence of the metal atom could complicate matters and it must be considered. Suddenly, the reason for our early observations of low homopolymer molecular weights,[31,37] and the lack of increase in molecular weight with a decrease in initiator concentration made sense. The polymer chain self-terminated. Furthermore, the more rapid copolymerizations that led to higher molecular-weight polymers made sense. The internal electron transfer that quenched the growing chain occurred far less often because it was not competitive with addition of the comonomer (e.g., k_{12} was faster than internal termination). While molecular weights were impacted, the effect was less pronounced.

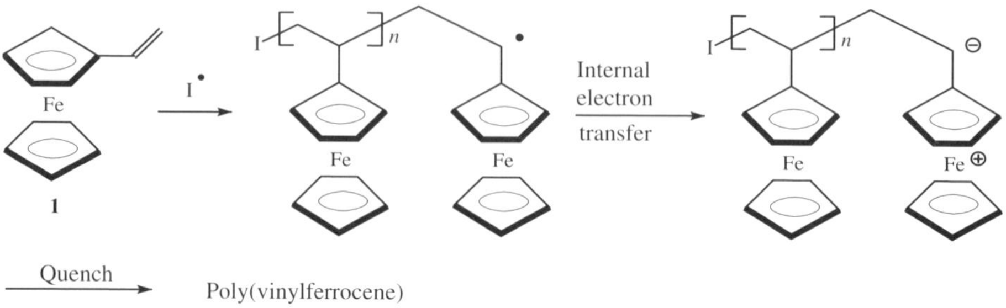

In these early years, nobody had a way to predict the composition of copolymers containing vinylferrocene or any other vinyl organometallic monomer. A better understanding of the reactivity was needed. A major effort to classify vinylferrocene copolymerization reactivity (and that of many other organometallic monomers) within the Alfrey–Price *Qe* scheme was begun.[42,43] This scheme, reminiscent of the Hamment σρ approach. It classifies monomers according to two parameters: *Q*, a resonance parameter, and *e*, related to the vinyl group's polarity. If the *Q* and *e* values of both comonomers are known, the copolymerization reactivity ratios, r_1 and r_2, can be calculated. This allows the use of the integrated form of the copolymer equation[44,45] to calculate the composition of the copolymer at any conversion from any M_1/M_2 starting composition.

The expressions relating *Q* and *e* to the reactivity ratios are:

$$r_1 = (Q_1/Q_2)\ \text{exp-}e_1\ (e_1 - e_2)$$

$$r_1 = (Q_1/Q_2)\ \text{exp-}e_2\ (e_2 - e_1)$$

$$r_1\ r_2 = \text{exp-}(e_1 - e_2)^2$$

where the subnumbers 1 and 2 refer to monomer M_1 and M_2 in the copolymerization.

To avoid electron transfer/quenching termination, the propagating radical center was moved away from and out of conjugation with the iron. Acrylic monomers **2** and **3** were soon made, and characterized.[25,32] They underwent "normal" radical addition polymerization kinetics: $r = k[\text{M}]^1\ [\text{I}]^{1/2}$ and higher molecular weights were achieved. The radical center, no longer adjacent to the cyclopentadienyl (Cp) ring, was far enough away from the iron atom to avoid electron transfer from iron. The ferrocene moieties in these polymers were, however, easily oxidized by the addition of strongly electron-attracting quinones such as 2,3-dichloro-5, 6-dicyanobenzoquinone (DDQ). Similar oxidations occurred with poly(vinylferrocene) to generate systems where all, or a portion of, the ferrocene groups in the polymers were converted to ferricenium units (see **4** and **5**). This work eventually led us to prepare a variety of designed mixed oxidation state systems. Mössbauer spectroscopy was developed as a superb analytical tool to define the oxidation stoichiometry with the assistance of Mary Good, then at Louisiana State University in New Orleans.[25,32]

The *Q* and *e* values for monomers are particularly valuable because, at a glance, they classify structure/reactivity features of monomers in a useful way that chemists can grasp. For example, monomers with a high resonance stability have larger *Q* values, while electron-rich

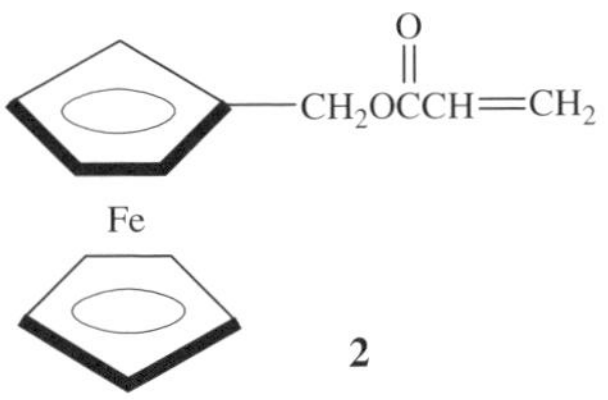

2

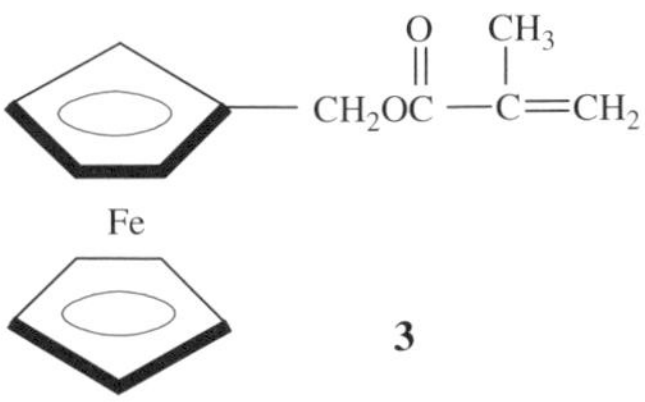

3

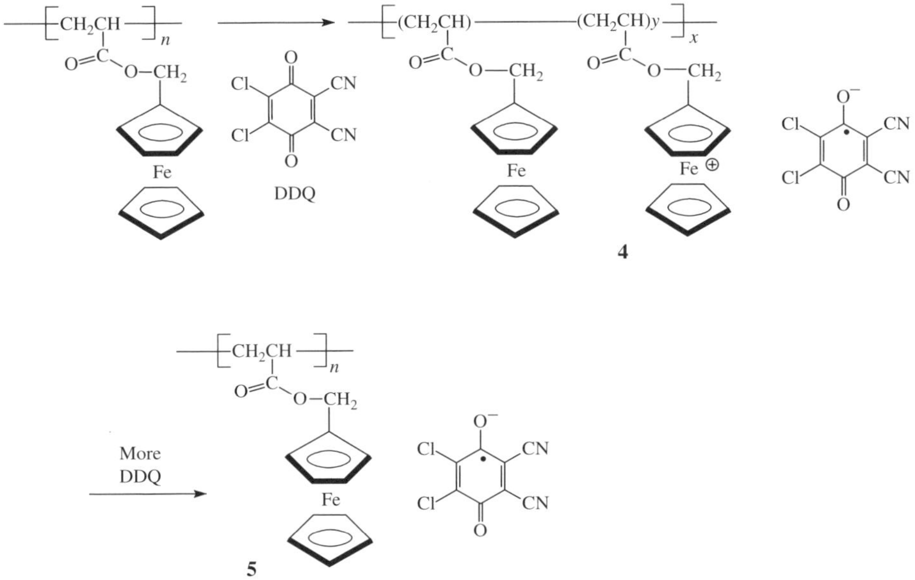

4

5

monomers have negative values of e and electron-poor monomers have a positive value of e. In the first application of physical organic approaches to organometallic polymers, the Q and e values of many organometallic monomers were determined by obtaining the relative reactivity ratios of organometallic monomers with numerous common organic monomers in over a decade of work. This established that vinylferrocene and all the other cyclopentadienyl metal monomers are extremely electron rich and have high resonance stabilization[46,47] (see Table 1.1). Many new copolymer systems were characterized and a general understanding of these addition polymerizations emerged during this program. This effort was augmented by studies of mixed oxidation state metal-containing polymers.[48]

The great stability of the ferrocenylmethyl cation suggested that the polymers of **2** and **3** would be susceptible to hydrolytic instability in polar ionizing solvents. Therefore, their 2-ferrocenylethyl analogs **6** and **7** were synthesized, homopolymerized, and their polymerization rate laws were established.[49] The copolymerizations of all four acrylates **2**, **3**, **6**, and **7**, with a large variety of common organic monomers, were established and the reactivity ratios were obtained.[33,36,49,50] The ferrocene moiety sharply increased the glass transition temperature of acrylate and methacrylate polymers.[51] The homopolymer examples are listed in Table 1.2. Apparently, ferrocene's high molecular weight and steric bulk strongly retard segmental motion, much like an anchor attached to a jump rope. Likewise, poly(vinylferrocene) exhibits a T_g = 184–194°C far above that of its nonorganometallic analog polystyrene.

TABLE 1.1 Alfrey Price *Q*, *e* Values For Selected Vinylcyclopentadienyl Metal-Containing Monomers Versus Common Organic Monomers

Monomer	*Q*	*e*
Vinylferrocene, **1**	1.03	−2.1
η^5-(Vinylcyclopentadienyl)tricarbonylmanganese, **8**	1.1	−1.99
η^5-(Vinylcyclopentadienyl)dicarbonylnitrosylchromium, **10**	3.1	−1.98
η^5-(Vinylcyclopentadienyl)dicarbonylnitrosylmolybdenum, **11**	3.1	−1.98
η^5-(Vinylcyclopentadienyl)tricarbonylmethyltungsten, **13**	1.66	−1.98
η^5-(Vinylcyclopentadienyl)dicarbonyliridium, **12**	4.1	−2.08
Styrene	1.00	−0.80
Propene	0.002	−0.78
Maleic anhydride	0.23	+2.25

Source: Reference 46.

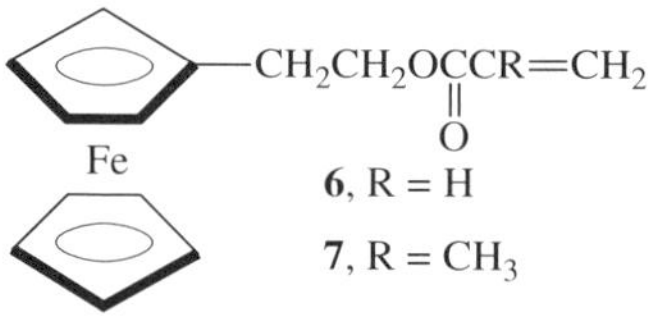

6, R = H

7, R = CH_3

TABLE 1.2 Glass Transition Temperatures, T_g, of Acrylic Ferrocene Polymers [30]

Polymer	T_g (°C)
Poly(methyl acrylate)	3
Poly(ferrocenylmethyl acrylate)	197–210
Poly(2-ferrocenylethyl acrylate)	157
Poly(methyl methacrylate)	57–68
Poly(ferrocenylmethyl methacrylate)	185–195
Poly(2-ferrocenylethyl methacrylate)	209

Source: Reference 30.

A. n^5-Cyclopentadienylmetal and n^6-Phenylmetal Carbonyl Monomers

The chemistry of organometallic metal carbonyl compounds exploded in the early 1960s. With the advent of compounds like η^5-(cyclopentadienyl)manganesetricarbonyl and many other related species, the obvious questions became, could the vinyl derivatives of these compounds function like styrene? Would addition polymerizations be possible, or would the metal atom interfere? Would carbonyl groups and other substituents modify the reactivity of such monomers and how might they compare with vinylferrocene? Where might such monomers lie on the *Q*, *e* scheme? These were key questions we sought to answer. A vigorous program was mounted to prepare a variety of metal carbonyl-containing monomers, study their homopolymerizations, their polymerization kinetics, and their copolymerizations with common organic monomers. The structures of several of these monomers (**8–32**), references to their preparation, and where available, polymerizations and their *Q* and *e* values, are displayed in Scheme 1.1.

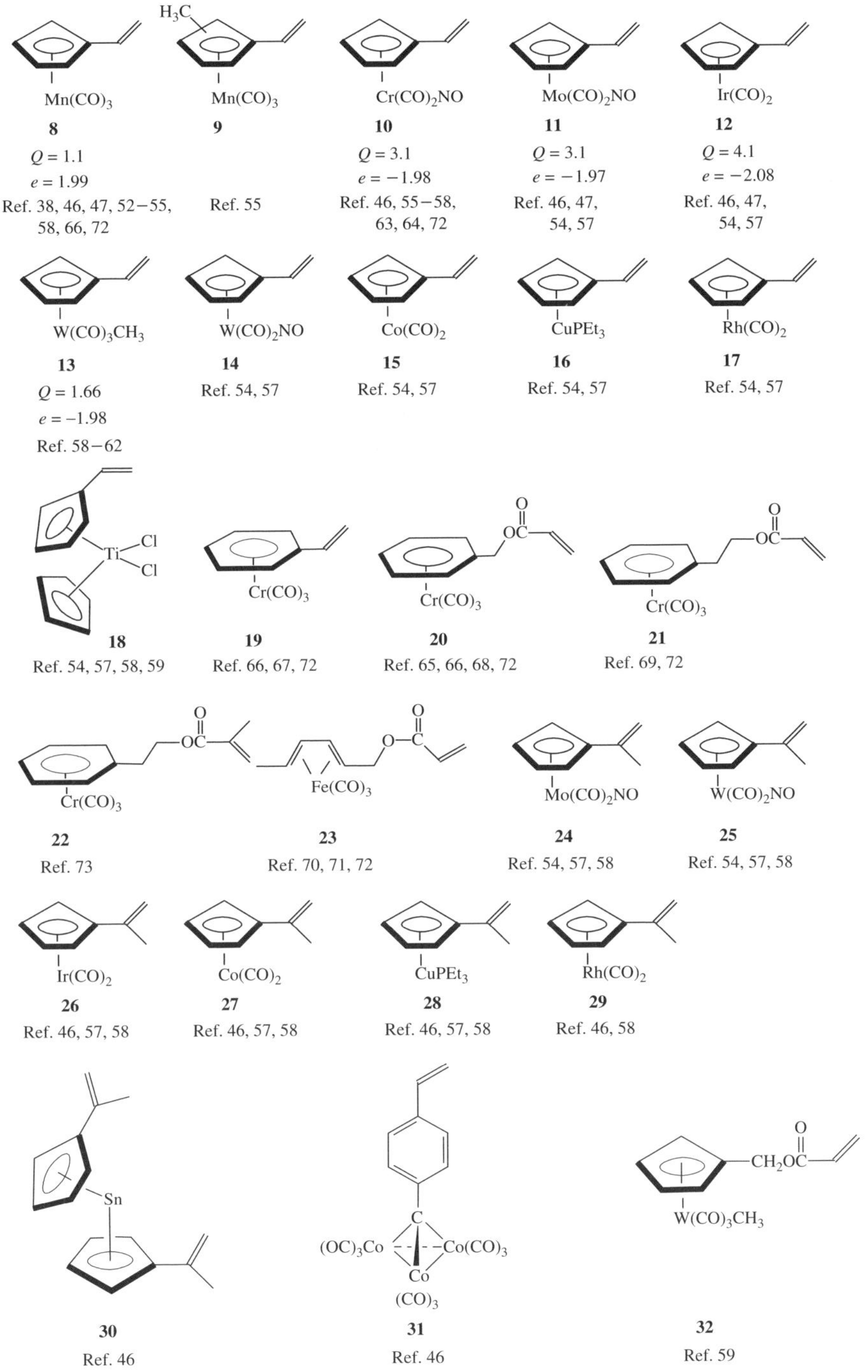
H3C
Mn(CO)3
8
Q = 1.1
e = 1.99
Ref. 38, 46, 47, 52−55, 58, 66, 72
Mn(CO)3
9
Ref. 55
Cr(CO)2NO
10
Q = 3.1
e = −1.98
Ref. 46, 55−58, 63, 64, 72
Mo(CO)2NO
11
Q = 3.1
e = −1.97
Ref. 46, 47, 54, 57
Ir(CO)2
12
Q = 4.1
e = −2.08
Ref. 46, 47, 54, 57
W(CO)3CH3
13
Q = 1.66
e = −1.98
Ref. 58−62
W(CO)2NO
14
Ref. 54, 57
Co(CO)2
15
Ref. 54, 57
CuPEt3
16
Ref. 54, 57
Rh(CO)2
17
Ref. 54, 57
Ti
Cl
Cl
18
Ref. 54, 57, 58, 59
Cr(CO)3
19
Ref. 66, 67, 72
O
OC
Cr(CO)3
20
Ref. 65, 66, 68, 72
O
OC
Cr(CO)3
21
Ref. 69, 72
O
OC
Cr(CO)3
22
Ref. 73
O
O−C
Fe(CO)3
23
Ref. 70, 71, 72
Mo(CO)2NO
24
Ref. 54, 57, 58
W(CO)2NO
25
Ref. 54, 57, 58
Ir(CO)2
26
Ref. 46, 57, 58
Co(CO)2
27
Ref. 46, 57, 58
CuPEt3
28
Ref. 46, 57, 58
Rh(CO)2
29
Ref. 46, 58
Sn
30
Ref. 46
C
(OC)3Co
Co(CO)3
Co
(CO)3
31
Ref. 46
O
CH2OC
W(CO)3CH3
32
Ref. 59

Scheme 1.1

B. η^5-(Cyclopentadienyl)tricarbonylmanganese, 8

We synthesized **8**[52,53,55] and its methyl analogs, **9**,[54] in a manner similar to that of vinylferrocene, **2**. Freidel Crafts acylation proceeds readily on η^5-(cyclopentadienyl)tricarbonylmanganese due to the aromatic nature of the cyclopentadienyl ring when complexed to the metal. Sodium borohydride reduction of the acetyl derivative and dehydration over $KHSO_4$ at 180–190°C generated **8**. The η^5-(methylcyclopentadienyl)tricarbonylmanganese precursor of **9**, was available at low cost in the late 1960s from Ethyl Corporation, since it was being used as an experimental cetane additive to gasoline. The high-temperature dehydration leading to **8** was somewhat surprising in view of an earlier report by Kozikowski and Cais[74] that a thermal decomposition residue was formed during the $KHSO_4$-catalyzed dehydration.

The radical-initiated copolymerization of **8** was demonstrated with styrene, methyl methacrylate, acrylonitrile, and vinyl acetate, in benzene or ethyl acetate using azobisisobutyronitrile (AIBN) as the initiator.[52a] Copolymerizations of **8** with *N*-pyrrolidone generated water-soluble materials.[52b,c] Molecular weights and the Mark Houwink *K* and *a* values were defined.[38] Monomer **8**, like vinylferrocene, proved to be extraordinarily electron-rich, exhibiting an *e* value of -1.99 and a larger resonance parameter ($Q = 1.1$) than styrene (0.80).[52a] Upon heating the *N*-vinyl-2-pyrrolidone copolymers of **8** to 260°C under nitrogen, gas evolution (CO) and weight loss occurred. The products became increasingly insoluble with time. Water-soluble copolymers from *N*-vinyl-2-pyrrolidone with excellent film adhesion to glass surface.[52b,c]

Copolymers of **8** and vinylferrocene, containing two different transition metals as shown in **33**, were readily prepared by radical addition polymerization.

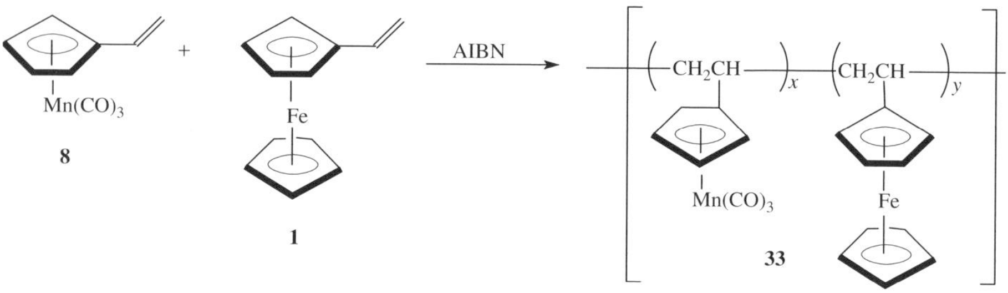

Would the tricarbonyl manganese function in monomer **8** prevent homopolymerization, or undergo electron transfer from manganese to the propagating radical center? Homopolymerization kinetic studies in benzene, benzonitrile, and acetone demonstrated that the effect of manganese was different than that of iron in **1**. The rate law was three halves order in the concentration of monomer **8** and half-order in initiator concentration.[53]

$$r = k\,[\mathrm{M}]^{1.5}\,[\mathrm{I}]^{0.5}$$

Molecular weight measurements confirmed that the degree of polymerization followed the relationship: $\mathrm{DP} = [k_p/(2\,f'\,k_t k_d)^{0.5}][[\mathrm{M}]/[\mathrm{I}]]^{0.5}$, which is predicted by this rate law. The measured rate law is consistent with a very low initiation efficiency, where $f' = f\,[\mathrm{M}]$.

We demonstrated that the observed rate law was not due to rapid initiator decomposition catalyzed by **8**. Instead, low initiation efficiency is a fundamental property of **8**, along with its very electron-rich vinyl group and its ability to provide a significant resonance interaction to the chain radical α to the ring.

C. η^5-(Benzene)tricarbonylchromium Acrylates and Methacrylates

The acrylate- and methacrylate-derivatized η^5-(benzene)tricarbonylchromium monomers **20**,[65,66,68,72] **21**,[69,72] and **22**[73] (Scheme 1.2) were synthesized from benzyl alcohol or 2-phenylethanol when reacted with $Cr(CO)_6$. The alcohols were esterified with either acrylyl or methacrylyl chloride in ether/pyridine and purified by multiple recrystallizations from CS_2. Homopolymerizations proceeded in classic fashion with no special electronic effects from the π-complexed $Cr(CO)_3$ moiety.[65,73] Acrylate **20** was copolymerized with styrene and methyl methacrylate and the reactivity ratios were obtained.[65] Acrylate **21** and methacrylate, **22**, copolymerized readily with styrene, methyl acrylate, acrylonitrile, and 2-phenylethyl acrylate to give bimodal molecular-weight distributions using AIBN initiation.[69] Copolymerization of **20** with ferrocenylmethyl acrylate, **2**, generates copolymers with varying mole ratios of two transition metals, Cr and Fe (see structure **34**).[65]

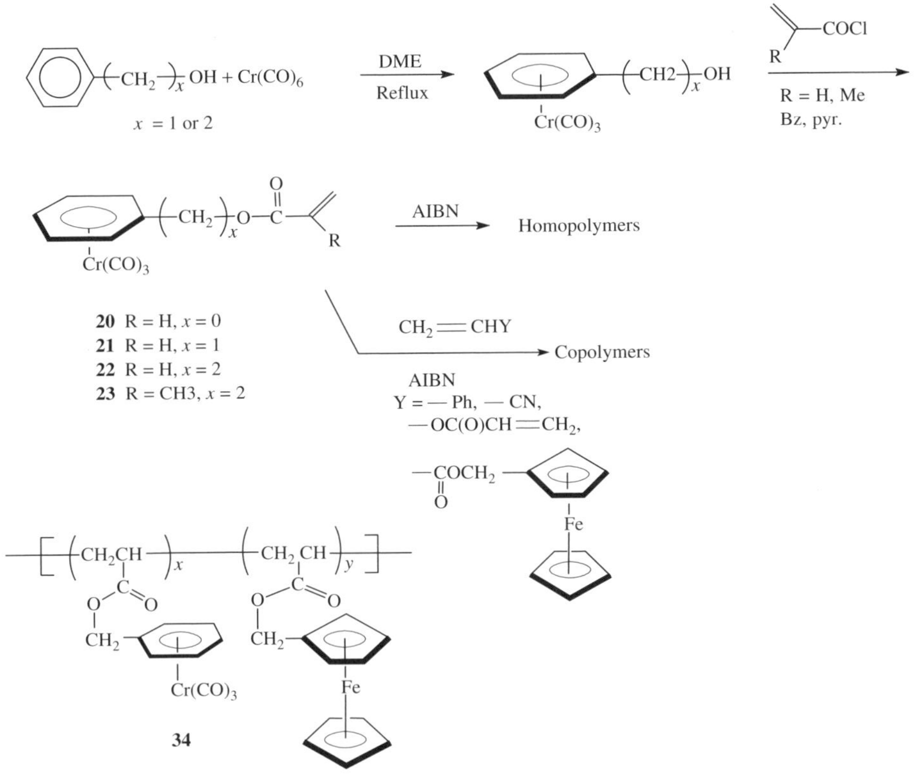

Scheme 1.2

η^6-(Phenyl)tricarbonylchromium-containing polymers are pastel yellow, but they become greenish blue on exposure to ultraviolet light due to formation of Cr_2O_3 particles as the organometal function decomposes. Thermal decompositions also form Cr_2O_3 with some polymer cross-linking. A bottom-up synthetic approach to generate nanocomposites of Cr_2O_3 was sought in the early 1970s. Cross-linked resins of **22** with diacrylate monomers were made, followed by either photochemical or thermal decomposition of the π-complexed $Cr(CO)_3$ moiety. This approach was also tried with resins containing **8** and other systems. In the early 1970s, our group did not have access to transmission electron microscopy (TEM), X-ray diffraction (XRD), or other methods to fully characterize the particle-size distribution of the resulting polymer-entrapped oxide particles. Thus, the potential of this bottom-up approach was never realized.

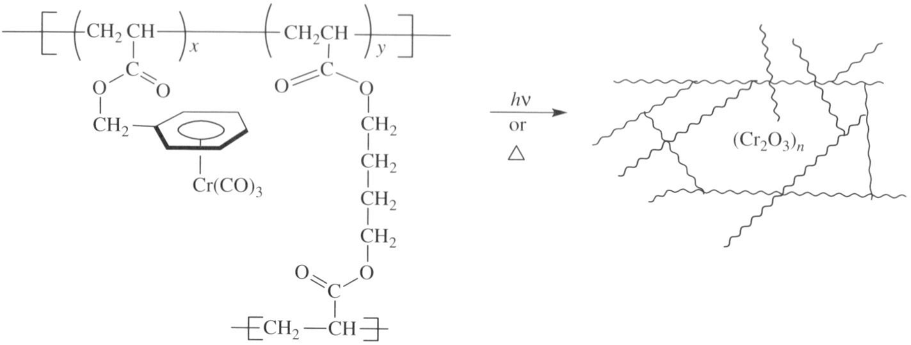

D. η^4-(Diene)tricarbonyliron Monomers

Novel iron carbonyl monomer, η^4-(2,4-hexadien-1-yl acrylate)tricarbonyl-iron, **23**, was prepared and both homopolymerized and copolymerized with acrylonitrile, vinyl acetate, styrene, and methyl methacrylate using AIBN initiation in benzene.[70,71,72] The reactivity ratios obtained demonstrated that **23** was a more active acrylate than ferrocenylmethyl acrylate, **2.** The thermal decomposition of the soluble homopolymer in air at 200°C led to the formation of Fe_2O_3 particles within a cross-linked matrix. This monomer raised the glass transition temperatures of the copolymers.[70] The η^4-(diene)tricarbonyliron functions of **23** in styrene copolymers were converted in high yields to π-allyltetracarbonyliron cations in the presence of HBF_4 and CO.[71] Exposure to nucleophiles gave 1,4-addition products of the diene group.[71]

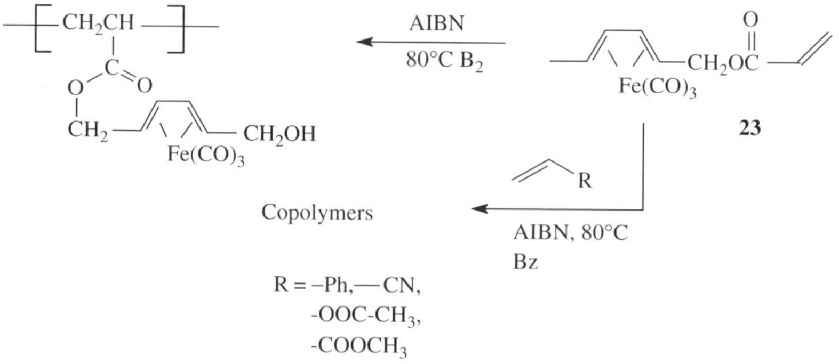

E. Other η^5-(Cyclopentadienyl)metal and η^6-(Phenyl)metal Containing Monomers and Their Polymers

An extremely valuable collaboration was forged with the respected synthetic organometallic chemist Mavin Rausch of the University of Massachusetts during the 1970s. Working with us, Rausch's group perfected the synthesis of many organometallic monomers for use in polymerization studies in our laboratory at the University of Alabama. This collaboration resulted in the preparation and polymerization of a large number of monomers, exemplified by **10–19**, **24–30**, and **32** (see Scheme 1.1, where references are listed).

Our first joint effort was the synthesis and polymerization of η[6]-(styrene)tricarbonylchromium, **19**.[66,67,72] This synthesis could not be achieved using the direct route from styrene and $Cr(CO)_6$. However, replacing $Cr(CO)_6$ with the more reactive $(NH_3)_3Cr(CO)_3$ and heating with styrene in dioxane led to **19**.[67] Free radical–initiated homopolymerization of **19** failed. However, radical–initiated copolymerizations with styrene and methyl acrylate succeeded. Manganesetricarbonyl monomer **8** copolymerized with **19** to give **35** successfully. Defining **19** as M_1, the reactivity ratios were $r_1 = 0$, $r_2 = 1.39$ in styrene copolymerizations, and $r_1 = 0$, $r_2 = 0.75$ with methyl acrylate. Thus, a chain radical of **19** cannot add to another **19** to give **36**, but can add to either styrene or methyl acrylate. This suggests the 3-dimensional steric bulk of **19** may decrease the k_{11} value close to zero. Also, **19**/styrene copolymers were prepared with different sequence distributions by the direct reaction of polystyrene with $Cr(CO)_6$.[67]

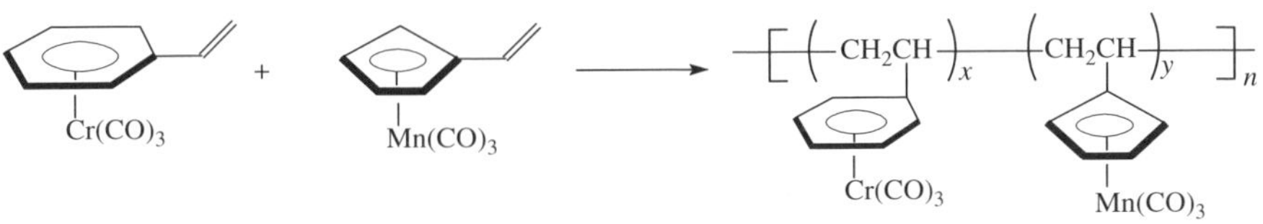

19 **8** **35**

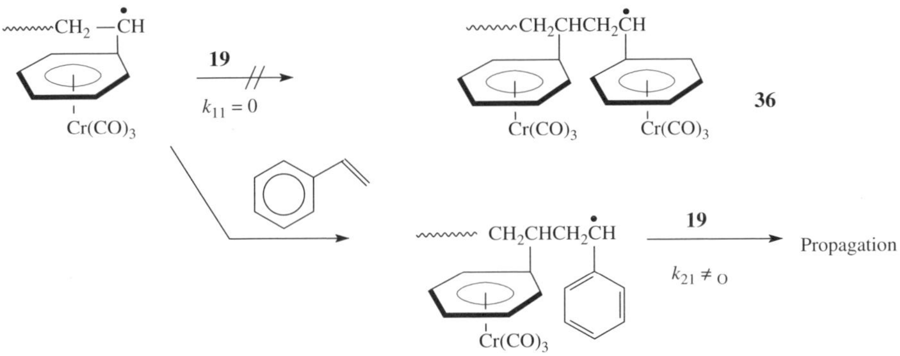

Copolymer **35**, generated from **19** and **8**, was the first transition metal containing organometallic polymer including one metal as a η[6]-complex and the other as a η[5]-complex.[67] The remarkable terpolymer, **37**, which contains η[4]-(diene)iron, η[5]-(cyclopentadienyl)manganese, and η[6]-(styryl)chromium species in each chain, was made. Thermal decomposition in air was conducted to see if mixed metal oxide particles of novel composition could be generated, but characterization of these products was beyond our capabilities in 1972.

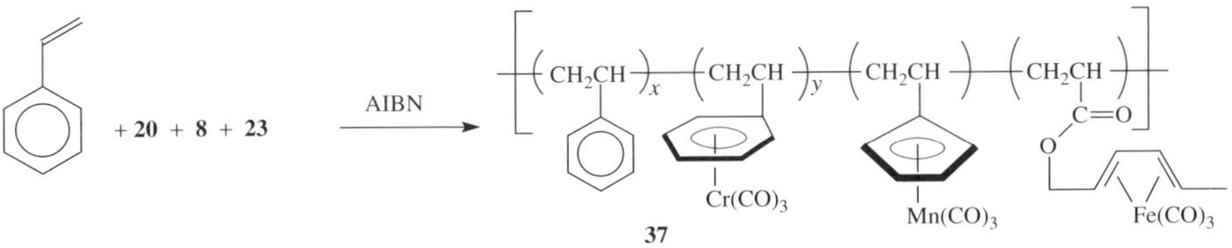

37

The preparation[64] and addition homo- and copolymerization[56] of η^5-(vinylcyclopentadienyl)dicarbonylnitrosylchromium, flowed from our cooperation with Rausch's group. Copolymerizations with styrene (M_2) gave values of $r_1 = 0.30$ and $r_2 = 0.82$, from which the values of $Q = 3.1$ and $e = -1.98$ were deduced for **10**. Similar results were obtained in *N*-vinyl-2-pyrrolidone copolymerizations. Thus, **10** resembles **1** and **8** as an exceptionally electron-rich vinyl monomer. The very large resonance parameter for **10** ($Q = 3.1$) resembles that of 2-vinylthiophene (2.86). A dual metal-containing copolymer **38** of **10** and **8** was readily obtained by radical initiated addition.

The homopolymers of **10** were branched and exhibited broad GPC molecular weight distributions. Studies of the homopolymers' molecular weights from polymerization at different monomer concentrations while (1) holding the [**10**]/[I] ratio constant, and (2) employing different [**10**]/[I] ratios confirmed that major differences existed in homopolymerizations of **10** versus vinylferrocene.[56] In ethylacetate the rate law was $r = k\,[M]^1\,[I]^{0.5}$. Polymerizations in benzene exhibited low initiator efficiencies. The rate was three halves order in the concentration of **10**, similar to that found for **8**.[53] Polymers incorporating **10** were able to catalyze the selective 1,4-hydrogenation of methyl sorbate, but not terminal or internal olefins.[56] This resembled the catalytic behavior of styrene/η^6-(styrene)tricarbonylchromium copolymers in hydrogenation.[75]

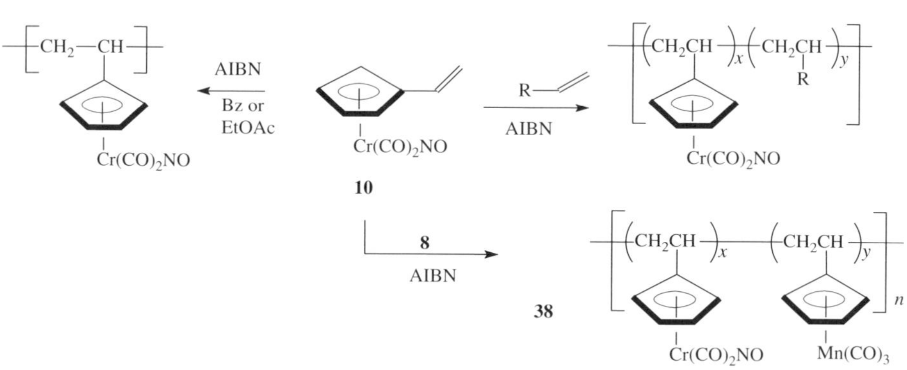

A new general route to a large variety of η^5-(vinylcyclopentadienyl) organometallic monomers was achieved in cooperation with Rausch.[46,54,57,76] The lithium salts of vinylcyclopentadienide and isopropenylcyclopentadienide were prepared (respectively), from which a large variety of new monomers, including **2–18** and **24–30** and **39** were generated (see Schemes 1.3 and 1.4). This led to many polymerizations; a few are mentioned here.

Vinyltitanocene dichloride, **18**, made as shown in Scheme 1.3, was copolymerized with styrene. Attempts to reduce the dichlorotitanocene to isolated titanocene moieties failed. Instead, the bridged dimer **40** formed, causing some cross-linking in the system. Structural analysis of this system was very difficult.

η^5-(Vinylcyclopentadienyl) monomers **11** (Mo), **12** (Ir), **14** (W), **15** (Co), and **16** (Cu) (Scheme 1.3) were subjected to homo- and copolymerizatons.[46,54,57,76–79] From reactivity ratios obtained in copolymerizations with styrene, the Q and e values of molybdenum and iridium-containing monomers, **11** and **12** (Scheme 1.4), were obtained.[57] Both exhibited large negative e values (**11**, -1.97, and **12**, -2.08) and large resonance stabilization Q parameters (**11,** 3.1, and **12**, 4.1). Combined with the data from vinylferrocene, **1**, and monomers **8** and **10**, it is clear that, regardless of the presence of carbonyl and nitrosyl substituents, all of these

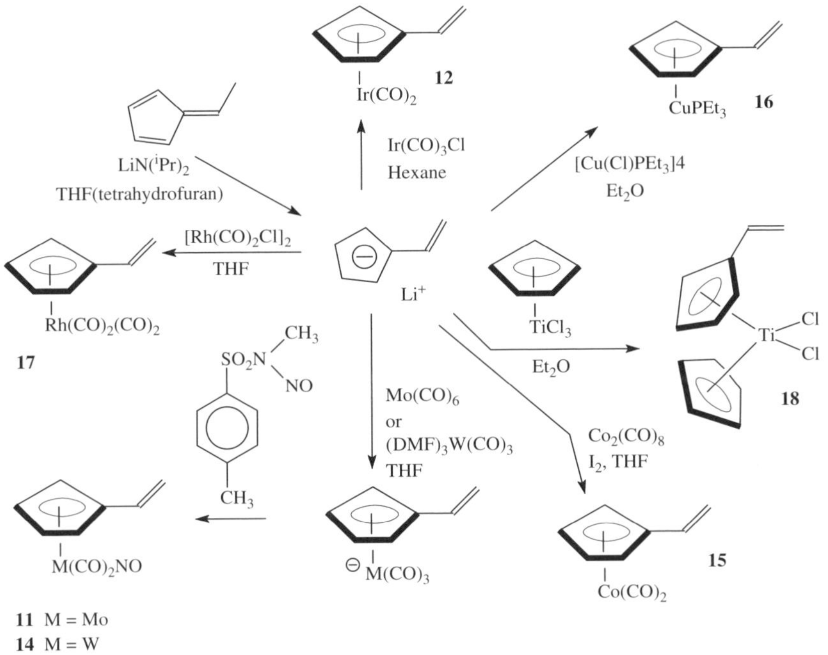

Scheme 1.3

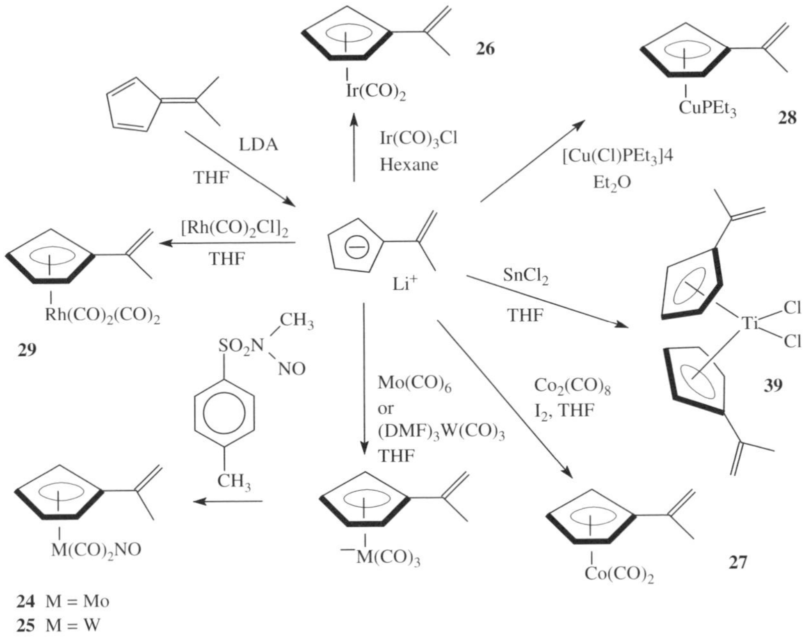

Scheme 1.4

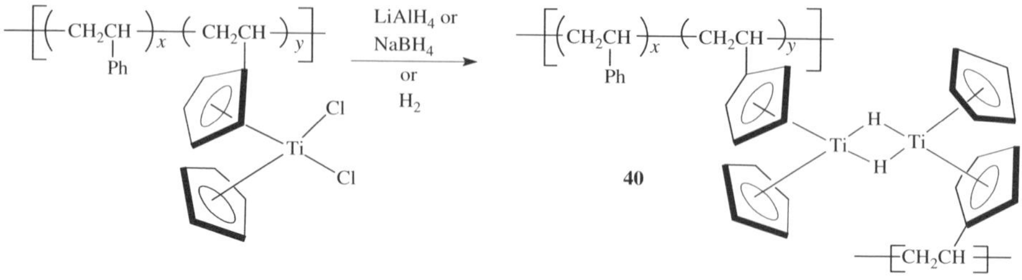

40

η^5-(vinylcyclopentadienyl) monomers are exceptionally electron-rich and provide stronger resonance stabilization of their α-radicals than is exhibited by styrene. Thus, they can be approximated as the resonance hybrids **41a–c**.

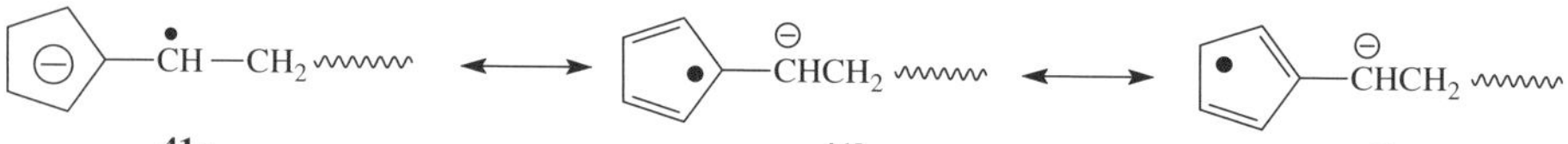

41a **41b** **41c**

The first η^5-(cyclopentadiene) tungsten monomer, **13**[59] (Scheme 1.5) was prepared from formylcyclopentadienylsodium, **42**. Salt **42** had been prepared first by Hafner[80] and later by Rausch.[81] Treatment of **42** with $W(CO)_6$ gave the η^5-(formylcyclopentadienyl)tricarbonyl-tungstate sodium salt, which was methylated by methyl iodide to give **43**. A phase-transfer Wittig reaction on **43** generated monomer **13**. Aldehyde **43** was also converted to the tungsten acrylate monomer **32** via $NaBH_4$ reduction to the alcohol, followed by esterification with acrylyl chloride.[59]

Monomer **13** homopolymerization was very sluggish, but it copolymerized in good yields with acrylonitrile, methyl methacrylate, and *N*-vinyl-2-pyrrolidone (Scheme 1.6).[59,61] However, styrene copolymerizations required several subsequent reinitiations to get good yields of copolymers. The reactivity ratios obtained in **13**/styrene copolymerizations were $r_1 = 0.16$ and $r_2 = 1.55$ (when M_1 = **13**),[61] giving values of $Q = 1.66$ and $e = -1.98$ for monomer **13** in direct accord with the *Qe* values found for monomers **1**, **8**, **10**, **11**, and **12**, as discussed earlier.[61]

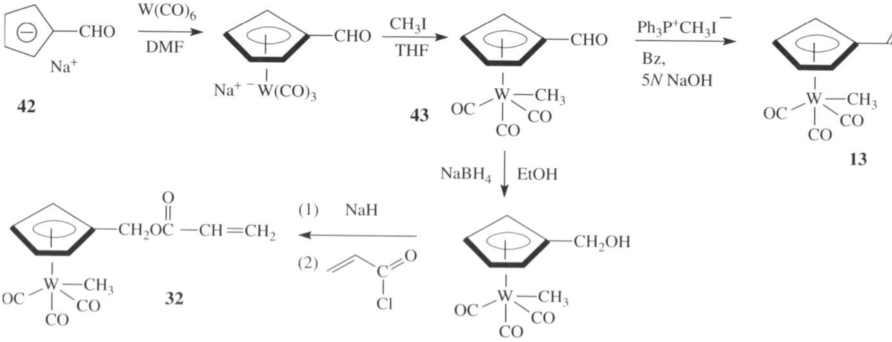

42 **43** **13** **32**

Scheme 1.5

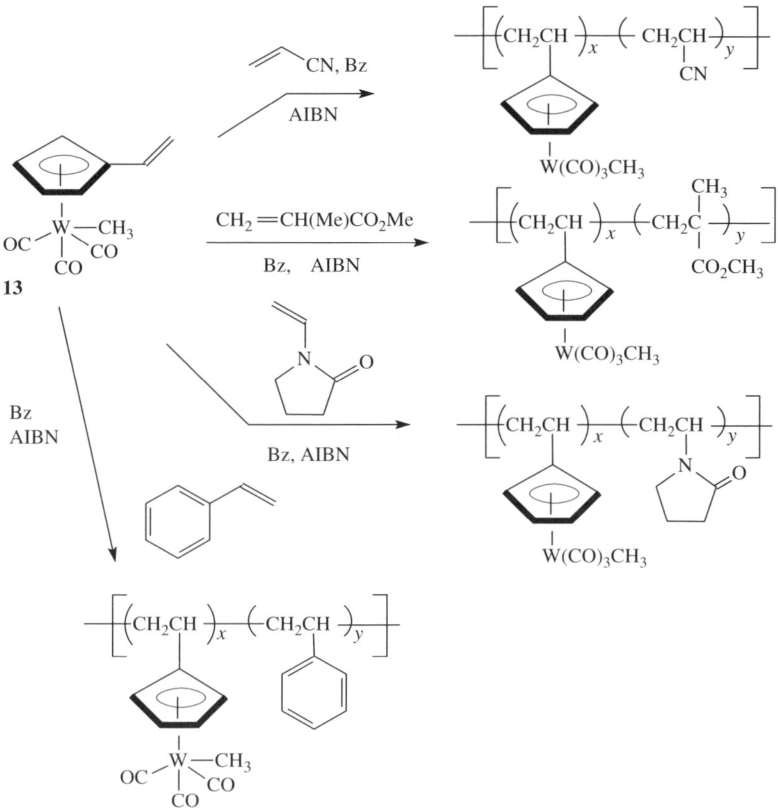

Scheme 1.6

The low homo- and copolymerization rates of **13** suggested that the methyl group, bound to tungsten, might stop chain growth by hydrogen-atom donation. If this methyl served as a chain transfer site, the molecular weight could be depressed. The molecular weights of polystyrene samples, produced by radical initiation in the presence of various concentrations of the model compound, η^5-(cyclopentadienyl)tricarbonylmethyltungsten, however, did not vary appreciably.[61] Furthermore, adding preformed **13**/styrene copolymers to styrene polymerizations did not depress the polystyrene molecular weights as the copolymer concentration was increased. Thus, chain transfer from the W—CH_3 function was not occurring.[61]

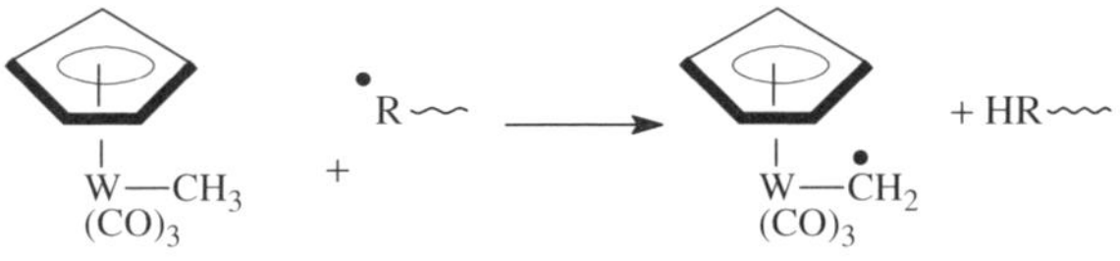

The homopolymerization kinetics of **13** were studied to determine its rate law.[62] The sluggish rates could have been due to low initiator efficiencies, as observed with **8**. The two most likely homopolymerization routes were those outlined in Scheme 1.7. Normal 1,2-addition polymerization to make **44** might be occurring. Alternatively, an internal H-atom abstraction by chain radical **45** from the tungsten-bound methyl group to give **46**, followed by addition to more **13**, might occur to generate polymer **47**. The rate law in benzene (AIBN initiation)[62] was $r_p = 1.13 \times 10^{-2}$ $[M]^{0.8}$ $[I]^{2.3}$. This rate law provided no clear evidence of the mechanism. The activation energy of propagation (determined from $E_p = E_o - 0.5(E_d - E_t)$ was very high (80.0 kcal/mol^{-1}), suggesting that standard kinetic analysis may not be applicable to homopolymerizations of **13**. However, structural analyses of the homopolymer by NMR clearly established its structure as **44**. No segments with structure **47** could be detected.[62]

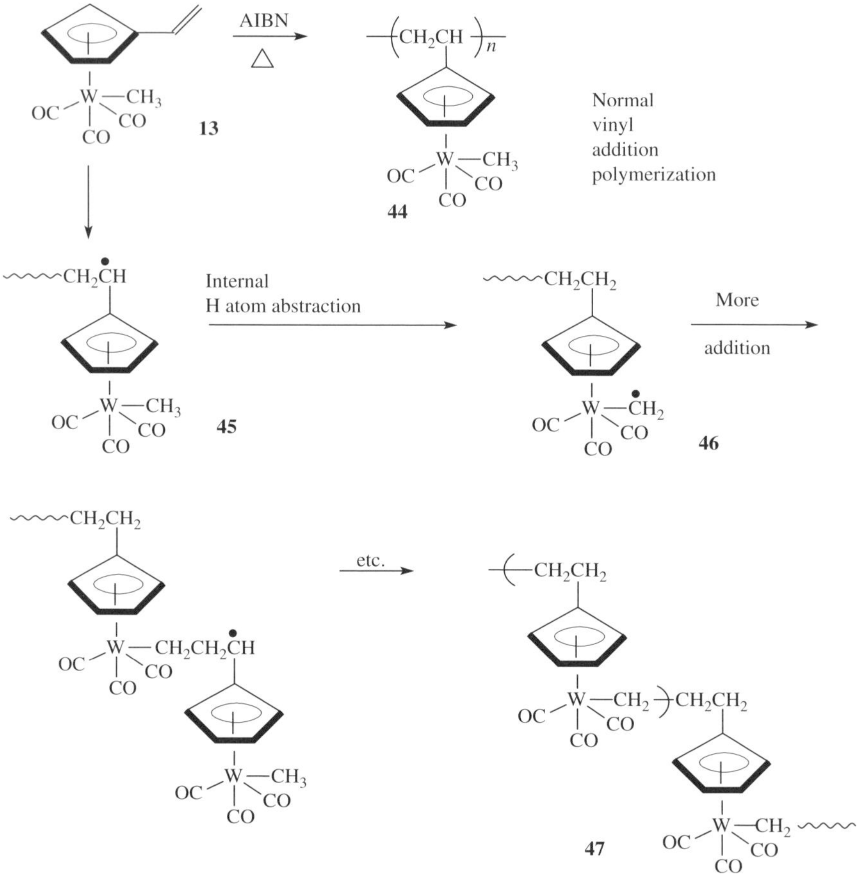

Scheme 1.7

F. Mixed-Valence Semiconducting Metallocene Polymers

In 1970 Cowan and Kaufman reported that the mixed-valent biferrocene [Fe(II)/Fe(III)] picarate, **48**, exhibited a single-crystal conductivity that was six orders of magnitude greater than either of its components[82,83] and a new electronic transition at 1900 nm, assigned to an electronic-transfer band. We had previously thought that it might be possible to alter the electrical properties of ferrocene polymers by partial oxidation to create mixed-valent materials. The Cowan–Kaufman report spurred us into action. In a joint effort with Dwain Cowan's group at Johns Hopkins University and Tapan Mukherjee's group at the Air Force Cambridge Research Labs, poly(vinylferrocene), a ferrocene-*o*-anisaldehyde condensation polymer, **49**, and poly(ferrocenylene), **50** were oxidized to their corresponding mixed-valence polymers using benzoquinone/HBF_4, 2,3-dichloro-5,6-DDQ, and *o*-chloranil.[84] This generated blue-green or black solids due to the ferricenium 620 nm $^2E_{2g} \rightarrow {}^2E_{1u}$ transition. Scheme 1.8 shows some of these mixed oxidation-state polymers, **51–53**.

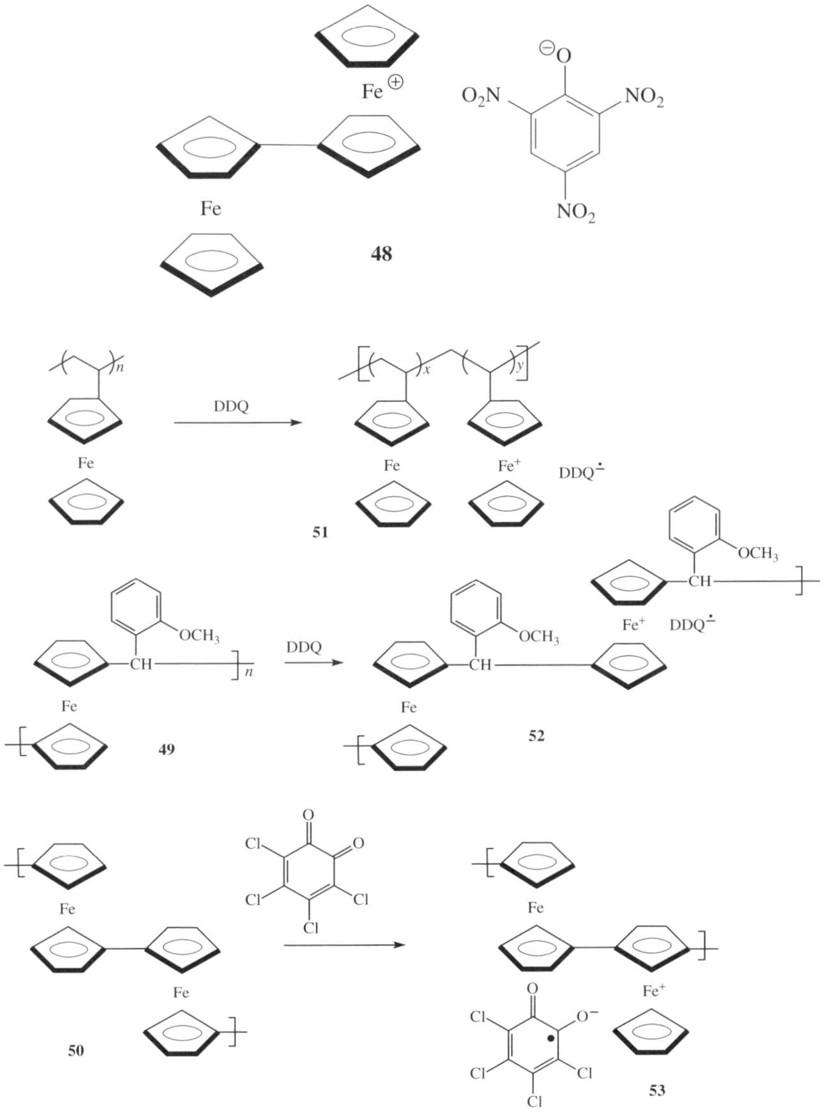

Scheme 1.8

Compressed pellet samples of these materials were semiconductors (conductivities of 10^{-11} to $10^{-6}\ \Omega^{-1}$) and no noticeable change in conductivity was detected after prolonged passage of current.[84] A weak transition occurred in the 1700–1900-nm region of **53** and the other poly(ferrocenylene) mixed-valent polymers, but not for **51** or **52**, where the ferrocene cyclopentadienyl rings were not directly bonded to one another. The presence of the 1700–1900-nm transition did not appear to be related to the insulator-to-semiconductor change in conductivity behavior. The conductivities appeared to be very sensitive to the percentage of the ferrocene rings in the polymers that had been oxidized to ferrocinium moieties. The conductivities reached a maximum at about 50% oxidation.[84] The oxidation stoichiometry was precisely established by Mössbauer spectroscopy.

Cooperation with Mukerjee continued under support from the Office of Naval Research. At this time Heeger and MacDiarmid were involved with the early, but groundbreaking, work on conducting stacked tetrathiafulvalene/tetracyanoquinodimethane (TCNQ) and related salts,[85] as well as the first studies of poly(acetylene). Not long before, Rosenberg had published studies of the synthesis of poly(ferrocenylene) **50**, which had helped in our joint work with Cowan. Efforts to understand mixed-valent semiconducting polymers continued in the hope that high conductivities might be achieved.

The structure of poly(ferrocenylene), **50**, involved conformations where the directly bonded cyclopentadienyl rings were canted (rotated) to create sizable dihedral angles, which interfered with extended conjugation. We synthesized poly(ethynylferrocene), **54**.[86] This poly

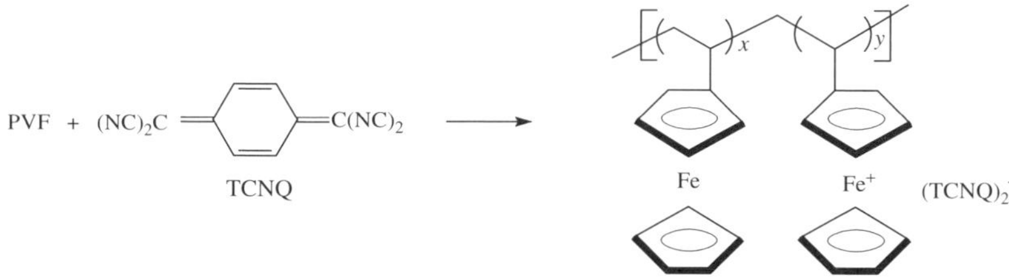

55

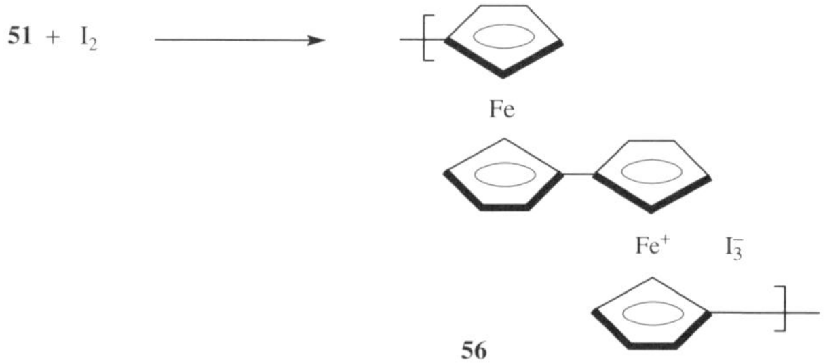

56

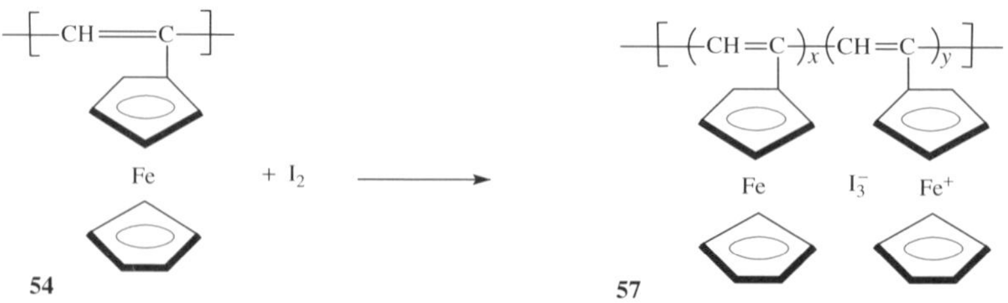

54

57

Scheme 1.9

(ethynylferrocene) (Scheme 1.9) did not, in retrospect, have a perfect poly(ethynylferrocene) structure over long persistence lengths, but it was the best we could do at the time. Mixed-valent polymers **55**, **56**, and **57** were then prepared from poly(vinylferrocene), poly(ferrocenylene), and poly(ethynylferrocene), respectively. Each was oxidized with DDQ, iodine, and TCNQ to generate nine series of polymers having a large range of Fe(II)/Fe(III) ratios.[87] Some examples are shown in both Schemes 1.8 and 1.9. Then the compressed-pellet conductivities were measured on samples across the range of Fe(II)/Fe(III) ratios.[87] The results were similar across the entire series. Conductivity was largely independent of the anion (or radical anion) present and always maximized between 30 and 70% Fe(III) content. When no (or small amounts) of ferricinium [Fe(III)] moieties were present or when 90% or more Fe(III) sites were present, the conductivities approached those of the starting unoxidized polymers, which were insulators (10^{-16} to $10^{-14}\,\Omega^{-1}$). The highest conductivities were $\sim 10^{-7}$ to $10^{-6}\,\Omega^{-1}$. The structure of the starting polymer also played only a minor role in the magnitude of the conductivity. In summary, these systems were charge-hopping semiconductors as compressed pellets.

In 1972 Mueller-Westerhoff at IBM[88] and Cowan at Johns Hopkins[89] reported the very high conductivity of powder samples of the bisfulvalenediiron $(TCNQ)_2$ salt, **58** ($\sigma \sim 10\,\Omega^{-1}\,cm^{-1}$). These reports immediately attracted our attention because the two ferrocene moieties in **58** must be coplanar, allowing resonance interactions between rings and the iron atoms. The compound differs sharply from poly(ferrocenylene) and its mixed-valence salts in this respect. Furthermore, single crystals of **58** would be expected to exhibit significantly higher conductivities. Therefore, synthesis and polymerization of 3-vinylbisfulvalenediiron, **61**, was undertaken in order to react the polymer with TCNQ and measure its conductivity.

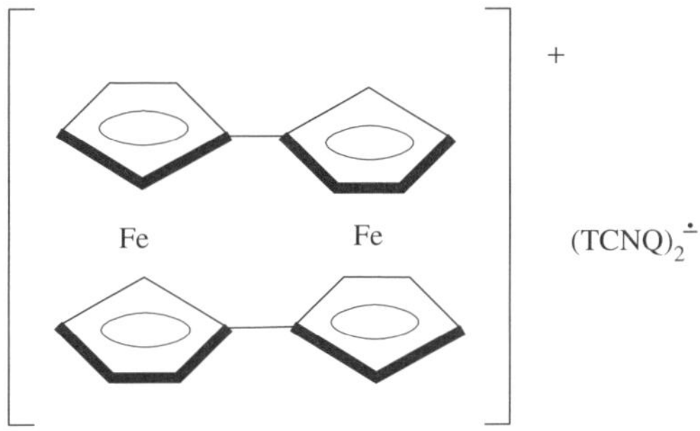

58

This synthesis proved very challenging, but it was eventually accomplished as outlined in Scheme 10.[90,91] Bisfulvalenediiron, **59**, could only be obtained in low (14% max) yields, even after improving the IBM synthesis. Acetylation produced several compounds, including several disubstituted isomers. The separation and purification of 3-acetylbisfulvalenediiron, **60**, was tedious, and yields of only 8–10% were obtained in this step. $NaBH_4$ reduction of the carbonyl group (78% Y) was followed by a low-yield vacuum pyrolytic dehydration (14% Y) to 3-vinylbisfulvalenediiron, **61**. Thus, the logistics of this process were daunting. Solubilities were low, compounding the problem. Once available, **61** homopolymerization was studied.

Monomer **61** has a very large steric demand and it is subject to the same internal electron-transfer/termination process discovered earlier for vinylferrocene polymerizations. Only modest molecular weights were achieved and reinitiation sequences were required to get, in our hands, a 31% maximum yield. Copolymerizations with styrene were also successfully performed.[90,91]

Homopolymer, **62**, and styrene copolymers, **63**, were converted to their $(TCNQ)_2^{\dot{-}}$ salts, **64** and **65**, respectively.[90,91] Mössbauer spectra only exhibited one type of iron in **58**, suggesting that each iron is in the +2.5 oxidation state. This is in accord with a broad near-IR absorption at 1000–2000 nm (maximizing at 1400–1700 nm), which is due to a photon-assisted intramolecular intervalence exchange. If an equilibrium between Fe(II)/Fe(III) and Fe(III)/Fe(II) was occurring, its rate had to be fast on the Mössbauer timescale ($\sim 10^7\ s^{-1}$).

Because only small amounts of polymer **62** were available, we reacted samples first with small stochiometric amounts of TCNQ to create samples of **64** with small *y*/*x* ratios (see Scheme 1.10). The conductivities of these samples were measured (Cambridge AFL) and returned to us to use again for conversion to larger *y*/*x* ratios. Homopolymer **62** had a conductivity of $10^{-14}\ \Omega^{-1}\ cm^{-1}$. The conductivities of 10 samples of **64** with different *y*/*x* ratios rose steadily to $6–9 \times 10^{-3}\ \Omega^{-1}\ cm^{-1}$, when 71% of the bisfulvalenediiron had been oxidized.[90,91] Unfortunately, our sample stocks were then accidentally destroyed while at Cambridge AFL.

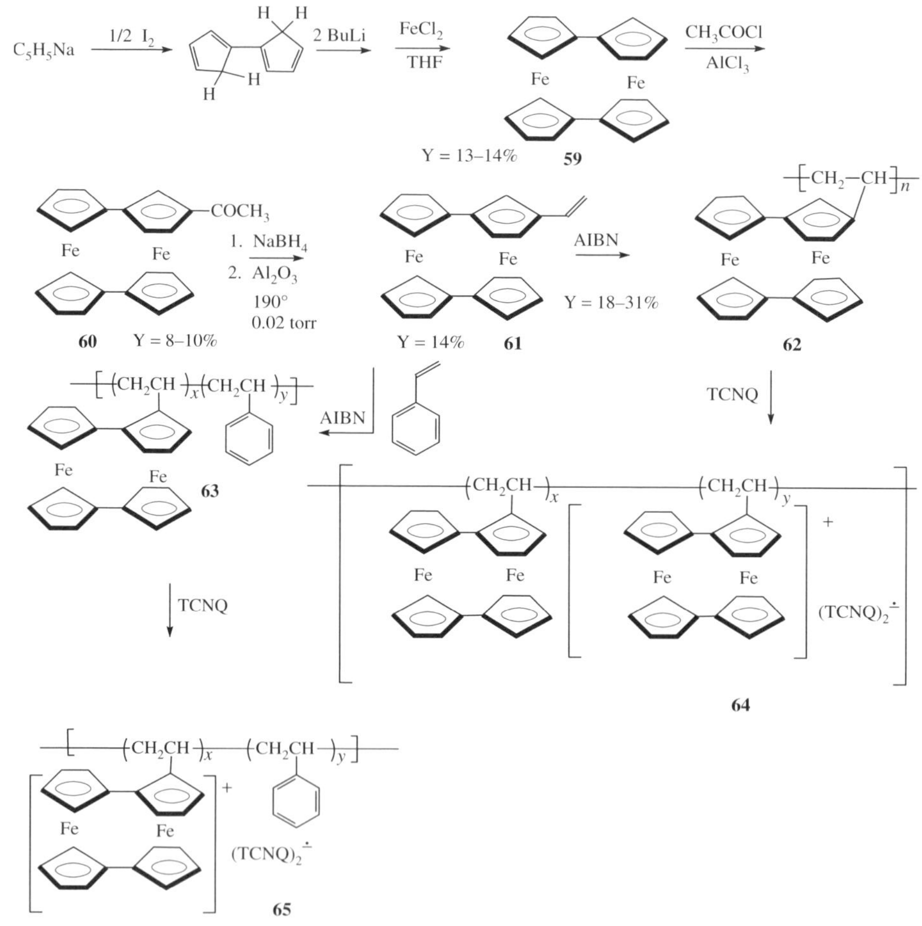

Scheme 1.10

The logistic challenges prevented resynthesis of **62** in order to study 80, 90, and 100% levels of bisfulvalenediiron monoxidation. Extrapolating the data obtained to a 100% conversion to the $(TCNQ)_2^-$ complexes, suggested that the conductivity of such a sample might be 0 to $10\,\Omega^{-1}\,cm^{-1}$.

Vinylruthenocene, **66**, and vinylosmocene, **67**, are the ruthenium and osmium analogs of vinylferrocene. They can be polymerized neat or in benzene using AIBN initiation at 60–80°C.[92] These monomers were provided to our group by John Sheats of Rider College in very small amounts.[93,94,95] Copolymerization of **66** with styrene at a 5/95 molar feed ratio produced copolymers with 3 mole % vinylruthenocene and molecular weights of 1 to 2×10^5.[54] Increasing the **66**/styrene feed ratio to 40/60 resulted in 25–30 mole % incorporation of **66** and lower molecular weights ($8–20 \times 10^4$) (Scheme 1.11). These copolymers, **67**, were oxidized with DDQ and TCNE to generate ruthenocenium salts, **68**, but their conductivities were never measured.

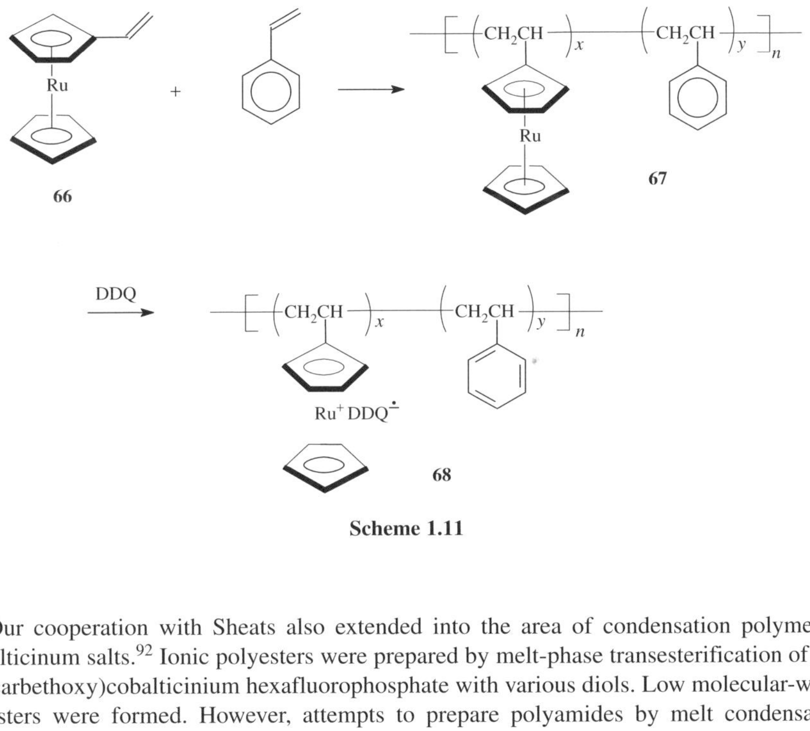

Scheme 1.11

Our cooperation with Sheats also extended into the area of condensation polymers of cobalticinum salts.[92] Ionic polyesters were prepared by melt-phase transesterification of 1,1′-bis(carbethoxy)cobalticinium hexafluorophosphate with various diols. Low molecular-weight polysters were formed. However, attempts to prepare polyamides by melt condensations of alkylene diammonium salts of 1,1′-bis(carboxy)cobalticinium PF_6^- salts resulted in decomposition.[92]

Another cooperative effort arose between Samuel McManus of the University of Alabama in Huntsville and our group. This cooperation produced results in pure heterocyclic organic chemistry, oxazoline polymerizations,[96] and siloxane polymers.[97] Oxysilane ferrocene-containing polymers were explored.[98,99] Several bis(dimethylamino)silanes were made and reacted with 1,1′-bis(hydroxymethyl)ferrocene to give the early ferrocene oxysilane polymers **69** and **70** (Scheme 1.12).[98]

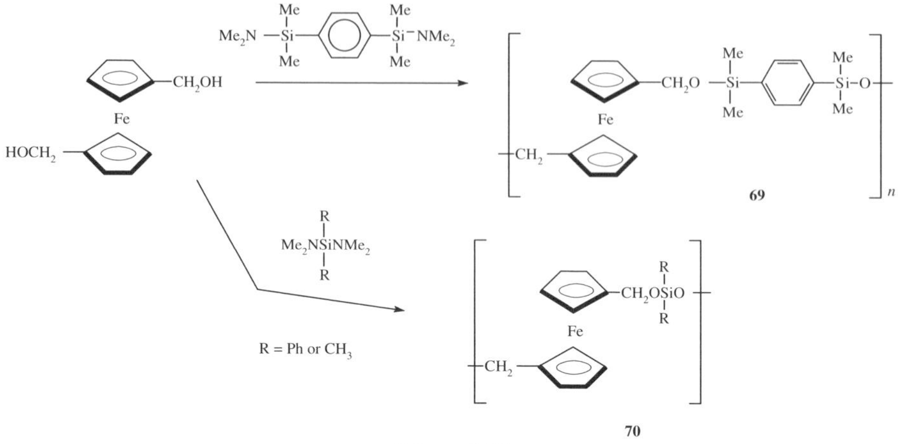

Scheme 1.12

Later this work was expanded when 1,1′-bis(dimethylaminodimethylsilyl)ferrocene **71** was synthesized and polymerized with three aryl disilanols to give polymers **72–74** (Scheme 1.13).[99] Melt polymerizations carried out at 1 torr and 100°C gave very high molecular weights. Tough flexible films could be cast, fibers were drawn from their melts, and these polymers were hydrolytically stable.

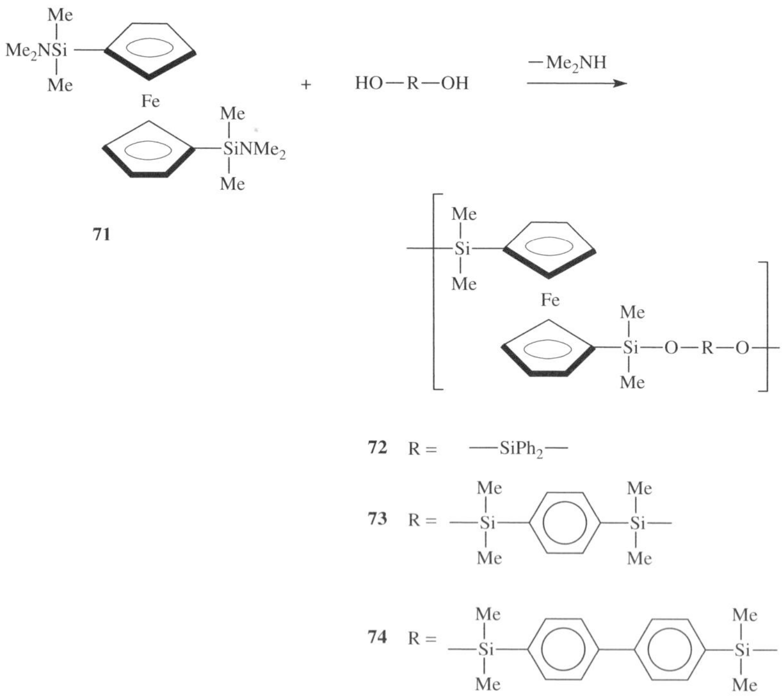

Scheme 1.13

G. Metal Carbonyl Anion Substitutions on Polymers

Synthesis with metal carbonyl anions was developed as a key route in organometallic chemistry in the 1960s.[100] We capitalized on this rapid progress by employing metal carbonyl anions in nucleophilic displacements onto preexisting polymers. For example, treating linear and cross-linked chloromethylated polystyrenes with $NaMo(CO)_5$-generated pendant η^1-benzylpentacarbonylmanganese functions throughout the polymer (Scheme 1.14).[101] Thermal decomposition of these polymers, **75**, resulted in cross-linking to **76** through benzyl radical

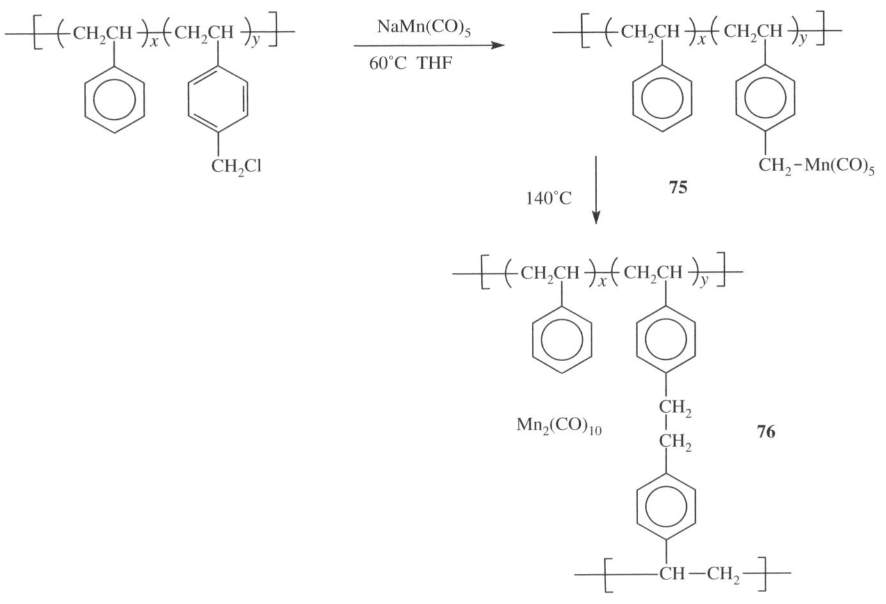

Scheme 1.14

dimerizations and the formation of liberated $Mn_2(CO)_{10}$. Using polymers of appropriate cross-link densities enabled the trapping of liberated $Mn_2(CO)_{10}$ within the polymer. This work foreshadowed the use of metal carbonyl chemistry in polymers that are photochemically degraded (recently reviewed by Tyler).[102]

This work was expanded through the use of the series of η^5-(cyclopentadienyl)metalcarbonyl anion nucleophiles as outlined in Scheme 1.15. Parallel work was carried out on the monomeric benzyl analogs that confirmed the generality of the synthetic work and elucidated differences in thermal decompositions of the polymeric versus monomeric species. In this manner, Fe-, Mo-, and W-containing polymers **77–79** were achieved, which contained with metal-to-carbon sigma bonds and η^5-Cp-to-metal coordination.[103] Neat polymer **77** was more thermally stable than its monomeric benzylic analog. When swollen in a solvent, **77** underwent thermolysis to form **80–82** (Scheme 1.15). A major pathway in the thermal reactions of both **78** and **79** involved migration of the benzyl group from the metal to the cyclopentadienyl ring and dimerization by metal–metal bond formation. Thus, structures such as **83** appeared in the products after heating.

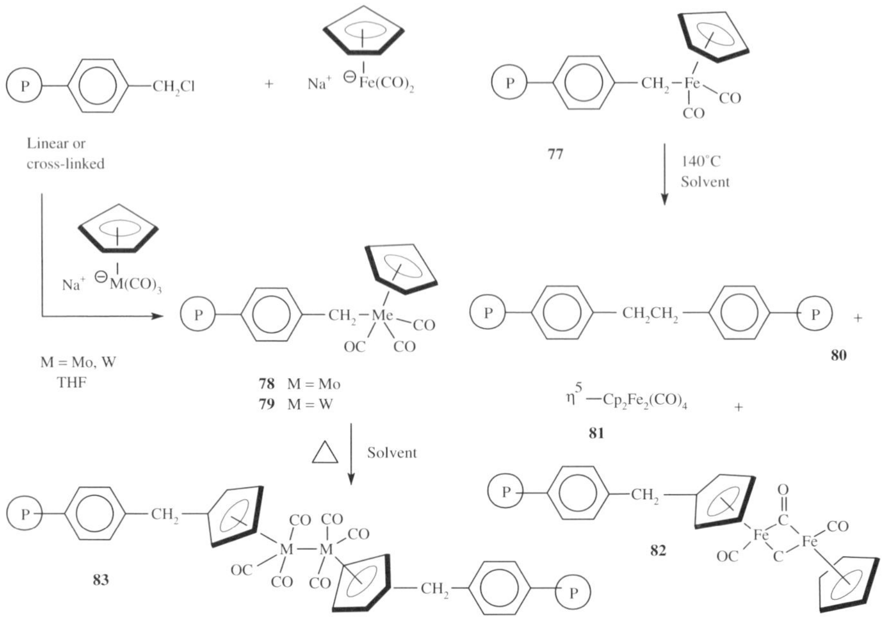

Scheme 1.15

H. Anionic Initiation

The analogy between styrene and the η^5-(vinylcyclopentadienyl)metal monomers vinylferrocene, **1**, vinylruthenocene, **66**, vinylosmocene, and the η^5-(vinylcyclopentadienyl)metal carbonyl monomers in radical-initiated polymerizations summarized in Scheme 1.1 no longer exists for anionically initiated addition polymerizations. Styrene is readily initiated by such anionic species as BuLi and $Na^+Naphth^-$. Living anionic styrene homopolymerizations and block copolymerizations have been extensively commercialized for many years (e.g., Kraton thermoplastic elastomers). However, the exceptionally electron-rich vinyl metal-containing monomers **1**, **8–18**, **24–30**, and **66** were never successfully initiated by anionic systems in our laboratory despite many attempts. In these systems, the α-carbocations are very stable, but the α-carbanions are quite unstable. Thus, the addition of an anion to the vinyl function of these monomers is unfavorable.

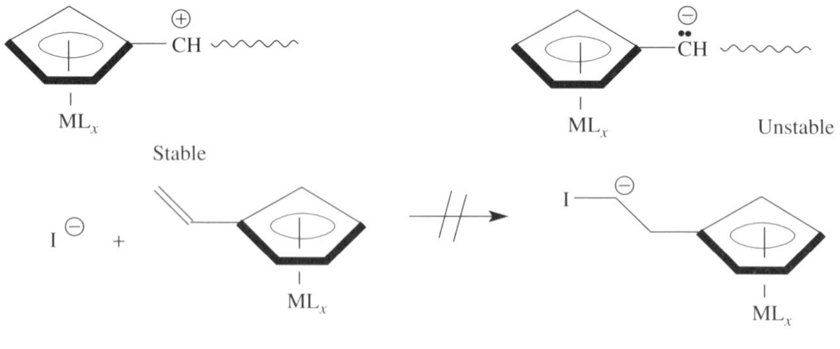

The acrylate and methacrylate monomers **2**, **3**, **6**, **7**, **20**, **23**, and **32** can undergo anionic initiation. These reactions are still not well studied at the present time. Akira Hirao (currently at Tokyo Institute of Technology), while working as a postdoctoral associate in my laboratory, demonstrated that the anionic polymerization of ferrocenylmethyl methacrylate, **3**, could be initiated by *n*-BuLi, Na^+Napht^-, K^+Napht^-, EtMgBr, PhMgBr, and $LiAlH_4$ using high vacuum techniques.[104,105] The molecular weight was controlled by varying the monomer-to-initiator ratio when $LiAlH_4$ was employed in THF (tetrahydrofuran). Molecular weights to 277,000 (M_n) were obtained with relatively narrow distributions, suggesting that quasi-living polymerizations were involved. The Mark-Houwink *K* and *a* values were determined in THF.[104]

Hirao, a glassblowing wizard, readily constructed a plethora of break-seal vacuum-line glassware. He then developed a $LiAlH_4$–tetramethylethylenediamine (TMEDA) initiation systems that formed excellent living polymers of **2** (Scheme 1.16). The addition of methyl methacrylate or acrylonitrile to the living poly(ferrocenylmethyl methacrylate) **84** (e.g., $PFMMA^-$) led to block copolymers **85**.[104,105] The $PFMMA^-$ anions **84** were not nucleophilic enough to initiate styrene. Therefore, living polystyrene was prepared by Na^+Napht^- initiation in THF (−78°C) and styrene–block–FMMA copolymers were formed upon subsequent addition of **2** or **3**.[105] These were the first true block copolymers and living anionic polymers made of transition metal organometallic species.

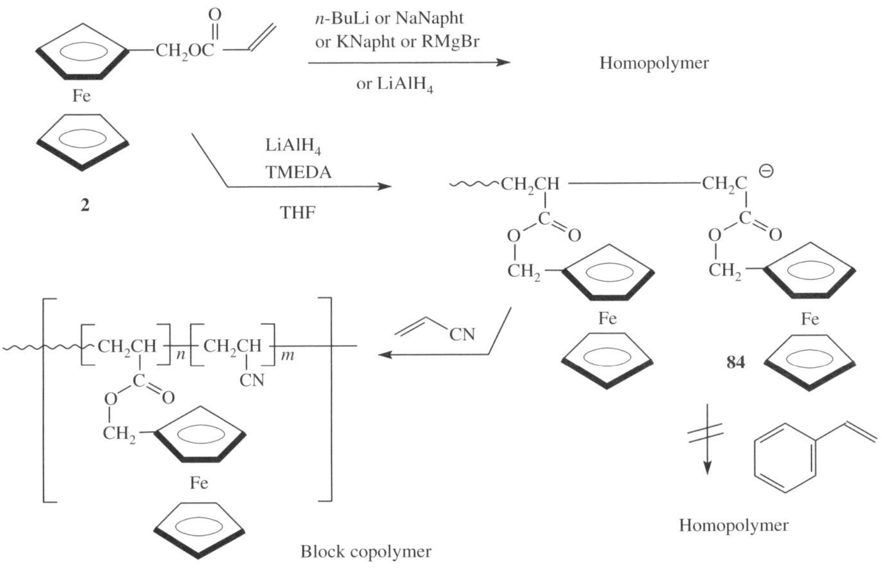

Scheme 1.16

I. Polymer-Anchored Organometallic Catalysts

In the period from 1967 to 1980, homogeneous organometallic and metal complex catalysts were just being developed. Examples like Wilkinson's hydrogenation catalyst, $Rh(PPh_3)_3$ $(PPh_3)_3$ Cl, or the rhodium hydroformylation catalyst $Rh(PPh_3)_3Cl$, were being employed mostly in batch or semibatch processes and the catalyst had to be recovered. A few groups in industry, R. Grubbs' group at Michigan State University and Pittman at Alabama[106] began to attach these catalysts to both soluble and cross-linked polymer matrices. Soon the effects of the polymer matrix, site isolation, size and diffusion effects, multistep reactions with several anchored catalysts in the same reactor, and ligand equilibria within the matrices came under intense study. This combination of polymer and homogeneous organometallic catalystic chemistry proved exciting. Pittman's laboratory was intensely involved, and the reader is referred to references 106–145 to gain a view of these early contributions. The emphasis on that area was mainly on catalysis versus that of polymeric materials and will not be covered further here.

II. CONDENSATION POLYMERS

Carraher's group was the predominant force in the early development of metal condensation polymers. They synthesized a large series of transition metal-containing polymers based on the analogy between acid halides and organometal halides. The simple idea was that typical Lewis bases such as diamines, diols, and dithiols would react to form polymers with organometallic dihalides (Scheme 1.17). Biscyclopentadienyl metal dichlorides were employed as the metallic diacid chloride analog in many of these studies. Thus, titanocene dichloride,[146–151,153] zirconocene dichloride,[146,148,149,150,152,153] and hafnacene dichloride[146,149,150,153] were reacted with a variety of bis-Lewis bases. These reactions can be thought of as analogs of polyester and polyamide synthesis, as shown below. This idea actually worked in some situations, but the early learning curve was steep. The presence of water caused hydrolysis products, rather than polymers, to dominate in many cases.

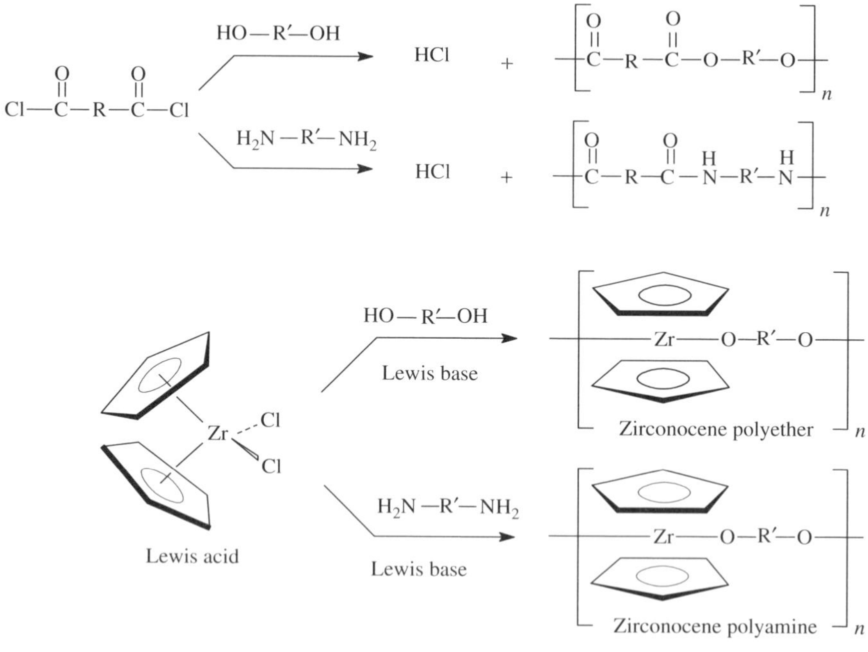

Scheme 1.17

How did this work originate? In 1965–1968, Zilkha et al. reported low molecular-weight tin polyesters formed from the reaction of a dialkyl(or aryl)dichlorotin or a tetraalkyldichlorotin with the sodium salt of bis[*o*-(carboxymethyl)phenyl]- or bis[*p*-(carboxymethyl)phenyl] dimethylsilanes.[154,155,156]

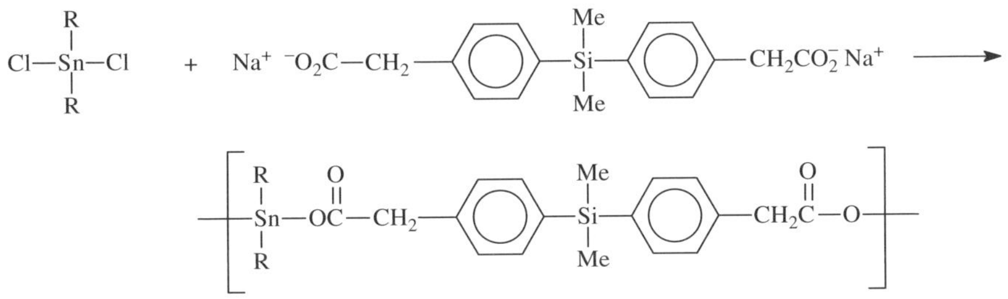

Carraher was a young assistant professor at the University of South Dakota who was interested in interfacial (IF) polymerization processes. Zilkha's results suggested that the more nucleophilic bisalkoxides might polymerize with dichlorosilanes. Thus, a series of organodichlorosilanes were reacted with ethylene glycol in two phase systems consisting of water and an organic solvent (pentane, cyclohexane, CCl_4 etc).[157] This led to a series of fairly low molecular-weight polymers, according to the limiting viscosity numbers obtained. No other molecular-weight measuring instrumentation was available at that time. The molecular weights were higher using dialkyldichlorosilanes than when employing diphenyldichlorosilane.

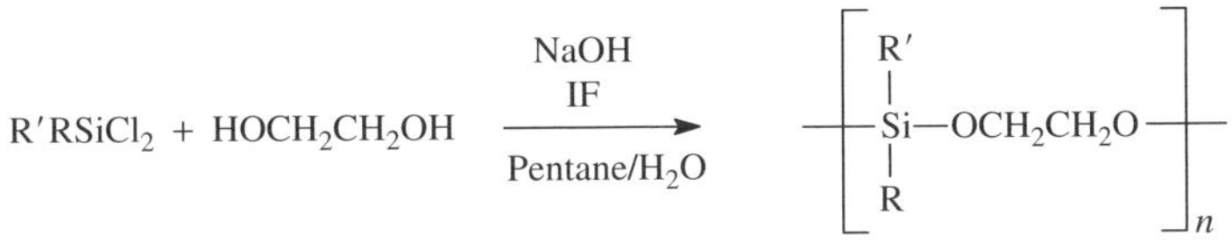

The stirred IF production of polyoxyethyleneoxy(diphenylsilylene), where R′,R = Ph above, was 5/3 order and dependent on the concentrations of ethylene glycol and diphenyldichlorosilane.[158]

$$r = k\,[\text{ethylene glycol}]^{2/3}[Ph_2SiCl_2]$$

This agreed with a model where ethylene glycol resided in the organic phase and where the rate depended upon the organic spherical surface area.[158] At that time, few rate studies of IF polymerizations were known due to the problem of removing the polymer from the reaction zone at a constant or known rate. The heterogeneous nature of these systems complicated the gathering and interpretation of the data.

This initial formation of silicone polymers by IF processes led to the key thought that many organometallic metal dihalides might be successfully polymerized by applying these methods. After all, everyone knew that dichlorosilanes rapidly hydrolyzed in water. However, it was not apparent that polymers would ever form from R_2MCl_2 plus diols in the presence of water. By and large, chemists routinely avoided the presence of water in the reactions of dichlorosilanes with organic reagents. The initial success with dichlorosilanes suggested that the door may be open to expand the number of elements that could be incorporated into polymer chains. With the explosive growth that had taken place in organometallic chemistry, an array of new metal-containing analogs of dichlorosilanes were now available. Examples included the Group IV dichlorometallocenes (of Ti, Zr, Hf) and dichlorotin compounds.

III. INTERFACIAL POLYMERIZATION

Many of these polycondensations employed the IF polycondensation reaction popularized by Paul Morgan and co-workers at Dupont.[159] The topic of IF synthesis has been reviewed.[159–162] Polycondensations via IF occur rapidly and at room temperature, allowing less thermally stable reactants to be employed. The equipment required to carry out these reactions is generally only a simple blender assembly. Early studies demonstrated stirred reactions were much faster (1 minute and less compared with a week) and more reproducible.

The formation of five- and six-member rings versus polymerization is always a problem when Lewis bases such as ethylene diamine and ethylene glycol were used with dihalometal compounds. Thus, product the original yields were low for these systems. If internal cyclization did not occur during the first condensation step, then additional chain growth could occur because cyclization of the longer chains was slower. The IF system allowed the use of a relatively high concentration of both reactants. This encouraged the second condensation step to occur with another Lewis acid rather than internal cyclization.

Several modifications to the water-containing IF process were developed,[159–162] including those that did not require the reactant to be dissolved and did not require the presence of water. Briefly the three most used IF systems are the

1. Classic system where the Lewis base, generally along with an added base, are dissolved in water and the second phase consists of the Lewis acid, here the organometallic reactant, dissolved in a suitable organic liquid.
2. Nonaqueous two-phase systems employ two largely immiscible liquids. Examples include dissolving the Lewis base in acetonitrile, nitrobenzene, or 2,5-hexadione and the organometallic acid chloride in a nonpolar organic liquid such as hexane, decane, or carbon tetrachloride.
3. Nonorganic solvent systems where the Lewis base is dissolved in water and a liquid Lewis acid is used neat.

Reaction occurs at or near the interface. Morgan showed that most organic acid chloride systems react with diols and diamines in the organic layer. Carraher found organometallic systems employing neutral Lewis bases, such as diols and diamines, react within the organic phase. But for charged Lewis bases, such as salts of diacids, reaction occurs even closer to the interface and possibly at the interface, due to the poor solubility of the ionized salt in the organic phase. The reaction site also varies with the Lewis acid. Thus, organotin chlorides are highly hydrophobic so there is less tendency for them to enter the aqueous phase. In contrast, Group IVB metallocene dihalides readily hydrolyze, forming species that react to place metallocene moieties into the polymer. Acid salts react in the aqueous phase, but reactions with diols and diamines still occurs in the organic phase.

IF polymerization is often a race between the instability of the developing polymer to an added base and product capture before the polymer degrades. Added bases were often required to accomplish polymerization. Most metal-containing condensation polymers exhibit decent stability to acids, but undergo rapid degradation in the presence of base. Addition of dilute acid to neutralize excess base was often sufficient to allow polymer retrieval.

The Lewis base at the droplet surface that is holding the Lewis acid, encounters a high Lewis acid concentration. This may be why polymer formation occurs instead of hydrolysis. The enormous water stoichiometry dwarfs the amount of Lewis base within the entire phase, but not necessarily at the interface surface. Thus, polymerization can proceed instead of hydrolysis. Polymer formed almost instantaneously in rapidly stirred systems. The reactions were completed within several seconds to minutes. In general, IF surface area increases with increasing stirring rate and up to some limit. A limit is reached where the amount of reactant to coat the droplets is too small to accomplish this objective. Rapid stirring, in the range of 18,000 rpm (several times the speed of a model airplane propeller) was easily accomplished by most commercial blenders. This stirring rate is about 70% of the speed where large molecules are sheared.

Base degradation of polymer is prevented by simple addition of dilute acid or by using base systems where the pH is low. Slightly soluble bases such as calcium hydroxide often work. Sterically hindered bases are often used since the amines compete with the desired Lewis base reactant. Bases complete with polymerization and causes chain termination. The associated base is an end group. Organotin reactions employing triethylamine, for example, generally have amine end groups. The pH control can also be accomplished through the use of phosphate salt buffer systems. Rapid polymer precipitation from solution probably assists in protecting especially base sensitive products, allowing them to be recovered.

A. Group IVB (Ti, Zr, Hf) Polyethers

Titanocene dichloride has an intense off-red color. Upon polymerization or hydrolysis the color changes from red to yellow. Thus, color change is a convenient indicator of rate.

A number of Group IVB polyethers, some of which are illustrated below, have been synthesized utilizing both the classic and inverse IF systems, as well as a totally aqueous system.[146–153,163–179] In the inverse system, the metallocene is contained in water and the diol contained in the organic liquid. From 30 to above 60% yields result using inverse interfacial systems. In classic IF systems, the metallocene is contained in the organic liquid, and the diol and added base are present in an aqueous phase. Product yields varied from about 15% to over 60%, similar to that of analogous inverse IF polymerizations.

Squaric acid formed polymer **87** with titanocene dichloride.[172–175] IR and Raman spectra of the systems were very similar to the starting spectra with the exception of the formation of Ti-O and loss of —OH stretching bands.

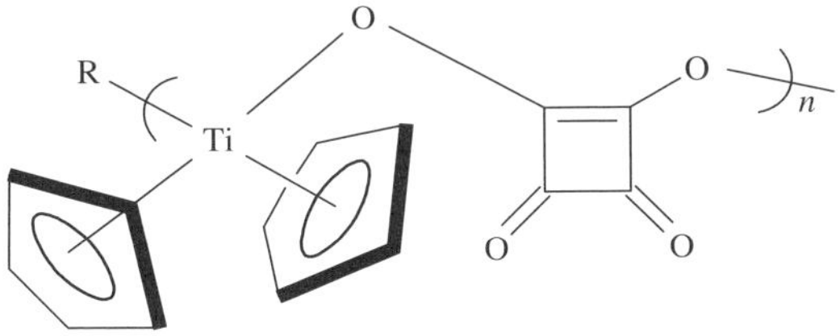

87

Titanocene polymers, **88**, were also synthesized from polyethylene glycol in hopes of producing water-soluble products.[168–171] Instead, the products were insoluble in all solvents. More recent work has shown that analogous organotin polyethers from poly(ethylene glycol) were initially insoluble, but on heating became soluble in a wide variety of liquids, including water.

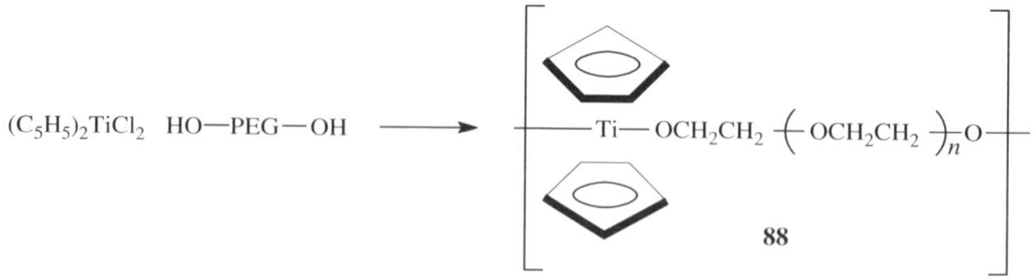

88

Insoluble

Titanocene polyferrocene ethers, **89**, were more recently synthesized from reaction from 1,1′-dihydroxymethylferrocene (HMF), 1,1′-bis(α-hydroxyisopropyl)ferrocene (HIF), and 1,1′-bis(α-hydroxyethyl)ferrocene (HEF), upon reaction with titanocene dichloride using the classic IF process.[171] The less sterically hindered HEF and HMF reactions were complete after 15 seconds, whereas for the more sterically hindered HIF, reaction continued to 30 seconds. Yields were 50 to 70%. The presence of inorganic base (sodium hydroxide) or organic base (triethylamine) was required. While yield varied, it was still decent, greater than 30%, for systems ranging from a ferrocene to titanocene ratio of 2:1 to 1:2. IF systems are far less sensitive than traditional polycondensation, due to the need to have a 1:1 mole ratio for high molecular-weight polymers to form (Crauthers equation). These Ti/Fe polymers retained ~50% of their original mass when heated to 900°C in air or nitrogen.

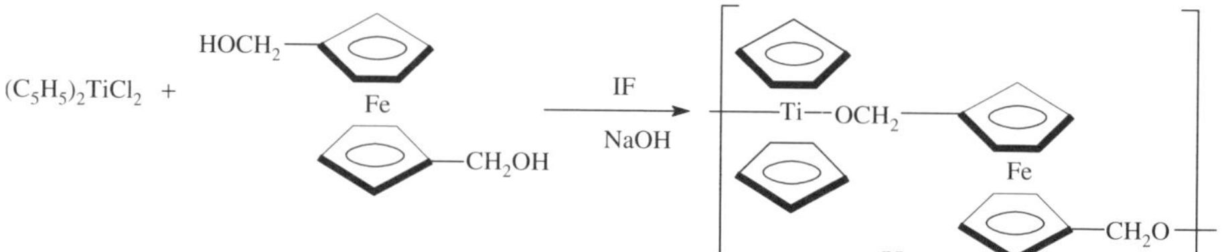

89

B. Polyesters

Analogous Group IVB–containing polyesters were synthesized using salts of both aliphatic or aromatic dicarboxylic acids employing both IF and aqueous-solution systems.[146,180–196] Weight-average molecular weights were in the range of 10^4 to 10^6. Over a short timespan (5 to 60 seconds), yield increases with time. The aqueous solution syntheses were faster than the interfacial syntheses. The reaction of diacid salts with titanocene dichloride are believed to proceed through the carboxylate anion reacting with $(C_5H_5)_2Ti^{+2}$ within the aqueous layer.[181,183] The rate followed pseudo-first-order kinetics in the condensation with disodium teraphthalate ($R = k[Cp_2TiCl_2]$).

The reaction between titanocene dichloride and salts of diacids were studied.[181,183] The reactive species in both the aqueous solution and IF systems are believed to be $R\text{-}COO_2^-$ and Cp_2Ti^{+2} with reaction occurring within the aqueous layer.[181,183] The condensation of titanocene dichloride with disodium terephthalate in the IF systems occurs via a pseudo-first-order reaction: $R = k[Cp_2TiCl_2]$

Diffusion of the titanocene dichloride into the aqueous layer and/or hydrolysis of the titanocene dichloride to its active form, Cp_2Ti^{+2}, is rate determining.[181,183]

Both ferrocene[188,189] and cobalticinum-containing[184,190] polyesters were generated using zirconocene dichloride the corresponding dicarboxylate (see structure **90**). Yield varies greatly with pH, consistent with the need to have the carboxylate salt. These polymers rapidly degrade at high pH. Product yield can be increased by addition of HCl to lower pH and sodium chloride to "salt out" the Lewis base.

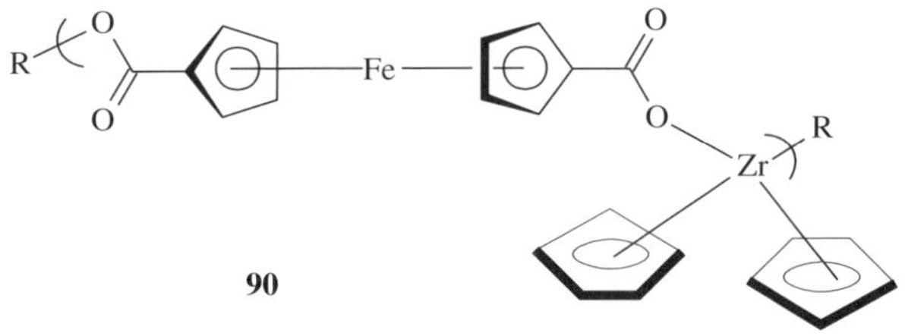

90

Cobalticinum-containing products, **91**, were synthesized utilizing 1,1′-dicarboxycobalticinium containing counteranions.[184,190] The results as a function of various reaction variables are consistent with other studies previously reported. Product yields increase as stirring rate increases to 15,700 rpm.

The hexafluorophosphate counterion gave much better yields compared with the bromide and nitrate counteranions.

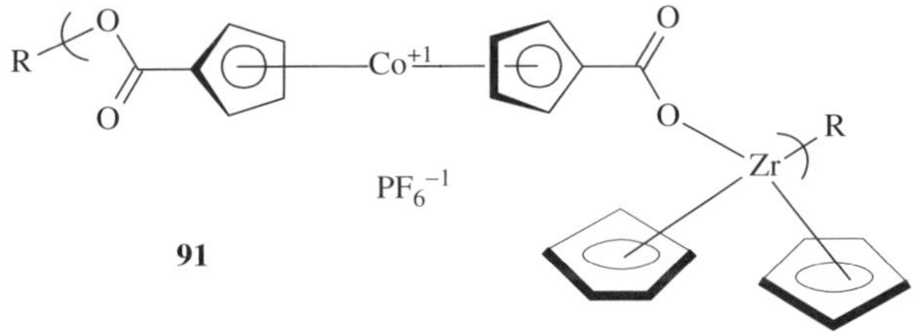

91

Product yields also decrease on the aging of the cobalticinum acid, apparently due to exchange of the hexafluorophosphate with another counteranion if the reaction is not conducted immediately after addition of water.

Many of the more rigid Group IVB polyesters form fibers or fibrils.[146] The products were often a mixture of these fibers and what appeared to be nonfibrous polymer. Fibers were examined under microscopy to 3200 magnification. The fibers are strong and flexible and retain some flexibility to over 1000°C. They have small "hooks" that come off the main shaft. The fibers are about 2×10^{-4} mm in diameter and reach 30 mm in length, giving an aspect ratio of 150,000. The fibrous materials go into a gel state but do not dissolve, whereas the nonfiber portion is soluble in dimethyl sulfoxide (DMSO). The IR spectra show little change on dissolution. Vanadocene and niobocene polyesters give such fibrils after physical agitation or drying. None of these fibers have yet been further studied or exploited.

C. Other Group IVB Products

Group IVB metallocene polythioethers[188,191–196] have been prepared from dithiols. Similar condensation polymers were constructed using both amidoximes and oximes.[197–210] Polyamines have been made from a variety of diamines.[211,212,213] The activation energy for addition of the reaction of Cp_2Ti^{+2} with diamines is about 10 kcal/mol (40 kJ/mol). Studies are consistent with the active diamine species being the un-ionized RNH_2. Increases in yield occurred as the amine to titanocene dichloride ratio increased. To achieve high product yields, relatively high concentrations of diamine should be employed. Cross-linked polyamines were also made using a variety of tetraamines.[214]

Vitamin K3, progesterone, and androstadienedione were converted to their dioximes. These dioximes were reacted with Group IVB metallocene dichlorides forming products with the general repeating unit **92**.[207,208] These polyoximes are soluble in dimethylformamide (DMF) and 1-methylimidazole. They exhibit degrees of polymerization from 20 to 600. Solubility is independent of chain length. Copolymers were synthesized to increase the solubility, but no regular trend was evident.

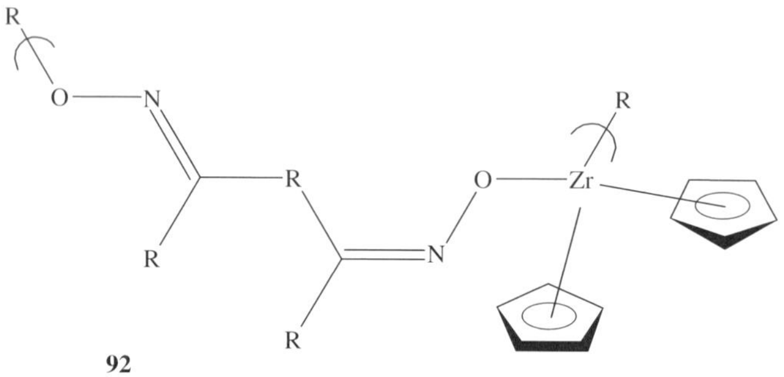

92

The electrical properties of a number of semiconductor polyoximes was studied.[209,210] These structures offer "whole-chain resonance," where the polymer backbone consists entirely of alternating double bonds with the exception of the metal atom. Values for D (dissipation factor) ranged from 3 to 10 (at 10^3 hertz), bulk resistivity from 0.05 to 0.6 (at 10^3 hertz), and bulk capacitance from 0.2 K to 0.9 K.

D. Mixed Functional Lewis Bases

New porphyrin-containing Group IVB polymers were made using ricinoleic acid[215] or hematoporphine IX.[216,217,218] Polymer **93** is a representative example that was used as a selective chelating agent. Macrocyclic "tetrapyrrole" has four methyl, two ethanolic, and two propionic acid side chains. The polymers are cross-linked because of the four reactive functions (two acid and two alcohol groups). The adsorption of Ni^{+2} by porphyrin moieties was observed for the titanium polymer.

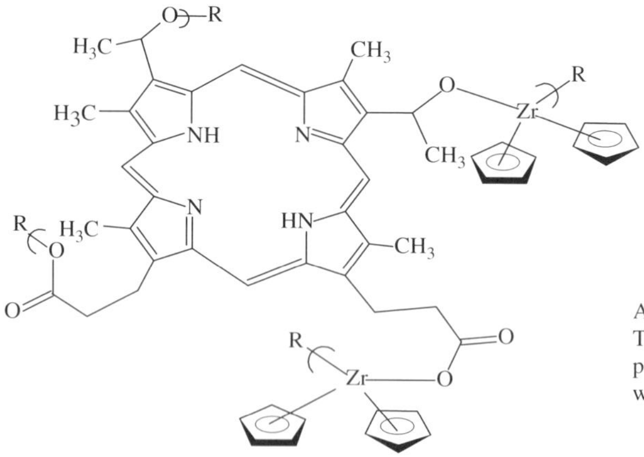

Also
Ti and Hf
polymers
were made.

93

E. Organotin and Lead Polymers

Early tin-modified polymers were produced by Carraher's group by reacting R_2SnCl_2 with polyacrylonitrile,[219,220] polyethyleneimine,[221] and poly(acrylic acid).[222] R_2SnCl_2 reactions with dialcohols, diamines, diacid salts, dithiols, urea, and thiourea generated such polymers as **94–99** shown below, together with early references from our laboratory.

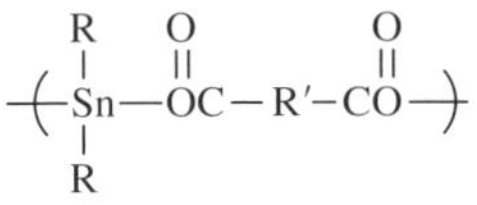

94[223,224,225]

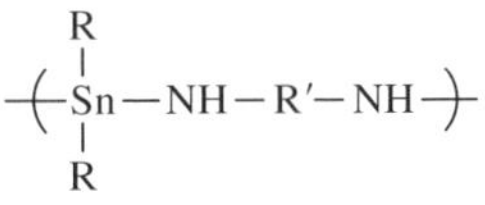

95[226, 227]

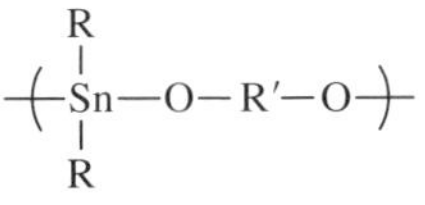

96[228]

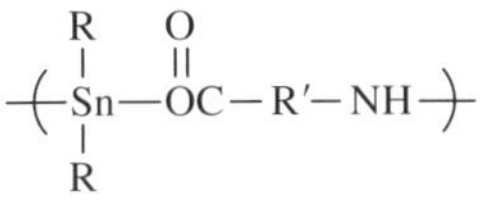

97[229]

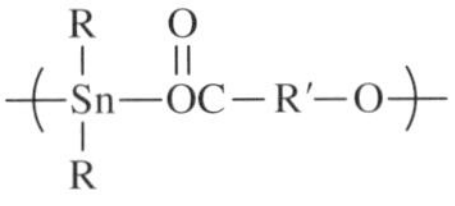

98[228, 230]

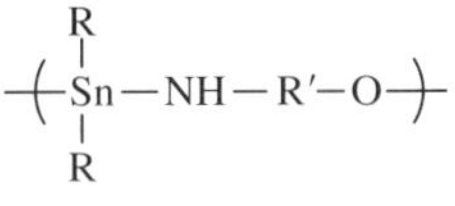

99[231]

Fluorescein has a carboxyl and phenol hydroxyl group. These groups were used to polycondense fluorescein with Me_2SnCl_2 via IF reactions to generate polydye, **100**.[230]

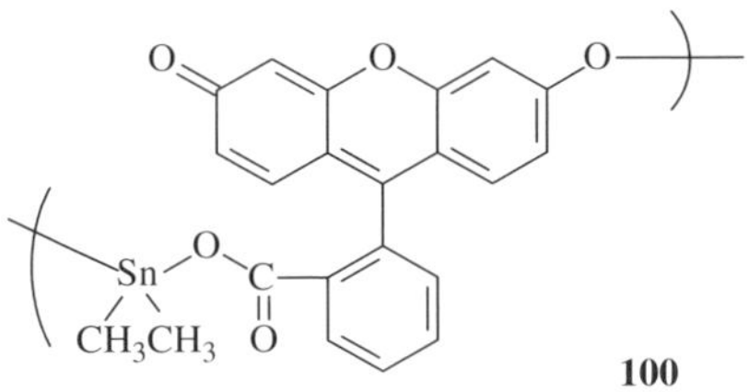

100

A more complete review of the entire field of organic tin polymers, including a more detailed list of the contributions from Carraher's laboratories, has recently appeared.[232]

Condensation of dicarboxylic acids and both diaryl- or dialkyllead dichlorides generated lead-containing polymers, **101**.[233,234,235] The IF polymerization method was used and higher yields resulted using the bis-carboxylate salts. The lead centers in these polymers can exist as tetrahedral or octahedral geometries. The octahedral lead centers result from further coordination of two carbonyl oxygens to the lead atom. IR studies demonstrated both lead structures exist within the polymers.

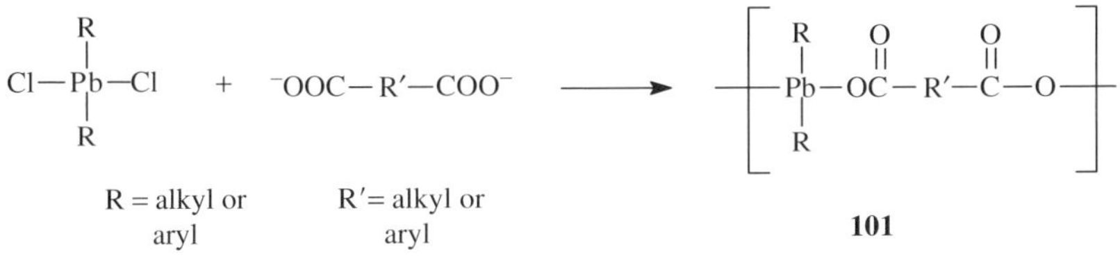

101

F. Ruthenium, Platinum, and Uranium Polymers

Ruthenium complexes are of interest in light harvesting and electron transfer. Therefore many groups have been incorporating ruthenium complexes into a variety of macromolecular systems (for a review, see reference 236). We have successfully prepared ruthenium-containing polythiols, **102**, and polyamines, **103**, from bpy ruthenium complexes.[237,238,239]

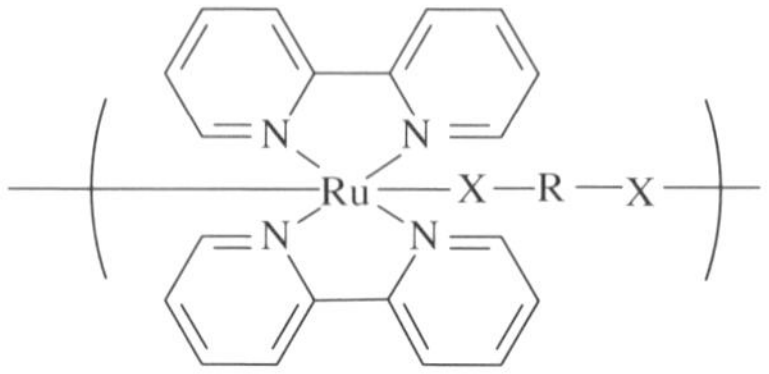

102 X = S[237,238]

103 X = NH_2[239]

Polymeric derivatives of *cis*-diaminodichloroplatinum have been prepared[240,241,242] because of the wide use of the anticancer monomeric drug *cis*-platinum. Polymeric platinum systems may function as medicinal agents. These polymers were prepared using diamines, substituted hydrazines, ureas, and thioureas and K_2PtCl_4. Structure **104** is representative of this coordination polymer class.

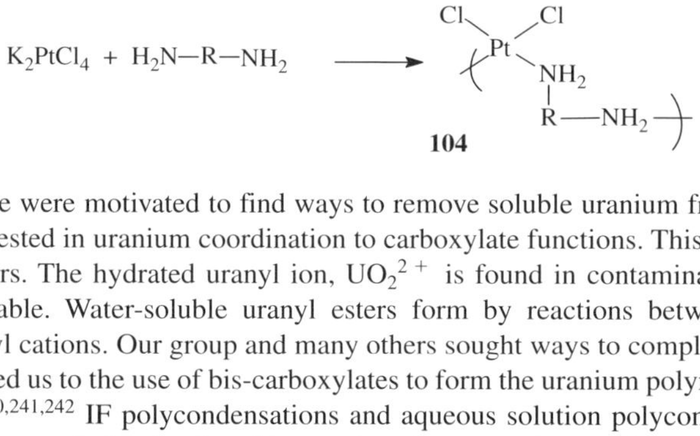

104

Originally, we were motivated to find ways to remove soluble uranium from water. Thus, we became interested in uranium coordination to carboxylate functions. This led to studies of uranium polymers. The hydrated uranyl ion, UO_2^{2+} is found in contaminated waters. It is exceptionally stable. Water-soluble uranyl esters form by reactions between carboxylate anions and uranyl cations. Our group and many others sought ways to complex uranyl ions to polymers. This led us to the use of bis-carboxylates to form the uranium polymers represented by **105** below.[240,241,242] IF polycondensations and aqueous solution polycondensations were used. Another approach involved making carboxylate complexes of uranyl ions with the sodium salt of poly(acrylic acid).[243]

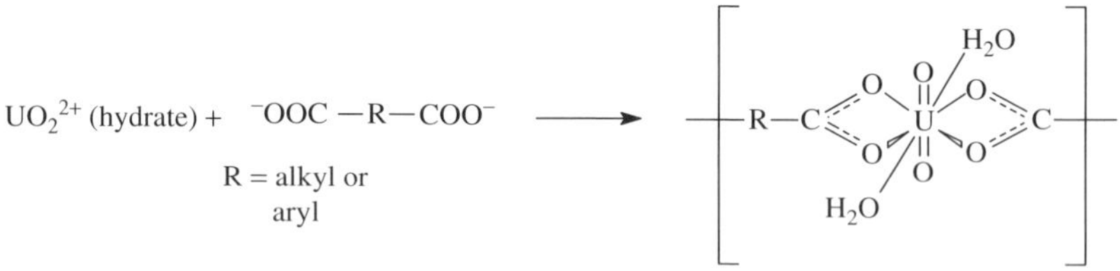

105

IV. RECENT DIRECTIONS

Interfacial polycondensations of organometal dihalides have now expanded the periodic table for polymer chemists. After these early years demonstrated this capability, the Carraher group has concentrated on applications of these polymers. Slow-release plant growth-hormone polymers were made, such as **106**, which contains kinetin in the backbone.[244] Others contained Gibberillic acid (GA 3).[245] These polymers have the potential to help re-seed portions of the Everglades using boat or aircraft casting of seeds plus slow-release growth hormone.

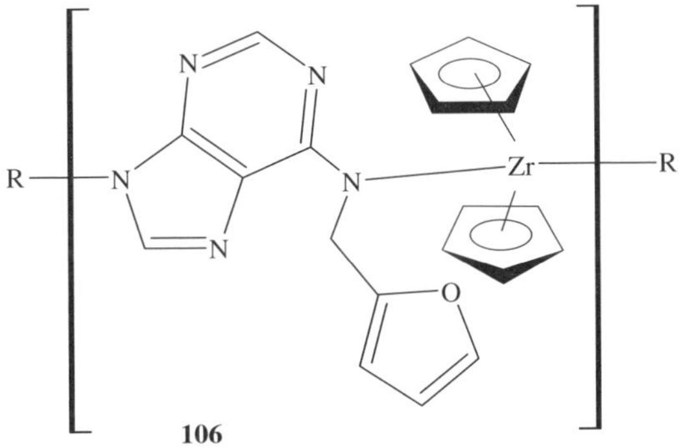

106

Polymeric organometallic drug candidates represent another current applied thrust. This topic has been recently reviewed.[246] Group IVB metallocene dichlorides polycondense with functional groups on the drug and may later hydrolyze to release the active agent. In most cases, the metallocene units act simply as carriers for eventual drug release. However, anticancer activity exists for the polymer formed from titanocene dichloride and cephalexin, **107**. Cephalexin polymers were formed with zirconocene and hafnacene dichlorides in yields of ~55%, but only ~5% with titanocene dichloride. The degree of polymerization (DP) ranged from ~70 with Hf to ~3,000 for Ti. Ciprofloxacin was also incorporated in **108** as part of an effort to make antibacterial agents.[247]

107

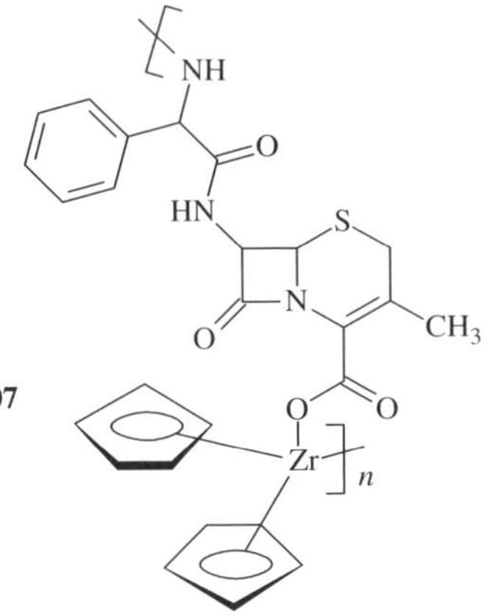

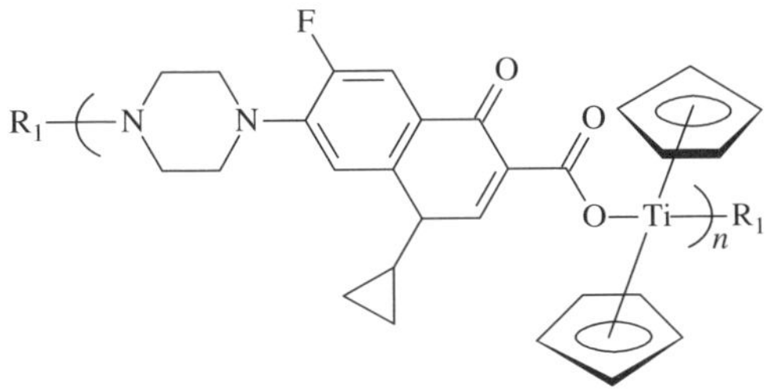

108

Vanadocene-containing polyesters and polyethers are now under active study.[248,249,250] Vanadocene dichloride has known anticancer activity, so polymers, such as **109**, will be evaluated.

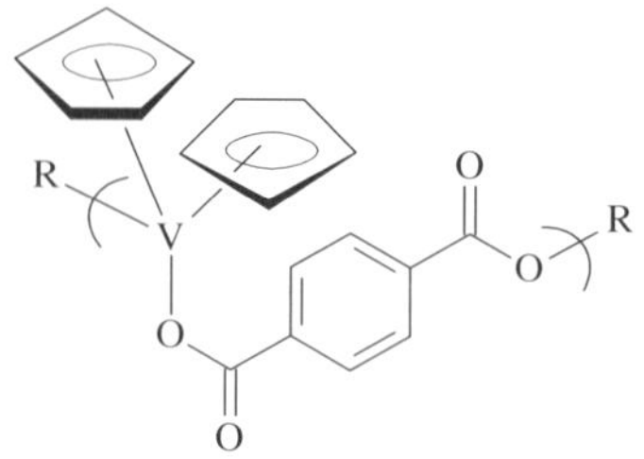

109

Polydyes have also figured prominently in our recent applications of Group IV B polymers.[251] Fluorescein, sulphonaphathalein, erythrosine B, azodye orange B, flavazine L, and Eriochrome black T have all been polycondensed with Group IV B dichlorometallocenes. The stability of these polydyes to photodegradation was studied using ultraviolet (UV) lamps and both argon ion and carbon dioxide lasers. The polydyes were more photostable than the original dyes. In the IR region, these polydyes protected materials they coated. This work and other contributions were recently reviewed.[252]

V. CLOSING REMARKS

This chapter has reviewed work from the early days of organometallic polymers as seen by two of the early contributors. The work of others from the 1950s to the 1970s can be found in various books[3,8,9,10,18,252–262] and reviews.[49,145,163,232,236,251,262–265] The discovery process has been fun. It was so much fun that it is a bit surprising to suddenly find ourselves being considered as the "old guard" in the field. The time passed so quickly! So many wonderful students contributed so much! Bailor, Marvel, Korshak, Rosenburg, etc., are gone. Nevertheless, from our perspective, the field is in good hands. The torch has been passed. Around the world talented and energetic chemists (including the editors of this text) are driving the field in a myriad of directions. An unprecedented growth is now occurring. Just as organometallic chemistry was merged with polymer chemistry in the late 1960s, organometallic polymer science is now blending with electronics, nanomaterials, ceramic sciences, thin-film physics, pharmaceutical sciences, coatings science, electrochemistry, and on and on. As interesting as our ride has been, we both believe the excitement has just begun. Some of that can be seen in the following chapters.

REFERENCES

1. E. Rochow, *Chem. Eng. News*, **23**, 612 (1945).
2. E. G. Rochow, *An Introduction to the Chemistry of Silicones*, Wiley, New York, 1947.
3. F. G. A Stone and W. A. G. Graham, *Inorganic Polymers*, Academic Press, New York, 1962.
4. C. D. West, *Z. Kristallogr.*, **90**, 555 (1935).
5. K. A. Jensen, *Z. Anorg. Chem.*, **252**, 227 (1944).
6. C. S. Marvel, J. H. Rassweiler, *J. Am. Chem. Soc.*, **80**, 1197 (1958).
7. W. C. Drinkard, J. C. Bailar, Jr., *J. Am. Chem. Soc.*, **81**, 4795 (1959).
8. A. Abd-El-Aziz, C. E. Carraher, Jr., C. U. Pittman, Jr., and M. Zeldin (eds.), *Macromolecules Containing Metal and Metal-Like Elements*, Vol. 5: *Metal-Coordination Polymers*, Wiley Interscience, New York, 2005.

9. R. Archer, *Inorganic and Organometallic Polymers*, Wiley, New York, 2001.
10. A. Abd-El-Aziz, C. E. Carraher, C. U. Pittman, Jr., J. Sheats, M. Zeldin, *Macromolecules Containing Metal and Metal-Like Elements*, Vol. 1: *A Half Century of Metal-Containing Polymers*, Wiley, New York (2003).
11. T. J. Kealy, P. L. Pauson, *Nature*, **168**, 1039 (1951).
12. G. Wilkinson, M. Rosenblum, H. C. Whiting, R. B. Woodward, *J. Am. Chem. Soc.*, **74**, 2125 (1952).
13. R. B. Woodward, M. Rosenblum, H. C. Whiting, *J. Am. Chem. Soc.*, **74**, 3458 (1952).
14. R. Dagani, "Fifty Years of Ferrocene Chemistry," *Chem. Eng. News*, **79**(49), 37–38 (Dec. 3, 2001); also see "50th Anniversary of the Discovery of Ferrocene," R. D. Adams, Ed., special edition of *J. Organometal. Chem.*, **637–639**, 1–875 (2001).
15. F. S. Arimoto, A. C. Haven, Jr., *J. Am. Chem. Soc.*, **77,** 6295 (1955).
16. A. C. Haven, Jr., US Patent 2,821,512 (January 28, 1958).
17. *J. Inorg. Organometal. Polym. Mater.*, **15**(1), 1 (March 2005).
18. E. W. Neuse, H. Rosenberg, *Metallocene Polymers*, Chapter 2, Marcel Dekker, New York, 1970; also see N. Bilow, A. L. Landis, H. Rosenberg, *J. Polym. Sci., Part A-1*, **7,** 2719 (1969).
19. V. V. Korshak, S. L. Sosin, "Synthesis and Polymerization of Some Ferrocene Derivatives," in *Organometallic Polymers*, C. E. Carraher, Jr., J. E. Sheats, and C. U. Pittman, Jr., Eds, pp. 25–38 and references therein, Academic Press, New York, 1978.
20. C. U. Pittman, Jr., *J. Paint Technol.*, **39**(513), 585 (1967). (This journal is now the *J. Coatings Technol.*)
21. C. U. Pittman, Jr., J. C. Lai, D. P. Vanderpool, "Polymers of Ferrocenylmethyl Acrylate and Ferrocenylmethyl Methacrylate and Their Ferricinium Salts," in *Polymer Characterization: Interdisciplinary Approaches*, C. D. Craver, Ed., pp. 97–124, Plenum Press, New York, 1971.
22. F. Rodriguez, *Principles of Polymer Systems*, Second Edition, McGraw-Hill, New York, 1970 and 1982, p. 107.
23. C. U. Pittman, Jr., *Polym. Lett.*, **6**, 19 (1968).
24. H. Benoit, Z. Grubisic, R. Rempp, *J. Polym. Sci., Part B*, **5,** 753 (1967).
25. C. U. Pittman, Jr., J. C. Lai, D. P. Vanderpool, M. Good, R. Prado, *Macromolecules*, **3**, 746 (1970).
26. D. O. Cowan, F. Kaufman, *J. Am. Chem. Soc.*, **92**, 219 (1970).
27. L. H. Caveny, C. U. Pittman, Jr., *AIAA J.*, **6**(8), 1461 (1968).
28. C. U. Pittman, Jr., *AIAA J.*, **7**(2), 328 (1969).
29. (a) C. U. Pittman, Jr., J. A. Eikenberry, *J. Ala. Acad. Sci.*, **38** (1967); (b) C. U. Pittman, Jr., *Polymer Lett.*, **6,** 18 (1968).
30. (a) C. U. Pittman, Jr., *J. Polym. Sci., Part A-1*, **5**, 2927 (1967); (b) C. U. Pittman, Jr., *J. Polym. Sci., Part A-1*, **6**, 1687 (1968).
31. J. C. Lai, T. Rounsefell, C. U. Pittman, Jr., *J. Polym. Sci., Part A-1*, **9**, 651 (1971).
32. C. U. Pittman, Jr., J. C. Lai, D. P. Vanderpool, *Macromolecules*, **3**(1), 105 (1970).
33. C. U. Pittman, Jr., J. C. Lai, D. P. Vanderpool, M. Good, R. Prados, *Macromolecules*, **3**(6), 746 (1970).
34. J. C. Lai, T. D. Rounsefell, C. U. Pittman, Jr., *Macromolecules*, **4**(2), 155 (1971).
35. C. U. Pittman, Jr., R. L. Voges, J. Elder, *Polymer Lett.*, **9**, 191 (1971).
36. C. U. Pittman, Jr., R. L. Voges, W. R. Jones, *Macromolecules*, **4**(3), 291 (1971).
37. C. U. Pittman, Jr., R. L. Voges, W. R. Jones, *Macromolecules*, **4**(3), 298 (1971).
38. Y. Sasaki, L. L. Walker, E. L. Hurst, C. U. Pittman, Jr., *J. Polym. Sci., Polym. Chem. Ed.*, **11,** 1213 (1973).
39. M. George, G. Hayes, *Polymer*, **15**, 397 (1974).

40. M. George, G. Hayes, *J. Polym. Sci. Polym. Chem. Ed.*, **13**, 1049 (1975).
41. M. George, G. Hayes, *J. Polym. Sci. Polym. Chem. Ed.*, **14**, 475 (1976).
42. C. C. Price, *J. Polym. Sci.*, **3,** 772 (1948).
43. T. Alfrey, Jr., J. J. Bohrer, H. Mark, *Copolymerization*, Vol. VIII: *High Polymers*, Interscience Publishers, New York, 1952.
44. F. R. Mayo, F. M. Lewis, *J. Am. Chem. Soc.*, **66**, 1954, (1944).
45. D. R. Montgomery, C. E. Fry, *J. Polym. Sci., Part C*, **25**, 59 (1968).
46. C. U. Pittman, Jr., M. D. Rausch, *Pure Appl. Chem.*, **58**(4), 617 (1986), and references therein.
47. C. U. Pittman, Jr., T. D. Rounsefell, *Macromolecules*, **9**(6), 936 (1976).
48. C. U. Pittman, Jr., Y. Sasaki, *Chemistry Lett.* (Japan), 383 (1975).
49. C. U. Pittman, Jr., *Chem. Technol.* **1**, 416 (1971).
50. O. E. Ayers, S. P. McManus, C. U. Pittman, Jr., *J. Polym. Sci., Polym. Chem. Ed.*, **11**, 1201 (1973).
51. C. U. Pittman, Jr., J. C. Lai, D. P. Vanderpool, M. Good, R. Prados, "Polymers of Ferrocenylmethyl Acylate and Ferrocenylmethyl Methacrylate and Ferricinium Salts," in *Polymer Characterization: Interdisciplinary Approaches*, C. D. Carver, Ed., pp. 97–124, Plenum Press, New York, (1971).
52. (a) C. U. Pittman, Jr., G. V. Marlin, T. D. Rounsefell, *Macromolecules*, **6**(1), 1 (1973); (b) C. U. Pittman, Jr., P. L. Grube, *J. Polym. Sci., Part A-1*, **9**, 3175 (1971); (c) C. U. Pittman, Jr., P. L. Grube, *J. Appl. Polym. Sci.*, **18**, 2269 (1974).
53. C. U. Pittman, Jr., C.-C. Lin, T. D. Rounsefell, *Macromolecules*, **11**(5), 1022 (1978).
54. C. U. Pittman, Jr., T. V. Jayaraman, R. D. Priester, Jr., M. D. Rausch, D. W. Macomber, W. P. Hart, *Polymer Preprints*, **23**(2), 73 (1982).
55. C. U. Pittman, Jr., J. Saad, J. Chapman, J. C. Lai, T. D. Rounsefell, Joint Southeast-Southwest Regional Meeting of the American Chem. Soc., New Orleans, LA, Dec. 2–4, 1970, papers 627, 664, and 666; see *Abstracts*, pp. 187, 199–200.
56. C. U. Pittman, Jr., T. D. Rounsefell, E. A. Lewis, J. E. Sheats, B. H. Edwards, M. D. Rausch, E. A. Mintz, *Macromolecules*, **11**, 560 (1978).
57. (a) D. W. Macomber, W. P. Hart, M. D. Rausch, R. D. Priester, Jr., C. U. Pittman, Jr., *J. Am. Chem. Soc.*, **104,** 884 (1982); (b) D. W. Macomber, W. P. Hart, M. D. Rausch, *J. Am. Chem. Soc.*, **104**, 884 (1982).
58. M. D. Rausch, D. W. Macomber, K. Gonsalves, F. G. Fang, Z.-R. Lin, C. U. Pittman, Jr., "The Synthesis and Polymerization of Vinyl Organometallic Monomers," in *Metal-Containing Polymeric Systems*, J. E. Sheats, C. E. Carraher, C. U. Pittman, Jr., Eds., pp. 43–57. Plenum Publishing, New York, 1985.
59. D. W. Macomber, W. P. Hart, M. D. Rausch, R. D. Priester, C. U. Pittman, Jr. *J. Organometallic Chem.*, **205**, 353 (1981).
60. M. D. Rausch, D. W. Macomber, C. U. Pittman, Jr., R. D. Priester, Jr., and T. V. Jayaraman, *J. Organometal. Chem.*, **205**, 353 (1981).
61. C. U. Pittman, Jr., T. V. Jayaraman, R. D. Priester, Jr., S. Spenser, M. D. Rausch, D. W. Macomber, *Macromolecules*, **14**, 237 (1981).
62. C. U. Pittman, Jr., R. D. Priester, Jr., T. V. Jayaraman, *J. Polym. Sci., Polym. Chem. Ed.*, **19**, 3351 (1981).
63. C. U. Pittman, Jr., T. D. Rounsefell, J. E. Sheats, B. H. Edwards, M. D. Rausch, E. A. Mintz, *Organometallic Polymers*, Academic Press, New York, 1978, pp. 67–78.
64. E. A. Mintz, M. D. Rausch, B. H. Edwards, J. E. Sheats, T. D. Rounsefell, C. U. Pittman, Jr., *J. Organometal. Chem.*, **137**, 199 (1977).
65. C. U. Pittman, Jr., R. L. Voges, J. Elder, *Macromolecules*, **4**(3), 302 (1971).
66. C. U. Pittman, Jr., *J. Paint Technol.*, **43**(561), 29 (1971).
67. C. U. Pittman, Jr., P. L. Grube, O. E. Ayers, S. P. McManus, M. D. Rausch, G. A. Moser, *J. Polymer Sci., Part A-1*, **10**, 379 (1972).

68. C. U. Pittman, Jr., R. L. Voges, "Polymerization of π-(Benzyl acrylate)-Chromium Tricarbonyl," in *Macromolecular Syntheses*, Vol. 4, W. J. Bailey, Ed., p. 175, Wiley, New York, 1972.
69. C. U. Pittman, Jr., G. V. Martin, *J. Polym. Sci., Polym. Chem. Ed.*, **11**, 2753 (1973).
70. C. U. Pittman, Jr., O. E. Ayers, S. P. McManus, *J. Macromol. Sci. Chem.*, **A7**(8), 1563 (1973).
71. C. U. Pittman, Jr., *Macromolecules*, **7**(3), 396 (1974).
72. C. U. Pittman, Jr., P. L. Grube, R. M. Hanes, *J. Paint Technol.*, **46**(597), 35 (1974).
73. C. U. Pittman, Jr., O. E. Ayers, S. P. McManus, *Macromolecules*, **7**(6), 737 (1974).
74. J. Kozikowski, M. Cais, US Patent 3,290,337 (1966).
75. C. U. Pittman, Jr., B. T. Kim, W. M. Douglas, *J. Org. Chem.*, **40**, 590 (1975).
76. M. D. Rausch, C. U. Pittman, Jr., in *New Monomers and Polymers*, B. M. Culbertson, C. U. Pittman, Jr., Eds., pp. 243–267, Plenum Publishing, New York, 1984.
77. D. W. Macomber, M. D. Rausch, *J. Organometal. Chem.*, **205**, 311 (1983).
78. M. D. Rausch, D. W. Macomber, K. Gonsalves, F. G. Fang, Z.-R. Lin, C. U. Pittman, Jr., in *Metal-Containing Polymeric Systems*, J. E. Sheats, C. E. Carraher, Jr., C. U. Pittman, Jr., Eds., pp. 43–57, Plenum Publishing, New York, 1985.
79. M. D. Rausch, D. W. Macomber, K. Gonsalves, F. C. Fan, Z. R. Lin, C. U. Pittman, Jr., *Polymeric Materials: Sci. Eng. Preprints* (ACS), **49**, 358 (1983).
80. K. Hafner, G. Schultz, K. Wagner, *Justus Liebigs Ann. Chem.*, **678**, 39 (1964).
81. W. P. Hart, D. W. Macomber, M. D. Rausch, *J. Am. Chem. Soc.*, **102**, 1196 (1980).
82. D. O. Cowan, F. Kaufman, *J. Am. Chem. Soc.*, **92**, 219 (1970).
83. D. O. Cowan and F. Kaufman, *J. Am. Chem. Soc.*, **92**, 6198 (1970).
84. D. O. Cowan, J. Park, C. U. Pittman, Jr., Y. Sasaki, T. K. Mukherjee, N. A. Diamond, *J. Am. Chem. Soc.*, **94**, 5110 (1972).
85. A. F. Garito, A. J. Heeger, *Accounts Chem. Res.*, **7**, 232 (1974).
86. C. U. Pittman, Jr., Y. Sasaki, P. L. Grube, *J. Macromol. Sci.-Chem.*, **A-8**(5), 923 (1974).
87. C. U. Pittman, Jr., Y. Sasaki, *Chem. Lett.* (Japan), 383 (1975).
88. U. T. Mueller-Westerhoff, P. Eilbracht, *J. Am. Chem. Soc.*, **94**, 9272 (1972).
89. D. O. Cowan, C. LeVanda, *J. Am. Chem. Soc.*, **94**, 9271 (1972).
90. C. U. Pittman, Jr., B. Surynarayanan, *J. Am. Chem. Soc.*, **96**, 7916 (1974).
91. C. U. Pittman, Jr., B. Surynarayanan, Y. Sasaki, "Mixed Valence, Semiconducting Ferrocene-Containing Polymers," in *Advances in Chemistry Series, Number 150: Inorganic Compounds with Unusual Properties*, R. B. King, Ed., pp. 46–55, ACS Publishers, Washington DC, 1976.
92. C.U. Pittman, Jr., O.E. Ayers, B. Suryanarayanan, S.P. McManus, J.E. Sheats, *Die Makromolekulare Chemie*, **175**, 1427 (1974).
93. J. E. Sheats, T. C. Willis, *Org. Coat. Plast. Preprints*, **41**(2), 33 (1979).
94. J. E. Sheats, T. C. Willis, *J. Polym. Sci., Polym. Chem. Ed.*, **22**, 1077 (1984).
95. J. E. Sheats, F. Hessel, L. Tsarouhas, K. G. Podejko, T. Porter, L. B. Kool, R. L. Nolen, Jr., *Polymeric. Mat. Sci. Eng.*, **49**(2), 363 (1983).
96. S. P. McManus, W. J. Patterson, C. U. Pittman, Jr., *J. Polym. Sci. Polym. Chem. Ed.*, **13**, 1721 (1975).
97. (a) S. P.. McManus, W. J. Patterson, C. U. Pittman, Jr., *J. Polym. Sci., Polym. Chem. Ed.*, **12**, 825 (1974); (b) C. U. Pittman, Jr., W. J. Patterson, S. P. McManus, *J. Polym. Sci., Polym. Chem. Ed.*, **14**, 1715 (1976).
98. C. U. Pittman, Jr., W. J. Patterson, S. P. McManus, *J. Polym. Sci., Part A-1*, **9**, 3187 (1971).
99. W. J. Patterson, S. P. McManus, C. U. Pittman, Jr., *J. Polym. Sci., Polym. Chem. Ed.*, **12**, 837 (1974).
100. R. B. King, *Accounts Chem. Res.*, **3**, 417 (1970).
101. C. U. Pittman, Jr., and R. F. Felis, *J. Organometal. Chem.*, **72**, 389 (1974).

102. D. R. Tyler, "Mechanistic Aspects of the Photodegradation of Polymers Containing Metal-Metal Bonds Along Their Backbones," Chapter 4 in *Macromolecules Containing Metals and Metal-Like Elements*, Vol. 6, *Transition Metal Containing Polymers*, Eds., A. Abd-El-Aziz, C. E. Carraher, C. H. Pittman, Jr., and M. Zeldin, pp. 77–109, Wiley, Hoboken, NJ, 2006.
103. C. U. Pittman, Jr., R. F. Felis, *J. Organometal. Chem.*, **72**, 399 (1974).
104. C. U. Pittman, Jr., A Hirao, *J. Polym. Sci., Polym. Chem. Ed.*, **15**, 1677 (1977).
105. A. Hirao, C. U. Pittman, Jr., *Polymer Preprints, Japan*, **26**(4), 634 (1977).
106. C. U. Pittman, Jr., G. O. Evans, *Chemtech*, 560 (Sept. 1973).
107. G. O. Evans, C. U. Pittman, Jr., R. McMillan, R. T. Beach, R. Jones, *J. Organometal. Chem.*, **67**, 295 (1974).
108. C. U. Pittman, Jr., R. M. Hanes, *Ann. New York Acad. Sci.*, **239**, 76 (1974).
109. C. U. Pittman, Jr., L. R. Smith, *J. Am. Chem. Soc.*, **97**, 341 (1975).
110. C. U. Pittman, Jr., L. R. Smith, R. M. Hanes, *J. Am. Chem. Soc.*, **97**, 1742 (1975).
111. C. U. Pittman, Jr., L. R. Smith, *J. Am. Chem. Soc.*, **97**, 1749 (1975).
112. C. U. Pittman, Jr., L. R. Smith, "Sequential Multistep Reactions Catalyzed by Polymer-Bound Homogeneous Ni, Rh and Ru Catalysts," in *Organotransition-Metal Chemistry*, Y. Ishii, M. Tsutsui, Eds., pp. 143–156, Plenum Publishing, New York, 1975.
113. C. U. Pittman, Jr., L. R. Smith, S. E. Jacobson, "Sequential Multistep Reactions Catalyzed by Polymer-Anchored Homogeneous Ni, Rh, Ru and Ir Catalysts," in *Catalysis: Heterogeneous and Homogeneous*, B. Delmon and G. Jannes, Eds., pp. 393–406, Proceedings Internat. Symp. Relations between Heterogeneous and Homogeneous Catalytic Phenomena, Brussels, Belgium, Oct. 23–25, 1974, Elsevier Scientific, Amsterdam, 1975.
114. J. Haggin, *Chem. Eng. News*, 11–17 (Nov. 15, 1982).
115. *Science*, **194**, 1261–1263 (Dec. 17, 1976).
116. S. E. Jacobson, C. U. Pittman, Jr., *J. Chem. Soc. Chem. Commun.*, 187 (1975).
117. C. U. Pittman, Jr., S. E. Jacobson, H. Hiramoto, *J. Am. Chem. Soc.*, **97**, 4774 (1975).
118. C. U. Pittman, Jr., S. E. Jacobson, L. R. Smith, W. Clements, H. Hiramoto, "Polymer-Anchored Homogeneous Hydrogenation Catalysts and Their Use in Multistep Synthetic Reactions," in *Catalysis in Organic Syntheses*, P. N. Rylander, H. Greenfield, Eds., pp. 161–180, Academic Press, New York, 1976.
119. S. Jacobson, W. Clements, H. Hiramoto, C. U. Pittman, Jr., *J. Mol. Catal.*, **1**, 73 (1975/76).
120. C. U. Pittman, Jr., R. M. Hanes, *J. Am. Chem. Soc.*, **98**, 5402 (1976).
121. C. U. Pittman, Jr., S. K. Wuu, S. E. Jacobson, *J. Catalysis*, **44**, 87 (1976).
122. C. U. Pittman, Jr., R. M. Hanes, *J. Org. Chem.*, **42**, 1194 (1977).
123. C. U. Pittman, Jr., A. Hirao, C. Jones, R. M. Hanes, Q. Ng, *Ann. New York Acad. Sci.*, **295**, 15–35 (Nov. 25, 1977).
124. C. U. Pittman, Jr., *Polymer News*, **4**(1), 5–15 (1977).
125. C. U. Pittman, Jr., *J. Mol. Catal.*, **3**, 293 (1977/78).
126. C. U. Pittman, Jr., A. Hirao, *J. Org. Chem.*, **43**, 640 (1978).
127. C. U. Pittman, Jr., Q. Ng, *J. Org. Chem.*, **153**, 85 (1978).
128. C. U. Pittman, Jr., C.-C. Lin, *J. Org. Chem.*, **43**, 4928 (1978).
129. C. U. Pittman, Jr., Q. Ng, A. Hirao, W. Honnick, R. Hanes, in *Colloques Internat. du Centre National de la Recherche Scientifique: No 281, Relations Entre Catalyse Homogéne Et Catalyse Hétérogene*, pp. 4–100, Du Centre National De La Recherche Scientifique, Lyon, France, Nov. 3–6, 1977.
130. R. D. Sanner, R. G. Austin, M. S. Wrighton, W. D. Honnick, C. U. Pittman, Jr., *Inorg. Chem.*, **18**, 928 (1979).
131. C. U. Pittman, Jr., W. D. Honnick, M. S. Wrighton, R. D. Sanner, R. G. Austin, "Photogeneration of Polymer-Anchored Catalytic Species from Iron Carbonyls," in *Fundamental Research in Homogeneous Catalysis*, Vol. 3, M. Tsutsui, Ed., pp. 603–619, Plenum Publishing, New York, 1979.

132. C. U. Pittman, Jr., and Q. Ng, US Patent 4,258,206 (March 24, 1981).

133. C. U. Pittman, Jr., Y. F. Liang, *J. Org. Chem.*, **45**, 5048 (1980).

134. C. U. Pittman, Jr., W. D. Honnick, J. J. Yang, *J. Org. Chem.*, **45**, 684 (1980).

135. C. U. Pittman, Jr., G. Wilemon, *Ann. New York Acad. Sci.*, **333**, 67 (1980).

136. C. U. Pittman, Jr., "Catalysis by Polymer-Supported Transition Metal Complexes," Chapter 5 in *Polymer-Supported Reactions in Organic Synthesis*, P. Hodge, D. C. Sherrington, Eds., pp. 249–291, Wiley, New York, 1980.

137. R. D. Sanner, R. G. Austin, M. S. Wrighton, W. D. Honnick, C. U. Pittman, Jr., "Photoactivation of Polymer-Anchored Catalysts. Iron Carbonyl Catalyzed Reactions of Alkenes," Chapter 2 in *Interfacial Photoprocesses: Energy Conversion and Synthesis*, M. S. Wrighton, Ed., pp. 13–26, Advances in Chemistry Series 184, ACS Publishers, 1980.

138. C. U. Pittman, Jr., W. D. Honnick, *J. Org. Chem.*, **45**, 2132 (1980).

139. C. U. Pittman, Jr., G. M. Wilemon, Q. Y. Ng, L. I. Flowers, *Polymer Preprints*, **22**(1), 153 (1981).

140. Y. Kawabata, C. U. Pittman, Jr., R. Kobayashi, *J. Mol. Catal.*, **12**, 113 (1981).

141. C. U. Pittman, Jr., G. Wilemon, *J. Org. Chem.*, **46**, 1901 (1981).

142. C. U. Pittman, Jr., W. D. Honnick, in *Catalysis of Organic Reactions*, W. R. Moser, Ed., pp. 353–380, Marcel Dekker, New York, 1981.

143. C. U. Pittman, Jr., Y. Kawabata, L. I. Flowers, *J. Chem. Soc. Chem. Commun.*, 473 (1982).

144. G. Consiglio, P. Pino, L. I. Flowers, C. U. Pittman, Jr., *J. Chem. Soc. Chem. Commun.*, 612 (1983).

145. C. U. Pittman, Jr., "Polymer-Supported Catalysts," Chapter 55 in *Comprehensive Organometallic Chemistry*, Vol. 8, G. Wilkinson, F. G. A. Stone, E. W. Abel, Eds., pp. 553–611, Pergamon Press, Oxford, 1982.

146. C. Carraher, Jr., *Chem. Tech*, 741 (1972).

147. C. Carraher, Jr., P. Lessek, *Eur. Polym. J.*, **8**, 1339 (1972).

148. C. Carraher, Jr., "Modification of Polymers," in *Reactions on Polymers*, J. Moore, Ed., pp. 1–12, Plenum Publishing, New York, 1972.

149. C. Carraher, Jr., J. Persma, *J. Macromol. Sci.*, **A7**, 913 (1973).

150. C. Carraher, Jr., L. Wang, *Angew. Makromolekular Chemie*, **25**, 121 (1972).

151. C. Carraher, Jr., R. Nordin, *J. Polym. Sci., Part A-1*, **10**, 3367 (1972).

152. C. Carraher, Jr., *Eur. Polym. J.*, **8**, 215 (1972).

153. C. Carraher, Jr., *Makromolekulare Chemie*, **166**, 31 (1973).

154. S. Migdal, D. Gether, A. Zilkha, *J. Organometal. Chem.*, **11**, 441 (1968).

155. M. Frankel, G. Gerther, D. Wagner, and A. Zilkha, *J. Organomet. Chem.*, **9**, 83 (1967).

156. M. Frankel, G. Gerther, D. Wagner, A. Zilkha, *J. Appl. Polym. Sci.*, **9**, 3383 (1965).

157. C. E. Carraher, Jr., *J. Polym. Sci., Part A-1*, **7**, 2351 (1969).

158. C. E. Carraher, Jr., *J. Polym. Sci., Part A-1*, **7,** 2359 (1969).

159. P. W. Morgan, *Condensation Polymers by Interfacial and Solution Methods*, Interscience, New York, 1965.

160. F. Millich, C. Carraher, *Interfacial Synthesis*, Vol. I: *Fundamentals*, Marcel Dekker, New York, 1977.

161. F. Millich, C. Carraher, *Interfacial Synthesis,* Vol. II: *Polymer Applications and Technology*, Marcel Dekker, New York, 1977.

162. C. Carraher, J. Preston, *Interfacial Synthesis*, Vol. III: *Recent Advances*, Marcel Dekker, New York, 1982.

163. C. Pittman, C. Carraher, J. Reynolds, *Encyclopedia of Polymer Science and Engineering*, Second Edition, Vol. 10, Wiley, New York, 1987.

164. C. Carraher, S. Bajah, *Polym. J.* (Br.), **7**, 155 (1975).

165. C. Carraher, S. Bajah, *Polymer* (Br.), **14**, 42 (1973).

166. C. Carraher, S. Bajah, *Polymer* (Br.), **15**, 9 (1974).
167. C. Carraher, S. Bajah, *Org. Coat. Plast. Chem.*, **33**(1), 624 (1973).
168. C. Carraher, L. Reckleben, in "Synthesis and Structural Characterization of Titanocene-Containing Polyethers Based on Reaction with Ethylene Oxide-Containing Diols, Including Poly(ethylene glycol)," *Polymer Modification*, G. Swift, C. Carraher, C. Bowman, Eds., p. 171–177, Plenum Publishing, New York, 1997.
169. C. Carraher, L. Reckleben, *Polym. Mater. Sci. Eng.*, **75**, 184 (1996).
170. C. Carraher, L. Reckleben, *Polym. Mater. Sci. Eng.*, **69**, 314 (1993).
171. C. Carraher, G. Burrish, *J. Macromol. Sci.-Chem.*, **A10**(8), 1457 (1976).
172. M. Williams, C. Carraher, F. Medina, M. Aloi, "Comparative Raman and Infrared Vibrational Study of the Polymer Derived from Titanocene Dichloride and Squaric Acid," in *Structure-Property Relations in Polymers*, M. Urban, C. Craver, Eds., p. 769–776, American Chemical Society, Washington, DC, 1993.
173. M. Williams, C. Carraher, "Comparative Infrared and Raman Spectroscopy of the Condensation Product of Squaric Acid and Bis(cyclopentadienyl)titanium Dichloride," in *Inorganic and Metal-Containing Polymeric Materials*, J. Sheats, C. Carraher, C. Pittman, Jr., M. Zeldin, B. Currell, Eds., pp. 295–318, Plenum Publishing, New York, 1995.
174. M. Williams, C. Carraher, F. Medina, M. Aloi, *Polym. Mater. Sci. Eng.*, **61**, 227 (1989).
175. M. Williams, C. Carraher, F. Medina, M. Aloi, *Polym. Mater. Sci. Eng.*, **64**, 6 (1991).
176. C. Carraher, L. Jambaya, *J. Macromol. Sci.-Chem.*, **A8**(7), 1249 (1974).
177. C. Carraher, L. Jambaya, *Org. Coat. Plast. Chem.*, **34**(2), 484 (1974).
178. C. Carraher, L. Jambaya, *Angew. Makromolekulare Chemie*, **39**, 69 (1974).
179. C. Carraher, L. Jambaya, *Angew. Makromolekular Chemie*, **52**, 111 (1976).
180. C. Carraher, *Org. Coat. Plast. Chem.*, **31**(2), 338 (1971).
181. C. Carraher, J. Lee, *Org. Coat. Plast. Chem.*, **34**(2), 478 (1974).
182. C. Carraher, J. Sheats, *Org. Coat. Plast. Chem.*, **33**(1), 634 (1973).
183. C. Carraher, J. Lee, *J. Macromol. Sci.-Chem.*, **A9**, 191 (1975).
184. C. Carraher, J. Sheats, *Makromolekulare, Chemie*, **166**, 23 (1973).
185. C. Carraher, *Europ. Polym. J.*, **8**, 215 (1972).
186. C. Carraher, *Angew. Makromolekular Chemie*, **28**, 145 (1973).
187. C. Carraher, L. Jambaya, S. Bajah, *Polymer Prep.*, **18**(2), 403 (1977).
188. C. Carraher, J. Reimer, *J. Polymer Sci., Polym. Chem. Ed.*, **10**, 3367 (1972).
189. C. Carraher, J. Reimer, *Polymer* (Br.), **13**, 153 (1972).
190. J. Sheats, C. Carraher, D. Bruyer, M. Cole, *Org. Coat. Plast. Chem.*, **34**(2), 474 (1972).
191. C. Carraher, R. Nordin, *Makromolekulare Chemie*, **164**, 87 (1973).
192. C. Carraher, R. Nordin, *J. Polym. Sci., Part A-1*, **10**, 521 (1972).
193. C. Carraher, R. Nordin, *J. Applied. Polym. Sci.*, **18**, 53 (1974).
194. C. Carraher, *Polymer* (Br.), **17**, 231 (1976).
195. C. Carraher, *Org. Coat. Plast. Chem.*, **33**(1), 629 (1973).
196. C. Carraher, R. Nordine, *J. Macromol. Sci.-Chem.*, **A15**(1), 143 (1981).
197. C. Carraher, L. Torre, *Org. Coat. Plast. Chem.*, **42**, 18 (1980).
198. C. Carraher, R. Frary, *Br. Polym. J.*, **6**, 255 (1974).
199. C. Carraher, R. Frary, *Makromolekulare Chemie*, **175**, 2307 (1974).
200. C. Carraher, L. Torre, H. M. Molloy, in *Interfacial Synthesis*, Vol. III: *Recent Advances*, Chapter 5, Marcel Dekker, New York, 1982.
201. C. Carraher, M. Christensen, *Angew. Makromolekulare Chemie*, **69**, 61 (1978).
202. C. Carraher, L. Torre, H. M. Molloy, *J. Macromol. Sci.-Chem.*, **A15**, 757 (1981).

203. C. Carraher, M. Christensen, *Polym. Prepr.*, **17**(2), 530 (1976).

204. C. Carraher, M. Christensen, *J. Macromol. Sci.-Chem.*, **A11**, 2021 (1977).

205. C. Carraher, R. Frary, *J. Polym. Sci., Polym. Chem. Ed.*, **12**, 799 (1974).

206. C. Carraher, R. Frary, *Org. Coat. Plast. Chem.*, **33**(2), 433 (1973).

207. C. Carraher, L. Torre, "Solution Behavior of Group IVB Polyoximes," Chapter 6 in *Macromolecular Solutions: Solvent-Property Relationships*, R. Seymour, G. Stahl, Eds., Pergamon, New York, pp. 61–69, 1982.

208. C. Carraher, L. Torre, *Org. Coat. Plast. Chem.*, **45**, 252 (1981).

209. C. Carrher, T. Manek, R. Linville, J. R. Taylor, L. Torre, W. Venable, *Org. Coat. Plast. Chem.*, **43**, 753 (1979),

210. C. Carraher, R. Linville, T. Manek, H. Blaxall, J. R. Taylor, L. Torre, "Electrical Properties of Group IVB Metallocene Polyoximes," in *Conductive Polymers*, R. Seymour, Ed., pp. 77–84, Plenum, Publishing, New York, 1981.

211. C. Carraher, S. Jorgensen, *Polym. Prepr.*, **16**(1), 671 (1975).

212. C. Carraher, P. Lessek, *Europ. Polym. J.*, **8**, 1339 (1972).

213. C. Carraher, S. Jorgensen, *J. Polym. Sci., Polym. Chem. Ed.*, **16**, 2965 (1978).

214. C. Carraher, R. Feiffer, P. Fullenkamp, *J. Macromol. Sci.-Chem.*, **A10**, 1221 (1976).

215. C. Carraher, E. Frankel, *Polym. Mater. Sci. Eng.*, **73**, 396 (1995).

216. C. Carraher, A. Rivalta, J. Haky, *Polym. Mater. Sci. Eng.*, **74**, 149 (1996).

217. C. Carraher, J. Haky, A. Rivalta, D. Sterling, *Polym. Mater. Sci. Eng.*, **70**, 329 (1993).

218. C. Carraher, J. W. Louda, D. Sterling, A. Rivalta, Q. Zhang, E. Baker, *Polym. Mater. Sci. Eng.*, **71**, 386 (1994).

219. C. Carraher, J. Piersma, L.-S. Wang, *Org. Coat. Plastics Chem.*, **31**, 254 (1971).

220. C. Carraher, L.-S. Wang, *Makromolekulare Chemie*, **54**, 119 (1976).

221. C. Carraher, M. Fedderson, *Angew. Makromolekulare Chemie*, **54**, 119 (1976).

222. C. Carraher, J. Piersma, *J. Appl. Polym. Sci.*, **16**, 1851 (1972).

223. C. Carraher, *Angew. Makromolekulare Chemie*, **31**, 115 (1973).

224. C. Carraher, R. Dammier, *Polym. Prepr.*, **11**, 606 (1970).

225. C. Carraher, R. Dammier, *J. Polym. Sci., Part A-1*, **8**, 3367 (1970).

226. C. Carraher, D. Winter, *Makromolekulare Chemie,* **141**, 237, 259 (1971).

227. C. Carraher, D. Winter, *Makromolekulare Chemie,* **152**, 55 (1972).

228. C. Carraher, R. Venkatachalam, T. Tiernan, M. Taylor, *Org. Coat. Appl. Polym. Sci.*, **47**, 119 (1982).

229. C. Carraher, G. Scherubel, *J. Polym. Sci., Part A-1*, **9**, 983 (1971).

230. C. Carraher, L. Tisinget, G. Solimine, M. Williams, S. Carraher, R. Strother, *Polym. Mater. Sci. Eng.*, **55**, 469 (1986).

231. C. Carraher, A. Taylor, F. Medina, R. Linville, E. Randolph, J. Kloss, D. Stevison, *Polym. Prepr.*, **34**, 166 (1993).

232. C. E. Carraher, "Organotin Polymers," Chapter 10 in *Macromolecules Containing Metals and Metal-Like Elements*, Vol. 4, *Group IV A Polymers*, A. Abd-El-Aziz, C. E. Carraher, Jr., C. U. Pittman, Jr., M. Zeldin, Eds., pp. 263–310, Wiley, Hoboken, NJ, 2005.

233. C. Carraher, C. Deremo Reese, *Org. Coat. Plastics* Chem., **37**, 162 (1977).

234. C. Carraher, C. Deremo Reese, *Angew. Makromolekulare Chemie*, **65**, 95 (1977).

235. C. Carraher, C. Deremo Reese, *J. Polym. Sci.*, **16**, 491 (1978).

236. C. Carraher, A. Murphy, "Ruthenium-Containing Polymers for Solar Energy Conversion," Chapter 13 in *Macromolecules Containing Metal and Metal-Like Elements*, Vol. 5, *Metal Coordination Polymers*, A. Abd-El-Aziz, C. E. Carraher, Jr., C. U. Pittman, Jr., M. Zeldin, Eds., pp. 325–342, Wiley, Hoboken, NJ, 2005.

237. C. Carraher, Q. Zhang, C. Parkanyi, *PMSE*, **71**, 398, 505 (1994).
238. C. Carraher, Q. Zhang, C. Parkanyi, *PMSE*, **73**, 398, (1995).
239. C. Carraher, Q. Zhang, "The Use of Ruthenium-Containing Polythiols for Solar Energy Conversion," in *Metal-Containing Polymeric Materials*, pp. 109–118, Plenum Publishing, New York, 1996.
240. C. Carraher, Jr., D. J. Giron, I. Lopez, D. R. Cerutis, W. J. Scott, *Org. Coat. Plastics Chem.*, **44**, 120 (1981).
241. C. Carraher, W. J. Scott, J. A. Schroeder, D. J. Giron, *J. Macromol. Sci.-Chem.*, **A15**(4), 625 (1981).
242. C. Carraher, *Org. Coat. Plastics Chem.*, **42**, 428 (1980).
243. C. Carraher, S. Tsuji, W. Field, J. Dinunzo, *Org. Coat. Appl. Polym. Sci.*, **46**, 254 (1982).
244. C. Carraher, H. Stewart, W. Soldani, J. delaTorre, B. Pandya, L. Reckleben, *Polym. Mater. Sci. Eng.*, **71**, 783 (1994).
245. C. Carraher, D. Chamely, *Polym. Mater. Sci. Eng.*, **85**, 358 (2001).
246. D. W. Siegmann-Louda, C. Carraher, "Polymeric Platnium-Containing Drugs in the Treatment of Cancer," Chapter 7 in *Macromolecules Containing Metal and Metal-Like Elements*, Vol. 3, *Biomedical Applications*, A. Abd-El-Aziz, C. E. Carraher, Jr., C. U. Pittman, Jr., M. Zeldin, Eds., pp. 119–192, Wiley, Hoboken, NJ, 2004.
247. C. Carraher, L. Lanz, *Polym. Mater. Sci. Eng.*, **81**, 243 (2002).
248. J. Peterson, C. Carraher, A. Salamone, A. Francis, *Polym. Mater. Sci. Eng.*, **81**, 149 (1999).
249. C. Carraher, E. Randolph, *Polym. Mater. Sci. Eng.*, **79**, 50 (1998).
250. D. Sigemann-Louda, C. Carraher, unpublished results.
251. C. E. Carraher, Jr., *J. Inorg. Organomet. Polym. Mater.*, **15**(1), 121 (2005).
252. "The 50th Anniversary of Metallocene-Based Polymers," a special issue of *J. Inorg. Organometal. Polym. Mater.*, **15**(1), 1–196 (2005).
253. C. E. Carraher, Jr., J. E. Sheats, C. U. Pittman, Jr. (eds.), *Organometallic Polymers*, Academic Press, New York, 1978.
254. C. E. Carraher, Jr., J. E. Sheats, C. U. Pittman, Jr. (eds.), *Advances in Organometallic and Inorganic Polymer Science*, Marcel Dekker, New York, 1982.
255. J. E. Sheats, C. E. Carraher, Jr., C. U. Pittman, Jr. (eds.), *Metal-Containing Polymeric Systems*, Plenum Press, New York, 1985.
256. J. E. Sheats, C. E. Carraher, Jr., C. U. Pittman, Jr., M. Zeldin, B. Currell (eds.), *Inorganic and Metal-Containing Polymeric Materials*, Plenum Press, New York, 1990.
257. C. U. Pittman, Jr., C. E. Carraher, Jr., M. Zeldin, J. E Sheats, B. M. Culbertson (eds.), *Metal-Containing Polymeric Materials*, Plenum Press, New York, 1996.
258. M. Zeldin, K. J. Wynne, H. R. Allcock (eds.), *Inorganic and Organometallic Polymers*, ACS Symp. Series, Vol. 360, ACS Publishers, Washington, DC, 1988.
259. K. A. Andrianov, *Metallorganic Polymers*, Wiley, New York, 1965.
260. A. D. Pomogailo, *Catalysis by Polymer Immobilized Metal Complexes*, Gordon and Breach Science Publishers, Amsterdam, 1998.
261. A. Abd-El-Aziz, C. E. Carraher, Jr., C. U. Pittman, Jr., M. Zeldin, Eds., *Macromolecules Containing Metal and Metal-Like Elements*, Volumes 1–7, Wiley, Hoboken, NJ. (This comprehensive series has appeared from 2003–2006.)
262. C. U. Pittman, Jr., "Vinyl Polymerization of Organic Monomers Containing Transition Metals," in *Organometallic Reactons*, Vol. 6, E. I. Becker, M. Tsutsui, Eds. pp. 1–62, Marcel Dekker, New York, 1977.
263. C. U. Pittman, Jr., C. E. Carraher, Jr., J. R. Reynolds, "Organometallic Polymers" in the *Encyclopedia of Polymer Science and Engineering*, Vol. 10, H. Mark, N. Bikales, C. Overberger, J. Menges, Eds., pp. 541–594, Wiley, New York, 1987.
264. C. E. Carraher, Jr., *Inorg. Macromol. Rev.*, **1**, 271 (1972).
265. A. Abd-El-Aziz, *Coord. Chem. Rev.*, **233–234**, 177 (2002).

CHAPTER 2

Recent Developments in Organometallic Polymers

ALAA S. ABD-EL-AZIZ AND PATRICK O. SHIPMAN

University of British Columbia Okanagan, Kelowna, British Columbia, Canada

I. INTRODUCTION

Since in early 1990s, organometallic compounds have been at the forefront of chemical research. While the start of the twentieth century showed vast developments in polymer chemistry, it was not until halfway through the decade that organometallic polymers really began to appear in the literature.[1,2,3] Shortly after the discovery of ferrocene in the 1950s, the first organometallic polymer was reported by Arimoto and Haven.[4] Since that time there have been significant developments in this field. There are a number of books and reviews that cover the early developments of this field.[5–33] While the first chapter of this book by Pittman and Carraher is dedicated to their early days of research in this field, this chapter focuses on the recent developments from the early 1990s to the present. The past 10 years have seen an exponential increase in the number of publications in this field, with topics as diverse as the synthesis of new macromolecules containing metallic moieties to the thorough investigations of the chemical and physical properties of these new and exciting materials. This chapter classifies these polymeric materials based on the bonding of the metals to the polymeric chain. The synthetic methodologies of the materials as well as their properties and applications are highlighted.

II. METALLIC MOIETIES IN THE POLYMER BACKBONE

A. Metals Directly Incorporated into the Polymer Chain

The research area of organometallic polymers containing transition metals σ–bonded in the polymer backbone is dominated by poly(metal acetylides).[34,35] Since their discovery in the 1970s,

Frontiers in Transition Metal-Containing Polymers, edited by Alaa S. Abd-El-Aziz and Ian Manners.

many researchers have described the incorporation of various transition metals into this class of materials. These organometallic polymers contain rigid rod structures, making them ideal for electrical and optical applications. The incorporation of transition metals into polymers containing triple bonds enhances the metal $d\pi$–$p\pi^*$ back bonding, which contributes to π-electron delocalization.[35]

i. *Pt, Pd, and Au Containing Rigid-Rod Polymers.* The 1990s was an exciting time for transition metal containing polymers. During this decade, Lewis and co-workers developed a new innovative method for the preparation of rigid rod transition metal containing polyynes. The reaction of bis-trimethylstannyl(acetylide) with $(PBu_3)_2MCl_2$ resulted in metal-containing polymers with molecular weights around 1×10^5.[36–47] Cross-linked polymers **5** and **4** in Scheme 2.1 were prepared using trialkyne-substituted derivative (1).[44] This particular type of platinum polyyne has been known to exhibit third-order optical nonlinearity.[48–51]

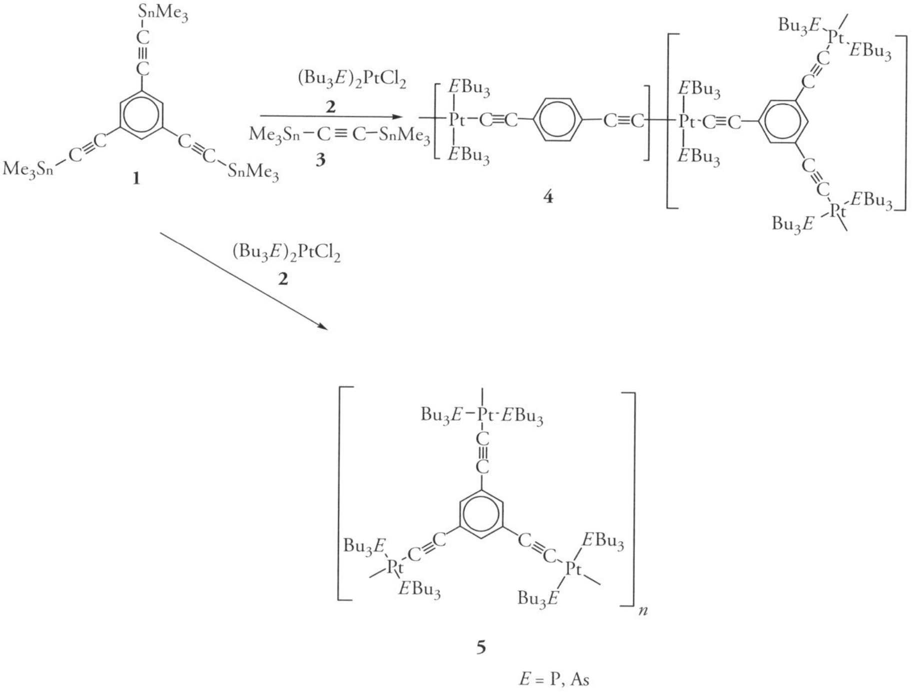

Scheme 2.1

Platinum and palladium acetylides containing cyclodiborazane moieties in the backbone have been prepared.[52] Hydroboration reactions between cyano-functionalized platinum- or palladium-containing monomers and tripylborane resulted in the isolation of polymers **6**. The weight-average molecular weights of the platinum- and palladium-containing polymers **6** were 9700 and 10,400, respectively. The polymers also displayed good solubility in organic solvents. Excitation of the polymers at 400 nm showed that the platinum-based polymer gave a weak fluorescence emission at 460 nm, whereas the palladium-based polymers showed a stronger fluorescence emission. The palladium-containing polymer also fluoresced at 530 nm when excited at 450 nm.[52]

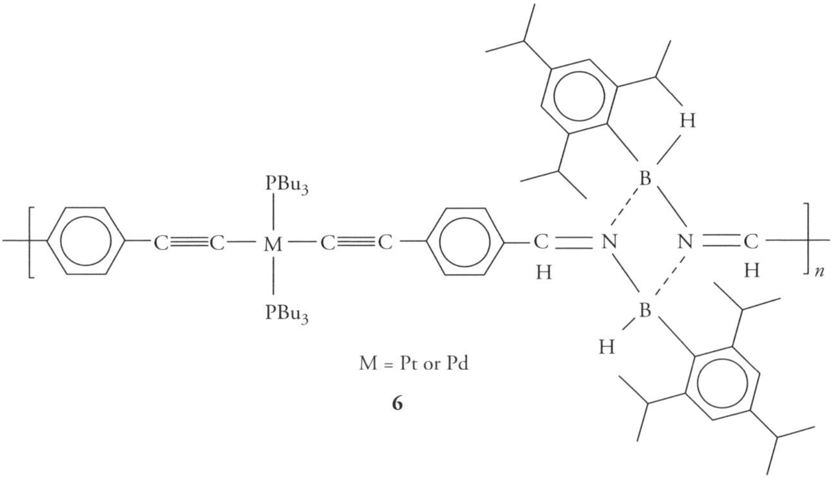

M = Pt or Pd

6

Vicente and co-workers have recently synthesized platinum(II) **7** and mixed platinum(II)/gold(I) σ-alkynyl polymers **8**.[53] The monomers were prepared via dehydrohalogenation between *cis*-[$PtCl_2(PR_3)_2$] and various alkynes. Polymerization occurred via the reaction of the platinum complex with PPN[$Au(acac)_2$], giving polymers with mixed metal backbones.

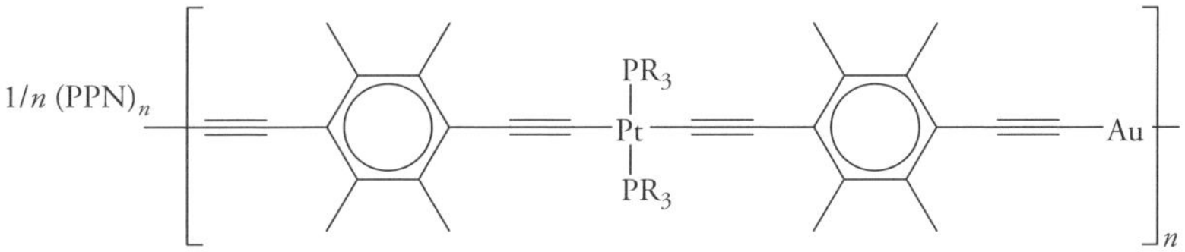

7

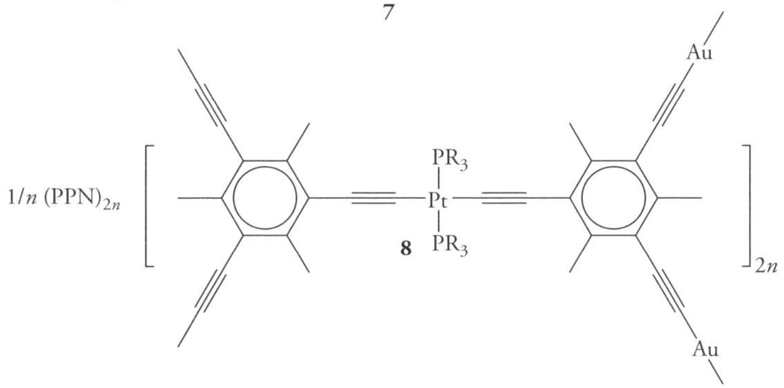

8

Rigid-rod organometallic polymers containing platinum **9** also have been reported by Fratoddi et al.[54] These polymers had the formula [PtL_2—(C≡C)—*p*—$C_6H_4(2,5\text{-}R)_2$—(C≡C)—]$_n$, where L═PPh_3, PBu_3 R═H, OC_4H_9, or $OC_{16}H_{33}$. Two synthetic procedures were compared for these syntheses: a dehydrohalogenation reaction and a modified Stille coupling reaction. Polymers of {[2,5-bis(ethynyl)benzene]-*trans*-bis(tributylphosphine) platinum(II)}(Pt-B-DEB$_{(n)}$) with approximately 80 repeating units were obtained from dehydrohalogenation reactions. An oligomeric material (poly-{[1,4-bis(ethynyl)-2,5-di(butoxy)benzene]-*trans*-bis(tributylphosphine) platinum(II)}) with approximately 12 repeating units was prepared from dehydrohalogenation reactions, whereas the modified Stille coupling only gave dinuclear and tetranuclear compounds. However, using a Pt-B-HDOB$_{(n)}$ monomer in the Stille coupling gave a polymeric material with approximately 20 repeating units. A number of polymers with varying chain lengths (2–7) were studied for absorption and emission optics. The π–π* absorption bands of the organometallic compounds were found to be red shifted in comparison to the organic monomers and were found to increases as the chain length increase from 360 nm to 390 nm.

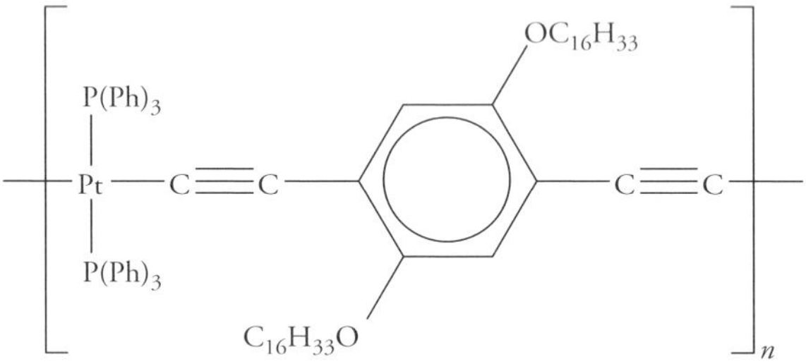

9

Pt-B-BOB$_n$

Fratoddi and co-workers synthesized organometallic polymers containing platinum(II) or palladium(II).[55] Polymers **10a**,**b** were prepared from the dehydrohalogenation reactions between bisphosphine dichloride complexes of Pt(II) and Pd(II) and 2,6-diethynyl-4-nitroaniline (DENA). The resulting polymers contained Pt(II) or Pd(II) in a square planar geometry in the polymer backbones and formed a helical pattern. Molecular weights for the polymers were 600 and 14,000 amu for the palladium and platinum polymers, respectively. Thermal studies showed that the platinum-containing polymer had higher thermal stability than the palladium polymer, with the first weight losses occurring at 267°C and 210°C, respectively.

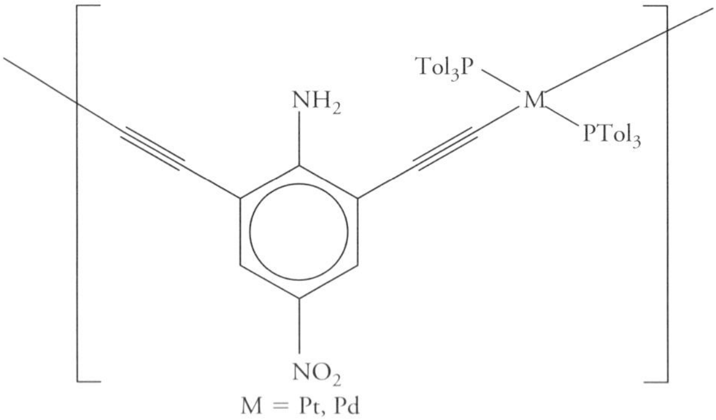

M = Pt, Pd

10a,b

Khan and co-workers reported the preparation of platinum-containing organometallic polyynes. The reaction of *trans*-[(^{n}Bu$_3$P)$_2$PtCl$_2$] with various conjugated bis-ethynyl ligands gave the organometallic polymers **11** in 85–90% yield.[56] These polymers possessed weight-averaged molecular weights between 28,000 and 41,000 with polydispersity indices (PDI)s between 1.8 and 1.9. The number of repeating units in the polymer greatly depended on the size of the spacers used. As the spacer size increased, the number of repeating units decreased due to steric effects caused by the orientation of the spacer. For example, the napthalene and anthracene spacers adopted an orientation orthogonal to the orientation of the platinum molecule. Optical studies showed that the transition between the ligand π and Pt $5d$-orbitals and the ligand π^* and Pt $6d$-orbitals shifted to lower energies as the size of the spacer increased. This shift is indicative of increased donor–acceptor behavior between the Pt and the ligand. The polyyne bands in the absorption spectra were red shifted compared to the analogous diynes; however, the red shift decreased as the size of the spacer increased.

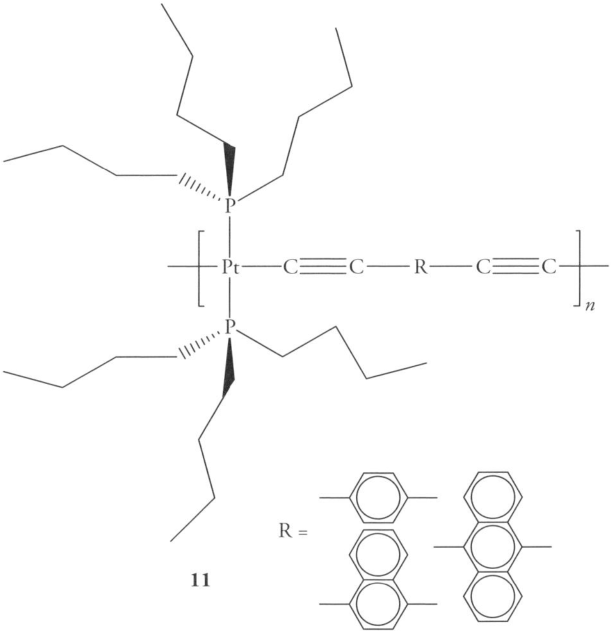

11

Low solubilities are a major problem that commonly occurs in the preparation of metal-containing acetylide polymers. Puddephatt and co-workers found that altering the phosphine ligands around the Au centers could control the solubility of gold acetylide polymers.[34,57,58,59] Polymer **14** possessed molecular weights estimated to range between 13,700 and 27,000.[23]

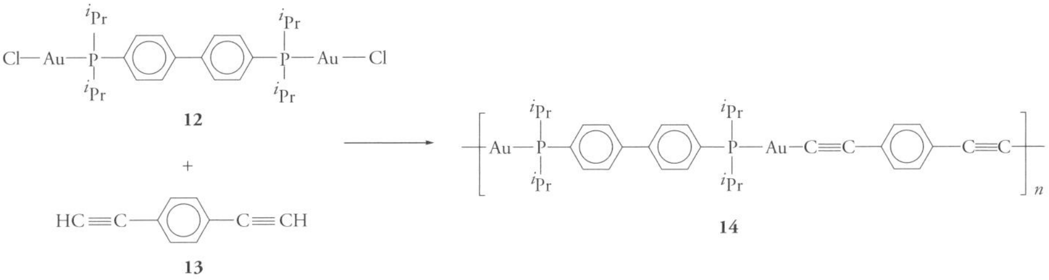

Over the past 10 years there has been a focus on the incorporation of thiophene and pyridine rings into acetylide polymers containing transition metals.[60–68] Kohler and co-workers have studied the effect that the conjugation level in transition metal containing polyacetylides has on photoluminescence.[69] Altering the conjugation of the spacer units in polymers **15–17** resulted in the alteration of the triplet energies. The triplet energies varied between 1.3 and 2.5 eV, and the polymers were found to possess lower quantum yields than their corresponding monomers.

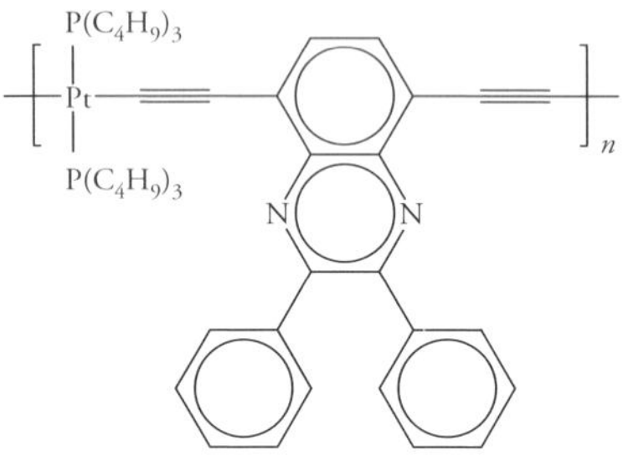

15

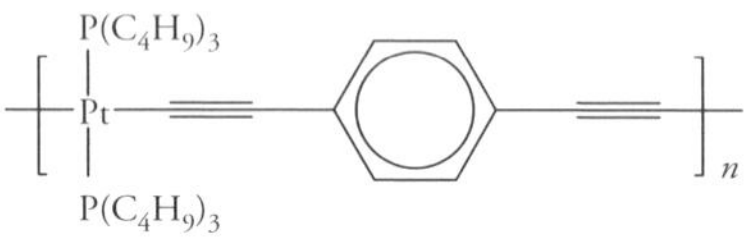

16

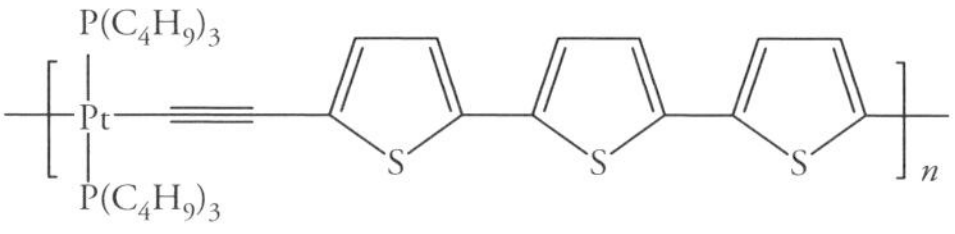

17

The incorporation of heterocyclic units into the backbone of acetylide-containing polymers allows for the alteration of the electronic and optical properties of the polymers.[61–66,70,71] Platinum-containing rigid-rod polymers have been reported by Wong and co-workers.[70,72] Polymers **19** and **21** were prepared by reacting monomers **18** and **20** with *trans*-[$PtCl_2(PBu_3)_2$], respectively (Scheme 2.2).[72] Increasing the conjugation in the polymers resulted in a decrease in the optical gap and the intersystem crossing from the singlet to the triplet state.

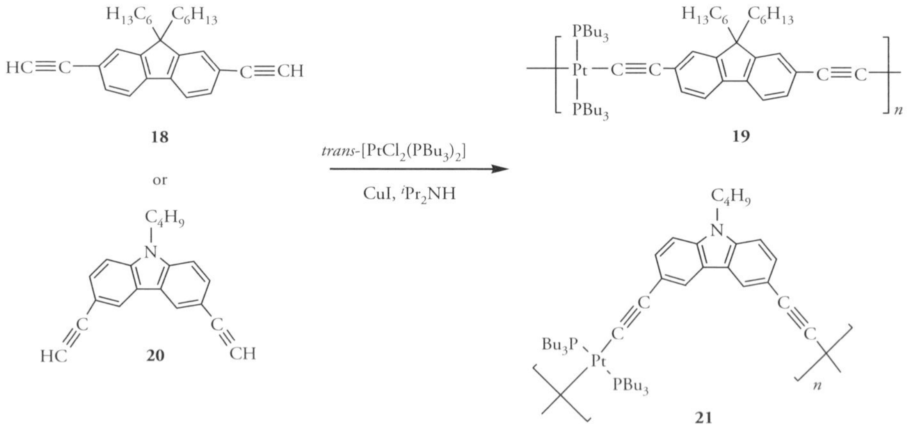

Scheme 2.2

Soluble platinum polyynes containing bithiazole units in the backbone were prepared by Wong et al.[71] Polymer **22** was prepared via dehydrohalogenation reactions and contains electron-withdrawing and electron-donating groups on the bithiazole ring. Polymer **22** displayed a M_w of 47,500 and a M_n of 40,390 with a degree of polymerization of 44. Polymer **22** displayed a luminescence peak at 539 nm when excited at 425 nm. This polymer was also photoresponsive with, its photocurrent increasing with increasing bias voltage.

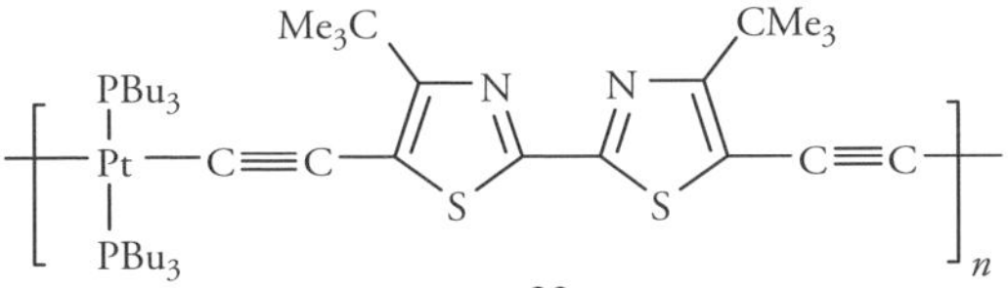

22

ii. Fe and Ru Containing Rigid-Rod Polymers. A variety of metal-containing rigid-rod polyynes have been reported by Lewis et al.[36,37,39,40,46,47,73] The reaction of metal-containing dihalide complexes with bis-trimethylstannyl-alkynyl compounds and catalytic amounts of CuI led to the isolation of high molecular weight polyynes containing a number of different metals (Scheme 2.3).[47]

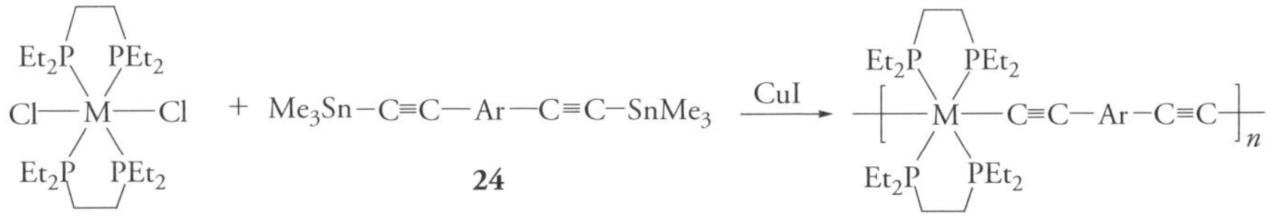

23 **24** **25**

M = Fe, Ar = p-$C_6H_2(CH_3)_2$
M = Ru, Ar = p-$C_6H_2(CH_3)_2$, p-C_6H_4-C_6H_4-p

Scheme 2.3

iii. Metal–Metal Bond. Puddephatt and co-workers have reported the synthesized polymers with main-chain Pt–Pt bonds via the reaction of a platinum complex with various diacetylides and diisocyanides (Scheme 2.4).[74] The reaction of **26** with diacetylides formed oligomers possessing low solubility. Polymers isolated from the reactions of **26** with diisocyanides did not possess any uncoordinated isocyanide groups.

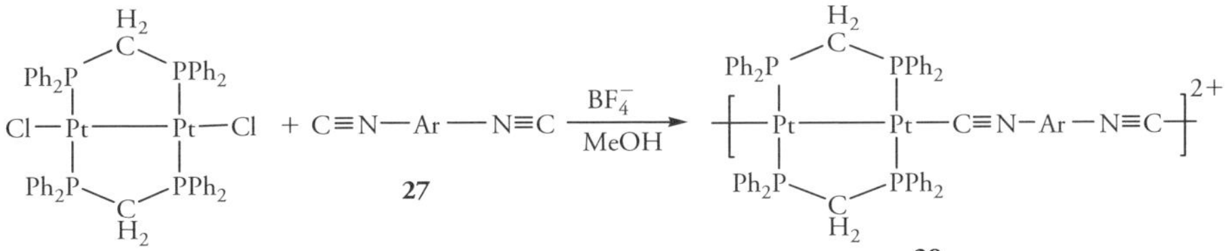

26 **27** **28**

Ar = C_6H_4, $C_6H_2Me_2$,
$C_6H_2{}^tBu_2$, C_6Me_4

Scheme 2.4

Tanase et al. described the synthesis of platinum-containing organometallic rigid-rod polymers **29**.[75] These polymers were prepared by reacting $[Pt_3(\mu\text{-dpmp})_2(XylNC)_2](PF_6)_2$ with π-conjugated bisisocyanides through an axial ligand exchange to give rigid-rod polymers with linear triplatinum moieties in the backbone. The molecular weights of the polymers, obtained by vapor-pressure osmometric analysis, ranged between 10,000 and 12,000 for the soluble portions of the polymers. Due to the rigidity of the polymers, a 2-electron oxidation-reduction couple that was seen in the monomeric unit was not present in the cyclic voltammagram of the polymer (Figure 2.1). The polymers prepared were further reacted with various small molecules and ions to result in a zigzag structure.

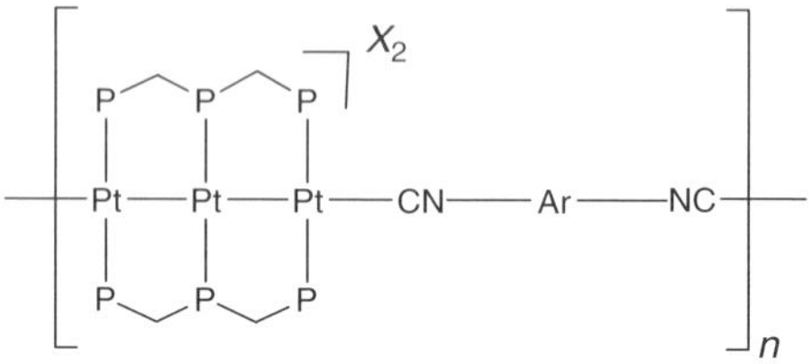

29

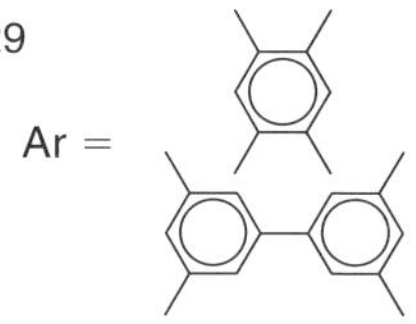

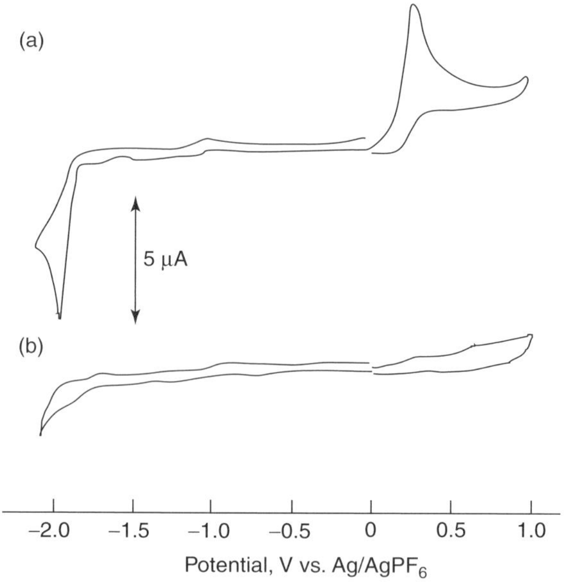

Figure 2.1 Cyclic voltammograms of (a) the monomer of polymer **29**, and (b) polymer **29** in acetonitrile containing 0.1 M [n-Bu$_4$N][PF$_6$]. (Reprinted from *Organometallics*, 2004, *23*, 5975–5988. Copyright © 2004, American Chemical Society.)

Tyler and co-workers isolated photochemically reactive polymers **30** containing Fe–Fe or Mo–Mo bonds.[76–80] Photolysis of polymer **30** resulted in the decomposition of the polymer due to cleavage of the metal–metal bond, yielding **31**. The photolysis of polymer **30** could also result in the formation of a metal–metal triple bond to give polymer **32** (Scheme 2.5).[76]

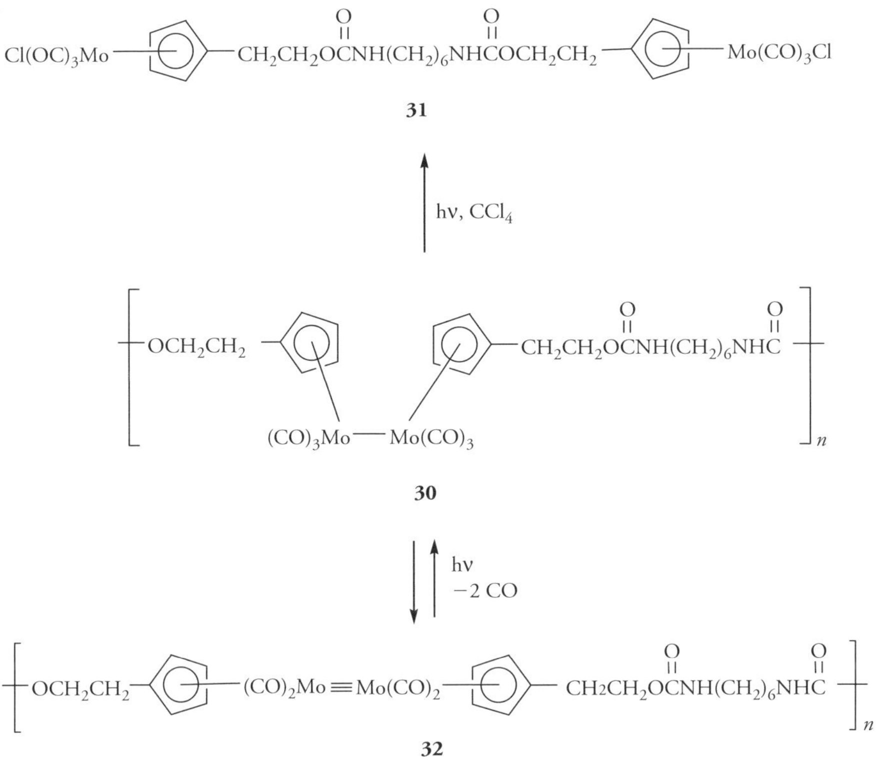

Scheme 2.5

Polymers containing all metal backbones of Ru–Ru or Os–Os bonds have been prepared via the electrochemical reduction of ruthenium and osmium complexes containing *trans*-chloride ligands.[81,82] Scheme 2.6 shows the synthesis of polymers with their backbones comprised solely of metal–metal bonds. The polymers were prepared by reducing $[M^{II}(trans\text{-}Cl_2)(bipy)(CO)_2]$ (M = Ru, Os), **33**, to M^0 complexes and forming the polymer after the loss of the chloride ligands. In both cases, the polymers were selective for the reduction of carbon dioxide.

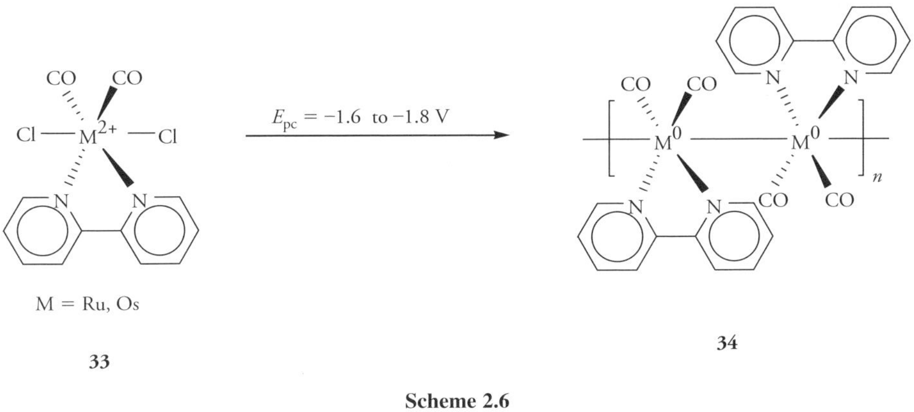

Scheme 2.6

Scheme 2.7 shows the synthesis of organomolybdenum polymers containing metal–metal bonds.[83] Polymer **36** was prepared via reaction of *trans*-$Mo(CO)_3Cl$ containing η^5-Mo complexes with MeMgCl. They displayed poor solubilities.

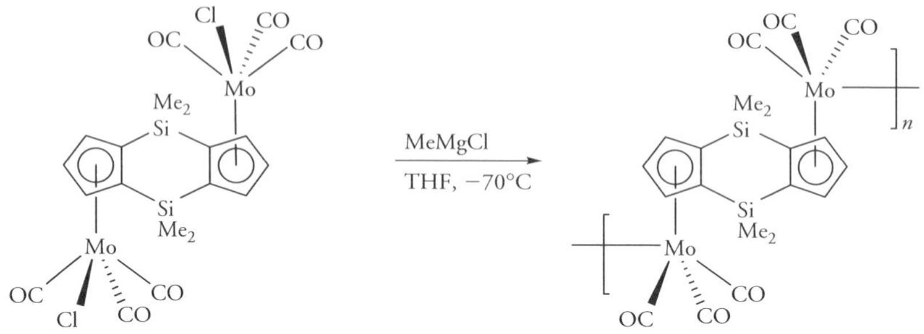

35 **36**

Scheme 2.7

B. π-Coordinated Metals

i. Coordination of Transition Metals to Aromatic Rings. Ferrous chloride reacts with the dilithium salt of dicyclopentdienyldimethylsilane (Scheme 2.8), permitting the isolation of low molecular-weight poly(ferrocenylsilane) and cyclic [1.1]ferrocenophane.[84]

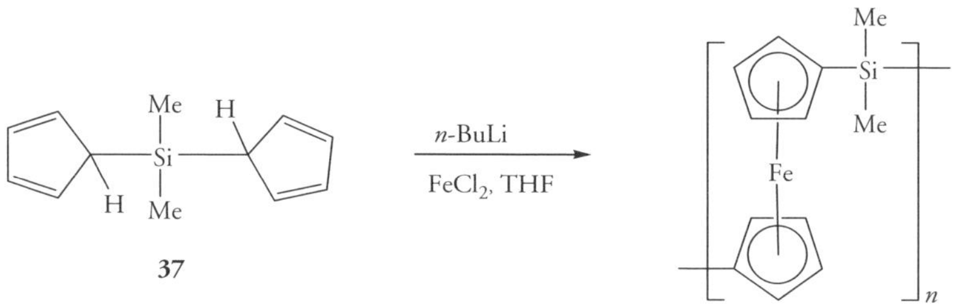

37 **38**

Scheme 2.8

Kohler et al. have prepared linear and cyclic polyferrocenes containing bridging dimethylsilane groups.[85] Polyferrocenes containing 6 to 18 ferrocene units in the backbone, **39**, have been prepared. Decreasing the reaction temperature caused an increase in the mean ring size from 8.1 to 10.8.

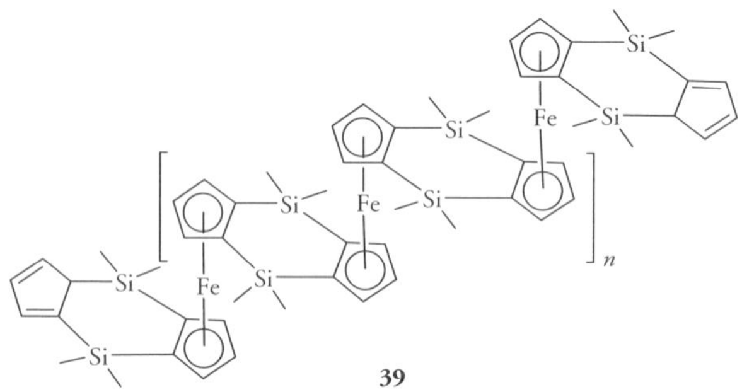

39

Polyferrocenes substituted with hexyl groups have been isolated in oligomeric and polymeric forms from the reaction of $[FeCl_2(THF)_2]$ (THF-tetrahydrofuran) with a dihexylfulvalene dianion (Scheme 2.9).[86] Due to the steric bulk of the hexyl groups high molecular weights could not be obtained even though the solubility of the polymers was increased by the alkyl groups. Electrochemical studies of the polymers showed that there were interactions between the iron centers.

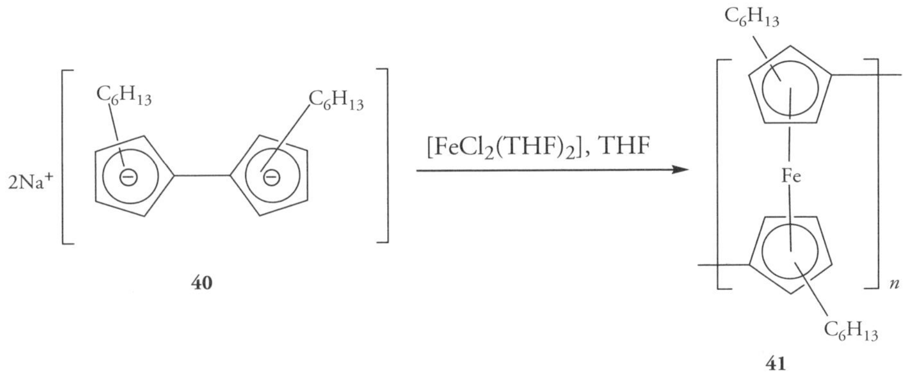

Scheme 2.9

Southard and Curtis reported the synthesis of conjugated polymers by reacting ferrous iodide with an isomeric mixture of bis(alkylcyclopentadienide)arenes.[87,88,89] This reaction generated high molecular-weight polymers with high polydispersities; oxidation of polymers **42a, b** resulted in electrical conductivities between 10^{-10} and 10^{-7} S/cm.

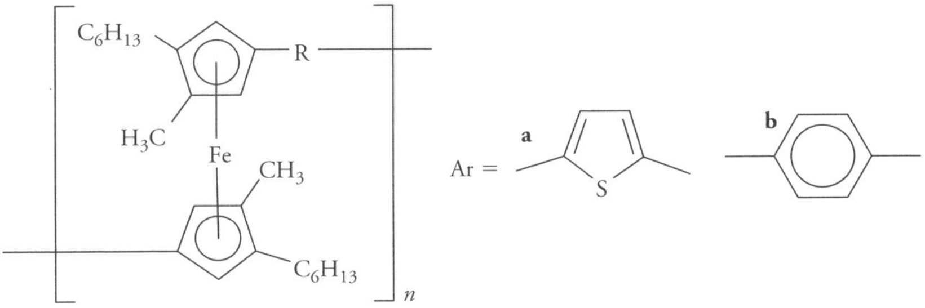

42a, b

Similarly, Park, Chang, and co-workers produced low molecular-weight poly(ferrocenylsilanes) from the condensation of $FeCl_2$ with the dilithium salts of dicyclopentadienyldimethylsilanes.[90] Electrochemical studies of these polymers showed that there were strong interactions between the iron centers. Scheme 2.10 shows the synthesis of face-to-face polymetallocenes using methodology developed by Rosenblum et al.[91,92,93] This methodology has been used to prepare paramagnetic mixed-metal polymers containing nickel/iron or nickel/cobalt with magnetic moments of 5.3 and 5.2 μ_B, respectively.

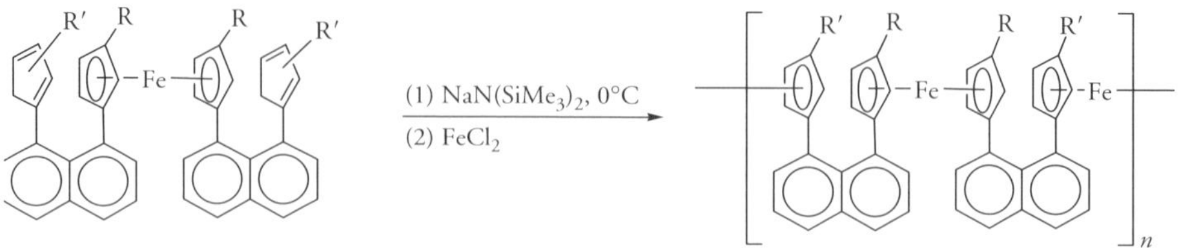

43a–d **44a–d**

R = 2-octyl, R′ = 2-octyl,
R = 2-octyl, R′ = H
R = 3,7-dimethyloctyl, R′ = H
R = decyl, R′ = H

Scheme 2.10

ii. Ring-Opening Polymerization of Strained Metallocenes. The ring-opening metathesis polymerization (ROMP) of ferrocenophanes containing bridging olefinic units generated conjugated polyferrocenes.[94,95,96] Homopolymers of *ansa*-(vinylene)ferrocene and copolymers with norbornene have been reported.[94] Homopolymerization gave insoluble polymers possessing conductivities in the range of $10^{-3}\,\Omega^{-1}\text{cm}^{-1}$. The copolymerization resulted in polymers with weight-average molecular weights of 21,000. Grubbs and Lee functionalized the bridging units with alkyl groups, enhancing the solubility of this class of polyferrocenes.[95] Polymers with weight-average molecular weights greater than 300,000 were obtained using this method (Scheme 2.11).

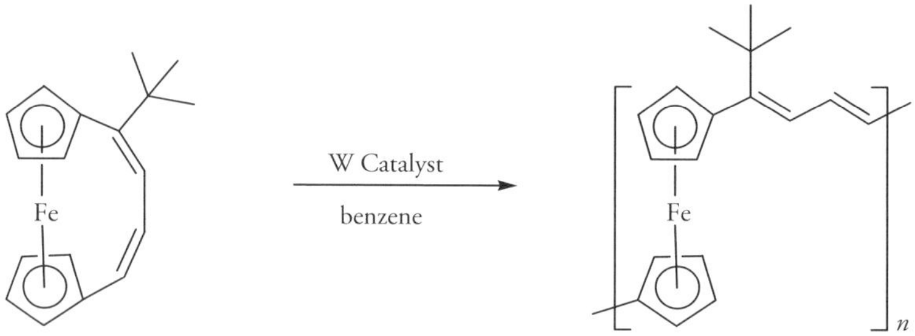

45 **46**

W catalyst = W(=NPh)-(=CHPh(2-OMe))(OC$(CF_3)_2(CH_3)_2$)(THF)

Scheme 2.11

Manners and co-workers have also described the thermal and anionic ring-opening polymerization (ROP) of [1]thia- and [1]selea-ferroceneophanes. The majority of the products of these reactions were insoluble; however, functionalization of the cyclopentadienyl rings with methyl groups resulted in the formation of low molecular-weight soluble poly(ferrocenyl sulfide)s.[95] The polymers displayed two reversible oxidation processes at −0.07 and +0.24 V, indicating strong electron communication between the metal centers in the polymers.[95] Manners and co-workers synthesized polymers containing carbon–phosphorus, –silicon, and –sulfur bridges that have been prepared from [2]ferrocenophanes.[97,98] The ferrocenophanes containing phosphorus or sulfur could be polymerized through thermal ROP, whereas the C-Si bridged [2]ferrocenophane was resistant to thermal, anionic, and transition-metal-catalyzed ROP.[99]

[2]Carbathioferrocenophane, **47**, underwent cationic ROP with methyl triflate and boron trifluoride etherate, but was resistant to both anionic and transition-metal-catalyzed ROP.[100,101] However, cationic ROP initiated with methyl triflate caused oxidation of the iron centers in the polymers . Me^+ and H^+ were determined to initiate cationic ROP for the methyl triflate and boron trifluoride, respectively.[101] Scheme 2.12 shows a proposed mechanism of the methyl triflate, **48**, initiated ROP of [2]carbathioferrocenophane, **47**.

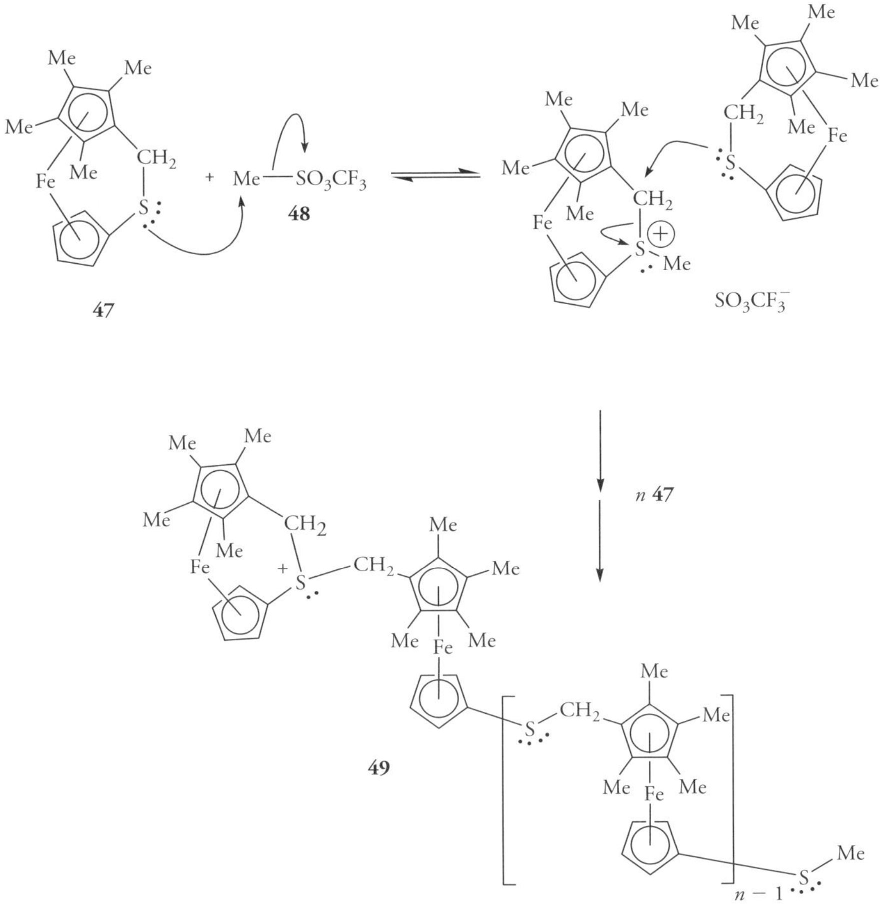

Scheme 2.12

The ROP of [3]-trithia- or [3]-triselenaferrocenes followed by de-sulfurization or de-selenization led to the isolation of polyferrocenylenes containing S or Se bonds in the backbone (Scheme 2.13). The reaction of PBu_3 with trithiaferrocenes yielded poly(ferrocene persulfide)s as high molecular-weight linear polymers as well as network polymers.[93–96,102–111] It was determined that these polymers could undergo decomposition. The addition of the reducing agent $LiBHEt_3$ to the polymers resulted in decomposition, whereas addition of iodine caused the regeneration of the polymer.

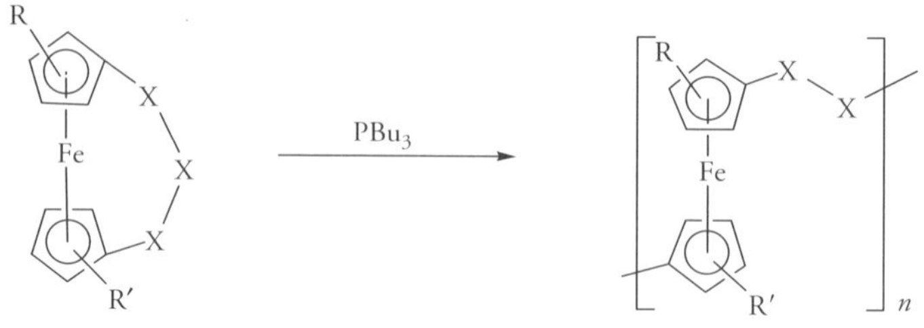

X = S, Se R, R′ = H, n-C_4H_9, $^tC_4H_9$

50

51

Scheme 2.13

High molecular-weight mixed-metal polyferrocenes have been prepared by Miyoshi et al. via the photochemical ROP of phosphorus-bridged [1]ferrocenophanes containing organometallic units on the phosphorus ligand (Scheme 2.14).[111]

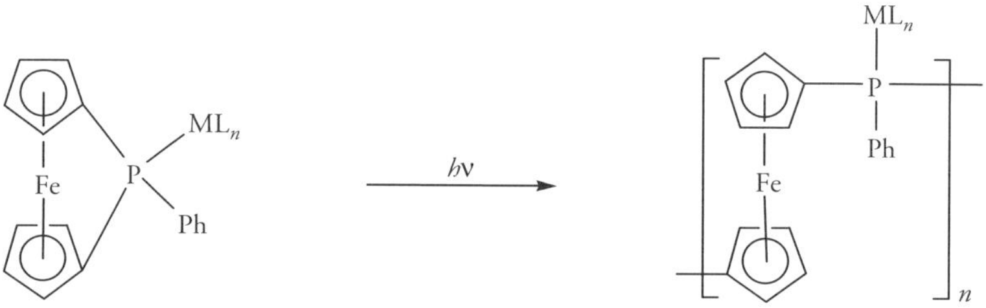

52

53

ML_n = $Mn(C_5H_4Me)(CO)_2$
$Mn(C_5H_5)(CO)_2$
$W(CO)_5$

Scheme 2.14

Homo- and block copolymers from phosphorus-bridged [1]ferrocenophanes were isolated from living anionic ring opening polymerizations.[113] Scheme 2.15 shows preparation of borane adducts of poly(ferrocenylphosphine)s by thermal ROP or addition of BX_3 (X = H, Cl).[114]

Scheme 2.16 displays the thermal ROP of a number of phosphorous-bridged ferrocenophanes, resulting in the isolation of phosphorus(III)- and (V)-containing polymers **59a–c**.[115] Reaction of poly(ferrocenylphosphines) **59a–c** with elemental sulfur generated poly(ferrocenylphosphine sulfides) **60a–c**. Polymers **60a–c** could not be directly prepared from thermal ROP of phosphine sulfide–bridged ferrocenophanes due to partial decomposition of the products. However, ferrocenophane **58a** polymerized via anionic ROP and reacted with sulfur.[116] The resultant polymers had M_n = 3600–32,000 for monomer–catalyst ratios that were varied from 11:1 to 100:1.

Many research groups have studied the synthesis and properties of polyferrocenophanes. The Manners group has been at the forefront of polyferrocenophane research for many years and has produced many major contributions to this field.[97–101,113,114,115,117–146] Scheme 2.17 shows the preparation of high molecular-weight poly(ferrocenyldimethylsilane), **38**, by thermal,

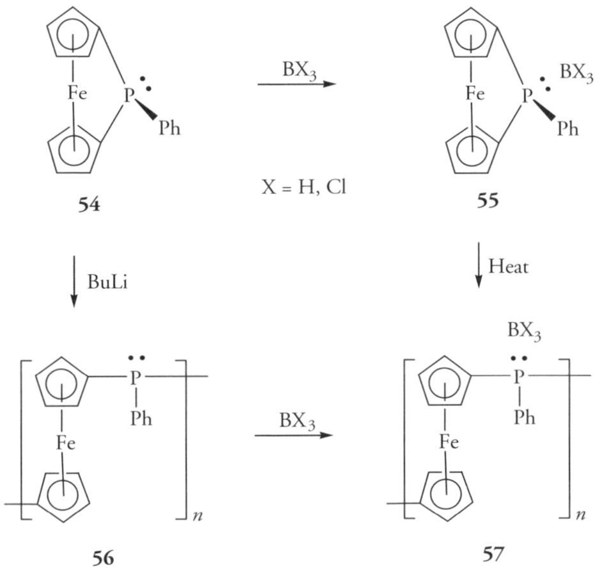

Scheme 2.15

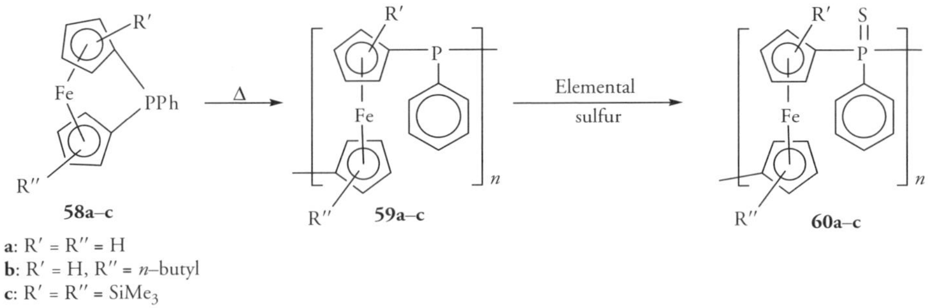

Scheme 2.16

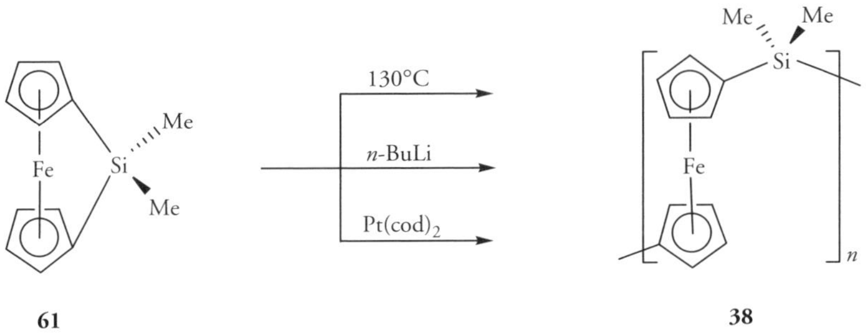

Scheme 2.17

anionic, and transition metal-catalyzed ROP.[120,121,122] Solid-state ROP of $Fe(\eta\text{-}C_5H_4)_2SiMePh$ using a ^{60}C γ-ray gave high-molecular weight stereoregular polymers.[123,124]

Polychromarenylsilanes containing bis(benzene)chromium units in the main chain were prepared using Karstedt's catalysts, as shown in Scheme 2.18.[147]

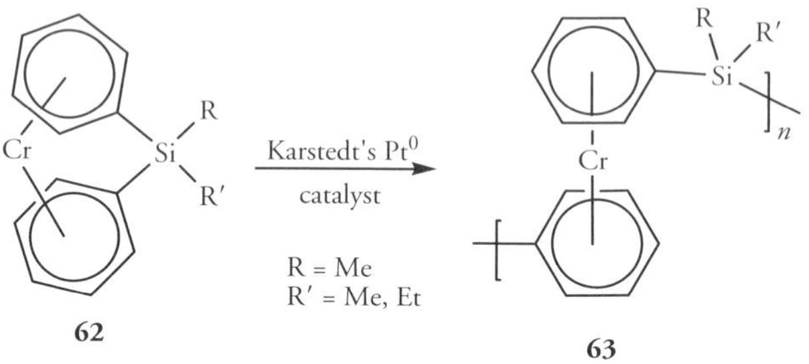

Scheme 2.18

A number of poly(dimethylsiloxane) and poly(ferrocenylsilane) copolymers have been prepared using thermal, anionic, and transition metal catalyzed ROP.[97,98,121,124–131] Manners et al.[132] described the ring-opening of [1]thia- and [1]selenaferrocenophanes using anionic initiators. Electrochemical studies of poly(ferrocenyl sulfide)s prepared by these methods elicited two reversible oxidation processes, indicating that strong Fe–Fe interactions exist in the polymers.

Manners and co-workers recently described the preparation of polyferrocenylsilane.[137] The polymerization occurred through living anionic ring-opening of ethylmethylsila[1]ferrocenophane with *n*-BuLi. The polymers obtained possessed molecular weights (M_n) between 4000 and 41,400, with PDI = 1.01–1.02.

Ruthenium and iron containing [2]metallocenophanes have undergone thermally initiated ROP to yield polymers, **65**, containing insulating bridges (Scheme 2.19).[140,141,142] The use of monomers containing methyl-functionalized cyclopentadienyl rings yielded soluble poly(ferrocenylethylene)s and poly(ruthenocenylethylene)s with high molecular weights. Electrochemical studies of the ruthenium-based polymers displayed irreversible oxidations, whereas the iron-containing polymers displayed reversible oxidations.[141] Antiferromagnetic interactions were detected in the poly(ferrocenylethylene)s when oxidized with tetracyanoethylene.[142]

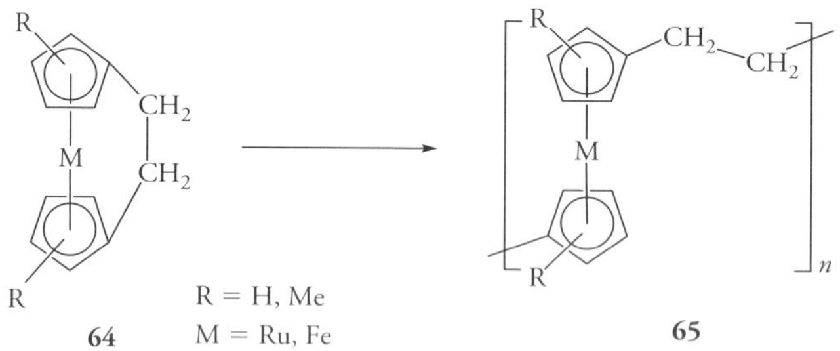

Scheme 2.19

Originally, Seyferth et al. reported the anionic ring-opening of phosphorus-bridged ferrocenophanes in 1982.[115] Subsequently, Manners reported the preparation of high molecular-weight polyferrocene via the thermal ring-opening of phosphorus-bridged [1]ferrocenophanes.[143] Manners and co-workers also described the ROP of boron-bridged [1]ferrocenophanes. The thermal ROP of highly strained boron-bridged [1]ferrocenophanes is shown in Scheme 2.20.[143] The resultant polymers had low molecular weights, determined by dynamic light scattering.

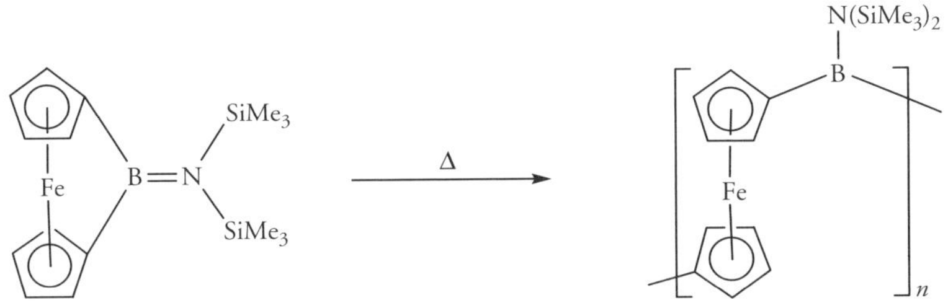

66 **67**

Scheme 2.20

Thermal ring opening of [1]germaferrocenophanes-generated high molecular-weight ($M_w = 10^6$), easily isolated poly(ferrocenyl germanes), **68**.[99,146,148,149,150] Manners and co-workers have also reported the platinum- and palladium-catalyzed ROP of [1]germaferrocenophanes.[146,148]

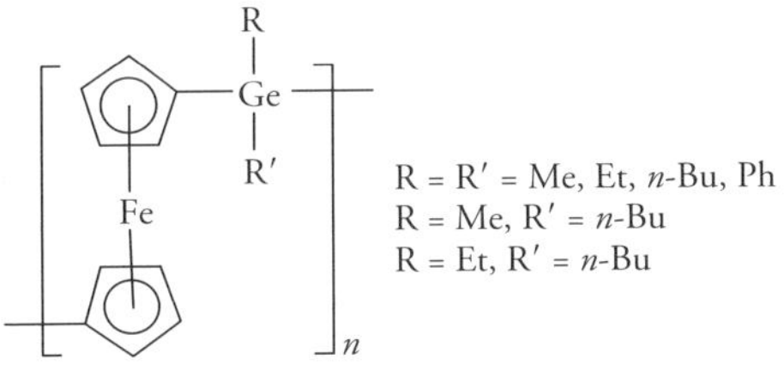

68

High molecular-weight polymers have been prepared via the thermal ROP of ferrocenophanes containing tin bridges.[116,151,152,153] These polymers could also be prepared at room temperature. The rate of the room temperature ROP was increased by the addition of amines (Scheme 2.21).[152,153,154] This was also seen in the room-temperature ROP of silicon- and germanium-bridged [1]ferrocenophanes. However, the silicon-containing monomers displayed low reactivity in the room temperature ROP, whereas the germanium and tin monomers showed

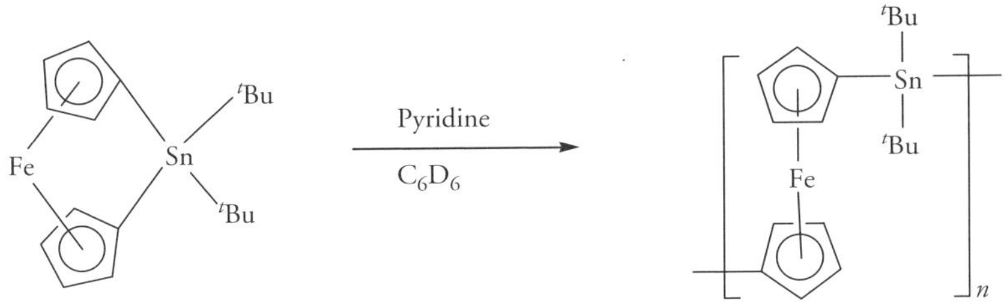

69 **70**

Scheme 2.21

good reactivity.[152] Tin-bridged ferrocenophanes also underwent cationic ROP in the presence of small amounts of Lewis acids.[154]

High molecular-weight polyferrocenes have been prepared from the thermal ROP of stained [1]stannaferrocenophanes at high temperatures.[152] The ROP temperature greatly affected the polymer molecular weights. High temperatures gave high molecular weights, whereas room temperature gave low molecular weights. Investigating the reaction mechanisms showed that the ROP of [1]stannaferrocenophanes occurs in the presence of catalytic amounts of nucleophilic species.[153,154,155] The ROP of silicon- and germanium-bridged [1]ferrocenophanes was examined to determine their reactivity toward nucleophilic species. Germanium monomers showed similar reactivities to the tin monomers, whereas the silicon monomers were much more resistant to polymerization under these conditions.[155] Scheme 2.22 shows the mechanism proposed for the nucleophilically assisted ROP.[153] Further studies of these reactions showed that the cationic ROP could also be initiated by electrophilic initiators.[153]

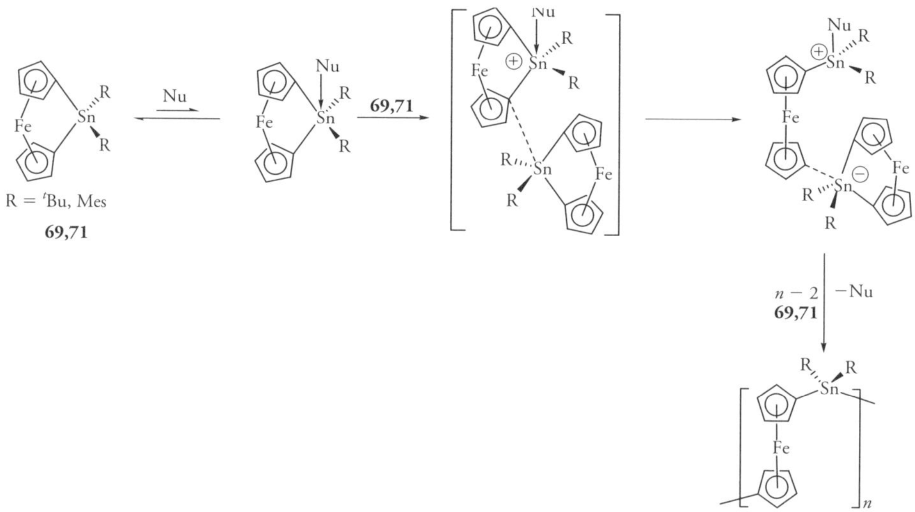

Scheme 2.22

Manners and co-workers prepared high molecular-weight poly(ruthenocenylstannanes), **73**, via the thermal ROP of [1]ruthenocenophanes. These polymers possessed a broad molecular-weight distribution with $M_n = 270{,}000$ and PDI = 2.28.[156]

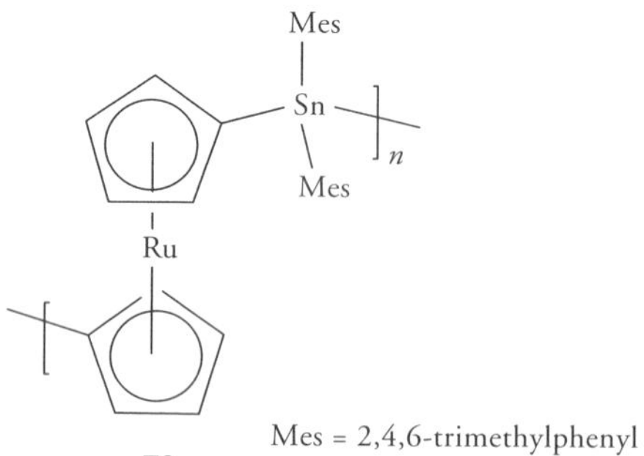

73

In 1995 Manners and co-workers prepared polyferrocenes containing boron bridges in their backbone.[143,144] The thermal ring opening of boron-bridged ferrocenophanes, **74a–c**, is shown in Scheme 2.23. This reaction formed both soluble and insoluble materials. The soluble polymer, **75b**, consisted of a mixture of low molecular-weight linear and cyclic oligomers, while polymers **75a** and **75c** were insoluble.[143] Similar polymers incorporating Lewis acid boron centers were prepared by ROP of [1]ferrocenophanes using boron halides as initators.[99] Capping poly(ferrocenylsilane)s with ferrocenylboranes produced polymers containing boron-functionalized end groups.[99]

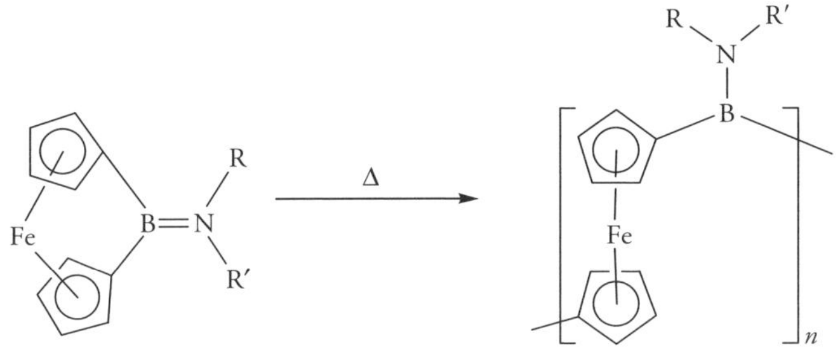

74a–c **75a–c**

a: R = R′ = $SiMe_3$
b: R′ = R″ = ^iPr
c: R′ = ^tBu, R″ = $SiMe_3$

Scheme 2.23

The thermal ROP of [2]ferrocenophanes generated poly(ferrocenylethylene)s.[139,141] Soluble polymers with bimodal molecular-weight distributions were prepared by functionalization of the cyclopentadienyl ring of the [2]ferrocenophanes, with methyl groups followed by ROP at 300°C. The isolated polymers possessed number-average molecular weights of 86,000 and 3500. Two reversible oxidation couples were observed at −0.25 and −0.16 V by electrochemical studies.[141] Ferromagnetic iron carbide ceramics were isolated from the pyrolysis of these polymers at 600°C. Oxidation of the polymers with tetracyanoethylene gave soluble and insoluble materials. Magnetic susceptibility studies on the soluble material showed significant antiferromagnetic interactions.[141]

Manners first proposed that the transition-metal-catalyzed ROP occurred via a homogenous mechanism.[157] However, a heterogenous catalytic cycle has been reported.[158] The proposed mechanism for the $Pt(1,5\text{-cod})_2$ (cod-cyclooctadiene) catalyzed reaction is shown in Scheme 2.24. The $Pt(1,5\text{-cod})_2$ forms a [2]platinasilaferrocenophane through oxidative addition to the zero-valent Pt complex via elimination of a 1,5-cod ligand. Platinum colloids are then formed by the elimination of the second 1,5-cod ligand; these platinum colloids are proposed to be the active catalysts. The polymers are then formed by subsequent oxidative addition and reductive eliminations at the colloid surface.

Polyferrocenes containing functionalized side chains have been the focus of several studies. A number of reports have appeared on the preparation of polyferrocenylsilanes containing long-chain alkyl, chloroalkyl, chloro, and alkoxy side chains.[159–162] Scheme 2.25 describes the synthesis of high molecular-weight polymers containing acetylenic substituents via transition-metal-catalyzed ROP.[163] Magnetic ceramics were generated from the pyrolysis of polymer **82**.

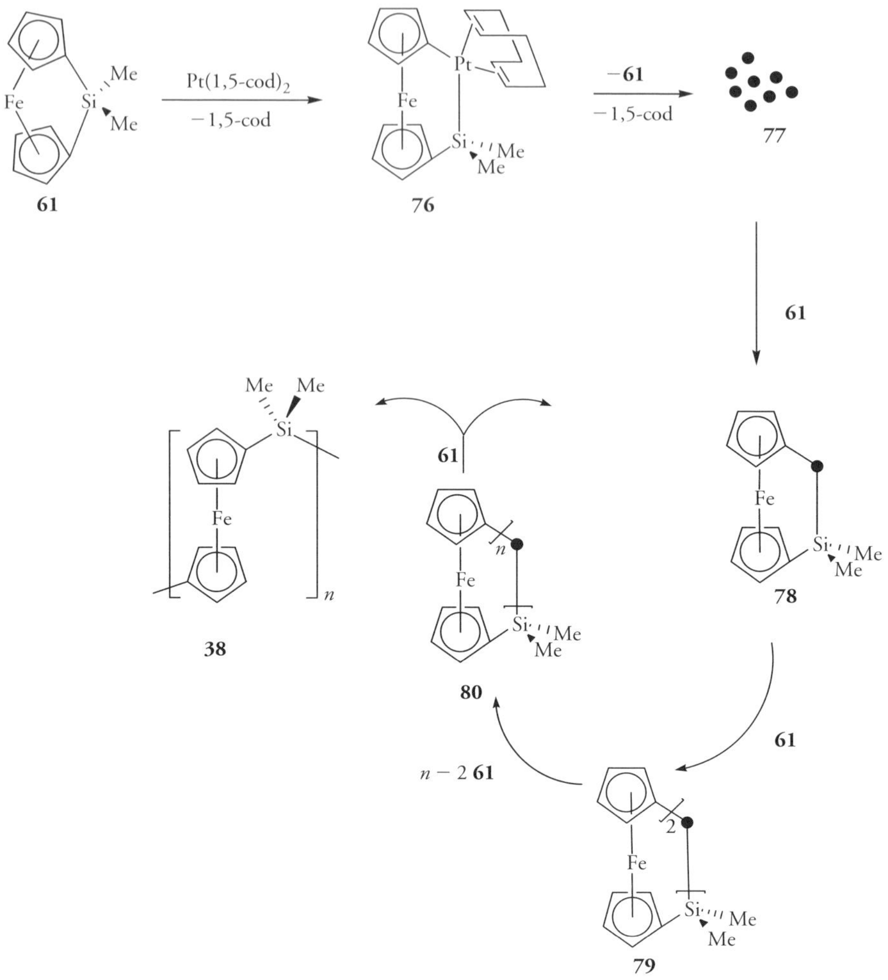

Scheme 2.24

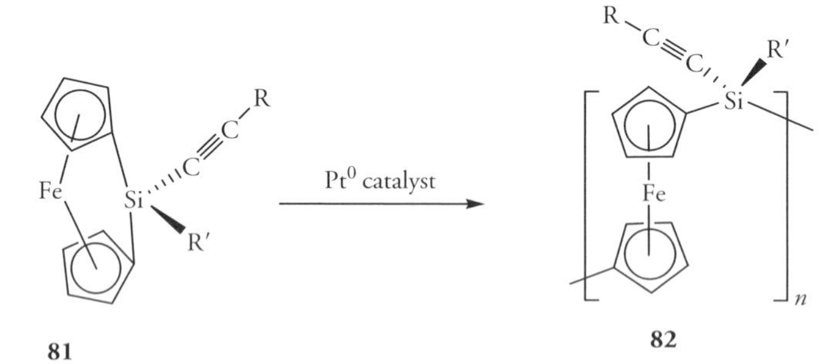

R = Ph, *n*-Bu
R′ = Me, CCPh

Scheme 2.25

The ROP of [1]germaferrocenophanes has been reported.[116,146,150,161,164] High molecular-weight polymers **68** were formed from the facial thermal ROP of [1]germaferrocenophane, **83** (Scheme 2.26). This reaction occurred in milder conditions than similar reactions using silicon-containing [1]ferrocenophanes.[146] The thermal ROP of ferrocenophane, **83**, occurred at temperatures as low as 90°C; however, the strained [1]germaferrocenophanes were moisture sensitive and decomposed to give the bis(ferrocenyl)germoxanes, **84**. In 2002, Pannell et al. produced polyferrocenes containing germanium in the backbone. These polyferrocenes, **68**, displayed insulator properties with conductivities of $10^{-14}\ \Omega^{-1}cm^{-1}$.[116] However, upon doping with iodine, these polymers became semiconducting, with conductivities between 10^{-6} to $10^{-4}\ \Omega^{-1}cm^{-1}$.

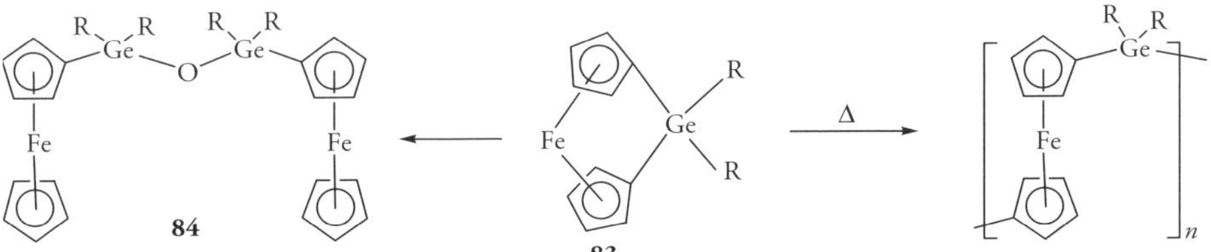

84 **83** R = Me, Et **68**

Scheme 2.26

Polymers formed from the ROP of [1]ferrocenophanes have been investigated for a variety of uses. Vancso et al. has tested them for nanolithographic and plasma etch-resistant materials.[165,166] Pannel and co-workers have reported ferrocenylsilylene polymers for use as tapered optical-fiber sensors in the detection of ammonia and carbon dioxide.[167]

iii. Radical and Condensation Polymerization. New classes of organometallic polymers have been isolated from the reaction of chloro-substituted polymers with alcohols or amines.[168,169,170] The reaction of the acetylide, **86**, with polymer **85** is shown in Scheme 2.27. Polymer **88** was isolated from the reaction of polymer **87** with hydrazine.[169,170]

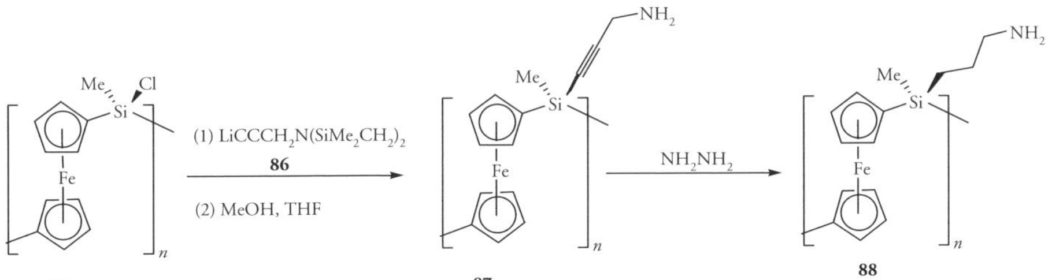

85 **87** **88**

Scheme 2.27

Over the past 10 years, research into water-soluble ferrocene-based polymers and polyelectrolytes has increased.[168–175] Vancso and co-workers have reported the preparation and self-assembly of anionic and cationic polyelectrolytes.[171,172] Polyanionic, **89**, and polycationic, **90**, ferrocene-based electrolytes are two examples of such polyelectrolytes.[170]

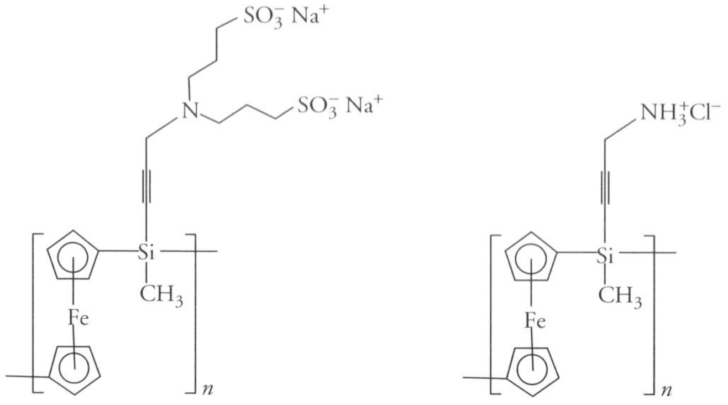

89 **90**

Plenio and co-workers synthesized and polymerized of 1-iodo-2-methoxymethyl-3-ethynylferrocene and 1-iodo-2-(*N*,*N*-dimethylamino methyl)-3-ethynylferrocene to give 1,3-linked ferrocene-acetylene polymers.[176,177] These polymers were synthesized by Sonogahira coupling reactions. These polymers showed optical activity[176] or possessed functionalized side chains.[177,178] Scheme 2.28 shows the reaction of diiodoferrocenes with diethynyl monomers.[178,179] The palladium-catalyzed reactions led to polymers **93a,b**, which when doped with iodine displayed semiconducting properties. The iodine-oxidized polymer **93a** displayed an electrical conductivity of 1.3×10^{-4} S/cm.

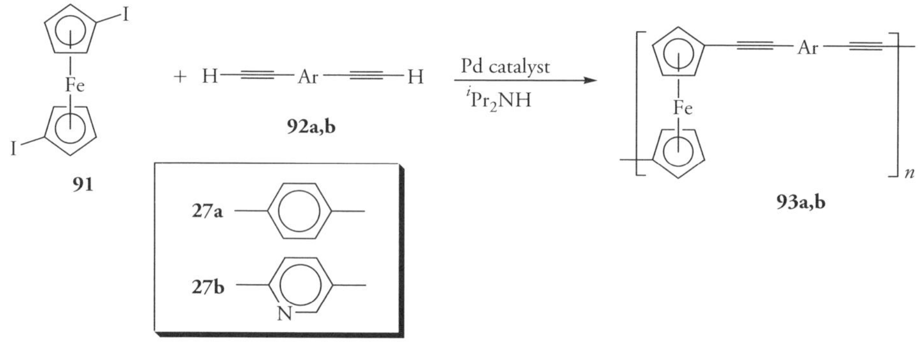

Scheme 2.28

Cuadrado and co-workers prepared polysiloxanes with ferrocene and cobalticenium moieties in their backbone.[180,181] Polymetallocene, **94**, was synthesized by reacting 1,3-bis(3-aminopropyl)-1,1,3,3-tetramethyldisiloxane with 1,1′-bis(chlorocarbonyl)ferrocene or 1,1′-bis(chlorocarbonyl)cobalticenium hexafluorophosphate. The organoiron polymer displayed a $M_n = 10{,}600$,[180] whereas the organocobalt polymer possessed very low solubility in polar organic solvents.[181] An alternative preparative route to this class of polymer is the reaction of 1,1′-bis(β-aminoethyl)ferrocene with dimethylbis(4-chlorocarbonylphenyl) silane.[180]

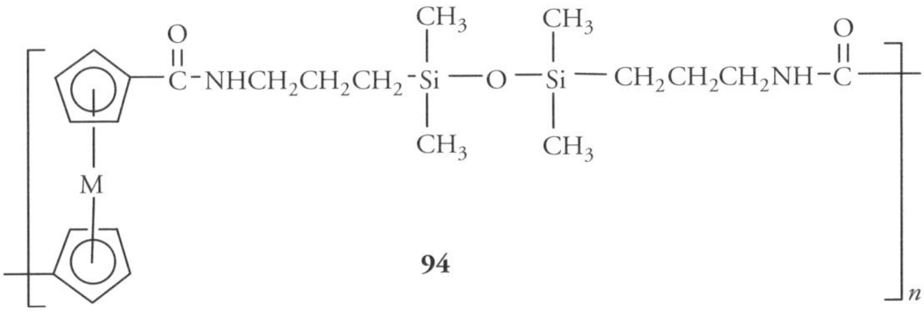

94

Gao et al. have recently prepared a number of thermotropic liquid crystalline main-chain ferrocenes exhibiting fluorescent properties.[182] The reaction of 1,1′-bis(functionalized)ferrocenes, **95**, with 1,4-dibromobutane, **97a**, or α,α′-dibromo-*p*-xylene **97b** generated polyelectrolytes **97a,b** with M_n ranging from 5400–14,700 (Scheme 2.29). These polymers were highly thermally stable with the first weight loss occurring between 172 and 330°C. The liquid crystalline properties showed these materials formed smectic liquid-crystal phases and displayed batonnet textures.

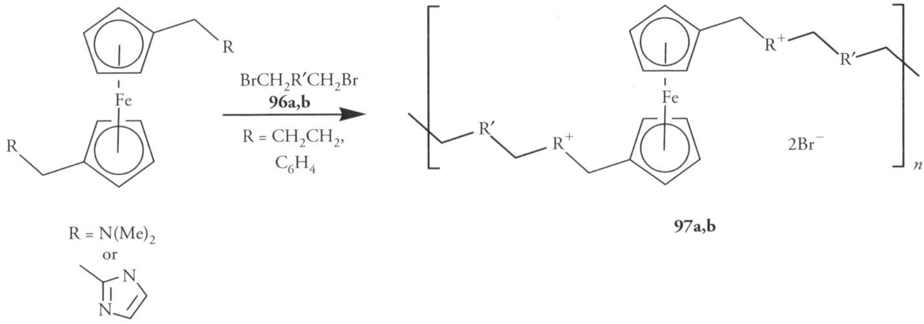

Scheme 2.29

Khan et al. recently polymerized 1,1′-ferrocene dicarbonyl chloride with various aromatic diamines in solution at low temperatures.[183] The resulting polyaramides, **100**, possessed glass transition temperatures above 350°C in the range of 360°–390°C and 10% decomposition temperatures between 300 and 430°C (Scheme 2.30).

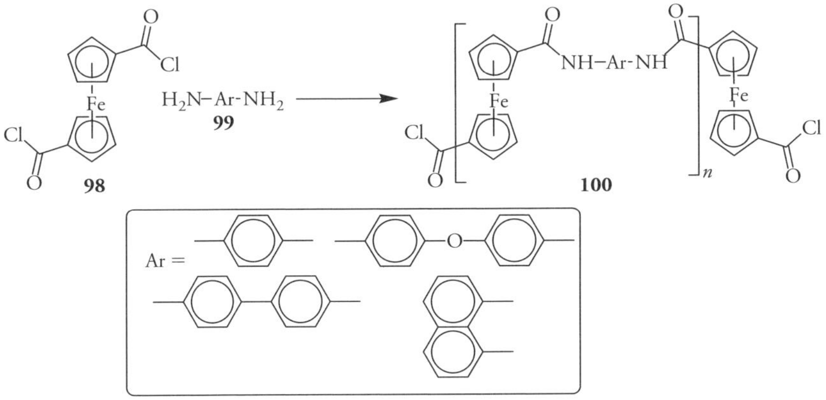

Scheme 2.30

Organometallic polyurethanes containing backbone polyether units have been reported.[184,185] The incorporation of ferrocenyl segments into the backbone of polyurethanes enhanced thermal stability.[186] There have been a number of studies on the incorporation of ferrocenyl units into the backbone of polyesters **101**.[187–191] A number of polymers of this class exhibited liquid crystalline properties.[189–192] The thermal stability and glass transition temperatures of these polymers decreased with an increase in the length of the alkyl chains.

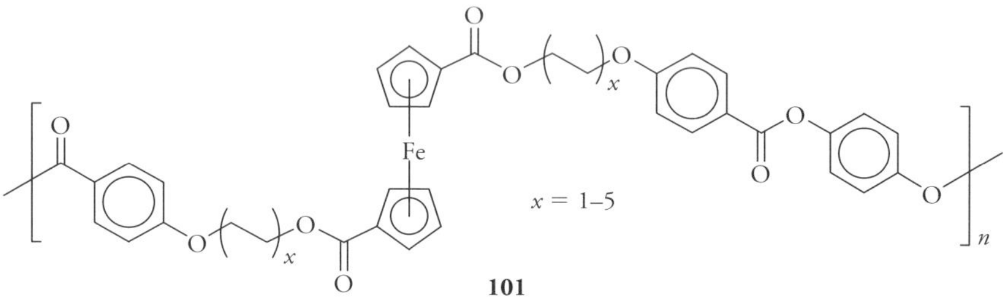

101

Senthil and Kannan prepared ferrocene-based liquid crystalline polymers containing phosphate groups in the backbone.[191,192] The majority of these polymers showed grainy nematic textures, whereas polymers with 8 and 10 methylene groups in the backbone displayed clear nematic textures. Liquid crystalline behavior was not observed for the polymer incorporating both ethylene spacers and a pendent phenyl group.

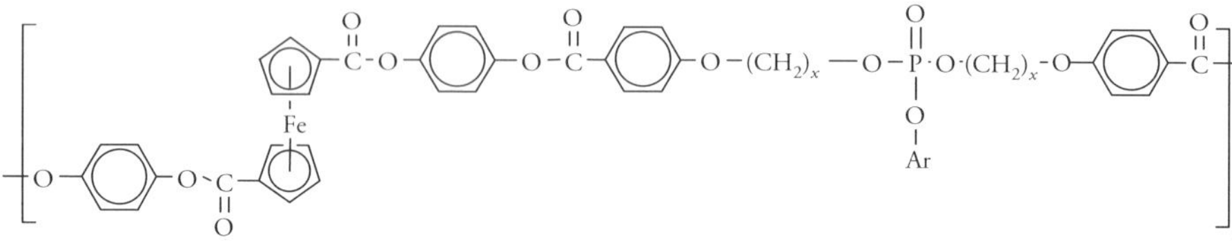

102

Polyferrocenes containing benzimidazole units in the backbone **105** were made from the reaction of 1,1′-ferrocenedimethanol **103** with 3,3′,4,4′-biphenyltetraamine **104** in the presence of [$RuCl_2(PPh_3)_3$] (Scheme 2.31).[193] Polymer **105** possessed a weight-average molecular weight of 45,000, and contained 20% oxidized iron centers due to the polymerization conditions. Electrochemical studies showed that these polymers displayed irreversible oxidation.

An organometallic elastomeric thermoset polymer was made by heating polymer **106** to 350°C under an inert atmosphere.[190] IR analysis confirmed the cross-linking of the butadiyne groups. Subsequent heating to 1000°C gave hard ceramics that displayed magnetic properties. Boron in the polymer backbone rendered the ceramic materials stable toward oxidation. Similar low molecular-weight ferrocene-based polymers containing diacetylene–organosilyl units have been reported.[194]

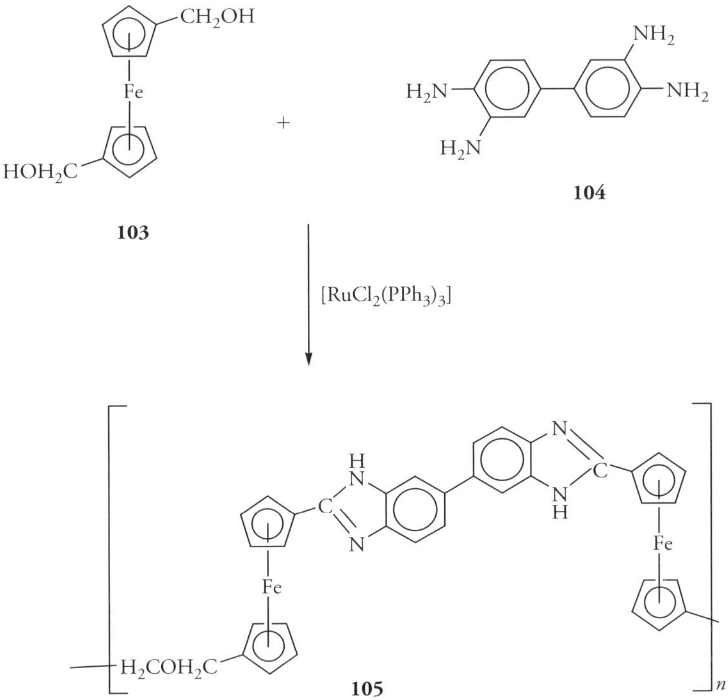

Scheme 2.31

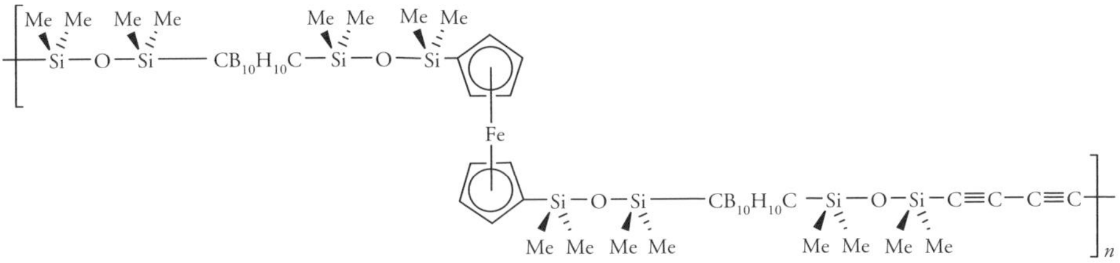

106

Abd-El-Aziz and co-workers recently synthesized main-chain polyferrocenes containing cationic η^5-cyclopentadienyliron groups complexed pendent to η^6-arene rings along the backbone **107**.[195] These polymers were obtained in approximately 90% yield and displayed good solubility in polar organic solvents. Analogous neutral polyferrocenes **108** were also prepared by the photolytic cleavage of the cationic iron moieties. The weight-average molecular weights of the cationic polyferrocene **107** ranged from 8700 to 13,500 with PDI ranging from 1.1 to 2.2. Electrochemical studies showed that the cationic polymers displayed two redox couples between −1.28 and −1.02 V and 0.891 and 1.03 V for the cationic and neutral iron moieties, respectively. The neutral polyferrocenes displayed one redox couple at 0.69 V for the oxidation of the ferrocene groups. Thermal studies of the cationic polymers displayed two degradative steps, one starting at 210°C corresponding to the cationic iron groups and another between 343 and 465°C due to the degradation of the backbone. The neutral polymers only showed one degradation step between 290 and 433°C due to the decomposition of the backbone. The neutral polymers possessed lower T_g than their cationic analogs. Viscosity measurements indicated that the cationic polymers displayed the polyelectrolyte effect (Figure 2.2).

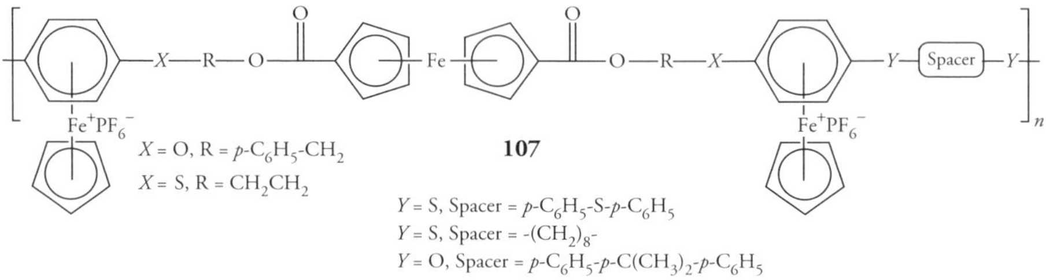

107

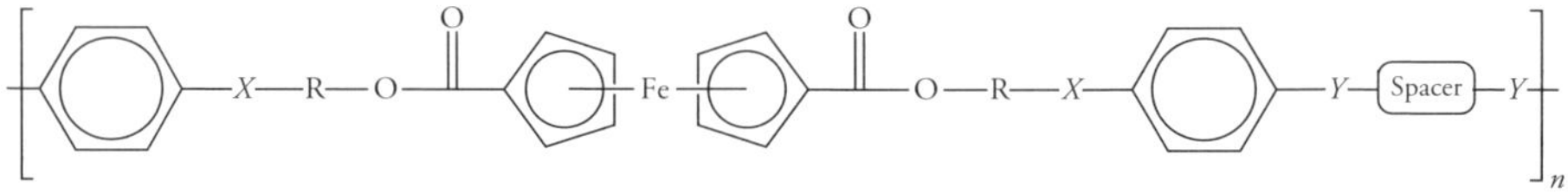

108

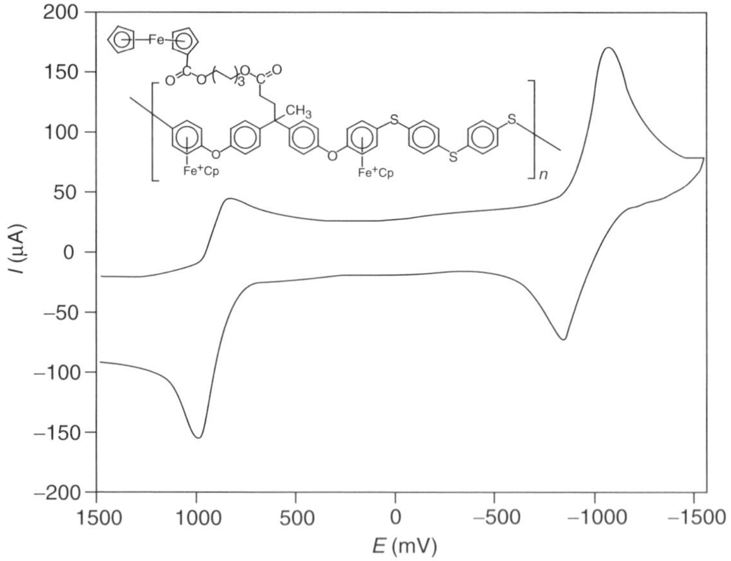

Figure 2.2 Cyclic voltammogram of polymer **214** in 0.1 M TBAP in propylene carbonate at −20°C.

Nishihara and co-workers generated main-chain organoruthenium polymers via the reaction of [(η^5-C_5H_4Hex)RuBr(cod)] with 4,4′-diethylbiphenyl in DCM.[196] The polymer was air sensitive and possessed low thermal stability. The molecular weight of the polymer was as high as 20,000. Cyclic voltammetry showed a 1-electron reduction of the ruthenacyclopentatriene at $E = -1.01$ V (Figure 2.3).

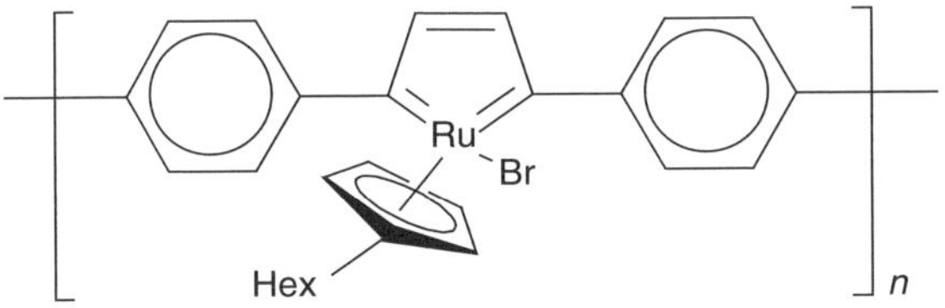

109

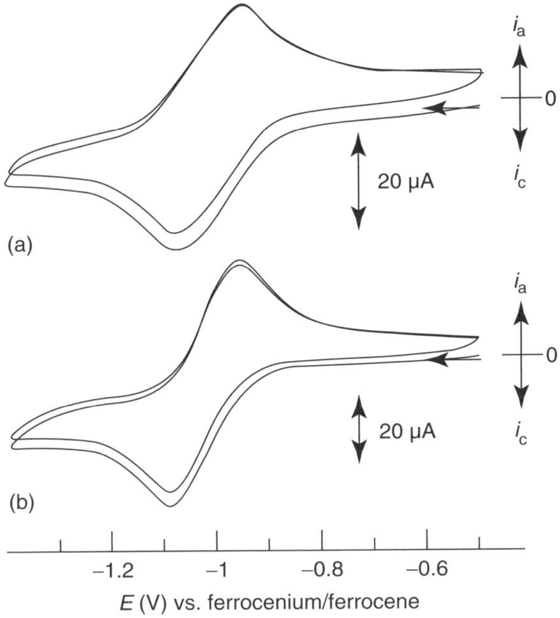

Figure 2.3 Cyclic voltammograms of (a) the monomer **109**, and (b) polymer **109** on a glassy carbon disk in 0.1-M Bu_4NClO_4-CH_2Cl_2. (Reprinted from *J. Am. Chem. Soc.*, 2003, *125*, 12420–12421. Copyright © 2003, American Chemical Society.)

C. σ- and π-Coordinated Metals

The reaction of bis-(pentamethylcyclopentadienyl) zirconium(IV) dichloride, **110**, with dilithioacetylene or dilithiodiacetylene, **111**, formed zirconocene acetylene polymers (Scheme 2.32).[197] Polymers **112** showed limited solubility in organic solvents other than *n*-hexane and possessed weight-average molecular weights between 55,000 and 68,000.

Lidrissi et al. synthesized air-sensitive, water-soluble coordination polymers containing two different organometallic moieties.[198] After reacting a cyclopentadienylruthenium complex containing water-soluble phosphane groups with AgOTf, polymers were isolated containing both ruthenium and silver (Scheme 2.33).

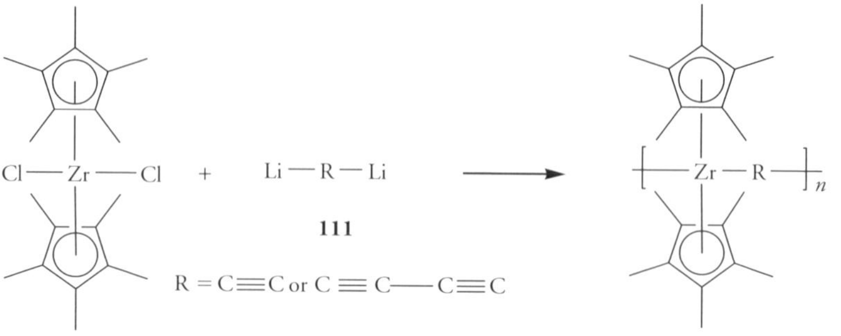

Scheme 2.32

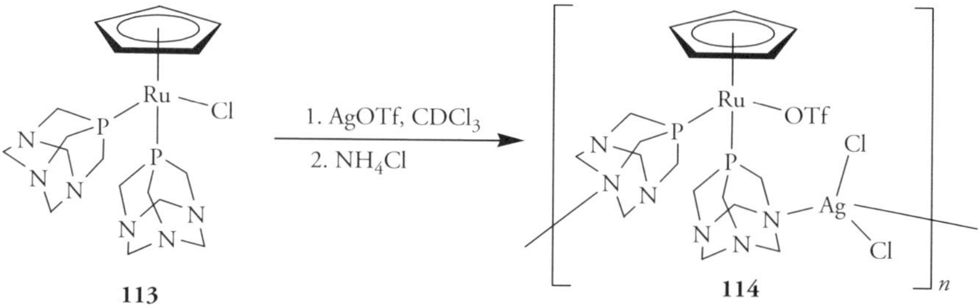

Scheme 2.33

Wang and co-workers reported the polycondensation of ferrocene with aromatic diazonium salts to give polymers containing iron in their backbones **115**.[199] The polymers were obtained in 38–69% yield and were insoluble in all organic solvents. The polymers did not display T_g and decomposed between 100 and 510°C. All the polymers showed ferromagnetism except for one that displayed paramagnetism.

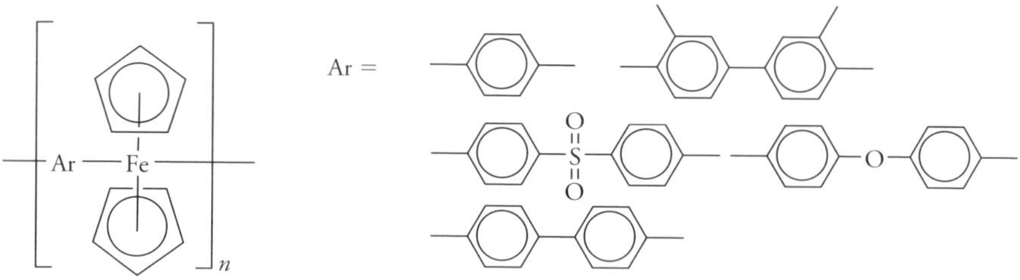

115

Chujo and co-workers made conjugated organometallic polymers containing platinum and tricarbonyl(arene)chromium moieties.[200] (η^6-1,4-Diethynylbenzene)tricarbonylchromium reacted with *trans*-$(PBu_3)_2PtCl_2$ in the presence of CuI produced copolymer **116** with M_w = 31,700 and PDI = 1.9. Analysis of the polymer showed 70% incorporation of chromium blocks and 30% uncoordinated phenylene units. Thermal analysis showed two degradative steps, the first occurring at 230°C due to the loss of carbon monoxide. The second decomposition step, at 310°C, was due to the degradation of the polymer backbone. Electrochemistry studies showed two broad redox couples at 0.52 V and 1.07 V for the oxidation of the chromium center and the platinum center, respectively. Strong interactions existed between the two metal centers due to the conjugated rigid-rod bridge.

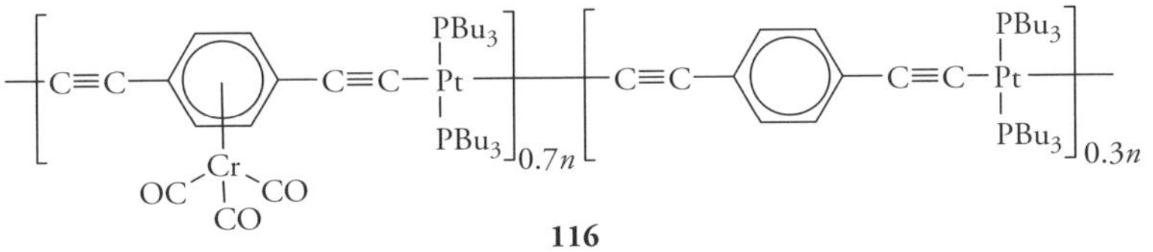

116

Pannel and co-workers isolated an organometallic polymer containing (η^5-C_5H_4)Fe in its backbone **117**.[201] (η^5-C_5H_4)Fe$(CO)_2SiMe_2SiMe_2Cl$ reacted with lithium diisopropylamide formed a polymer with a molecular weight of 6400 and PDI = 2.0 by light scattering. Thermal studies showed two weight losses, one between 50 and 75°C and another between 200 and 300°C. At a temperature of 500°C, 50% of the polymer remained.

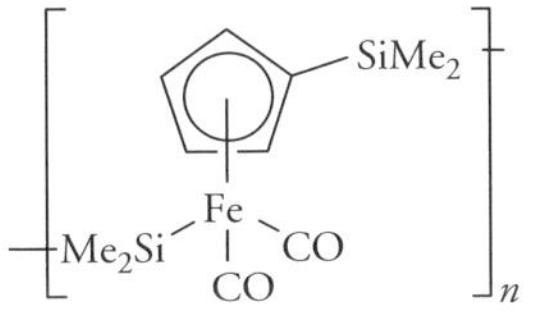

117

Polymers containing zirconacyclopentadiene rings in their backbone **119** have been prepared (Scheme 2.34).[202–206] Polymer **119** was isolated with a weight-average molecular weight of 13,000 and number-averaged molecular weight of 4600.[202] Low molecular-weight trimeric macrocycles **120** were generated when the polymerization took place at high temperatures or when the polymer was refluxed in THF. The metallacycle rings in the macrocycle could be converted to *trans*-dienyl units with hydrolysis by HCl.[202,203,204]

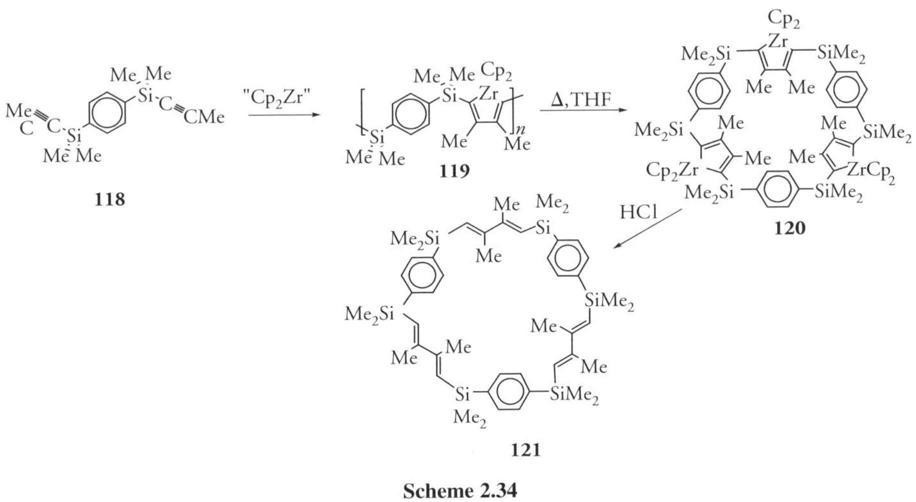

Scheme 2.34

A variety of different functionalities were introduced into this class of polymers by conversion of the zirconacyclopentadiene units (Scheme 2.35).[205,206] The reaction of polymer **124** with dimethylaceteylene dicarboxylate **125** in the presence of CuCl/LiCl produced the phenylene-based polymer **126** as an example. Polymers containing thiophene and phosphole units **127** were also prepared.

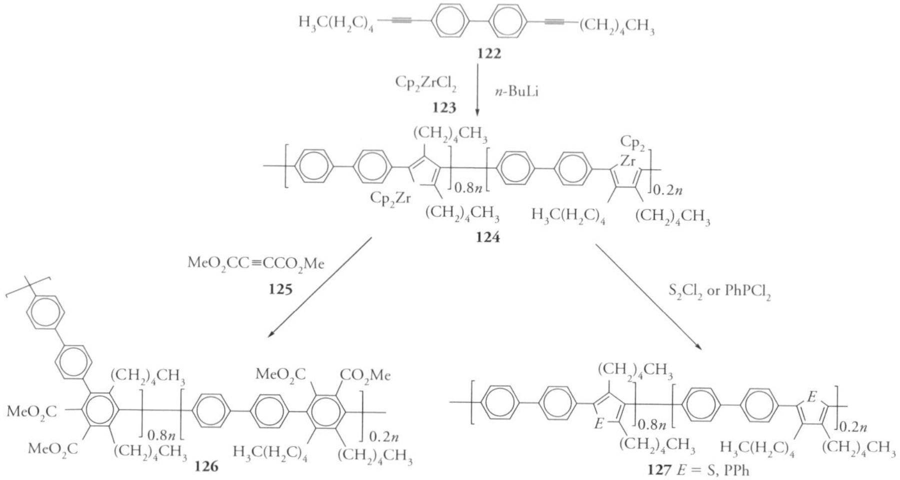

Scheme 2.35

Nishihara and Endo used a similar methodology for the synthesis of organocobalt polymers.[207–212] Scheme 2.36 shows the reaction of the organocobalt complex **128** with **129** to give the cobaltacyclopentadiene-containing polymer **130**.[212] Further reaction of polymer **130** with the isocyanate **131** gave a 2-pyridone ring **132**. Organic polymers containing pyridine, thiophene, selenophene, dithiolactone, phenylene, and diketone moieties were obtained from the reaction of polymer **130** with nitriles, sulfur, selenium, carbon disulfide, acetylenes, and oxygen, respectively.[212,213,214]

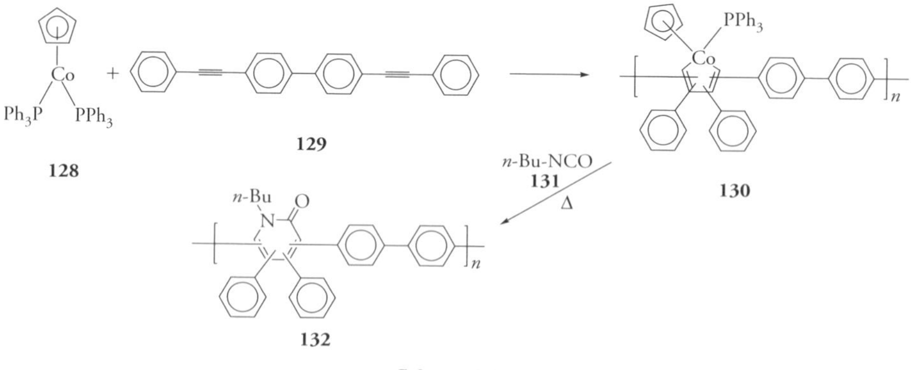

Scheme 2.36

Tomita and co-workers have reported the isolation of polymers containing titanocene.[215,216] The reaction of Cp^*_2TiCl with a number of acetylene containing monomers resulted in the formation of the titanocene containing polymer **133**. Polymers **133** were determined to possess weight-average molecular weights ranging from 2000 to 5500 (PDI = 2.0). Due to partial decomposition of the polymers, however, the molecular weights may have been underestimated.

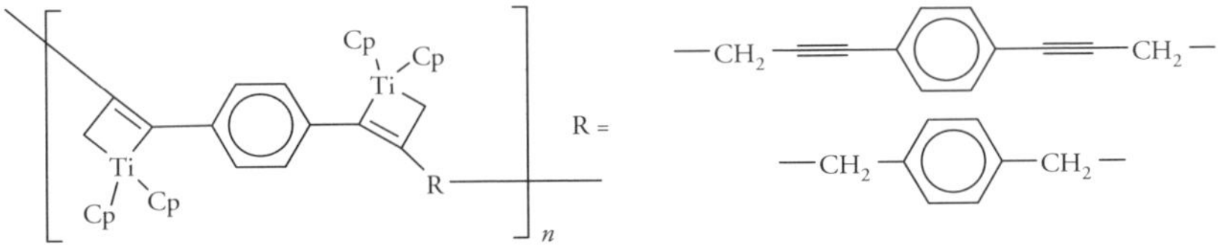

133

III. METALLIC MOIETIES PENDENT TO THE POLYMER BACKBONE OR IN THE SIDE CHAINS

A. σ-Coordinated Metals

The organocobalt polymer **134** was prepared from the reaction of $Co_2(CO)_8$ and an acetyl containing polysiloxane.[217] Mixed-metal polymers via the reaction of chromium tricarbonyl arene complexes with $Mo_2Cp_2(CO)_6$ or $Co_2(CO)_8$ were also reported.

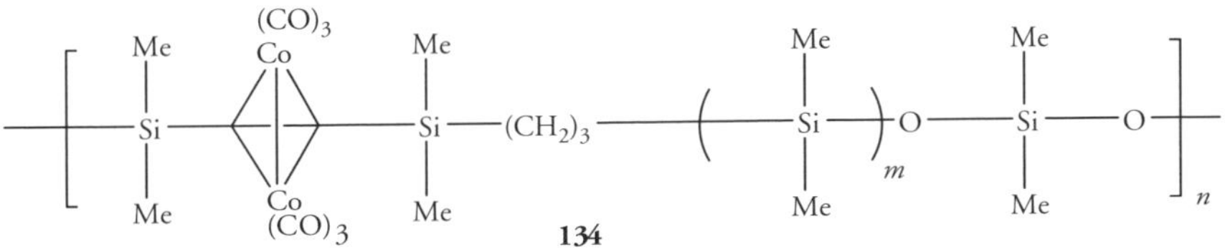

134

Swager and co-workers reported the preparation of conjugated poly(phenylene)s **135** containing cyclometallated platinum(II).[218] The Suzuki cross-coupling of dibromophenylpyridine-ligated Pt(II) with a fluorene diboronic ester resulted in the formation of a polymer containing square planar Pt pendent groups. The molecular weight of the polymer was 12,000 g/mol with a polydispersity index of approximately 2.5. Photophysical studies showed that this polymer was phosphorescent with a strong emission band at 585 nm, and a stokes shift of 200 nm.

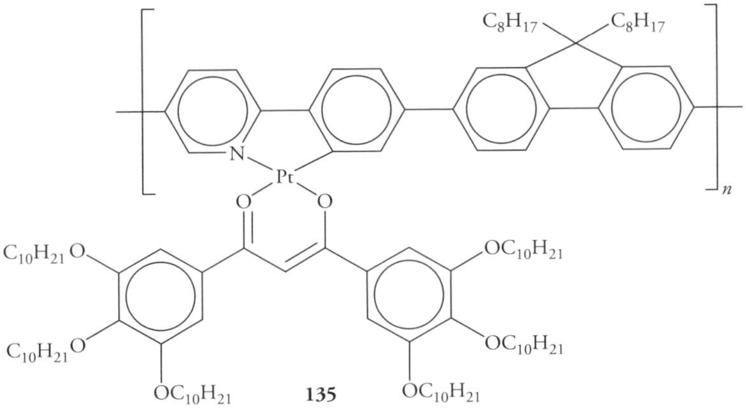

135

Diaz and co-workers isolated organometallic copolymeric phosphazenes containing a number of different metals (Fe, Ru, Cr, Mn, W).[219] These polymers were synthesized by first preparing the organic copolymers and then coordinating the metal unit. The organic copolymers $\{[NP(O_2C_{12}H_8)]_x[NP(OC_5H_4N)_2]_{1-x}\}_n$ were prepared via the reaction of $[NPCl_2]_n$ with 2,2′-dihydroxybiphenyl followed by the reaction with 4-hydroxypyridine. The copolymers had weight-averaged molecular weights of 696,000 and 550,000. Incorporation of W into the polymer proceeded via the reaction of the copolymers with $[W(CHOMe)(CO)_5]$. These organotungsten polymers were shown to be of higher thermal stability, but possessed lower T_g than the organic copolymers. The coordination of the metal moiety to these polymers resulted in the cross-linking of the polymers. Organometallic copolymers **136** containing Fe, Ru, and Mn units were also prepared using CpFe(dppe), $CpRu(PPh_2)_2Cl$, and $(\eta^5\text{-}CH_3C_5H_4)Mn(CO)_3$, and displayed 32–69% metal incorporation, with Fe showing the highest. The incorporation of the metal moieties resulted in some cross-linking of copolymers as well as an increase in the thermal stability.

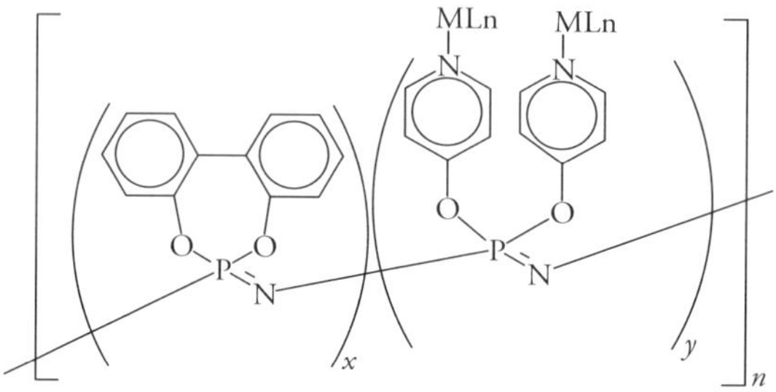

MLn = $W(CO)_5$, $Mn(\eta^5\text{-}CH_3C_5H_4)(CO)_2$, $Fe(\eta^5\text{-}C_5H_5)$dppe, or $Ru(\eta^5\text{-}C_5H_5)(PPh_3)_2$

136

Roviello et al. generated a number of polymers and copolymers containing cyclopalladated dinuclear complexes.[220] The homopolymers **137** were prepared by the reaction of the dinuclear acetato-bridged complexes with the polymeric ligand. Thermal studies showed that the poling temperatures ranged between 120 and 160°C. Liquid crystalline textures indicative of a smectic phase were observed for the polymer bearing R = H and R′ = Me groups. Decomposition occurred above 150°C and T_g ranged up to 110°C. Incorporation of the metallic unit was determined to be no less than 90%.

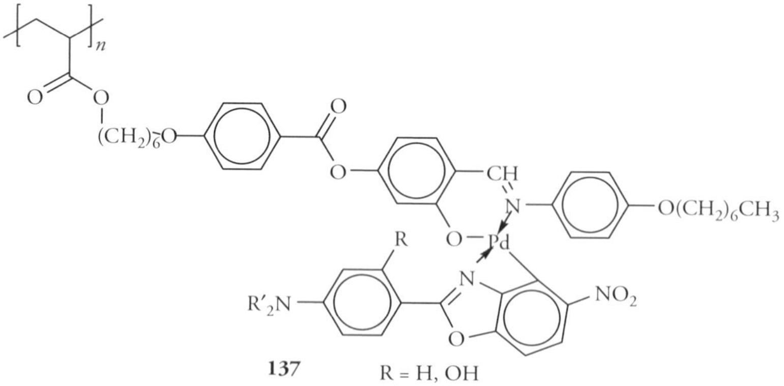

137

R = H, OH
R′ = Me, Et

B. π-Coordinated Metals

i. Metallocene-Containing Polymers. Aly has reported a new class of organometallic polyketones and copolyketones prepared via Friedel–Crafts reactions between 2,6-[bis(2-ferrocenyl)methylene]cyclohexanone(II) with various diacid chlorides (Scheme 2.37).[221] A series of Lewis acids such as $FeCl_3$, $SbCl_5$, and $AlCl_3$ were tested to produce the polymers. $AlCl_3$ was determined to give the highest yield and best degree of polymerization with a catalyst-to-reactant ratio of approximately 20. Polymers and copolymers of bis-ferrocenylidene cyclohexanone, **138**, and bis-benzylidenecyclohexanone, **141**, have also been reported.[221] The polymers and copolymers showed poor solubility in common organic solvents, but displayed good solubility in strong acids. The copolymers showed higher solubility than the homopolymers due to the increased flexibility from the cyclohexane unit. Thermal analysis of the polymers showed that degradation of 50% of the polymers occurred between 300 and 500°C. These materials had electrical behavior close to semiconductor behavior with conductivities around 3.5×10^{-10}–$2.9 \times 10^{-9}\,\Omega^{-1}\,cm^{-1}$.

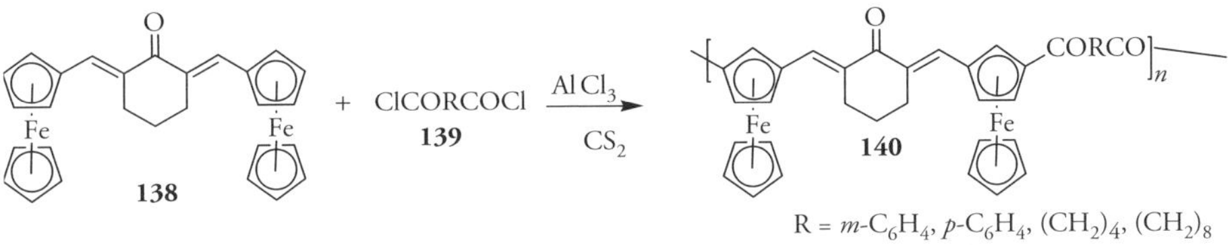

Scheme 2.37

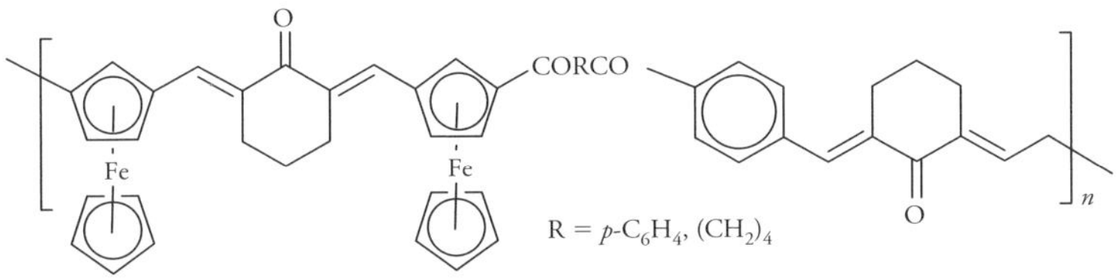

141

The generation of homo- and copolymers prepared by Friedel–Crafts reactions of 2,7-bis [(2-ferroceneyl)methylene]cycloheptanone with various diacid chlorides has been reported.[222] These polyketones were insoluble in most common organic solvents, but were quite soluble in concentrated H_2SO_4. The homopolymers were thermally stable with two decompositions between 220 and 520°C, and the copolyketones showed decomposition at temperatures between 250 and 600°C. In both cases the first weight loss was determined to depend on the nature of the polymers and occurred at a faster rate than the second degradation.

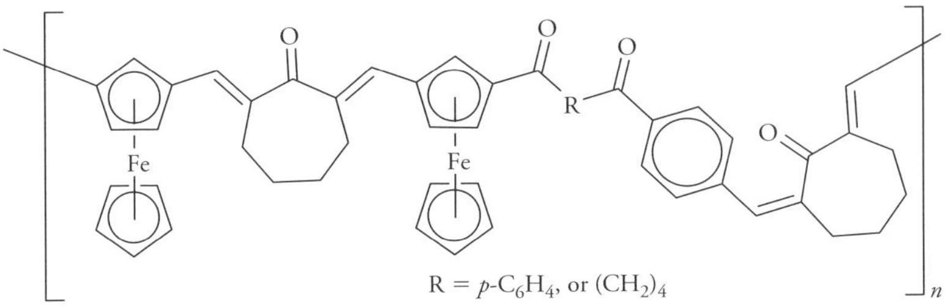

142

Durkee and co-workers have prepared polyferrocene block copolymers for use in catalysis.[223] The reported polymers were synthesized via sequential anionic polymerization of vinylferrocene and isoprene, followed by oxidation using silver triflate. The fraction of ferrocene converted to ferrocenium was directly proportional to the amount of Ag^+ added. These materials were tested for their catalytic activity toward the Michael addition reaction of ethyl-2-oxycyclopentane barboxylate and methylvinylketone. These materials showed k values similar in magnitude to the rates of iron(III) chloride.

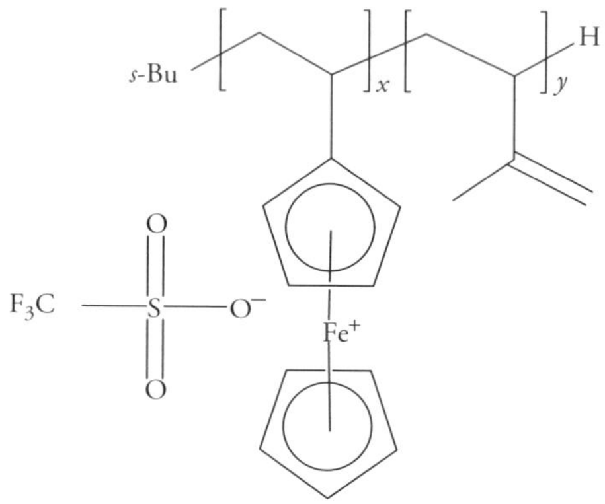

143

Plenio and co-workers isolated ferrocene-containing polymers with conjugated spacers in their backbones.[176,177] Low molecular-weight polymers **145** were synthesized via Sonogashira coupling of a diasteromeric mixture of monomer **146** (Scheme 2.38). Reacting monomer **144** with MeI and aza-12-crown-4 allowed for the isolation of **146**. Polymerization of monomer **146** under the same conditions as monomer **145** resulted in the isolation of the high molecular-weight polymer **147**. The molecular weight of polymer **147** may be overestimated, however, due to polymer aggregation, as was indicated by trails in the GPC traces.[176]

In 2000, Plenio et al. reported the synthesis of an enantiomerically pure (ee $>$ 98%) planar ferrocene complex **148**.[176] The palladium-catalyzed coupling reaction of monomer **148** gave polymer **149** with an optical purity of $[\alpha]_D^{20} = -198.0$ per ferrocene unit and a molecular weight of 10,000 as determined from GPC-LALLS (gel permeation chromatography with a low-angle laser light scattering detector) (Scheme 2.39).[176] Electrochemical studies on oligomer model complexes containing 1,1′- and 1,3-linked ferrocene units indicated that the homoannularly substituted polymers should give better electronic communication than the heteroannularly substituted polymers. This result was supported by the longer bathochromic shifts in these materials.[176,177]

The homopolymerization of ethynylferrocene **150a** and ruthenium **150b** (Scheme 2.40) produced polymers **151a** and **151b** containing pendent metallocenyl moieties.[224] Polymer **153** prepared by the copolymerization of **150a** with **152** using the Schrock molybdenum metathesis catalyst ($Mo(N\text{-}2,6\text{-}i\text{-}Pr_2C_6H_3)(CHCMe_2Ph)(OCMe(CF_3)_2)_2$). Buchmeiser and co-workers reported soluble living polymer systems with up to 50 double bonds along the backbone.[224,225,226] These polymers were in a head-to-tail configuration with *trans* double bonds.

The radical homo- and copolymerization of the ferrocene-containing monomer **154** allowed for the isolation of polymers containing pendent ferrocene moieties (Scheme 2.41).[228] High molecular-weight polymers were obtained from AIBN-initiated reactions, whereas other cationic initiators gave only low molecular-weight materials. Electrochemical studies showed that homopolymer **155** underwent two oxidations at $E_{1/2} = -0.13$ V and $E_{1/2} = +0.05$ V in

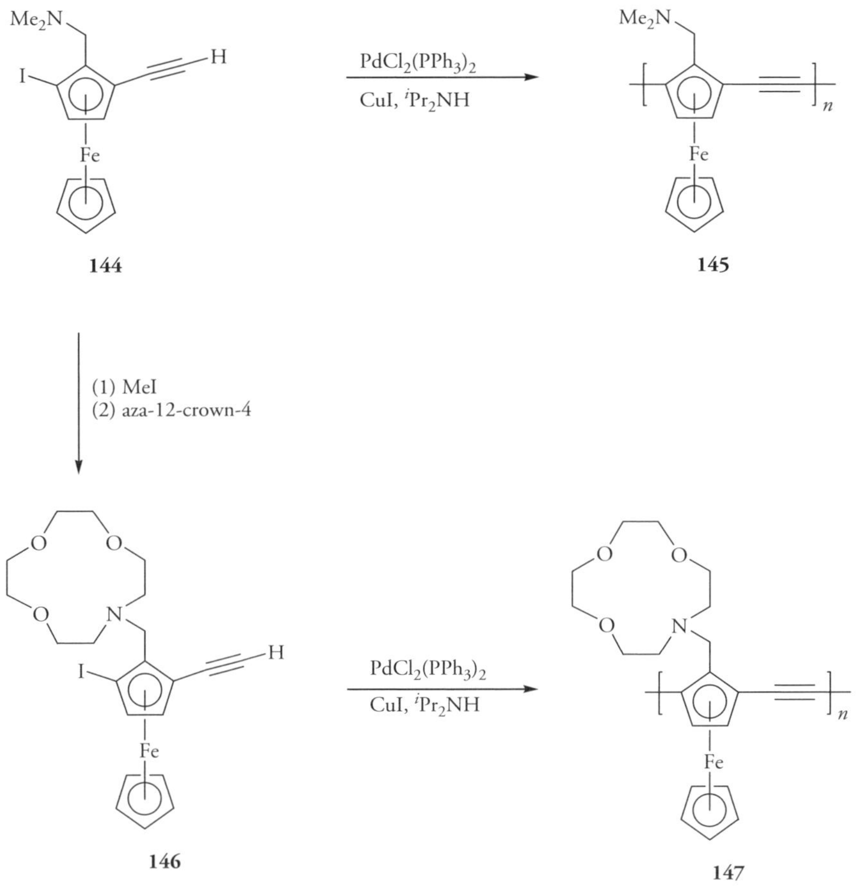

Scheme 2.38

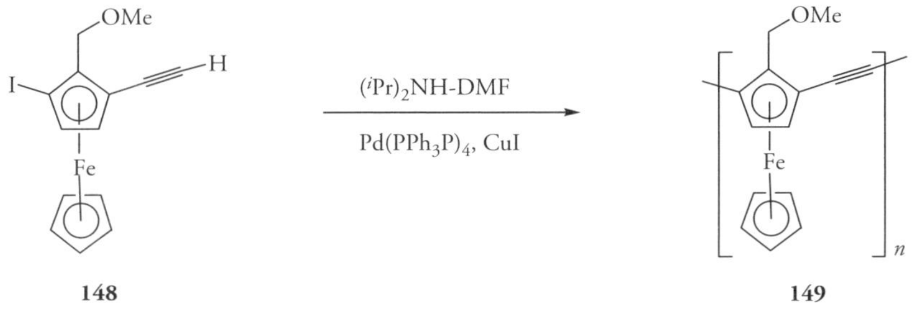

Scheme 2.39

dichloromethane. Copolymer **157**, however, showed only one oxidation at $E_{1/2} = -0.03$ V. Doping of the homo- and copolymers with iodine under argon gave conductive polymers with conductivities of 7.6×10^{-6} S/cm and 9.5×10^{-5} S/cm, respectively. Oxidation of the homopolymer **155** with quinone under vacuum resulted in lower conductivities than when doped with iodine, with conductivities ranging from 10^{-10} to 10^{-9} S/cm by increasing the oxidation level from 8% to 56%; however, under ambient conditions, the conductivity was 10^{-6} S/cm.

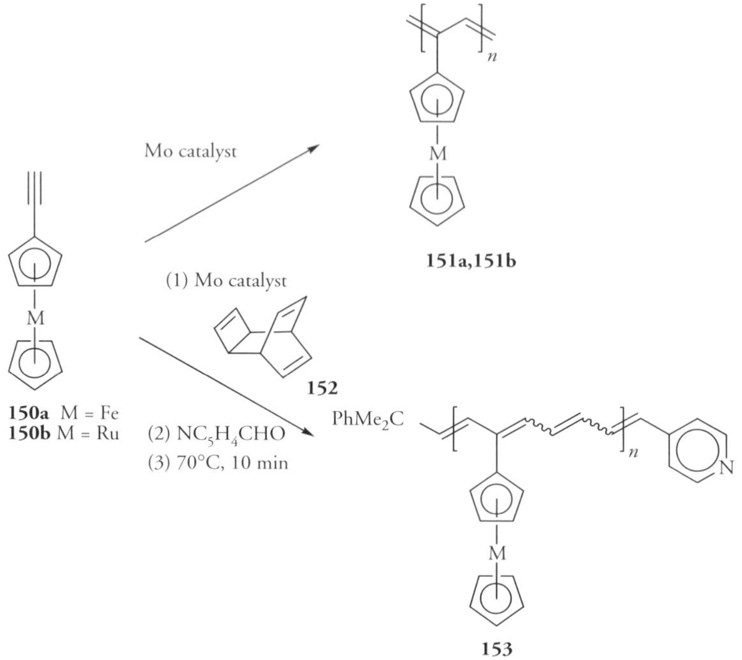

Scheme 2.40

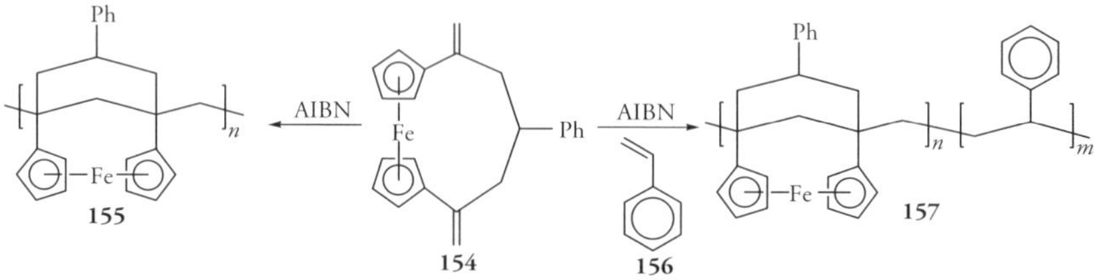

Scheme 2.41

The polycarbosilane $[(CH_2)_3Si(\eta^5\text{-}C_5H_4)_2ZrCl_2]_n$, **159**, was prepared from the ring-opening polymerization of the spirocyclic silacyclobutane-bridged monomer $(CH_2)_3Si(\eta^5\text{-}C_5H_4)_2ZrCl_2$, **158**, with Karstedt's catalyst (Scheme 2.42).[229] The ring opening of the zirconocene ring was also attempted, but polymerization could not occur under the reaction conditions. Both **158** and **159** showed some catalytic activity for ethylene polymerization.

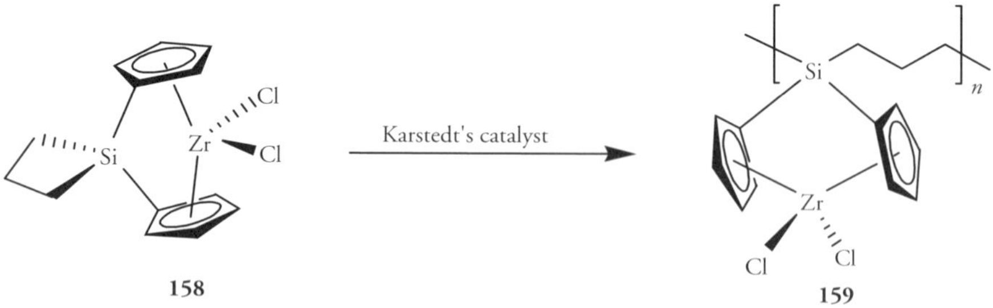

Scheme 2.42

Le Floch and co-workers isolated ferrocene-containing polythiophenes for use in deoxyribonucleic acid (DNA) detection.[230] The reaction between thiophene **160** and the functionalized ferrocene **161** gave the ferrocene functionalized polythiophene in a quantitative yield (Scheme 2.43). Polymerization in the presence of $FeCl_3$ led to a water-soluble polymer containing a cationic side chain and an electroactive ferrocene moeity. These polymers showed a DNA detection limit of 5×10^{-10} M.

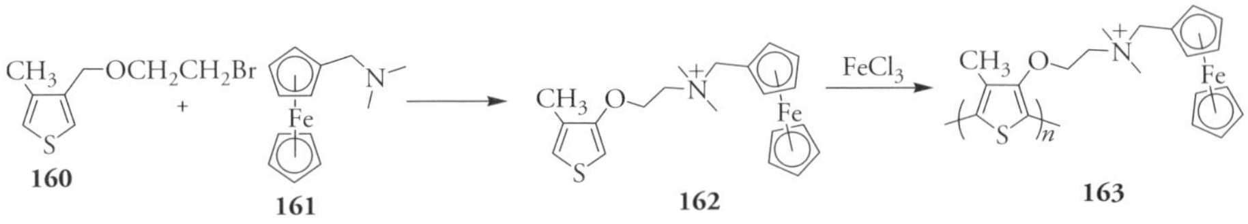

Scheme 2.43

Yang, Xie, and Wu reported the syntheses of 2-ferrocenylethyl methacrylate, acrylate, methacrylamide, and acrylamide.[231] Scheme 2.44 shows the preparation of the water-soluble polymer **167** by the copolymerization of 2-ferrocenylacrylamide with isopropylacrylamide. Depending on the ratio of organometallic-to-organic units the copolymer possessed low critical solution temperatures (LCST) of 26–29°C. Comparatively the pure organic poly(*N*-isopropylacrylamide) possessed an LCST of 32°C.

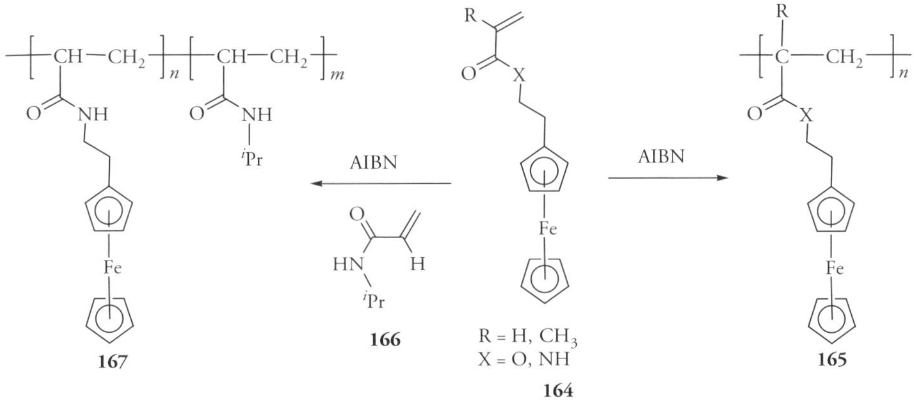

Scheme 2.44

The AIBN-initiated copolymerization of vinyl ferrocene with *N*-ethyl- or *N,N*,-diethylacrylamide gives thermosensitive polymers with less than 3% incorporation of vinyl ferrocene.[232] Increasing the amount of organometallic comonomer results in a decreased LCST value, Wright and co-workers generated methyl methacrylate copolymers containing ferrocene in their side chains, that display nonlinear optical properties (Scheme 2.45).[233,234] These polymers were prepared by the copolymerization of 5 mol % ferrocenyl functionalized monomer **168** with methyl methacrylate **169**. The isolated polymer **170** displayed second-harmonic-generation activity. Thermal studies indicated that the glass transition and melting temperatures of the organometallic polymers were similar to those of poly(methyl methacrylate). The T_g and T_m of the organometallic polymers were 120°C and 225°C, respectively.

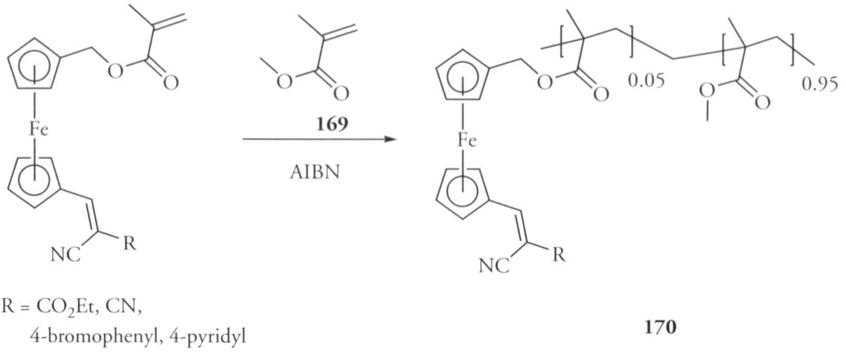

168

170

Scheme 2.45

Liquid crystalline polymers containing ferrocene in their side chains have been reported.[235–240] Deschenaux used free-radical synthesis to prepare thermotropic liquid crystalline polymethacrylates containing ferrocene (Scheme 2.46).[235] Polarized light microscopy showed that monomer **171** and its corresponding polymer **172** exhibited enantiotropic smectic A and C phases.

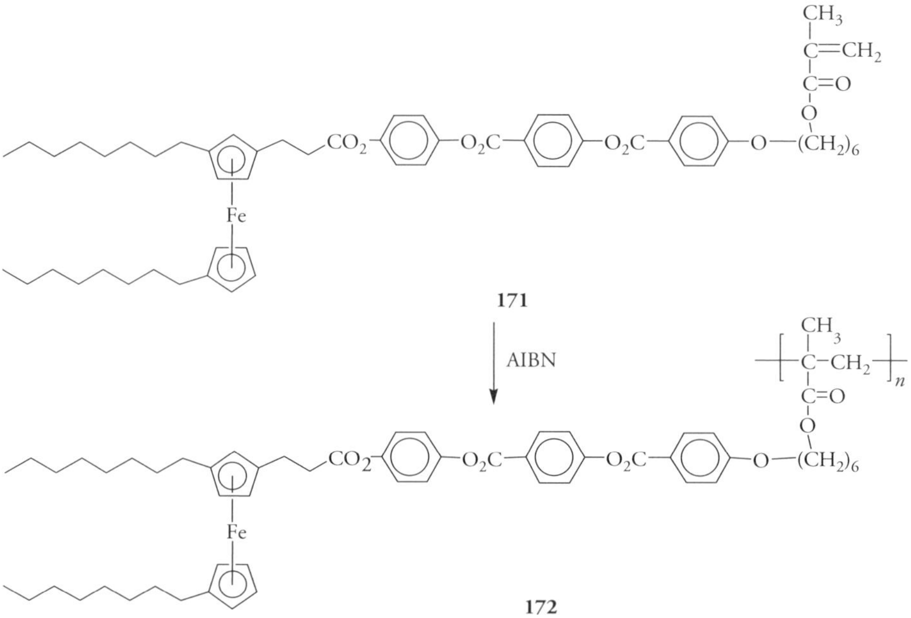

171

172

Scheme 2.46

Deschenaux and co-workers also isolated liquid crystalline polysiloxanes containing ferrocene in their side chains.[236] These polymers contained either 1,3- or 1,1′-disubstituted ferrocene and were synthesized from the reaction of preformed polysiloxanes and vinyl organoiron monomers.

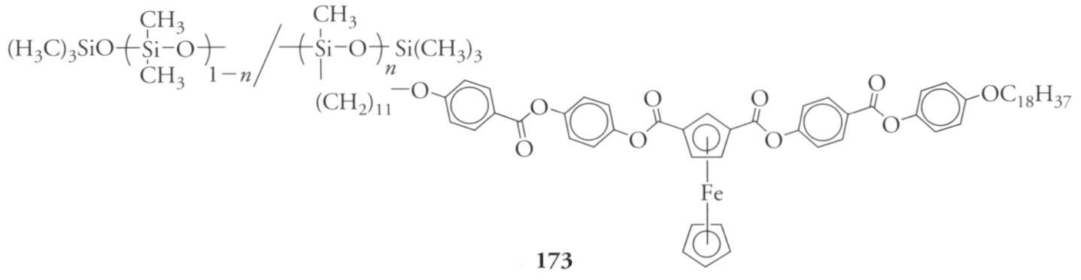

173

Liquid crystalline copolymers containing ferrocene in their side chains, **174**, have been prepared by the methodology outlined in Scheme 2.47.[240] Ionomers **175** and **176** were generated by the oxidation of polymer **174** with copper(II) perchlorate or benzoquinone, respectively. Using ferrocene derivatives with four alkyl substituents on the cyclopentadienyl rings, the stability of the ferrocene-containing polymers increased compared to similar polymers containing monoalkylated cyclopentadienyl rings. Thermal analysis showed that ionomer **175** was unstable above 100°C, whereas the polymers with sulfate counterions were much more stable. Thermal studies also determined that the transition temperatures of the polymers were not significantly affected by up

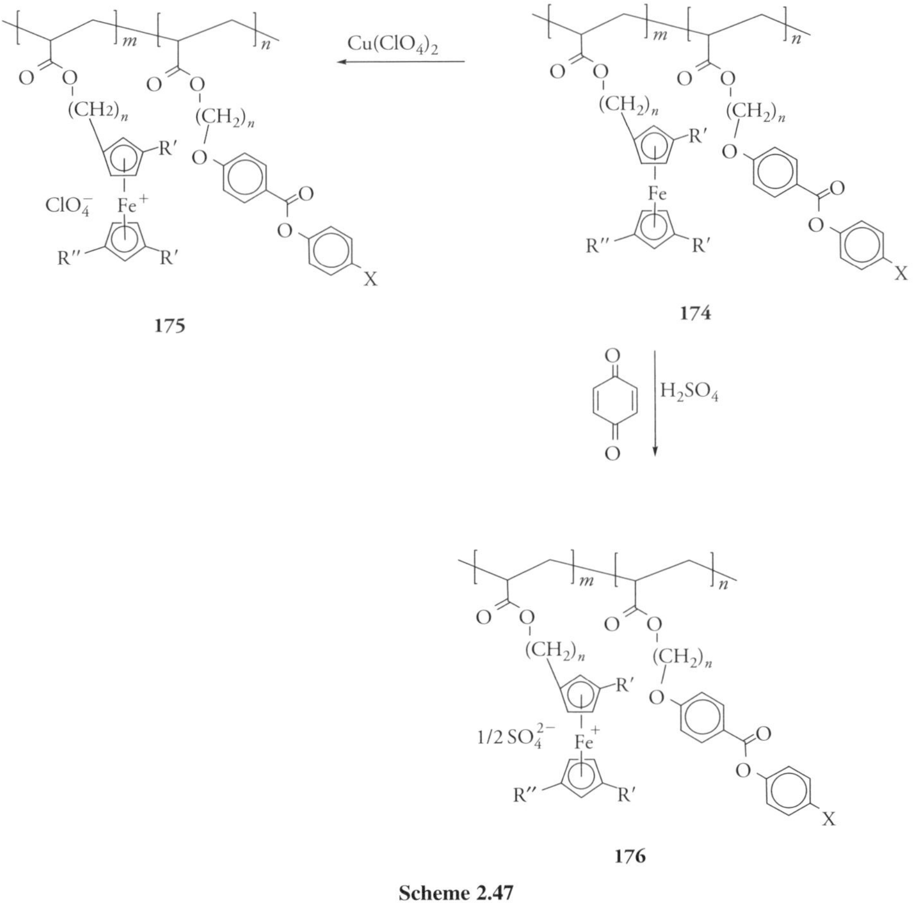

Scheme 2.47

to 10% incorporation of ferrocene derivatives. The ferrocenium complexes formed from the oxidation of the polymers were found to form clusters and segregate from the liquid crystalline phase.

The electropolymerization of metallocene-functionalized pyrrole and thiophene has been utilized for the creation of organometallic polypyrroles and polythiophenes.[181,241,242,243] Zotti et al. reported the electrochemical homo- and copolymerization of ferrocene-functionalized pyrrole, **177**.[242] Polymer **179** was prepared via the copolymerization of monomer **177** with pyrrole (Scheme 2.48). This polymer possessed an electrical conductivity of 1.5×10^{-2} S/cm.

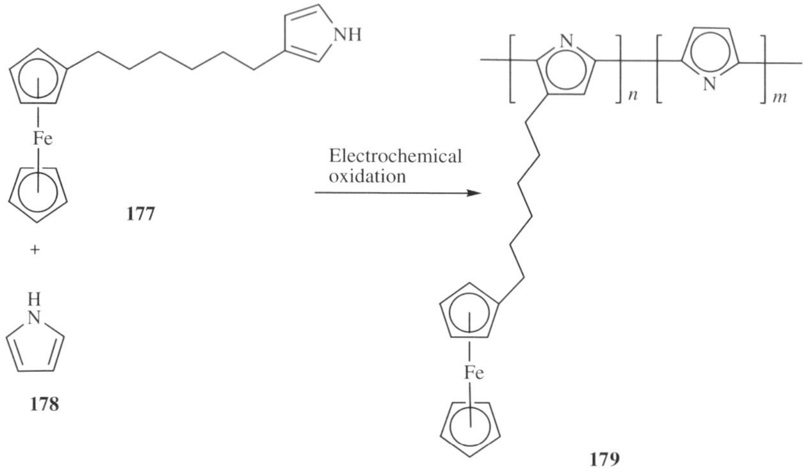

Scheme 2.48

Polythiophenes with ferrocene-containing side chains have been isolated from the electrochemical polymerization of monomers **180–182**.[242,243] The polymerization of monomer **180** gave a polymer with a redox potential identical to that of *n*-hexylferrocene. However, polymers obtained from monomers **181** possessed redox potentials 0.03 V to 0.15 V higher than that of the polymer of **180**. This increase in redox potential was due to the stronger interactions between the ferrocene moieties and the polymer backbone. Hence, by increasing the alkyl chain length connecting the thiophene to the ferrocene, the redox potential decreased.[242,242] The redox potential of the polymer obtained from monomer **180** increased through copolymerization with 4-hexylcyclopentadithiophene and gave a conductivity of 1 S/cm. The homopolymer of monomer **180** prepared by electrochemical oxidation displayed an in situ conductivity of 40 S/cm.

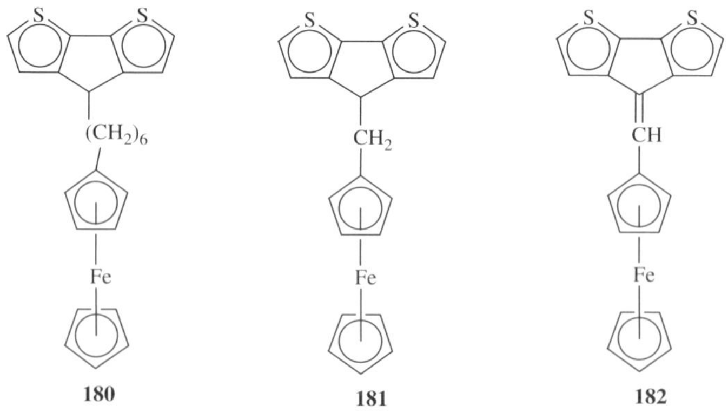

Takahashi and co-workers have performed the living polymerization of ferrocene derivatives to obtain helical, chiral polyisocyanides **183**.[244] Using 100 equivalents of the monomer and a 20-hour reaction time, polymers with number-average molecular weights ranging between 9800 and 22,000 and PDI = 1.15 were obtained in 90% yield.

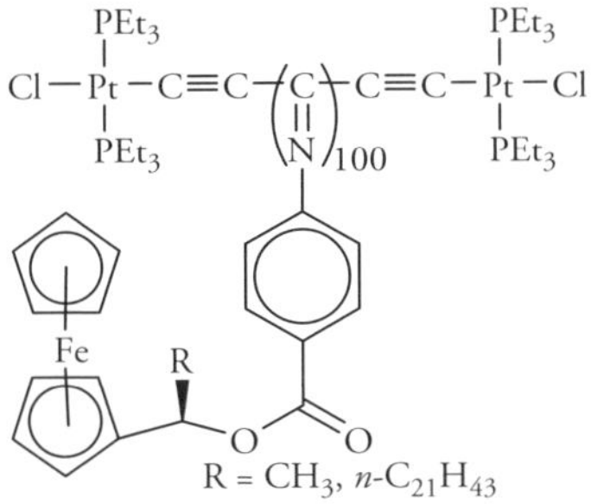

183

Ferrocene-containing polymers were examined for potential applications in cancer research.[244–249] Water-soluble ferrocene-containing polymers such as **184** possess interesting antiproliferative properties. These polymers contain easily cleavable ferrocenyl groups; ferrocenium salts can trap free radicals that can play a role in cancer progression.

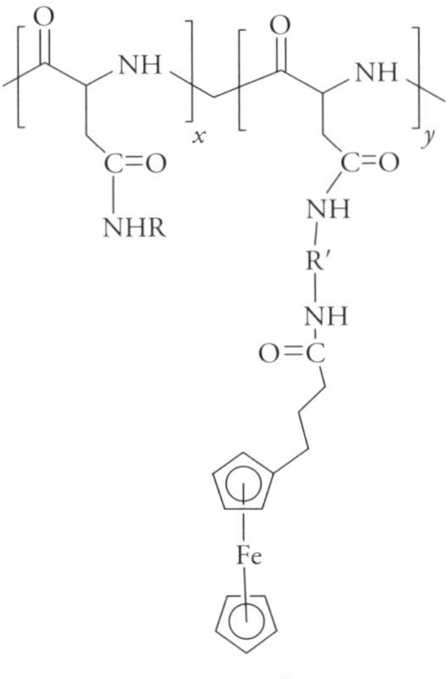

184

Ferrocene-functionalized norbornene monomers have been copolymerized with norbornene attached to a gold nanoparticle through alkanethiol bonds (Scheme 2.49).[250] This polymerization was a three-step synthesis where the first step was the metathesis of the norbornene on the gold nanoparticle with one equivalent of the catalyst. This step was then followed by the addition of monomer **186,** and the addition of monomer **187** 20 minutes later. This methodology could be used to prepare nanoparticles with varying layers of polynorbornene.

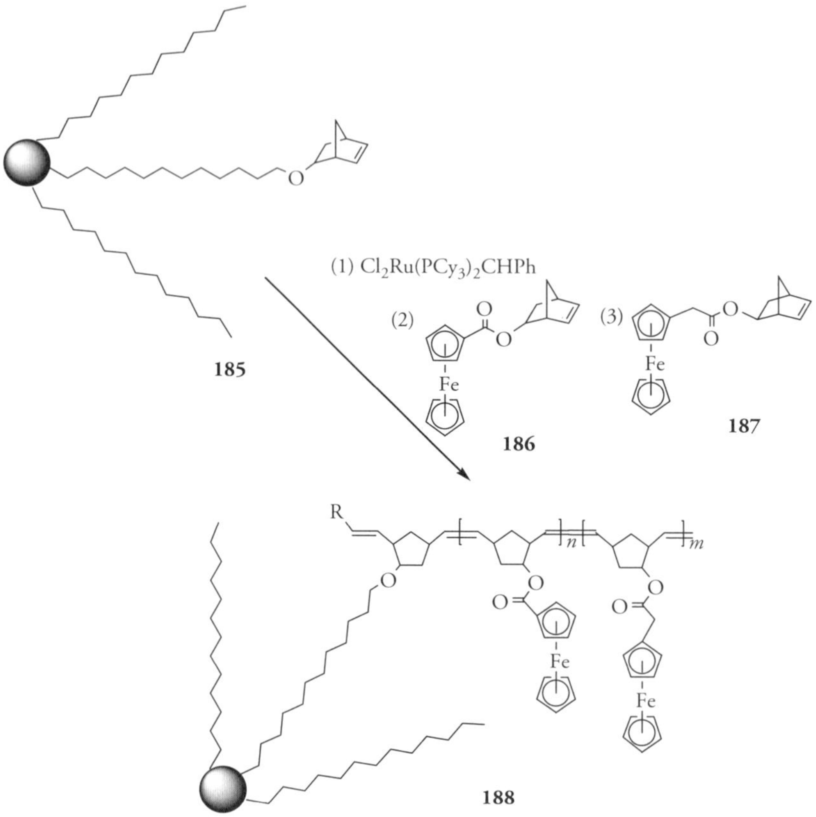

Scheme 2.49

ii. Cyclopentadienyl–Metal Arene Complex Containing Polymers. Abd-El-Aziz and co-workers have reported the synthesis of polyether/imines containing cationic cyclopentadienyliron moieties.[251] The reaction of the dialdehyde cyclopentadienyliron complex, **189**, with aliphatic and aromatic diamines, **190**, led to the isolation of soluble polymers **191** (Scheme 2.50). These polymers displayed two thermal decomposition steps due to the decoordination of the cation iron moieties and the degradation of the polymer backbone. Interestingly, increasing the length of the aliphatic spacer (4–12 CH_2 groups) increased the temperature required to cleave the cationic cyclopentadienyl groups from 250°C for rigid spacers to 350°C.[251,252]

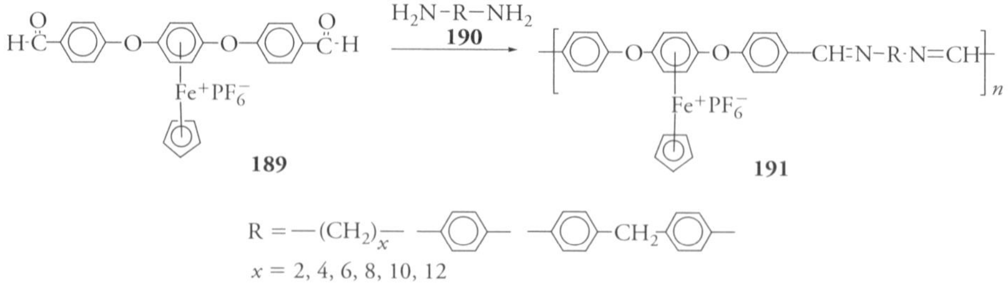

Scheme 2.50

The stepwise synthesis of aromatic ether complexes containing up to 35 pendent organoiron moieties in their backbones has been reported.[253] Abd-El-Aziz and co-workers have prepared high molecular-weight organoiron polymers containing aliphatic or aromatic bridges using this stepwise method (Scheme 2.51).

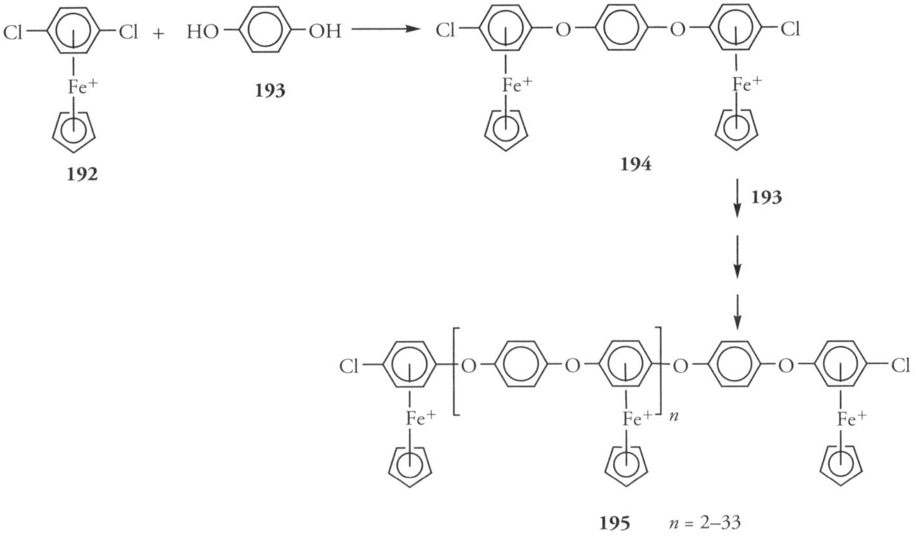

Scheme 2.51

The reaction of the cationic polyether, **196**, with sodium cyanide led to the isolation of the neutral adduct, **197**.[254] Demetallation of these adducts via chemical oxidation with 2,3-dichloro-5,6-dicyano-1,4-benzoquinone (DDQ) gave the nitrile functionalized polyaromatic ethers **198** (Scheme 2.52).

Abd-El-Aziz and co-workers also reported S_NAr reactions of dichloroarene cyclopentadienyliron complexes with sulfur-, oxygen-, and nitrogen-based nucleophiles to prepare soluble polyaromatic ethers, thioethers, and amines.[255,256] The reaction of the η^6-1,4-dichlorobenzene-η^5-cyclopentadienyliron complex, **192**, with dinucleophiles, **199**, in the presence of K_2CO_3 gave organoiron polymers **200** containing pendent cyclopentadienyliron moieties (Scheme 2.53).[256] Polymers **200** possessed good solubility in polar organic solvents. Photolytic demetallation of these polymers in acetonitrile gave their organic analogs, **201**. Irradiation of the polymers caused the excitation of the cationic cyclopentadienyliron moieties, converting them from η^6 to η^4. This new η^4 complex coordinates to the solvent, forming $[CpFe(NCCH_3)_3]^+$, which then decomposes into ferrocene and various iron salts. The organic analogs displayed lower solubilities than their corresponding cationic polymers; however, the presence of aliphatic groups in the backbone enhanced solubility. The weight-average molecular weights of the organic polymers were determined to be between 7300 and 21,400. Using 1,2- and 1,3-dichlorobenzene complexes was also found to increase solubility.

Electrochemical studies of these polymers showed a reversible reduction couple.[257] The cyclic voltammogram of an organometallic polythioether is shown in Figure 2.4 where the $E_{1/2} = -0.85$ V.

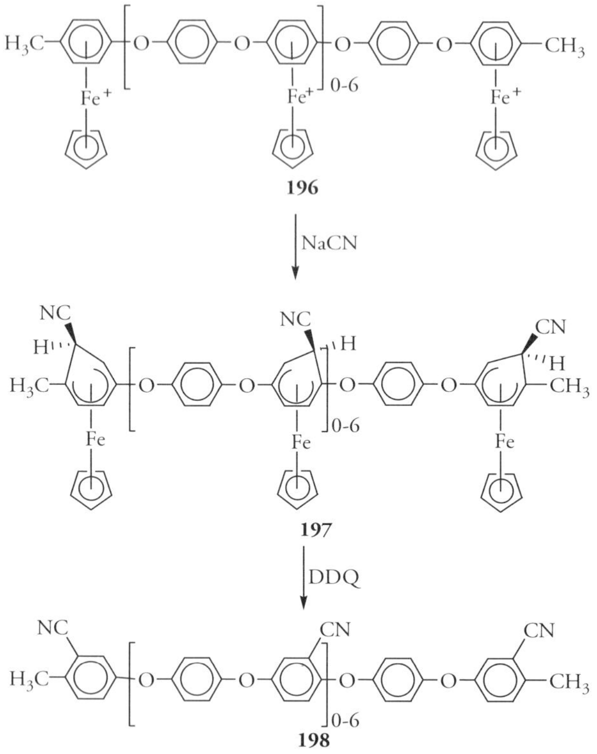

Scheme 2.52

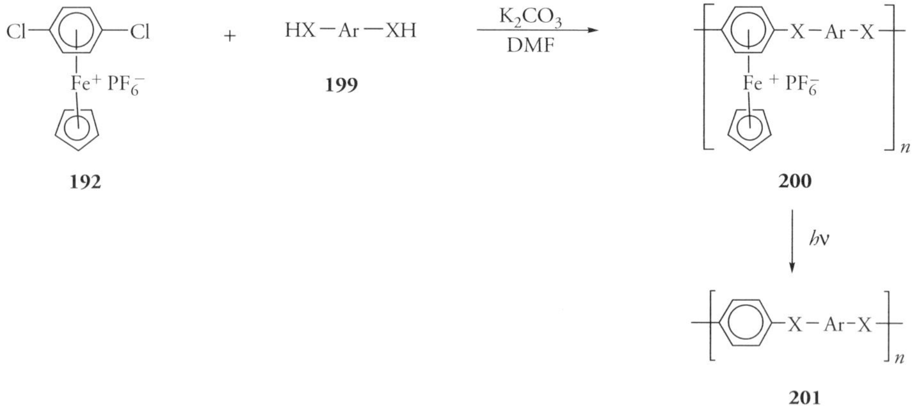

Scheme 2.53

Aromatic and aliphatic spaces have been incorporated into organoiron polymers to determine their affect on the thermal stability and glass transition temperature of the polymers.[256] The thermal stability of these polymers increased after the loss of the metallic moieties. The metallated polymer shows two decomposition steps at 245°C, corresponding to decomposition of the cyclopentadienyliron moieties and a second weight loss beginning at 511°C, corresponding to

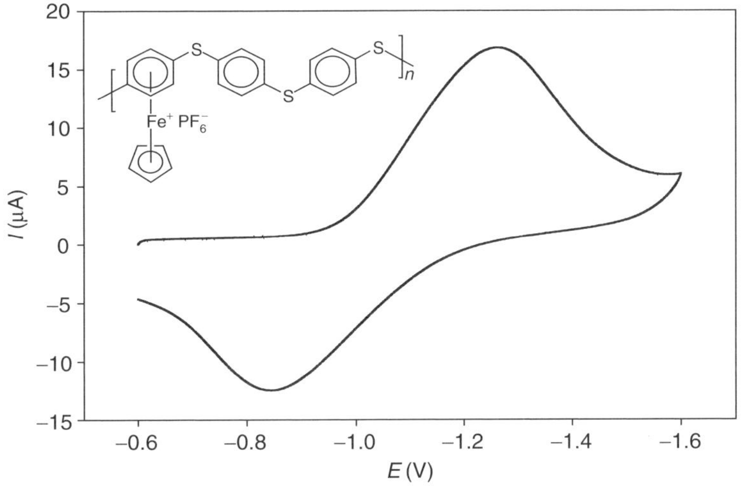

Figure 2.4 Cyclic voltammogram of organoiron polyphenylene sulfide obtained at −30°C in propylene carbonate with a scan rate of 2 V/s.

degradation of the polymer backbone. The organic analogue of this polymer only shows one weight-loss step beginning at 512°C. Differential scanning calorimetry (DSC) indicated that the more rigid polymers with oxygen-containing spacers had the higher glass transition temperatures and the polymers with aliphatic sulphur-containing spacers possessed the lower T_g.

Polymers containing alternating ether–thioether or thioether–amine bridges were also prepared.[256] Scheme 2.54 shows the synthesis of polymers containing alternating ether–thioether bridges. The organometallic polymers all displayed good solubility in polar organic solvents; however, the removal of the iron moieties by photolysis gave an insoluble analog of polymer **204** and a soluble analog of polymer **206**. The organic analog of polymer **206** was determined to possess a weight-average molecular weight of 21,700 and a PDI of 2.4.

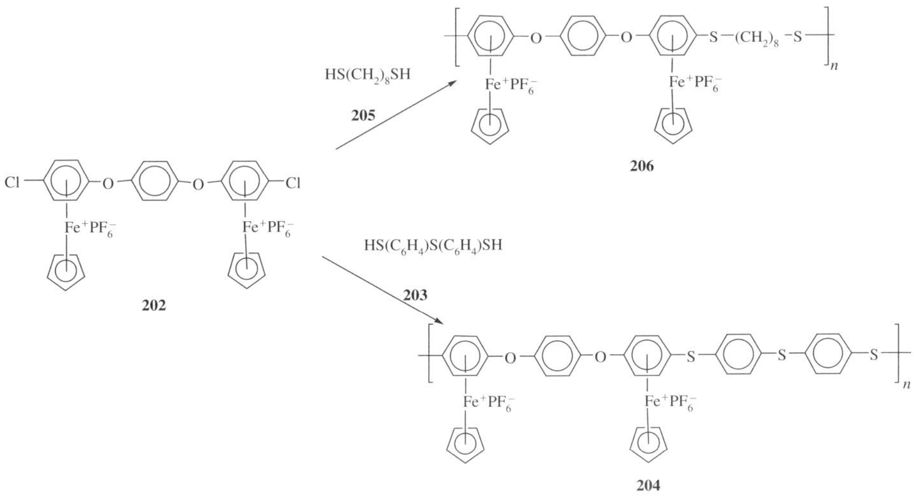

Scheme 2.54

Recently, Abd-El-Aziz and co-workers generated organoiron polymers containing azo dyes in their side chains.[258,259] The reaction of an azo dye containing bimetallic organoiron complex, **206**, with various S- and O-containing dinucleophiles gave organoiron polymers **208** with azobenzene chromophores in their side chains (Scheme 2.55). Polymers **208** were red in color and absorbed between 417 and 489 nm. Decomplexation of the cyclopentadienyliron moieties by photolysis produced their organic analogs, **209**. Polymers **209** underwent decoloration upon irradiation in the presence of hydrogen peroxide.

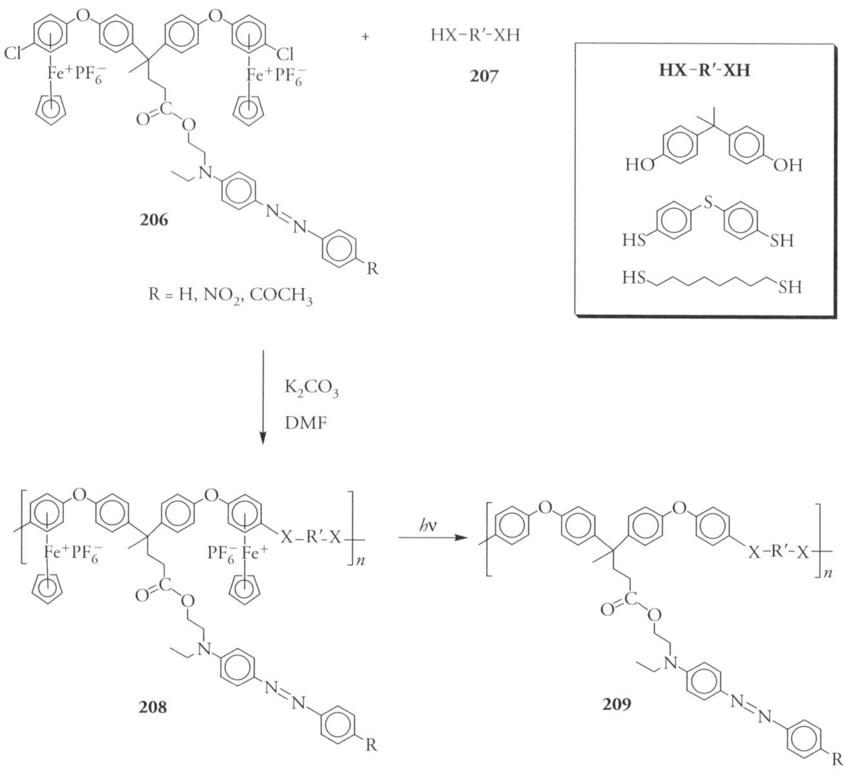

Scheme 2.55

At low concentrations (0.025 M) the photolysis of these polymers resulted in the decoloration due to the degradation of the azo dye.[260] After the cationic cyclopentadienyliron moieties are cleaved from the polymer, they form complexes with the azo chromophore, which then leads to photodegradation of the azo chromophore.

Cationic polyferrocenes containing pendent cyclopentadienyliron moieties as well as azo dyes in their backbone, **210**, have been prepared.[261] These polymers displayed excellent solubility in polar organic solvents and exhibited λ_{max} around 419 nm. Electrochemical studies showed that these polymers underwent two redox couples corresponding to the two different iron centers. The oxidation of the ferrocene occurred at $E_{1/2} = 0.89$ V and the reduction of the cationic cyclopentadienyliron complex occurred at $E_{1/2} = -1.42$ V. Photolytic cleavage of the cationic cyclopentadienyliron moieties produced the neutral polyferrocene analogs.

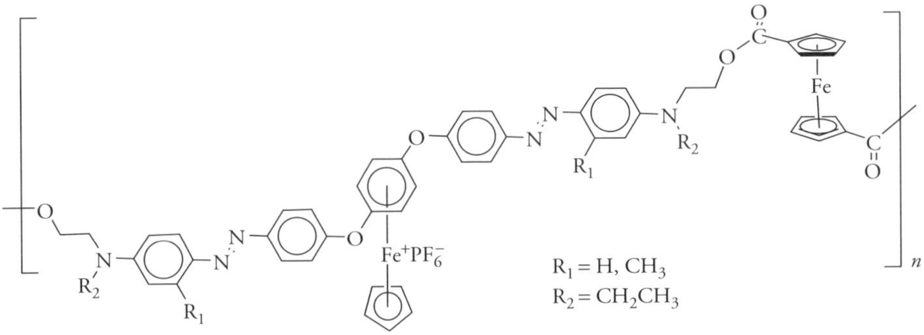

210

Polymers containing pendent mixed metal groups, **212**, have been isolated from the reaction of cyclopentadienyliron-containing complexes with cyclopentadienylruthenium-containing complexes (Scheme 2.56).[262,263] Cleavage of the metallic moieties of polymer **213** allowed the determination of the weight average molecular weight of the organic analog (M_w = 12,600, PDI = 1.9), which then gave an estimate of the weight-average molecular weight of the metallated polymer (M_w = 19,000).

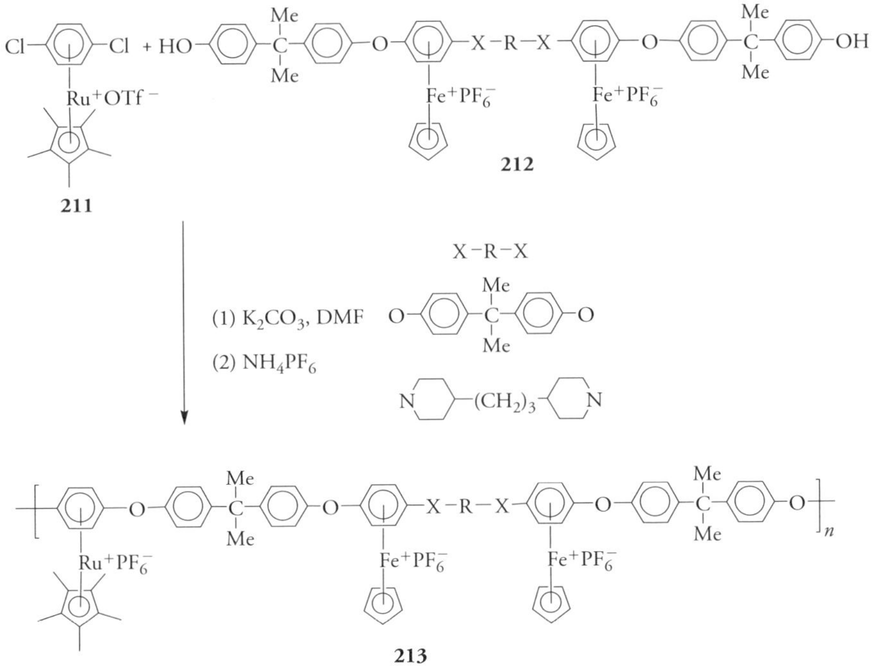

211 **212** **213**

Scheme 2.56

A series of side-chain polyferrocenes containing cationic iron groups pendent to the polymer backbone, **214**, have been reported by Abd-El-Aziz and co-workers.[264] Prepared by nucleophilic aromatic substitution reactions, these polymers displayed good solubilities in polar solvents, and M_w ranged from 11,100 to 56,200, with PDIs between 1.2 and 3.1. Electrochemical analysis showed that the cationic iron moiety was reduced at −1.1 V and the neutral iron group oxidized at 0.7 V (Figure 2.5). Photolytic cleavage of the cationic iron moieties gave neutral polymers containing ferrocene in their side chains, **215**.

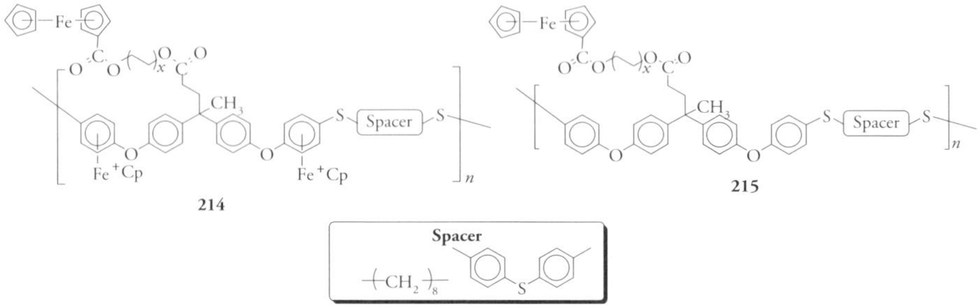

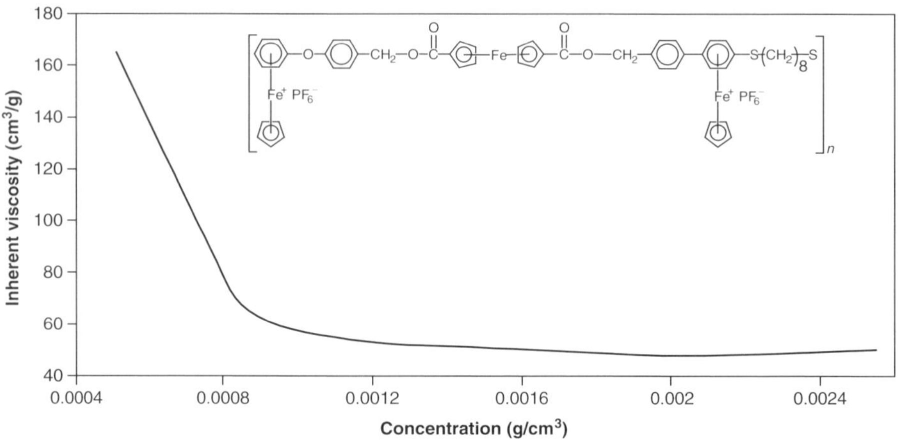

Figure 2.5 Plot of inherent viscosity versus concentration.

The synthesis of cationic organoiron containing polymethacrylates and polystyrenes has been reported. The radical polymerization of methacrylate and styrene monomers containing cationic cyclopentadienyliron complexes produced polymers with cationic cyclopentadienyliron complexes in their side chains (Scheme 2.57).[265,266] Monomer **216** and polymer **217** were both redox active and underwent reversible reduction processes. An organic analog of the cationic organoiron polymer was obtained by photolytic cleavage of the cationic cyclopentadienyliron moieties. The organic analog possessed a weight-average molecular weight between 48,000 and 68,000.

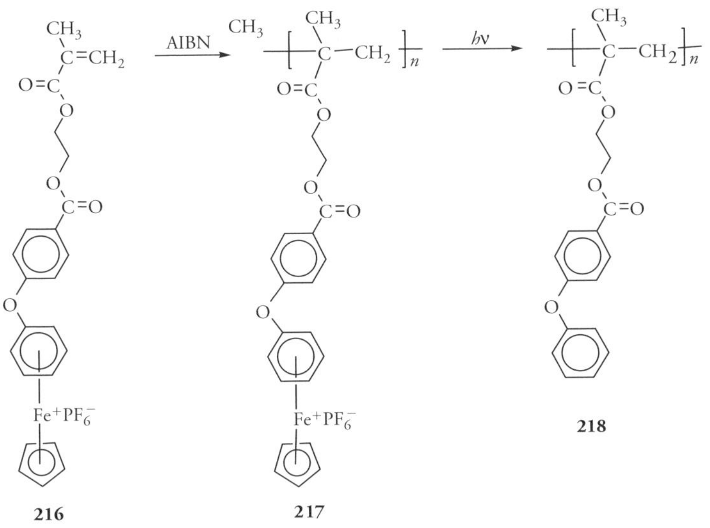

Scheme 2.57

Similar polymethacrylates containing two cationic groups in the side chain **219a, b** were also prepared.[266] Electrochemical studies on polymers **219a, b** showed a reversible reduction of the iron centers between $E_{1/2} = -1.1$ and -1.2 V.

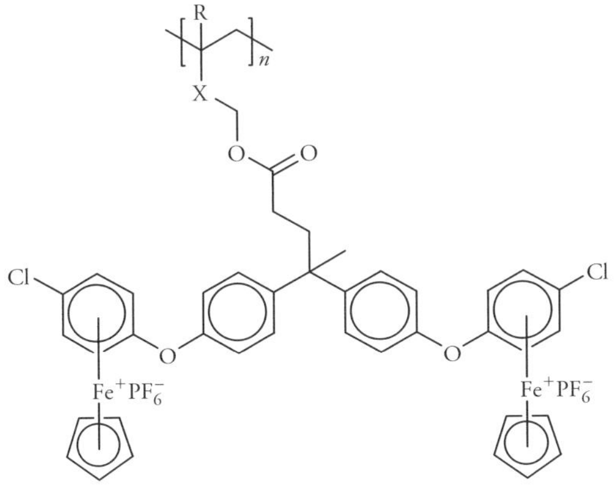

(a) R = CH_3, X = CH_2OOC
(b) R = H, X = p-C_6H_4

219a,b

Organoiron polynorbornenes have been prepared from the ROMP of cationic cyclopentadienyliron-functionalized norbornene monomers in the presence of Grubbs' catalyst (Scheme 2.58).[267–271] Photolysis of the organometallic polymers, **221**, resulted in organic analogs, **222**, with weight-average molecular weights between 10,600 and 18,600 (PDI 1.3–1.6). Electrochemical studies showed the organometallic polymers were reduced between $E_{1/2} = 1.2$ and $E_{1/2} = -1.4$ V.

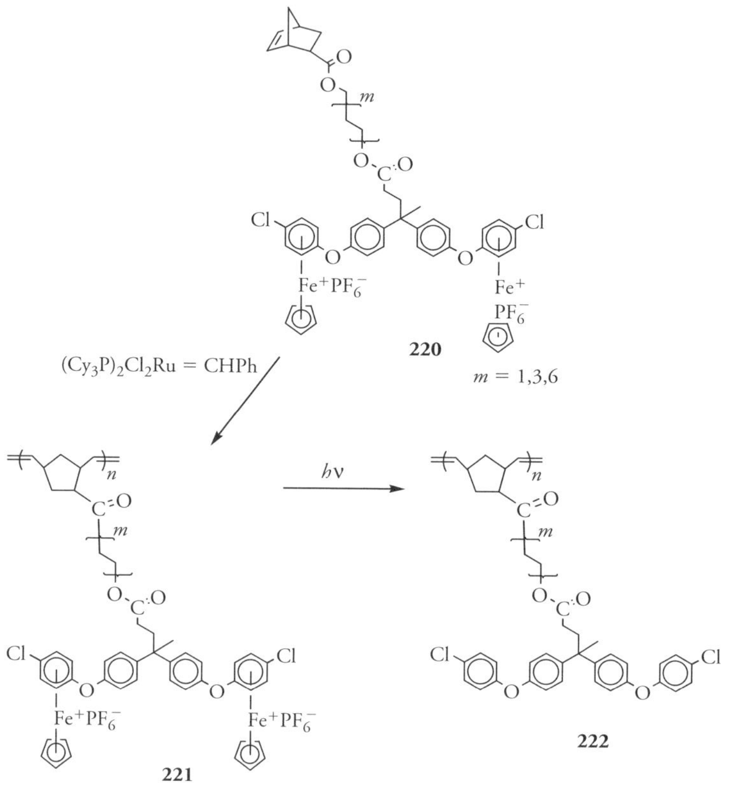

Scheme 2.58

Abd-El-Aziz and co-workers have also isolated organometallic polynorbornenes incorporating azobenzene chromophores.[272] The ROMP polymerization of azo dye functionalized organoiron norbornene-containing monomers **223a,b** with Grubbs' catalyst resulted in colored polynorbornenes (Scheme 2.59). Polymers **224a,b** were dark orange in color and displayed λ_{max} values between 420 and 430 nm in DMF. The addition of HCl to the polymer solution caused a bathochromic shift to λ_{max} values between 510 and 520 nm. Cyclic voltammetry showed that polymers **224a,b** underwent a reversible reduction between −1.2 and −1.4 V. Organic analogs were obtained via photolysis and possessed M_w values ranging from 9200 to 21,800.[272]

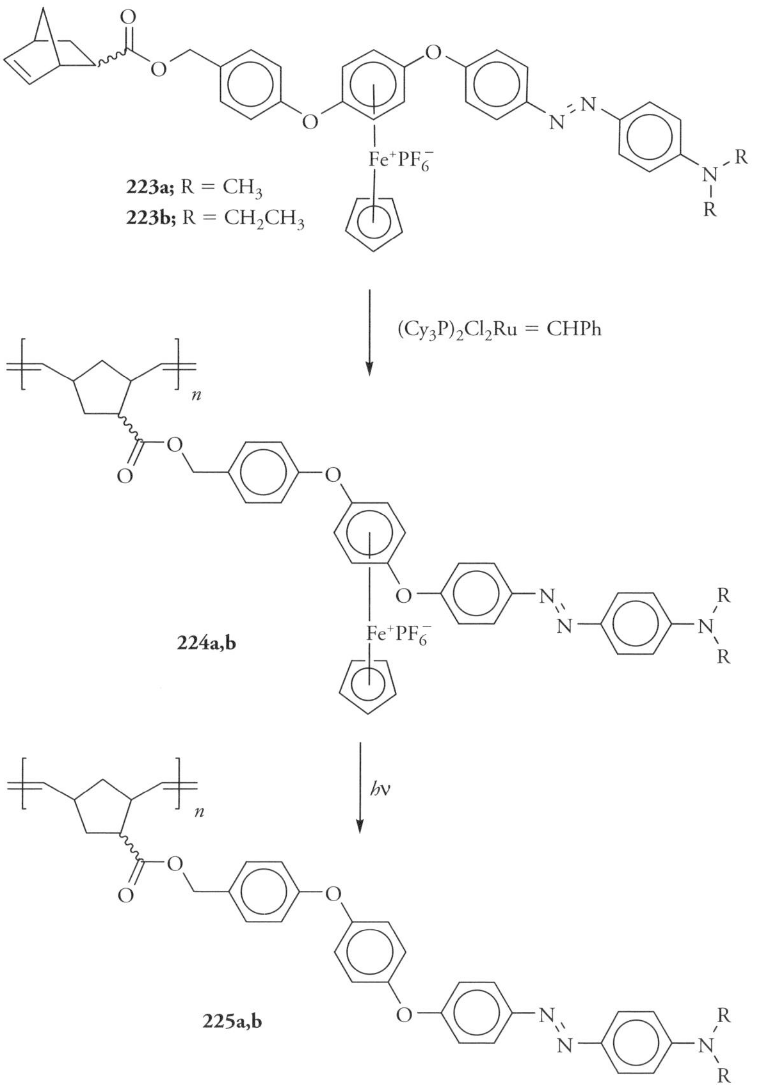

Scheme 2.59

Organoiron polynorbornenes incorporating benzothiazole dyes into the side chains have also been prepared.[273] These polymers displayed $\lambda_{max} = 511$ nm in DMF and $\lambda_{max} = 608$ nm, with a shoulder at 574 nm upon addition of a hydrochloric acid solution.

The coordination of tricarbonylmaganese and pentamethylcyclopentadienylcobalt to poly(9-hexylfluorene) was examined (Scheme 2.60).[274] Polymer **226** contained both 5- and 6-membered aromatic rings that can coordinate the transition metal complexes; however, studies indicated that the coordination to the 5-membered ring was preferred. Reaction of **226** with $MnBr(CO)_5$ yielded polymer **227** with a ratio of 1:1.5 for the complexed to the uncomplexed rings; however, reaction of **226** with BuLi followed by $(C_5Me_5)Co(acac)$ and then oxidation of the solution with HCl gave polymer **228**. Polymer **228** was determined to be a mixture of

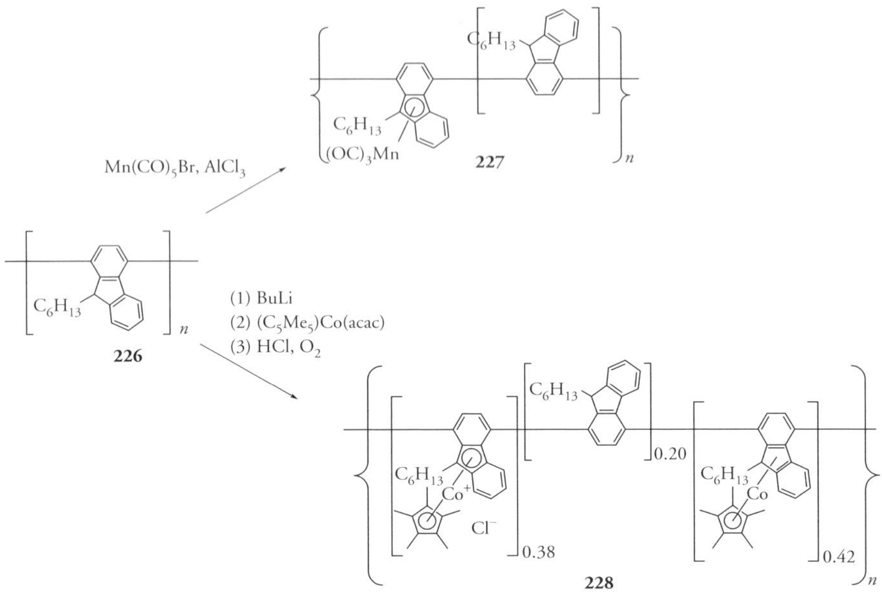

Scheme 2.60

cobalt-coordinated η^5-fluorenyl and η^4-fluorene rings. Conductivity studies on the polymers indicated that polymer **227** had higher conductivities than polymer **226**, and that polymer **228** possessed slightly lower conductivities then polymer **226**.

Bunz and co-workers reported the preparation of polymers containing cyclopentadienyl-cobalt moieties coordinated to cyclobutadiene rings, using diethylnyl and dipropynyl cyclobu-

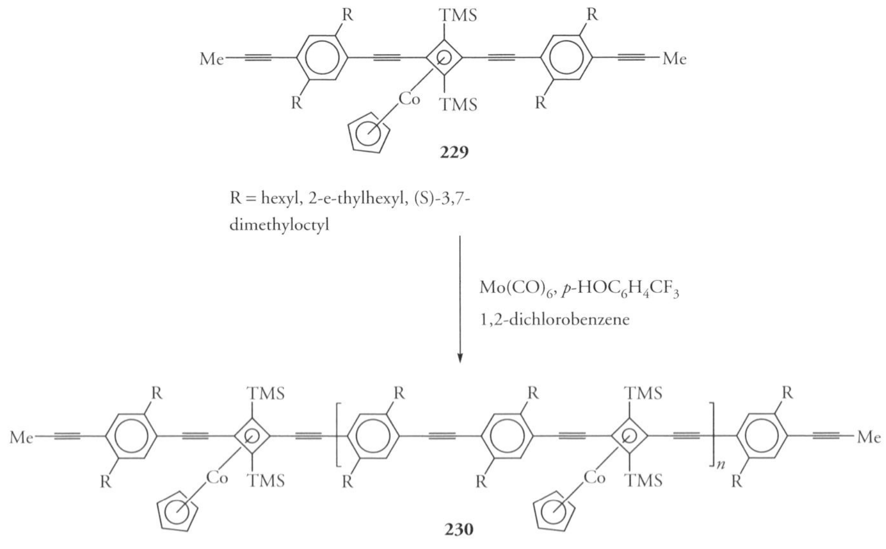

Scheme 2.61

tadiene complex.[275–279] The acyclic alkyne metathesis polymerization of monomer **229** yielded the conjugated organocobalt polymer **230** (Scheme 2.61).[279] Some of these polymers displayed liquid crystalline behavior, and photophysical studies showed that the cyclopentadienylcobalt moieties quenched fluorescence.[278,279,280] Liquid crystalline polymers containing cyclopentadienylcobalt moieties coordinated to cyclobutadiene rings also have been reported by Tomita, Endo, and co-workers.[279,280]

iii. Metal Carbonyl–Containing Polymers. Polysiloles containing iron carbonyl moieties coordinated to the silole rings, **231**, were prepared by Ohshita.[281] The molecular weights of the organometallic polymers were between 11,000 and 7100, and thermal properties showed that 5% of the polymers degraded at temperatures between 214 and 220°C. Ohshita also found that unless they contained two sterically bulky substituents at the silicon atom, 2,5-diethynyldilole units can be used as coordination sites for transition metal moieties. It was also discovered that by coordinating $Fe(CO)_3$ units to the polysilole unit, the conjugation of the system was decreased.

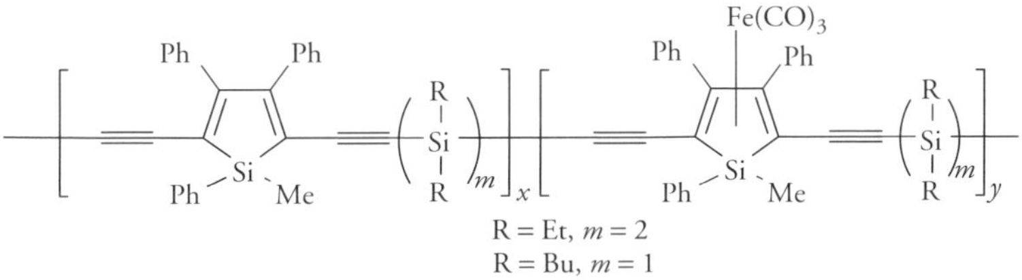

231

Ligand exchange reactions between poly(*n*-hexylphenylene) (PHP) and various organometallic compounds have given polymers containing coordinated molybdenum tricarbonyl and cyclopentadienyliron (Scheme 2.62).[282–287] These displayed better solubility than the metallated poly(*n*-propylphenylene) (PPP) analogs in organic solvents due to the flexible alkyl chains in the polymer backbone. Elemental analysis of the $Mo(CO)_3$-containing polymer indicated that 1 in every 4.8 aromatic rings was coordinated to a metal.[284,285] The cyclopentadienyliron containing polymer contained 1 in 1.6 coordinated arene rings.[284] Further analysis on the iron-containing polymers showed the formation of a network between aromatic rings on neighboring polymer chains after the cationic centers were reduced to neutral radicals. Nishihara and co-workers also reported the synthesis of a $Cr(CO)_3$ coordinated poly(*n*-butylphenylene) (PBP); however, this reaction was less efficient than when $Mo(CO)_3$ was used.[284] Analysis of the organic and organometallic polymers showed that the conductivity of the polymers increased with metal coordination.

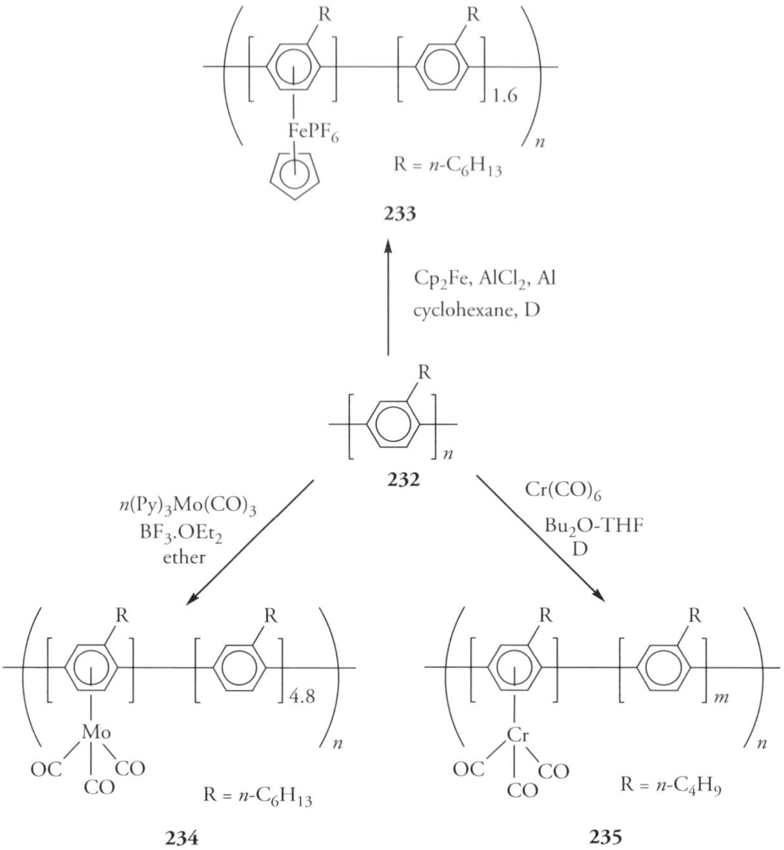

Scheme 2.62

Chromium tricarbonyl–containing polymers were prepared from the reaction of a 1,4-diethynylbenzene complex **236** with a number of different 3-alkyl-2,5-dibromothiophenes **237** (Scheme 2.63).[288] The resultant polymers possessed weight-average molecular weights of 13,500 to 24,400 with PDIs ranging from 3.2 to 3.6. Thermal studies displayed two weight losses corresponding to the cleavage of the chromium tricarbonyl and decomposition of the polymer backbone. Spectral studies of these polymers showed that they absorbed between 375 and 386 nm, and when excited at 370 nm, there was a weak emission at 450 nm. Electrochemical studies on the dodecyl-substituted polymer showed one irreversible oxidation at 1.5 V, and an electrical conductivity of 8.1×10^{-6} S/cm.[288]

Chromium tricarbonyl has been incorporated into polyimine via the reaction of η^6-terephthaldialdehyde-$Cr(CO)_3$ with 1,3-phenylenediamine.[289] Due to the rigidity of the backbone, the resulting conjugated polymers were insoluble in common organic solvents. Hay[290] and Suzuki[291] coupling reactions were utilized in the synthesis of polymers containing pendent manganese tricarbonyl moieties. High molecular-weight polymers were prepared from the reaction of the diiodocyclopentadienylmanganese complex **240** with the diboronic acid monomers **241** (Scheme 2.64). The polymers were determined to possess number-average molecular weights ranging from 1.1 to 1.2×10^4 with broad polydispersities (PDI = 6–6.4).

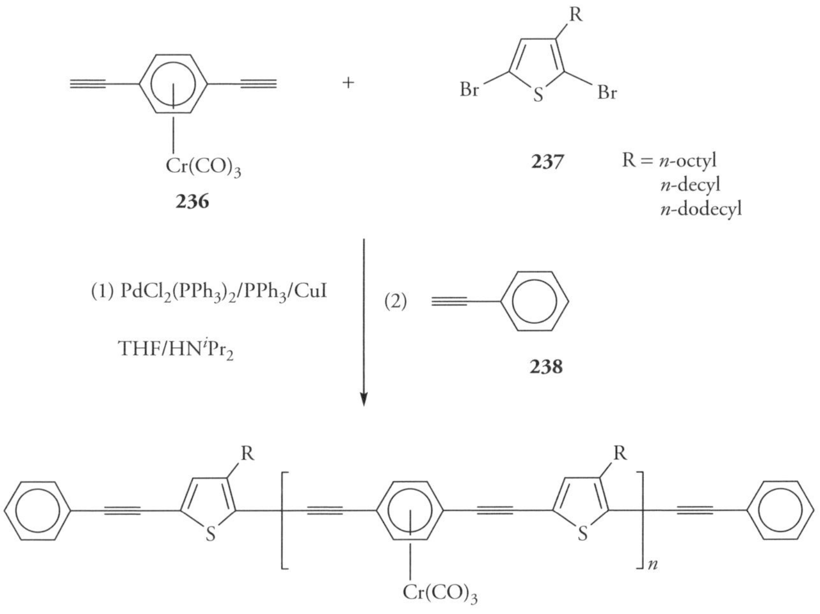

Scheme 2.63

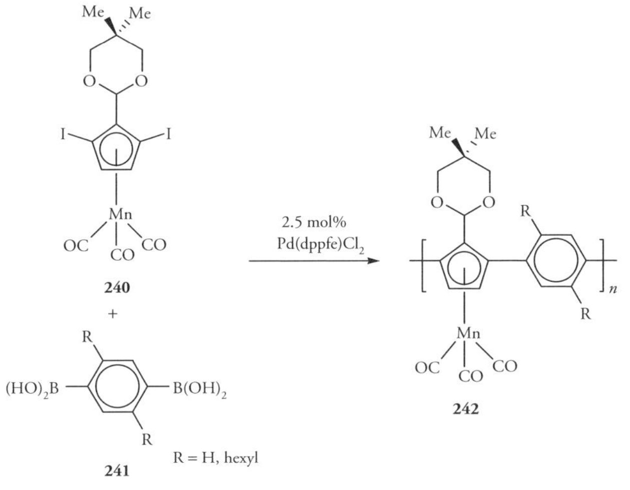

Scheme 2.64

Irradiation of polymer **243** in an excess of iron pentacarbonyl led to the production of the iron tricarbonyl-containing polymer **244** (Scheme 2.65).[292] Polymer **244** possessed a 2:1 ratio of coordinated to noncoordinated silole rings. Conductivity studies on the $FeCl_3$ doped polymers showed that polymer **244** had a conductivity of 10^{-8} S/cm and polymer **244** had a conductivity of 10^{-5} S/cm. X-ray studies on model compounds indicated that $Fe(CO)_3$-coordinated silole rings have enhanced π-conjugation.

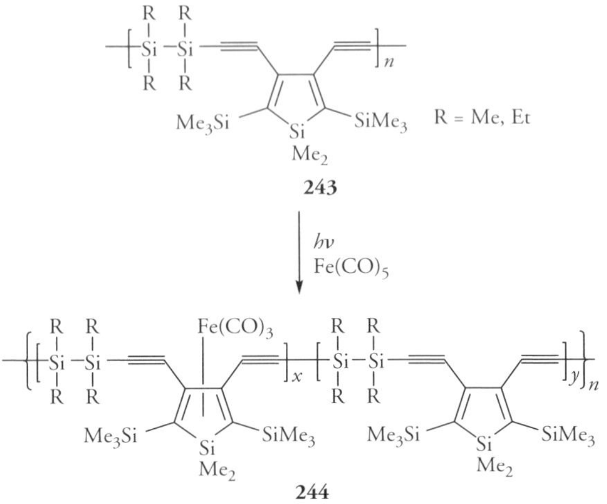

Scheme 2.65

Mapolie and co-workers have reported the synthesis of polymers containing cyclopentadienylmetal complexes in their side chains.[293,294] The polymers were prepared by homopolymerization of iron- or tungsten-containing vinyl monomers or copolymerization with styrene.[293] Low molecular weight polyesters containing rhenium and iron complexes in their side-chains have also been prepared via polycondensations (Scheme 2.66).[294] It was determined that by increasing the reaction time of the polymerization, the amount of insoluble material recovered increased.

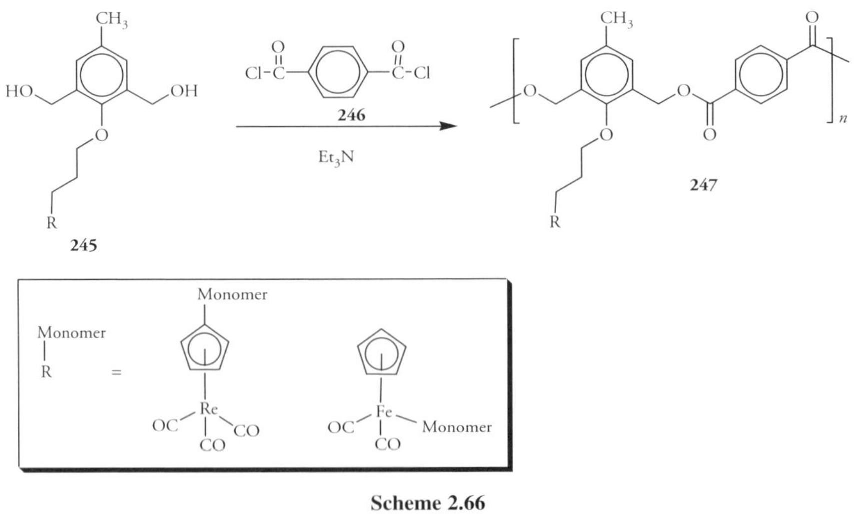

Scheme 2.66

IV. STARS, DENDRIMERS, AND HYPERBRANCHED POLYMERS

A. Stars and Hyperbranched Polymers

Cationic star-shaped molecules containing cyclopentadienyliron complexes, where the organometallic cations are evenly distributed throughout the molecule branches, **250**, have been reported (Scheme 2.67).[295,296,297] Electrochemical studies of these materials indicated that the iron centers underwent reversible reductions. For the hexametallic star-shaped molecule ($n = 1$), two redox potentials were observed at $E_{1/2} = -1.20$ and -1.30 V.

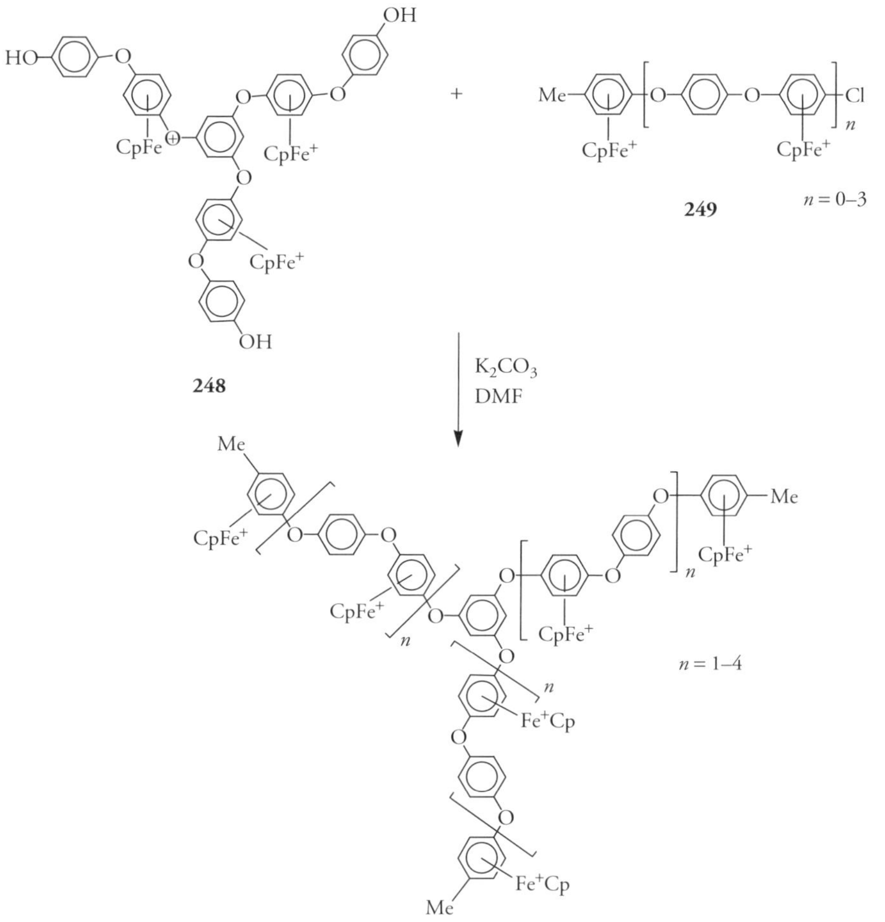

Scheme 2.67

Various first-generation dendrimers and star-shaped molecules were generated from the nucleophilic aromatic substitution of chlorosubstituted arenes coordinated to cyclopentadienyliron complexes (Schemes 2.68 and 2.69). These compounds contained either ether or ester bridges. The star-shaped molecules prepared with ether bridges showed a decrease in solubility with an increase in star size, whereas the incorporation of ester bridges resulted in a decrease

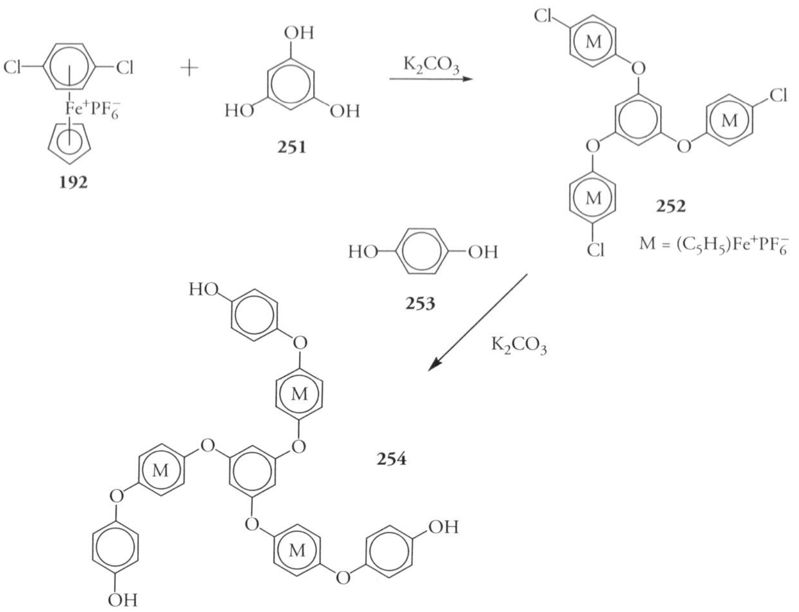

Scheme 2.68

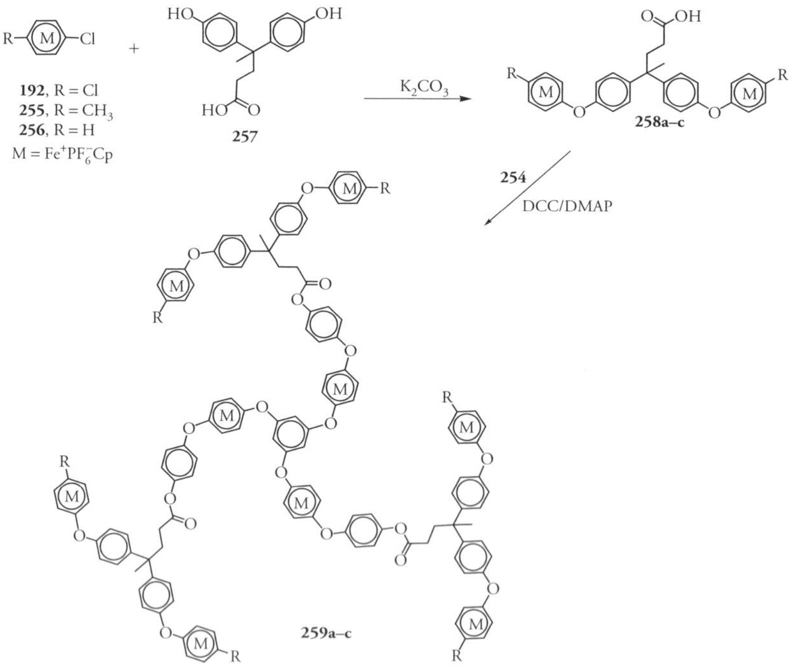

Scheme 2.69

in solubility independent of star size. Thermal studies of these polymers indicated that the ester-containing stars displayed T_g between 150 and 195°C, whereas the ether-containing stars displayed T_g between 120 and 200°C.

Star-shaped molecules containing cationic arene complexes of iron and ruthenium have been reported by Astruc and co-workers.[298] Utilizing the activating nature of the cyclopentadienyliron moiety on the complexed arene, **260** was converted into **262** via bromobenzylation. The photolysis of **262** gave **263**, which was subsequently reacted with **264** to give the hexametallic complex **265**. Further nucleophilic aromatic substitution reactions with phenol **266** gave the allyl-substituted complex **267** (Scheme 2.70).

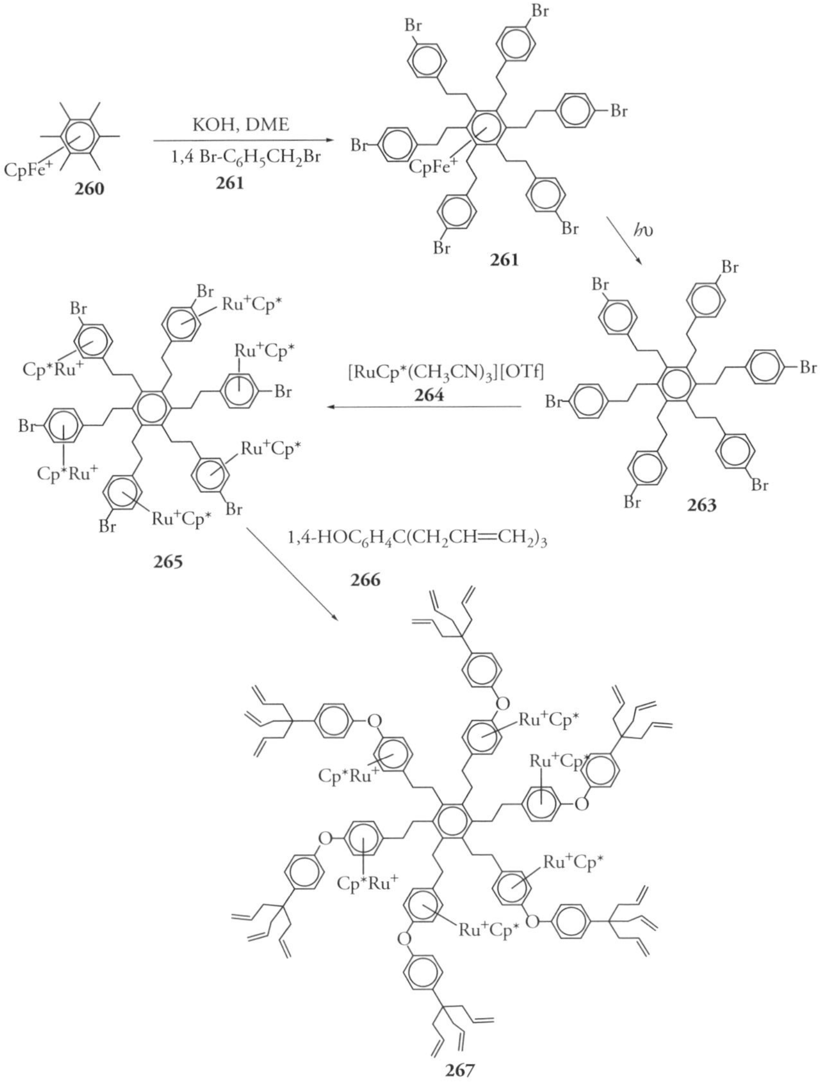

Scheme 2.70

Magnetic ceramics were isolated from the pyrolysis of the hyperbranched polymer **268**.[299] Polymer **268** was prepared from the reaction of 1,1′-dilithioferrocene with methyltrichlorosilane and possessed a weight-average molecular weight of 6300 (PDI = 3.8). Thermal studies showed that polymer **268** possessed a glass transition at −54.5°C and showed a 5% weight loss at 306°C. The ceramic material was prepared by heating polymer **268** to 1000°C under nitrogen or to 1200°C under argon. This was followed by calcination or sintering at temperature for 1 hour. Ceramics prepared from the hyperbranched polymers showed higher yields than those prepared from linear analogs.[299] The ceramic formed under argon had a high saturation magnetization with a M_s of approximately 49 emu/g and a near zero coercivity and remanence, indicating that it functions as an excellent soft ferromagnetic material.

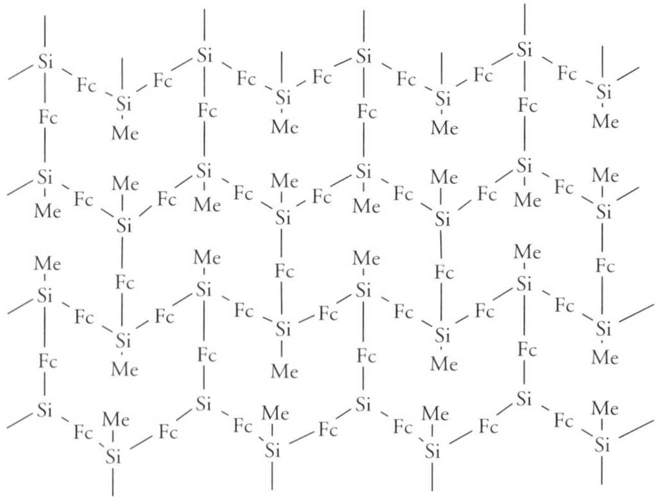

Fc = 1,1′-ferrocenylene

268

A number of research groups have reported the preparation of a large number of star-shaped molecules and dendrimers containing ferrocenyl or arene cyclopentadienyliron complexes at the core or the peripheries.[300–320] A number of these dendrimers were prepared via cyclopentadienyliron-mediated per-alkylation, -benzylation and -allylation reactions of cationic tri-, tetra- and hexamethylbenzene complexes. These dendrimers were multifunctional materials and have been used in the synthesis of branched organic and organometallic polymers.

Scheme 2.71 shows the preparation of an octametallic star-shaped molecule containing cyclopentadienyliron arene complexes at the peripheries. The reaction of the organic core, **269**, with the organometallic complex, **270**, gave the organometallic star-shaped molecule, **271**. A previously reported synthesis of **271** occurred via deprotonation of a permethylated iron complex. Astruc et al. have also reported silicon-containing star-shaped molecules, **272**.[312]

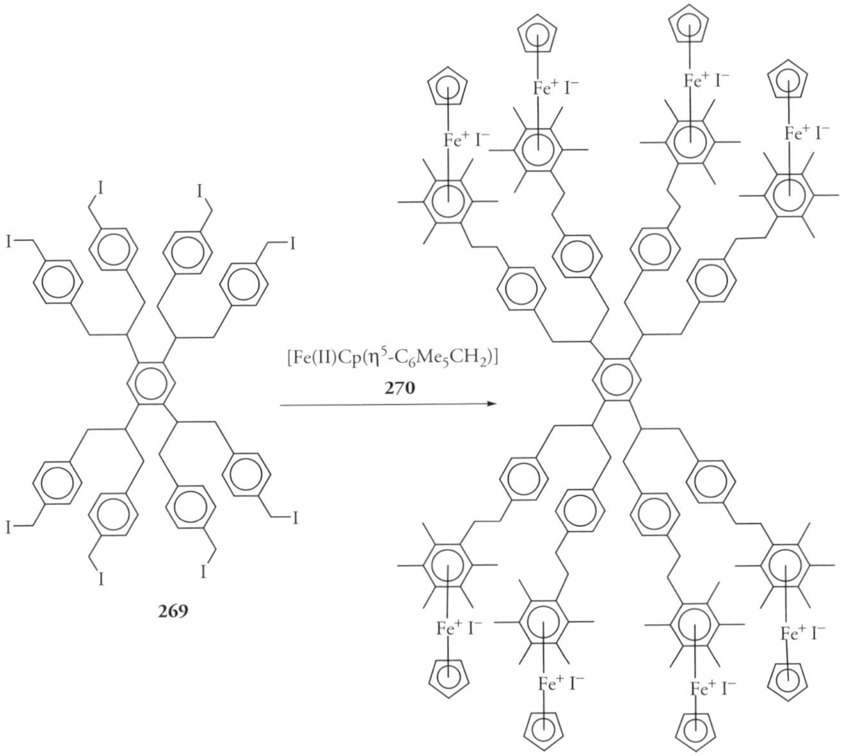

269

271

Scheme 2.71

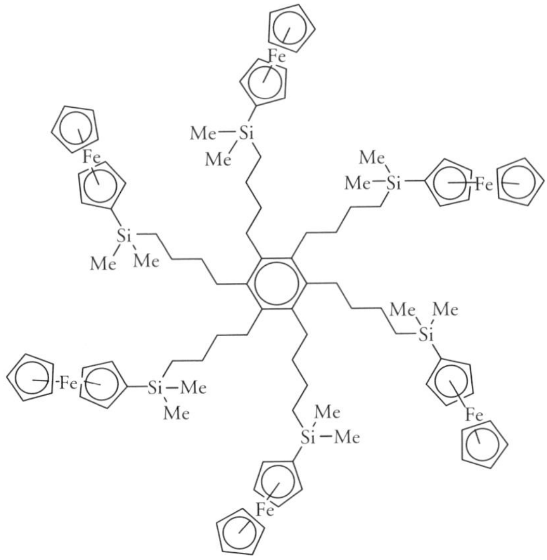

272

Tang and co-workers have recently reported the synthesis of hyperbranched poly(ferrocenylenesilyne)s **273**.[320] These polymers were prepared via the one pot coupling reactions of dilithioferrocene with trichlorosilanes. The organometallic hyperbranched polymers possessed higher solubility in organic solvents than poly(methylsilyne) networks and were obtained in 71% yield. The solubility of the hyperbranched networks was greatly enhanced by increasing the length of the aliphatic chain on the silane. An increase in the chain length to 8 carbons caused the hyperbranched networks to become completely soluble. Increasing to 12 carbons resulted in a lower T_g and the isolation of elastomers. The molecular weight of the hyperbranched polymer with the 18 carbon chain was 50,0000 Da determined by light scattering.

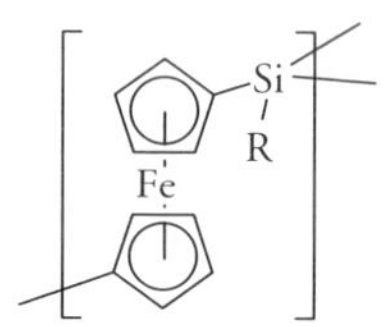

273

Tang and co-workers have also prepared cobalt-containing hyperbranched polymers by the reaction of cobalt complexes with a preformed hyperbranched polymer **274** containing acetylene groups.[321] The percent of cobalt incorporated into the polymers ranged from 27.8 to 36.7 wt %, and upon high temperature pyrolysis, the polymers formed ceramics with magnetic properties.

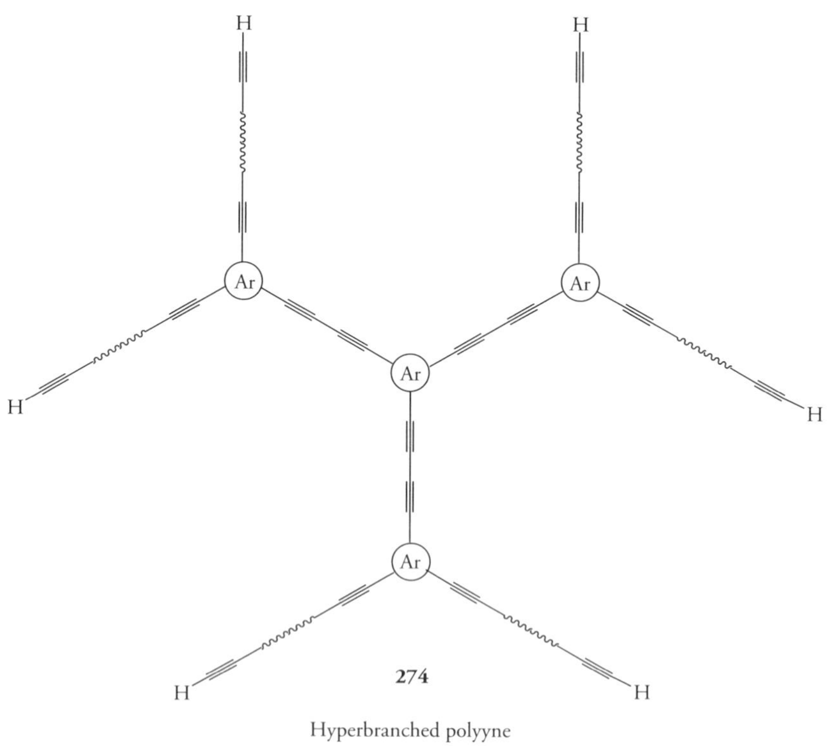

274

Hyperbranched polyyne

Abd-El-Aziz and co-workers have recently reported the isolation of iron-containing hyperbranched polymers.[296] Using an $A_2 + B_3$ methodology, a number of different diols, dithiols, or organoiron complexes were reacted with star-shaped molecules to produce hyperbranched polymers (Scheme 2.72). Viscometric studies showed that the polymers possessed low viscosities ranging from 0.175 dl/g to 0.300 dl/g. Thermal studies indicated that the cationic cyclopentadienyliron moieties were cleaved between 230 and 280°C, and the polymeric backbone degraded between 390 and 567°C. The polymers possessed glass transition temperatures between 60 and 134°C and melting points in the range of 155–190°C. Polymer networks were also obtained from these reactions, giving materials with higher viscosities and thermal stability; however, these networks possessed lower T_g.

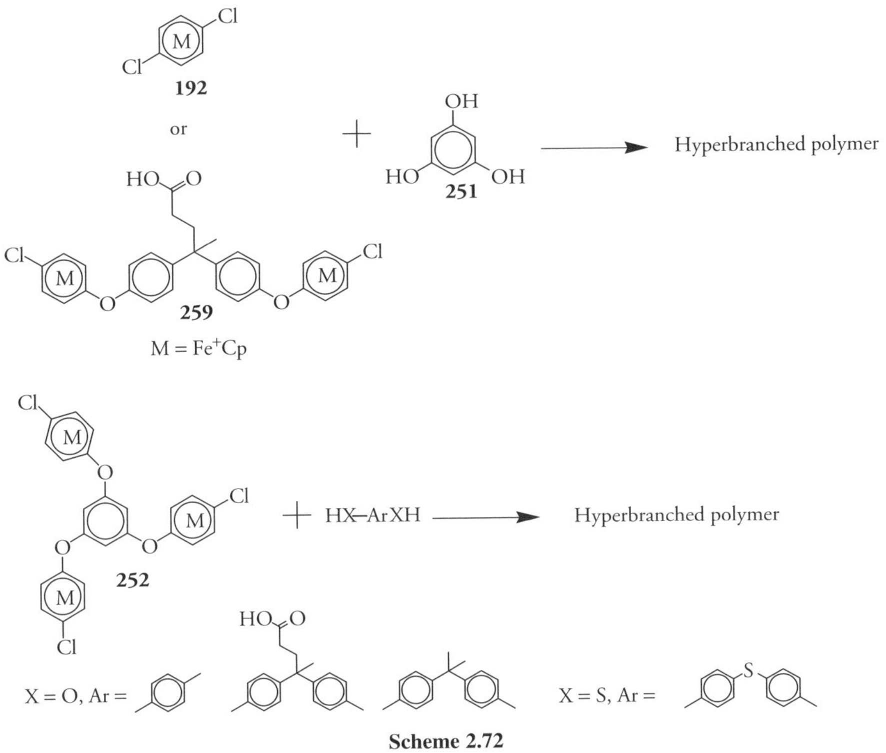

Scheme 2.72

B. Dendrimers

Dendrimers containing cyclopentadienylruthenium complexes have been reported by Moss and co-workers via a convergent method.[322] This reaction has led to a fourth-generation dendrimer, **275**, containing 48 organometallic moieties. Low generation dendrimers containing six terminal ruthenium complexes were also reported.

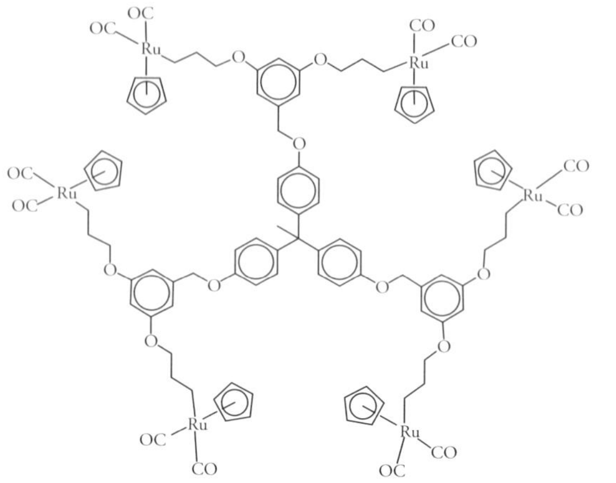

275

Dendrimers containing σ- and metal-bipyridine–bonded platinum and palladium complexes have been prepared.[323–327] Puddephatt and co-workers have generated dendrimers containing both palladium and platinum complexes coordinated to bidentate ligands, **276**.[325] Dendrimer **276** was synthesized via oxidative-addition of a C—Br bond to a platinum complex to form the dendrimer core. Subsequent reaction with metal complexes gave homo- or heterometallic materials.

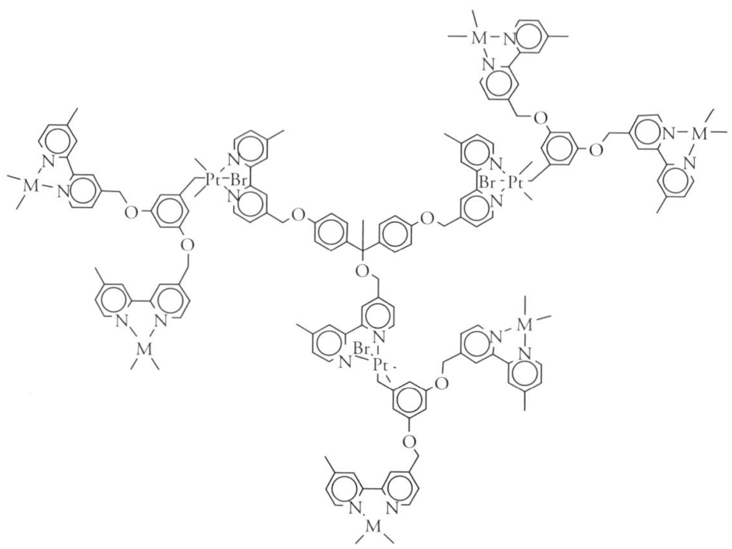

276 M = Pt, Pd

Dendrimers with periphery carbosilane containing palladium complexes were prepared by van Koten et al. through the insertion of $Pd^{(0)}$ into carbon–iodine bonds.[327,328] The dendrimers prepared by this method were based on a first-generation dendrimer, **277**, containing 12 palladium complexes.[328]

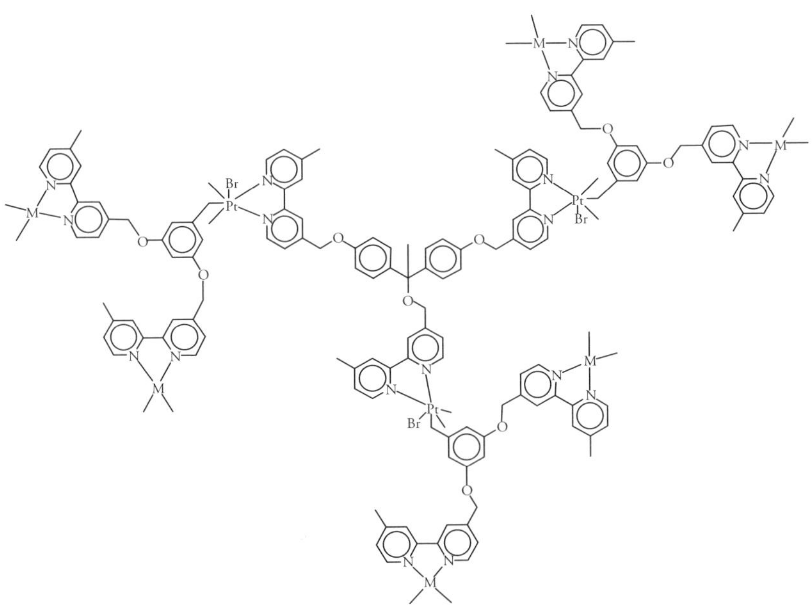

M = Pt, Pd

276

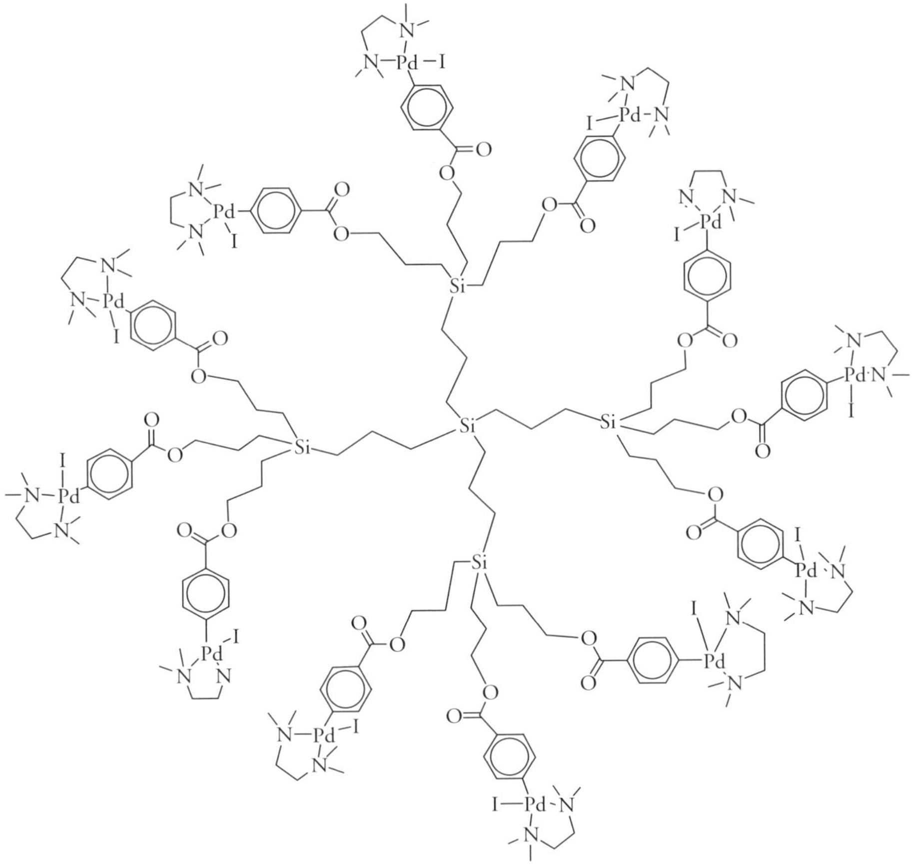

277

A variety of metal acetylide–containing stars and dendrimers have been synthesized.[329–332] Dendrimer **280** was prepared via reaction of a chloro-substituted platinum complex **278** with ethynyl functionalized dendron **279** in the presence of CuI (Scheme 2.73).[329] This method generated dendrimers with up to 45 platinum units.

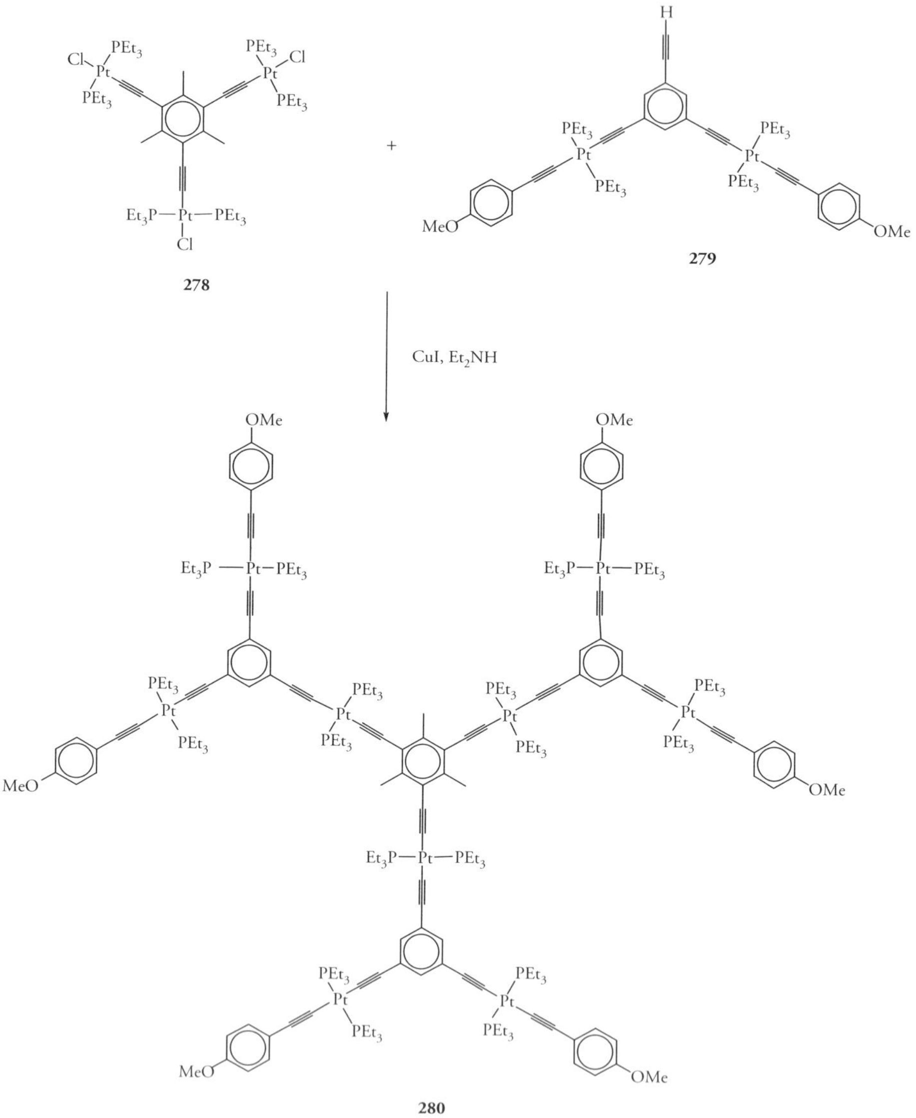

Scheme 2.73

Star and dendrimer core molecules were prepared by the peralkylation or allylation of cyclopentadienyliron complexes containing methyl-substituted arenes.[298,301,302,304–311,333] The preparation of water-soluble metallodendrimers containing six cationic cyclopentadienyliron moieties, **281**, has also been reported.[301] Dendrimer **281** was tested for potential use as a redox catalyst for the cationic reduction of nitrates and nitrites to ammonia.

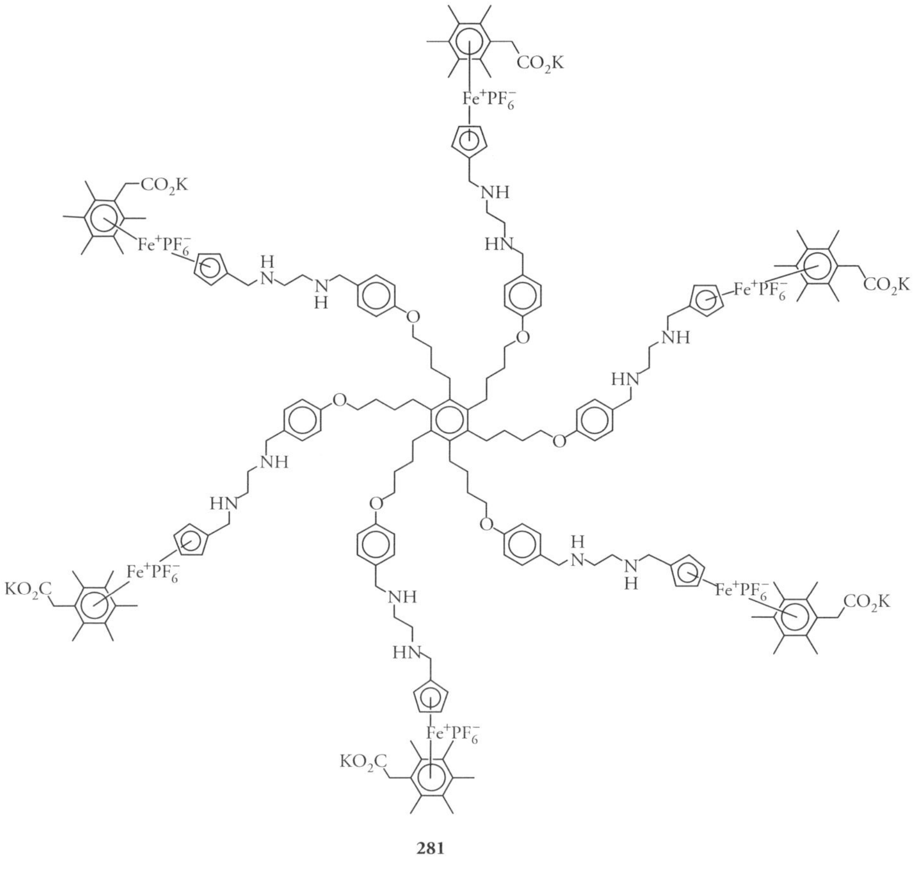

281

Cuadrado and co-workers have reported the synthesis of chromium-containing organosilicon dendrimers.[334] These dendrimers, **282**, have chromium tricarbonyl units incorporated pendent to the terminal aromatic rings.[334] Synthesized via the reaction of the silane dendrimer precursor with chromium hexacarbonyl, complete complexation was not possible due to steric hindrance at higher generations. Electrochemical studies showed that the oxidation of the chromium atoms occurred reversibly in the absence of a nucleophilic species and that the chromium tricarbonyl units behaved as isolated redox centers.

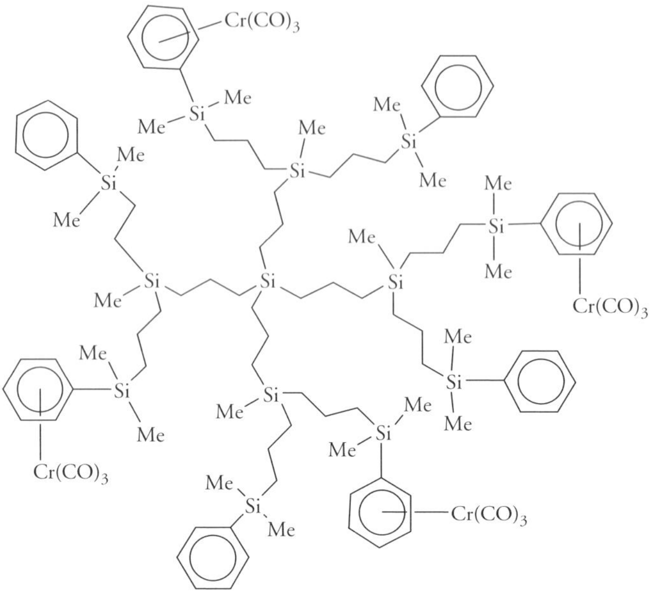

282

Tilley and co-workers have also reported organosilane dendrimers containing transition metal moieties.[335] Tilley et al. reported the synthesis of ruthenium containing dendrimers via the methodology outlined in Scheme 2.74. This synthesis allowed for the isolation of dendritic molecules containing 12, 24, 36, and 72 metallic moieties. Mass spectrometry showed that complex **285** with 72 metallic moieties was prepared; however, as with the dendrimers reported by Cuadrado,[334] not all the aromatic rings on the dendrimer could be complexed to metals due to steric hindrance at higher generations.

Scheme 2.75 shows the functionalization of a polyamine dendrimer **286** with the tricarbonylpentadienylmethyl–tungsten derivative, **287**.[336] Dendrimer **288** was subjected to photolysis and the resulting radical complexes with two and four tungsten moieties were found to cleave plasmid DNA in a double-stranded manner. While these low-generation materials were effective at DNA scission, the higher-generation dendrimers precipitated from solution and formed aggregates.

Ferrocene-functionalized star-shaped and dendritic compounds have received much attention.[307,319,337–347] The synthesis and properties of various ferrocene-containing star-shaped and dendritic molecules have been reported by Astruc et al.[307,337,338] The synthesis and electrochemical properties of a dendrimer containing 54 ferrocene moieties were reported and showed that all 54 peripheral ferrocene groups underwent reversible oxidation.[337] The iron centers could also be oxidized chemically using $NOPF_6$. The reaction of amine-functionalized dendrimers with the acid chlorides of ferrocene or cobaltocenium resulted in dendrimers containing nine peripheral organoiron[338] or organocobalt[307] groups, **289**. The ferrocene-functionalized polymers have been used as supramolecular redox sensors for the recognition of small inorganic anions.[338]

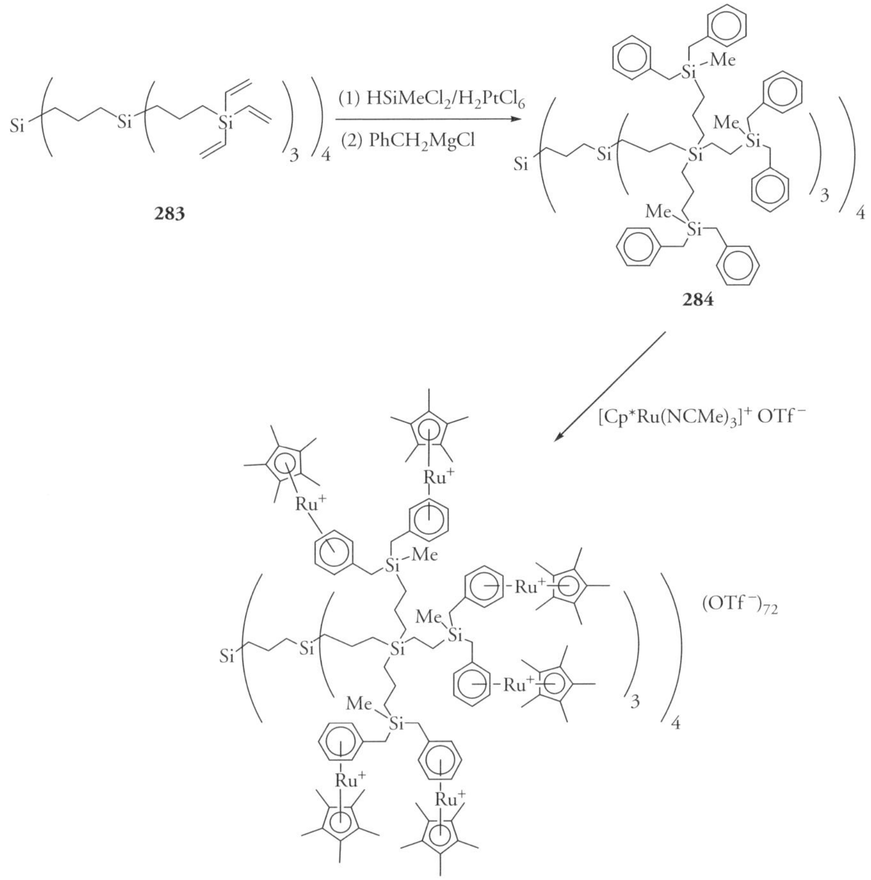

Scheme 2.74

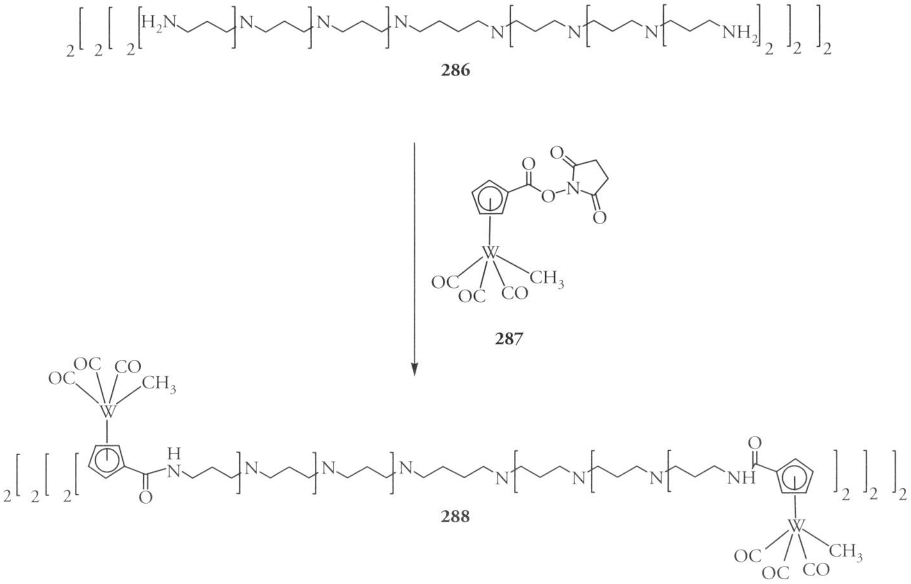

Scheme 2.75

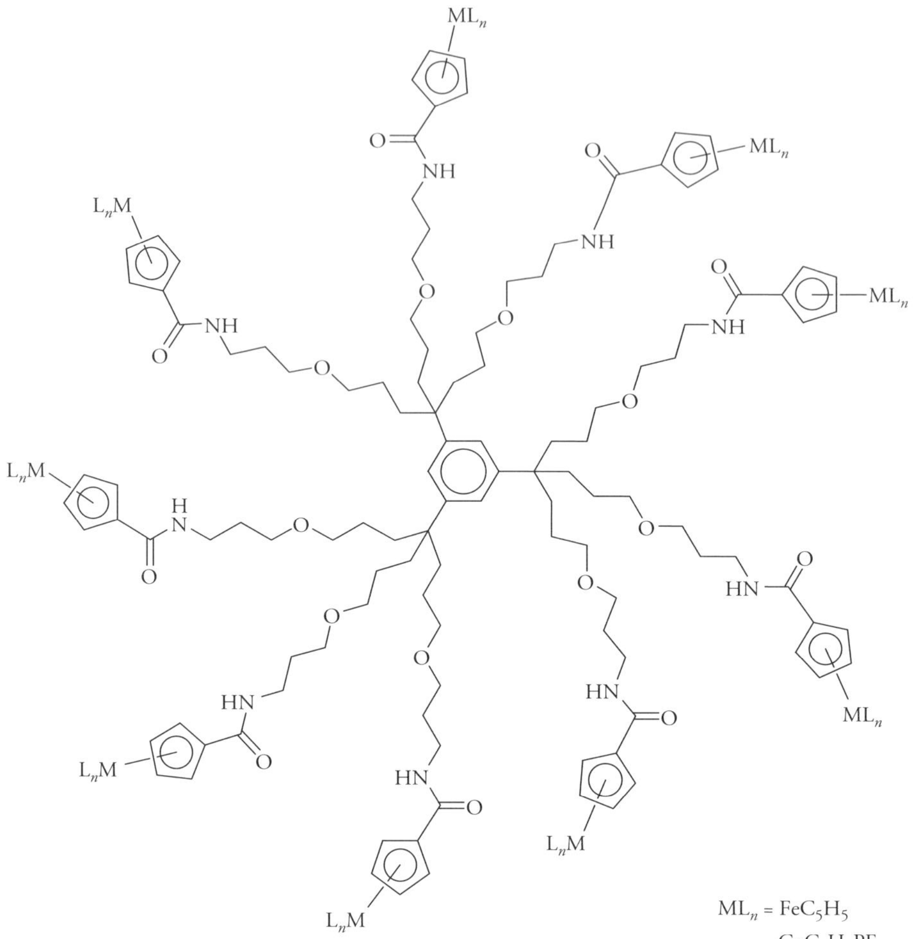

289

The guest-host properties of some low-generation dendrimers with cyclodextrins have been analyzed.[340] Complete complexation of β-cyclodextrin occurred with dendrimers containing four to eight peripheral ferrocene units; however, higher-generation dendrimers (16 ferrocene moieties) showed only partial complexation. The complexation of the dendrimers led to an increased solubility in aqueous solutions. The complexation with cyclodextrins could be made reversible by the addition of 2-napthalenesulfonate. Silicon-based ferrocenyl dendrimers possessing electrochemical communication between the iron centers were also synthesized (Scheme 2.76).[346] Dendrimers containing ferrocene and cobaltocenium units at their periphery were also examined.[348]

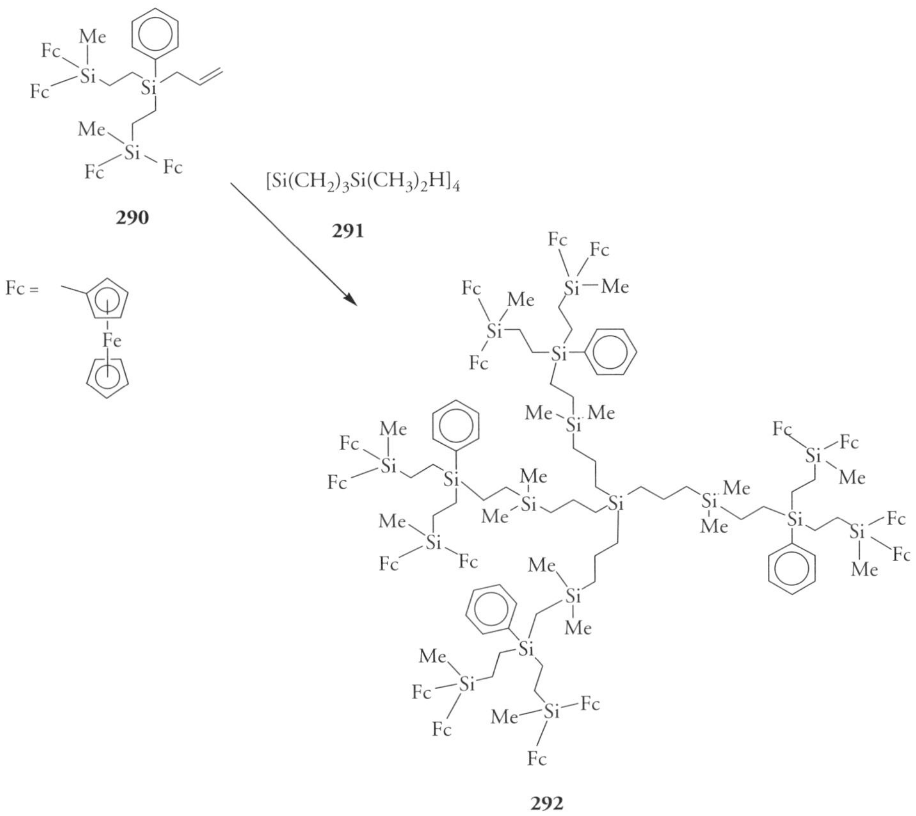

Scheme 2.76

A number of dendrimers showing potential as sensors for inorganic anions have been prepared containing 4,8,16, or 32 peripheral ferrocene or ferrocenyl–urea units.[339,343–347,349,350] These dendrimers were generated by the reaction of 1-chlorocarbonylferrocene or isocyanatoferrocene with the amine-terminated dendrimers (Scheme 2.77). Dendrimers **297** have shown potential uses as sensors for inorganic anions such as $H_2PO_4^-$ and HSO_4^-. The second- and third-generation dendrimers showed the strongest binding of $H_2PO_4^-$, whereas the first and fourth generations displayed the weakest. Electrochemical investigation of dendrimer **292** showed no interaction between the ferrocenyl redox centers. This class of dendrimer is also capable of electrochemically recognizing anionic compounds.

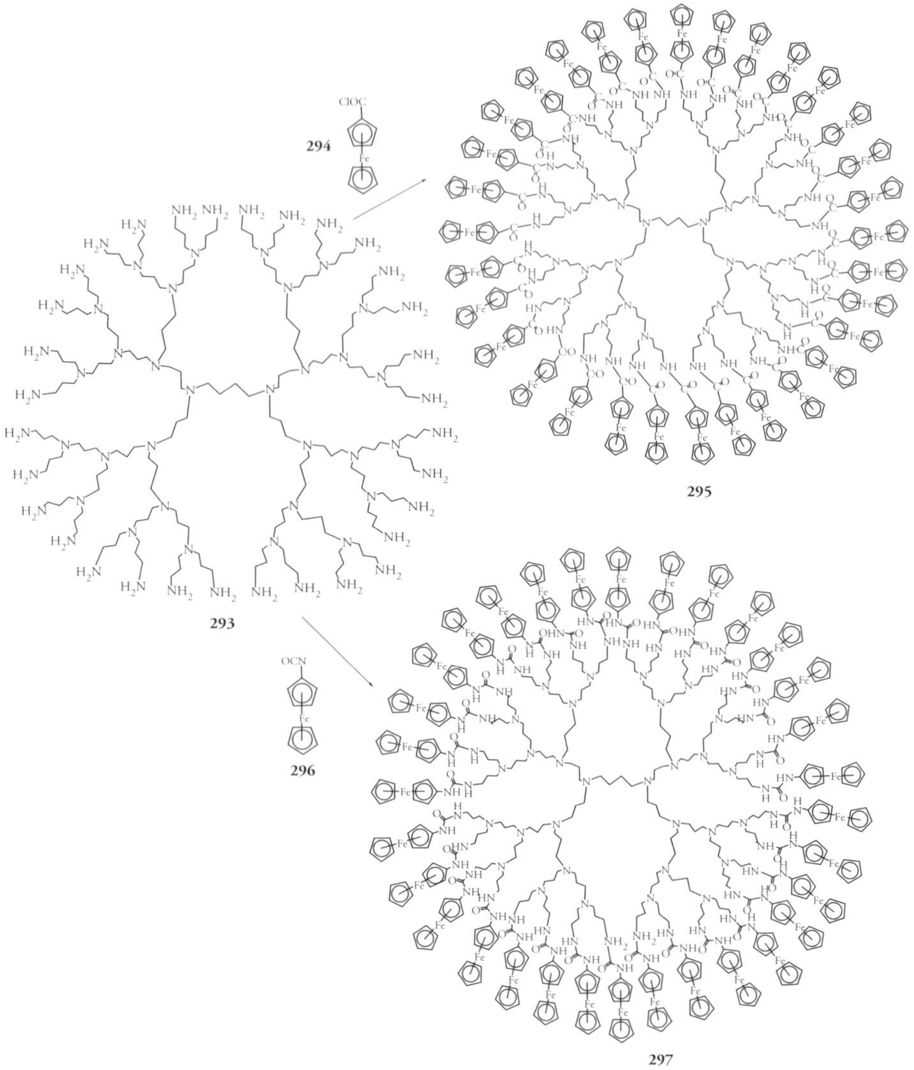

Scheme 2.77

The convergent synthesis of dendrimers has allowed for the preparation of hexa-functionalized organometallic dendrimers containing olefinic groups, **298**, as well as dendrimers containing up to 57 silylferroceneyl moieties.[310,319,327,351,352,353] However, due to competing reactions this method cannot be used to prepare larger dendrimers.

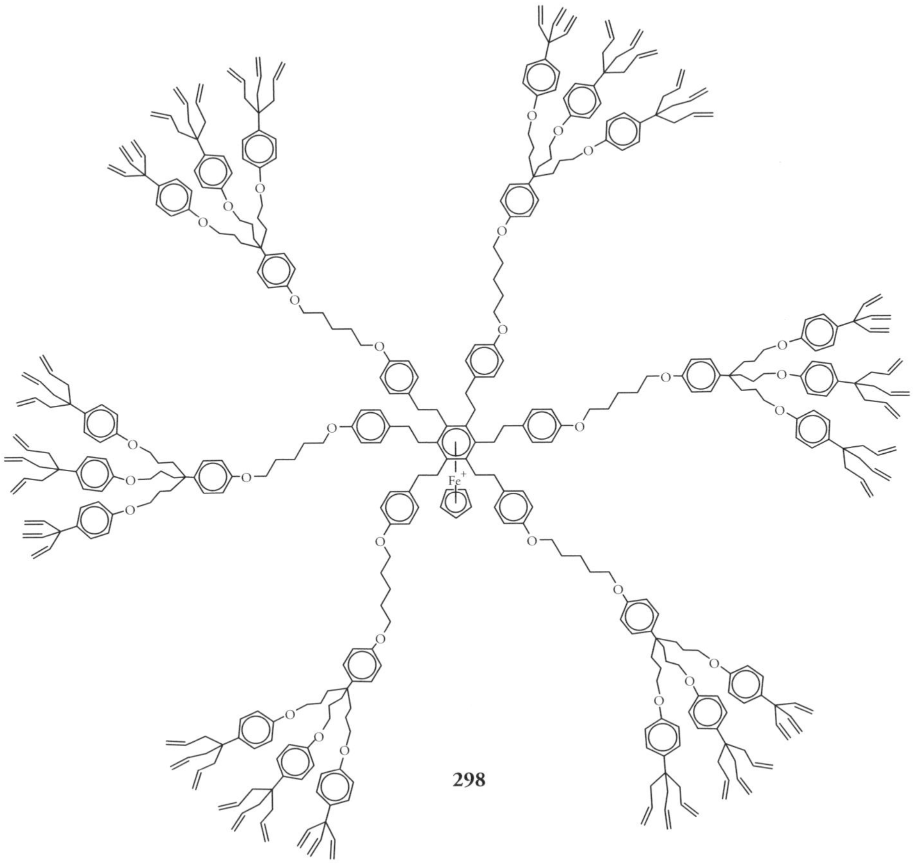

298

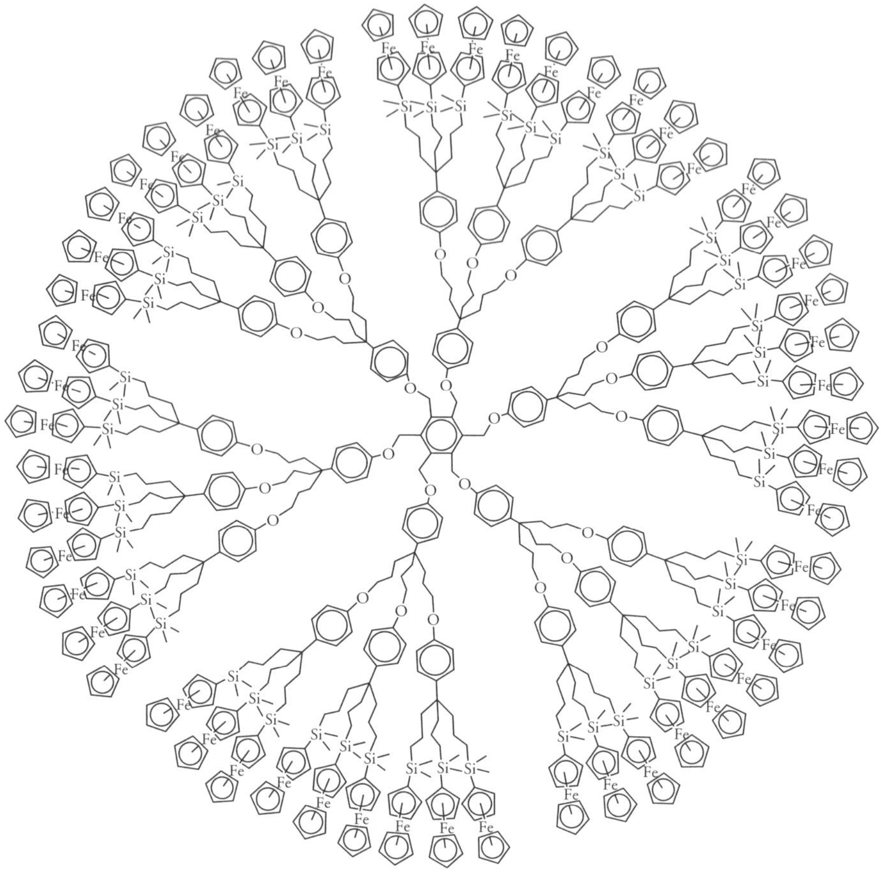

299

The reaction of allyl-terminated dendrimers with ferrocenyldimethylsilane using Karstedt's catalysts resulted in dendrimers containing up to 243 ferrocenyl groups, as reported by Astruc and co-workers. Electrochemical studies showed that these dendrimers underwent chemically and electrochemically reversible oxidation processes.[319,337,348]

Recently, Astruc and co-workers have reported the synthesis of dendrimers based on cluster cores.[354,355] The reaction of an octahedral molybdenum cluster $[n\text{-}\mu_4N]_2[Mo_6Br_8(CF_3SO_3)_6]$ with the dendritic $p\text{-NaO-}_6H_4C\{CH_2CH_2CH_2Si(Me)_2Fc\}_3$ led to the isolation of an air-sensitive Mo_6-cluster-cored octadecylferrocenyl dendrimer, **300**.[354] This dendrimer was found to be useful as a detector for ATP^{2-}, and the may be advantageous over amido-ferrocenyl containing dendrimers, since the completely oxidized ferrocenium form is stable.[354]

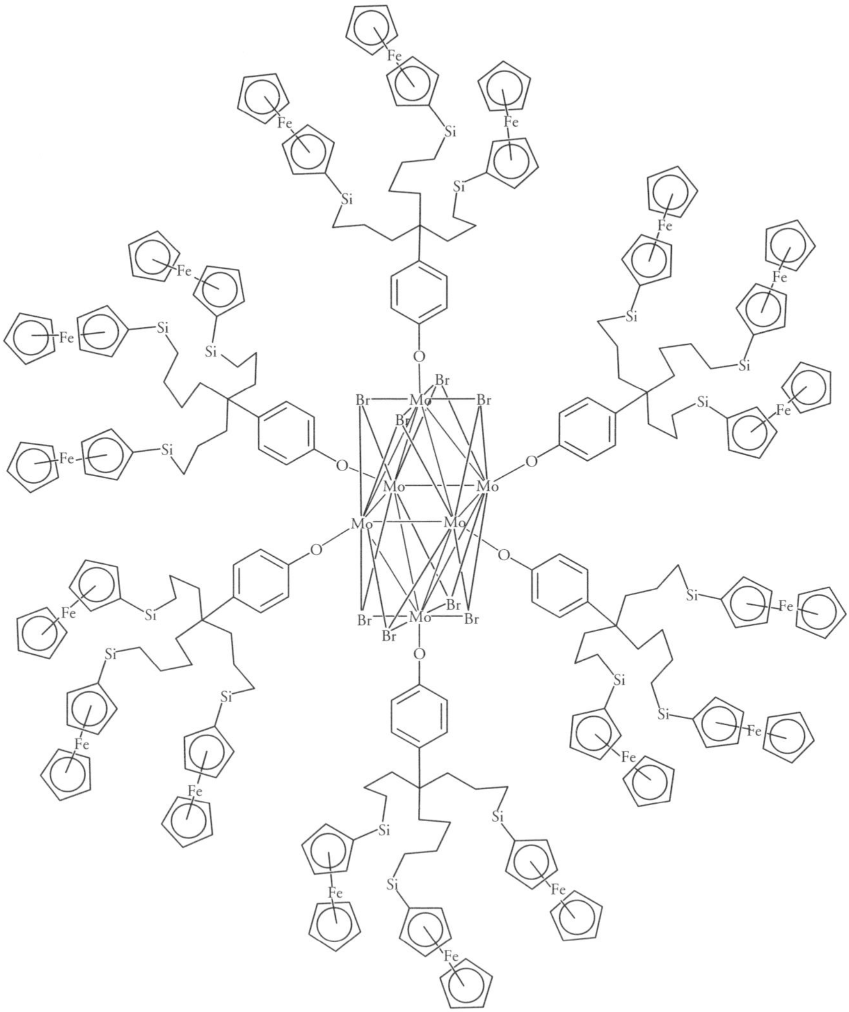

300

Astruc and co-workers have also isolated dendrimers using a tetrairon cluster [{CpFe(μ_3-CO)}$_4$], **302**.[355] Dendrimer **303**, with nine organoiron clusters at the peripheries, was prepared via the reaction of the tetrairon cluster with dendrimer **301** (Scheme 2.78). These types of dendrimers were air-stable, and electrochemistry studies showed that the iron clusters were separated enough from each other to prevent any electrostatic interaction. These dendrimers are potential exoreceptors, which are generally confined to dendrimers containing metallocene units.[355]

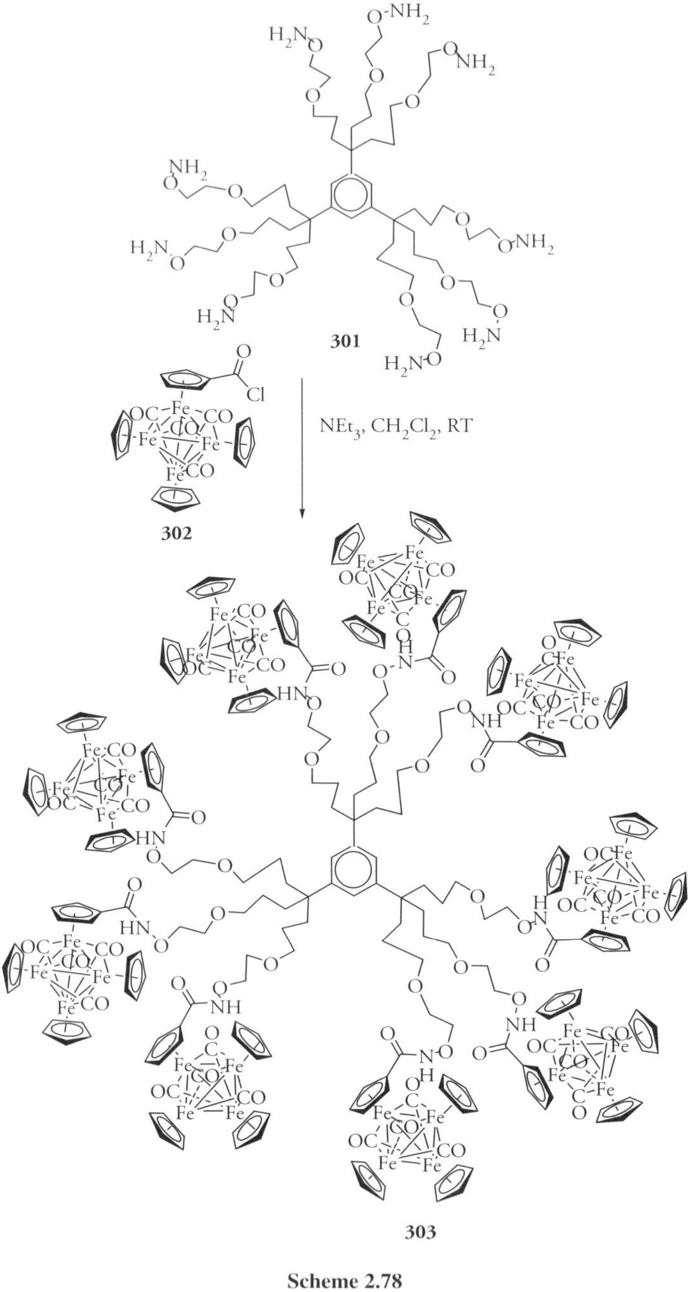

301

302

303

Scheme 2.78

The synthesis of a ferrocene-capped dendrimer with a stealth cyclotriphazene core was reported by Sengupta.[320] Electrochemical studies showed that this dendrimer contained equivalent iron centers. A liquid crystalline ferrocene-containing six-arm star, **304**, has been reported.[344] This orange star showed an enantiotropic smectic A phase.[349]

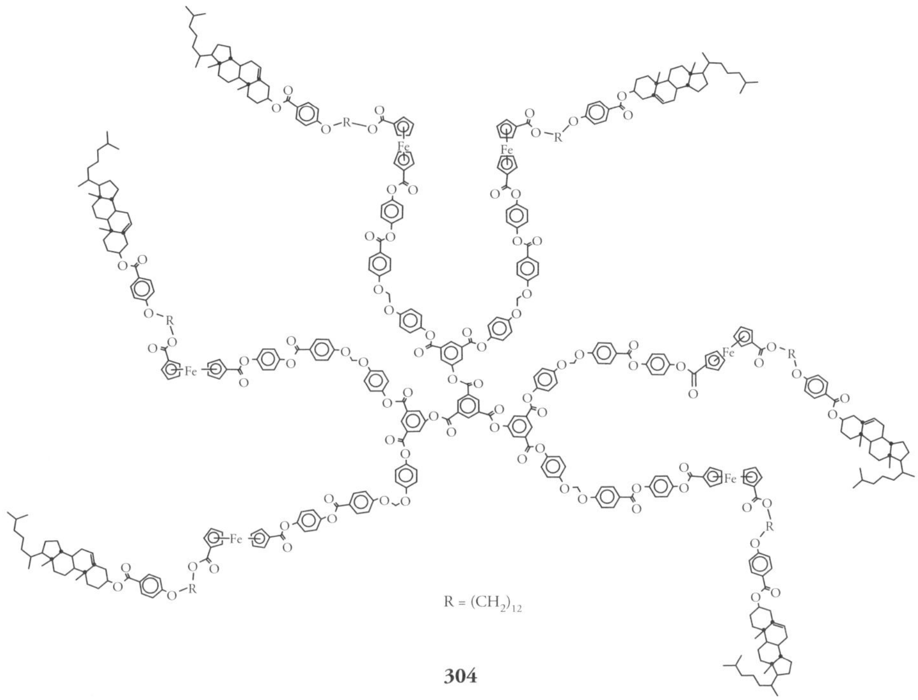

304

A water-soluble ferrocene-containing dendrimer, **305**, has been prepared via the incorporation of carbohydrate groups.[356] This dendrimer was found to mimic redox active proteins and formed different conformers in aqueous solution.

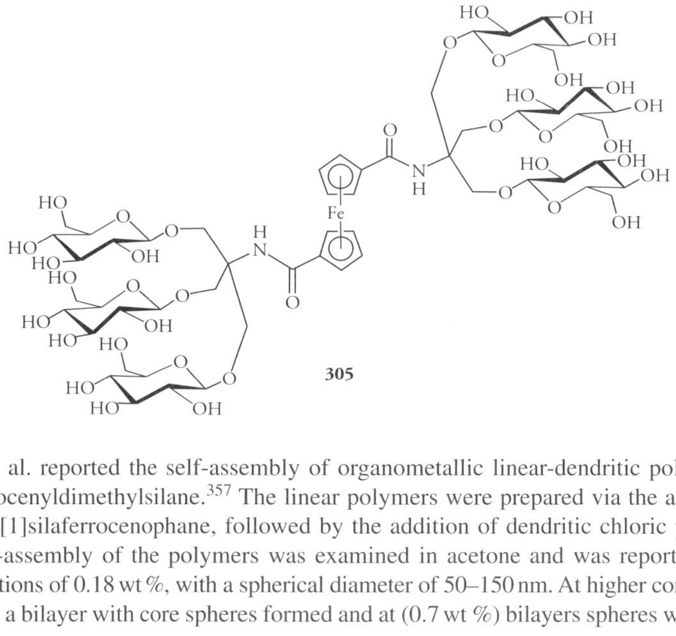

305

Wang et al. reported the self-assembly of organometallic linear-dendritic polymers containing ferrocenyldimethylsilane.[357] The linear polymers were prepared via the anionic ring-opening of [1]silaferrocenophane, followed by the addition of dendritic chloric poly(benzyl ether). Self-assembly of the polymers was examined in acetone and was reported to occur at concentrations of 0.18 wt %, with a spherical diameter of 50–150 nm. At higher concentrations (0.35 wt %), a bilayer with core spheres formed and at (0.7 wt %) bilayers spheres were formed.

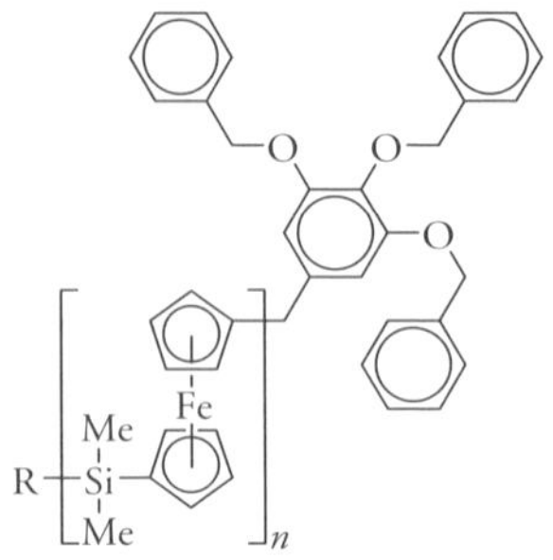

306

Majoral et al. has recently synthesized organometallic phosphorus dendrimers up to the fourth generation.[358] The core was prepared using a Staudinger reaction between bis-di-phenylphosphinoferrocene with $N_3P(S)(OC_6H_4CHO)_2$. The first dendrimer generation was prepared by reacting the core with $H_2NNMeP(S)Cl_2$. The second generation was synthesized via nucleophilic substitution of the chlorine groups by a hydroxybenzaldehyde sodium salt. The sequential use of these reactions allowed for the isolation of fourth-generation dendrimer.

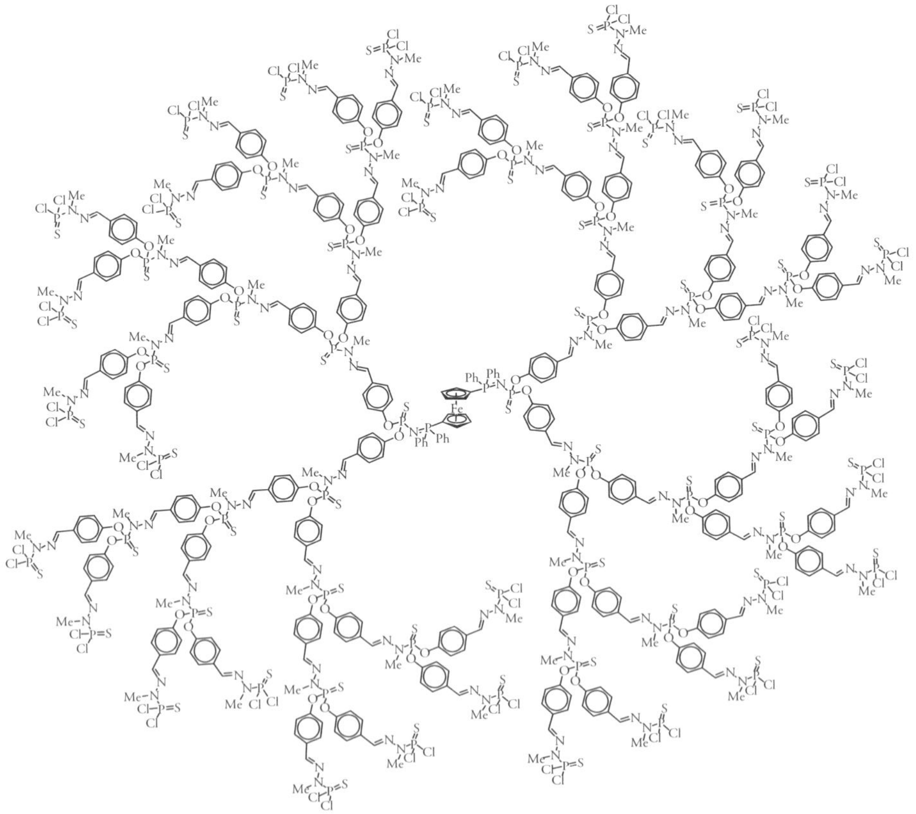

307

V. SUMMARY

Ever since the discovery of the first organometallic polymer in 1956, the field has been expanding exponentially, especially since 1991. The unique electrical, optical, catalytic, and magnetic properties that these organometallic polymers possess have attracted many research groups around the globe to be involved in designing new polymeric materials that contain metallic moieties, and in examining their properties and potential applications. The metallic moieties can either reside as part of the macromolecular backbone, in side-chains, coordinated to the backbone, or as integral parts of dendrites, stars, and rods. In this chapter, we briefly presented the various classes of organometallic polymers and highlighted their properties. We believe that many of tomorrow's new materials will contain metals in their framework and that many of the discoveries so far will be the bases of strong foundations for new and commercially important materials.

ACKNOWLEDGMENTS

This work is supported by the Natural Sciences and Engineering Research Council of Canada (NSERC), Manitoba Hydro, and The University of Winnipeg.

REFERENCES

1. *Ferrocenes: Heterogenousn Catalysis. Organic Synthesis. Materials Science*, A. Togni, T. Hayashi, Eds. VCH, Wienheim, 1995.
2. E. W. Neuse, H. Rosenburg, *Metallocene Polymers*, Marcel Dekker, New York, 1970.
3. K. A. Andrianov, *Metallorganic Polymers*, Wiley, New York, 1965.
4. F. S. Arimoto, A. C. Haven, Jr., *J. Am. Chem. Soc.* **77**, 6295 (1955).
5. C. E. Carraher, Jr., J. E. Sheats, C. U. Pittman, Jr. (eds.), *Organometallic Polymers*, Academic Press, New York, 1978.
6. C. E. Carraher, Jr., J. E. Sheats, C. U. Pittman, Jr. (eds.), *Advances in Organometallic and Inorganic Polymers Science*, Marcel Dekker, New York, 1982.
7. J. E. Sheats, C. E. Carraher, Jr., C. U. Pittman, Jr. (eds.), *Metal-Containing Polymeric Systems*, Plenum Press, New York, 1985.
8. J. E. Sheats, C. E. Carraher, Jr., C. U. Pittman, Jr., M. Zeldin, B. Currell (eds.), *Inorganic and Metal-Containing Polymeric Materials*, Plenum Press, New York, 1990.
9. C. U. Pittman, Jr., C. E. Carraher, Jr., M. Zeldin, J. E. Sheats, B. M. Culbertson (eds.), *Metal-Containing Polymeric Materials*, Plenum Press, New York, 1996.
10. M. Zeldin, K. J. Wynne, H. R. Allcock (eds.), *Inorganic and Organometallic Polymers*, ACS Symposium Series Vol. 360, ACS Publishers, Washington, DC, 1988.
11. M. Zeldin, "Inorganic and Organometallic Polymers," in *Encyclopedia of Physical Science and Technology*, Third Edition, Vol. 12, pp. 675–695, Wiley, New York, 2002.
12. K. A. Andrianov, *Metallorganic Polymers*, Wiley, New York, 1965.
13. E. W. Neuse, H. Rosenburg, *Metallocene Polymers*, Marcel Dekker, New York, 1970.
14. A. D. Pomogailo, *Catalysis by Polymer Immobilized Metal Complexes,* Gordon and Breach Science Publishers, Amsterdam, 1998.
15. R. D. Archer, *Inorganic and Organometallic Polymers*, Wiley-VCH, New York, 2001.
16. A. S. Abd-El-Aziz, *Coord. Chem. Rev.,* **233–234**, 177 (2002).

17. A. S. Abd-El-Aziz, "Metal-Containing Polymers," in *Encyclopedia of Polymer Science and Technology*, Third Edition, J. I. Kroschwitz, Ed., Wiley, New York, 2002.
18. A. S. Abd-El-Aziz, E. K. Todd, *Polymer News*, **5**, 26 (2001).
19. A. S. Abd-El-Aziz, *Macromol. Rapid Commun.*, **23**, 995 (2002).
20. P. Nguyen, P. Gomez-Elipe, I. Manners, *Chem. Rev.*, **99**, 1515 (1999).
21. I. Manners, *Angew. Chem. Int. Ed. Engl.*, **35**, 1603 (1996).
22. F. G. A. Stone, W. A. G. Graham, *Inorganic Polymers*, Academic Press, New York, 1962.
23. C. E. Carraher, Jr., C. U. Pittman, Jr., "Inorganic Polymers," in Ullmann's *Encyclopedia of Industrial Chemistry*, Fifth Edition, Vol. A14, pp. 241–262, VCH Press, 1989.
24. B. P. Block, *Inorganic Macromol. Rev.* **1**(2), 115 (1970).
25. M. D. Rausch, C. U. Pittman, Jr., et al. in *New Monomers and Polymers*, B. M. Culbertson, C. U. Pittman, Jr., Eds., pp. 243–267, Plenum Publishing, New York, 1984.
26. C. U. Pittman, Jr., in *Organometallic Reactions*, E. I. Becker, M. Tsutsui, Eds., Vol. 6, pp. 1–62, Marcel Dekker, New York, 1977.
27. C. U. Pittman, Jr., C. E. Carraher, Jr., J. R. Reynolds, "Organometallic Polymers," in *Encyclopedia of Polymer Science and Engineering*, H. Mark, N. Bikales, C. Overberger, J. Menges, Eds., Vol. 10, pp. 541–594, Wiley, 1987.
28. A. D. Pomogailo, V. S. Savostianov, *J. Macromol. Sci. Rev.* in *Macromol. Sci. Chem. Phys.*, **25**(3), 375–479 (1985).
29. C. U. Pittman, Jr., in *Comprehensive Organometallic Chemistry*, G. Wilkinson, F. G. A. Stone, E. W. Abel, Eds., Vol. 8, pp. 553–608, Pergamon Press, Oxford, UK, 1982.
30. C. E. Carraher, Jr., *Inorg. Macromol. Rev.*, **1**, 271 (1972).
31. R. G. Jones, W. Ando, J. Chojnowski (eds.), *Silicon-Containing Polymers: The Science and Technology of Their Synthesis and Applications*, Kluwer Academic Publishers, Dordrecht/Boston/London, 2000.
32. J. M. Zeigler, F. G. Fearon (eds.), *Silicon-Based Polymer Science: A Comprehensive Resource*, Advances in Chemistry, 224, American Chemical Society, Washington, DC, 1990.
33. S. J. Clarson, J. J. Fitzgerald, M. J. Owen, S. D. Smith (eds.), *Silicones and Silicone-Modified Materials*, ACS Symposium Series 729, American Chemical Society, Washington, DC, 2000.
34. R. J. Puddephatt, *Chem. Commun*, 1055 (1998).
35. M. H. Chisholm, *Angew. Chem. Int. Ed. Engl.*, **30**, 673 (1991).
36. S. J. Davies, B. F. G. Johnson, M. S. Khan, J. Lewis, *J. Chem. Soc., Chem. Commun.*, 187 (1991).
37. M. S. Khan, S. J. Davies, A. K. Kakkar, D. Schwartz, B. Lin, B. F. G. Johnson, J. Lewis, *J. Organomet. Chem.*, **424**, 87 (1992).
38. J. Lewis, M. S. Khan, A. K. Kakkar, B. F. G. Johnson, T. B. Marder, H. B. Fyfe, F. Wittmann, R. H. Friend, A. E. Dray, *J. Organomet. Chem.*, **425**, 165 (1992).
39. N. Chawdhury, A. Kohler, R. R. Friend, M. Younus, N. J. Long, P. R. Raithby, J. Lewis, *Macromolecules*, **31**, 722 (1998).
40. R. D. Markwell, I. S. Butler, A. K. Kakkar, M. S. Khan, Z. H. Al-Zakwani, J. Lewis, *Organometallics*, **15**, 2331 (1996).
41. M. Younus, A. Kohler, S. Cron, N. Chawdhury, M. R. A. Al-Mandhary, M. S. Khan, J. Lewis, N. J. Long, R. H. Friend, P. R. Raithby, *Angew. Chem. Int. Ed.*, **37**, 3036 (1998).
42. A. E. Dray, R. Rachel, W. O. Saxton, J. Lewis, M. S. Khan, A. M. Donald, R. H. Friend, *Macromolecules*, **25**, 3473 (1992).
43. M. S. Khan, A. K. Kakkar, N. L. Long, J. Lewis, P. Raithby, P. Nguyen, T. B. Marder, F. Wittmann, R. L. Friend, *J. Mater. Chem.*, **4**, 1227 (1994).
44. M. S. Khan, D. J. Schwartz, N. A. Pasha, A. K. Kakkar, B. Lin, P. R. Raithby, J. Lewis, *Z. Anorg. Allg. Chem.*, **616**, 121 (1992).

45. S. J. Davies, B. F. G. Johnson, J. Lewis, P. R. Raithby, *J. Organomet. Chem.*, **414**, C51 (1991).

46. B. F. G. Johnson, A. K. Kakkar, M. S. Khan, J. Lewis, *J. Organomet. Chem.*, **409**, C12 (1991).

47. Z. Atherton, C. W. Faulkner, S. L. Ingham, A. K. Kakkar, M. S. Khan, J. Lewis, N. J. Long, P. R. Raithby, *J. Organomet. Chem.*, **462**, 265 (1993).

48. C. C. Frazier, S. Guha, W. P. Chen, M. P. Cockerham, P. L. Porter, E. A. Chauchard, C. H. Lee, *Polymer*, **28**, 553 (1987).

49. S. Guha, C. C. Frazier, P. L. Porter, K. Kang, S. E. Finberg, *Optics Lett.*, **14**, 952 (1989).

50. H. B. Fyfe, M. Mlekuz, G. Stringer, N. J. Taylor, T. B. Marder, in *Inorganic and Organometallic Polymers with Special Properties*, R. M. Laine, Ed., NATO ASI Series, Series E, Applied Sciences, Vol. 206, pp. 331–344, Kluwer Academic, Dordecht, 1992.

51. M. Yang, L. Zhang, Z. Lei, P. Ye, J. Si, Q. Yang, Y. Wang, *J. Appl. Polym. Sci.*, **70**, 1165 (1998).

52. F. Matsumoto, N. Matsumi, Y. Chujo, *Polym. Bull.*, **48**, 119 (2002).

53. J. Vicente, M.-T. Chicote, M. M. Alvarez-Falcón, P. G. Jones, *Organometallics*, **24**, 2764 (2005).

54. I. Fratoddi, C. Battocchio, A. Furlani, P. Mataloni, G. Polzonetti, M. V. Russo, *J. Organomet. Chem.*, **674**, 10 (2003).

55. R. D'Amato, I. Fratoddi, A. Cappotto, P. Altamura, M. Delfini, C. Bianchetti, A. Bolasco, G. Polzonetti, M. V. Russo, *Organometallics*, **23**, 2860 (2004).

56. M. S. Khan, M. R. A. Al-Mandhary, M. K. Al-Suti, F. R. Al-Battashi, S. Al-Saadi, B. Aherns, J. K. Bjernemose, M. F. Mahon, P. R. Raithby, M. Younus, N. Chawdhury, A. Köhler, E. A. Marseglia, E. Tedesco, N. Feeder, S. J. Teat, *Dalton Trans.*, 2377 (2004).

57. G. Jia, R. J. Puddephatt, J. D. Scott, J. J. Vittal, *Organometallics*, **12**, 3565 (1993).

58. G. Jia, N. C. Payne, J. J. Vittal, R. J. Puddephatt, *Organometallics*, **12**, 4771 (1993).

59. M. J. Irwin, L. Manojlovic-Muir, K. W. Muir, R. J. Puddephatt, D. S. Yufit, *Chem. Commun.*, 219 (1997).

60. M. Younus, A. Kohler, S. Cron, N. Chawdhury, M. R. A. Al-Mandhary, M. S. Khan, J. Lewis, N. J. Long, R. H. Friend, P. R. Raithby, *Angew. Chem. Int. Ed.*, **37**, 3036 (1998).

61. M. S. Khan, M. R. A. Al-Mandhary, M. K. Al-Suti, A. K. Hisahm, P. R. Raithby, B. Ahrens, M. F. Mahon, L. Male, E. A. Marseglia, E. Tedesco, R. H. Friend, A. Kohler, N. Feeder, S. J. Teat, *J. Chem. Soc., Dalton Trans.*, 1358 (2002).

62. M. S. Khan, M. R. A. Al-Mandhary, M. K. Al-Suti, N. Feeder, S. Nahar, A. Kohler, R. H. Friend, P. J. Wilson, P. R. Raithby, *J. Chem. Soc., Dalton Trans.*, 2441 (2002).

63. J. S. Wilson, N. Chawdhury, M. R. A. Al-Mandhary, M. Younus, M. S. Khan, P. R. Raithby, A. Kohler, R. H. Friend, *J. Am. Chem. Soc.*, **123**, 9412 (2001).

64. G. Polzonetti, G. Lucci, P. Altamura, A. Ferri, G. Paolucci, A. Goldoni, P. Parent, C. Laffon, M. V. Russo, *Surf. Interface Anal.*, **34**, 588 (2002).

65. A. Buttinelli, E. Viola, E. Antonelli, C. Lo Sterzo, *Organometallics*, **17**, 2574 (1998).

66. P. Altamura, G. Giardina, C. Lo Sterzo, M. V. Russo, *Organometallics*, **20**, 4360 (2001).

67. M. Yang, L. Zhang, Z. Lei, P. Ye, J. Si, Q. Yang, Y. Wang, *J. Appl. Polym. Sci.*, **70**, 1165 (1998).

68. W.-Y. Wong, S.-M. Chan, K.-H. Choi, K.-W. Cheah, W.-K. Chan, *Macromol. Rapid Commun.*, **21**, 453 (2000).

69. J. S. Wilson, N. Chawdhury, M. R. A. Al-Mandhary, M. Younus, M. S. Khan, P. R. Raithby, A. Kohler, R. H. Friend, *J. Am. Chem. Soc.*, **123**, 9412 (2001).

70. W.-Y. Wong, G.-L. Lu, K.-H. Choi, J.-X. Shi, *Macromolecules*, **35**, 3506 (2002).

71. W.-Y. Wong, S.-M. Chan, K.-H. Choi, K.-W. Cheah, W.-K. Chan, *Macromol. Rapid Commun.*, **21**, 453 (2000).

72. L. Liu, W.-Y. Wong, S.-Y. Poon, K.-W Cheah, *J. Inorg. Organomet. Polym. Mater.*, xxxx-xxxx, 1, (2006).

73. E. Dray, R. Rachel, W. O. Saxton, J. Lewis, M. S. Khan, A. M. Donald, R. H. Friend, *Macromolecules*, **25**, 3473 (1992).

74. M. J. Irwin, G. Jia, J. J. Vittal, R. J. Puddephatt, *Organometallics*, **15**, 5321 (1996).
75. T. Tanase, E. Goto, R. A. Begum, M. Hamaguchi, S. Zhan, M. Iida, K. Sakai, *Organometallics*, **23**, 5975 (2004).
76. D. R. Tyler, J. J. Wolcott, G. W. Nieckarz, S. C. Tenhaeff, Chapter 36, in *Inorganic and Organometallic Polymers II*, P. Wisian-Neilson, H. R. Allcock, K. J. Wynne, Eds., ACS Symposium Series 572, pp. 481–496, American Chemical Society, Washington, DC, 1994.
77. S. C. Tenhaeff, D. R. Tyler, *Organometallics,* **10**, 1116 (1991).
78. S. C. Tenhaeff, D. R. Tyler, *Organometallics,* **10**, 473 (1991).
79. S. C. Tenhaeff, D. R. Tyler, *Organometallics,* **11**, 1466 (1992).
80. J. L. Male, M. Yoon, A. G. Glenn, T. J. R. Weakley, D. R. Tyler, *Macromolecules*, **32**, 3898 (1999).
81. S. Chardon-Noblat, A. Deronzier, D. Zsoldos, R. Ziessel, M. Haukka, T. Pakkanen, T. Venalainen, *J. Chem. Soc., Dalton Trans.,* 2581 (1996).
82. S. Chardon-Noblat, A. Deronzier, F. Hartl, J. van Slageren, T. Mahabiersing, *Eur. J. Inorg. Chem.,* 613 (2001).
83. F. Amor, E. de Jesus, A. I. Perez, P. Royo, A. V. de Miguel, *Organometallics*, **15**, 365 (1996).
84. J. Park, Y. Seo, S. Cho, D. Whang, K. Kim, T. Chang, *J. Organomet. Chem.*, **489**, 23 (1995).
85. F. H. Kohler, A. Schell, B. Weber, *Chem. Eur. J.*, **8**, 5219 (2002).
86. T. Hirao, M. Kurashina, K. Aramaki, H. Nishihara, *J. Chem. Soc., Dalton Trans.*, 2929 (1996).
87. G. E. Southard, M. D. Curtis, *Organometallics*, **16**, 5618 (1997).
88. G. E. Southard, M. D. Curtis, *Organometallics*, **20**, 508 (2001).
89. G. E. Southard, M. D. Curtis, *Synthesis*, **9**, 1177 (2002).
90. J. Park, Y. Seo, S. Cho, D. Whang, K. Kim, T. Chang, *J. Organomet. Chem.*, **489**, 23 (1995).
91. H. M. Nugent, M. Rosenblum, P. Klemarczky, *J. Am. Chem. Soc.*, **115**, 3848 (1993).
92. M. Rosenblum, H. M. Nugent, K.-S. Jang, M. M. Labes, W. Cahalane, P. Klemarczyk, W. M. Reiff, *Macromolecules*, **28**, 6330 (1995).
93. R. D. A. Hudson, B. M. Foxman, M. Rosenblum, *Organometallics*, **18**, 4098 (1999).
94. M. A. Buretea, T. D. Tilley, *Organometallics,* **16**, 1507 (1997).
95. C. E. Stanton, T. R. Lee, H. R. Grubbs, N. S. Lewis, J. K. Pudelski, M. R. Callstrom, M. S. Erickson, M. L. McLaughlin, *Macromolecules,* **28**, 8713 (1995).
96. R. W. Heo, F. B. Somoza, T. R. Lee, *J. Am. Chem. Soc.,* **120**, 1621 (1998).
97. R. Rulkens, Y. Ni, I. Manners, *J. Am. Chem. Soc.* **116**, 12121 (1994).
98. J. A. Massey, K. Temple, L. Cao, Y. Rharbi, J. Raez, M. A. Winnik, I. Manners, *J. Am. Chem. Soc.*, **122**, 11577 (2000).
99. A. Berenbaum, H. Braunschweig, R. Dirk, U. Englert, J. C. Green, F. Jakle, A. J. Lough, I. Manners, *J. Am. Chem. Soc.*, **122**, 5765 (2000).
100. R. Resendes, P. Nguyen, A. J. Lough, I. Manners, *Chem. Commun.*, 1001 (1998).
101. R. Resendes, J. M. Nelson, A. Fischer, F. Jakle, A. Bartole, A. J. Lough, I. Manners, *J. Am. Chem. Soc.*, **123**, 2116 (2001).
102. M. Bochmann, J. Lu, R. D. Cannon, *J. Organomet. Chem.* **518**, 97 (1996).
103. S. L. Ingham, M. S. Khan, J. Lewis, N. J. Long, P. R. Raithby, *J. Organomet. Chem.*, **470**, 153 (1994).
104. K. Naka, T. Uemura, Y. Chujo, *Macromolecules*, **33**, 6965 (2000).
105. N. J. Long, A. J. Martin, R. Vilar, A. J. P. White, D. J. Williams, M. Younus, *Organometallics*, **18**, 4261 (1999).
106. T. Yamamoto, T. Morikita, T. Maruyama, K. Kubota, M. Katada, *Macromolecules*, **30**, 5390 (1997).
107. T. Morikita, T. Maruyama, T. Yamamoto, K. Kubota, M. Katada, *Inorg. Chim. Acta*, **269**, 310 (1998).
108. P. F. Brandt, T. B. Rauchfuss, *J. Am. Chem. Soc.*, **114**, 1926 (1992).
109. C. P. Galloway, T. B. Rauchfuss, *Angew. Chem. Int. Ed. Engl.*, **32**, 1319 (1993).

110. D. L. Compton, T. B. Rauchfuss, *Organometallics*, **13**, 4367 (1994).

111. D. L. Compton, P. F. Brandt, T. B. Rauchfuss, D. F. Rosenbaum, C. F. Zukoski, *Chem. Mater.*, **7**, 2342 (1995).

112. T. Mizuta, M. Onishi, K. Miyoshi, *Organometallics,* **19**, 5005 (2000).

113. C. E. B. Evans, A. J. Lough, H. Grondey, I. Manners, *New J. Chem.*, **24**, 447 (2000).

114. T. J. Peckham, J. A. Massey, C. H. Honeyman, I. Manners, *Macromolecules*, **32**, 2830 (1999).

115. H. P. Withers, Jr., D. Seyferth, J. D. Fellmann, P. E. Garrou, S. Martin, *Organometallics,* **1**, 1283 (1982).

116. L. Espada, K. H. Pannell, V. Papkov, L. Leites, S. Bukalov, I. Suzdalev, M. Tanaka, T. Hayashi, *Organometallics*, **21**, 3758 (2002).

117. I. Manners, *Can. J. Chem.*, **76**, 371 (1998).

118. I. Manners, *Chem. Commun.*, 857 (1999).

119. K. Kulbaba, I. Manners, *Macromol. Rapid. Commun.*, **22**, 711 (2001).

120. D. A. Foucher, B. Z. Tang, I. Manners, *J. Am. Chem. Soc.*, **114**, 6246 (1992).

121. R. Rulkens, A. J. Lough, I. Manners, *J. Am. Chem. Soc.*, **116**, 797 (1994).

122. Y. Z. Ni, R. Rulkens, J. K. Pudelski, I. Manners, *Macromol, Rapid Commun.*, **16**, 637 (1995).

123. J. Rasburn, D. A. Foucher, W. F. Reynolds, I. Manners, G. J. Vancso, *Chem. Commun.*, 843 (1998).

124. J. Rasburn, R. Petersen, T. Jahr, R. Rulkens, I. Manners, G. J. Vancso, *Chem. Mater.*, **7**, 871 (1995).

125. R. G. H. Lammertink, M. A. Hempenius, G. J. Vancso, K. Shin, M. H. Rafailovich, J. Sokolov, *Macromolecules*, **34**, 942 (2000).

126. J. Raez, I. Manners, M. A. Winnik, *J. Am. Chem. Soc.*, **124**, 10381 (2002).

127. J. Raez, I. Manners, M. A. Winnik, *Langmuir*, **18**, 7229 (2002).

128. R. Resendes, J. Massey, H. Dorn, K. N. Power, M. A. Winnik, I. Manners, *Angew. Chem. Int. Ed. Engl.* **38**, 2570 (1999).

129. R. Resendes, J. A. Massey, K. Temple, L. Cao, K. N. Power-Billard, M. A. Winnik, I. Manners, *Chem. Eur. J.*, **7**, 2414 (2001).

130. J. Massey, K. N. Power, I. Manners, M. A. Winnik, *J. Am. Chem. Soc.*, **120**, 9533 (1998).

131. J. A. Massey, M. A. Winnik, I. Manners, *J. Am. Chem. Soc.*, **123**, 3147 (2001).

132. R. Rulkens, D. P. Gates, D. Balaishis, J. K. Pudelski, D. F. McIntosh, A. J. Lough, I. Manners, *J. Am. Chem. Soc.*, **119**, 10976 (1997).

133. K. Temple, S. Dziadek, I. Manners, *Organometallics*, **21**, 4377 (2002).

134. R. Resendes, J. Massey, H. Dorn, M. A. Winnik, I. Manners, *Macromolecules*, **33**, 8 (2000).

135. X.-S. Wang, M. A. Winnik, I. Manners, *Macromol. Rapid Commun.*, **23**, 210 (2002).

136. X.-S. Wang, M. A. Winnik, I. Manners, *Macromolecules*, **35**, 9146 (2002).

137. D. A. Rider, K. A. Cavicchi, K. N. Power-Billard, T. P. Russell, I. Manners, *Macromolecules*, **38**, 6931 (2005).

138. X. Wang, M. A. Winnik, I. Manners, *Macromolecules*, **38**, 1928 (2005).

139. I. Korczagin, M. A. Hempenius, G. J. Vancso, *Macromolecules*, **37**, 1686 (2004).

140. J. M. Nelson, H. Rengel, I. Manners, *J. Am. Chem. Soc.*, **115**, 7035 (1993).

141. J. M. Nelson, A. J. Lough, I. Manners, *Angew. Chem. Int. Ed. Engl.,* **33**, 989 (1994).

142. J. M. Nelson, P. Nguyen, R. Petersen, H. Rengel, P. M. Macdonald, A. J. Lough, I. Manners, N. P. Raju, J. E. Greedan, S. Barlow, D. O'Hare, *Chem. Eur. J.*, **3**, 573 (1997).

143. C. H. Honeyman, D. A. Foucher, F. Y. Dahmen, R. Rulkens, A. J. Lough, I. Manners, *Organometallics*, **14**, 5503 (1995).

144. H. Braunschweig, R. Dirk, M. Muller, P. Nguyen, R. Resendes, D. P. Gates, I. Manners, *Angew. Chem. Int. Ed. Engl.*, **36**, 2338 (1997).

145. F. Jakle, A. Berenbaum, A. J. Lough, I. Manners, *Chem. Eur. J.*, **6**, 2762 (2000).
146. D. A. Foucher, M. Edwards, R. A. Burrow, A. J. Lough, I. Manners, *Organometallics*, **13**, 4959 (1994).
147. A. Berenbaum, I. Manners, *Dalton Trans.*, 2057 (2004).
148. N. P. Reddy, H. Yamashita, M. Tanaka, *J. Chem. Soc., Chem. Commun.*, 2263 (1995).
149. T. J. Peckham, J. A. Massey, M. Edwards, I. Manners, D. A. Foucher, *Macromolecules*, **29**, 2396 (1996).
150. R. N. Kapoor, G. M. Crawford, J. Mahmoud, V. V. Dementiev, M. T. Nguyen, A. F. Diaz, K. H. Pannell, *Organometallics*, **14**, 4944 (1995).
151. H. K. Sharma, F. Cervantes-Lee, J. S. Mahmoud, K. H. Pannell, *Organometallics*, **18**, 399 (1999).
152. R. Rulkens, A. J. Lough, I. Manners, *Angew. Chem. Int. Ed. Engl.*, **35**, 1805 (1996).
153. F. Jakle, R. Rulkens, G. Zech, D. A. Foucher, A. J. Lough, I. Manners, *Chem. Eur. J.*, **4**, 2117 (1998).
154. T. Baumgartner, F. Jakle, R. Rulkens, G. Zech, A. J. Lough, I. Manners, *J. Am. Chem. Soc.*, **124**, 10062 (2002).
155. F. Jakle, R. Rulkens, G. Zech, J. A. Massey, I. Manners, *J. Am. Chem. Soc.*, **122**, 4231 (2000).
156. U. Vogel, A. J. Lough, I. Manners, *Angew. Chem. Int. Ed.*, **43**, 3321 (2004).
157. I. Manners, *Polyhedron*, **15**, 4311 (1996).
158. K. Temple, F. Jakle, J. B. Sheridan, I. Manners, *J. Am. Chem. Soc.*, **123**, 1355 (2001).
159. H. Yim, M. D. Foster, D. Balaishis, I. Manners, *Langmuir*, **14**, 3921 (1998).
160. K. N. Power-Billard, T. J. Peckham, A. Butt, F. Jakle, I. Manners, *J. Inorg. Organomet. Polym.*, **10**, 159 (2000).
161. D. L. Zechel, K. C. Hultzsch, R. Rulkens, D. Balaishis, Y. Ni, J. K. Pudelski, A. J. Lough, I. Manners, *Organometallics*, **15**, 1972 (1996).
162. P. Nguyen, G. Stojcevic, K. Kulbaba, M. J. MacLachlan, X.-H. Liu, A. J. Lough, I. Manners, *Macromolecules*, **31**, 5977 (1998).
163. A. Berenbaum, A. Lough, I. Manners, *Organometallics*, **21**, 4415 (2002).
164. M. J. Maclachlan, J. Zheng, K. Thieme, A. J. Lough, I. Manners, C. Mordas, R. LeSuer, W. E. Geiger, L. M. Liable-Sands, A. L. Rheingold, *Polyhedron*, **19**, 275 (2000).
165. R. G. H. Lammertink, M. A. Hempenius, V. Z.-H. Chan, E. L. Thomas, G. J. Vancso, *Chem. Mater.*, **13**, 429 (2001).
166. R. G. H. Lammertink, M. A. Hempenius, J. E. van den Enk, V. Z.-H. Chan, E. L. Thomas, G. J. Vancso, *Adv. Mater.*, **12**, 98 (2000).
167. L. I. Espada, M. Shadaram, J. Robillard, K. Pannell, *J. Inorg. Organomet. Polym.*, **10**, 169 (2000).
168. K. N. Power-Billard, I. Manners, *Macromolecules*, **33**, 26 (2000).
169. F. Jakle, Z. Wang, I. Manners, *Macromol. Rapid. Commun.*, **21**, 1291 (2000).
170. Z. Wang, A. Lough, I. Manners, *Macromolecules*, **35**, 7669 (2002).
171. M. A. Hempenius, G. J. Vancso, *Macromolecules*, **35**, 2445 (2002).
172. M. A. Hempenius, M. Peter, N. S. Robins, E. S. Kooij, G. J. Vancso, *Langmuir*, **18**, 7629 (2002).
173. M. A. Hempenius, N. S. Robins, R. G. H. Lammertink, G. J. Vancso, *Macromol. Rapid Commun.*, **22**, 30 (2001).
174. K. N. Power-Billard, I. Manners, *Macromol. Rapid Commun.*, **23**, 607 (2002).
175. J. Halfyard, J. Galloro, M. Ginzburg, Z. Wang, N. Coombs, I. Manners, G. A. Ozin, *Chem. Commun.*, 1746 (2002).
176. H. Plenio, J. Hermann, A. Sehring, *Chem. Eur. J.*, **6**, 1820 (2000).
177. H. Plenio, J. Hermann, J. Leukel, *Eur. J. Inorg. Chem.*, **12**, 2063 (1998).
178. T. Morikita, T. Mauyama, T. Yamamoto, K. Kubota, M. Katada, *Inorg. Chim. Acta*, **269**, 310 (1998).
179. T. Yamamoto, T. Morikita, T. Maruyama, K. Kubota, M. Katada, *Macromolecules*, **30**, 5390 (1997).
180. C. M. Casado, M. Moran, J. Losada, I. Cuadrado, *Inorg. Chem.*, **34**, 1668 (1995).

181. I. Cuadrado, C. M. Casado, F. Lobete, B. Alonso, B. Gonzalez, J. Losada, U. Amador, *Organometallics*, **18**, 4960 (1999).
182. Y. Gao, J. M. Shreeve, *J. Polym. Sci., Part A: Polym. Chem.*, **43**, 974 (2005).
183. Z. Akhter, M. A. Bashir, M. S. Khan, *Appl. Organometal. Chem.*, **19**, 848 (2005).
184. K. E. Gonsalves, M. D. Rausch, *J. Polym. Sci., Part A: Polym. Chem.*, **24**, 1599 (1986).
185. Q.-D. Shen, S.-L. Ni, H.-S. Xu, C.-Z. Yang, *J. Appl. Polym. Sci.*, **74**, 2674 (1999).
186. N. Najafi-Mohajeri, G. L. Nelson, R. Benrashid, *J. Appl. Polym. Sci.*, **76**, 1847 (2000).
187. M. M. Abd-Alla, M. F. El-Zohry, K. I. Aly, M. M. M Abd-El-Wahab, *J. Appl. Polym. Sci.*, **47**, 323 (1993).
188. P. Singh, M. D. Rausch, R. W. Lenz, *Polym. Bull.*, **22**, 247 (1989).
189. G. Wilbert, A. Wiesemann, R. Zentel, *Macromol. Chem. Phys.*, **196**, 3771 (1995).
190. P. Kannan, S. Senthil, R. Vijayakumar, R. Marimuthu, *J. Appl. Polym. Sci.*, **86**, 3494 (2002).
191. S. Senthil, P. Kannan, *J. Polym. Sci., Part A: Polym. Chem.*, **39**, 2396 (2001).
192. S. Senthil, P. Kannan, *J. Appl. Polym. Sci.*, **85**, 831 (2002).
193. I. Yamaguchi, K. Osakada, T. Yamamoto, M. Katada, *Bull. Chem. Soc. Jpn.*, **72**, 2557 (1999).
194. R. J. P. Corriu, N. Devylder, C. Guerin, B. Henner, A. Jean, *Organometallics*, **13**, 3194 (1994).
195. A. S. Abd-El-Aziz, E. K. Todd. R. M. Okasha, P. O. Shipman, T. E. Wood, *Macromolecules*, **38**, 9411 (2005).
196. M. Kurashina, M. Murata, T. Watanabe, H. Nishihara, *J. Am. Chem. Soc.*, **125**, 12420 (2003).
197. P. K. Sahoo, S. K. Swain, *J. Polym. Sci., Part A: Polym. Chem.*, **37**, 3899 (1999).
198. C. Lidrissi, A. Romerosa, M. Saoud, M. Serrano-Ruiz, L. Gonsalvi, M. Peruzzini, *Angew. Chem. Int. Ed.*, **44**, 2568 (2005).
199. C.-P. Chai, Y.-P. Wang, R.-M. Wang, H.-X. Ren, C.-J. Hao, *Polym. Adv. Technol.*, **15**, 55 (2004).
200. Y. Morisaki, H. Chen, Y. Chujo, *J. Organomet. Chem.* **689**, 2684 (2004).
201. H. K. Sharma, K. H. Pannel, *Chem. Commun.*, 2556 (2004).
202. S. S. H. Mao, T. D. Tilley, *J. Am. Chem. Soc.*, **117**, 5365 (1995).
203. S. S. H. Mao, T. D. Tilley, *Macromolecules*, **29**, 6362 (1996).
204. S. S. H. Mao, F.-Q. Liu, T. D. Tilley, *J. Am. Chem. Soc.*, **120**, 1193 (1998).
205. S. S. H. Mao, T. D. Tilley, *Macromolecules*, **30**, 5566 (1997).
206. B. L. Lucht, S. S. H. Mao, T. D. Tilley, *J. Am. Chem. Soc.*, **120**, 4354 (1998).
207. M. Murata, T. Hoshi, I. Matsuoka, T. Nankawa, M. Kurihara, H. Nishihara, *J. Inorg. Organomet. Polym.*, **10**, 209 (2000).
208. J.-C. Lee, A. Nishio, I. Tomita, T. Endo, *Macromolecules*, **30**, 5205 (1997).
209. I. Matsuoka, K. Aramaki, H. Nishihara, *J. Chem. Soc., Dalton Trans.*, 147 (1998).
210. T. Shimura, A. Ohkubo, K. Aramaki, H. Uekusa, T. Fujita, S. Ohba, H. Nishihara, *Inorg. Chim. Acta*, **230**, 215 (1995).
211. T. Shimura, A. Ohkubo, N. Matsuda, I. Matsuoka, K. Aramaki, H. Nishihara, *Chem. Mater.*, **8**, 1307 (1996).
212. I. Tomita, A. Nishio, T. Endo, *Macromolecules*, **28**, 3042 (1995).
213. J.-C. Lee, I. Tomita, T. Endo, *Polym. Bull.*, **39**, 415 (1997).
214. J.-C. Lee, I. Tomita, T. Endo, *Macromolecules*, **31**, 5916 (1998).
215. I. Tomita, M. Ueda, *Macromol. Symp.*, **209**, 217 (2004).
216. I. Tomita, M. Ueda, *J. Inorg. Organomet. Polym. Mater.*, **15**, 511–518 (2006).
217. T. Kuhnen, M. Stradiotto, R. Ruffolo, D. Ulbrich, M. J. McGlinchey, M. A. Brook, *Organometallics*, **16**, 5048 (1997).
218. S. W. Thomas III, S. Yagi, T. M. Swager, *J. Mater. Chem.*, **15**, 2829 (2005).
219. C. Diaz, M. L. Valenzuela, *Macromolecules,* **39**, 103–111 (2006).

220. I. Aiello, U. Caruso, M. Ghedini, B. Panunzi, A. Quatela, A. Roviello, F. Sarcinelli, *Polymer*, **44**, 7635 (2003).
221. K. I. Aly, *J. App. Polym. Sci.*, **94**, 1440 (2004).
222. K. Aly, M. I. Abdel Monem, *J. Appl. Polym. Sci.*, **98**, 2394 (2005).
223. D. A. Durkee, H. B. Eitouni, E. D. Gomez, M. W. Ellsworth, A. T. Bell, N. P. Balsara, *Adv. Mater.*, **17**, 2003 (2005).
224. M. Buchmeiser, R. R. Schrock, *Macromolecules*, **28**, 6642 (1995).
225. M. R. Buchmeiser, *Macromolecules*, **30**, 2274 (1997).
226. M. R. Buchmeiser, N. Schuler, G. Kaltenhauser, K.-H. Ongania, I. Lagola, K. Wurst, H. Scottenberger, *Macromolecules*, **31**, 3175 (1998).
227. K. Eder, E. Reichel, H. Scottenberger, C. G. Huber, M. R. Buchmeiser, *Macromolecules*, **34**, 4334 (2001).
228. C. J. Neef, D. T. Glatzhofer, K. M. Nicholas, *J. Polym. Sci., Part A: Polym. Chem.*, **35**, 3365 (1997).
229. T. Peckham, P. Nguyen, S. C. Bourke, Q. Wang, D. G. Harrison, P. Zoricak, C. Russell, L. M. Liable-Sands, A. L. Rheingold, A. J. Lough, I. Manners, *Organometallics*, **20**, 3035 (2001).
230. F. Le Floch, H.-A. Ho, P. Harding-Lepage, M. Bédard, R. Neagu-Plesu, M. Leclerc, *Adv. Mater.*, **17**, 1251 (2005).
231. Y. Yang, Z. Xie, C. Wu, *Macromolecules*, **35**, 3426 (2002).
232. N. Kuramoto, Y. Shishido, K. Nagai, *J. Polym. Sci., Part A: Polym. Chem.*, **35**, 1967 (1997).
233. M. E. Wright, E. G. Toplikar, R. F. Kubin, M. D. Seltzer, *Macromolecules*, **25**, 1838 (1992).
234. M. E. Wright, B. B. Cochran, E. G. Toplikar, H. S. Lackritz, J. T. Kerney, in *Inorganic and Organometallic Polymers II*, P. Wisian-Neilson, H. R. Allcock, K. J. Wynne Eds., ACS Symposium Series 572, American Chemical Society, Washington, DC, 1994.
235. R. Deschenaux, F. Turpin, D. Guillon, *Macromolecules*, **30**, 3759 (1997).
236. R. Deschenaux, I. Jauslin, U. Scholten, F. Turpin, D. Guillon, B. Heinrich, *Macromolecules*, **31**, 5647 (1998).
237. A. Wiesemann, R. Zentel, T. Pakula, *Polymer*, **33**, 5315 (1992).
238. G. Wilbert, R. Zentel, *Macromol. Chem. Phys.*, **197**, 3259 (1996).
239. G. Wilbert, S. Traud, R. Zentel, *Macromol. Chem. Phys.*, **198**, 3769 (1997).
240. A. Wiesemann, R. Zentel, G. Lieser, *Acta Polymer.*, **46**, 25 (1995).
241. N. C. Foulds, C. R. Lowe, *Anal. Chem.*, **60**, 2473 (1988).
242. G. Zotti, S. Zecchin, G. Schiavon, A. Berlin, G. Pagani, A. Canavesi, *Chem. Mater.*, **7**, 2309 (1995).
243. G. Zotti, G. Schiavon, S. Zecchin, A. Berlin, G. Pagani, A. Canavesi, *Synth. Met.*, **76**, 255 (1996).
244. N. Hida, F. Takei, K. Onitsuka, K. Shiga, S. Asaoka, T. Iyoda, S. Takahashi, *Angew. Chem. Int. Ed.*, **42**, 4349–4352 (2003).
245. E. W. Neuse, M. G. Meirim, D. D. N''Da, G. Caldwell, *J. Inorg. Organomet. Polym.*, **9**, 221 (1999).
246. B. Schechter, G. Caldwell, E. W. Neuse, *J. Inorg. Organomet. Polym.*, **10**, 177 (2000).
247. E. W. Neuse, *Macromol. Symp.*, **172**, 127 (2001).
248. M. G. Meirim, E. W. Neuse, G. Caldwell, *J. Appl. Polym. Sci.*, **73**, 2143 (1999).
249. G. Caldwell, M. G. Meirim, E. W. Neuse, K. Beloussow, W.-C. Shen, *J. Inorg. Organomet. Polym.*, **10**, 93 (2000).
250. K. J. Watson, J. Zhu, S. T. Nguyen, C. A. Mirkin, *J. Am. Chem. Soc.*, **121**, 462 (1999).
251. A. S. Abd-El-Aziz, T. H. Afifi, E. K. Todd, G. Z. Ma, *Polym. Prepr., Am. Chem. Soc. Div. Polym. Chem.*, **42**(2), 450 (2001).
252. A. S. Abd-El-Aziz, T. C. Corkery, E. K. Todd, T. H. Afifi, G. Ma, *J. Inorg. Organomet. Polym.*, **13**, 113, (2003).

253. A. S. Abd-El-Aziz, C. R. de Denus, M. J. Zaworotko, L. R. MacGillivray, *J. Chem. Soc., Dalton Trans.*, 3375, (1995).
254. A. S. Abd-El-Aziz, D. A. Armstrong, S. Bernardin, H. M. Hutton, *Can. J. Chem.*, **74**, 2073 (1996).
255. A. S. Abd-El-Aziz, E. K. Todd, K. M. Epp, *J. Inorg. Organomet. Polym.*, **8**, 127 (1998).
256. A. S. Abd-El-Aziz, E. K. Todd, G. Z. Ma, *J. Poly. Sci., Part A: Polym. Chem.*, **39**, 1216 (2001).
257. A. S. Abd-El-Aziz, E. K. Todd, R. M. Okasha, unpublished results.
258. A. S. Abd-El-Aziz, T. Afifi, W. Budakowski, K. Friesen, E. K. Todd, *Macromolecules*, **35**, 8929 (2002).
259. A. S. Abd-El-Aziz, N. M. Pereira, W. Boraie, E. K. Todd, W. Budakowski, K. Friesen, *J. Inorg. Organomet. Polym. Mater.*, **15**, 497–509 (2005).
260. A. S. Abd-El-Aziz, B. Elmayergi, B. Asher, T. A. Afifi, K. J. Friesen, *Inorg. Chim. Acta*, **359**, 3007–3013 (2006).
261. A. S. Abd-El-Aziz, R. M. Okasha, P. O. Shipman, T. H. Afifi, *Macromol. Rapid Commun.*, **25**, 1497 (2004).
262. A. S. Abd-El-Aziz, E. K. Todd, C. R. de Denus, A. A. Dembek, P. J. Fagan, *Polym. Prepr., Am. Chem. Soc. Div. Polym. Chem.*, **40**(2), 926 (1999).
263. C. R. de Denus, L. M. Hoffa, E. K. Todd, A. S. Abd-El-Aziz, *J. Inorg. Organomet. Polym.*, **10**, 189 (2000).
264. A. S. Abd-El-Aziz, E. K. Todd, R. M. Okasha, P. O. Shipman, T. E. Wood, *Macromolecules*, **38**, 9411 (2005).
265. A. S. Abd-El-Aziz, E. K. Todd, G. Z. Ma, J. DiMartino, *J. Inorg. Organomet. Polym.*, **10**, 265 (2000).
266. A. S. Abd-El-Aziz, E. K. Todd, *Polym. Mater. Sci. Eng.*, **86**, 104 (2002).
267. A. S. Abd-El-Aziz, L. May, A. L. Edel, *Macromol. Rapid Commun.*, **21**, 598 (2000).
268. A. S. Abd-El-Aziz, A. L. Edel, L. May, K. M. Epp, H. M. Hutton, *Can. J. Chem.*, **77**, 1797 (1999).
269. A. S. Abd-El-Aziz, L. J. May, J. A. Hurd, R. M. Okasha, *J. Polym. Sci., Part A: Polym. Chem.*, **39**, 2716 (2001).
270. A. S. Abd-El-Aziz, R. M. Okasha, J. Hurd, E. K. Todd, *Polym. Mater. Sci. Eng.*, **86**(1), 91 (2002).
271. A. S. Abd-El-Aziz, L. Edel, A. K. M. Epp, H. M. Hutton, *New J. Chem.*, **23**, 569 (1999).
272. A. S. Abd-El-Aziz, R. M. Okasha, T. H. Afifi, E. K. Todd, *Macromol. Chem. Phys.*, **204**, 555 (2003).
273. A. S. Abd-El-Aziz, R. M. Okasha, T. H. Afifi, *J. Inorg. Organomet. Polym.*, **14**, 269 (2004).
274. J. Matsuda, K. Aramaki, H. Nishihara, *J. Chem. Soc. Faraday Trans.*, **91**, 1477 (1995).
275. M. Altmann, V. Enkelmann, F. Beer, U. H. F. Bunz, *Organometallics*, **15**, 394 (1996).
276. M. Altmann, U. H. F. Bunz, *Angew. Chem. Int. Ed. Engl.*, **34**, 569 (1995).
277. W. Steffen, B. Kohler, M. Altmann, U. Scherf, K. Stitzer, H.-C. zur Loye, U. H. F. Bunz, *Chem. Eur. J.*, **7**, 117 (2001).
278. W. Steffen, U. H. F. Bunz, *Macromolecules*, **33**, 9518 (2000).
279. Y. Sawada, I. Tomita, T. Endo, *Polym. Bull.*, **43**, 165 (1999).
280. Y. Sawada, I. Tomita, I. L. Rozhanskii, T. Endo, *J. Inorg. Organomet. Polym.*, **10**, 221 (2000).
281. J. Ohshita, H. Arase, T. Sumida, N. Mimura, K. Yoshimoto, Y. Tada, Y. Kunugi, Y. Harima, A. Kunai, *Inorg. Chim. Acta*, **358**, 4156 (2005).
282. S. I. Yaniger, D. J. Rose, W. P. McKenna, E. M. Eyring, *App. Spectrosc.*, **38**, 7 (1984).
283. S. I. Yaniger, D. J. Rose, W. P. McKenna, E. M. Eyring, *Macromolecules*, **17**, 2579 (1984).
284. H. Nishihara, H. Funaki, T. Shimura, K. Aramaki, *Synth. Met.*, **55–57**, 942 (1993).
285. H. Funaki, K. Aramaki, H. Nishihara, *Chem. Lett.*, 2065 (1992).
286. H. Funaki, K. Aramaki, H. Nishihara, *Synth. Met.*, **74**, 59 (1995).
287. H. Nishihara, in *Handbook of Organic Conductive Molecules and Polymers*, H. S. Nalwa, Ed., Vol. 2, p. 799, Wiley, New York, 1997.
288. Y. Morisaki, H. Chen, Y. Chujo, *Polym. Bull.*, **48**, 243 (2002).

289. M. E. Wright, C. K. Lowe-Ma, *Inorg. Chim. Acta.*, **232**, 223 (1995).
290. U. H. F. Bunz, V. Enkelmann, F. Beer, *Organometallics*, **14**, 2490 (1995).
291. S. Setayesh, U. H. F. Bunz, *Organometallics*, **15**, 5470 (1996).
292. J. Ohsita, T. Hamaguchi, E. Toyoda, A. Kunai, K. Komaguchi, M. Shiotani, M. Ishikawa, A. Naka, *Organometallics*, **18**, 1717 (1999).
293. S. F. Mapolie, J. R. Moss, G. S. Smith, *J. Inorg. Organomet. Chem.*, **7**, 233 (1997).
294. S. F. Mapolie, I. J. Mavunkal, J. R. Moss, G. S. Smith, *Appl. Organomet. Chem.*, **16**, 307 (2002).
295. A. S. Abd-El-Aziz, E. K. Todd, T. H. Afifi, *Macromol. Rapid Commun.*, **23**, 113 (2002).
296. A. S. Abd-El-Aziz; S. A. Carruthers, P. M. Aguiar, S. Kroeker, *J. Inorg. Organometal. Polym. Mater.*, **15**, 349, (2005).
297. A. S. Abd-El-Aziz, S. A. Carruthers, E. K. Todd, T. A. Afif, J. M. A. Gavina, *J. of Polym. Sci. Part A: Polym. Chem.*, **43**, 1382 (2005).
298. B. Alonso, J.-C. Blais, D. Astruc, *Organometallics*, **21**, 1001 (2002).
299. Q. Sun, J. W. Y. Lam, K. Xu, H. Xu, J. A. K. Cha, P. C. L. Wong, G. Wen, X. Zhang, X. Jing, F. Wang, B. Z. Tang, *Chem. Mater.*, **12**, 2617 (2000).
300. D. Astruc, F. Chardac, *Chem. Rev.*, **101**, 2991 (2001).
301. S. Rigaut, M.-H. Delville, D. Astruc, *J. Am. Chem. Soc.*, **119**, 11132 (1997).
302. J.-L. Fillaut, D. Astruc, *New J. Chem.*, **20**, 945 (1996).
303. C. Valério, F. Moulines, J. Ruiz, J.-C. Blais, D. Astruc, *J. Org. Chem.*, **65,** 1996 (2000).
304. F. Moulines, D. Astruc, *J. Chem. Soc., Chem. Commun.*, 614 (1989).
305. F. Moulines, L. Djakovitch, R. Boese, B. Gloaguen, W. Thiel, J.-L. Fillaut, M.-H. Delville, D. Astruc, *Angew. Chem. Int. Ed. Engl.*, **32**, 1075 (1993).
306. C. Valerio, B. Gloaguen, J.-L. Fillaut, D. Astruc, *Bull. Soc. Chim. Fr.*, **133**, 101 (1996).
307. E. Alonso, C. Valerio, J. Ruiz, D. Astruc, *New J. Chem.*, **21**, 1139 (1997).
308. C. Valerio, E. Alonso, J. Ruiz, J.-C. Blais, D. Astruc, *Angew. Chem. Int. Ed.*, **38**, 1747 (1999).
309. J. Ruiz, C. Pradet, F. Varret, D. Astruc, *Chem. Commun.*, 1108, (2002).
310. S. Nlate, Y. Nieto, J.-C. Blais, J. Ruiz, D. Astruc, *Chem. Eur. J.*, **8**, 171 (2002).
311. C. Valerio, F. Moulines, J. Ruiz, J.-C. Blais, D. Astruc, *J. Org. Chem.*, **65**, 1996 (2000).
312. J. Ruiz, E. Alonso, J.-C. Clais, D. Astruc, *J. Organometal. Chem.*, **582**, 139 (1999).
313. S. Rigaut, M.-H. Delville, J. Losada, D. Astruc, *Inorg. Chim. Acta*, **334**, 225 (2002).
314. D. Astruc, in *Electron Transfer in Chemistry*, J. Matay, D. Astruc, Eds., Vol. II, pp. 714–803, Wiley, Weinheim, 2001.
315. F. Moulines, L. Djakovitch, M.-H. Delville, F. Robert, P. Gouzerh, D. Astruc, *J. Chem. Soc., Chem. Commun.*, 463 (1995).
316. F. Moulines, L. Djakovitch, D. Astruc, *New J. Chem.*, **20**, 1071 (1996).
317. V. Sartor, L. Djakovitch, J.-L. Fillaut, F. Moulines, F. Neveu, V. Marvaud, J. Guittard, J.-C. Blais, D. Astruc, *J. Am. Chem. Soc.*, **121**, 2929 (1999).
318. V. Sartor, S. Nlate, J.-L. Fillaut, F. Djakovitch, F. Moulines, V. Marvaud, F. Neveu, J.-C. Blais, *New J. Chem.*, **24**, 351 (2000).
319. S. Nlate, J. Ruiz, V. Sartor, R. Navarro, J.-C. Blais, D. Astruc, *Chemistry Eur. J.*, **6**, 2544 (2000).
320. S. Sengupta, *Tetrahedron Lett.*, **44**, 7281 (2003).
321. Q. Sun, K. Xu, H. Peng, R. Zheng, M. Häussler, B. Z. Tang, *Macromolecules*, **36**, 2309 (2003).
322. Y.-H. Liao, J. R. Moss, *Organometallics*, **15**, 4307 (1996).
323. S. Achar, R. J. Puddephatt, *Angew. Chem. Int. Ed. Engl.*, **33**, 847 (1994).
324. S. Achar, J. J. Vittal, R. J. Puddephatt, *Organometallics*, **15**, 43 (1996).
325. G.-X. Liu, R. J. Puddephatt, *Organometallics*, **15**, 5257 (1996).
326. S. Achar, C. E. Immoos, M. G. Hill, V. J. Catalano, *Inorg. Chem.*, **36**, 2314 (1997).

327. N. J. Hovestad, J. L. Hoare, J. T. B. H. Jastrzebski, A. J. Canty, W. J. J. Smeets, A. L. Spek, G. van Koten, *Organometallics*, **18**, 2970 (1999).

328. J. L. Hoare, K. Lorenz, N. J. Hovestad, W. J. J. Smeets, A. L. Spek, A. J. Canty, H. Frey, G. van Koten, *Organometallics*, **16**, 4167 (1997).

329. K. Onitsuka, A. Iuchi, M. Fujimoto, S. Takahashi, *Chem. Commun.*, 741 (2001).

330. S. Leininger, P. J. Stang, S. Huang, *Organometallics*, **17**, 3981 (1998).

331. K. Onitsuka, M. Fujimoto, N. Ohshiro, S. Takahashi, *Angew. Chem. Int. Ed.*, **38**, 689 (1999).

332. A. M. McDonagh, M. G. Humphrey, M. Samoc, B. Luther-Davies, *Organometallics*, **18**, 5195 (1999).

333. S. K. Hurst, M. P. Cifuentes, M. G. Humphrey, *Organometallics*, **21**, 2353 (2002).

334. F. Lobete, I. Cuadrado, C. M. Casado, B. Alonso, M. Moran, J. Losada, *J. Organomet. Chem.*, **509**, 109 (1996).

335. J. W. Kriesel, S. Konig, M. A. Freitas, A. G. Marshall, J. A. Leary, T. D. Tilley, *J. Am. Chem. Soc.*, **120**, 12207 (1998).

336. A. A. Hurley, D. L. Mohler, *Org. Lett.*, **2**, 2745 (2000).

337. S. Nlate, J. Ruiz, J.-C. Blais, D. Astruc, *Chem. Commun.*, 417 (2000).

338. C. Valerio, J.-L. Fillaut, J. Ruiz, J. Guittard, J.-C. Blais, D. Astruc, *J. Am. Chem. Soc.*, **119**, 2588 (1997).

339. I. Cuadrado, M. Moran, C. M. Casado, B. Alonso, F. Lobete, B. Garcia, M. Ibisate, J. Losada, *Organometallics*, **15**, 5278 (1996).

340. R. Castro, I. Cuadrado, B. Alonso, C. M. Casado, M. Moran, A. E. Kaifer, *J. Am. Chem. Soc.*, **119**, 5760 (1997).

341. I. Cuadrado, C. M. Casado, B. Alonso, M. Moran, J. Losada, V. Belsky, *J. Am. Chem. Soc.*, **119**, 7613 (1997).

342. C. M. Casado, B. Gonzalez, I. Cuadrado, B. Alonso, M. Moran, J. Losada, *Angew. Chem., Int. Ed.*, **39**, 2135 (2000).

343. B. Dardel, R. Deschenaux, M. Even, E. Serrano, *Macromolecules*, **32**, 5193 (1999).

344. C.-F. Shu, H.-M. Shen, *J. Mater. Chem.*, **7**, 47 (1997).

345. P. Gomez-Elipe, R. Resendes, P. F. Macdonald, I. Manners, *J. Am. Chem. Soc.*, **120**, 8348 (1998).

346. B. Alonso, I. Cuadrado, M. Morán, J. Losada, *J. Chem. Soc., Chem. Commun.*, 2575, (1994).

347. B. Alonso, C. Casado, I. Cuadrado, M. Morán, A. Kaifer, *Chem. Commun.*, 1778 (2002).

348. J. Ruiz, E. Alonso, J.-C. Clais, D. Astruc, *J. Organometal. Chem.*, **582**, 139 (1999).

349. R. Deschenaux, E. Serrano, A.-M. Levelut, *Chem. Commun.*, 1577 (1997).

350. K. Takada, D. J. Diaz, H. D. Abruna, I. Cuadrado, C. Casado, B. Alonso, M. Moran, J. Losada, *J. Am. Chem. Soc.*, **119**, 10763 (1997).

351. F. Moulines, L. Djakovitch, D. Astruc, *New J. Chem.*, **20**, 1071 (1996).

352. V. Sartor, L. Djakovitch, J.-L. Fillaut, F. Moulines, F. Neveu, V. Marvaud, J. Guittard, J.-C. Blais, D. Astruc, *J. Am. Chem. Soc.*, **121**, 2929 (1999).

353. V. Sartor, S. Nlate, J.-L. Fillaut, F. Djakovitch, F. Moulines, V. Marvaud, F. Neveu, J.-C. Blais, *New J. Chem.*, **24**, 351 (2000).

354. D. Méry, C. Ornelas, M.-C. Daniel, J. Ruiz, J. Rodrigues, D. Astruc, S. Cordier, K. Kirakci, C. Perrin, *C. R. Chimie*, **8**, 1789 (2005).

355. J. R. Aranzaes, C. Belin, D. Astruc, *Angew. Chem. Int. Ed.*, **45**, 132 (2005).

356. P. Ashton, B. Vincenzo, M. Clemente-León, B. Colonna, A. Credi, N. Jayaraman, F. Raymo, F. Stoddart, M. Venturi, *Chem. Eur. J.*, **8**, 673 (2002).

357. T. Chen, L. Wang, G. Jiang, J. Wang, X. Wang, J. Zhou, W. Wang, *Eur. Polym. J.*, **42**, 687–693 (2006).

358. C.-O. Turrin, B. Donnadieu, A.-M. Caminade, J.-P., Majoral, *Z. Anorg. Allg. Chem.*, **631**, 2881 (2005).

CHAPTER 3

Block Copolymers with Transition Metals in the Main Chain

DAVID A. RIDER

University of Toronto, Toronto, Ontario, Canada

IAN MANNERS

University of Bristol, Bristol, United Kingdom

I. INTRODUCTION

A substantial fraction of the valuable physical and chemical properties and applications of solid-state and biological materials can be attributed to the presence of metallic elements. Examples include magnetic materials used in data storage, electrical conductors and superconductors, electrochromic materials, and catalysts, including metalloenzymes. It has long been recognized that the incorporation of metal atoms into synthetic polymer chains can also lead to desirable properties, and thereby generate a new and versatile class of functional materials capable of enhanced processability.

Until recently, difficulties with syntheses associated with the creation of macromolecular chains possessing metal atoms as a key structural component have held back progress in the field. Since about 1996, these obstacles to synthesis have been, in part, overcome through many creative procedures to prepare new materials.[1–9] The new approaches that are now available have led to macromolecular structures in which metals are not only incorporated via the use of traditional covalent bonds but also by potentially reversible coordination interactions. Of comparable importance to these developments in synthesis, these new materials have shown that they can possess a diverse range of interesting and useful properties and potential applications that complement those of organic macromolecules.[1–9]

The additional challenge of preparing well-defined metallopolymer structures has become a key goal since the early 1990s. Metallopolymers with precisely controlled chain length, or with block, star, or dendritic architectures, are attracting rapidly expanding attention as a result of their properties and potential applications. This chapter focuses on metal-containing block copolymers where the metal atoms are present in the main chain. These materials represent an area of rapidly growing interest as a result of their self-assembly into phase-separated metal-rich

Frontiers in Transition Metal-Containing Polymers, edited by Alaa S. Abd-El-Aziz and Ian Manners.

nanodomain structures in thin films and micelles in block-selective solvents. The resulting nanostructured materials have a wealth of potential applications, and recent breakthroughs in this area are discussed. The subject matter is divided into subsections covering block copolymers with metallo-linkers, metal-centered star structures, and polyferrocenylsilane-based materials that have been extensively developed by ourselves and our collaborators, as well as by other groups.

II. METALLO-LINKED BLOCK COPOLYMERS

Schubert and co-workers have reported the preparation of terpyridine (terpy) end-group functionalized polystyrene (PS) and polyethylene oxide (PEO) from a nitroxide-mediated terpy functionalized initiator and OH-terminated PEO transformation, respectively (Scheme 3.1,

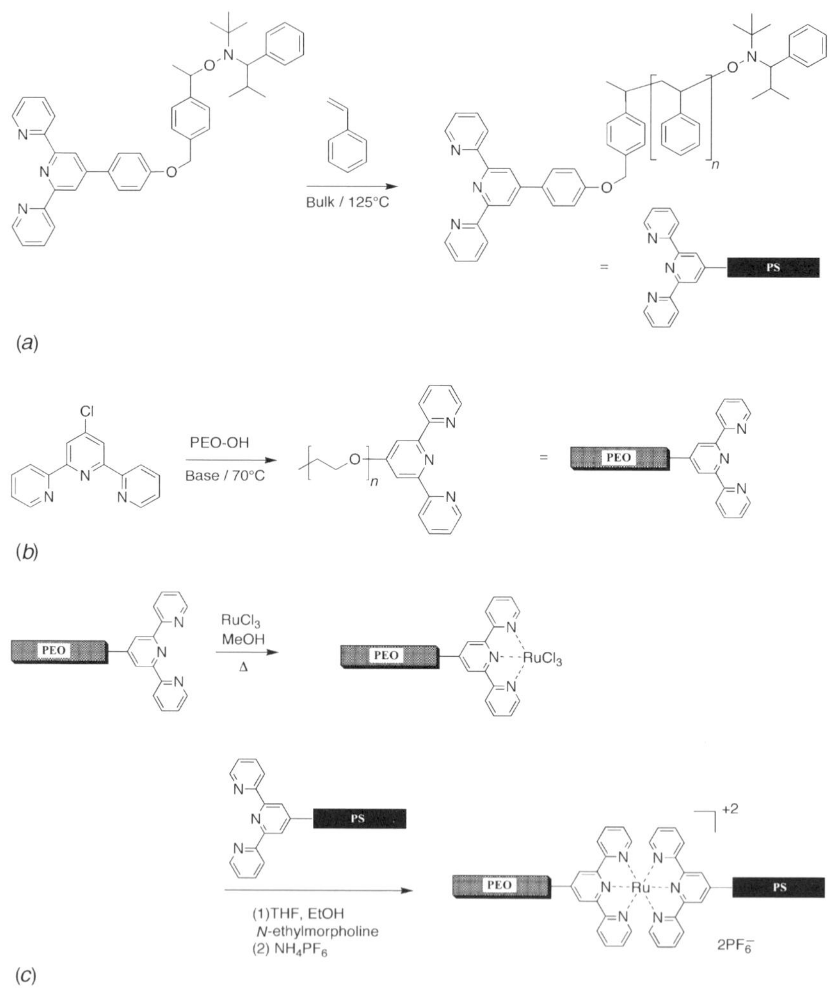

Scheme 3.1 Synthesis of terpyridine-functionalized polystyrene (a), terpyridine-functionalized polyethyleneoxide (b), and ruthenium(II)-connected PEO-*b*-PS diblock copolymer (c).

top and middle).[10] These polymer segments act as the building blocks for the coupling reaction between terpy moieties and a transition metal, namely ruthenium(II). The AB block copolymerization process involves two simple steps (Scheme 3.1, bottom). First, the terpy end-group functionalized polystyrene or polyethylene can be complexed with $RuCl_3$ to selectively afford a monocomplex (PS-$RuCl_3$ or PEO-$RuCl_3$). Second, the monocomplex is reacted with an additional terpy-containing polymer segment under reducing conditions, forming the diblock copolymer linked by the transition metal. Molecular-weight estimates for these AB metal-linked block copolymers from gel permeation chromatography (GPC) were not possible, as the ruthenium center was reported to interact unfavorably with the size exclusion media in the GPC. Although the metal content in these materials is relatively low, important features are introduced with the redox active link. First, the link permits copolymer characterization and manipulation. Second, easily cleaved diblock copolymers are available, permitting the formation of nanoporous thin films (see below). Recently, Harruna et al. have extended this methodology to include terpy-containing reversible addition fragmentation termination (RAFT) polymerization initiators for the controlled polymerization of styrene (M_n = 4500–13,900; PDI = 1.04–1.18) and *N*-isopropylacrylamide (M_n = 1900–2900; PDI = 1.06–1.11).[11,12]

In 2005, Schubert et al. applied their supramolecular diblock copolymers toward the generation of a nanoporous thin film.[13] The diblock copolymer studied was PS_{375}-[Ru]-PEO_{225} with a Ru(II) junction point (where [Ru] denotes bis(terpyridine)Ru complex). It was found that by casting a thin film, ca. 74 nm thick, spontaneous orientation of cylindrical microdomains of PEO occurred within a PS matrix (atomic force microscopy (AFM) phase image shown in Scheme 3.2, bottom left). The center-to-center distance was 63 nm, whereas the cylindrical diameter was 33 nm. The film was then exposed to deep ultra violet (UV) radiation so as to stabilize the PS matrix via cross-linking. Next, a simple 1h immersion in an acidic (pH = 1) aqueous solution of $Ce(SO_4)_2$, followed by a water rinse and N_2 dry afforded a nanoporous PS film. The mechanism for removal of the PEO cylindrical domains involves an oxidation of the junction Ru(II) by the dissolved $Ce(SO_4)_2$. This involved (1) swelling of the PEO phase so as to permit an oxidation of the Ru(II) by the PEO infiltrated Ce(IV) ions, (2) formation of various cleaved copolymer chains made up of PS and PEO segments, and (3) selective dissolution of the PEO chains and Ru(III) species into the aqueous phase. The resulting film (AFM phase image shown in Scheme 3.2, bottom right) exhibits a phase contrast that is opposite to that of the precursor film, consistent with its nanoporosity. The generation of a porous PS thin film was further confirmed by X-ray photoelectron spectroscopy and X-ray reflectivity experiments on both PS_{375}-[Ru]-PEO_{225} and nanoporous PS films.

III. METALLO-CENTERED STAR-BLOCK COPOLYMERS

The Fraser group has recently investigated the cationic polymerization of 2-ethyl-2-oxazoline (EOX) using di-, tetra-, and hexafunctional ruthenium and iron tris(bipyridine) metalloinitiators.[14] The resulting starlike poly(2-ethyl-2-oxazoline) $(Ru\text{-}PEOX)_{6-2x}$ illustrates a rare example of well-controlled cationic polymerization to afford a transition metal–containing polymer system. It was found that a Ru-center was required for these metalloinitiators, as the resulting polymers were more resilient against arm dismemberment (e.g., loss of $bipy(PEOX)_2$). For polymerization, the Ru-metalloinitiator and EOX were dissolved in acetonitrile and heated to 80°C for 1.5 days. For deliberate termination, a small amount of dipropylamine was added after cooling to room temperature. The $Ru\text{-}(PEOX)_{6-2x}$ was isolated in moderate yields as a red-orange glassy solid (for $x = 0$; M_n = 6300–22,700; PDI = 1.24–1.09). Near-linear plots of molecular weight versus conversion and ln(conversion) versus time were found for conversions up to 60%. Also reported was the synthesis of the Fe(II)-centered polyoxazoline block copolymers from 2-ethyl-2-oxazoline and 2-undecyl-2-oxazoline (UOX) monomers (Scheme 3.3a). The resulting

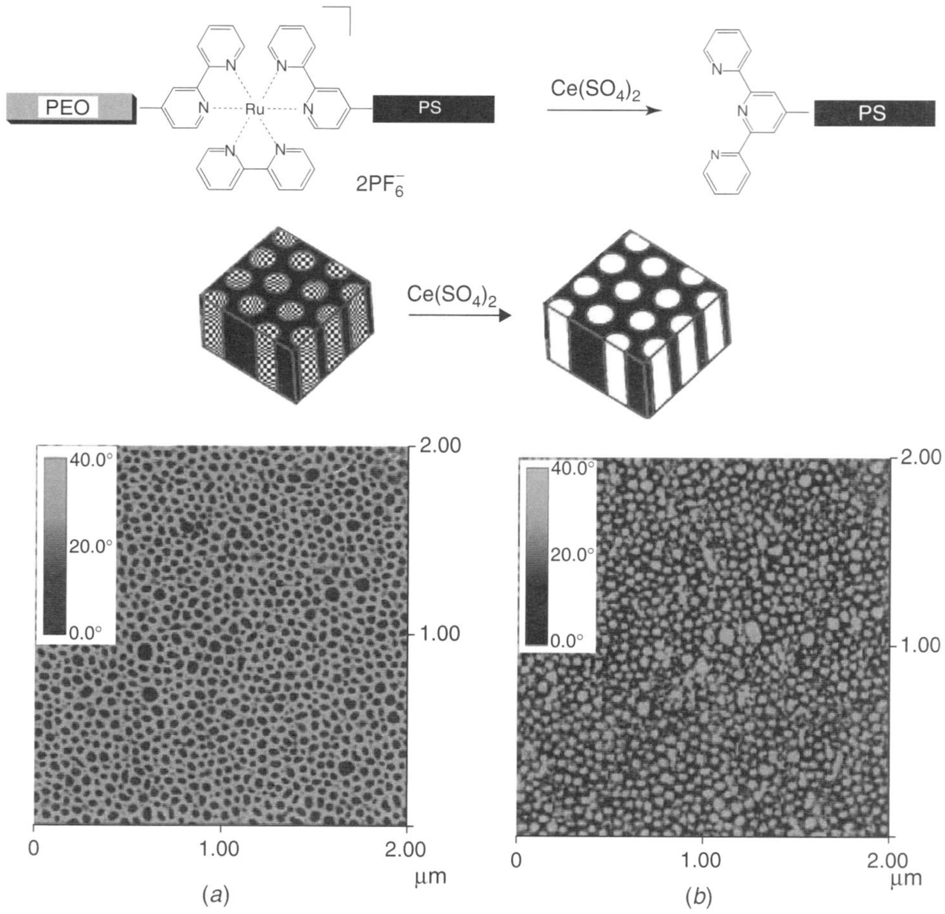

Scheme 3.2 (Top and middle) General schematic for the generation of nanoporous PS film. (Bottom) Atomic force microscopy (AFM) phase images of (a) PS_{375}-[Ru]-PEO_{225} thin film, and (b) nanoporous PS film from oxidation-induced cleavage of the [Ru] junction.

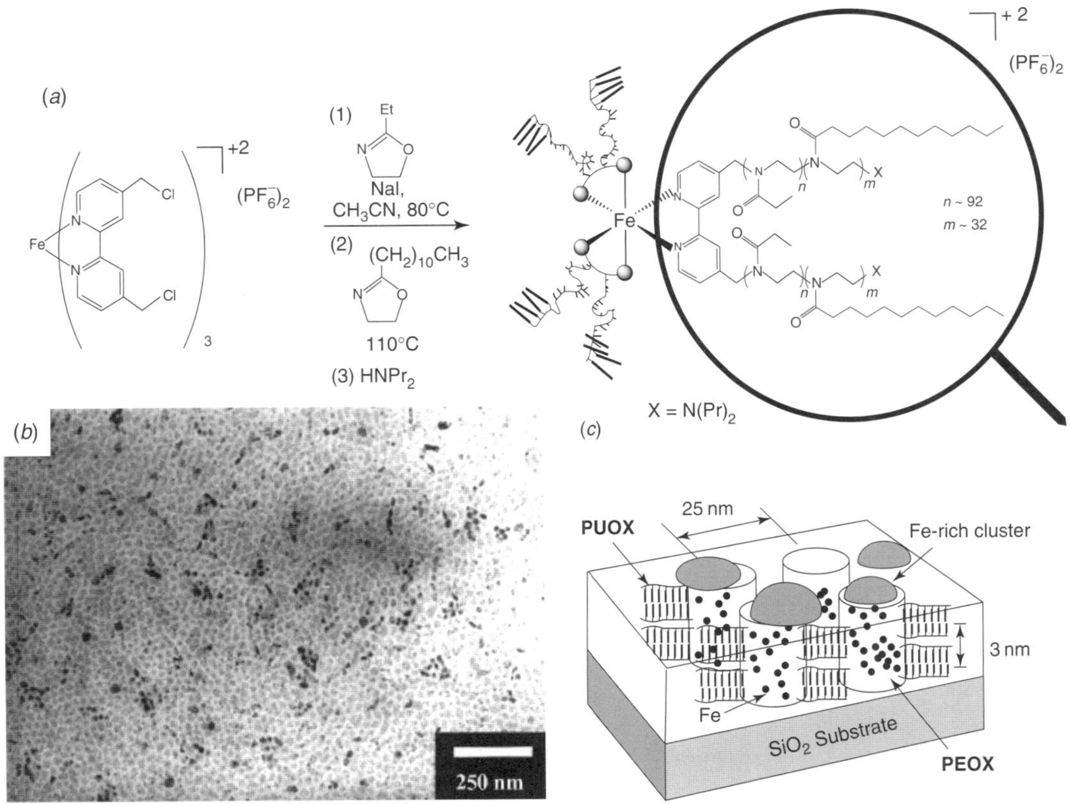

Scheme 3.3 (a) Sequential cationic polymerization of 2-ethyl-2-oxazoline and 2-undecyl-2-oxazoline from a metalloinitiator to produce $[Fe\{bipy(PEOX\text{-}b\text{-}PUOX)_2\}_3]^{+2}$. (b) Stained, annealed (160°C) sample showing 20–40-nm Fe clusters (black) and cylindrical PEOX (gray) in a PUOX matrix (white). (c) Diagram of the microstructure of $[Fe\{bipy(PEOX\text{-}b\text{-}PUOX)_2\}_3]^{+2}$.

$[Fe\{bipy(PEOX\text{-}b\text{-}PUOX)_2\}_3]^{+2}$ materials were estimated to have controlled molecular weights of up to 120,000. When these materials were cast into thin films and annealed for 2 days at 160°C, 20–40-nm-diameter iron nanoclusters decorating a phase-separated film of vertically oriented cylindrical PEOX domains in a PUOX matrix were observed (Scheme 3.3b). It was suggested that the formation of nanoclusters was due to kinetically unstable bonding of the iron atom to the $bipy(PEOX\text{-}b\text{-}PUOX)_2$ macroligands. When heated, the iron ions became mobile, aggregated, and oxidized to give rise to the nanoparticles with diameters in the range of 20–40 nm (Scheme 3.3c).

IV. POLYFERROCENYLSILANE (PFS) BLOCK COPOLYMERS

A. Introduction

High molecular-weight polyferrocenylsilane (PFS) macromolecules were discovered in the early 1990s and represent an interesting class of metal-containing polymers.[15] Depending on the substituent groups on silicon, PFS materials can be semicrystalline or amorphous.[16] In this chapter, PFS refers to semicrystalline dimethyl substituted polyferrocenylsilane (R=R′=Me; see structure PFS) unless otherwise indicated. The presence of iron in PFSs imparts unique properties that are otherwise not found or are difficult to acquire with traditional organic polymers. Some of these properties include interesting redox-activity (associated with the reversible Fe(II)/Fe(III) couple), semi- and photoconductivity, and an ability to act as a magnetic ceramic or catalyst precursor.[2,8,17]

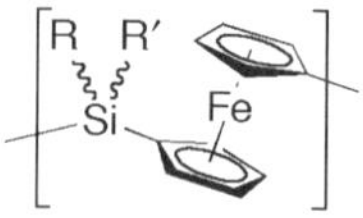

PFS

Due to the availability of controlled polymerization routes for PFS monomers, well-defined architectures with organic and inorganic coblocks are available. The incorporation of PFS segments into self-organizing motifs, such as block copolymers, provides further possibilities for supramolecular chemistry and the development of functional nanomaterials.[18–23] This section summarizes recent developments in the synthesis and self-assembly of PFS block copolymers, as well as their applications in material science.

B. Synthesis of PFS Block Copolymers

In the mid-1990s, the living anionic ring-opening polymerization (ROP) of silicon-bridged [1]ferrocenophanes (denoted fcSiRR′; see structure **1**) using initiators such as BuLi was reported (see Scheme 3.4).[24] This permitted the synthesis of PFS with controlled molecular weights and narrow molecular-weight distributions (polydispersities < 1.2). The comparative reactivity of living PFS anionic chain-end groups follows the sequence of PS $\backsim$ PI $>$ PFS $>$ PDMS (PS: polystyrene; PI: polyisoprine; PDMS: polydimethylsiloxane).[24] Therefore, by sequential addition of monomers in an order of decreasing end-group reactivity, block copolymers with skeletal transition metal atoms were successfully prepared. The first "prototypical" block copolymers synthesized by consecutive addition of monomers in a one-pot anionic polymerization were the organic-organometallic (PS-*b*-PFS) (see Scheme 3.5) and the organometallic-inorganic (PFS-*b*-PDMS) block copolymers (see Scheme 3.6).[5] Simply extending this method to other monomers has allowed the synthesis of PI-*b*-PFS,[25] PFS-*b*-PMVS[26] (PMVS: polymethylvinylsiloxane), and PFP-*b*-PFS-*b*-PDMS (PFP: polyferrocenylphosphine) triblock copolymers.[27] In principle, this sequential anionic polymerization technique can be applied to integrate a PFS block with any polymers compatible with the anionic polymerization mechanism. The exceptions are methacrylate-type polymers as side reactions, such as nucleophilic attack of the carbanionic chain ends at the carbonyl groups of methacrylates may occur if the monomers are added directly at the anionic chain end derived from structure **1**. Until recently, this type of block copolymer could therefore not be obtained in high yields with well-defined molecular weights and architectures.

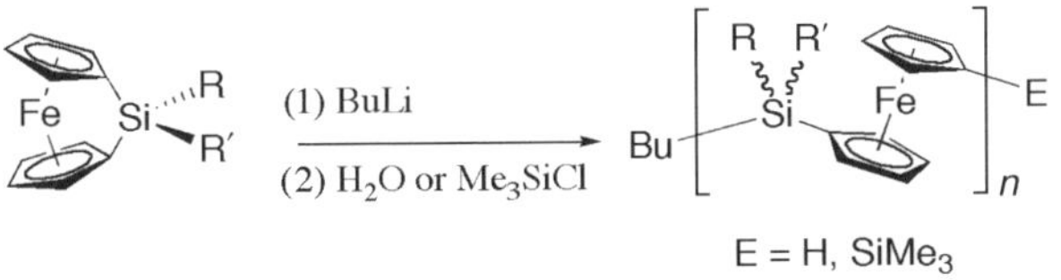

1 **PFS**

Scheme 3.4 Anionic polymerization of dimethylsila[1]ferrocenophane.

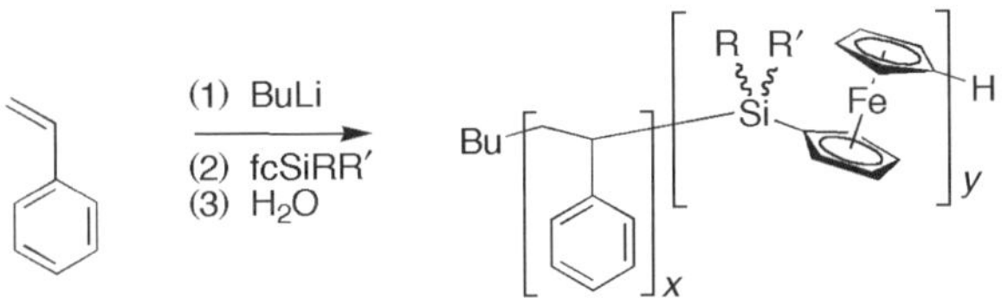

PS-*b*-PFS

Scheme 3.5 Sequential anionic polymerization of styrene and dimethylsila[1]ferrocenophane.

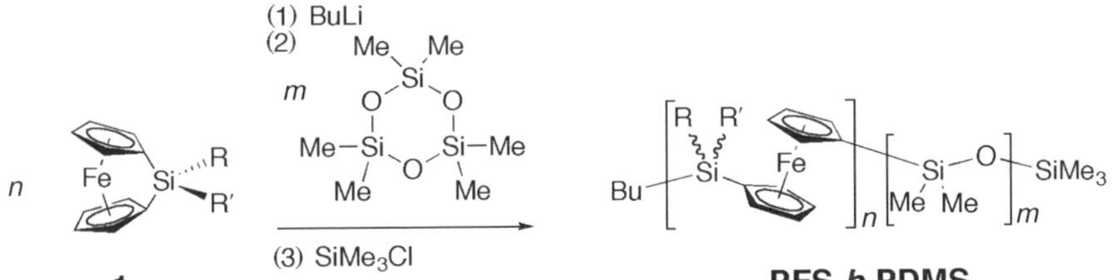

1 **PFS-*b*-PDMS**

Scheme 3.6 Sequential anionic polymerization of dimethylsila[1]ferrocenophane and hexamethylcyclotrisiloxane.

The first PFS-*block*-polymethacrylate copolymer was reported in 2002.[28] The procedure involved a two-step anionic polymerization (Scheme 3.7). Hydroxyterminated polyferrocenylsilane (PFS-OH) was initially synthesized using *t*-butyldimethylsilyloxy-1-propyllithium as an initiator bearing a protected alcohol functionality. Once isolated, PFS-OH was deprotonated using potassium hydride to afford the alkoxy chain end that initiates dimethylaminoethyl methacrylate (DMAEMA) for anionic polymerization. The PFS-*b*-PDMAEMA block copolymer was obtained in a high yield (M_n = 11,000; PDI = 1.3) and with a low polyferrocenylsilane content (PFS:PDMAEMA = 1:5).

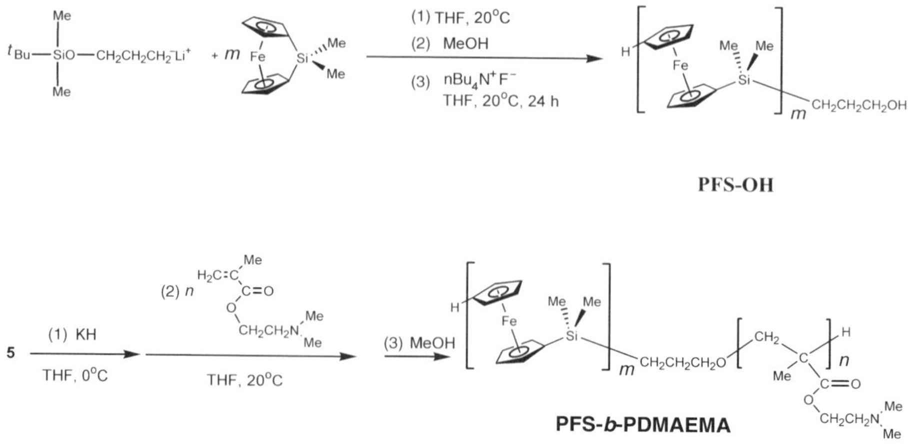

Scheme 3.7 Synthesis of polyferrocenyldimethylsilane-*block*-polydimethylaminoethylmethacylate. (From Wang et al.[28] Reproduced with permission.)

Alternatively, Vancso and co-workers recently investigated the anionic ROP of dimethylsila[1]ferrocenophane followed by end-group transformations for the isolation of a polyferrocenylsilane-based atom-transfer radical polymerization (ATRP) macroinitiator for methyl methacrylate (MMA) polymerization.[29] By quenching the living anionic polyferrocenylsilane chain end with siloxypropylchlorosilane, a homopolymer bearing a protected alcohol chain end was generated (Scheme 3.8). Subsequent chain-end hydrolysis and derivatization with 2-bromoisobutyric anhydride afforded a PFS macroinitiator for the ATRP of MMA. The macroinitiators were isolated as well-defined homopolymers (M_n = 8000–12,000; PDI = 1.02–1.07). Following introduction of MMA monomer in the presence of an ATRP Ru-based catalyst (*p*-cymeneruthenium(II)chloride-tricyclohexylphosphine), PFS-*b*-PMMA diblock copolymers were produced. The metal-containing diblock copolymers were available with tunable polyferrocenylsilane content (between 7 and 26 wt %) via variations in the macroinitiator concentration with respect to monomer concentration. Molecular weights were also varied from 40,500 to 101,000 (PDI = 1.06–1.18) by maintaining constant monomer to macroinitiator ratios. MMA conversion was 45–60% in all cases.

Last, Rehahn et al. successfully synthesized PFS-*b*-PMMA without homopolymer isolation and various chain-end transformations.[22] To circumvent the undesired reaction of anionic PFS chain ends with the carbonyl functionality of MMA that prevented sequential anionic polymerization, mediation of the reactivity of the living anionic PFS chain end was conducted. A PFS chain end mediated by diphenylethylene (DPE) was investigated. First, to ensure optimal coupling of DPE, living PFS chain ends were used for the ring-opening reaction of dimethylsilacyclobutane (Scheme 3.9). By matrix-assisted laser ionization–time of flight (MALDI–TOF) analysis of oligomeric model homopolymers for PFS, it was found that 1–3 units of the silacyclobutane are incorporated in the process required for efficient anionic chain-end transfer to DPE (>95% by ^{1}H-NMR). Having sufficiently reduced the reactivity of the chain via incorporation of DPE, subsequent anionic polymerization of MMA was completed. Unfortunately, approximately 5% of the chain ends were reported to deactivate due to undesired spontaneous chain termination or homocoupling. Nevertheless, following selective precipitation PFS-*b*-PMMA diblock copolymers were isolated with tunable molecular weights (M_n = 27,000–84,500; PDI = 1.03–1.08) and polyferrocenylsilane content (27–47 wt %).

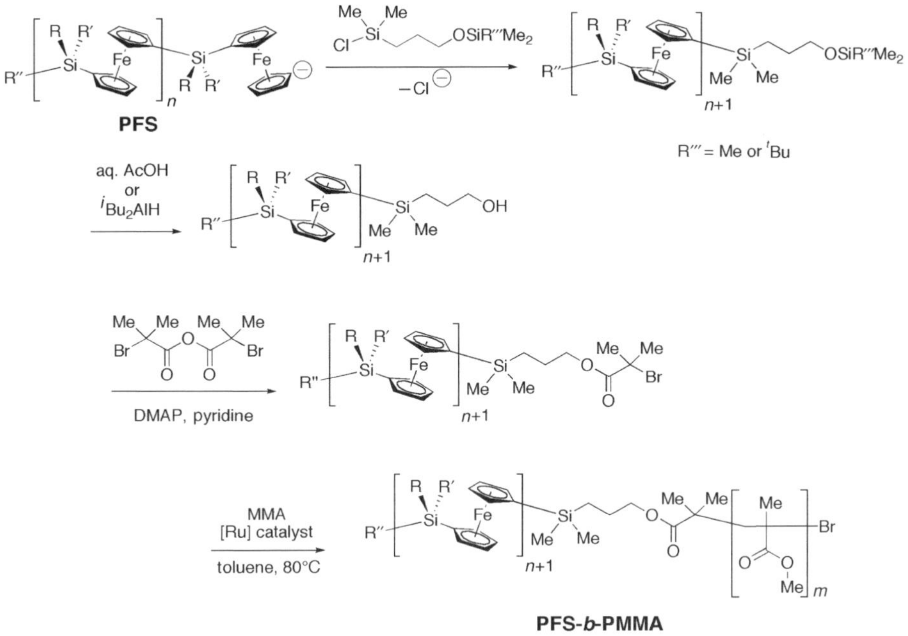

Scheme 3.8 Synthesis of polyferrocenyldimethylsilane-*block*-polymethylmethacrylate via a chain-end switch from anionic to ATRP.

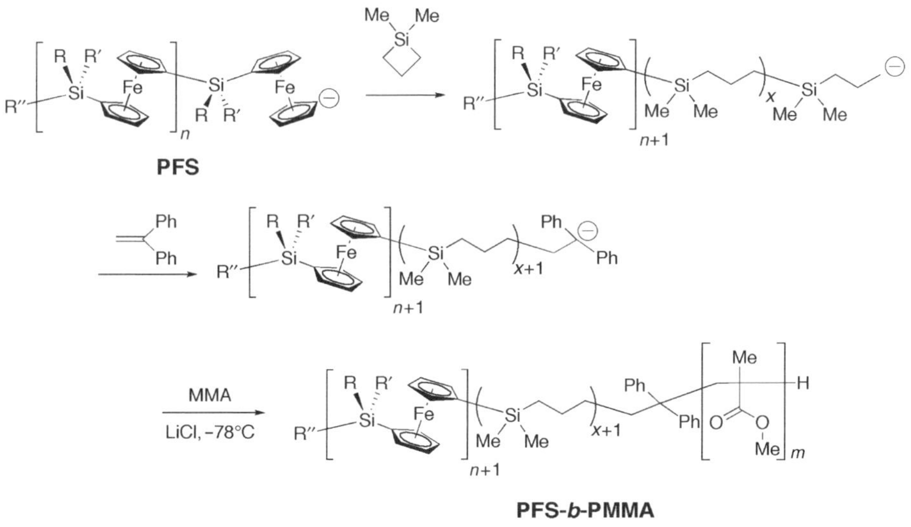

Scheme 3.9 Synthesis of polyferrocenyldimethylsilane-*block*-polymethylmethacrylate via a chain-end mediation using DPE.

We have also observed that, at elevated temperature (50°C), DPE could effectively cap the living anionic polyferrocenylsilane directly.[30] Therefore, the addition of DMSB to "pump up" the activity of PFS anionic chain ends can potentially be omitted, but the yield of the isolated PFS-*b*-PMMA is significantly reduced.[31] The living DPE-capped PFS polymers also can react with chloromethyl functionalities of poly(styrene-*co*-chloromethylstyrene) (PS-*co*-PCMS), leading to the first PS-*g*-PFS graft copolymers (see Scheme 3.10).[30]

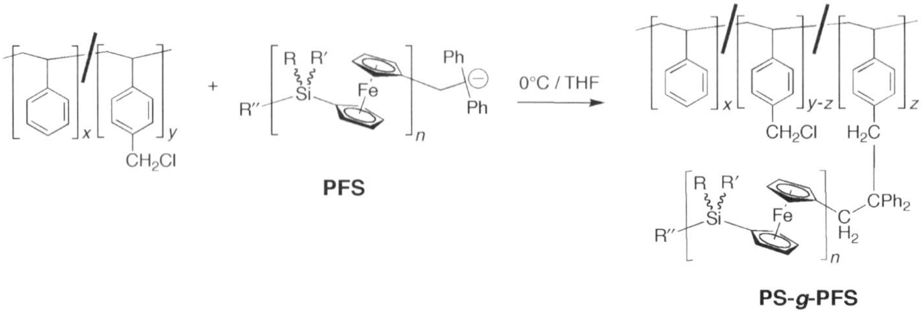

Scheme 3.10 Synthesis of polystyrene-*graft*-polyferrocenyldimethylsilane using DPE.

The first organometallic miktoarm star copolymer, PFS(PI)$_3$, with PDI of 1.04 was synthesized through an anionic polymerization by using $SiCl_4$ as a coupling agent, as shown in Scheme 3.11.[32] The well-defined structure was confirmed by the characterizations of GPC and ^{1}H nuclear magnetic resonance (NMR) spectroscopy. The PFS(PI)$_3$ miktoarm star copolymer was obtained in a moderate yield after size-exclusion column purification (M_n = 21,300; PDI = 1.05) and with a composition ratio of PFS:PI = 1:9.5.

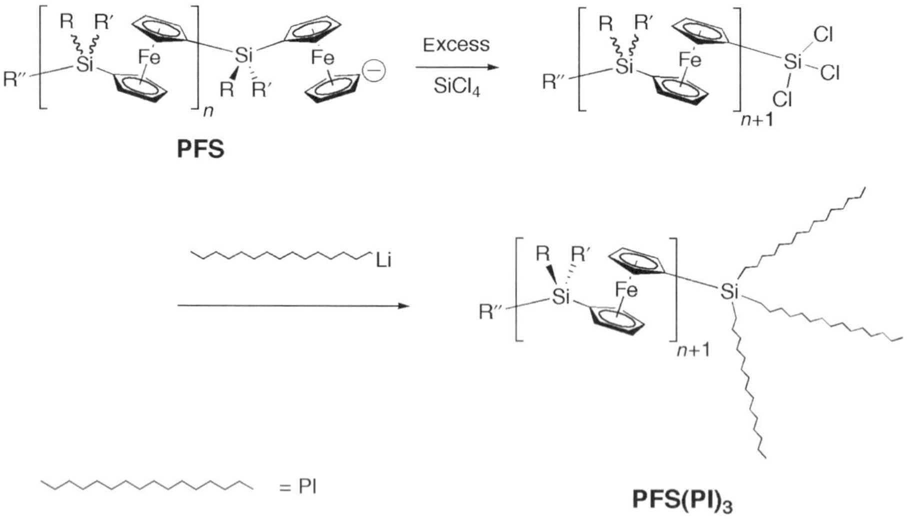

Scheme 3.11 Synthesis of the PFS(PI)$_3$ miktoarm star copolymer.

In this context, it is worthwhile to note a recently reported photolytic living anionic ROP of structure **1** in Scheme 3.4. The polymerization was initiated by ultraviolet-visible (UV-vis) irradiation in the presence of anionic initiator $Na[C_5H_5]$ and propagated via the cleavage of the Fe-η^5-C_5H_4 bonds in the monomer (see Scheme 3.12).[33] This method differs fundamentally from the previously reported living anionic ROP (illustrated in Scheme 3.4) where the organolithium-initiated anionic polymerization proceeds in the absence of UV-vis irradiation and involves Si-Cp bond cleavage. Significantly, the propagating centers for the new photolytic methodology are silyl-substituted cyclopentadienyl anions, which are less basic than iron coordinated Cp anions. Variation of the ratio of monomer to $Na[C_5H_5]$ from 25:1 to 200:1 under UV-vis irradiation at 5°C, followed by chain termination with water, afforded samples of polyferrocenylsilane with controlled molecular weights (M_n = 9300–70,000) and narrow polydispersities (PDI = 1.04–1.21). This method not only provides a new concept for living anionic polymerization, but also offers opportunities to synthesize PFS block copolymers under mild conditions.

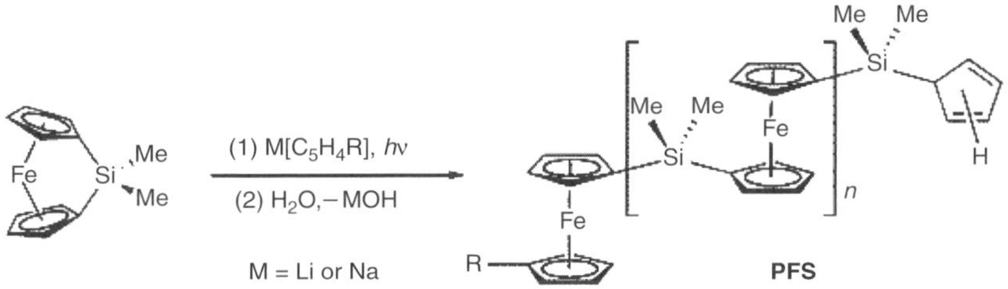

Scheme 3.12 *Photolytic* anionic ring opening polymerization of dimethylsila[1]ferrocenophane (From Tanabe and Manners.[33] Reproduced with permission.)

The hybridization of PFS with polypeptides recently produced a new organometallic–organic block copolymer offering opportunities for biological applications for PFS.[34] As shown in Scheme 3.13, a PFS-*b*-PBLG (PBLG: poly(γ-benzyl-L-glutamate)) block copolymer was synthesized through two steps: preparing amino-terminated PFS (PFS-NH_2, M_n = 9330, PDI = 1.10) through anionic ROP of ferrocenylsilane monomer (**1**), followed by a ROP of γ-benzyl-L-glutamate-*N*-carboxyanhydrides initiated by PFS-NH_2 macroinitiator.[34] PFS-*b*-PBLG samples with controlled molecular weights (M_n = 15,300–30,200) and narrow polydispersities (PDI = 1.13–1.21) were isolated with tunable polyferrocenylsilane contents (PFS:PBLG = 1:1.8–1:3.5). Moreover, amphiphilic block copolymers were easily isolated through hydrogenation of the PBLG block with H_2 and Pd/C in DMF. The resulting polyferrocenylsilane-*b*-poly(L-glutamic acid) (PFS-*b*-PGA) materials were soluble in base (0.1 M NaOH).

C. Solution Self-Assembly of PFS Block Copolymers

Solution self-assembly of PFS block copolymers allows the generation of discrete supramolecular organometallic nanomaterials with a range of one-dimensional micellar morphologies including cylinders,[18] tubes,[35] fibers,[36] and tapes.[25] Typically, block selective solvents are used to induce aggregation of the diblock copolymer macromolecules where there is sequestering of the incompatible blocks within a core stabilized by solvent-swollen corona-forming blocks. In most cases the solvent-swollen corona, with a dramatically increased volume, imparts a significant driving force for curvature of the segment–segment interface in the diblock copolymer micelle. It follows, then, when a bulk sample of PFS_{50}-*b*-$PDMS_{300}$ (block ratio of PFS:PDMS = 1:6) was heated in

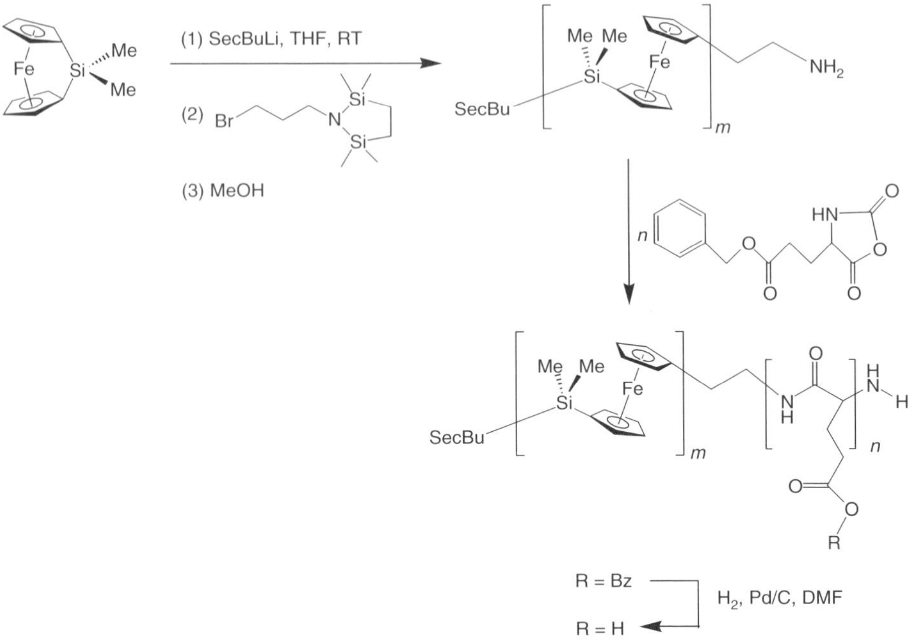

Scheme 3.13 Synthesis of PFS-*b*-PBLG block copolymer.

n-hexane, a selective solvent for PDMS, micellar aggregates became solubilized and could be visualized by TEM (see Figure 3.1a). TEM imaging of these morphologies was effective, as the high electron density of iron-rich, core-forming PFS blocks was sufficient for contrast in the electron-microscope.[18] As schematically illustrated in Figure 3.1b, the cylinders possess an iron-rich, organometallic core of PFS surrounded by an insulating sheath of PDMS. The cylindrical morphology observed in these systems was found to be a result of semicrystallinity in the core-forming PFS blocks, a characteristic that was confirmed by differential scanning calorimetry and wide-angle X-ray scattering.[17,40]

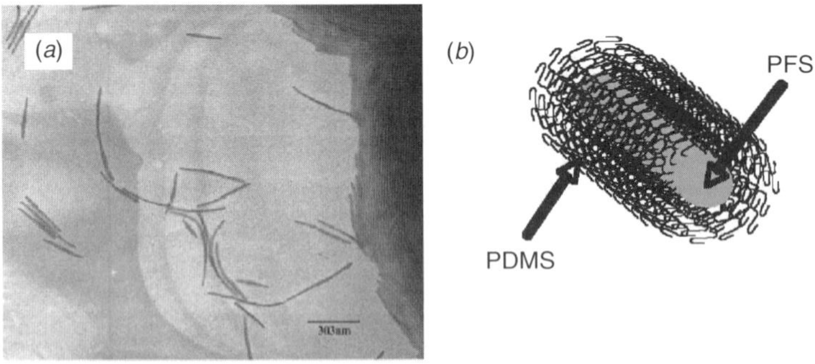

Figure 3.1 (a) Transmission electron microscopy (TEM) image, and (b) scheme for PFS nanocylinders self-assembled from PFS_{50}-*b*-$PDMS_{300}$ (PFS:PDMS = 1:6) in hexane. (From Massey et al.[18] Reproduced with permission.)

Apart from cylinders, PFS-*b*-PDMS block copolymers can also self-assemble into apparently hollow tubular structures when the block ratio of PFS:PDMS reaches 1:12.[35] As shown in Figure 3.2, PFS_{40}-*b*-$PDMS_{480}$ (PFS:PDMS = 1:12) were self-organized in hexane into nanotubes, in which the PFS blocks aggregate to form a shell with a cavity in the middle of the tube, while the PDMS blocks form the corona (see Figure 3.2b).[35] Several additional TEM techniques, including dark-field TEM and energy-filtered TEM, also supported the nanotube structures.[39] The presence of the hollow cavities was further supported by trapping tetrabutyllead in the voids and performing energy-dispersive X-ray measurements on the resulting structure.

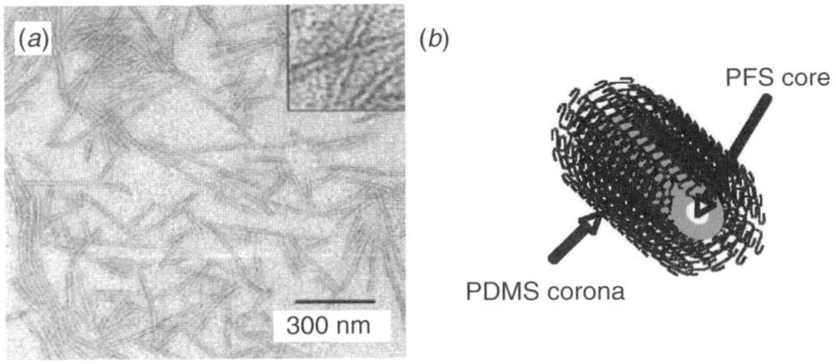

Figure 3.2 (a) TEM image, and (b) scheme for PFS nanotubes self-assembled from PFS_{40}-*b*-$PDMS_{480}$ (PFS:PDMS = 1:12) in hexane. (From Raez et al.[35] Reproduced with permission.)

The essential role of the block ratio in determining the micellar morphologies of PFS block copolymers was also discerned in studies on the PI-*b*-PFS system.[25] PI_{30}-*b*-PFS_{60} and PI_{320}-*b*-PFS_{53} have similar PFS length, but differ significantly in PI length, with block ratios (PI:PFS) of 1:2 and 6:1, respectively. As shown in Figure 3.3, the block copolymers of PI_{320}-*b*-PFS_{53} with longer PI chains, cylinders were observed (see Figure 3.3a), while the tapelike micelles, as shown in Figure 3.3b, self-assembled from PI_{30}-*b*-PFS_{60}.

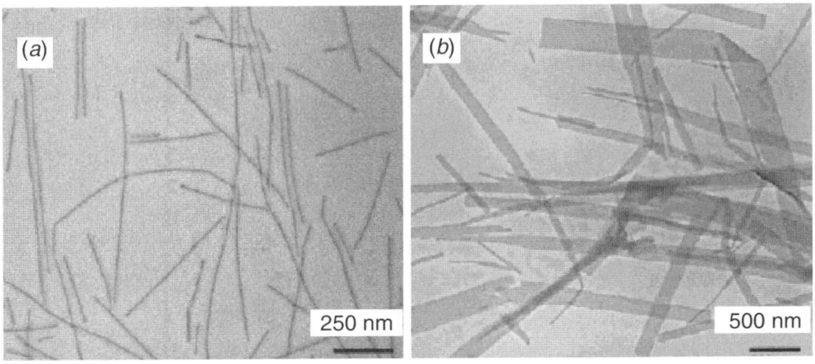

Figure 3.3 TEM images for the micelles self-assembled from (a) PI_{320}-*b*-PFS_{53} in THF/hexane [2/8, volume-to-volume ratio (v/v)], and (*b*) PI_{30}-*b*-PFS_{60} THF/hexane (3/7, v/v). (From Cao et al.[25] Reproduced with permission.)

Under certain circumstances, spherical aggregates were also found. For example, when PFS cylindrical micelles were heated above the T_m of PFS blocks (ca. 120–145°C),[40] or an amorphous poly(ferrocenylmethylphenylsilane)[40] block was used to replace semicrystalline PFS in the block copolymers, spherical aggregates became the only morphology observed by TEM. These experiments suggested that the crystallization of PFS chains could be a possible reason for the formation of observed one-dimensional supramolecular assemblies. PFS crystallization was indeed detected in both PFS nanocylinders[40] and nanotubes[35] by wide-angle X-ray diffraction (WAXD) experiments on the solid micellar samples dried from the corresponding solutions.

PFS block copolymers are also able to self-assemble in water when hydrophilic blocks are incorporated.[41,42] Following the successful synthesis of PFS_9-*b*-$PDMAEMA_{50}$ (see Scheme 3.7), we performed a reaction of the block copolymer with methyl iodide, obtaining a corresponding amino-quaternized PFS_9-*b*-$qPDMAEMA_{50}$.[42] We further investigated the micellization behavior of these two block copolymers. A range of intermediate morphologies, such as vesicles, rods, or cylinders, were captured in the case of pristine PFS_9-*b*-$PDMAEMA_{50}$ during the course of our investigation, though spheres appear to be the thermodynamic stable structures for the aggregates of both polymers in water. Nevertheless, alcohol solvents such as EtOH and ^{i}PrOH are able to induce PFS_9-*b*-$PDMAEMA_{50}$ to self-assemble into cylindrical micelles by simply dissolving the block copolymers in the solvent. Therefore, water-soluble cylindrical micelles could then be prepared via the dialysis of PFS_9-*b*-$PDMAEMA_{50}$ in ethanol against water.[42]

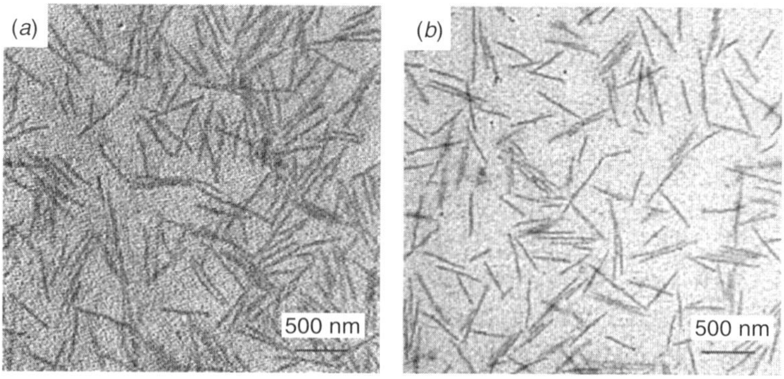

Figure 3.4 TEM for the micelles prepared by (a) mixing PFS_9-*b*-$PDMAEMA_{50}$ with ethanol, and (b) replacing ethanol with water by dialysis. (From Wang et al.[42] Reproduced with permission.)

One potential application of these well-defined aggregates is as etch resists for semiconducting substrates, such as GaAs or Si, which would offer potential access to magnetic or semiconducting nanoscopic patterns on various substrates.[43] In a collaboration with J. Spatz and M. Möller at the University of Ulm, Germany, the PFS cylinders have been positioned on the surface of a GaAs resist by capillary forces along grooves, which were previously formed from electron-beam etching of the surface. Subsequent reactive ion plasma etching generated connected ceramic lines of reduced size (see Figure 3.5).[44]

Despite the success in the fabrication of nanoceramic lines by plasma etching of aligned micellar precursors, our further exploitations of PFS nanostructures were often frustrated by the labile nature of the self-assembled micelles in that they easily undergo dissociation in a common solvent or at an elevated temperature.

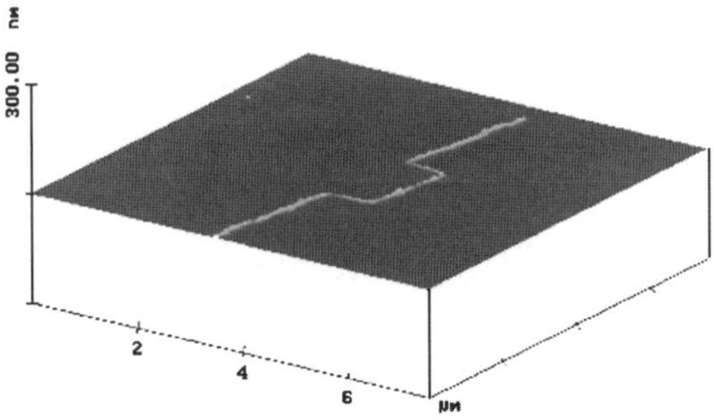

Figure 3.5 Scanning force micrograph for the ceramic line derived from aligned PFS-*b*-PDMS cylindrical micelles by H_2 plasma etching. (From Massey et al.[44] Reproduced with permission.)

D. Shell-Cross-Linked Nanocylinders and Nanotubes

Several groups have demonstrated that the stability of self-assembled micelles can be improved by performing a shell-cross-linking reaction.[45–48] The synthesis of PFS-*b*-PMVS and PI-*b*-PFS block copolymers through the sequential anionic polymerization, followed by the self-assembly in hexane allowed access either nanocylinders or nanotubes with pendent vinyl groups attached to the corona chains. By taking advantage of pendent vinyl groups, we performed a Pt-catalyzed hydrosilylation cross-linking reaction with tetramethyldisiloxane as a cross-linker, leading to shell-cross-linked nanocylinders and nanotubes.[26,49]

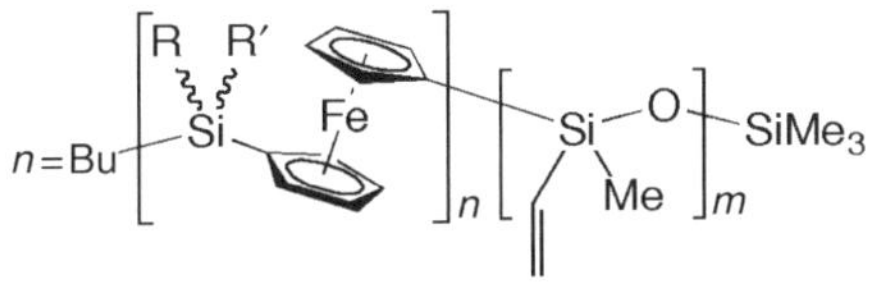

PFS-*b*-PMVS

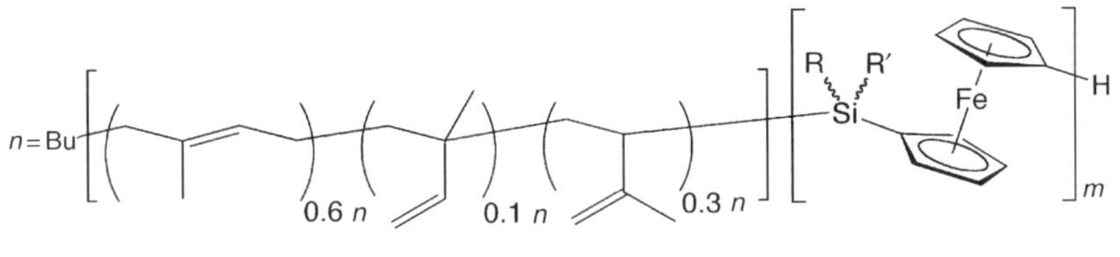

PI-*b*-PFS

As a result of shell-crosslinking, the one-dimensional micellar structures were locked-in and preserved even if transferred from hexane to a common solvent for both blocks.[26,49] Figure 3.6 illustrates a TEM image for PI_{320}-*b*-PFS_{53} shell-cross-linked micelles. The TEM sample was prepared from a micellar solution in THF, a common solvent for both PFS and PI blocks.

We have further demonstrated that the shell-cross-linking reaction is essential in making PFS nanoceramics. Upon heating up to 600°C with a temperature ramp of 1°C/min under N_2, we pyrolyzed the shell-cross-linked micelles. The resulting ceramics with excellent shape retention were characterized by TEM (see Figure 3.7). In a control experiment, pyrolysis of uncross-linked PI_{320}-*b*-PFS_{53} cylindrical micelles was found to lead to the destruction of the structure. This comparison indicated that the shell-cross-linking plays an essential role in the

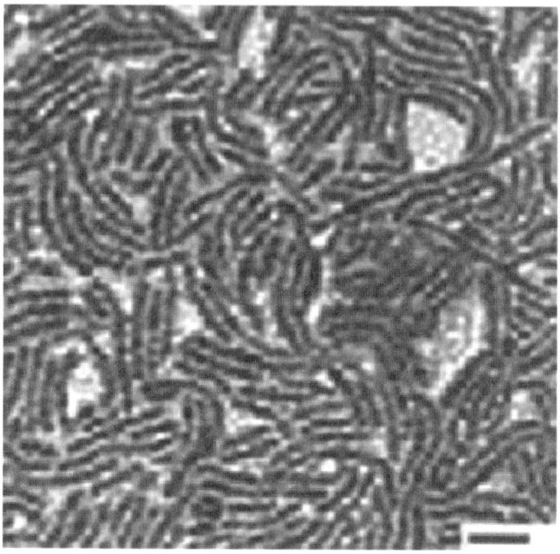

Figure 3.6 TEM image for shell-cross-linked PI_{320}-*b*-PFS_{53} cylindrical micelle solid samples prepared by drying a drop of THF solution on a carbon-coated copper grid. THF is a common solvent for both blocks. Scale bar = 250 nm. (From Wang et al.[49] Reproduced with permission.)

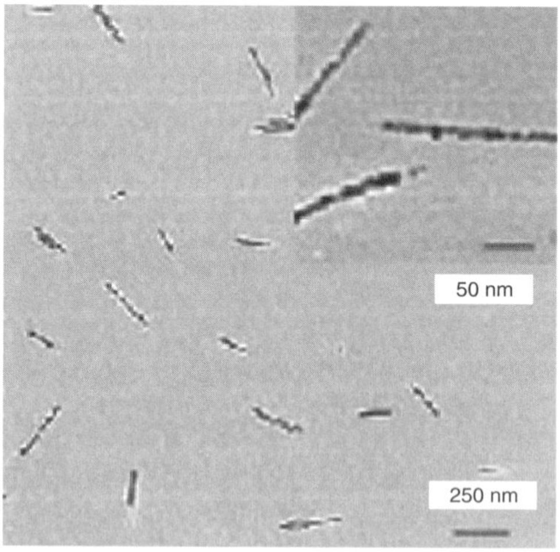

Figure 3.7 TEM image for cylindrical nanoceramic replica derived from PI_{320}-*b*-PFS_{53} shell-cross-linked micelles through a pyrolysis process under N_2 with a temperature ramp of 1°C/min up to 600°C. (From Wang et al.[49] Reproduced with permission.)

shape retention and permits the formation of a ceramic replica, presumably due to the high ceramic yield resulted from cross-linked structure.[26,49]

Shell-cross-linked micelles have also been aligned and patterned on a flat silicon substrate by a microfluidic technique (see Figure 3.8). Ordered magnetic nanoceramic arrays derived from a pyrolysis process may be of interest as magnetic memory materials or as catalysts for the growth of carbon nanotubes.[49]

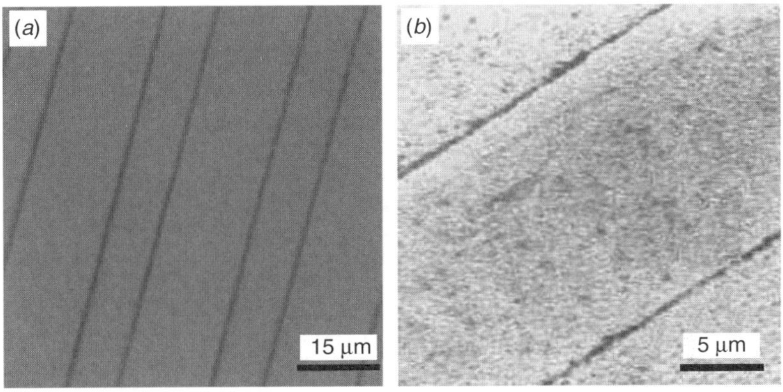

Figure 3.8 (a) Optical, and (b) scanning electron microscope (SEM) images for microfluidically aligned shell-cross-linked PI_{320}-*b*-PFS_{53} nanocylinders. (From Wang and Manners.[26] Reproduced with permission.)

As for shell-cross-linked nanotubes (see Figure 3.9a), the conservation of redox activity due to PFS chains has been illustrated by performing cyclic voltammetry experiments (see Figure 3.9b).[26] This type of nanotube represents a new type of reactive nanostructure and provides a chance to encapsulate guest compounds or particles in the cavities through an in situ redox reaction using PFS as a reductant.[50]

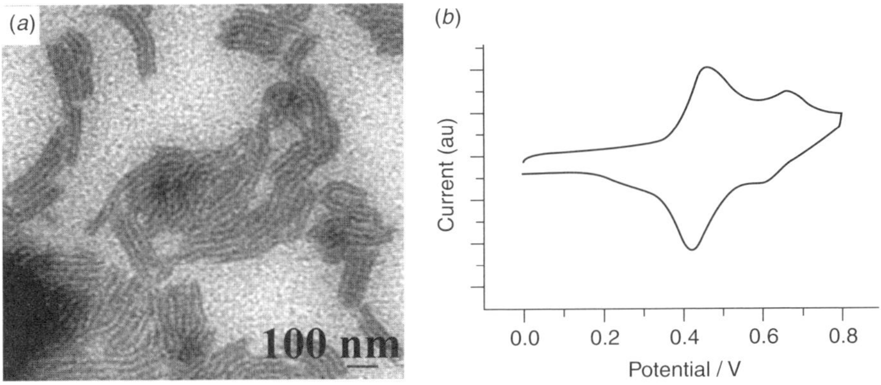

Figure 3.9 Shell-cross-linked nanotubes of PFS_{48}-*b*-$PMVS_{300}$ block copolymers: (a) TEM image for the sample prepared for THF solution; (b) cyclic voltammetry in dichloromethane:benzonitrile (2:1) with 0.1 M $[Bu_4N][PF_6]$ as supporting electrolyte. (From Wang and Manners.[26] Reproduced with permission.)

As illustrated in Scheme 3.14, Ag nanoparticles were prepared within the tubes via a reaction with $Ag[PF_6]$. We found that partial preoxidation of the PFS domains with an organic oxidant such as tris(4-bromophenyl)aminium hexachloroantimonate is a key step for the efficient formation of one-dimensional arrays of silver nanoparticles confined within the nanotubes (Figure 3.10).[50] Further attempts to synthesize metal nanowires through this encapsulation method are currently in progress.

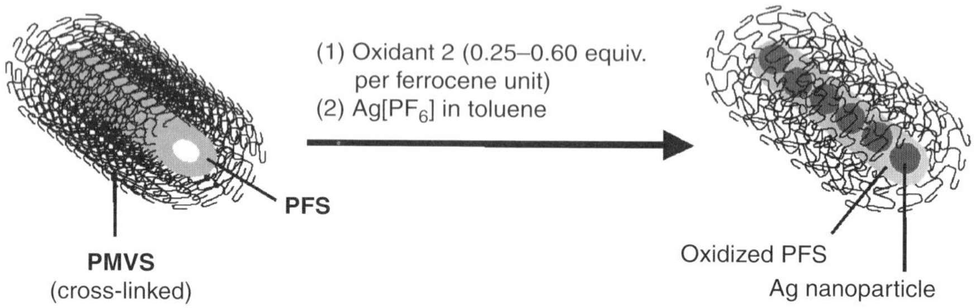

Scheme 3.14 Generation of Ag nanoparticles within self-assembled PFS-*b*-PMVS.

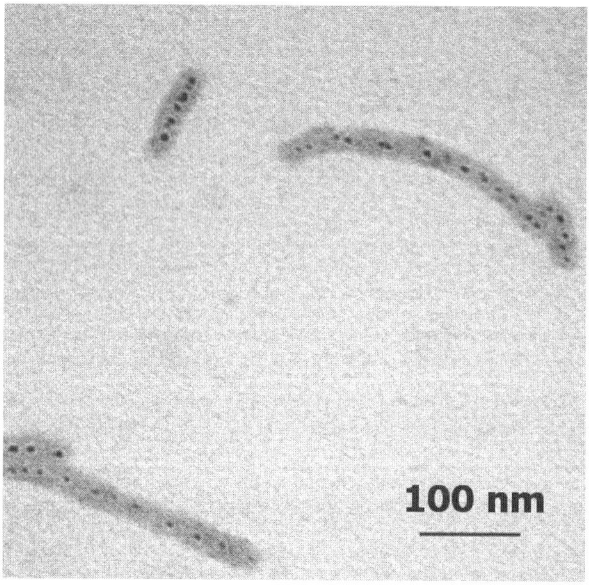

Figure 3.10 TEM images for shell-cross-linked nanotubes of PFS_{48}-*b*-$PMVS_{300}$ block copolymers encapsulated with Ag nanoparticles. (From Wang et al.[50] Reproduced with permission.)

PFS-*b*-PBLG block copolymers, which are soluble in hot toluene, were found to form optically transparent gels upon cooling to ambient temperature.[51] The gelation was thermally reversible in all cases. The critical concentration for gelation was found: (1) to strongly depend on the chain length of the PBLG block, suggesting that the interaction between the PBLG helices is a crucial factor in the gelation process, and (2) to be related to the chain-length ratio between the PFS and the PBLG blocks. Moreover, when the chain-length ratio between the PFS and the PBLG exceeded 1:3, incomplete gelation was observed (macroscopic phase separation into a dense block copolymer gel and the free solution phase). Hydrogen bonding was also shown to be a negligible factor in the gelation process. These PFS-*b*-PBLG block copolymer gels, with PBLG-based helices with a random coil of PFS at one end, formed fibrous nanoribbons assembled from the one-dimensional antiparallel stacking of the building blocks in a monolayer fashion (Figure 3.11). It was proposed that strong dipolar interactions between the PBLG helices stabilized the stacked structure where the PFS blocks protruded outside of the ribbon into the toluene-rich environment, thereby preventing aggregation of nanoribbons.

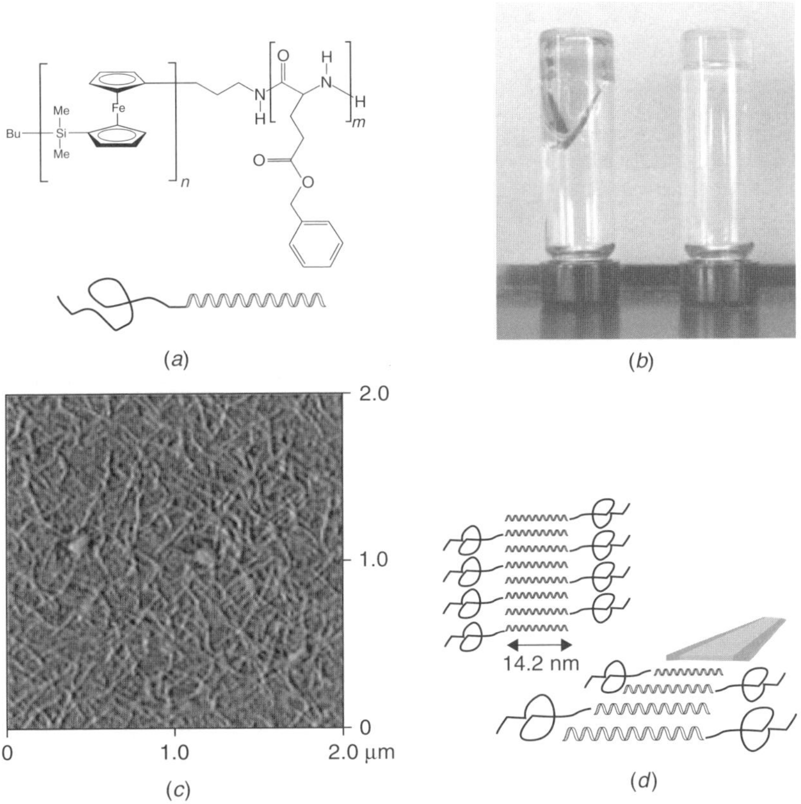

Figure 3.11 (a) PFS-*b*-PBLG. (b) Optical photograph of toluene swollen gel of PFS-*b*-PBLG. (c) AFM of PFS-*b*-PBLG gel. (d) Schematic for nanoribbon mechanism. (From Kim et al.[51] Reproduced with permission.)

E. Self-Assembly of PFS Block Copolymers in the Solid State

Polyferrocenylsilane block copolymers also phase separate in solid state to generate periodic, nanoscopic iron-rich domains that can be observed by TEM without resorting to staining techniques.[24] This bulk self-assembly behavior has been studied with various block copolymers, such as PS-*b*-PFS,[24,52] PI-*b*-PFS,[49] and PFS-*b*-PMMA.[22] Additional self-assembly complications arising from the semicrystallinity of the PFS can be avoided through synthesis of unsymmetrically substituted polyferrocenylsilanes as in the case of polyferrocenylethylmethylsilane (PFEMS). Shown in Figure 3.12 are TEM micrographs of solid-state bulk morphologies for PS-*b*-PFEMS,[53] PFS-*b*-PMMA,[54] (blended with homopolymer), and PMMA-*b*-PFS-*b*-PS-*b*-PFS-*b*-PMMA.[55] Furthermore, with the PS-*b*-PFS system, oxidation of PFS block was reported as having influence on the order–disorder transition temperature[21] of the phase-separated morphologies. Exploitation of the high-etch resistance of PFS nanodomains with respect to its surrounding matrix has also been used to nanopattern surfaces.[23]

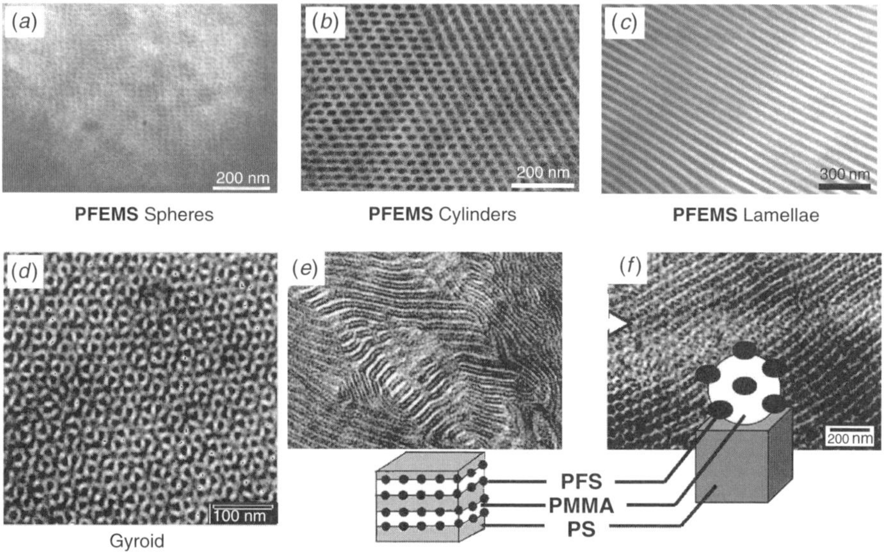

Figure 3.12 TEM images for bulk state morphologies of PS-*b*-PFEMS (a)–(c)[53], PFS-*b*-PMMA (d)[54], and PMMA-*b*-PFS-*b*-PS-*b*-PFS-*b*-PMMA (e)–(f)[55]. (Reproduced with permission.)

Additionally, the periodic PFS domains can be converted to iron-containing nanoclusters that can be used as heterogeneous catalysts.[56,57,58] Spontaneous perpendicular-to-surface orienting of cylindrical PFS microdomains was observed throughout a thin-film sample of PS_{374}-*b*-PFS_{45}. Shown in Figure 3.13a is a TEM image of the film where cylindrical domains were found to traverse the entire film thickness. Following pyrolysis above 600°C, an array of iron-rich ceramic nanodomain replicas was produced (Figure 3.13b).[52]

In ongoing collaborations with Lu at Agilent Laboratories, Russell at the University of Massachusetts, Liu at Duke University, Ajayan and Ryu at Rensselaer Polytechnic Institute, and Winnik at Toronto, we have used the iron-rich ceramics from films PS-*b*-PFEMS as catalysts for the growth of single-walled carbon nanotubes (SWCNT). Typically, thin films were self-assembled and treated with UV-ozone to remove organic materials and subsequently used for simple one-step chemical vapor deposition (CVD) growth of SWCNTs. The resulting SWCNTs were characterized by AFM, SEM, Raman spectroscopy, and X-ray photoelectron spectroscopy (XPS). Shown in Figure 3.14 is a SEM of a SWCNT array generated from lithographic patterning of a PS-*b*-PFEMS film.[56,57] In an analogous study by Hinderling et al., it was also found that multiwalled CNTs could also be grown from PFS-derived catalysts using a plasma

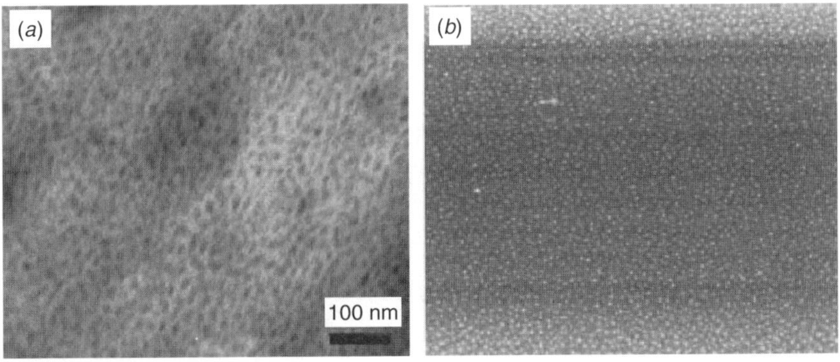

Figure 3.13 (a) TEM image for PS_{374}-*b*-PFS_{45} thin-film microtomed parallel to the substrate. (b) AFM image (2 μm × 2 μm) of pyrolyzed sample (600°C, 2 h, N_2) of a PS_{374}-*b*-PFS_{45} thin film in which the PS matrix has been cross-linked by UV radiation. (From Temple et al.[52] Reproduced with permission.)

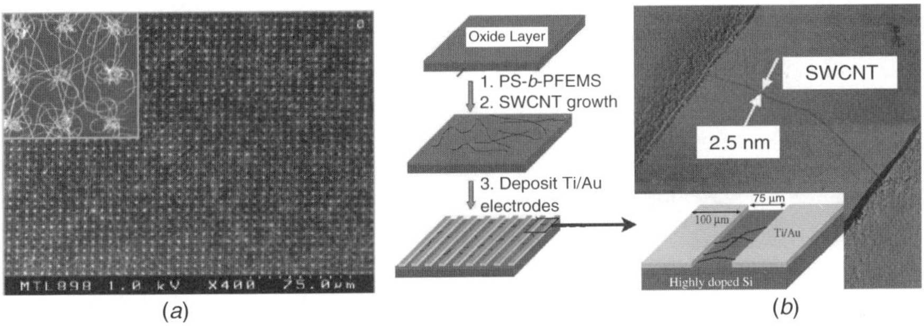

Figure 3.14 (a) SEM images for single-walled carbon nanotubes (SWCNT) synthesized by chemical vapor deposition (CVD) with iron-rich ceramic particles derived from patterned PS-*b*-PFEMS islands as catalyst. (From Lu et al.[57] Reproduced with permission.) (b) Schematic and AFM height image (11 μm^2) of high-throughput field-effect transistors (FETs) from SWCNTs afforded from pyrolyzed PS-*b*-PFEMS films. (From Lastella et al.[59] Reproduced with permission.)

treatment approach.[58] More recently, Lastella el al. have shown that the SWCNTs from pyrolyzed PS-*b*-PFEMS films could be used in high-throughput field-effect transistors (FETs).[59] Shown in Figure 3.14 is a schematic for the high device yield (96%) preparation of individually addressable FETs. Up to 160 devices on a 15 × 15 mm^2 chip were easily achieved.[59]

As the interest in nanolithography grows, efficient routes to nanotemplates are required. An emerging method involves films of phase-separated diblock copolymers. As an exciting application of transition metal–containing block copolymers, we discuss the lithographic template produced using PS-*b*-PFS. Ross and co-workers demonstrated that by spin coating a film of a PS-*b*-PFS, a material that affords plasma etch-resistant PFS spheres in a PS matrix, direct transfer of the pattern to underlying layers was possible.[60] Using oxygen reactive ion etching (RIE), much of the PS matrix is eliminated so as to permit the CF_3H etch of an underlying silica layer (Scheme 3.15g). Further, CF_4 and O_2 RIE followed by ashing permits the patterned etching of tungsten. Last, an ion beam etch was used to transfer the pattern into the cobalt layer. The end result is an array of tungsten-capped cobalt dots that have dimensions on the nanometer scale whose pattern has been inherited from the PS-*b*-PFS film (Scheme 3.15h).

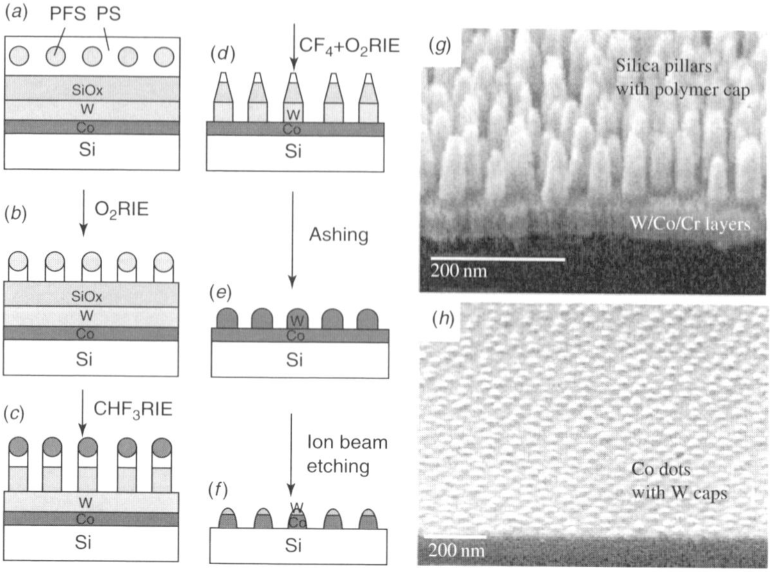

Scheme 3.15 (a) Thin film of PS-*b*-PFS (PFS spheres) on silicon substrate with Co, W, and SiO_x layers. (b) Oxygen reactive ion etch of PS-*b*-PFS. (c) CHF_3 reactive ion etch of SiO_x layer. (d) $CF_4 + O_2$ reactive ion etch of PS-*b*-PFS derived lithographic mask and W layer. (e) Ashing to remove all material above W. (f) Ion-beam etching to produce cobalt dots with W caps. (g) SEM of part (c). (h) SEM of part (f). (From Cheng et al.[60] Reproduced with permission.)

A new type of nanotextured Ag surface that was generated from a metal-containing diblock copolymer was found to exhibit high and uniform surface-enhanced Raman spectroscopy (SERS) activity.[61] First, the metal-containing diblock copolymer, PS-*b*-PFEMS, was self-assembled in a thin film into hexagonally close-packed standing cylindrical PFEMS-based nanostructures surrounded by a PS matrix. UV-ozonation was then used to selectively remove the PS matrix and convert PFEMS into iron-containing silicon oxide cylinders. A silver nanotextured surface was subsequently generated by sputtering a thin layer of silver over these inorganic cylinders (Scheme 3.16). Uniformly enhanced Raman signals on the silver nanotextured surfaces have been observed for absorbed benzenethiol, with enhancement factors up to 10^6 (Scheme 3.16, bottom). This metal-containing block-copolymer-template approach provides a simple and straightforward method to fabricate SERS active substrates with great manufacturing potential. Furthermore, since the size and spacing of the cylinders can be adjusted by tailoring the polymer chain lengths, the electromagnetic field can potentially be tuned to achieve even higher SERS activity. More importantly, this block-copolymer approach has established the potential for reproducible fabrication of uniformly enhancing SERS-active substrates.

Most studies[18–23,62] indicated that the self-assembly of PFS block copolymers in bulk state follows the theory that the longer block forms the continuous phase, whereas the shorter block forms the imbedded structures. In a recent study, an interesting exception was discovered.[63] As illustrated in Figure 3.17, the morphology of PFS_{90}-*b*-$PDMS_{900}$ film cast from toluene shows somewhat disordered hexagonal packed objects that are formed by the PDMS blocks (white areas), while the continuous phase appears gray. Scanning TEM (STEM) in the dark field mode to view an image of the electrons that are elastically scattered by the PFS domains also

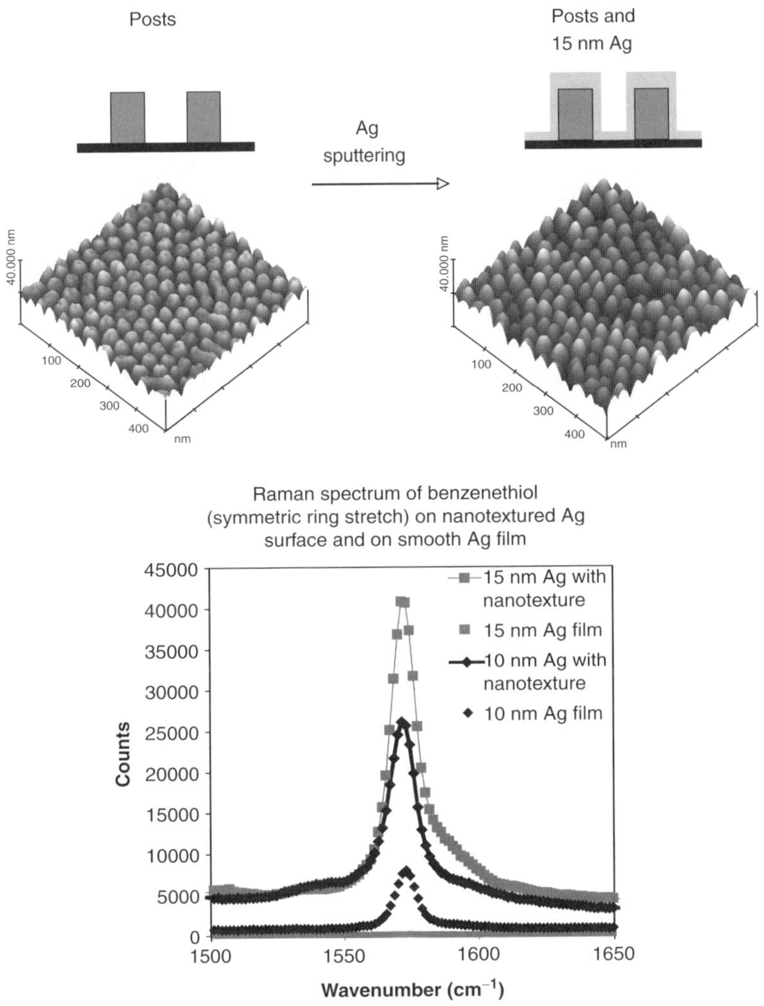

Scheme 3.16 Generation of nanotextured Ag film (AFM height image, top right) from sputtering on UV-ozone etched PS-*b*-PFEMS film (AFM height image, top left). Surface-enhanced Raman spectra (SERS) (bottom) of benzenethiol on nanotextured Ag films (10 and 15 nm: -◆- and -■-, respectively) and control experiments with smooth Ag films (10 and 15 nm : ◆ and ■, respectively).

confirmed this morphology. As shown in the inset of Figure 3.17, the presence of PFS rings (white) in the midst of the relatively electron-poor PDMS domains (black and gray) was observed. The small-angle X-ray scattering (SAXS) pattern of the film using a CuK source shows broad peaks with relative positions of $3^{1/2}$, $7^{1/2}$, and $21^{1/2}$, suggesting hexagonally packed cylinders. Therefore, based on preliminary experiments, it was confirmed that an asymmetric PFS_{90}-*b*-$PDMS_{900}$ copolymer undergoes phase segregation where the longer block forms isolated cylinders surrounded by a shell of PFS, with the remaining PDMS filling the interstitial spaces.

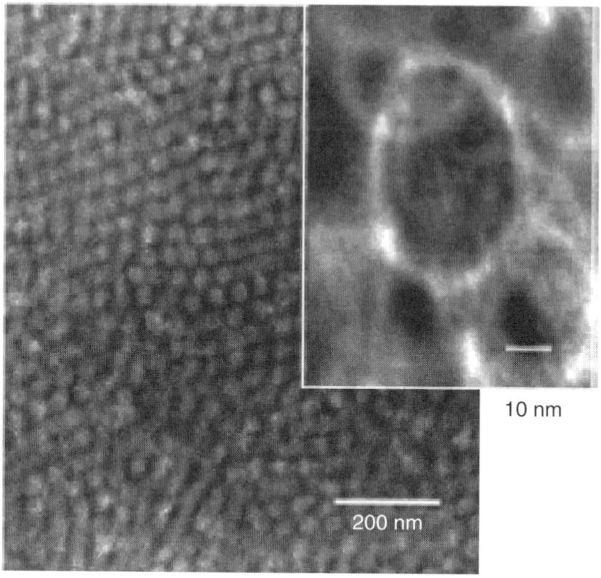

Figure 3.17 TEM image for a cryomicrotomed PFS_{90}-*b*-$PDMS_{900}$ film cast from toluene. *Inset*: High-resolution STEM image of this sample obtained in the dark-field mode. (From Raez et al.[63] Reproduced with permission.)

SUMMARY

The study of metal-containing block copolymers represents one of the most active areas in the metallopolymer field. As illustrated in this chapter, exciting possibilities exist for the development of new functional nanostructured materials via simple self-assembly approaches. Despite the obvious potential, many challenges still remain in the field. One of the most important is the development of new living polymerization procedures that are tolerant of functionality and a wide range of transition metal centers. The recently discovered living photolytic polymerization approach for ferrocenophanes holds significant promise in this area if the procedure can be extended to a range of strained metallo-ring systems.

REFERENCES

1. R. P. Kingsborough, T. M. Swager, *Prog. Inorg. Chem.*, **48**, 123 (1999).
2. I. Manners, *Science*, **294**, 1664 (2001).
3. G. R. Newkome, E. F. He, C. N. Moorefield, *Chem. Rev.*, **99**, 1689 (1999).
4. P. Nguyen, P. Gómez-Elipe, I. Manners, *Chem. Rev.*, **99**, 1515 (1999).
5. U. S. Schubert, C. Eschbaumer, *Angew. Chem., Int. Ed. Engl.*, **41**, 2893 (2002).
6. B. J. Holliday, T. M. Swager, *Chem. Commun.*, **1**, 23 (2005).
7. R. D. Archer, *Inorganic and Organometallic Polymers*, Wiley VCH, Weinheim, 2001.
8. I. Manners, *Synthetic Metal-Containing Polymers*, Wiley-VCH, Weinheim, 2004.
9. V. Chandrasekhar, *Inorganic and Organometallic Polymers*, Springer: New York, 2005.
10. B. G. G. Lohmeijer, U. S. Schubert, *J. Polym. Sci., Part A, Polym. Chem.*, **42**, 4016 (2004).

11. G. Zhou, I. I. Harruna, *Macromolecules*, **37**, 7132 (2004).
12. G. Zhou, I. I. Harruna, *Macromolecules*, **38**, 4114 (2005).
13. C. A. Fustin, B. G. G. Lohmeijer, A.-S. Duwez, A. M. Jonas, U. S. Schubert, J.-F. Gohy, *Adv. Mater.*, **17**, 1162 (2005).
14. J. E. McAlvin, C. L. Fraser, *Macromolecules*, **32**, 6925 (1999).
15. D. A. Foucher, B.-Z. Tang, I. Manners, *J. Am. Chem. Soc.*, **114**, 6246 (1992).
16. R. Rulkens, A. J. Lough, I. Manners, S. R. Lovelace, C. Grant, W. E. Geiger, *J. Am. Chem. Soc.*, **118**, 12683 (1996).
17. K. Kulbaba, I. Manners, *Macromol. Rapid Commun.*, **22**, 711 (2001).
18. J. A. Massey, K. N. Power, M. A. Winnik, I. Manners, *Adv. Mater.*, **10**, 1559 (1998).
19. R. G. H. Lammertink, M. A. Hempenius, E. L. Thomas, G. J. Vancso, *J. Polym. Sci., Part B, Polym. Phys.*, **37**, 1009 (1999).
20. W. Li, N. Sheller, M. D. Foster, D. Balaishis, I. Manners, B. Annis, J. S. Lin, *Polymer*, **41**, 719 (2000).
21. H. B. Eitouni, N. P. Balsara, *J. Am. Chem. Soc.*, **126**, 7446 (2004).
22. C. Kloninger, M. Rehahn, *Macromolecules*, **37**, 1720 (2004).
23. R. G. H. Lammertink, M. A. Hempenius, J. E. van den Enk, V. Z.-H. Chan, E. L. Thomas, G. J. Vancso, *Adv. Mater.*, **12**, 98 (2000).
24. Y. Ni, R. Rulkens, I. Manners, *J. Am. Chem. Soc.*, **118**, 4102 (1996).
25. L. Cao, I. Manners, M. A. Winnik, *Macromolecules*, **35**, 8258 (2002).
26. X. Wang, M. A. Winnik, I. Manners, *Angew. Chem., Int. Ed. Engl.*, **43**, 3703 (2004).
27. X. S. Wang, M. A. Winnik, I. Manners, *Macromolecules*, **35**, 9146 (2002).
28. X. S. Wang, M. A. Winnik, I. Manners, *Macromol. Rapid Commun.*, **23**, 210 (2002).
29. I. Korczagin, M. A. Hempenius, G. J. Vancso, *Macromolecules*, **37**, 1686 (2004).
30. K. N. Power-Billard, P. Wieland, M. Schafer, O. Nuyken, I. Manners, *Macromolecules*, **37**, 2090 (2004).
31. L. Vanderark, I. Manners, unpublished results, 2006.
32. X. S. Wang, M. A. Winnik, I. Manners, *Macromol. Rapid Commun.*, **24**, 403 (2003).
33. M. Tanabe, I. Manners, *J. Am. Chem. Soc.*, **126**, 11434 (2004).
34. K. T. Kim, G. W. M. Vandermeulen, M. A. Winnik, I. Manners, *Macromolecules*, **38**, 4958 (2005).
35. J. Raez, I. Manners, M. A. Winnik, *J. Am. Chem. Soc.*, **124**, 10381 (2002).
36. J. Raez, I. Manners, M. A. Winnik, *Langmuir*, **18**, 7229 (2002).
37. P. Nguyen, G. Stojcevic, K. Kulbaba, M. J. MacLachlan, X.-H. Liu, A. J. Lough, I. Manners, *Macromolecules*, **31**, 5977 (1998).
38. J. K. Pudelski, D. A. Foucher, C. H. Honeyman, P. M. MacDonald, I. Manners, S. Barlow, D. O'Hare, *Macromolecules*, **29**, 1894 (1996).
39. D. J. Frankowski, J. Raez, I. Manners, M. A. Winnik, S. A. Khan, R. J. Spontak, *Langmuir*, **20**, 9304 (2004).
40. J. A. Massey, K. Temple, L. Cao, Y. Rharbi, J. Raez, M. A. Winnik, I. Manners, *J. Am. Chem. Soc.*, **122**, 11577 (2000).
41. J.-F. Gohy, B. G. G. Lohmeijer, A. Alexeev, X. S. Wang, I. Manners, M. A. Winnik, U. S. Schubert, *Chem. Eur. J.*, **10**, 4315 (2004).
42. X. S. Wang, M. A. Winnik, I. Manners, *Macromolecules*, **38**, 1928 (2005).
43. L. Cao, J. A. Massey, M. A. Winnik, I. Manners, S. Riethmuller, F. Banhart, J. P. Spatz, M. Möller, *Adv. Funct. Mater.*, **13**, 271 (2003).
44. J. A. Massey, M. A. Winnik, I. Manners, V. Z.-H. Chan, J. M. Ostermann, R. Enchelmaier, J. P. Spatz, M. Möller, *J. Am. Chem. Soc.*, **123**, 3147 (2001).
45. K. B. Thurmond, T. Kowalewski, K. L. Wooley, *J. Am. Chem. Soc.*, **118**, 7239 (1996).
46. K. B. Thurmond, T. Kowalewski, K. L. Wooley, *J. Am. Chem. Soc.*, **119**, 6656 (1997).

47. J. Ding, G. Liu, *Macromolecules*, **31**, 6554 (1998).
48. V. Bütün, A. B. Lowe, N. C. Billingham, S. P. Armes, *J. Am. Chem. Soc.*, **121**, 4288 (1999).
49. X. S. Wang, A. C. Arsenault, G. A. Ozin, M. A. Winnik, I. Manners, *J. Am. Chem. Soc.*, **125**, 12686 (2003).
50. X. S. Wang, H. Wang, M. A. Winnik, I. Manners, *J. Am. Chem. Soc.*, **127**, 8924 (2005).
51. K. T. Kim, C. Park, G. W. M. Vandermeulen, D. A. Rider, C. Kim, M. A. Winnik, I. Manners, *Angew. Chem., Int. Ed. Engl.*, **44**, 7964 (2005).
52. K. Temple, K. Kulbaba, K. N. Power-Billard, I. Manners, K. A. Leach, T. Xu, T. P. Russell, C. J. Hawker, *Adv. Mater.*, **15**, 297 (2003).
53. D. A. Rider, K. A. Cavicchi, K. N. Power-Billard, T. P. Russell, I. Manners, *Macromolecules*, **38**, 6931 (2005).
54. C. Kloninger, M. Rehahn, *Macromolecules*, **37**, 8319 (2004).
55. U. Datta, M. Rehahn, *Macromol. Rapid Commun.*, **25**, 1615 (2004).
56. S. Lastella, Y. J. Jung, H. Yang, R. Vajtai, P. M. Ajayan, C. Y. Ryu, D. A. Rider, I. Manners, *J. Mater. Chem.*, **14**, 1791 (2004).
57. J. Q. Lu, T. E. Kopley, N. Moll, D. Roitman, D. Chamberlin, Q. Fu, J. Liu, T. P. Russell, D. A. Rider, M. A. Winnik, I. Manners, *Chem. Mater.*, **17**, 2227 (2005).
58. C. Hinderling, Y. Keles, T. Stöckli, H. F. Knapp, T. de los Arcos, P. Oelhafen, I. Korczagin, M. A. Hempenius, G. J. Vancso, R. Pugin, H. Heinzelmann, *Adv. Mater.*, **16**, 876 (2004).
59. S. Lastella, G. Mallick, R. Woo, S. P. Karna, D. A. Rider, I. Manners, Y. J. Jung, C. Y. Ryu, P. M. Ajayan, *J. Appl. Phys.*, **99**, 024302 (2006).
60. J. Y. Cheng, C. A. Ross, V. Z.-H. Chan, E. L. Thomas, R. G. H. Lammertink, G. J. Vancso, *Adv. Mater.*, **13**, 1174 (2001).
61. J. Q. Lu, D. Chamberlin, M. Liu, D. A. Rider, I. Manners, T. P. Russell, *Nanotechnology*. (available online October 2006.)
62. R. G. H. Lammertink, M. A. Hempenius, G. J. Vancso, K. Shin, M. H. Rafailovich, J. Sokolov, *Macromolecules*, **34**, 942 (2001).
63. J. Raez, Y. Zhang, L. Cao, S. Petrov, K. Erlacher, U. Wiesner, I. Manners, M. A. Winnik, *J. Am. Chem. Soc.*, **125**, 6010 (2003).

CHAPTER 4

Metal-Containing π-Conjugated Polymers

MARK J. MACLACHLAN
University of British Columbia, Vancouver, British Columbia, Canada

I. INTRODUCTION: BACKGROUND AND SCOPE

The discovery of conducting polyacetylene in 1977 ignited a revolution in materials chemistry, where researchers turned to organic materials as lightweight, tunable substitutes for existing technologies based primarily on silicon.[1] Materials that exhibit delocalization within π- or σ-conjugated frameworks are attractive for a wide variety of applications.[2] In the past two decades, we have seen organic materials applied as emissive layers in flexible displays [light-emitting diodes (LEDs)],[3] photoconductors,[4] and static dissipaters.[5] In addition, these materials are of interest for many underdeveloped applications, such as photovoltaic cells and field-effect transistors (FETs).[6] It is inevitable that conjugated organic polymers will play a pivotal role in future technologies.

The incorporation of metals into organic polymers is a much less widely explored area of research.[7] By integrating metals into a conjugated framework, researchers hope to influence electronic, optical, or magnetic properties of the polymer.[8] For example, the metal may act as an n- or p-dopant to alter the polymer's conductivity, provided the metal and polymer are electronically coupled. In addition to mediating electronic interactions within the polymer, the metal may play other roles, such as sensor, where coordination to a Lewis base modifies the properties of the polymer in a detectable way. Metals may also serve as structural or mechanical components in the conjugated polymer, just as σ-bonding to metals has been used widely to stabilize acetylenic polymers.[9] In Schiff base–containing polymers, the metal prevents the hydrolysis of the imine units that make up the backbone. Moreover, the metal may be used as a scaffold for supramolecular chemistry, such as in poly(rotaxane)s or supramolecular polymer grids and ladders.[10]

This chapter reviews research published on conjugated metal-containing polymers from late 1998 to mid-2005. It includes polymers where the metal is incorporated into the backbone or the side chain of a conjugated polymer. Kingsborough and Swager wrote an excellent review of the field up to late 1998, and we do not intend to repeat any of their efforts herein.[11] However,

Frontiers in Transition Metal-Containing Polymers, edited by Alaa S. Abd-El-Aziz and Ian Manners.

due to the expansive nature of this field and the presence of many specialized reviews, including ones in this book, we have decided to present a more focused review of the field. Topics not included in this review are σ-acetylenic polymers,[12] polymers that contain σ-bonded main-group elements that disrupt conjugation (e.g., polyferrocenylsilanes,[13] poly(silylzirconacyclopentadiene)s[14]), bis(terpyridine)metal polymers,[15] conjugated polymers where the metal complex is a charge-balancing ion,[16] oligomers (except in a few places),[17] dendrimers,[18] conjugated polymer/metal complex blends,[19] and insoluble coordination polymers that crystallize into extended structures.[20] The review also excludes conjugated polymers with metals linked via a nonconjugated tether.[21] We have tried to be comprehensive, but some articles have undoubtedly been missed, and we apologize for those omissions.

The chapter is primarily organized by the class of polymers: (1) metal-containing porphyrin and phthalocyanine polymers; (2) metal-containing polythiophene-based polymers; and (3) other metal-containing conjugated polymers. Section II includes all conjugated polymers with metalloporphyrins or metallophthalocyanines in the backbone. Section III includes all polythiophene derivatives, except those containing porphyrins or phthalocyanines. Section IV includes all remaining conjugated metallopolymers, for example, those with Schiff bases, bipyridines, or phenanthrolines in the backbone. We have tried to describe the synthesis and structure of each polymer presented, and summarize any related studies. Although many of the developments here are extensions of earlier work, this chapter focuses on results reported since 1998. References to the original work can usually be found in the papers discussed here. We close the chapter with a discussion of progress, challenges, and future directions for the field.

For your reference, here are some of the abbreviations used in this chapter:

bpy	bipyridine
CV	cyclic voltammetry
Da	dalton ($g\,mol^{-1}$)
EDOT	ethylenedioxythiophene
ee	enantiomeric excess
GPC	gel permeation chromatography
HOMO	highest occupied molecular orbital
LED	light-emitting diode
LUMO	lowest unoccupied molecular orbital
M_n	number-average molecular wt
M_w	wt-average molecular wt
phen	1,10-phenanthroline
PPE	poly(phenyleneethynylene)
PPV	poly(phenylenevinylene)
salen	*N,N′*-bis(salicylidene)ethylenediamine
salphen	*N,N′*-bis(salicylidene)-1,2-diaminobenzene
terpy	terpyridine
TOF	time-of-flight

II. METAL-CONTAINING PORPHYRIN AND PHTHALOCYANINE POLYMERS

Porphyrins and phthalocyanines continue to be attractive components for incorporation into π-conjugated frameworks.[22] Their high stabilities, well-developed syntheses, and open coordination sites at the metal center are some of the reasons chemists are drawn to them. Moreover, it is well known that some porphyrins and phthalocyanines are semiconductors in the solid state. It is notable that the direction of research in this area has shifted substantially since the 1990s. While phthalocyanine sheet polymers and one-dimensional (1-D) "shish-kebab" polymers were studied extensively, most recent research has focused on coupling porphyrins and phthalocyanines to the backbone of organic polymers such as PPE and PPV.

A. Metallophthalocyanine Polymers

A common route to phthalocyanine sheet polymers involves the condensation of 1,2,4,5-tetracyanobenzene with a metal salt, Scheme 4.1. Unfortunately, the sheet polymer is contaminated with polyisoindolenine and triazine linkages, which reduce its planarity and conductivity. Since these polymers are generally insoluble and not sublimable, it is difficult to remove the defects or impurities. Wöhrle has developed an improved route to **2** (M = Cu^{II}) by employing lithium pentanolate to reduce the density of defects in the resulting polymers.[23] Improved approaches for thin-film preparation were also explored.

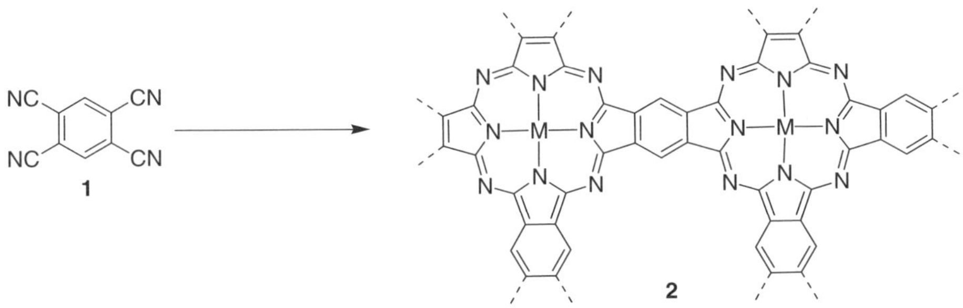

Scheme 4.1 Synthesis of phthalocyanine-based sheet polymer **2**.

Achar has synthesized cobalt, nickel, and copper phthalocyanine sheet polymers from the condensation of the metal phthalocyanine tetracarboxylic acids at 400°C (Scheme 4.2).[24] These

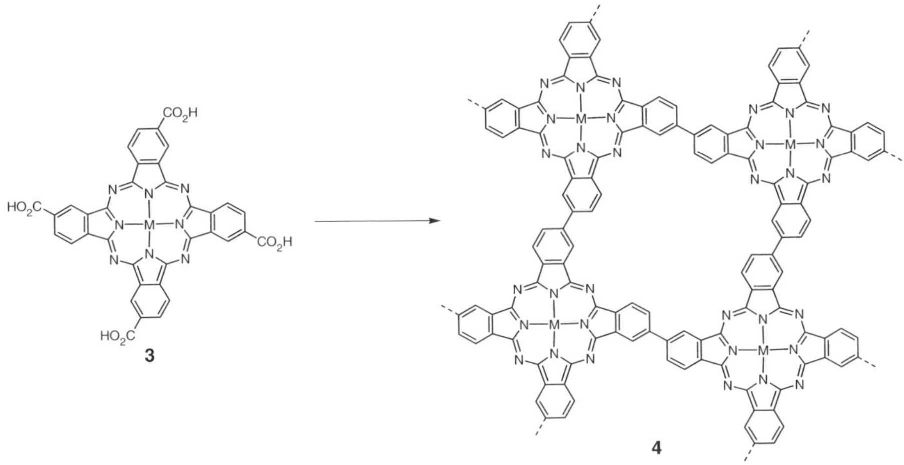

Scheme 4.2 Preparation of idealized polymer sheets **4** from **3** (M = Co^{II}, Ni^{II}, Cu^{II}).

dark blue to purple solids are soluble in 30 N sulfuric acid, but no molecular weights were determined. Although the authors refer to these as "sheet polymers," they are likely 3-D due to twisting of the biphenyl linkages between the phthalocyanines; in fact, an idealized planar structure would be very strained. These amorphous polymeric materials have high thermal stability (>500°C in air) with high ceramic yields (90–92% at 800°C) under nitrogen. The thermal decomposition mechanism of the Ni^{II} phthalocyanine polymer was studied by gas chromatography–mass spectrometry (GC–MS).[25] Temperature-dependent conductivity measurements of pressed pellets indicated that the Co and Ni polymers are semiconductors, while the Cu polymer is metallic at low temperature and semiconducting at high temperature. Achar concluded that room-temperature conductivity ($\sim 10^{-7}\,S\,cm^{-1}$) was much higher in these samples than in previously reported sheet polymers, possibly due to enhanced structural order in the sheets, and this system certainly merits further investigation.

1,4-Diazophenylene-bridged copper(II) phthalocyanine polymers **6** were prepared by reacting the diazonium salt of 1,4-phenylenediamine with copper(II) tetraaminophthalocyanine **5** (Scheme 4.3).[26] The polymer structure was determined by infrared (IR) and ultraviolet-visible (UV-vis) spectroscopies, and elemental analysis. Thermal analysis indicated that the polymer underwent a small decomposition at $T > 100°C$, presumably due to N_2 loss, but had a ceramic yield of 86% at 500°C under nitrogen. Although the bulk polymer was insoluble, a tetrahydrofuran (THF)-soluble fraction showed a molecular mass (M_n) of 22,000 Da by ebullioscopy. The viscosity of the polymer appeared qualitatively to be higher than that of the monomeric phthalocyanine complex. Four-probe conductivity measurements revealed conductivities as high as $3 \times 10^{-4}\,S\,cm^{-1}$ after doping with I_2.

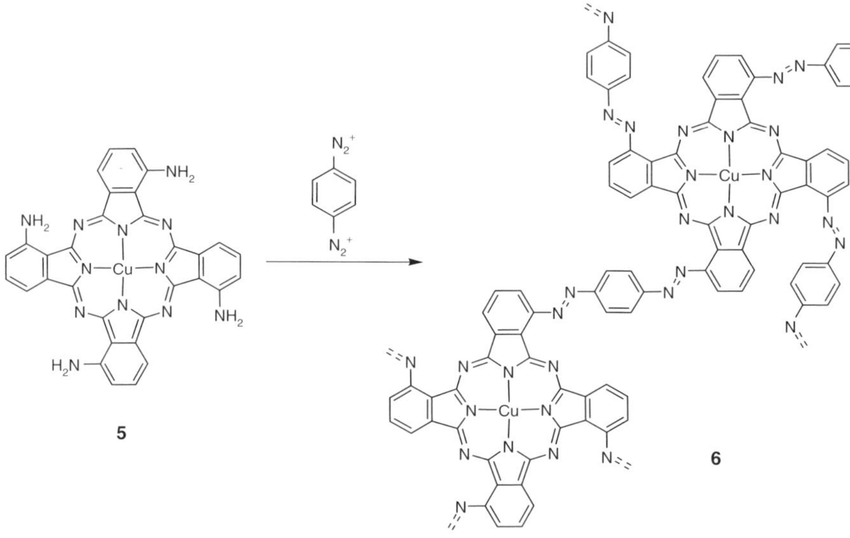

Scheme 4.3 Synthesis of diazophenylene-bridged phthalocyanine polymer **6**.

Swager has incorporated ethylenedioxythiophene (EDOT) substituents onto nickel(II) phthalocyanine, and electropolymerized to generate electroactive polythiophene–metallophthalocyanine hybrid materials **8** (Scheme 4.4).[27] Polymer films were formed between interdigitated microelectrodes by CV. CV of the polymer films showed one reversible reduction wave assigned to Ni^{I}/Ni^{II}, and a second assigned to the ligand. In the oxidative region, the polymer exhibited a broad wave assigned to the conjugated polymer. The polymer had a maximum conductivity ($\sim$30 S cm^{-1}) midway through this oxidation wave, suggesting that electrons were removed from the valence band to generate free charge carriers. The authors found that phthalocyanines functionalized on opposite isoindoles (e.g., **7**) underwent electropolymerization to afford films, while those functionalized on adjacent isoindoles (e.g., **9**) did not. Swager attributed this to the high solubility of poly-**9**, although the polymer was not isolated.

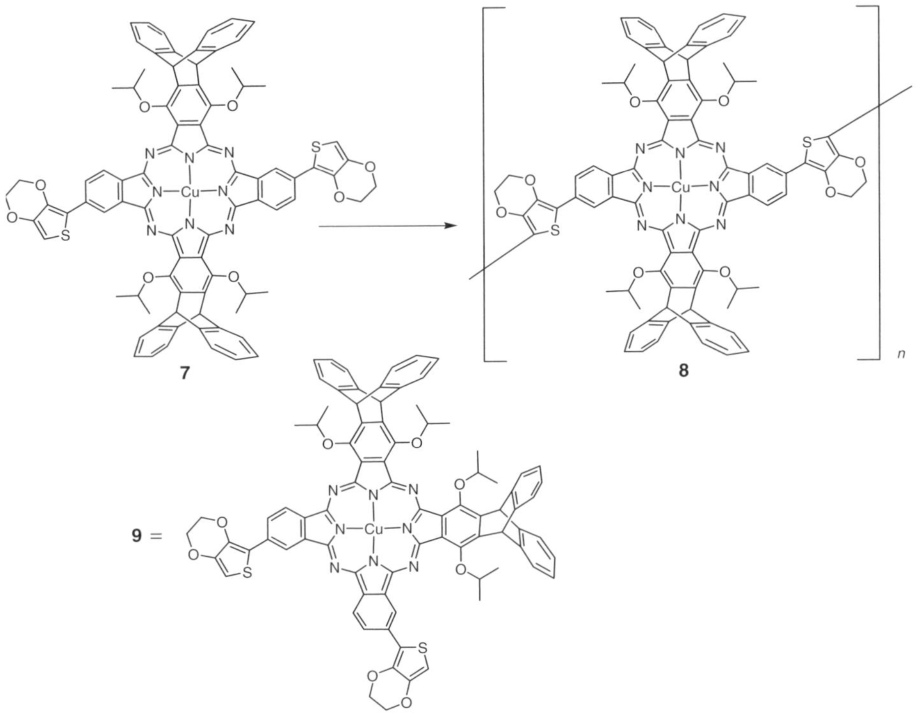

Scheme 4.4 Synthesis of polythiophene/nickel(II) phthalocyanine hybrid materials.

B. Metalloporphyrin Polymers

The main impetus for the incorporation of porphyrins into conjugated polymers is to develop photosynthetic and enzyme mimics. Porphyrins were recently incorporated into π-conjugated polymers (e.g., PPE, PPV), linear porphyrin tapes, and electropolymerized thin films.

In an effort to develop electrode-bound complexes as potentiometric sensors, Mori electropolymerized tetra(aminophenyl)porphyrin **10**, generating intensely colored, highly cross-linked polymer films with the structure suggested by **11** (Scheme 4.5).[28] In the case of Zn^{II} and Ni^{II}, the polymers exhibited low conductivity, presumably due to twisting of the polymer that limited conjugation to short polymer segments. In the case of Co^{II}, coupling of the redox states to the conduction path led to improved conductivity.

Osuka has published a beautiful study of conjugated porphyrin tapes prepared by scandium(III)-catalyzed oxidative coupling of meso-meso-linked zinc(II) porphyrin oligomers.[29] Although oligomeric, these molecules (e.g., **12**) are excellent models for conjugated porphyrin ladder polymers. The oligomers contain up to 10 Zn^{II} porphyrins and exhibit

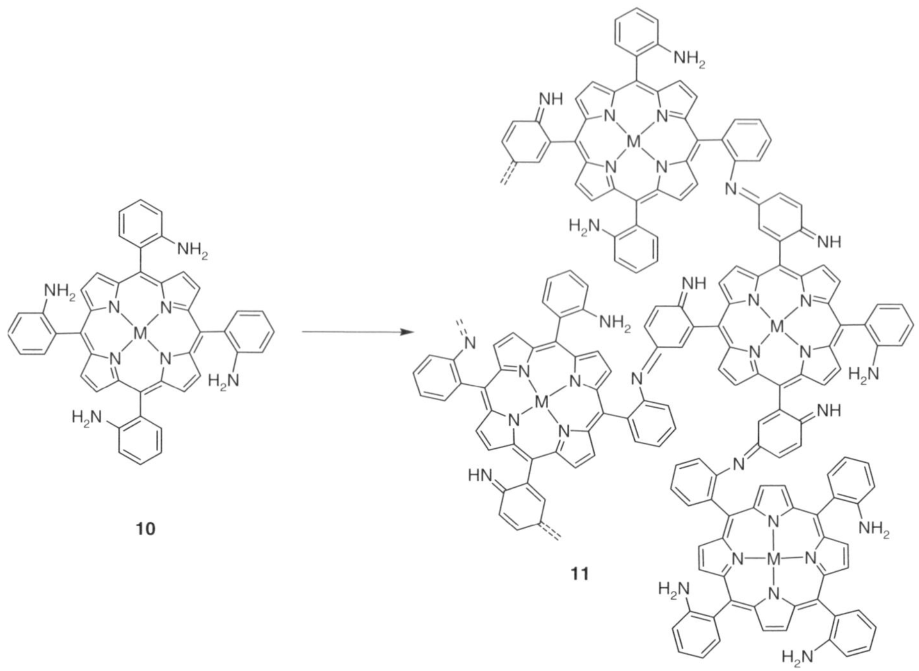

Scheme 4.5 Electropolymerization of porphyrin derivative **10** affords films **11** (M = Co^{II}, Ni^{II}, Zn^{II}).

very small HOMO to LUMO transitions that are found in the IR. Calculations show that the red-shifted absorption is a consequence of stabilization of the LUMO in these systems.[30] The calculations also indicate that 1-D zinc porphyrin tapes and related 2-D zinc porphyrin sheets will be metallic and semiconducting, respectively; these are challenging targets for synthetic chemists!

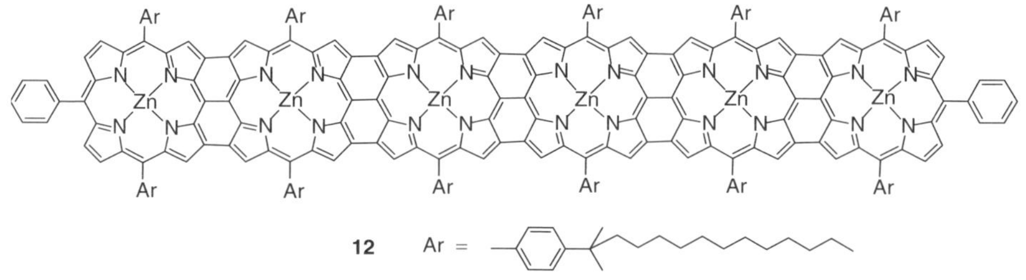

As with metallophthalocyanine-containing polymers, strong intermolecular interactions between porphyrins inhibit solubility. For this reason, many studies have been limited to oligomers. Anderson has studied poly(porphyrinbutadiyne) oligomers as building blocks for supramolecular materials and for nonlinear optical activity.[31] By employing very bulky substituents on the zinc porphyrin monomer **13**, they prepared soluble polybutadiynes **14** (Scheme 4.6), with molecular weights of ca. 60,000 Da (M_n, GPC).[32] On addition of 0.5 equiv. of 4,4′-bipyridine, the polymer forms a ladder structure with the metal centers of the two strands bridged by the bpy ligand (Scheme 4.7). Degenerate four-wave mixing (DFWM) experiments showed that ladder formation was accompanied by an order of magnitude enhancement of the real component of $\chi^{(3)}$ in solution. The enhanced nonlinear optical (NLO) properties arise from the improved conjugation when the molecule is in the ladder form. Extraordinarily high NLO properties were exhibited by the analogous Pb^{II} polymers, though they do not bind axial ligands to form ladders.[33] Anderson has also discovered that these polymers have very large 2-photon cross sections, which may make them ideal for photodynamic therapy, optical power limiting, and 3-D optical memory.

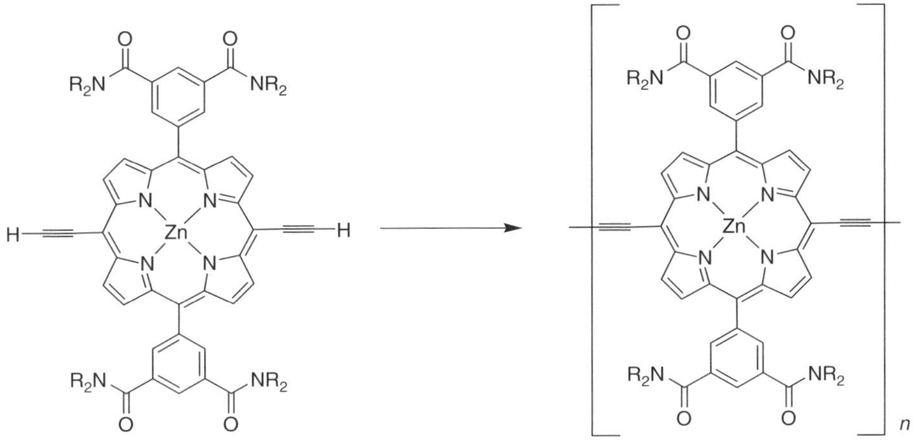

13 **14**

Scheme 4.6 Synthesis of poly(porphyrinbutadiyne) **14**.

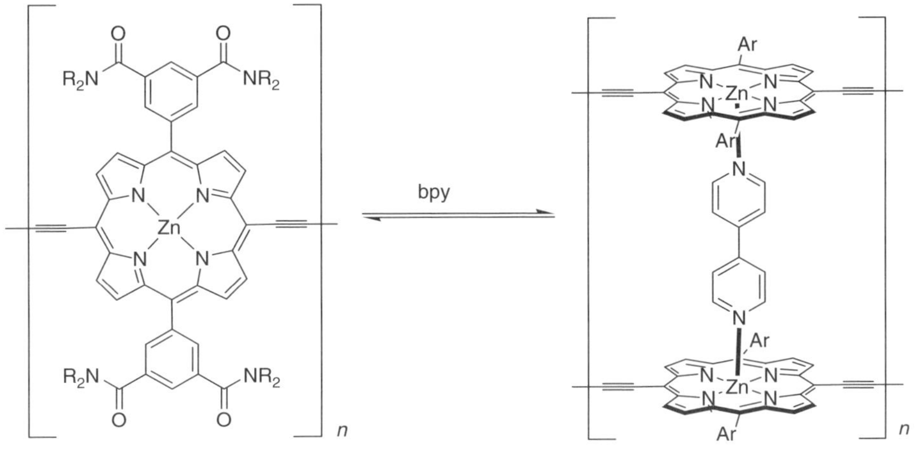

14 **15**

Scheme 4.7 Polymer **14** reversibly forms ladder polymers **15** in the presence of bpy.

There have been several studies of PPEs incorporating Zn^{II} porphyrins into the backbone. Using oligomeric models, Therien has demonstrated that the optical and potentiometric bandgaps of these polymers can be independently modulated.[34] High molecular-weight PPEs **17a** and **17b** incorporating Zn^{II} porphyrin in every repeat unit were prepared by Jones using a Sonogashira coupling strategy (Scheme 4.8).[35] It was necessary to use long substituents and mesomesityl groups in the monomers to render these polymers soluble. Molecular weights were determined by GPC ($M_n \sim 2.5 \times 10^4$), but could not be measured by light scattering due to absorption. Chen has also explored porphyrin-based PPE polymers, but utilized much larger groups to suppress interchain facial interactions that hinder solubility.[36] Using acyclic diyne metathesis (ADIMET) polymerization with a $Mo(CO)_6$ catalyst, they prepared PPE polymers incorporating 10–20% porphyrin; it is unclear whether or not metals were incorporated into the polymer before or after polymerization.

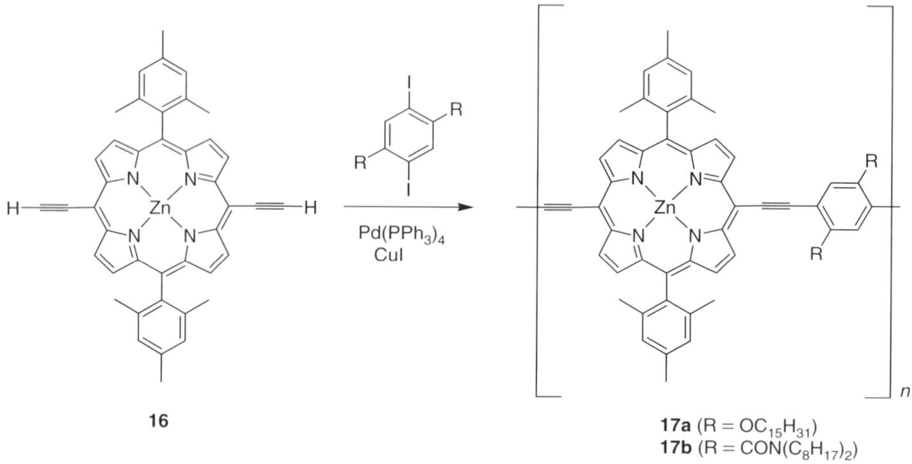

Scheme 4.8 Synthesis of PPEs containing Zn^{II} porphyrins.

Krebs has developed a method to incorporate a single Zn^{II} porphyrin into each polymer chain by coupling a dibromoporphyrin (**18**) with a bifunctional monomer (**19**), as shown in Scheme 4.9.[37] The porphyrins have very bulky *m*-isopropylbenzene substituents that prevent cofacial interactions between polymer strands and increase solubility. It was necessary to use a dibromometalloporphyrin rather than a dialkynylmetalloporphyrin to ensure that a single

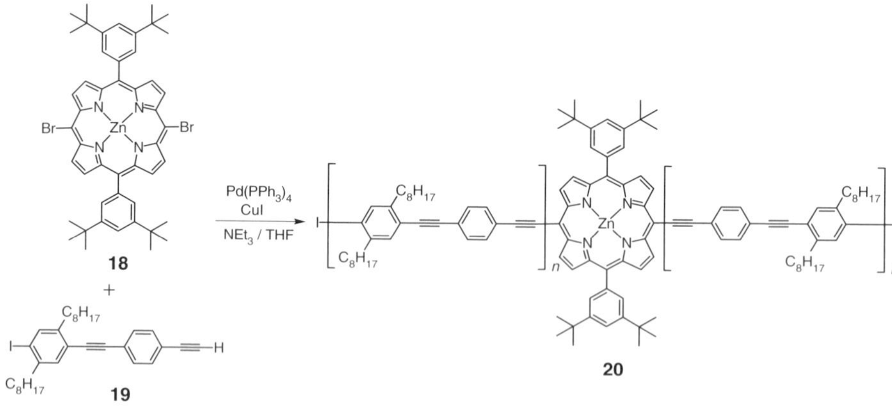

Scheme 4.9 Synthetic route to PPEs **20** containing only one porphyrin per polymer chain.

porphyrin was incorporated into the chain. If the dialkyne is used, butadiyne linkages form in the backbone, and multiple Zn^{II} porphyrins are incorporated into the backbone. This strategy was effective for forming medium molecular-weight ($M_w \sim 11{,}500$ Da; GPC) polymers. Fluorescence studies of **20** showed virtually no energy transfer to the Zn^{II} porphyrin center in solution. In the solid state, however, the energy transfer is nearly quantitative, with all of the emission coming from the dilute porphyrin species. Thin films of **20** electroluminesce in the near-IR, indicating effective exciton migration to the porphyrin centers.[38]

Surprisingly, in spite of the attractive electroluminescent properties of PPV and its derivatives, there have been few efforts to incorporate porphyrins into PPVs. In 1997, Jones reported the incorporation of 1–7 wt % Zn^{II} or Al^{III} porphyrin into a PPV backbone from the reaction of bis(halomethyl)porphyrins **21** and 2,5-bis(chloromethyl)-1,4-dihexyloxybenzene **22** (Scheme 4.10).[39] Lo and Burn have generated insoluble PPV-metalloporphyrin hybrids from a Zn^{II} porphyrin **24** (Scheme 4.11).[40] The porphyrins were incorporated as sidegroups decorated with

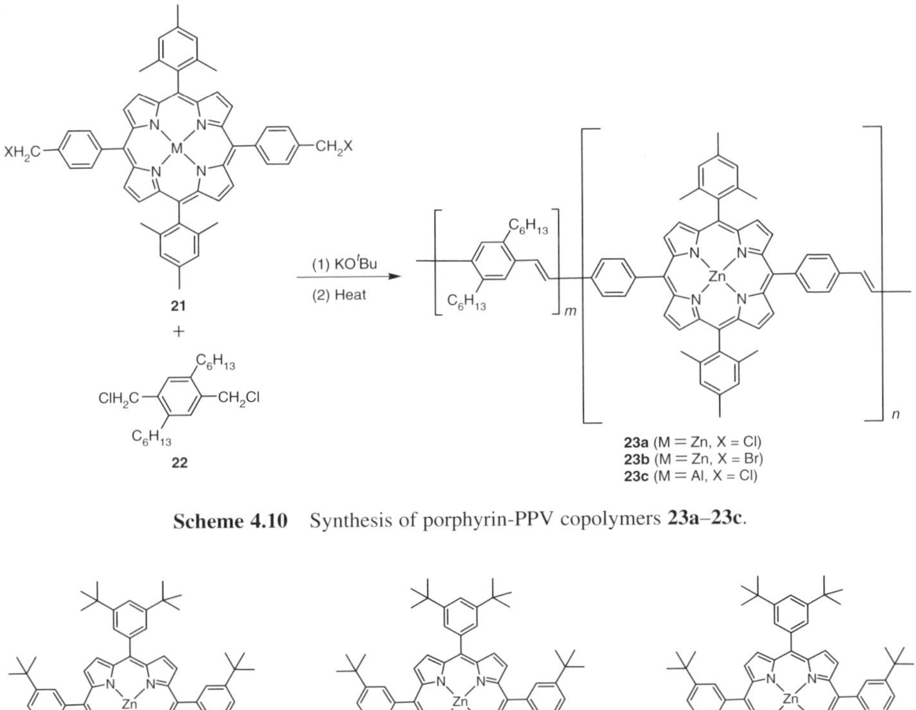

Scheme 4.10 Synthesis of porphyrin-PPV copolymers **23a–23c**.

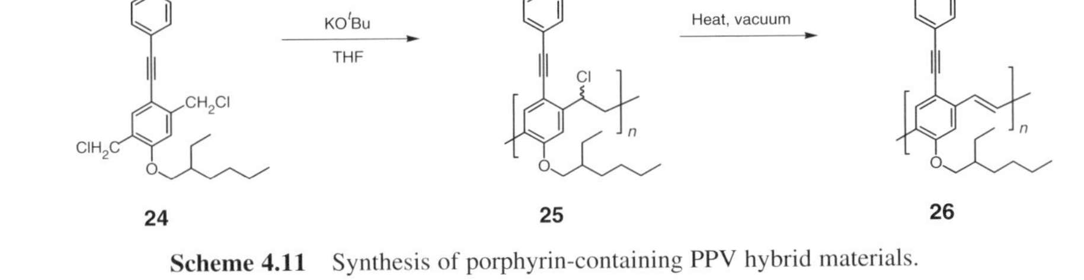

Scheme 4.11 Synthesis of porphyrin-containing PPV hybrid materials.

m-tert-butylphenyl substituents. The soluble precursor polymer **25** exhibited cooperative binding to bidentate ligands, suggesting the formation of a ladder polymer (analogous to PPEs prepared by Anderson). After heating to eliminate HCl, polymer **26** exhibited electroluminescence.

Lindsey has reported PPE **27** and polybutadiyne **28** incorporating Zn^{II} tetraphenylporphyrins, prepared by Sonogashira and Glaser coupling, respectively.[41] Perylene dye substituents were coupled via alkynyl linkages to the porphyrin to enhance the light harvesting ability of the polymers. Substantial molecular weights were obtained for **27** and **28**. Surprisingly, attempts to produce PPEs with metalloporphyrin moieties via Suzuki coupling failed to give even long oligomers. Preliminary studies of the optical properties of these polymers showed nearly quantitative energy transfer from the perylene to the Zn^{II} porphyrin.

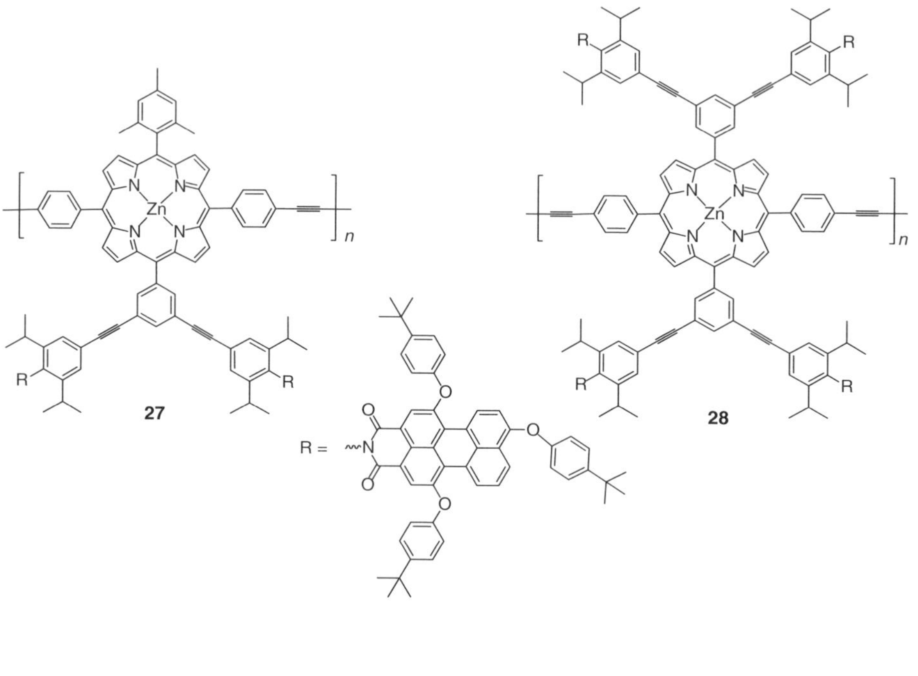

Yamamoto has synthesized a variety of zinc porphyrin–containing polymers (**29–35**) with arylene and aryleneethynylene linkers.[42] Arylene groups incorporated included 2,5,-dialkoxy-1,4-phenylene, 3-alkyl-2,5-thiophenyl, and 2-alkyl-3,6-pyridenyl species. Due to their long substituents, these polymers are soluble in THF and have molecular weights of 4600–38,000 Da (M_n, GPC). Powder X-ray diffraction revealed some degree of order in the thin films prepared from polymers with long alkyl substituents. Photoluminescence, thermal analysis, and CV of the new polymers were also reported.

Although a number of polymers incorporating porphyrins have been reported, much research remains in this area. Few of the studies have taken advantage of the metal for any purpose, such as catalysis. Moreover, the studies have been limited to Zn^{II} porphyrins (with the exception of Anderson's Pb and Cu porphyrin–containing polymers). Incorporating other metals with interesting electrochemistry and catalytic function will be a rich area of exploration.

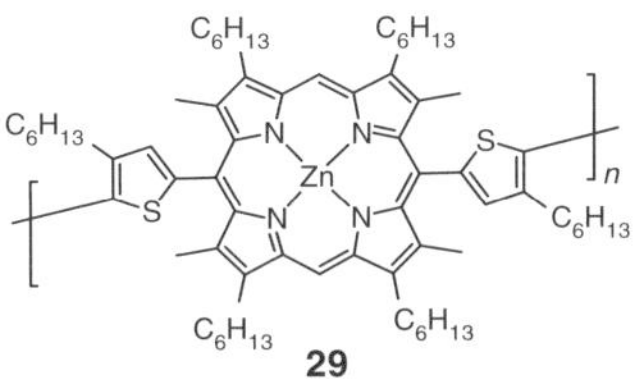

29

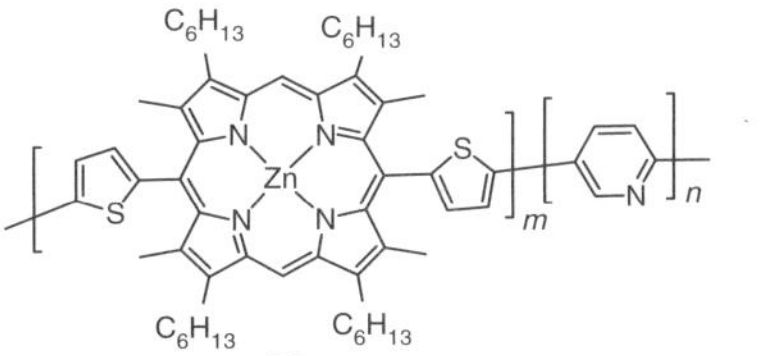

30

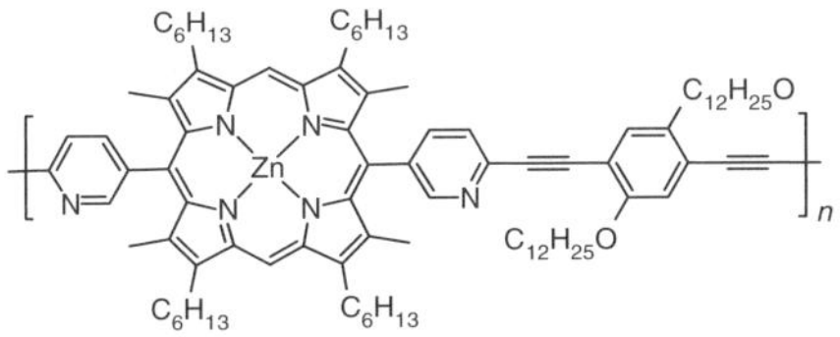

31

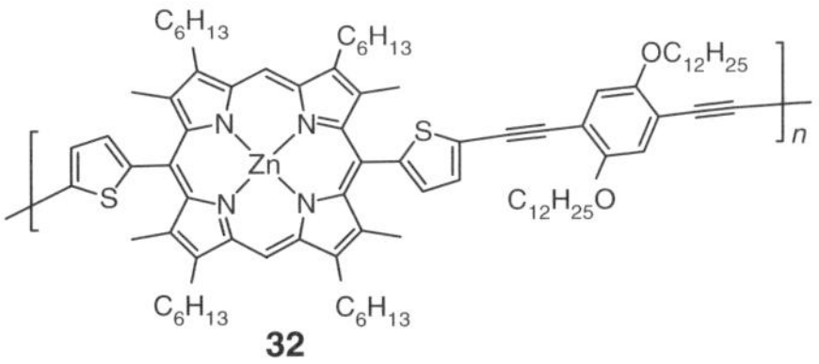

32

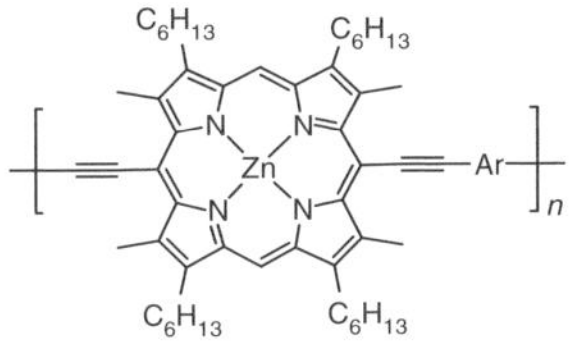

33–35

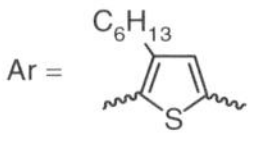

33

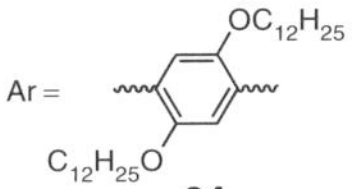

34

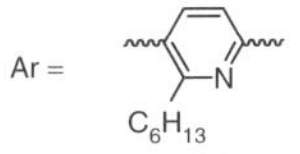

35

III. METAL-CONTAINING POLYTHIOPHENE-BASED POLYMERS

Polythiophenes are low-bandgap polymers that can be prepared by electropolymerization or C–C bond coupling reactions. These polymers are luminescent in their neutral state, and can exhibit high conductivities when oxidized. Metal-containing polythiophenes can be formed by polymerization of thiophene-derivatized metal complexes, or by incorporation of ligands into polythiophene, followed by coordination to metals. Each method has its own challenges: polymerization of metal-containing monomers often requires careful tuning of polymerization procedures, but guarantees complete metalation. Adding the metal after polymerization may result in less than complete incorporation. Both methods have been used to create metal-containing polythiophenes.[43]

As with other polymers, the metal complex may be part of the polymer backbone, or attached to the backbone by a conjugated linker. Investigations into the conjugated polymers have been assisted by studies of model complexes.[44] Polythiophenes incorporating organometallic complexes are discussed first, followed by those incorporating other coordination complexes with bipyridine, terpyridine, Schiff base, and other ligands.

A. Organometallic Polythiophenes

Ferrocene is an attractive and convenient organometallic species to incorporate into conjugated polymers—it is stable, easy to functionalize, has reversible electrochemistry, and its derivatives (both monomeric and polymeric) can be easily characterized [e.g., by nuclear magnetic resonance (NMR) spectroscopy]. Ferrocene has been incorporated into the backbone and side chains of polythiophene.[45] In 1999, both Higgins and Wolf independently reported the synthesis of polythiophenes with ferrocene in the backbone.[46,47] Electropolymerization of ferrocene substituted with two bithienyl, terthienyl, or dialkylterthienyl moieties (**36–38**, **41**) led to conjugated polythiophenes **39**, **40**, and **42** interrupted by ferrocene groups in the backbone (Scheme 4.12). Interestingly, both groups found that 1,1′dithienylferrocene **36** did not undergo electropolymerization. Polymer **39** grown on indium tin oxide (ITO) was investigated by spectroelectrochemistry. At −0.1 V [vs. standard calomel electrode (SCE)], the polymer is completely reduced and the polythiophene π–π* transition is observed at ca. 450 nm. Upon oxidation at 0.8 V, the oligothiophene segment's π–π* transition is red-shifted to ca. 500 nm, and a polythiophene→Fe^{III} charge-transfer band appears at 1395 nm. At higher potentials, the π–π* band is broadened and red-shifted and the charge-transfer band disappears, revealing that the oligothiophene segments and ferrocenyl groups are both oxidized.

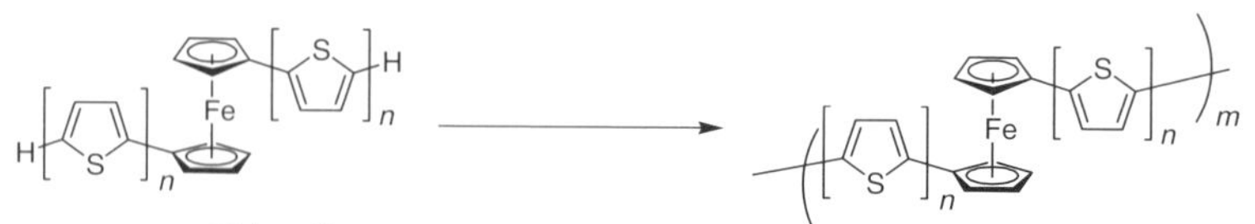

36 ($n = 1$)
37 ($n = 2$)
38 ($n = 3$)

39 ($n = 2$)
40 ($n = 3$)

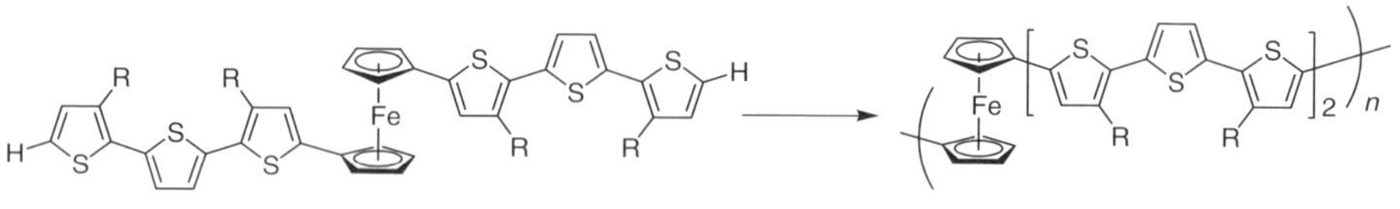

41 ($R = n = C_6H_{13}$)

42 ($R = n = C_6H_{13}$)

Scheme 4.12 Synthesis of ferrocene-containing polythiophenes **39**, **40**, and **42**.

Curtis has employed Ni^0 catalyzed coupling of bromothiophene derivatives **43** and **44** to synthesize polythiophene with pendant ferrocene groups (**44**, **45**) linked via vinylene and cyanovinylene bridges (Scheme 4.13).[48] The polymers were soluble in organic solvents and had molecular weights of ~16,000 to 22,000 Da (M_n, GPC). ^1H NMR studies suggest that the polymers are regiorandom, and include a mixture of E and Z isomers for the olefin. Upon p-doping the polymer with I_2, charge-transfer bands were observed at ca. 1700 nm. Fluorescence quenching of the polymer was attributed to electron transfer from the ferrocene substituents to the exciton in the conjugated polymer, though a Förster mechanism could not

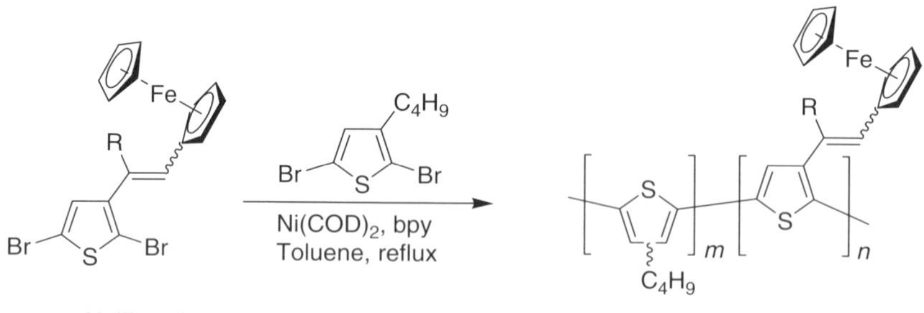

43 (R = H)
44 (R = CN)

45 (R = H)
46 (R = CN)

Scheme 4.13 Synthesis of polymers **45** and **46**.

be dismissed. The CV of the polymers revealed separate oxidation waves for the polymer and ferrocene substituents, with no evidence of metal–metal interactions. When incorporated into p/n photocells, these hybrid polymers exhibit higher photoconductivities than the parent polythiophene, likely due to efficient hole hopping between ferrocene groups.

Wolf and co-workers have synthesized phosphine-functionalized polythiophenes that are coordinated to Pd^{II} by P,C coordination. Electropolymerization of dimer **47** gave the first conductive ($\sim 10^{-3}$ S cm^{-1} when oxidized) polythiophene derivative with a metal σ-bonded to the conjugated backbone (Scheme 4.14).[49] Although the Pd^{II} dimers are intact during polymerization, they can be disrupted in **48** by adding an isocyanide ligand to increase the polymer's solubility. Using model complexes, it was demonstrated that the polymerization takes place exclusively at the α (2) positions of thiophene. Notably, the spectroscopic properties of **48** were not significantly different from polythiophene, suggesting there is little interaction between Pd and the conjugated polymer. This conclusion was supported by the observation that the addition of isocyanide ligands to **48** had little effect on its absorption spectrum.

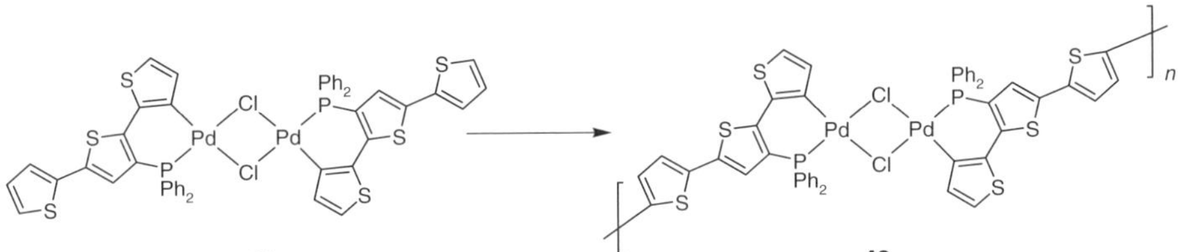

47 **48**

Scheme 4.14 Electropolymerization of Pd^{II} dimer **47** gives polymer **48**.

Continuing their investigations of conjugated polymers incorporating (η^5-cyclopentadienyl)(η^4-cyclobutadienyl)cobalt complexes, Endo has discovered a new route to polythiophenes that involves reaction of organocobalt polymers with S_8 (Scheme 4.15).[50] When polymer **49**

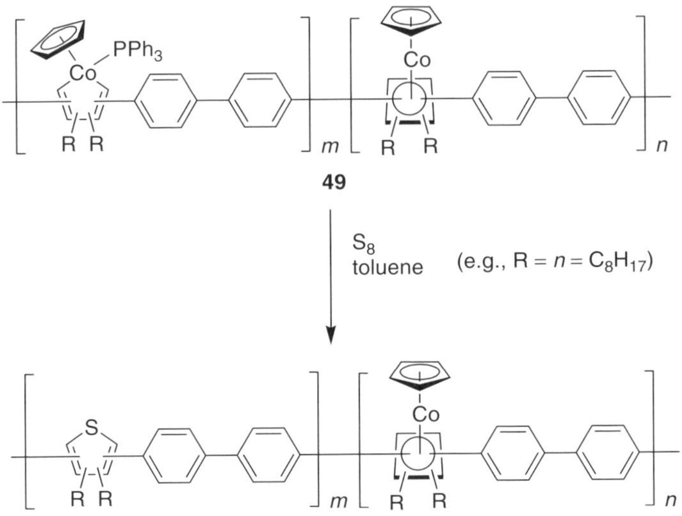

Scheme 4.15 Preparation of polymer **50**.

(various R groups were employed) was reacted with S_8, polythiophene/cobalt complex hybrid polymers formed. These polymers had molecular weights of 11,000–14,000 Da (M_n, GPC), showed no glass transition temperature [differential scanning calorimetry (DSC)], and were thermally stable to nearly 400°C. Although UV-vis spectra and cyclic voltammograms were measured for the polymers, no conclusions could be made about the extent of metal interaction with the polymer backbone.

Like the polymers just discussed, where the metal is coordinated to a conjugated ring in the backbone of the polymer, azulene groups have been incorporated into polythiophene for the purpose of coordinating to a metal carbonyl cluster.[51] Thus, azulene-containing polymers **53** and **54** with $M_n \sim$ 25,000–31,500 Da (GPC) were prepared by oxidative coupling of 1,3-bis(dialkylthienyl)azulene, **51**, or 1,3-bis(dialkoxythienyl)azulene, **52** (Scheme 4.16).[52] Subsequent reaction with $Ru_3(CO)_{12}$ gave polymers partially or fully metalated with ruthenium carbonyl cluster, depending on the stoichiometry of the reactants. Interestingly, coordination of ruthenium was accompanied by a blue shift in the absorption spectrum. This shift is attributed to the disruption of diatropicity of the azulene ring upon coordination to ruthenium, breaking the conjugation in the polymer, and thus increasing the HOMO–LUMO gap. Synthesis of a model oligomer showed that di-, tri-, and tetranuclear clusters were coordinated to the azulene. Nevertheless, crystallographic studies of the major product revealed a distorted tetrahedral ruthenium carbonyl coordinated by azulene. One of the Ru centers is coordinated in η^5 mode to the 5-membered ring, while two Ru centers are coordinated to the 7-membered ring.

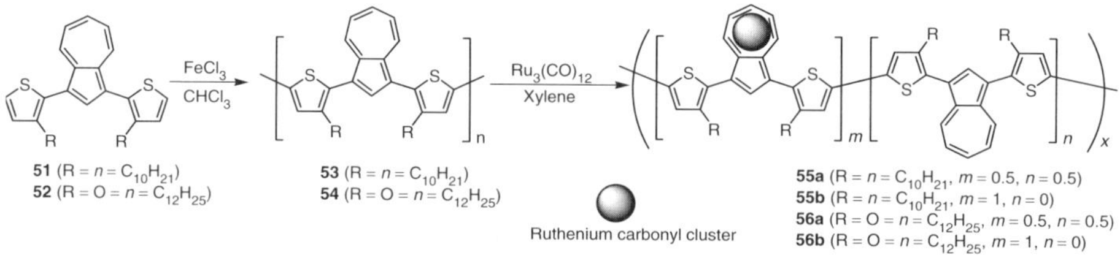

Scheme 4.16 Preparation of polymers **53–56**.

Although this may be the first conjugated polymer that has coordinated polynuclear metal complexes after polymerization, metal clusters have been incorporated into conjugated polymers in other ways. Shin has grafted clusters of both organomolybdenum oxide and $Co_3(CO)_9$ onto terthiophene (**57**, **59**), and demonstrated that these electropolymerize to metal-containing thin films of polythiophene (Scheme 4.17).[53,54] CV of polymers **58** and **60** reveal a strong electronic interaction between the metal clusters and the conjugated polymer backbone. Although the oxidation wave of the Mo_2 cluster could not be distinguished from that of the polythiophene in **58**, a reduction wave could be assigned to the metal. This polymer film exhibited strong electrochromicity, cycling between purple (undoped) and blue (doped), a color change different from that of polythiophene. The redox behavior of the polymer and cobalt carbonyl components of **60** were both different than for the separate components, indicating a substantial interaction. Organomolybdenum sulfide clusters **61** (R = H, Ph, butyl, thienyl, tolyl) appended with two terthiophene groups were also electropolymerized to afford electrochromic thin films (polymer **62**) that showed significant metal-conjugated polymer interactions by CV.[55]

B. Polythiophenes Containing $M(terpy)_2$ Groups

Although this chapter does not include the significant body of work on the condensation polymerization of bis(terpy) ligands with metal ions to form polymers with $M(terpy)_2$ linkages (e.g.,

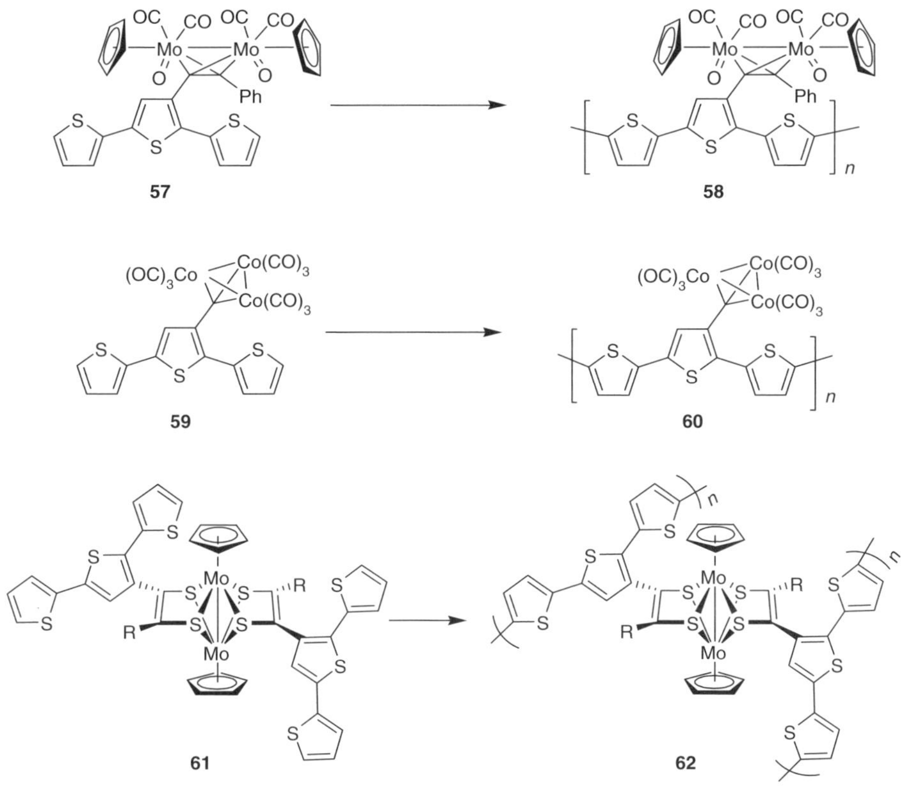

Scheme 4.17 Preparation of polythiophenes **58**, **60**, and **62** incorporating metal cluster side groups (R = H, Ph, butyl, thienyl, tolyl). Note that there is a lower degree of stereoregularity in **61** and **62** than indicated.

Rehahn, Schubert), it is appropriate to include examples here where M(terpy)$_2$ groups interrupt polythiophene. Ru(terpy)$_2$ functionalized with thienyl substituents can be electropolymerized to form a thin film (Scheme 4.18).[56] The metal-to-ligand charge transfer (MLCT) band of polymer **69** was observed at 521 nm, red-shifted from the MLCT transition observed for monomer **63**

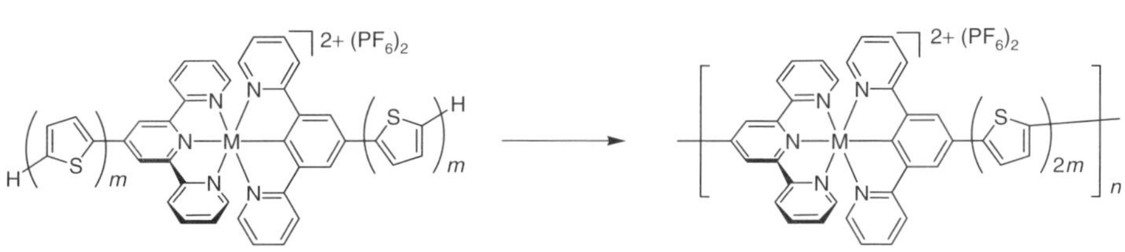

63 (M = Ru, *m* = 1)
64 (M = Ru, *m* = 2)
65 (M = Ru, *m* = 3)
66 (M = Os, *m* = 1)
67 (M = Os, *m* = 2)
68 (M = Os, *m* = 3)

69 (M = Ru, *m* = 1)
70 (M = Ru, *m* = 2)
71 (M = Ru, *m* = 3)
72 (M = Os, *m* = 1)
73 (M = Os, *m* = 2)
74 (M = Os, *m* = 3)

Scheme 4.18 Preparation polythiophene/M(terpy)$_2$ hybrid polymers incorporating Ru^{II} and Os^{II} centers.

(498 nm) due to extended conjugation. Electropolymerization of bithienyl-functionalized $Os(terpy)_2$ complexes gave thin films with conductivities of $0.5 - 1 \times 10^{-4}\,S\,cm^{-1}$.[57] Charge transport in interdigitated electrodes coated with polymer **69** was ~2 orders of magnitude greater than in structurally related systems with nonconjugated bridges. In addition to charge-transport dynamics, the authors have also studied the dynamics of electropolymerization of these complexes.[58]

Recently, Constable and Forster reported the polymerization of both $[Ru(terpy)_2]^{2+}$ and $[Os(terpy)_2]^{2+}$ complexes **63–68** functionalized with thienyl, bithienyl, and terthienyl substituents (Scheme 4.18).[59] Thin films of the quaterthienyl-bridged polymers had the highest conductivities ($\sigma \sim 1.6 \times 10^{-3}\,S\,cm^{-1}$) for both M = Ru^{II} (**70**) and Os^{II} (**73**). This may be due to a close match between the metal's redox potential and the electronic states of the thienyl-bridge revealed by conductivity measurements made during cyclic voltammetry of the films. These results indicate that redox conduction (charge hopping) is the dominant mechanism of charge transport.

C. Polythiophenes Incorporating Metalated Bipyridine Ligands

Ruthenium(II) bipyridine complexes are also attractive components of conducting polymers. The synthesis of such polymers is facilitated by the stability and well-understood molecular chemistry of $[Ru(bpy)_3]^{2+}$. In addition, the photochemical, photophysical, and redox properties of $[Ru(bpy)_3]^{2+}$ have been thoroughly investigated. It is no surprise that $[Ru(bpy)_3]^{2+}$ has been incorporated into a large number of conjugated polymers, including polythiophenes. These studies have given important information about the extent of coupling between metal-based and polymer energy states, as well as the mechanism of charge transport in conducting metallopolymers.

Swager has reported the synthesis and electropolymerization of $[Ru(bpy)_3]^{2+}$ complexes appended with multiple bithienyl groups.[60] The bithienyl substituents were placed at the 4,4′

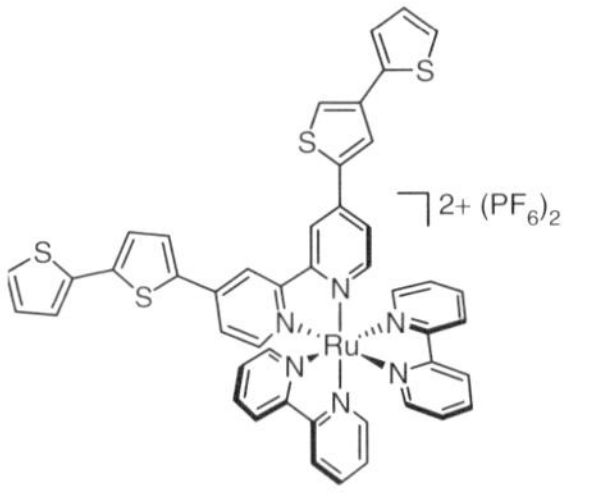

75

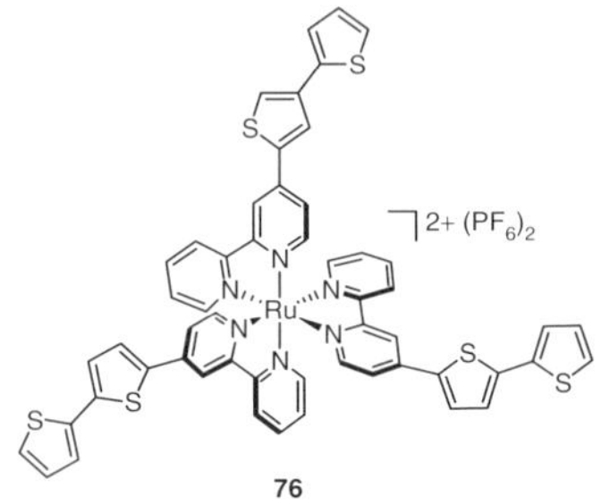

76

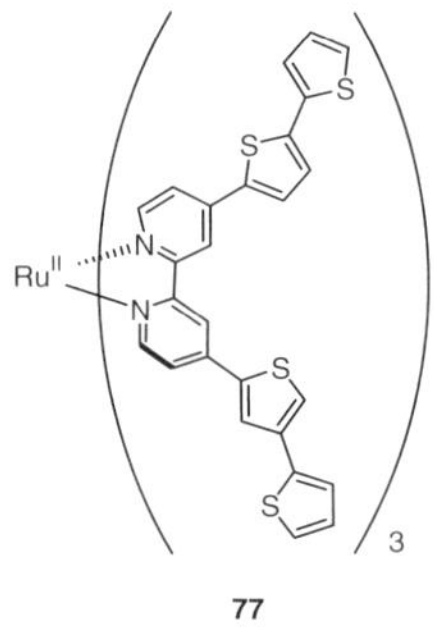

77

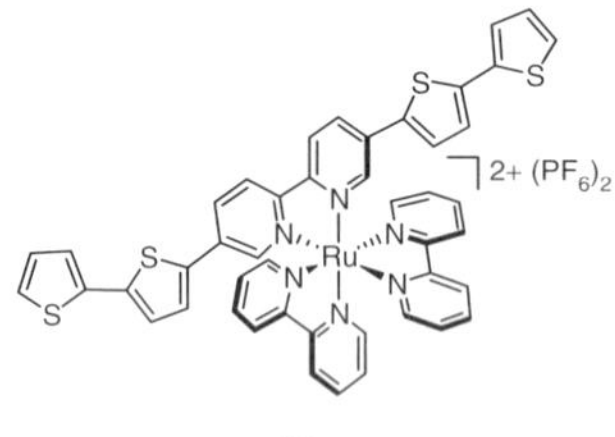

78

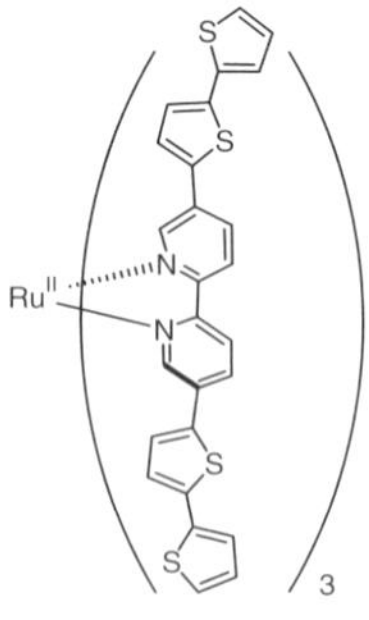

79

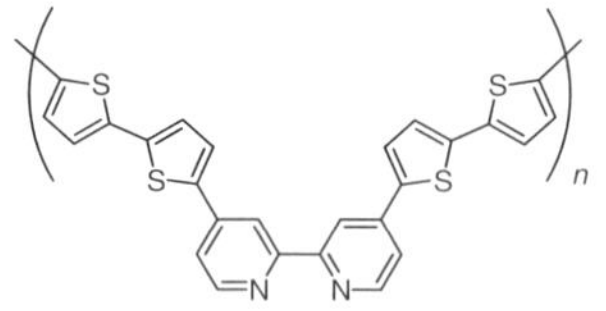

80

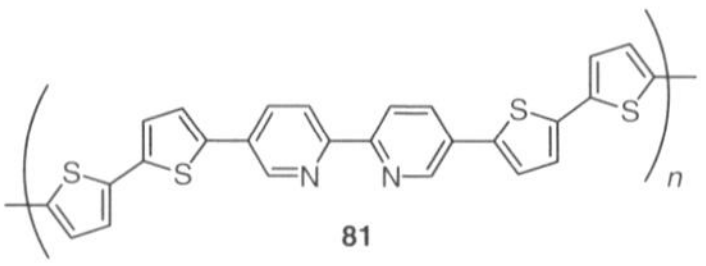

81

positions of the bpy ligand (as in **75–77**), or the 5,5′ positions (as in **78** and **79**). These complexes were electropolymerized and compared with metal-free analogs, **80** and **81**. The conductivity of the polymers was affected by both the substitution pattern of the metal complex and the degree of polymer cross-linking. Although the metal-free polymer **80** was insulating, polymer **77** metalated with $[Ru(bpy)_2]^{2+}$ showed conductivities up to $3.3 \times 10^{-3}\,S\,cm^{-1}$ during redox processes. Swager found that bpy's substituted at the 4,4′ positions were more effectively coupled to Ru^{II}, enhancing conductivity. Polymers prepared from **76**, where conjugation is interrupted by ruthenium centers, also exhibited moderate conductivities (up to $1.7 \times 10^{-3}\,S\,cm^{-1}$) due to redox coupling between the metal and the conjugated polymer.

Zotti and Berlin used electropolymerization to generate thin films of organic polymer **82** with a backbone of alternating bpy/bis(EDOT) groups.[61] Upon complexation with Fe^{II}, Co^{II}, Ni^{II}, and Cu^{II}, the polymers exhibit large ionochromic and redox changes, suggesting their application in ion sensors.

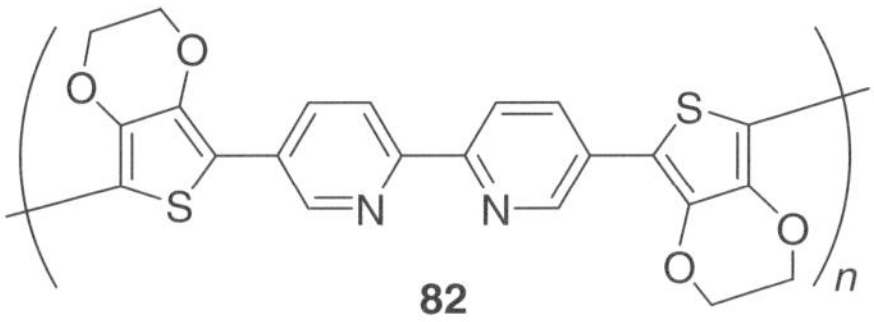

82

Figgemeier reported the synthesis and electropolymerization of a series of ruthenium complexes functionalized with ethenylthiophene groups, **83–85**.[62] Interestingly, the MLCT bands of the ruthenium complexes were shifted to shorter wavelengths after polymerization, suggesting conjugation is interrupted in the polymers. This may be due to free-radical coupling of ethenylthiophenes to give a reduced ethenyl group in the polymer.

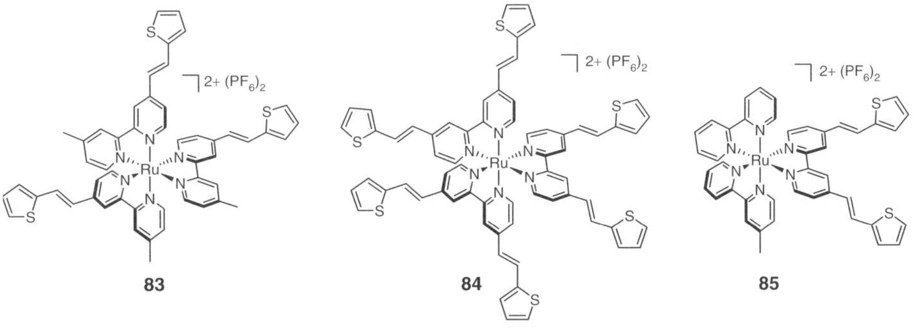

83 **84** **85**

Although most metal-containing polythiophenes have been synthesized by electropolymerization on an electrode surface, there are many reasons to chemically synthesize these polymers. Chemical synthesis may allow isolation of soluble polymers, enabling complete solution characterization (GPC, light scattering, NMR, etc.) and facilitating conductivity studies. Moreover, it can enable improved thin-film preparation and film deposition onto nonconducting substrates. Finally, monomers that are unsuitable for electropolymerization may be polymerized by chemical methods.

Guillerez synthesized polymer **86** with alternating bpy and 3-octylquaterthiophene units.[63] Subsequent reaction with $Ru(bpy)_2Cl_2$ or $[Ru(bpy)_2(EtOH)_2](PF_6)_2$ gave polymers **89** with incomplete metalation as shown by method A in Scheme 4.19. However, Stille coupling of 5,5‴-bis(trimethylstannyl)tetraoctylquaterthiophene **87** with [(5,5′-dibromo-2,2′-bipyridyl)Ru(bpy)$_2$](PF$_6$)$_2$ **88** gave polymer **89** in good yield. This polymer had a molecular weight of 70,000 Da (M_w, light scattering in DMSO) and was soluble in a variety of polar organic solvents. Notably, the absorption of the polymer was red-shifted relative to the metal-free polymer or to poly(3-octylthiophene). The authors attribute this red shift to an extension of the π-conjugation resulting from geometric factors—the two pyridyl rings of bpy remain coplanar when coordinated to a metal and therefore increase conjugation between the oligothiophene segments. By making a derivative with 6 octylthiophene bridges (rather than 4), the authors were able to identify three separate electrochemical events at the polymer and metal centers, and to confirm that complexation to Ru^{II} reduced the electrochemical gap of the polymer.

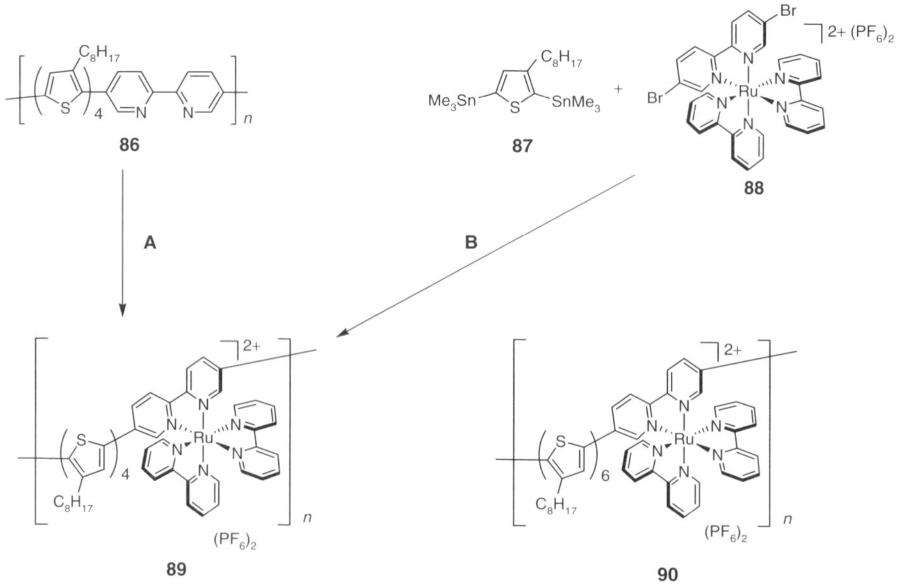

Scheme 4.19 Two synthetic routes to polymer **89**. Poly[hexa(3-octylthiophene)] analog **90** was prepared by method **B**.

A thorough electrochemical and photophysical investigation of Ru^{II} polymers **89** and **90** as well as the Os^{II} analog of polymer **89** was undertaken.[64] Luminescence lifetime measurements and transient absorption spectroscopy showed that luminescence from the MLCT transition of the metal complex was reduced by interaction with a low-lying $^3\pi,\pi^*$ state of the conjugated polymer. This result provides guidance to designing highly luminescent metal–organic materials: it is necessary to tune the polymer such that the energy of the $^3\pi,\pi^*$ state lies above the energy of the MLCT. Nevertheless, preliminary investigations of these polymers indicate that they are electroluminescent with a turn-on voltage around 10 V.[65] The transient absorption spectra of the oxidized polymers showed polarons on the poly(3-alkylthiophene) units.[63]

An electron spin-resonance (ESR)/UV-vis study of the Ru-containing polymer **89** has given insight into the charge carriers of the metalated polymers.[66] If the polymer was oxidized beyond the metal-based redox event, the polymer underwent irreversible decomposition. However, when the polymer was p-doped at lower potential, both the metalated and metal-free polymers exhibited sharp ESR lineshapes characteristic of mobile charge carriers (polarons) on the polymer. Upon further oxidation, the polarons in the poly(3-alkylthiophene) segments of **90** recombined to form bipolarons; these charge carriers were not observed when the $[Ru(bpy)_3]^{2+}$ complexes were bridged by shorter segments (polymer **89**). Clearly, the longer conjugation length in polymer **90** is important for stabilizing the bipolaron. In the reduced state (n-doped), the ESR signals of the metallopolymer are broadened due to spin-orbit coupling of the metal center, indicating that charge injection occurs preferentially on the bpy ligand. A SNIFTIRS (subtractively normalized interfacial Fourier-transform infrared spectroscopy) study of polymer **89** (and derivatives not shown here) indicated that $[Ru(bpy)_3]^{2+}$ behaves as an electronic gate for the injection of charge into the polythiophene.[67] Moreover, this technique provided evidence of communication between the conjugated polymer and the metal complex.

D. Rotaxanated Polythiophenes

Rotaxanes are molecules with a thread mechanically locked within a macrocycle. One common synthesis of these molecules involves the use of transition metal templates, and naturally leaves ligands (usually bpy or phenanthroline) appended to the structure.[68] By decorating these molecules with polymerizable thiophene or oligothiophene groups, researchers have made rotaxanated metal-conducting polymers.[69] This is a relatively new field and demonstrates that combining supramolecular chemistry with metal-containing conjugated materials offers unique opportunities to create materials.

Using a Cu^I templating strategy, Sauvage synthesized rotaxane **93** that combines macrocycle **92** with bithienyl-functionalized 1,10-phenantholine (phen), **91** (Scheme 4.20).[70] After electropolymerization, a thin red-orange film of polymer **94** was obtained. Electrochemical characterization of the polymer confirmed that the bithienyl groups were coupled, and that Cu^I was present in the polymer. Remarkably, the polymer could be demetalated in the presence of

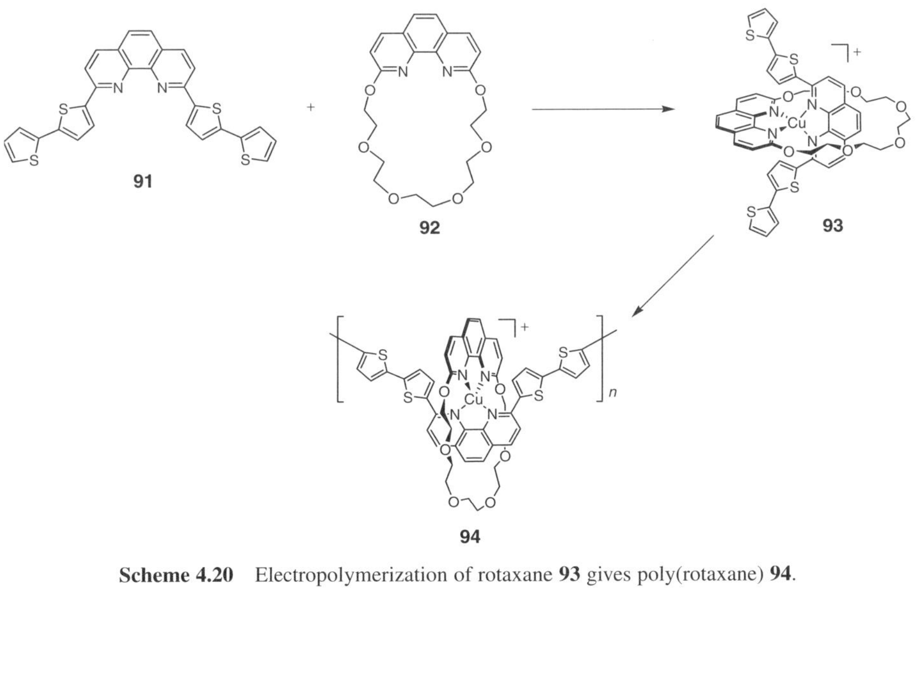

Scheme 4.20 Electropolymerization of rotaxane **93** gives poly(rotaxane) **94**.

Li^+ salts with conservation of the rotaxanated topography. This was proved when Cu^I was nearly quantitatively reincorporated into the polyrotaxane. An X-ray absorption spectroscopy (XAS) study confirmed that the Cu^I is in a similar environment to model compounds, but that the symmetry of the coordination sphere is reduced due to steric constraints of the polymer matrix.[71] Attempts to replace Cu^I with Co^{II} or Zn^{II} using the Li^+ intermediate, or by electropolymerizing rotaxane **93** with Co^{II} or Zn^{II} in the place of Cu^I failed.[72]

Sauvage also reported analogous rotaxane structures (**95**, **96**) where the bithienyl groups are incorporated into the 3,8 positions, rather than the 2,9 positions (i.e., **93**), of phen. These rotaxanes were also electropolymerized to generate thin films of polyrotaxanes **97** and **98** (Scheme 4.21). When the phen ligand was modified with methyl groups in the 2,9 positions, as in **96**, the stability of the polyrotaxane (**98**) was significantly enhanced.[73] This polymer was stable to multiple cycling of the Cu^I/Cu^{II} oxidation. Moreover, unlike polymers **94** and **97**, polymer **98** could be remetalated with Co^{II}, Zn^{II}, or Cu^I after demetalation in the presence of Li^+ (Scheme 4.22). Divisia-Blohorn et al. have undertaken in situ ESR and conductivity studies of polyrotaxanes **97–100**.[74] The individual components of the polymer showed widespread redox activity, indicating charge localization in the polymer; the metal introduced barriers to charge carrier mobility. The maximum conductivity of the metallopolymer was observed at the potential

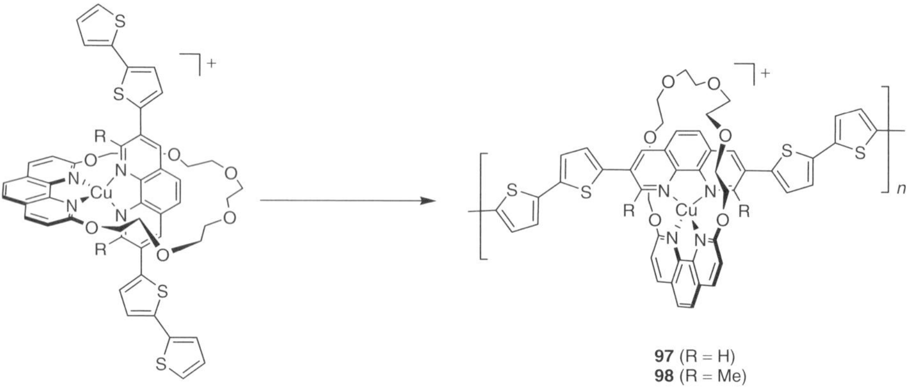

95 (R = H)
96 (R = Me)

97 (R = H)
98 (R = Me)

Scheme 4.21 Electropolymerization of rotaxanes **95** and **96** gives polyrotaxanes **97** and **98**, respectively.

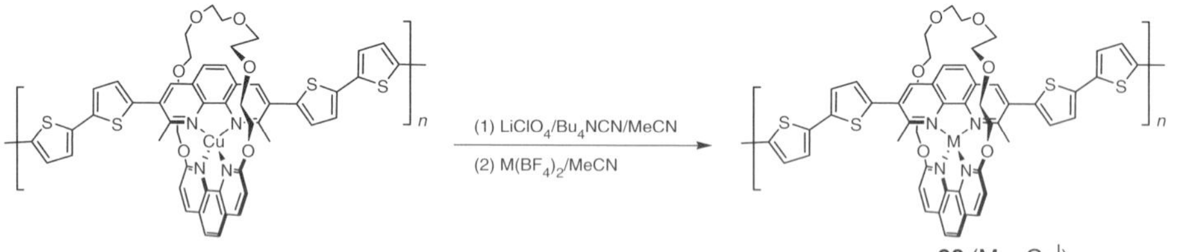

98

98 (M = Cu^I)
99 (M = Co^{II})
100 (M = Zn^{II})

Scheme 4.22 Lithium-mediated transmetalation of polyrotaxane **98** gives access to Co^{II} and Zn^{II} analogs **99** and **100**.

corresponding to the Cu^{I}/Cu^{II} couple, indicating that the mixed valence state was necessary for conduction. The maximum conduction for the metalated polymer was $2 \times 10^{-5}\,S\,cm^{-1}$.

Sauvage has also investigated the electropolymerization of thienyl-substituted phens **91** and **101** entwined around metal centers as shown in Cu^{I} complexes **102** and **103**.[75] Although the products of the electropolymerization are not rotaxanes, it is likely that macrocycles and catenanes are present within the cross-linked polymer matrix. Electropolymerization of the single thienyl-functionalized phen complex **102** gave only metal-free conjugated polymer. On the other hand, electropolymerization of the bithienyl-functionalized phen complex, **103**, generated a thin film containing Cu^{I}. The polymer showed separate redox waves for the Cu centers and for the conjugated polymer backbone. As removal of the metal from this polymer resulted in irreversible collapse of the structure, these researchers prepared the same polymer with dihexylterthienyl spacers.[76] Starting with precursor **104**, the authors prepared complex **105** incorporating Cu^{I}, then electropolymerized to form a stable film. In this case, the polymer could be demetalated without destruction of the polymer film, as the hexyl substituents prevent disentanglement of the polymer chains and retain the overall superstructure of the conjugated polymer. The binding environment of the Cu^{I} in the polymer was probed by XAS. In situ electrochemistry measurements revealed that poly(**103**) exhibits no metal–metal interactions, and the Cu^{I}/Cu^{II} redox process showed very little effect on the conductivity of the polymer. The

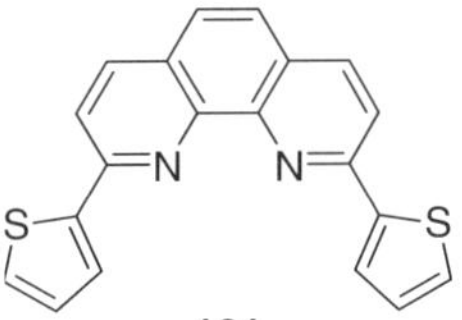

101

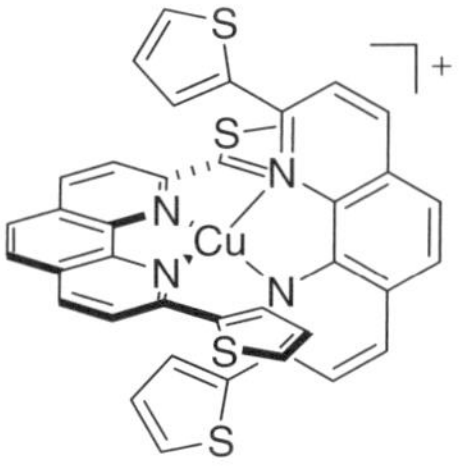

102

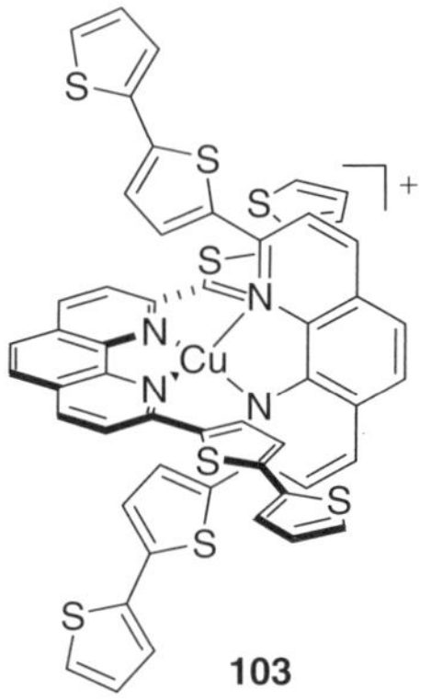

103

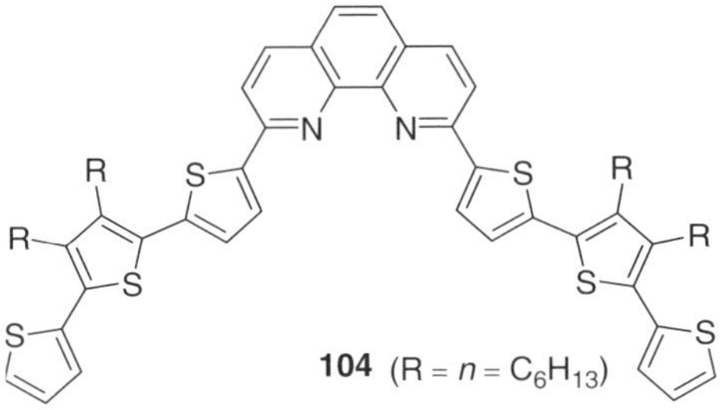

104 (R = n = C_6H_{13})

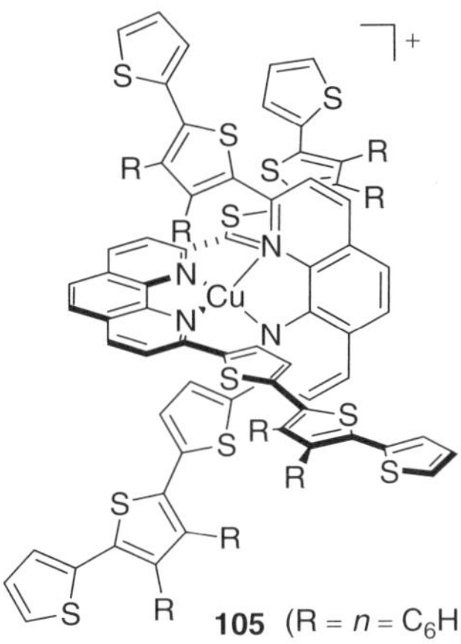

105 (R = n = C_6H_{13})

conductivity of poly(**105**), however, was strongly affected by the oxidation state of the metal, indicating a significant interaction between the metal complex and the π-conjugated polymer. At best, the conductivity of the polymer was $1.1 \times 10^{-4}\,S\,cm^{-1}$, indicating that in spite of the redox matching with electronic states of copper, poly(**105**) is a relatively poor conductor.

In all of the polyrotaxanes just discussed, the thread through the macrocycle formed part of the conjugated backbone. Buey and Swager made polyrotaxane ladder polymers, where one conjugated strand is formed by the thread and the other is formed with the macrocycle.[77] By placing EDOT groups on the macrocycle, these could be selectively polymerized before the threads were oxidatively coupled at higher potential. Thus, electropolymerization of pseudorotaxane, **106**, gave conjugated metallopolymer **107**, as shown in Scheme 4.23. Although molecular modeling suggested that a 3-strand ladder polymer could form, it is more likely that a disordered polymer matrix is formed. The use of spectroscopic techniques may help to probe the extent of order in these polymers. Nevertheless, the authors carried out conductivity and CV measurements on the polymeric components to support the proposed structure, and this is an elegant example of utilizing supramolecular chemistry to build sophisticated metal-containing polymeric materials.

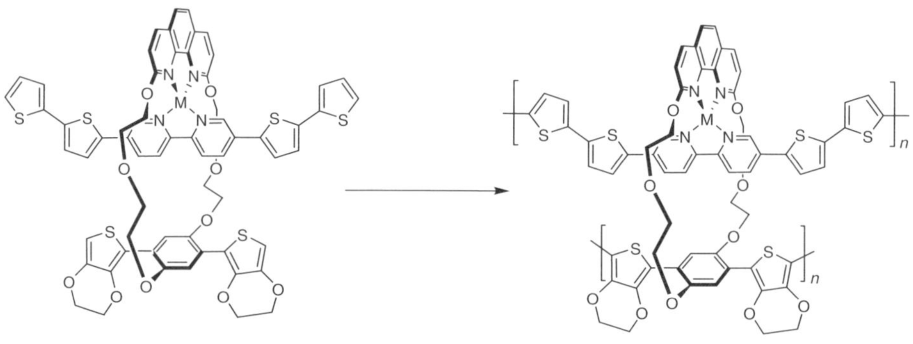

106a (M = Cu^{I})
106b (M = Zn^{II})

107a (M = Cu^{I})
107b (M = Zn^{II})

Scheme 4.23 Electropolymerization of pseudorotaxanes **106** gives ladder polyrotaxanes **107**.

E. Polythiophenes Incorporating Other Coordination Complexes

Polythiophene has been incorporated into π-conjugated polymers in other ways. In this section, we will discuss examples not previously covered in Sections A through D.

Araki and Ogawa electrodeposited a polymer with alternating sexithiophene and $[(phen)Ru(bpy)_2]^{2+}$ groups (**108**), and investigated its properties by spectroelectrochemistry.[78]

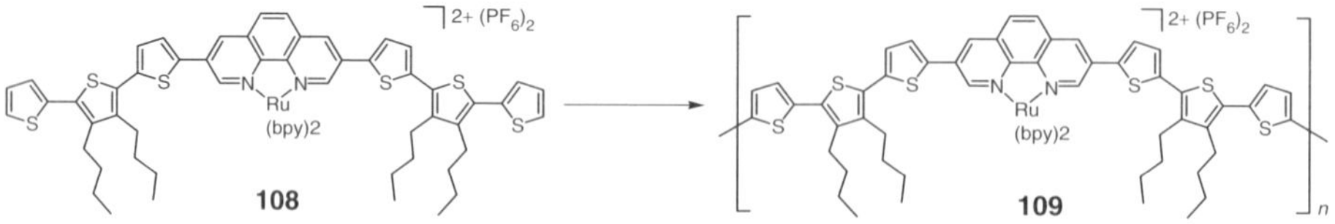

108

109

Scheme 4.24 Synthesis of polymer **109** from complex **108**.

Scanning electron microscopy (SEM) images showed that the polymer had a fibrous morphology. The metal-containing polymers were semiconducting and showed temperature-dependent I–V curves.

Swager has investigated the incorporation of tungsten-capped calixarenes **110–112** into conjugated polythiophene matrices, **113–115** (Scheme 4.25).[79] Conductivity studies showed that oxidation of the polymer led to high conductivities ($\sigma = 16\,\mathrm{S\,cm^{-1}}$). By attaching pendant adamantyl groups to the calixarene, the polymers exhibited molecular recognition of *p*-xylene. Polymers formed in the presence of *p*-xylene exhibited conductivities that were 10 times greater than those prepared in the absence of *p*-xylene. Moreover, films generated without *p*-xylene showed no increase in conductivity upon exposure to *p*-xylene, suggesting that the increased conductivity resulted from a molecular recognition event during polymerization. The adamantyl groups appear to be important for this recognition. The authors extended their studies to tungsten polymer **116**, which incorporates bithienyl units directly appended to the calixarene. This polymer exhibited conductivities similar to polymers **113–115**, but it was found that the conductivity depended on the guest coordinated to the base of the tungsten. Exposure to Lewis bases changed the conductivity, suggesting that these polymers intercalated neutral guests and could function as polymeric chemosensors.[80]

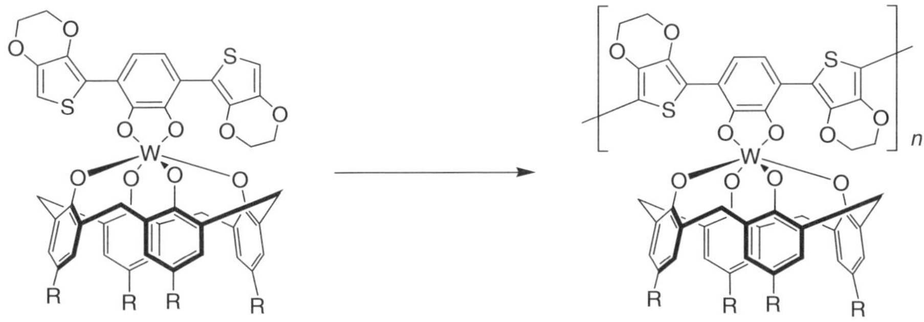

Scheme 4.25 Incorporation of tungsten-capped calixarenes into polythiophene matrices.

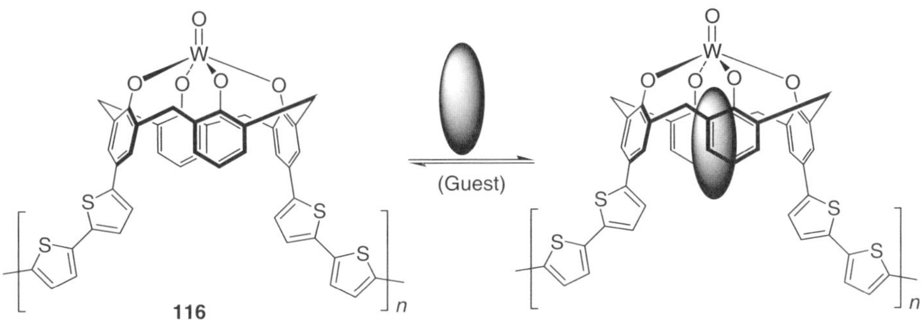

Scheme 4.26 Polymer **116** reversibly binds neutral guest molecules.

MacLean and Pickup have investigated the complexation of $[Ru(bpy)_2]^{2+}$ and $[Os(bpy)_2]^{2+}$ to bithienylbithiazole (**117–120**), and subsequent electropolymerization to generate polymers **121–124** (Scheme 4.27).[81] Although complexes **117–120** would not undergo electropolymerization themselves at the high potentials necessary for coupling, deactivation of residual water in the solvent with $BF_3 \cdot OEt_2$ facilitated growth of polymer **121**.[82] Films of **122–124** were grown in neat $BF_3 \cdot OEt_2$. The metallopolymers all exhibit reversible metal-based M^{III}/M^{II} electrochemistry. Further studies showed that the polymer oxidation is localized on the bithiophene segments, while reduction is localized on the bithiazole components. The lack of significant electronic interaction between metal sites in the polymer is attributed to the localization of the π-orbitals in the polymer backbone.

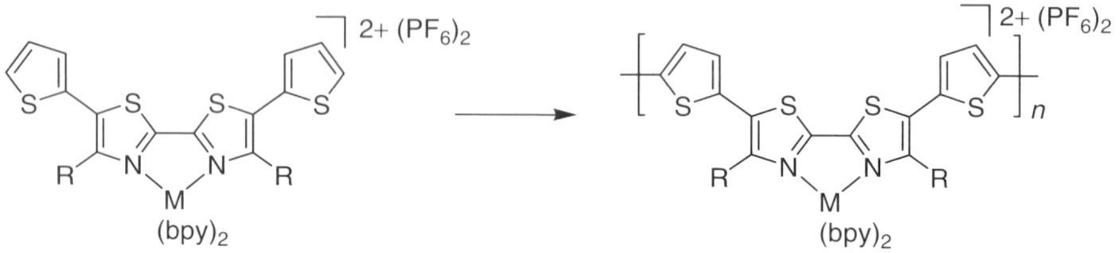

117 (M = Os, R = H)
118 (M = Ru, R = H)
119 (M = Os, R = Me)
120 (M = Ru, R = Me)

121 (M = Os, R = H)
122 (M = Ru, R = H)
123 (M = Os, R = Me)
124 (M = Ru, R = Me)

Scheme 4.27 Polymerization of osmium and ruthenium bithienylbithiazoles gives polymers **121–124**.

The Memorial University group has also explored low band-gap conjugated polymers **126** incorporating bis(dithiolene)nickel(II) cross-links.[83] These polymers, formed by the electropolymerization of the tetrathienyl-functionalized complex, **125**, (Scheme 4.28), displayed conductivities of ca. $10^{-4}\,S\,cm^{-1}$ when p-doped. Other analogs of polymer **126** incorporating nickel complexes, **127–129**, were also examined by CV and electronic spectroscopy.[84] In polymer **127**, where the polythiophene is interrupted by the Ni^{II} centers, the CV is very similar to that of isolated bithiophene segments, indicating that the metal centers do not mediate conductivity between dithiolene dimers in the polymer chain. On the other hand, polymer **129** exhibits electrochemistry characteristic of a conjugated polymer, but displays no interaction with the nickel centers. Polymer **128**, which has biphenyl linkages between the nickel dithiolene complexes, is nonconductive. Skabara has explored a related class of dithiolene polymers synthesized by electropolymerization of complexes incorporating Ni^{II}, Pd^{II}, and Au^{III}, **130–132**.[85] The polymers all showed modest conductivities ($\sigma \sim 10^{-6}$ to $10^{-5}\,S\,cm^{-1}$), as determined by impedance spectroscopy. Significantly, poly(**130**), where M = Ni^{II}, absorbs across the entire visible spectrum and into the near-IR (400–1100 nm), suggesting that these low band-gap polymers may be useful for light harvesting.

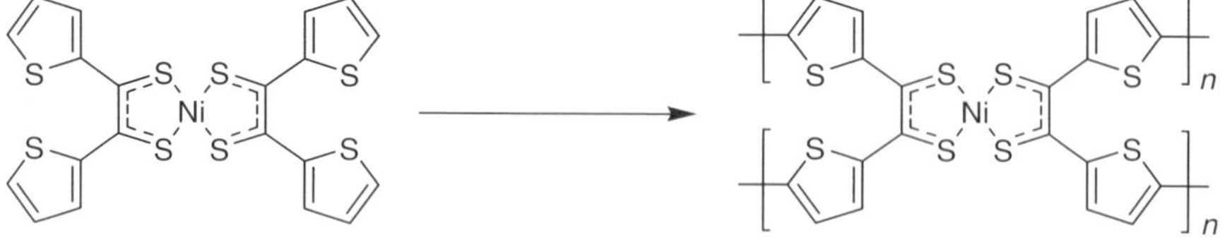

125

126

Scheme 4.28 Electropolymerization of **125** gives **126**.

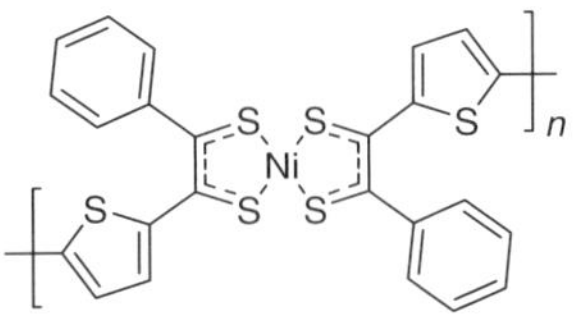

127

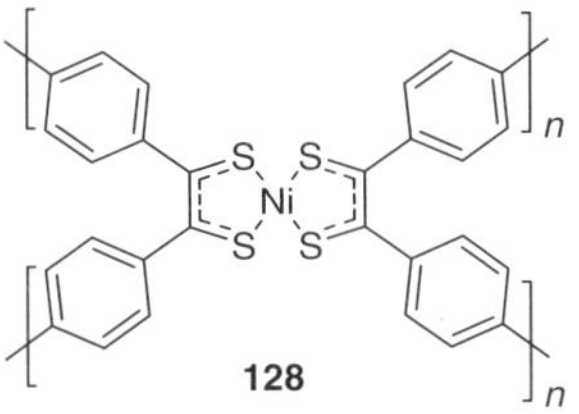

128

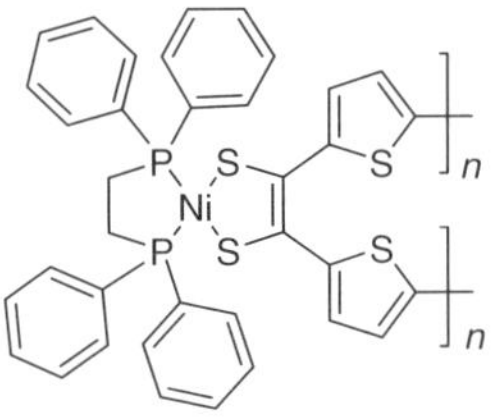

129

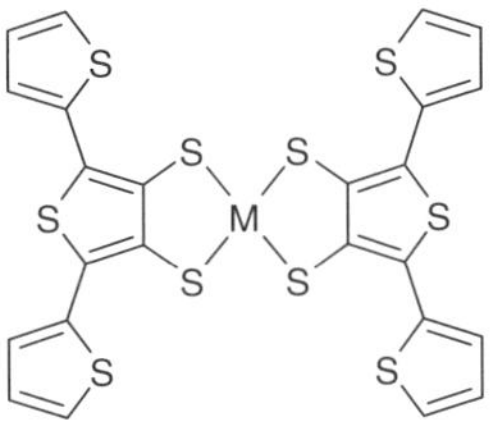

130 (M = Ni^{II})
131 (M = Pd^{II})
132 (M = Au^{III})

Seeber has probed the effect of Pd^{II} coordination to an alkylsulfanyl-substituted polythiophene, **133** (Scheme 4.29).[86] Complexation of $PdCl_2(MeCN)_2$ to the conjugated polymer was investigated by UV-vis spectroscopy. Upon addition of Pd^{II}, a large blue shift of the π–π^* transition (460 nm before metalation) occurred. This shift was attributed to conformational changes that occur upon metalation; crystallographic analysis of a model compound showed that adjacent thiophenes twist upon Pd^{II} coordination to alkylsulfanyl groups. A second factor contributing to the blue shift is the complexation of the lone pairs on sulfur that decreases the electron density on thiophene. Unfortunately, when attempts were made to completely metalate the polymer, the product was insoluble. A partially metalated polymer showed similar oxidation waves to the unmetalated polymer, but new metal-based reductive processes were observed.

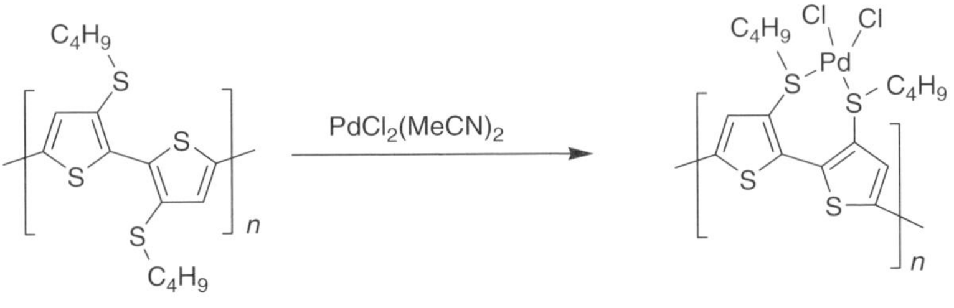

133 **134**

Scheme 4.29 Synthesis of partially metalated polythiophene, **134**.

Scheme 4.30 shows the synthesis of metal-containing conjugated polymer **136** from electropolymerization of Cu^{II} complex **135**, a terthiophene derivative substituted with oximato ligands.[87] Lacroix and co-workers used a multianalytical approach to investigate the properties of the new polymer, including the use of CV, UV-vis, IR, X-ray photoelectron spectroscopy (XPS), X-ray absorption near-edge structure (XANES), and extended X-ray absorption fine structure (EXAFS). Under an applied potential, small changes occurred to the Cu coordination

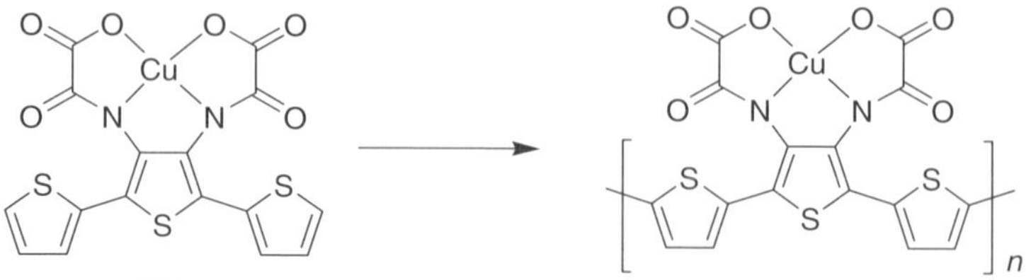

135 136

Scheme 4.30 Electropolymerization of **135** gives polymer **136**.

sphere, suggesting that these polymers act as molecular wires to modify the chemical and physical properties of the coordinated metals.

Swager and co-workers continued their investigations of polythiophenes containing metal salens (**137**) by systematically varying steric bulk at the diamine in Cu^{II} and Ni^{II} polymers.[88] All of the polymers exhibited high conductivities ($\sigma \sim$ 10–100 S cm^{-1}). Copper(II) polymers with bulky substituents that prevent interchain π-stacking exhibited two waves in the CV. For less bulky substituents, three or four waves were observed, suggesting interactions between polarons formed on adjacent polymer chains. In the case of the Ni^{II} polymers, the difference was less noticeable, suggesting that Ni–Ni interactions could take place between the polymer chains regardless of steric demands imposed by the salens, enhancing conductivity. The authors also employed a combination of spectroelectrochemistry and ESR to probe the oxidized polymers. Highly conductive polythiophene–cobalt salen hybrid polymer, **138**, was shown to be an effective electrocatalyst for the selective 4-electron reduction of O_2 to H_2O without formation of H_2O_2 by-products.[89] Moreover, this polymer could be used to sense nitric oxide in water.[90] In this case, binding of NO to Co resulted in a small change in the CV of the polymer. Exposure to a 7 mM NO solution resulted in a 30% increase in conductivity.

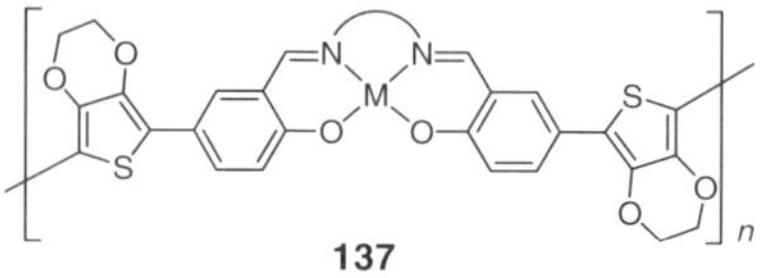

137

M = Cu, Ni, UO_2

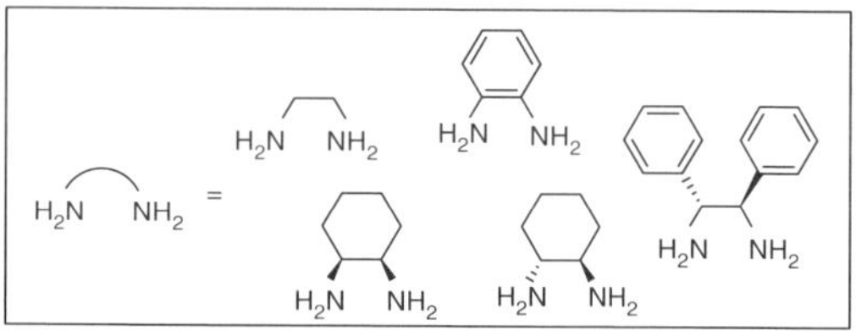

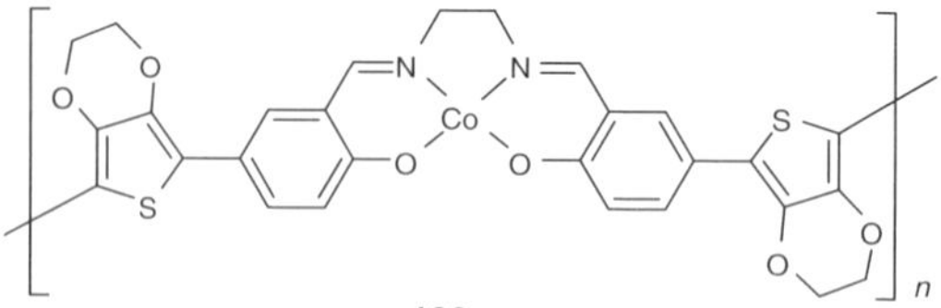

138

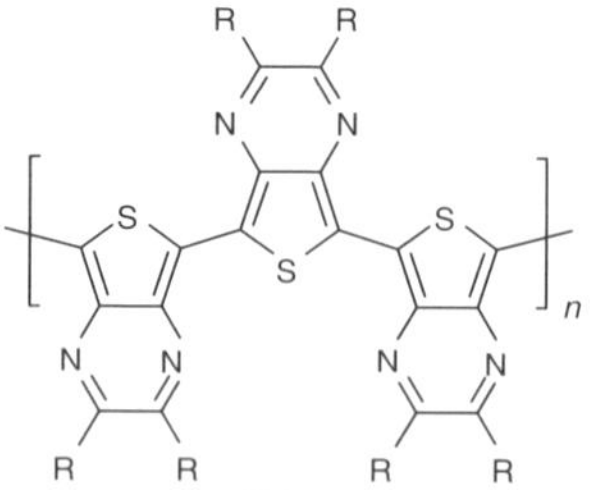

139

Rasmussen has prepared low band-gap polymers, **139**, by electropolymerization of thieno[3,4-*b*]pyrazine.[91] These polymers are anticipated to coordinate to metals via the *N*-donors of the pyrazines to offer environments akin to terpy ligands.[92]

To close this section, many interesting examples of polythiophene-based metallopolymers have been investigated. Although most have been synthesized to investigate the role of the metal complex in modifying the properties of the conjugated polymer, it is clear that other opportunities exist, including applications to potentiometric chemical sensing, electrocatalysis, and electroluminescent materials.

IV. OTHER METAL-CONTAINING CONJUGATED POLYMERS

Many of the principles of, and synthetic routes to, metal-containing π-conjugated polymers have been illustrated with porphyrin- and polythiophene-containing metallopolymers. Nevertheless, many examples lacking these groups remain, particularly metal-containing derivatives of PPV, PPE, and polyphenylene. Polymers incorporating bipyridines, phenanthrolines, and other ligands may also coordinate to transition metals. In this section, polymers are grouped into the following categories: (A) organometallic polymers; (B) salen- and salphen-based polymers; (C) polymers based on bipyridine and phenanthroline; and (D) other examples.

A. Organometallic Polymers

In addition to a few organometallic polymers that incorporate polythiophene (Section III.A), there have been many significant advances in connecting metals to the main chain of polymers by metal–carbon bonds. As was described in Section III.A, ferrocene is well suited for incorporation into conjugated polymers. In the past seven years, new methods have been developed to prepare ferrocene-containing conjugated polymers. Lee and co-workers prepared polymer **141** via ring-opening metathesis polymerization (ROMP) of ferrocenophane **140** (Scheme 4.31).[93] High molecular-weight ($M_w \sim$ 240,000 Da; GPC) **141** was soluble in common organic solvents, air stable, and exhibited high thermal stability (ca. 300°C). By varying the monomer-to-catalyst ratio, the molecular weight of the polymer could be modified. The authors attributed the shift of the absorption maximum from 240 nm in **140** to 320 nm in **141** to increased conjugation in the polymer.

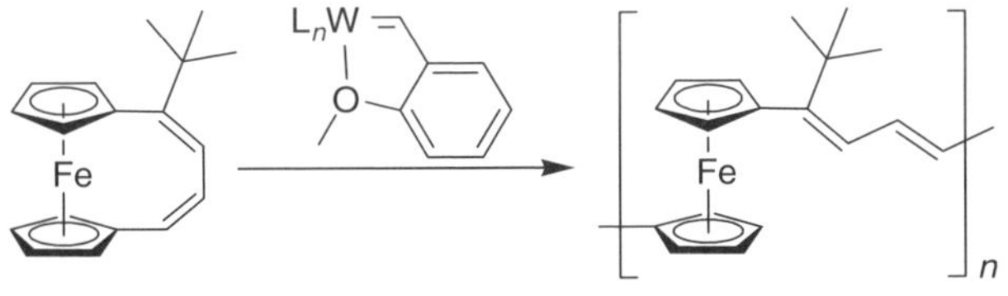

140 **141**

Scheme 4.31 Synthesis of poly(ferrocenylenebutadiene) **141**.

Plenio has reported poly(ferrocenylethynylene)s **145–147** synthesized by Sonogashira coupling of 1,3-disubstituted ferrocenes **142–144** (Scheme 4.32).[94,95] These polymers were of low molecular weight, although GPC analysis of **147** functionalized with a crown ether indicated it had a high molecular weight (M_n = 58,000 Da). The authors attributed this to aggregation in the mobile phase (THF). Two reversible oxidation waves (ΔE = 145 mV) were observed for **145** in CH_2Cl_2, indicating a metal–metal interaction in the polymer. This was further explained by comparison of the optical and redox properties of the polymer with those of a series of model oligomers.

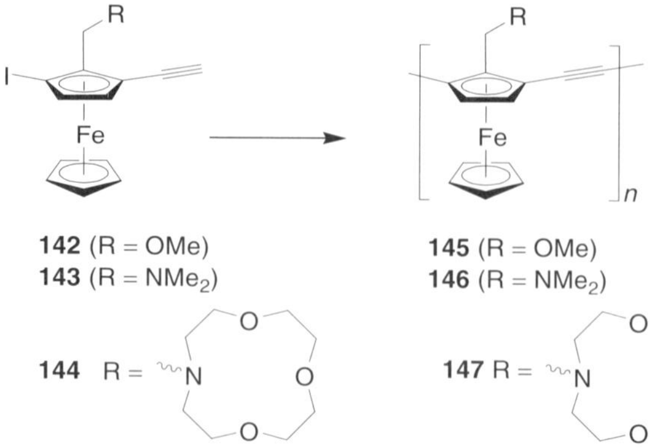

Scheme 4.32 Synthesis of poly(ferrocenylacetylene)s **145–147**.

Naka and Chujo have developed conjugated polymers with two different redox centers in the backbone: ferrocene and dithiafulvene.[96] The low molecular-weight polymer, **148** (M_n = 2020 Da, GPC), was characterized by several techniques, including ^{1}H and ^{13}C NMR spectroscopy. Polymer **148** showed only one broad oxidation wave in its CV, indicating strong interaction between the two donors in the polymer. When doped with TCNQ, **148** formed a charge-transfer salt that showed a 2-probe conductivity of $5.0 \times 10^{-6}\,S\,cm^{-1}$. A film of **148** doped with I_2 exhibited a higher conductivity of $3.6 \times 10^{-3}\,S\,cm^{-1}$.

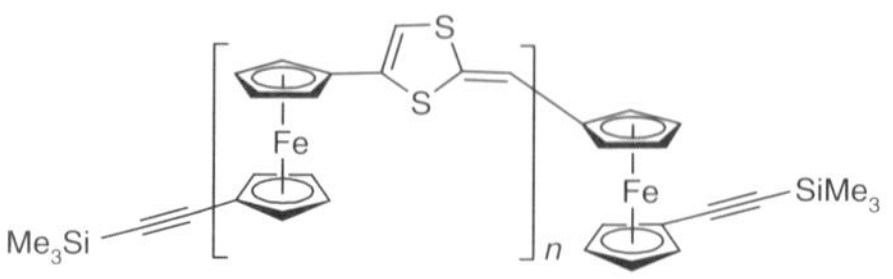

148

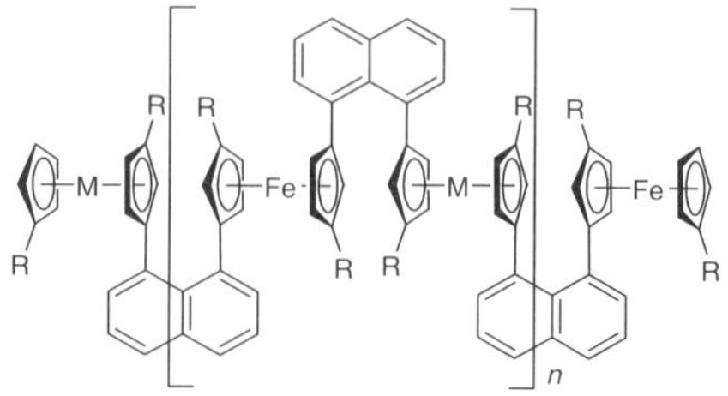

(M = Fe or Ni; R = H, 2-octyl, 4-NMe$_2$C$_6$H$_4$CH$_2$, etc.)

149

Foxman and Rosenblum prepared polymers **149** with various substituents attached to the cyclopentadienyl rings.[97] With alkyl substituents, the polymers were soluble and could be analyzed by GPC. Molecular weights (M_n) were between 1700 and 19,000 Da. Polymers could be prepared with all-Fe centers, or with alternating Fe^{II}/Ni^{II} metal centers. Although the redox and optical properties of these polymers were not investigated, they are sure to be interesting for this novel class of polymers.

Southard and Curtis have undertaken a thorough study of poly(ferrocenylarylenes) **152** and **153**.[98] These low molecular weight polymers (GPC, degree of polymerization ~ 7) were prepared by condensation of dicyclopentadienyl precursors, **150** and **151**, with Fe^{II} salts (Scheme 4.33) and contain substantial quantities of cyclic dimer and trimer. The authors also investigated the doping of the polymers with I_2 and studied comproportionation of the mixed-valence species. The oxidized polymers showed weak antiferromagnetic interactions as revealed by superconducting quantum interference device (SQUID) magnetometry. The metal–metal interaction in the polymers was also confirmed by CV, which revealed two reversible waves with $\Delta E = 170$ and 190 mV for **152** and **153**, respectively. Four probe conductivity measurements of thin films of **152** and **153** revealed conductivities $<10^{-10}\,S\,cm^{-1}$, but the conductivity increased to $6.5 \times 10^{-7}\,S\,cm^{-1}$ in doped polymer **153**. The polymers were also investigated by NMR, UV-vis, Mössbauer, IR, and ESR spectroscopies.

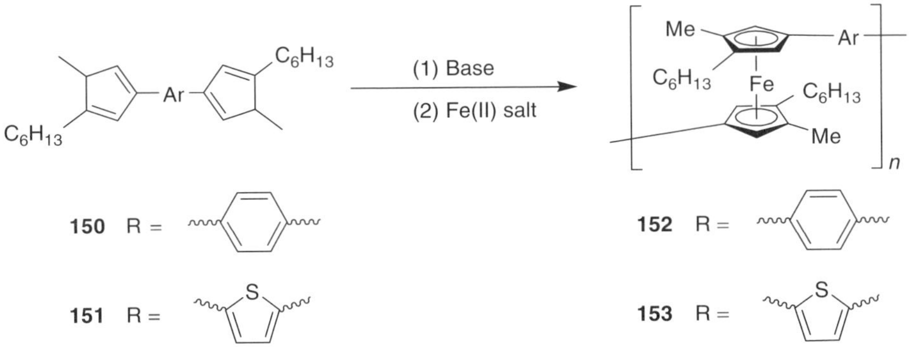

Scheme 4.33 Preparation of poly(ferrocenylarylene)s **152** and **153**.

Cobaltocene and related complexes have been incorporated into conjugated polymers. Reaction of cobaltocene-containing polymers **49** (R = aryl) with ^{t}butylisocyanide gave new metallopolymers, **154** (Scheme 4.34).[99] These polymers were soluble in organic solvents and had molecular weights of 6800–22,600 Da (M_n, GPC). Polymers **154** could be partially *N*-methylated with iodomethane.

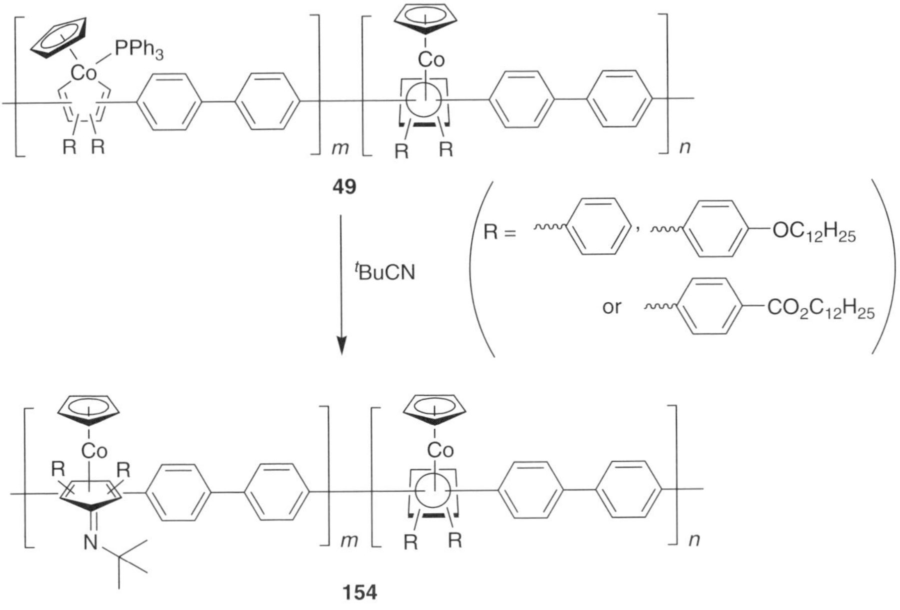

Scheme 4.34 Preparation of cobaltocene polymers **154**.

Bunz has explored the use of ADIMET to prepare conjugated organometallic polymers.[100] (η^4-C_4H_4)Co(η^5-C_5H_5) was incorporated into PPEs by Sonogashira coupling of **155** with **156** to give **157** (Scheme 4.35), or by ADIMET to give **159** (Scheme 4.36). While Sonogashira coupling gave polymers with molecular weights of 4300–24,000 Da (M_n, GPC), ADIMET polymerization generally gave much higher molecular weight (M_n = 37,000–85,000 Da) material. ADIMET also gave polymers in higher yield and purity. Bunz investigated the morphologies, aggregation, and supramolecular structures of the polymers. At 155°C, **157** (R_1 = $SiMe_3$, R_2 = hexyl) exhibited a Schlieren texture characteristic of a nematic liquid-crystalline state. Transmission electron microscopy (TEM) of **157** and **159** showed that they organized into irregular honeycomb architectures, or well-ordered lamellar structures, depending on the interactions of the organometallic backbone and alkyl substituents.

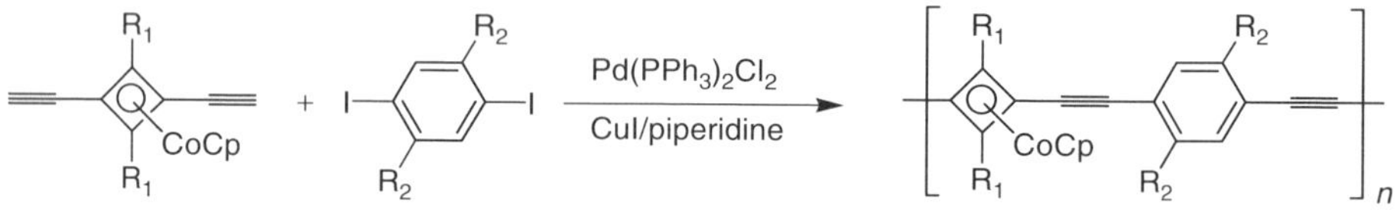

155 **156** **157**

(R_1 = H or $SiMe_3$; R_2 = H, C_6H_{13}, $C_{12}H_{25}$, or (*S*)-3,7-dimethyloctyl)

Scheme 4.35 Sonogashira coupling route to polymers **157**.

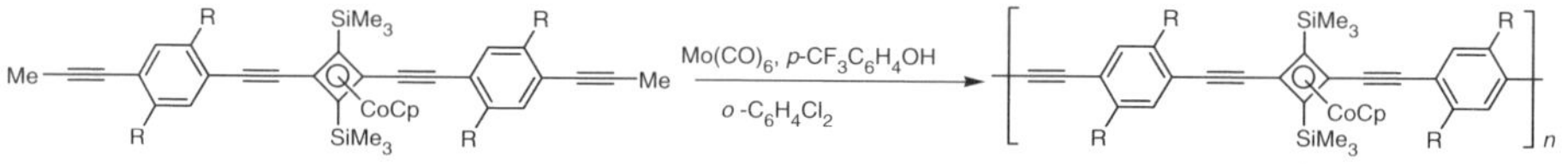

158 **159**

(R = hexyl, 2-ethylhexyl, or (S)-3,7-dimethyloctyl)

Scheme 4.36 ADIMET synthesis of polymers **159**.

Deck and Maiorana have reacted low molecular-weight (M_n = 3500 Da; GPC) polymer **160** with $Mn(CO)_5Br$ in THF to coordinate manganese to the anionic indenyl groups of the conjugated backbone (Scheme 4.37).[101] The incorporation of $[Mn(CO)_3]^+$ units in fluorinated **161** was confirmed by IR spectroscopy and elemental analysis of the polymer.

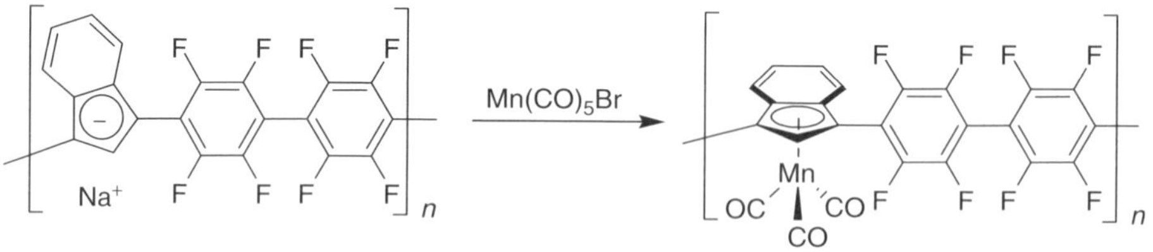

160 **161**

Scheme 4.37 Preparation of polymer **161**.

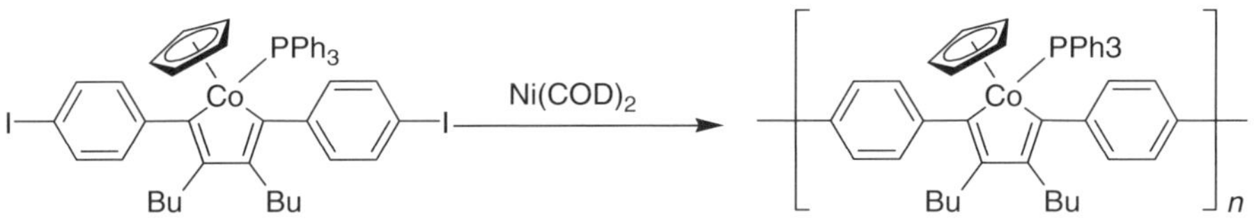

Scheme 4.38 Preparation of polymer **163**.

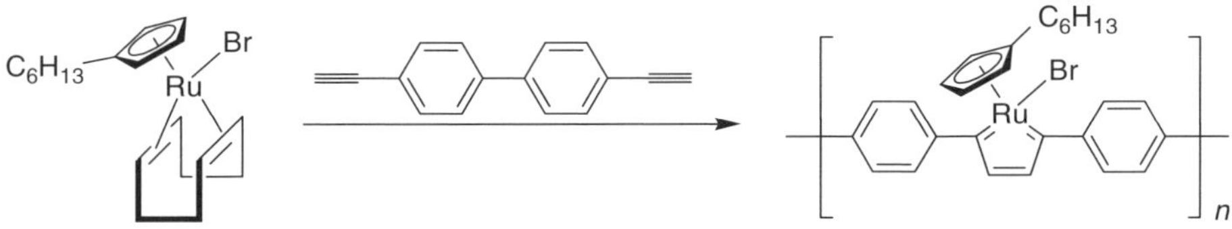

Scheme 4.39 Preparation of ruthenium-containing polymer **165** by MCP reaction.

Nishihara and Endo have developed an elegant route known as "metalacycling polymerization" (MCP) to incorporate metallocyclopentadiene complexes into conjugated polymers. Polymers similar to **49** can be prepared by this reaction. A related polymer, **163**, was recently prepared using Ni^0 catalyzed by cross-coupling of monomer **162**.[102] Attempts to couple the dibromo analog of **162** failed, but reaction of **162** with $Ni(COD)_2$ (COD = 1,5-cyclooctadiene) gave high molecular-weight polymer **163** (M_n 200,000 Da; GPC) (Scheme 4.38). The polymer exhibited a small red shift in the UV-vis spectrum relative to monomer **162**, indicative of extended conjugation in the polymer. Nishihara has utilized the MCP reaction to make polymer **165** from the reaction of dialkynylbiphenyl and **164** (Scheme 4.39).[103] Although the average molecular weight was only ca. 3400 Da (M_n, GPC), fractions with molecular weights up to 20,000 Da were observed. The π–π^* transition of the polymer is red-shifted by ca. 20–30 nm, indicating extended π-conjugation. Broad signals in the EPR spectrum of the reduced polymer were attributed to weak ferromagnetic coupling between metal centers.

Weder has examined the coordination of Pt^0 to unsaturated carbon–carbon bonds in PPE.[104] Reaction of $[Pt(\mu\text{-}Cl)Cl(PhCH{=}CH_2)]_2$, **167**, with polymer **166** afforded **168** with Pt complexes coordinated to the alkyne groups. Other Pt complexes (not shown) bridge the polymer chains, the styrene displaced by another alkyne. The complexation was studied by ^1H and ^{195}Pt NMR spectroscopy and compared to model complexes. Upon binding to Pt, the fluorescence of **166** is quenched due to spin-orbit coupling of the heavy Pt metal with the exciton of the conjugated polymer. The polymer also exhibits a small blue shift upon binding to Pt, suggesting a slight decrease in the extent of conjugation. Coordination of the Pt was reversible, and decomplexation occurred upon addition of a competing olefin, such as styrene, to **168**.

Reaction of polymer **166** with tristyreneplatinum(0), **169**, gave conjugated polymer network **170** (Scheme 4.41).[105] As in **168**, polymer **170** contains cross-links between the chains where a single Pt^0 is coordinated to two alkyne groups. TOF measurements indicated that the electron and hole mobilities in **166** were significantly enhanced upon Pt binding. The incorporation of metals into π-conjugated polymers may be useful for increasing charge-carrier mobility. Krebs has found that palladium nanoparticles present in PPVs coordinate to unsaturated bonds in the polymer, as shown for **170**, to quench luminescence.[106]

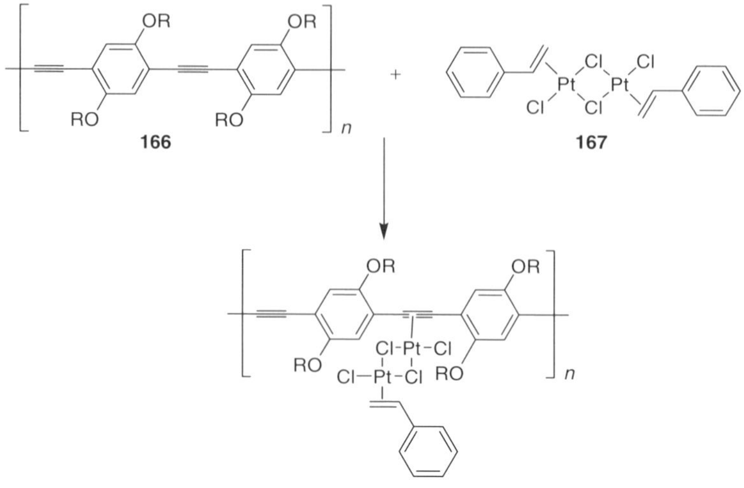

166

167

168 (R = alkyl)

Scheme 4.40 Preparation of Pt^0-containing polymer **168**.

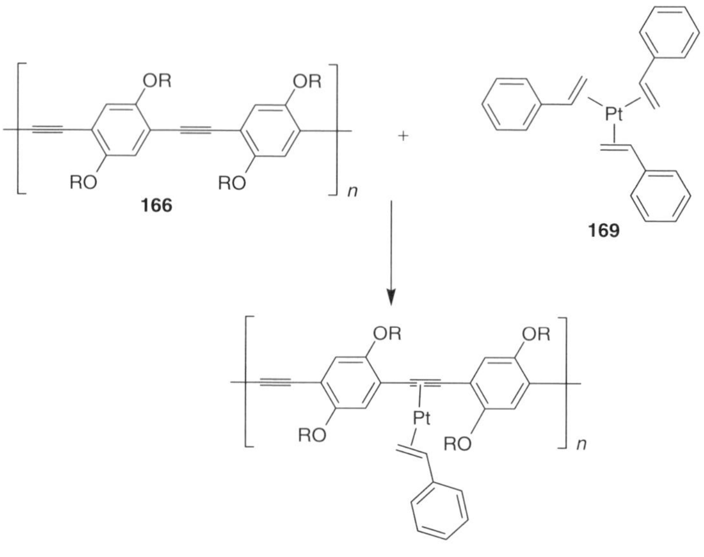

166

169

170 (R = alkyl)

Scheme 4.41 Preparation of polymer **170**.

Recently the Swager group at MIT described the Suzuki cross-coupling of cyclometalated Pt^{II} complex, **171**, with 2,7-fluorenediboronic ester, **172**, to give conjugated polymer **173** (Scheme 4.42).[107] The polymer ($M_n \sim 12{,}000$ Da; GPC) showed strong phosphorescence at 585 nm with a lifetime of 14.4 μs at room temperature in THF. This phosphorescence is assigned to a triplet π–π* transition, with mixing from an MLCT transition of the square planar Pt^{II} complex.

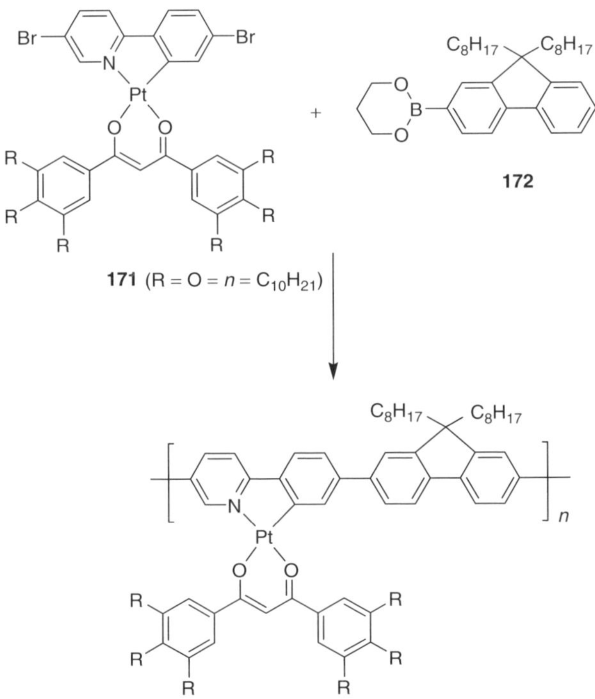

Scheme 4.42 Synthesis of cyclometalated Pd(II) polymer **173**.

Although its quantum yield is low (0.05), polymer **173** senses oxygen in solution, as measured by the decrease in excited state lifetime with increasing O_2 concentration in solution.

B. Salen- and Salphen-Based Polymers

A number of salen and salphen complexes have been incorporated into conjugated polymers by electropolymerization (including the use of thienyl-functionalized monomers, as discussed in Section III) or condensation polymerization.

Leung and MacLachlan have synthesized high molecular-weight ($M_w \sim$ 17,000–84,000 Da; GPC) poly(salphenyleneethynylene)s **176a–176c**, by Sonogashira coupling of diiodosalphens, **174a–174c**, with dialkoxydiethynylbenzene, **175** (Scheme 4.43).[108] It was necessary to employ long alkyl chains on both the spacer and the metal complex to obtain soluble polymers. Polymers **176** were cast into intensely colored red thin films. Polymer **176a** (M = Zn^{2+}) is weakly luminescent in solution, and forms highly viscous solutions in THF.

Building on the work of Katz,[109] two groups have prepared helical salen derivatives. Pu and co-workers explored two new types of potentially helical polymers.[110] Condensation of biphenol, **177**, with dialkylphenylenediamine, **178**, in the presence of nickel(II) acetate gave soluble oligomer **179** (Scheme 4.44). GPC analysis of the product indicated it had a molecular weight (M_n) of ca. 4900 Da. By reacting chiral binaphthol, **180**, with diamine, **179**, in the presence of $Ni(OAc)_2$, a low molecular-weight polymer ($M_n \sim$ 3600 Da, GPC) that was expected to adopt a helical conformation was also prepared. The authors speculated that the product was paramagnetic, since ^{1}H NMR spectra were broad, but EPR experiments were not reported.

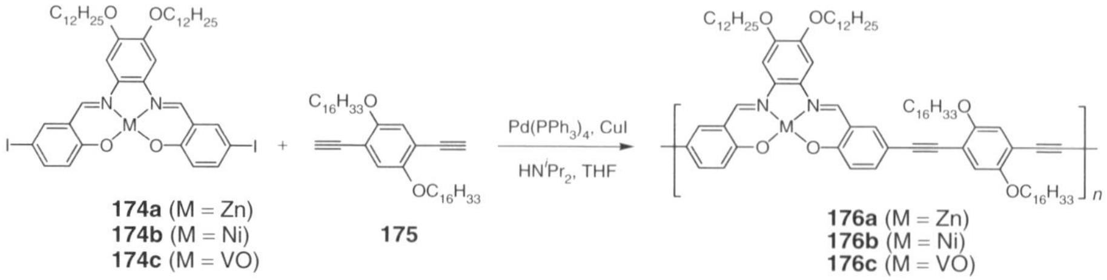

Scheme 4.43 Synthesis of poly(salphenyleneethynylene)s **176**.

Broadening may be due to aggregation in the NMR solvent. Copolymers were prepared by changing the ratio of precursors **177** and **178** used.

Takata and co-workers have also investigated the condensation of a binaphthyl derivative (**181**) with diamines **182a–182e** to give low molecular-weight polyimines ($M_n \sim$ 1400–17,000

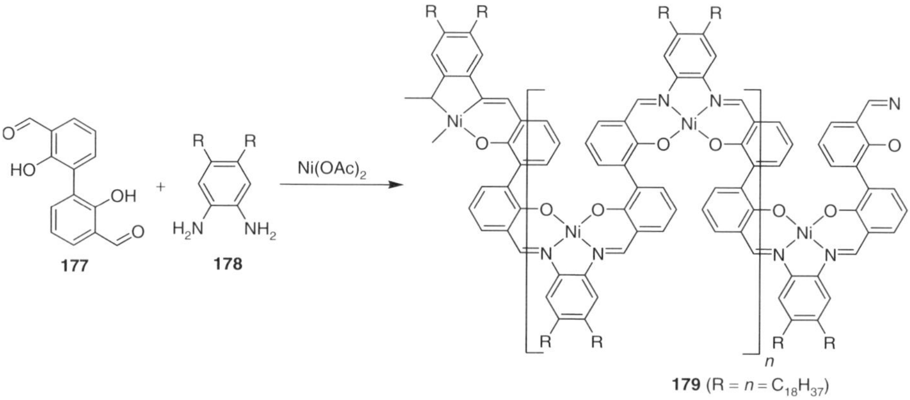

Scheme 4.44 Synthesis of polymer **179**.

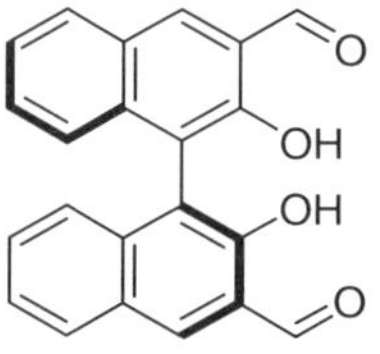

180

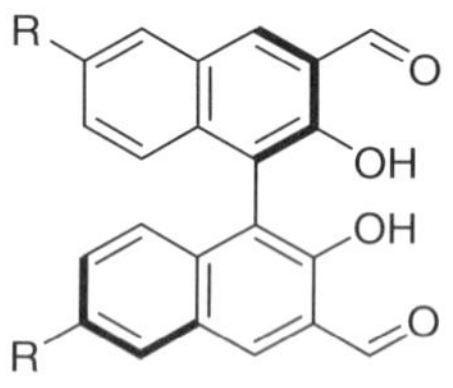

181 (R = n = C_8H_{17})

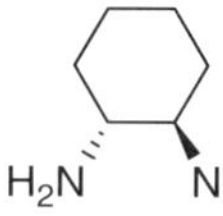

182a

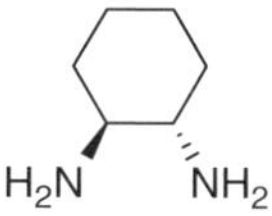

182b

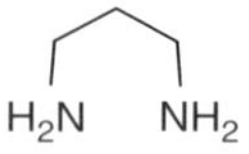

182c

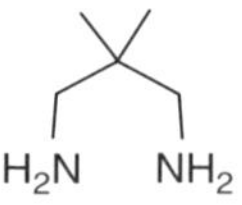

182d

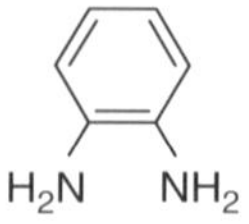

182e

Da, GPC).[111,112] Reaction with Et_2Zn gave conjugated Zn-containing polymers. Interestingly, the molecular weights of the metalated polymers were substantially lower than those measured for the metal-free precursor polymers by GPC. The authors attributed this to the helical superstructure of the metalated polymer, which would reduce the polymer's effective hydrodynamic volume. IR studies indicated that the backbone was not cleaved during the metalation step. The helical polymers catalyzed the addition of Et_2Zn to benzaldehyde, and significantly enhanced the enantioselectivity observed for this reaction compared with a molecular biphenylsalphen catalyst. Although the molecular weight of the polymer catalyst did not affect the ee for this reaction greatly, the ee was strongly temperature-dependent, with the highest ee's (but low yields) observed at −60°C. The authors postulate that the interaction of the helical polymer with the catalyst during the transition state is important for transfer of chirality to the product.

Electropolymerization of salen derivatives (e.g., **183**) gave thin films of poly(salen)s, which are believed to have the structure shown in **184** (Scheme 4.45).[113] Nickel polymer **184** was prepared as a yellow-green film on Pt electrodes, and exhibited high stability and good conductivity. Electrochemical kinetics were measured by CV and chronoamperometry. Iron salen could also be electropolymerized to give **185**, and the polymer was found to catalyze the reduction of oxygen and hydrogen peroxide.[114] It is notable that the poly(salen) could not be prepared in MeCN, and only thin films were deposited from CH_2Cl_2. Aubert has also investigated the electropolymerization of dinickel(II) bis(salphen) complexes, **186–188**.[115] In this case, the coordination complexes have two sites that undergo electropolymerization, generating cross-linked thin films. Attempts to make analogous copper or copper/nickel mixed-metal films failed due to insolubility of the precursors. The extent of doping, cross-linking, and the kinetics of electron transfer in the dinickel(II) polymers were investigated by the authors.

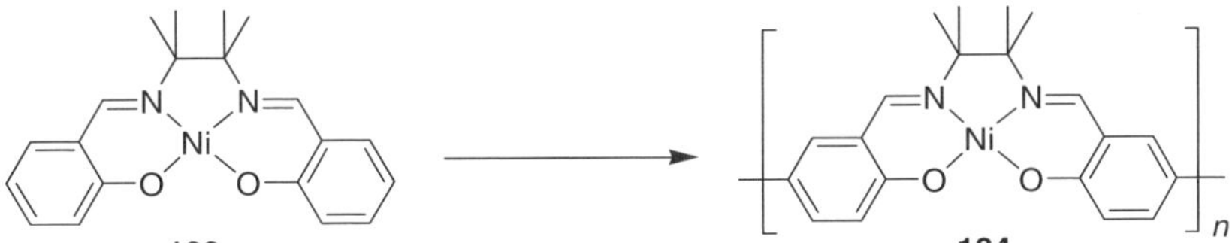

183 184

Scheme 4.45 Electropolymerization of **183** gives polymer **184**.

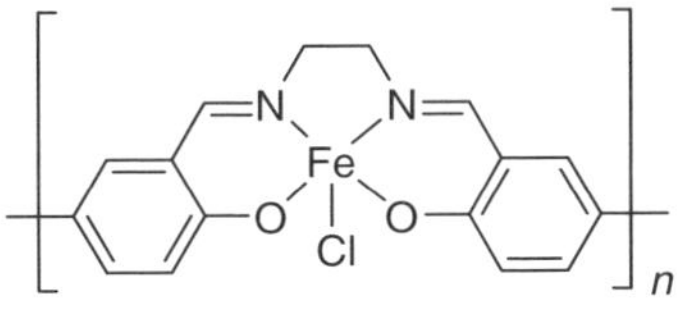

185

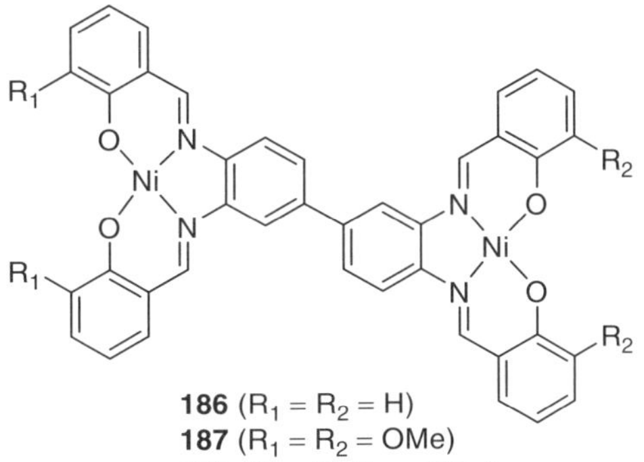

186 ($R_1 = R_2 = H$)
187 ($R_1 = R_2 = OMe$)
188 ($R_1 = H$, $R_2 = OMe$)

C. Polymers Based on Bipyridine and Phenanthroline

The past seven years have seen significant advances in the preparation of coordination polymers. A vast assortment of conjugated polymers has been prepared that incorporate bpy, phen, and other polydentate ligands. Complexation to metal ions affords new polymeric materials with properties modified by the presence of the transition metal.

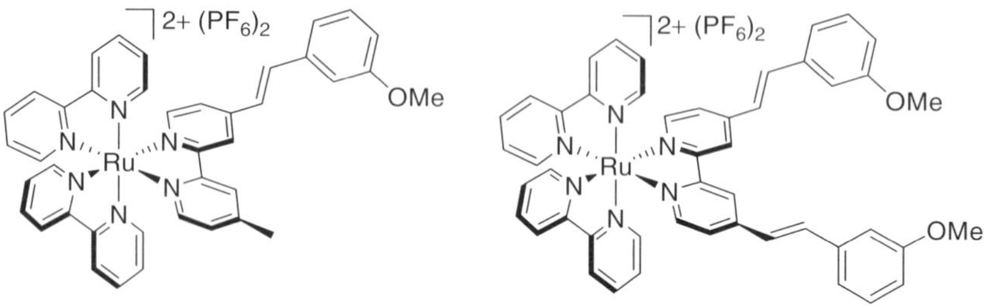

189 **190**

Grennberg et al. explored the electropolymerization of styryl-functionalized polybipyridyl complexes of Ru^{II}, such as **189** and **190**, in an effort to develop metal-containing conjugated polymers.[116] The electrodeposited thin films of these complexes and related structures showed blue-shifted absorption bands relative to the monomers, indicating that the extent of conjugation was reduced. The authors postulated that coupling of vinyl groups during the polymerization breaks the conjugation in the polymers.

Satyanarayana and Elsenbaumer attempted the Gilch polymerization of monomer **191** to afford poly(bipyridinevinylene), **192**, with $[Ru(bpy)_2]^{2+}$ complexed to each bpy in the conjugated backbone (Scheme 4.46).[117] Unfortunately, the polymer obtained from this reaction was insoluble, and thus difficult to characterize. A large red shift in the optical spectrum was observed, and thermogravimetric analysis and IR spectroscopy also supported the structure assigned. The conductivity of the blue-black product was 4.5×10^{-6} S cm^{-1}.

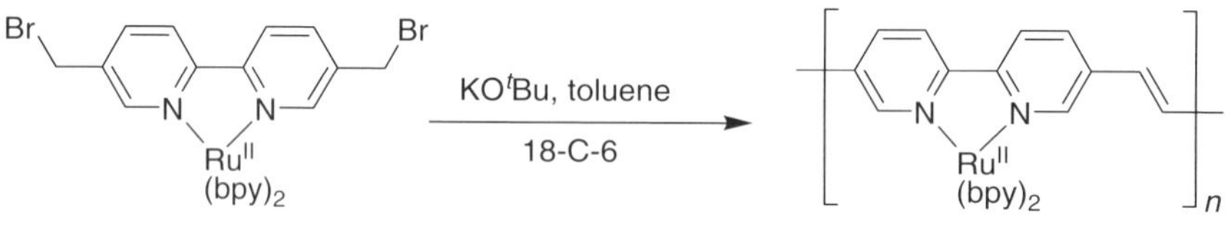

Scheme 4.46 Attempted synthesis of polymer **192**.

Yu and co-workers have incorporated $[Ru(bpy)_2]^{2+}$ groups into soluble PPV polymers in an effort to develop polymeric photorefractive materials.[118] As shown in Scheme 4.47, Heck coupling of divinylbenzene, monomer **193**, and 1% of Ru-containing monomer **194** gave metallopolymer **195** ($M_n \sim 18{,}000$ Da; GPC). In this polymer, it was anticipated that the MLCT band of the Ru^{II} complex would enhance photogeneration of charge carriers. However, extensive studies of the optical properties of the polymer (including 2-beam coupling, photoconductivity, optical gain coefficient, diffraction efficiency, and field-induced birefringence experiments) indicated that alignment of the dipoles in this polymer generated a screen that limited photocharge generation. As a result, the photorefractive properties of the polymer were poor, and it exhibited

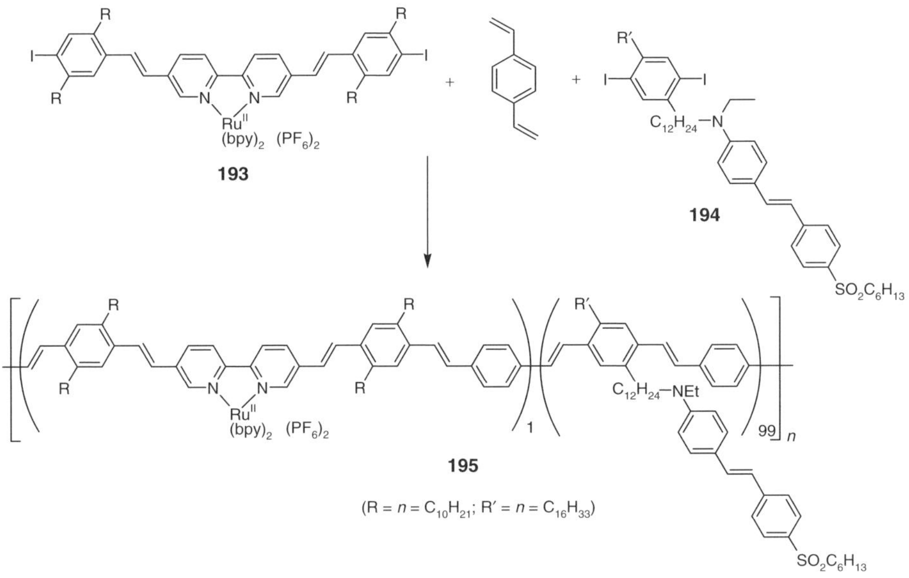

Scheme 4.47 Synthesis of polymer **195**.

saturation in the field dependence of photoconductivity. When a series of related polymers (**196–200**) incorporating mixed-ligand ruthenium complexes were investigated, the photosensitivity extended to longer wavelengths.[119] By varying the concentration of Ru in the polymer (*x*,*y* in **196–200**), the authors demonstrated that the photoconductivity increases with more Ru. When the charged Ru complexes in **195** were replaced by neutral Zn^{II} porphyrins or phthalocyanines, the PPV copolymers exhibited high photorefractive performance as a result of high photoconductivity and photogeneration quantum yield.[120] The presence of unpaired electrons in Cu^{II} porphyrin-containing polymers lowered photorefractive performance because

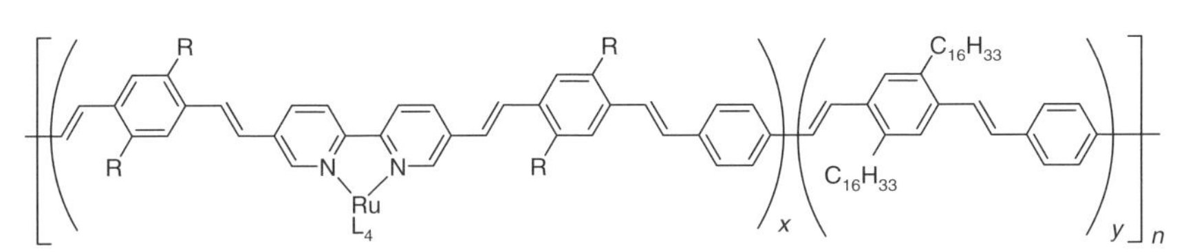

Polymers **196–200**

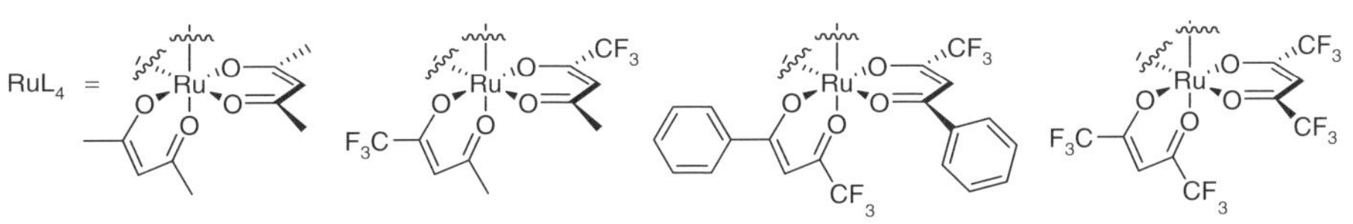

196 **197** **198** **199**

$Ru(bpy)_2(PF_6)_2$

200

of efficient excited-state relaxation. The optical and electrochemical properties of all the soluble polymers ($M_n \sim 3500–18,300$ Da, GPC) were investigated.

Lippard has investigated the use of a conjugated PPV/bipyridine polymer **201** as a substrate for NO sensing.[121] When complexed to Cu^{II}, the luminescence of polymer **201** is quenched. NO reduces Cu^{II} in the polymer to diamagnetic Cu^{I}, and the fluorescence is enhanced after most of the Cu^{II} centers have been reduced. Exciton migration in the polymer backbone reduces its sensitivity to NO, since its fluorescence will only increase substantially when most of the sites are reduced. Thus, **201** may be less sensitive to NO than a molecular analog due to exciton migration in the polymer.

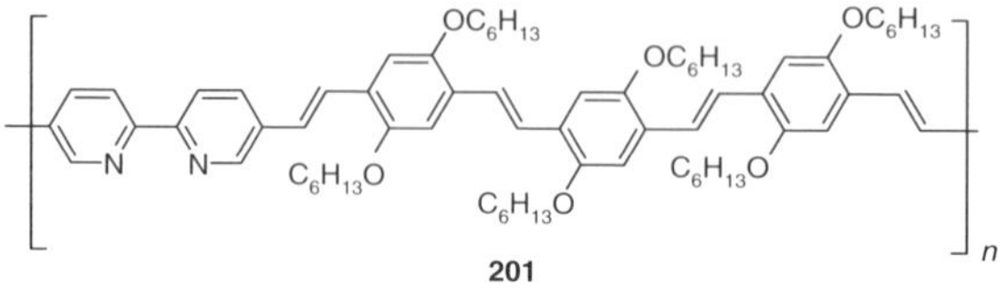

201

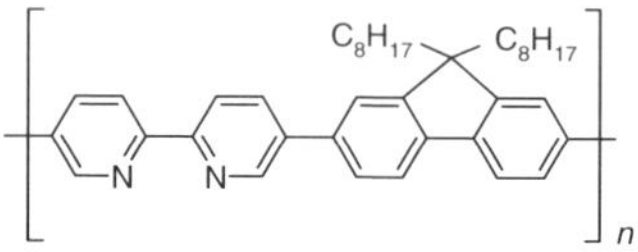

202

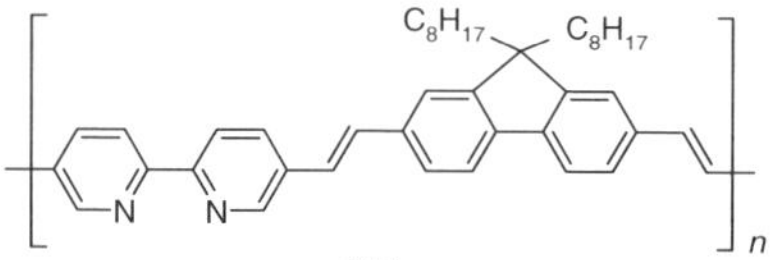

203

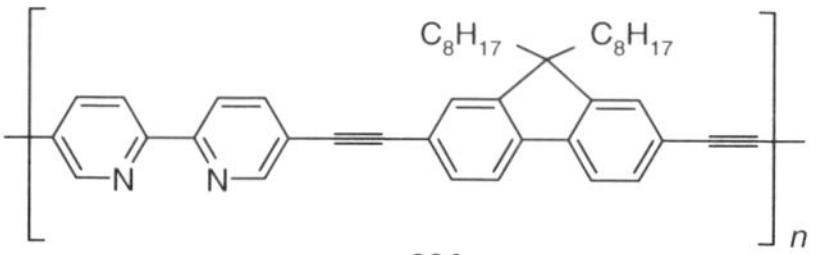

204

Lai and Huang have formed conjugated polymers **202–204** that incorporate alternating bpy and 9,9-dioctylfluorene moieties in the polymer backbone, linked by single bonds, vinylene groups, or alkynyl spacers.[122] The polymers were very sensitive to transition metal ions, showing 20–40-nm red shifts on complexation to some metals. In addition to the color change, some of the metals caused fluoresence quenching, which could also be used for sensing.

Pautzsch and Klemm have investigated conjugated PPEs containing substituted $[Ru(bpy)_3]^{2+}$ complexes.[123] Sonogashira coupling of monomers **205** and **207** with dialkynylbenzenes gave red polymers **206** and **208** (Scheme 4.48). In most cases (except where all substituents were H) the polymers were soluble and GPC analysis indicated molecular weights (M_n) of ~7000–25,000 Da. As these polymers are charged and rigid rods, GPC results calibrated to polystyrene standards are suspect. By measuring a series of model oligomers, the authors determined that GPC underestimates the molecular weights of the polymers by a factor of 2–2.4. The absorption spectra of the polymers show red-shifted π–π^* transitions associated with the conjugated polymer, but the transitions assigned to the $[Ru(bpy)_3]^{2+}$ complex are virtually unchanged. The polymers were found to be good photoconductors, and their fluorescence, electrochemical behavior, and thermal properties were also investigated.

Weder has made cross-linked metal–organic polymers by complexing bpy ligands in **209** with metal ions (Cu^{I}, Zn^{II}, Cd^{II}, Co^{II}, and Ni^{II}).[124] The polymers exhibit small red shifts in their absorption spectra, and fluorescence quenching upon metal-ion binding. Notably, there is little interaction between the metal sites in the partially metalated polymers; exciton migration is limited by the bpy "blocking" groups in the polymer chain, limiting the effective conjugation length of the polymer. When cast as a film, the metalated polymers became insoluble, consistent with cross-linking, but could be dissolved in the presence of excess competing ligand. The authors are investigating the Zn^{II} and Cd^{II} polymers for use in electroluminescent displays. If the metal is incapable of coordinating to two bpy sites, soluble metallopolymers are expected. Schanze has complexed $Re^{I}(CO)_3Cl$ moieties to poly(bipyridylethynylene)s to give polymer **210**.[125] This polymer exhibits a red-shifted absorption maximum compared with model oligomers reported.

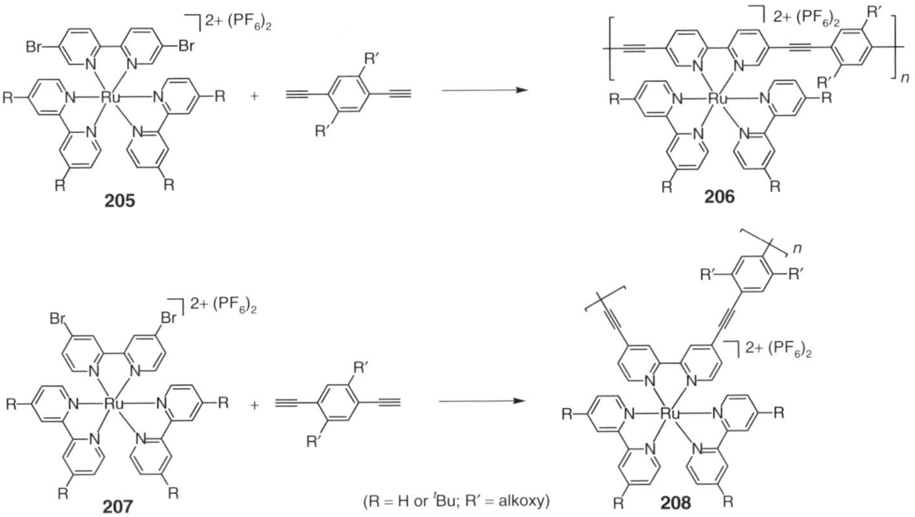

Scheme 4.48 Synthesis of polymers **206** and **208**.

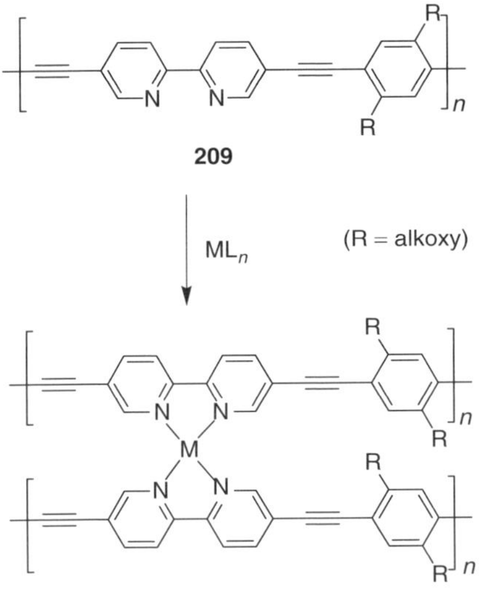

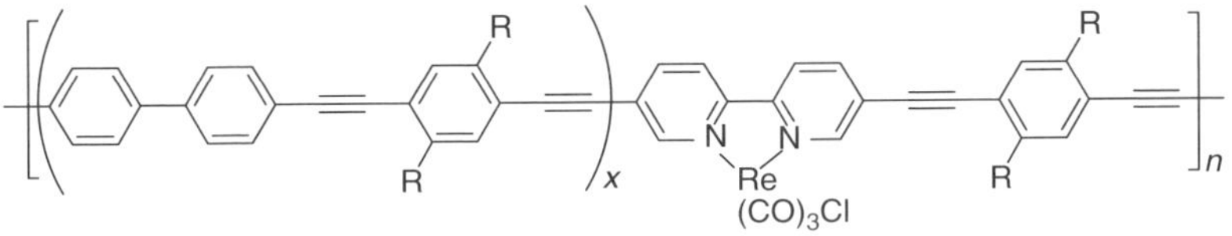

210

Naka et al. reported the preparation of low molecular-weight ($M_n \sim 1700$, GPC) copolymers of bpy and dithiafulvalene, **211**.[126] Reaction with $Ru(bpy)_2Cl_2$ in water gave a metallopolymer (**215**; Scheme 4.49), but only 29% of the bpy groups were complexed to Ru, probably limited by steric hindrance of the $[Ru(bpy)_3]^{2+}$ groups. CV of this metallopolymer revealed broad peaks, indicative of interaction between $[Ru(bpy)_3]^{2+}$ and dithiafulvalene. The conductivity of the metallopolymer was $1.1 \times 10^{-6}\ S\ cm^{-1}$, two orders of magnitude lower than that of the unmetalated polymer.

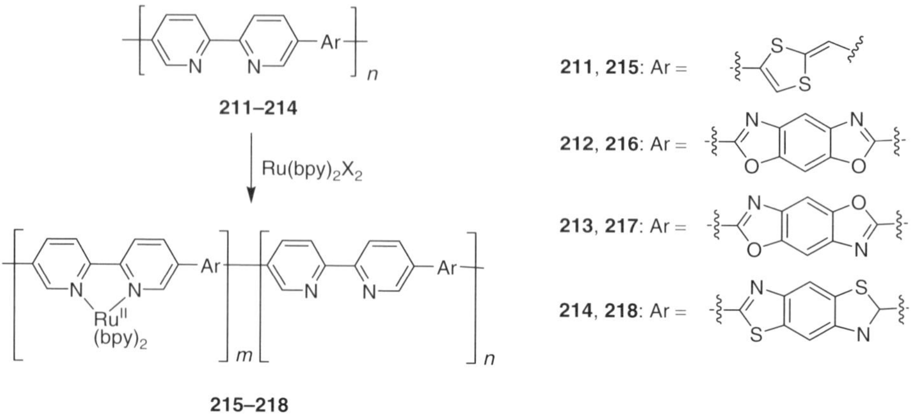

Scheme 4.49 Synthesis of polymers **215–218**.

Yu, Gong, and Chan reacted poly(bipyridinebenzobisthiazole)s **212–214** and **219–221** with $[Ru(bpy)_2(acetone)_2](OTf)_2$ to give partially metalated polymers **216–218** and **222–224**, respectively, each with 0.2–0.7 Ru per bpy (Schemes 4.49 and 4.50).[127] Upon metalation, the absorption spectra, fluorescence spectra, and cyclic voltammograms were changed and confirmed interaction of the Ru^{II} complexes with the polymer backbone. Moreover, the properties of the metalated polymers were compared with a series of model oligomers. The metallopolymers showed good thermal stability (>300°C). TOF measurements indicated that the drift velocity of charge carriers was enhanced by one to two orders of magnitude in the metallopolymer.

Yamamoto has investigated conjugated polymers **225** and **226**, incorporating phen in the backbone.[128,129] These polymers may function as chemosensors for transition metals as they

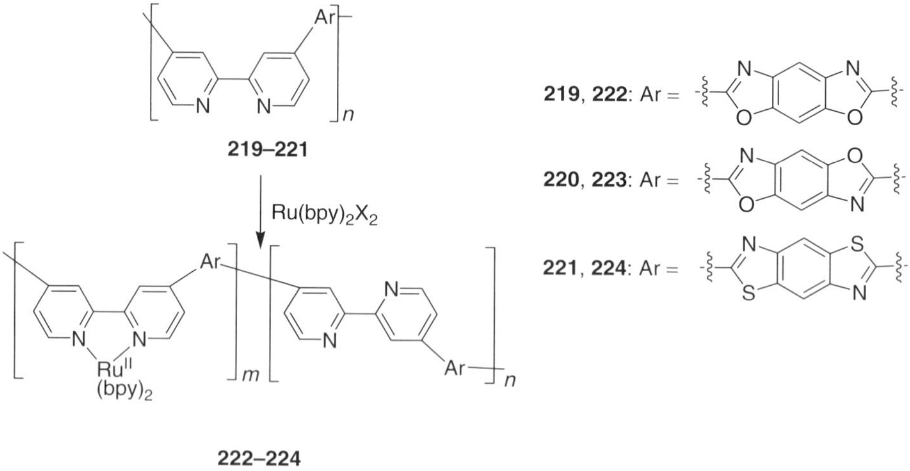

Scheme 4.50 Synthesis of polymers **222–224**.

undergo significant changes in color, as well as emission, upon binding metal ions. By coordinating Zn^{II}, Eu^{III}, and Ir^{III} complexes to the phen groups, the researchers produced highly luminescent polymers with tunable emission properties. Polymers **227** and **228** were reacted with $Ru(bpy)_2Cl_2$ to give polymers **229** and **230** with a high content of Ru (nearly 1 per phen) (Scheme 4.51).[130] Because the Ru polymer absorbs into the visible range, these polymers are attractive for optical applications.

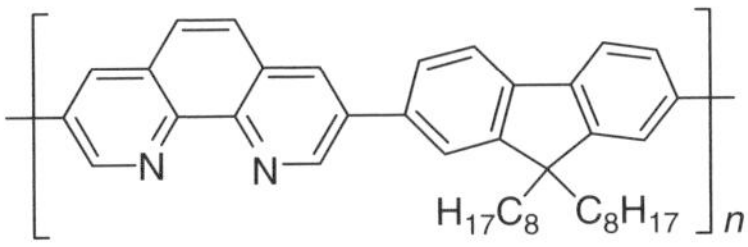

225

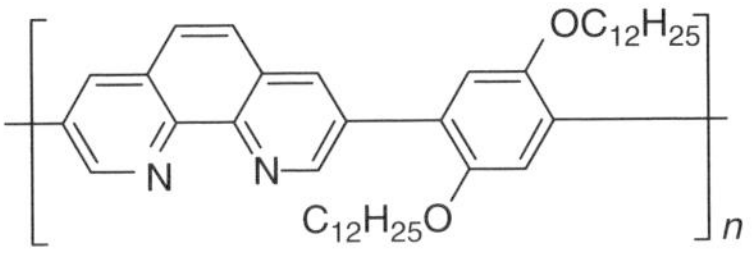

226

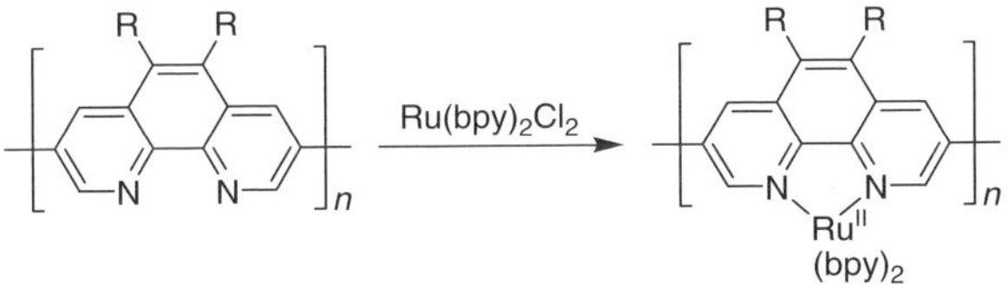

227 (R = H)
228 (R = Me)

229 (R = H)
230 (R = Me)

Scheme 4.51 Synthesis of polymers **229** and **230**.

Chan and co-workers have prepared several conjugated polymers incorporating rhenium(I) or ruthenium(II) dipyridophenazine complexes.[131] Palladium-catalyzed Heck coupling of metal-containing monomers, **231** or **232**, with **233** and *p*-divinylbenzene, gave metallopolymers **234** and **235** (Scheme 4.52). A series of copolymers was prepared where the metal content of the polymer was varied. Although no molecular weights were reported, the polymers showed inherent viscosities of 0.2–0.4 dL g^{-1}. In both polymers, the luminescence was partially quenched by energy transfer to the metal complex, with the completely metalated polymers showing almost no luminescence. Nevertheless, a simple LED device (ITO/polymer/Al) was prepared using polymer **234** ($x = y = 0.5$). This device emitted at 550–650 nm under forward bias (>10 V) with an external quantum efficiency of ca. 0.6%. TOF measurements indicated that electron-carrier mobility is higher than hole mobility in the polymers, in agreement with other electron-deficient quinoxaline-type polymers. The incorporation of metals into the polymer backbone was confirmed by TGA, CV, and UV-vis, NMR, and IR spectroscopies. Chan also reported conjugated polymers **236–239** that incorporate both metal phen complexes and aromatic oxadiazole units in the main chain.[132] These were made by a similar Heck coupling procedure to that in Scheme 4.52. Inherent viscosities of the polymers were in the range 0.30–0.71 dL g^{-1}. The proposed structures are based on the methods of characterization just listed. Notably, the authors demonstrated a 2-layer photovoltaic cell (ITO/polymer **237** ($x = 1$, $y = 0$)/C_{60}/Al) with short circuit current

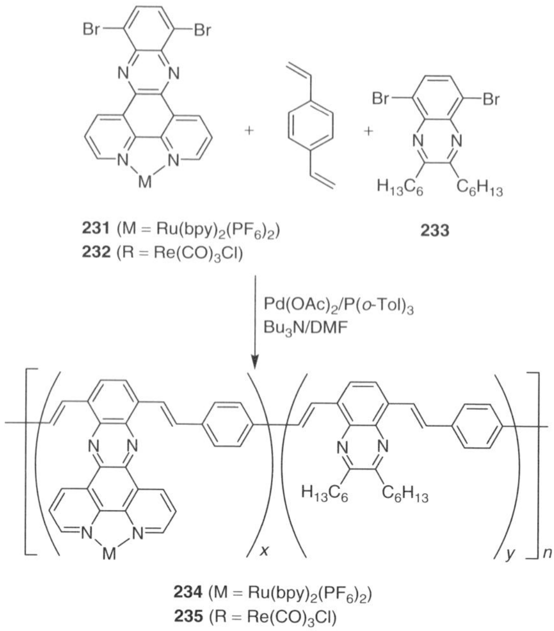

Scheme 4.52 Synthesis of metallopolymers **234** and **235**.

and open circuit voltage of 0.05 mA cm^{-2} and 0.35 V, respectively. The efficiency of the device was estimated at ca. 0.5% under a Xe arc lamp. Several of the polymers, **236–239**, were incorporated into single-layer LED devices, and showed luminescence between 570 and 710 nm (external quantum efficiencies ~0.1–0.2%).

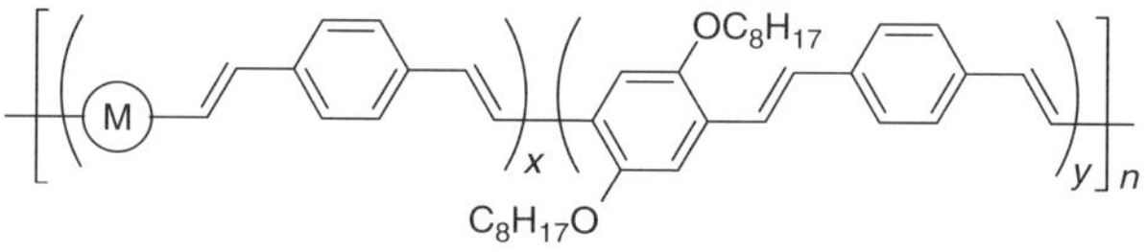

236–239

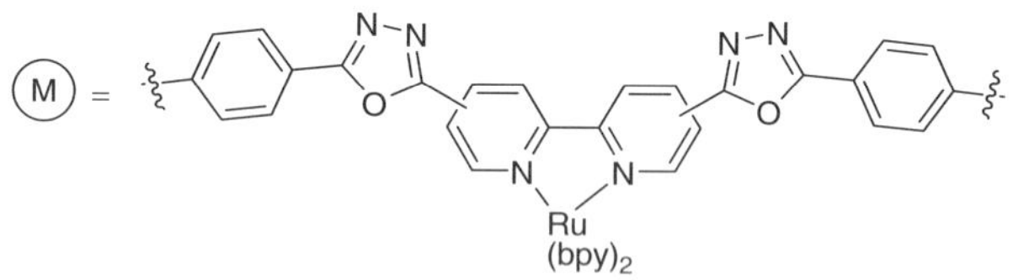

236 (4,4′-disubstituted bpy)
237 (5,5′-disubstituted bpy)

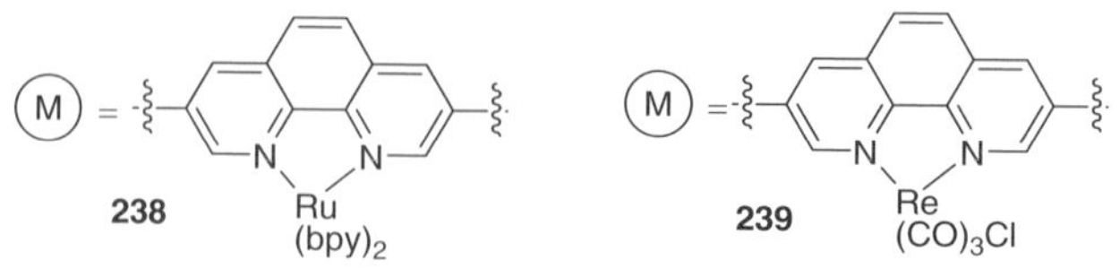

D. Other Examples

In addition to bpy and phen ligands, other chelating ligands have been incorporated into the backbone of conjugated polymers. Preformed Schiff-base polymers were reacted with $FeSO_4$ to yield polymers **240** and **241** that are partially soluble in DMSO.[133] Surprisingly, they claim that a related polymer with a nonconjugated spacer displays ferromagnetic interactions between the metal centers at 200 K. Polymers **242** and **243**, prepared by an analogous procedure, also display ferromagnetic coupling between the metal centers as revealed by a small hysteresis in the magnetization vs. field curves (5 K for Gd^{III} polymer **243**).[134] Sun has also reported the attachment of a conjugated bithiazole-containing polymer to functionalized single-walled carbon nanotubes, followed by coordination to Nd^{III}.[135] The resulting materials were characterized by IR spectroscopy and magnetic measurements. The authors suggest that, like polymers **242** and **243**, this single-walled nanotube (SWNT)–metallopolymer conjugate displays ferromagnetic interactions between the metal centers. More work is needed to substantiate these exciting claims of ferromagnetic polymers; it is possible that the results are due to small particles of metals in the polymer.

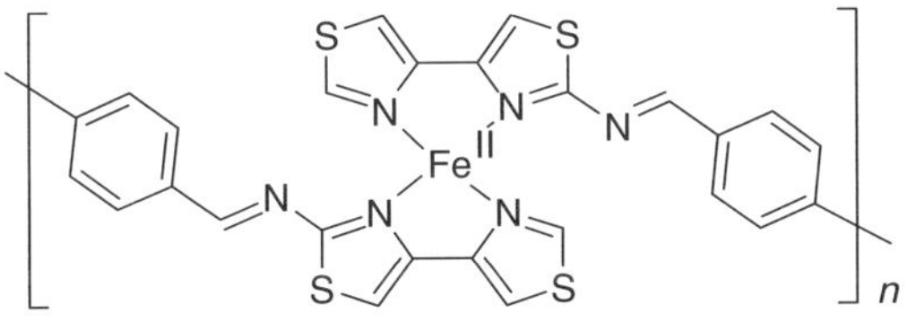

240

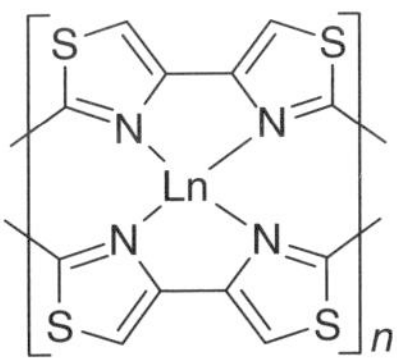

242 (Ln = Nd^{3+})
243 (Ln = Gd^{3+})

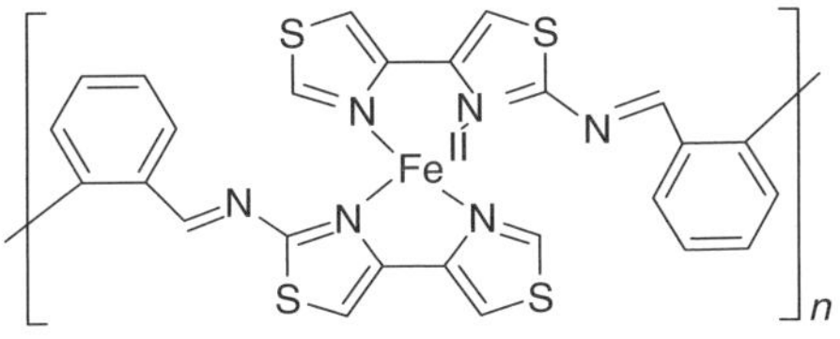

241

Cameron and Pickup investigated the addition of $Ru(bpy)_2Cl$ to poly(2-(2-pyridyl)bibenzimidazole) to afford polymer **244**.[136] This polymer was metalated to ca. 80% capacity, and gel permeation experiments indicated that it had a molecular weight >50,000 Da. An intervalence charge-transfer band for the partially oxidized (Ru^{II}/Ru^{III}) polymer was an indication of metal–metal communication in the polymer. Further, the researchers demonstrated that the conjugated polymer is an efficient conduit for electrons between Ru centers. The related Os polymer **247**, also prepared by metalation, showed strong coupling between the Os centers.[137] This coupling, mediated by the pyrazine linkage, was apparent in the cyclic voltammogram, which showed two Os^{III}/Os^{II} waves with a separation of 320 mV. Electrochemical characterization of the polymer film revealed fast electron hopping that was attributed to electron-type superexchange between the Os_2(pyrazine) moieties of the conjugated polymer. Pickup has reported extensive characterization of metalated polymers **244–248**.[138] Electron diffusion coefficients were determined using impedance spectroscopy and dual-electrode voltammetry.

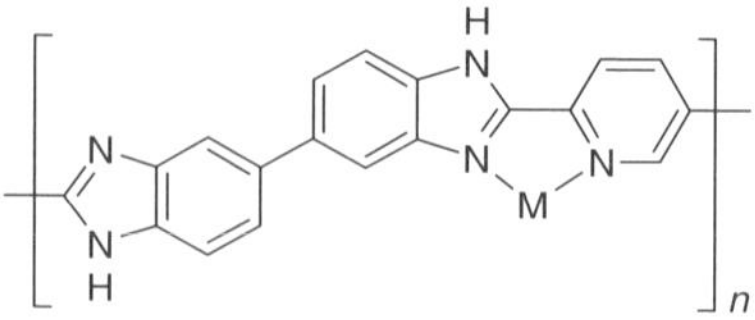

244 (M = Ru(bpy)$_2$)
245 (M = Os(bpy)$_2$)

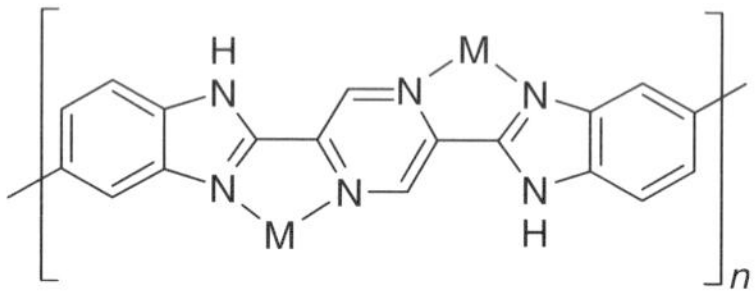

246 (M = Ru(bpy)$_2$)
247 (M = Os(bpy)$_2$)

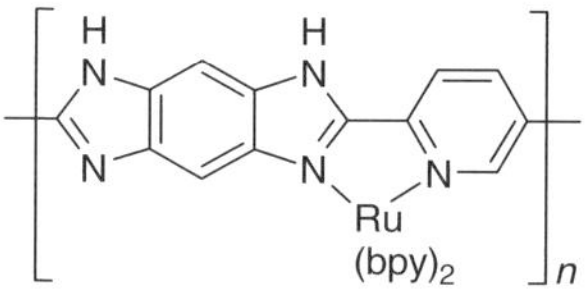

248

PPV polymers **249** containing bis(phenylimino)acenaphthalene rhenium(I) complexes were prepared by Heck coupling, as in Scheme 4.52.[139] The polymers had molecular weights of 9200–17,200 Da (M_n, GPC) and were characterized by a number of techniques. A multilayer photovoltaic device incorporating polymer **249** ($x = 0.3$, $y = 0.7$) showed a low conversion efficiency (~0.06%) with short circuit current and open circuit voltage of 380 μA cm^{-2} and 0.81 V, respectively.[140]

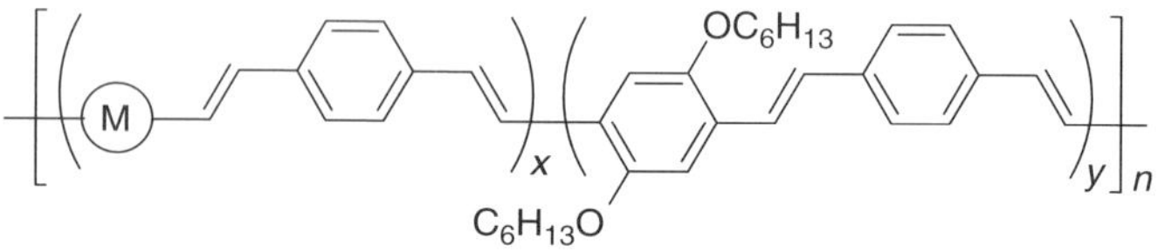

249

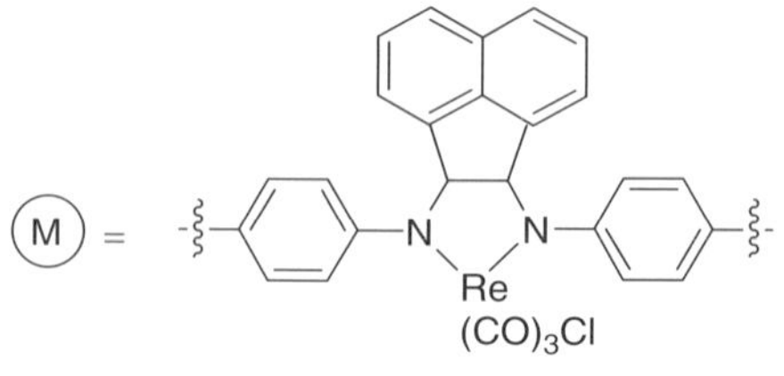

Polymers **250** and **251**, formed by coordination of Ru^{II} complexes to the preformed 2,6-poly(bis(benzimidazol-2-yl)pyridine), showed a strong enhancement of absorption and photosensitivity in the visible region relative to the unmetalated polymer.[141] The metalated polymers, with ca. 90% of the coordination sites occupied, showed an inherent viscosity of $1.8\,dL\,g^{-1}$. Although the polymers have low conductivities (ca. $10^{-10}\,S\,cm^{-1}$), they exhibited higher photocurrents than the organic precursor polymer.

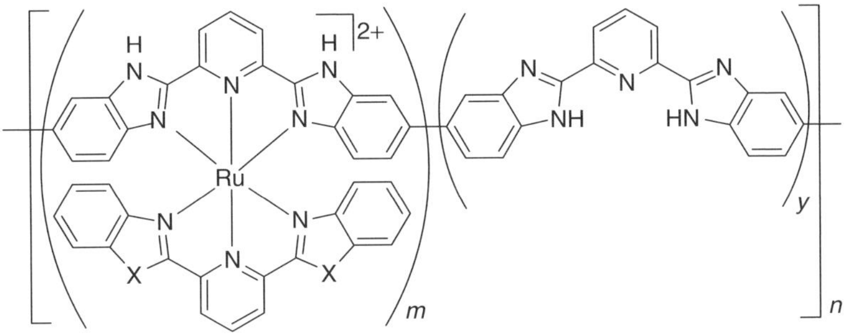

250 (X = NH)
251 (X = S)

Cheng and Euler have prepared polyazines functionalized with $[Ru(bpy)_2]^{2+}$ moieties (Scheme 4.53).[142] These polymers exhibit strong MLCT transitions assigned to the Ru complex. The authors prepared a number of model complexes for comparison of electrochemical and spectroscopic properties.

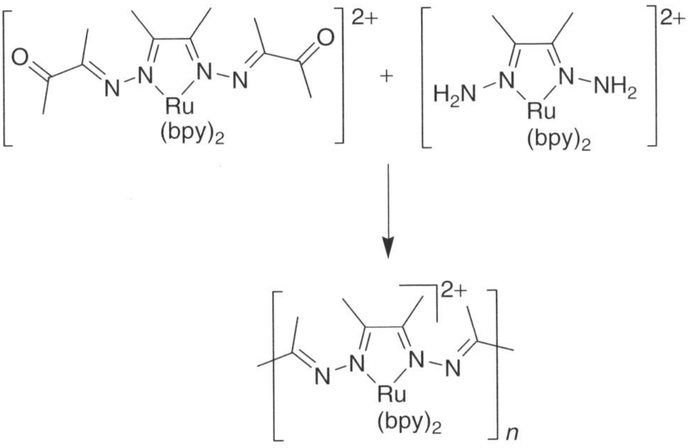

252

Scheme 4.53 Preparation of polymer **252** and related polymers.

Hanks has investigated the chemical polymerization of complexes **253–256**.[143,144] With appropriate oxidizing agents (e.g., phosphomolybdic acid or *t*-butyl ammonium persulfate), film-forming conductive (10^{-2} to $10^{-3}\,S\,cm^{-1}$) polypyrroles were generated. The polymer from Pd^{II} complex **255** was luminescent, while the other polymers were nonemissive. Dodecylbenzenesulfonate counterions improved the film quality of cast polymers.

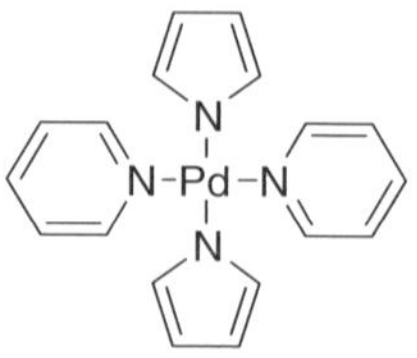

253

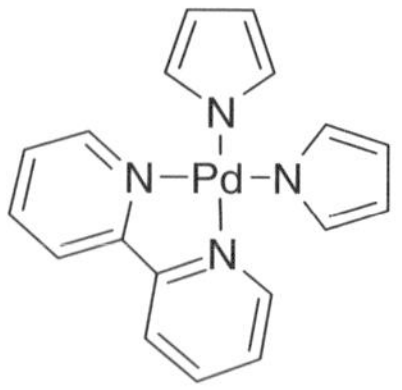

254

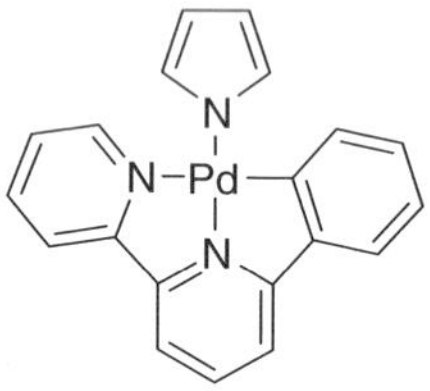

255

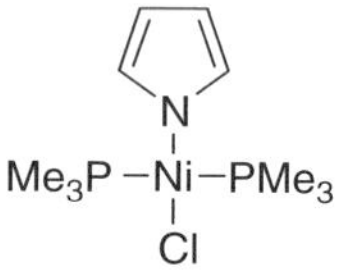

256

Conjugated polymers with O- and S-donor atoms have also been prepared. Sun reported a conjugated bithiazole-tetrathiapentalene polymer, **257**, complexed to Fe^{II} or Nd^{III}.[145] Vapor-pressure osmometry of the polymer (before metalation) indicated a molecular weight of 6700 (M_n). As with related polymers **242** and **243**, the authors found ferromagnetic coupling between the metal centers in polymer **257** metalated with Nd^{III}, while the ferrous polymer was antiferromagnetic.

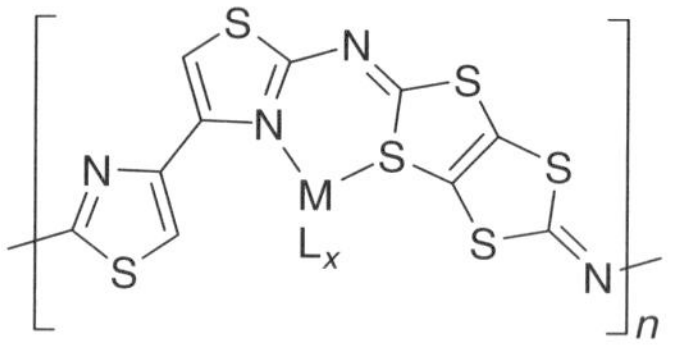

257

($ML_x = FeSO_4$ or $NdCl_3$)

Copper- and nickel-containing polymers **259a–259d** were prepared by the Schiffbase condensation of diamino metal complexes, **258a–258d**, with *p*-diformylbenzene (Scheme 4.54).[146] GPC analysis indicated that the polymers had low molecular weights (M_n) of 610–3500. Electronic spectroscopy and CV indicated that there was no communication between adjacent metal centers in the polymers.

A series of conjugated poly(phenylenefluorene)s with europium(III) complexes as side-groups (**260a–260c**; *x*:*y* 1:1, 1:2, 1:5) has been investigated.[147] These polymers were prepared by reaction of a Eu^{III} complex, ligands, and the preformed organic polymer. Incorporation of Eu^{III} was confirmed by IR analysis, TOF-SIMS (secondary-ion mass spectrometry), and NMR spectroscopy. The polymers had molecular weights of $1–4 \times 10^4$ Da (M_n; GPC), and thin films displayed the red emission characteristic of Eu complexes, indicating effective energy transfer from the polymer backbone to the metal center.

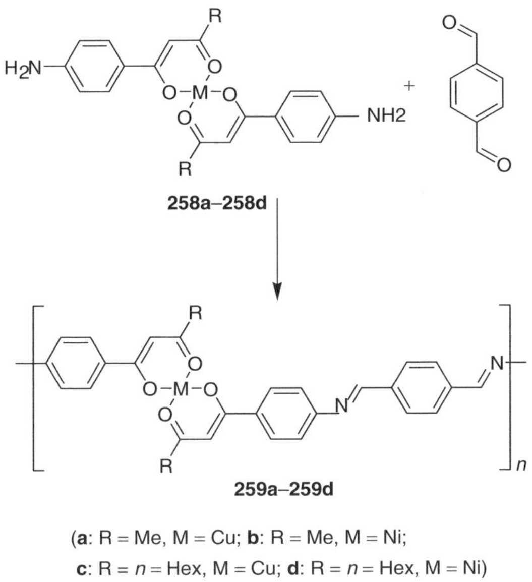

Scheme 4.54 Preparation of polymers **259a–259d**.

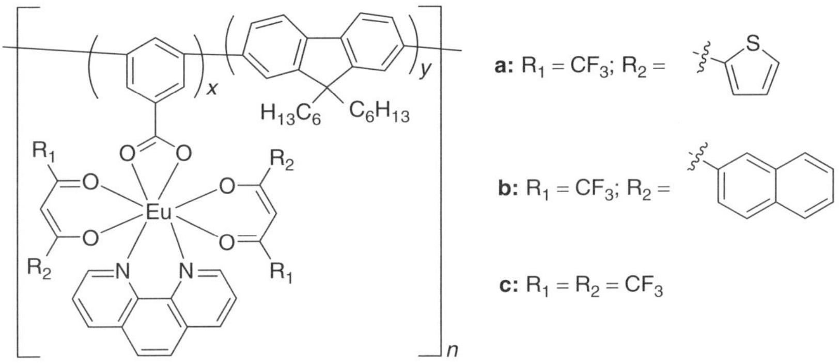

260a–260c

Peng and co-workers recently incorporated polyoxometalates into PPEs by Sonogashira coupling of cluster compound **261** with **262–263** (Scheme 4.55).[148] GPC analysis of polymers **264a**, **264b** indicated $M_n \sim 161{,}000$ Da, and light scattering indicated molecular weights of 91,000 Da. End-group analysis (NMR) of polymer **264b** indicated that the degree of polymerization was ~18, but end-groups were not observed for **264a**. In contrast to PPEs, these polymers are nonemissive, as the metal oxo clusters quench the luminescence. Photovoltaic devices of polymer **264b** sandwiched between ITO and a metal cathode gave a device with an open-circuit voltage and short-circuit current of 0.47 V and 1.12 mA cm^{-2}, respectively, and an overall efficiency of ca. 0.15%.

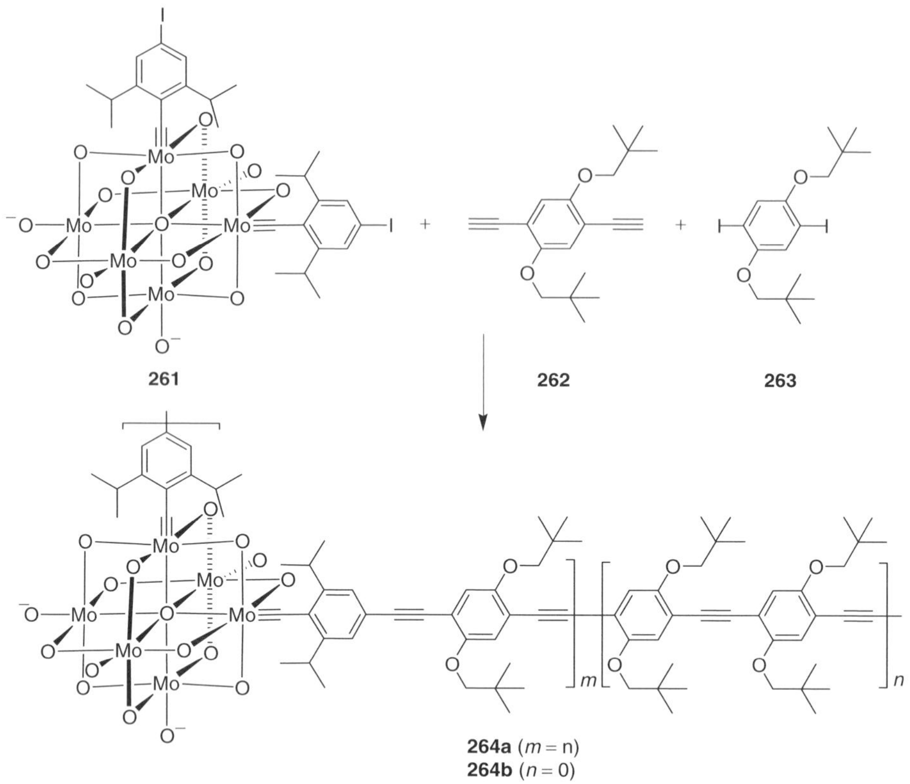

Scheme 4.55 Preparation of polymers **264a–b**.

V. PROGRESS, CHALLENGES, AND FUTURE DIRECTIONS

Over the past seven years, numerous π-conjugated metallopolymers have been synthesized and investigated. These polymers illustrate a broad range of synthetic routes, polymer architectures, and properties. Studies of these polymers have contributed to our understanding of charge-transport mechanisms, energy transfer, and conduction in π-conjugated materials. In spite of these developments, there are still no true applications of conjugated metallopolymers. Indeed, there are a number of challenges in the synthesis and characterization of these polymers that must be overcome to realize practical applications of π-conjugated metal-containing polymers.

Most studies indicate that the metal has the greatest effect on the properties of the polymer when it is part of the backbone, rather than a sidegroup. The direct coupling of the metal orbitals with the π-orbitals of the conjugated polymer may allow redox-matching to enhance the conductivity and perturb the band structure of the polymer. However, it is often synthetically easier to add the metal center as a side chain of a known polymer (e.g., on PPV) than to develop new polymerization procedures for the metal complex.

Polymerization of metal-containing monomers is often more complicated, but generally gives purer polymer than postpolymerization metalation. The former method allows 100% metalation of coordination sites in the polymer, whereas the latter method usually results in only partial metalation due to steric interactions between the sites. As this can affect the conformation of the polymer, and influence optical and electronic properties, direct polymerization of the metal-containing monomer is most desirable.

Although many polymers have been synthesized by step-growth or condensation methods, there are few examples of π-conjugated metallopolymers prepared by chain-growth mechanisms. The Gilch polymerization of porphyrin complex **21** is an example of this, although only a small amount of the complex was incorporated into the polymer. Most polymers that have been prepared by condensation methods have low molecular weights, with notable exceptions. Development of chain-growth procedures (e.g., free-radical, anionic, cationic, and ring-opening polymerization) may give access to conjugated metallopolymers with high molecular weights and controlled architectures (e.g., block copolymers).

Characterization of conjugated metallopolymers can be difficult. When the metals are paramagnetic, very broad NMR spectra are obtained. Also, many metal complexes, particularly porphyrins and phthalocyanines, aggregate in solution. This leads to broad NMR spectra and can lower solubility. Some researchers have used bulky substituents to disrupt cofacial stacking, but this can add several synthetic steps to the polymer preparation. It is often difficult to obtain clear, unequivocal characterization of the structures of metallopolymers.

Molecular-weight analysis of conjugated metallopolymers is perhaps the most frustrating characterization. Researchers have often used GPC and sometimes vapor-pressure osmometry, end-group analysis (NMR), and viscometry, but each of these techniques has drawbacks. Light-scattering analysis is often not possible due to strong absorption of the metal complex. Aggregation can lead to overestimation of molecular weights by light scattering and other techniques. Moreover, when ionic polymers are formed, the charged backbone and the spectator ions may impair analysis by many techniques. Most of the conjugated metallopolymers are rigid rod molecules. Analysis by GPC, which assumes a spherical polymer in solution, is qualitative since the instrument has been calibrated to polystyrene standards. In only a few cases have the molecular weights obtained by GPC of π-conjugated metallopolymers been corroborated by comparison with oligomeric model compounds. Additional studies of molecular-weight analysis of conjugated metallopolymers are required to advance this field.

This is still a young field, ripe with opportunities for fundamental investigations. For researchers contemplating entering this field, there are significant challenges posed by the incorporation of metals into polymer architectures—solubility is perhaps the most difficult to deal with, but one must also be concerned with the charge of the metal, counterions, "spectator" ligands on the metal, and stability of the metal complex. Characterization requires a multianalytical approach, since common techniques are often not useful. There are relatively few metal-containing moieties (e.g., bipyridines, salens, porphyrins, phthalocyanines) that have been used to build metal-containing polymers. This means that there are a large number of new metal-containing components (e.g., bisoxazoline and pincer complexes) that may be incorporated for various applications. Moreover, there is a critical need to develop improved synthetic methods to new and existing metallopolymers that will give access to block copolymers and other architectures.

A. Properties and Applications

π-Conjugated metallopolymers have been investigated for a diverse range of properties and anticipated applications. In nearly all the published articles, the authors indicate that addition of metals may affect the optical, electronic, or magnetic properties of the polymers. A survey of the references cited in this chapter gives the following applications for π-conjugated metallopolymers: chemical sensors for small molecules, ions, and organic vapors (luminescence sensors, optical sensors, potentiometric sensors, amperometric sensors), molecular electronics, electron donors, light harvesters/photosensitizers, semiconductor devices (LEDs), FETs, lasers, solar cells/other photoconductive materials, batteries (energy storage), electrochromic devices (e.g., smart windows), memory devices, photorefractive materials (gratings, image processing, optical-data storage), functionalized electrode surfaces, electrocatalysts (redox catalysts), magnetic materials, NLO devices (second-order, third-order), antistatic coatings, electro- or photochemically driven artificial muscles, substrates for photodynamic therapy, liquid crystals, and synthetic precursors to conjugated polymers. This range of applications takes advantage of the coordination ability of metals, the polarizability of metals, singlet–triplet conversion in metals, and the properties of the conjugated polymer.

We believe that commercially viable applications of conjugated metallopolymers are distant, primarily because they are much more difficult to synthesize and characterize than organic polymers. Significant advances in the synthesis and characterization of conjugated metallopolymers are necessary before they will be able to replace organic polymers for any applications. For example, although a large number of the cited articles have examined metallopolymers for application in LED devices, it is hard to imagine inorganic polymers outperforming organic analogs such that they would become competitive.

It is possible that conjugated metallopolymers could find application where they offer significant advances over organic polymers. For instance, they can be used for phosphorescent display technologies, where the metal is important for the singlet–triplet conversion in the polymer. π-Conjugated metallopolymers may also be useful for generating luminescent materials in the near-IR, a spectral region of interest for telecommunications. Still, it will be necessary to establish improved techniques for the preparation of well-defined, high molecular-weight polymers.

In addition to fundamental studies of the synthesis and characterization of metallopolymers, there are some additional areas that are worth exploring. The development of hybrid materials incorporating metallopolymers (e.g., nanotube/metallopolymer; nanoparticle/metallopolymer, etc.) is an emerging area of interest that is underdeveloped. Conjugated metallopolymers that form gels may be useful for designing artificial muscles and chemosensory gels that change properties when analytes bind to the metals. π-Conjugated metal-containing polymers may be useful as electrodes for supercapacitors. The additional charge-carrying capacity of the metals may enhance the performance of the supercapacitors. Further work in liquid-crystalline–conjugated metallopolymers is required. Although some work has been undertaken in this area (Bunz), there are enormous opportunities for forming new liquid-crystalline architectures from rigid, rodlike conjugated polymers. Along the same lines, the pyrolysis of π-conjugated metallopolymer fibers may give access to novel ceramic materials.

Clearly, many challenges remain for chemists interested in π-conjugated metal-containing polymers. Advances in the synthesis and characterization of these materials are critical to opening the door to practical and commercially viable applications. This field is a fertile research ground for young investigators.

REFERENCES

1. H. Shirakawa, E. J. Louis, A. G. MacDiarmid, C. K. Chiang, A. J. Heeger, *J. Chem. Soc, Chem. Commun.*, 578 (1997).
2. (a) D. T. McQuade, A. E. Pullen, T. M. Swager, *Chem. Rev.*, **100**, 2537 (2000); (b) R. D. Miller, J. Michl, *Chem. Rev.*, **89**, 1359 (1989).
3. R. H. Friend, R. W. Gymer, A. B. Holmes, J. H. Burroughes, R. N. Marks, C. Taliani, D. D. C. Bradley, D. A. Dos Santos, J. L. Brédas, M. Lögdlund, W. R. Salaneck, *Nature*, **397**, 121 (1999).
4. C.-T. Chen, *Chem. Mater.*, **16**, 4389 (2004).

5. L. B. Groenendaal, F. Jonas, D. Freitag, H. Pielartzik, J. R. Reynolds, *Adv. Mater.*, **12**, 481 (2000).

6. (a) T. Yamamoto, *Macromol. Rapid Commun.*, **23**, 583 (2002); (b) N. Stutzmann, R. H. Friend, H. Sirringhaus, *Science*, **299**, 1881 (2003); (c) K. Coakley, M. D. McGehee, *Chem. Mater.*, **16**, 4533 (2004).

7. I. Manners, *Synthetic Metal-Containing Polymers*, Wiley, Hoboken, New Jersey, 2002.

8. For some recent general reviews, see (a) W. E. Jones Jr., L. Hermans, B. Jiang, *Molecular Supramolecular Photochem.*, **4**, 1 (1999); (b) T. Hirao, *Coord. Chem. Rev.*, **226**, 81 (2002); (c) Y. Liu, Y. Li, K. S. Schanze, *J. Photochem. Photobiol., Part C: Photochem. Rev.*, **3**, 1 (2002); (d) B. J. Holliday, T. M. Swager, *Chem. Commun.*, 25 (2005); (e) P. G. Pickup, *J. Mater. Chem.*, **9**, 1641 (1999); (f) P. Nguyen, P. Gomez-Elipe, I. Manners, *Chem. Rev.*, **99**, 1515 (1999).

9. Wong, W.-Y., *J. Inorg. Organomet. Poly. Mater.*, **15**, 197 (2005).

10. M. Ruben, J. Rojo, F. J. Romero-Salguero, L. H. Uppadine, J.-M. Lehn, *Angew. Chem. Int. Ed.*, **43**, 3644 (2004).

11. R. Kingsborough, T. M. Swager, *Prog. Inorg. Chem.*, **48**, 123 (1999).

12. For examples, see (a) M. S. Khan et al. *New J. Chem.*, **27**, 140 (2003); (b) C. Battocchio, I. Fratoddi, M. V. Russo, G. Polzonetti, *Chem. Phys. Lett.*, **400**, 290 (2004); (c) N. J. Long, A. J. P. White, D. J. Williams, M. Younus, *J. Organomet. Chem.*, **649**, 94 (2002); (d) I. Fratoddi, C. Battocchio, A. Furlani, P. Mataloni, G. Polzonetti, M. V. Russo, *J. Organomet. Chem.*, **674**, 10 (2003); (e) R. D'Amato, I. Fratoddi, A. Cappotto, P. Altamura, M. Delfini, C. Bianchetti, A. Bolasco, G. Polzonetti, M. V. Russo, *Organometallics*, **23**, 2860 (2004); (f) N. Matsumi, Y. Chujo, O. Lavastre, P. H. Dixneuf, *Organometallics*, **20**, 2425 (2001); (g) J. S. Wilson, N. Chawdhury, M. R. A. Al-Mandhary, M. Younus, M. S. Khan, P. R. Raithby, A. Köhler, R. H. Friend, *J. Am. Chem. Soc.*, **123**, 9412 (2001); (h) W.-Y. Wong, L. Liu, J.-X. Shi, *Angew. Chem. Int. Ed.*, **42**, 4064 (2003); (i) W.-Y. Wong, *Coord. Chem. Rev.*, **249**, 971 (2005); (j) P. Altamura, G. Giardina, C. L. Sterzo, M. V. Russo, *Organometallics*, **20**, 4360 (2001).

13. (a) I. Manners, *Can. J. Chem.*, **76**, 371 (1998); (b) R. Jain, R. A. Lalancette, J. B. Sheridan, *Organometallics*, **24**, 1458 (2005).

14. S. S. H. Mao, F.-Q. Liu, T. D. Tilley, *J. Am. Chem. Soc.*, **120**, 1193 (1998).

15. For leading references, see (a) S. Kelch, M. Rehahn, *Chem. Commun.*, 1123 (1999); (b) K. Y. K. Man, H. L. Wong, W. K. Chan, C. Y. Kwong, A. B. Djurišić, *Chem. Mater.*, **16**, 365 (2004).

16. (a) T. Hirao, S. Yamaguchi, S. Fukuhara, *Electrochem. Proc.*, **15**, 96 (2000); (b) K. Yamamoto, S. Nakazawa, A. Matsufuji, *Mol. Cryst. Liq. Cryst.*, **342**, 255 (2000).

17. For examples, see (a) F. Zhao, X. Xu, S. B. Khoo, T. S. A. Hor, *Eur. J. Inorg. Chem.*, 69 (2004); (b) E. Solari, J. Hesschenbrouck, R. Scopelliti, C. Floriani, N. Re, *Angew. Chem. Int. Ed.*, **40**, 932 (2001); (c) K. D. Ley, Y. Li, J. V. Johnson, D. H. Powell, K. S. Schanze, *Chem. Commun.*, 1749 (1999); (d) S. Goeb, A. De Nicola, R. Ziessel, *J. Org. Chem.*, **70**, 6802 (2005); (e) Y. Zhu, M. O. Wolf, *J. Am. Chem. Soc.*, **122**, 10121 (2000).

18. (a) J. M. Lupton, I. D. W. Samuel, M. J. Frampton, R. Beavington, P. L. Burn, *Adv. Funct. Mater.*, **11**, 287 (2001); (b) E. C. Constable, "Coordination Polymers: Discrete Systems," in *Comprehensive Coordination Chemistry II*, Vol. 7, J. A. McCleverty, T. J. Meyer, Eds., pp. 263–302, Elsevier, Oxford, 2004; (c) K. Onitsuka, S. Takahashi, *Topics Curr. Chem.*, **228**, 39 (2003).

19. (a) J. Pei, X.-L. Liu, W.-L. Yu, Y.-H. Lai, Y.-H. Niu, Y. Cao, *Macromolecules*, **35**, 7274 (2002); (b) F. Laquai, C. Im, A. Kadashchuk, H. Bassler, *Chem. Phys. Lett.*, **375**, 286 (2003).

20. For some examples, see (a) M.-C. Brandys, R. J. Puddephatt, *J. Am. Chem. Soc.*, **123,** 4839 (2001); (b) M. B. Zaman, M. D. Smith, D. M. Ciurtin, H.-C. zur Loye, *Inorg. Chem.*, **41,** 4895 (2002); (c) E. Lozano, M. Nieuwenhuyzen, S. L. James, *Chem. Eur. J.*, **7,** 2644 (2001).

21. (a) M. Schäferling, P. Bäuerle, *J. Mater. Chem.*, **14**, 1132 (2004); (b) M. Schäferling, P. Bäuerle, *Synth. Met.*, **101**, 38 (1999); (c) M. Billon, B. Divisia-Blohorn, J.-M. Kern, J.-P. Sauvage, *J. Mater. Chem.*, **7,** 1169 (1997); (d) S. Hamar-Thibault, J.-C. Moutet, S. Tingry, *J. Organomet. Chem.*, **532**, 31 (1997).

22. (a) N. B. McKeown, *J. Mater. Chem.*, **10**, 1979 (2000); (b) D. Wöhrle, *Macromol. Rapid Commun.*, **22**, 68 (2001).
23. (a) D. Wöhrle, *J. Porphyrins Phthalocyanines*, **4**, 491 (2000); (b) D. Wöhrle, O. Hild, N. Trombach, R. Benters, G. Schnurpfeil, O. Suvorova, *Macromol. Symp.*, **186**, 99 (2002).
24. B. N. Achar, K. S. Lokesh, *J. Organometallic Chem.*, **689**, 2601 (2004).
25. B. N. Achar, G. M. Fohlen, K. S. Lokesh, *Polym. Degrad. Stab.*, **80**, 427 (2003).
26. C. Alkan, L. Aras, G. Gündüz, *e-Polymers*, No. *070*, 1 (2004).
27. R. P. Kingsborough, T. M. Swager, *Angew. Chem. Int. Ed.*, **39**, 2897 (2000).
28. E. M. Bruti, M. Giannetto, G. Mori, R. Seeber, *Electroanalysis*, **11**, 565 (1999).
29. (a) A. Tsuda, A. Osuka, *Science*, **293**, 79 (2001); (b) N. Aratani, A. Osuka, H. S. Cho, D. Kim, *J. Photochem. Photobiol., Part C: Photochem. Rev.*, **3**, 25 (2002).
30. Y. Yamaguchi, *J. Chem. Phys.*, **117**, 9688 (2002).
31. H. L. Anderson, *Chem. Commun.*, 2323 (1999).
32. T. E. O. Screen, J. R. G. Thorne, R. G. Denning, D. G. Bucknall, H. L. Anderson, *J. Am. Chem. Soc.*, **124**, 9712 (2002).
33. T. E. O. Screen, J. R. G. Thorne, R. G. Denning, D. G. Bucknall, H. L. Anderson, *J. Mater. Chem.*, **13**, 2796 (2003).
34. K. Susumu, M. J. Therien, *J. Am. Chem. Soc.*, **124**, 8550 (2002).
35. B. Jiang, S.-W. Yang, D. C. Barbini, W. E. Jones, Jr., *Chem. Commun.*, 213 (1998).
36. S. Chen, W. D. Michael, T. E. Johnson, *Polymeric Mater.: Sci. Eng.*, **86**, 45 (2002).
37. K. T. Nielsen, H. Spanggaard, F. C. Krebs, *Macromolecules*, **38**, 1180 (2005).
38. K. T. Nielsen, H. Spanggaard, F. C. Krebs, *Displays*, **25**, 231 (2004).
39. B. Jiang, W. E. Jones, Jr., *Macromolecules*, **30**, 5575 (1997).
40. S.-C. Lo, P. L. Burn, *Synth. Met.*, **102**, 1089 (1999).
41. R. S. Loewe, K. Tomizaki, W. J. Youngblood, Z. Bo, J. S. Lindsey, *J. Mater. Chem.*, **12**, 3438 (2002).
42. T. Yamamoto, N. Fukushima, H. Nakajima, T. Maruyama, I. Yamaguchi, *Macromolecules*, **33**, 5988 (2000).
43. For a review of metal-containing PT hybrid materials, see M. Wolf, *Adv. Mater.*, **13**, 545 (2001).
44. T. L. Stott, M. O. Wolf, *Coord. Chem. Rev.*, **246**, 89 (2003).
45. M. O. Wolf, Y. Zhu, *Adv. Mater.*, **12**, 599 (2000).
46. S. J. Higgins, C. L. Jones, S. M. Francis, *Synth. Met.*, **98**, 211 (1999).
47. Y. Zhu, M. O. Wolf, *Chem. Mater.*, **11,** 2995 (1999).
48. L. Tan, M. D. Curtis, A. H. Francis, *Macromolecules*, **35**, 4628 (2002).
49. (a) O. Clot, M. O. Wolf, B. O. Patrick, *J. Am. Chem. Soc.*, **122**, 10456 (2000); (b) O. Clot, M. O. Wolf, B. O. Patrick, *J. Am. Chem. Soc.*, **123**, 9963 (2001).
50. J.-C. Lee, I. Tomita, T. Endo, *Macromolecules*, **31**, 5916 (1998).
51. F. Wang, Y.-H. Lai, M. Y. Han, *Org. Lett.*, **5**, 4791 (2003).
52. F. Wang, Y.-H. Lai, *Macromolecules*, **36**, 536 (2003).
53. D. H. Kim, D.-S. Park, Y.-B. Shim, S. C. Shin, *J. Organomet. Chem.*, **608**, 133 (2000).
54. B. S. Kang, D. H. Kim, T. S. Jung, E. K. Jang, Y. Pak, S. C. Shin, D.-S. Park, Y.-B. Shim, *Synth. Met.*, **105**, 9 (1999).
55. D. H. Kim, J.-H. Kim, T. H. Kim, D. M. Kang, Y. H. Kim, Y.-B. Shim, S. C. Shin, *Chem. Mater.*, **15**, 825 (2003).
56. J. Hjelm, E. C. Constable, E. Figgemeier, A. Hagfeldt, R. Handel, C. E. Housecroft, E. Mukhtar, E. Schofield, *Chem. Commun.*, 284 (2002).
57. J. Hjelm, R. W. Handel, A. Hagfeldt, E. C. Constable, C. E. Housecroft, R. J. Forster, *J. Phys. Chem. B*, **107**, 10431 (2003).

58. J. Hjelm, R. W. Handel, A. Hagfeldt, E. C. Constable, C. E. Housecroft, R. J. Forster, *Electrochem. Commun.*, **6**, 193 (2004).
59. J. Hjelm, R. W. Handel, A. Hagfeldt, E. C. Constable, C. E. Housecroft, R. J. Forster, *Inorg. Chem.*, **44**, 1073 (2005).
60. S. S. Zhu, R. P. Kingsborough, T. M. Swager, *J. Mater. Chem.*, **9**, 2123 (1999).
61. G. Zotti, S. Zecchin, G. Schiavon, A. Berlin, M. Penso, *Chem. Mater.*, **11**, 3342 (1999).
62. V. Aranyos, A. Hagfeldt, H. Grennberg, E. Figgemeier, *Polyhedron*, **23**, 589 (2004).
63. L. Trouillet, A. De Nicola, S. Guillerez, *Chem. Mater.*, **12**, 1611 (2000).
64. K. A. Walters, L. Trouillet, S. Guillerez, K. S. Schanze, *Inorg. Chem.*, **39**, 5496 (2000).
65. L. Trouillet, M. Lapkowski, O. Stephan, S. Guillerez, *Synth. Met.*, **109**, 277 (2000).
66. F. Lafolet, F. Genoud, B. Divisia-Blohorn, C. Aronica, S. Guillerez, *J. Phys. Chem. B*, **109**, 12755 (2005).
67. S. Guillerez, M. Kalaji, F. Lafolet, D. N. Tito, *J. Electroanal. Chem.*, **563**, 161 (2004).
68. J.-M. Kern, J.-P. Sauvage, G. Bidan, B. Divisia-Blohorn, *J. Polym. Sci., Part A: Polym. Chem.*, **41**, 3470 (2003).
69. S. S. Zhu, P. J. Carroll, T. M. Swager, *J. Am. Chem. Soc.*, **118**, 8713 (1996).
70. P.-L. Vidal, M. Billon, B. Divisia-Blohorn, G. Bidan, J. M. Kern, J. P. Sauvage, *Chem. Commun.*, 629 (1998).
71. P.-L. Vidal, B. Divisia-Blohorn, G. Bidan, J.-M. Kern, J.-P. Sauvage, J.-L. Hazemann, *Inorg. Chem.*, **38**, 4203 (1999).
72. P. L. Vidal, B. Divisia-Blohorn, M. Billon, G. Bidan, J. M. Kern, J. P. Sauvage, *Synth. Met.*, **102**, 1478 (1999).
73. J.-P. Sauvage, J.-M. Kern, G. Bidan, B. Divisia-Blohorn, P.-L. Vidal, *New J. Chem.*, **26**, 1287 (2002).
74. B. Divisia-Blohorn, F. Genoud, C. Borel, G. Bidan, J.-M. Kern, J.-P. Sauvage, *J. Phys. Chem. B.*, **107**, 5126 (2003).
75. G. Bidan, M. Billon, B. Divisia-Blohorn, B. Leroy, P. L. Vidal, J. M. Kern, J. P. Sauvage, *J. Chim. Phys.*, **95**, 1254 (1998).
76. P.-L. Vidal, B. Divisia-Blohorn, G. Bidan, J.-L. Hazemann, J.-M. Kern, J.-P. Sauvage, *Chem. Eur. J.*, **6**, 1663 (2000).
77. J. Buey, T. M. Swager, *Angew. Chem. Int. Ed.*, **39**, 608 (2000).
78. K. Araki, H. Endo, G. Masuda, T. Ogawa, *Chem. Eur. J.*, **10**, 3331 (2004).
79. A. Vigalok, Z. Zhu, T. M. Swager, *J. Am. Chem. Soc.*, **123**, 7917 (2001).
80. A. Vigalok, T. M. Swager, *Adv. Mater.*, **14**, 368 (2002).
81. B. J. MacLean, P. G. Pickup, *Chem. Commun.*, 2471 (1999).
82. B. J. MacLean, P. G. Pickup, *J. Mater. Chem.*, **11**, 1357 (2001).
83. C. L. Kean, P. G. Pickup, *Chem. Commun.*, 815 (2001).
84. C. L. Kean, D. O. Miller, P. G. Pickup, *J. Mater. Chem.*, **12**, 2949 (2002).
85. C. Pozo-Gonzalo, R. Berridge, P. J. Skabara, E. Cerrada, M. Laguna, S. J. Coles, M. B. Hursthouse, *Chem. Commun.*, 2408 (2002).
86. A. Mucci, F. Parenti, L. Pigani, R. Seeber, C. Zanardi, M. I. Pilo, N. Spano, M. Manassero, *J. Mater. Chem.*, **13**, 1287 (2003).
87. C. Mangeney, J.-C. Lacroix, K. I. Chane-Ching, M. Jouini, F. Villain, S. Ammar, N. Jouini, P.-C. Lacaze, *Chem. Eur. J.*, **7**, 5029 (2001).
88. R. P. Kingsborough, T. M. Swager, *J. Am. Chem. Soc.*, **121**, 8825 (1999).
89. R. P. Kingsborough, T. M. Swager, *Chem. Mater.*, **12**, 872 (2000).
90. T. Shioya, T. M. Swager, *Chem. Commun.*, 1364 (2002).
91. D. D. Kenning, S. C. Rasmussen, *Macromolecules*, **36**, 6298 (2003).
92. D. D. Kenning, K. A. Mitchell, M. R. Funfar, S. C. Rasmussen, *Poly. Prepr.*, **42**, 665 (2001).

93. R. W. Heo, F. B. Somoza, T. R. Lee, *J. Am. Chem. Soc.*, **120**, 1621 (1998).
94. H. Plenio, J. Hermann, J. Leukel, *Eur. J. Inorg. Chem.*, 2063 (1998).
95. H. Plenio, J. Hermann, A. Sehring, *Chem. Eur. J.*, **6**, 1820 (2000).
96. K. Naka, T. Uemura, Y. Chujo, *Macromolecules*, **33**, 6965 (2000).
97. R. D. A. Hudson, B. M. Foxman, M. Rosenblum, *Organometallics*, **18**, 4098 (1999).
98. G. E. Southard, M. D. Curtis, *Organometallics*, **20**, 508 (2001).
99. I. Tomita, J.-C. Lee, T. Endo, *J. Organomet. Chem.*, **611**, 570 (2000).
100. W. Steffen, B. Köhler, M. Altmann, U. Scherf, K. Stitzer, H.-C. zur Loye, U. H. F. Bunz, *Chem. Eur. J.*, **7**, 117 (2001).
101. P. A. Deck, C. R. Maiorana, *Macromolecules*, **34**, 9 (2001).
102. I. Matsuoka, K. Aramaki, H. Nishihara, *J. Chem. Soc., Dalton Trans.*, 147 (1998).
103. M. Kurashina, M. Murata, T. Watanabe, H. Nishihara, *J. Am. Chem. Soc.*, **125**, 12420 (2003).
104. C. Huber, F. Bangerter, W. R. Caseri, C. Weder, *J. Am. Chem. Soc.*, **123**, 3857 (2001).
105. A. Kokil, I. Shiyanovskaya, K. D. Singer, C. Weder, *J. Am. Chem. Soc.*, **124**, 9978 (2002).
106. F. C. Krebs, R. B. Nyberg, M. Jørgensen, *Chem. Mater.*, **16**, 1313 (2004).
107. S. W. Thomas III, S. Yagi, T. M. Swager, *J. Mater. Chem.*, **15**, 2829 (2005).
108. A. C. W. Leung, J. H. Chong, B. O. Patrick, M. J. MacLachlan, *Macromolecules*, **36**, 5051 (2003).
109. Y. Dai, T. J. Katz, D. A. Nichols, *Angew. Chem. Int. Ed. Engl.*, **35**, 2109 (1996).
110. H.-C. Zhang, W.-S. Huang, L. Pu, *J. Org. Chem.*, **66**, 481 (2001).
111. Y. Furusho, T. Maeda, T. Takeuchi, N. Makino, T. Takata, *Chem. Lett.*, **30**, 1020 (2001).
112. T. Maeda, T. Takeuchi, Y. Furusho, T. Takata, *J. Polym. Sci., Part A: Polym. Chem.*, **42**, 4693 (2004).
113. M. Vilas-Boas, C. Freire, B. de Castro, A. R. Hillman, *J. Phys. Chem. B*, **102**, 8533 (1998).
114. F. Miomandre, P. Audebert, M. Maumy, L. Uhl, *J. Electroanal. Chem.*, **516**, 66 (2001).
115. P.-H. Aubert, P. Audebert, M. Roche, P. Capdevielle, M. Maumy, G. Ricart, *Chem. Mater.*, **13**, 2223 (2001).
116. V. Aranyos, J. Hjelm, A. Hagfeldt, H. Grennberg, *J. Chem. Soc., Dalton Trans.*, 1319 (2001).
117. S. Satyanarayana, R. L. Elsenbaumer, *Synth. Met.*, **102**, 1470 (1999).
118. Q. Wang, L. Wang, L. Yu, *J. Am. Chem. Soc.*, **120**, 12860 (1998).
119. Q. Wang, L. Yu, *J. Am. Chem. Soc.*, **122**, 11806 (2000).
120. Q. Wang, L. Wang, J. Yu, L. Yu, *Adv. Mater.*, **12**, 974 (2000).
121. R. C. Smith, A. G. Tennyson, M. H. Lim, S. J. Lippard, *Org. Lett.*, **7**, 3573 (2005).
122. B. Liu, W.-L. Yu, J. Pei, S.-Y. Liu, Y.-H. Lai, W. Huang, *Macromolecules*, **34**, 7932 (2001).
123. T. Pautzsch, E. Klemm, *Macromolecules*, **35**, 1569 (2002).
124. A. Kokil, P. Yao, C. Weder, *Macromolecules*, **38**, 3800 (2005).
125. K. D. Ley, K. A. Walters, K. S. Schanze, *Synth. Met.*, **102**, 1585 (1999).
126. K. Naka, T. Uemura, Y. Chujo, *J. Polym. Sci., Part A: Polym. Chem.*, **39**, 4083 (2001).
127. S. C. Yu, X. Gong, W. K. Chan, *Macromolecules*, **31**, 5639 (1998).
128. T. Yasuda, I. Yamaguchi, T. Yamamoto, *Adv. Mater.*, **15**, 293 (2003).
129. T. Yasuda, T. Yamamoto, *Macromolecules*, **36**, 7513 (2003).
130. T. Yamamoto, Y. Saitoh, K. Anzai, H. Fukumoto, T. Yasuda, Y. Fujiwara, B.-K. Choi, K. Kubota, T. Miyamae, *Macromolecules*, **36**, 6722 (2003).
131. W. K. Chan, P. K. Ng, X. Gong, S. Hou, *J. Mater. Chem.*, **9**, 2103 (1999).
132. P. K. Ng, X. Gong, S. H. Chan, L. S. M. Lam, W. K. Chan, *Chem. Eur. J.*, **7**, 4358 (2001).
133. J. Weng, W. Sun, L. Jiang, Z. Shen, *Macromol. Rapid Commun.*, **21**, 1099 (2000).
134. J. Tang, L. Jiang, W. Sun, Z. Shen, *React. Funct. Polym.*, **61**, 405 (2004).
135. B. He, W. Sun, M. Wang, Z. Shen, *Mater. Chem. Phys.*, **87**, 222 (2004).

136. C. G. Cameron, P. G. Pickup, *J. Am. Chem. Soc.*, **121**, 11773 (1999).
137. C. G. Cameron, P. G. Pickup, *J. Am. Chem. Soc.*, **121**, 7710 (1999).
138. C. G. Cameron, T. J. Pittman, P. G. Pickup, *J. Phys. Chem. B*, **105**, 8838 (2001).
139. W. K. Chan, C. S. Hui, K. Y. K. Man, K. W. Cheng, H. L. Wong, N. Zhu, A. B. Djurišić, *Coord. Chem. Rev.*, **249**, 1351 (2005).
140. C. S. Hui, H. L. Wong, C. Y. Kwong, A. B. Djurišić, W. K. Chan, *Poly. Prepr.*, **45**, 360 (2004).
141. S. C. Yu, S. Hou, W. K. Chan, *Macromolecules*, **32**, 5251 (1999).
142. M. Cheng, W. B. Euler, *Inorg. Chem.*, **42**, 5384 (2003).
143. T. W. Hanks, M. Mathis, W. Harsha, *Synth. Met.*, **102**, 1792 (1999).
144. M. Mathis, W. Harsha, T. W. Hanks, R. D. Bailey, G. L. Schimek, W. T. Pennington, *Chem. Mater.*, **10**, 3568 (1998).
145. W. Sun, L. Jiang, J. Weng, B. He, D. Cen, Z. Shen, *React. Funct. Polym.*, **55**, 249 (2003).
146. K. Naka, E. Horii, Y. Chujo, *Polym. J.*, **32**, 316 (2000).
147. Q. D. Ling, E. T. Kang, K. G. Neoh, W. Huang, *Macromolecules*, **36**, 6995 (2003).
148. M. Lu, B. Xie, J. Kang, F.-C. Chen, Y. Yang, Z. Peng, *Chem. Mater.*, **17**, 402 (2005).

CHAPTER 5

Metal Coordination Polymers for Nanofabrication

WAI KIN CHAN AND KAI WING CHENG

The University of Hong Kong, Hong Kong, China

I. INTRODUCTION

The quest for ordered and well-defined nanosized patterns composed of organics, metals, or semiconductors has been one of the major challenges for physicists, chemists, and materials scientists in the last sixty years. Not only are the synthesis and fabrication of these materials in nanoscale of fundamental scientific interest but the materials are also highly promising in various important technological areas, such as photonics, electronics, catalysis, sensing, energy conversion, and biomedical applications. Various chemical and physical techniques have been developed in the preparation and manipulation of nanosized materials. For example, scanning probe microscopy allows us to move a single atom/molecule or to write a pattern on a substrate in the dimension of a molecule. Also, the electron beam can be used to write a regular pattern in nanometer scale. However, these techniques have very low throughput and efficiency. In order to fabricate a nanoscaled device more effectively, techniques such as soft lithography and self-assembly have been proposed.[1] These techniques are more rapid fabrication processes that can yield large areas of nanosized patterns in a relatively short period of time. Soft lithography is a technique that generates micro- to nanopatterns of self-assembled monolayer by contact printing.[2,3] For the direct self-assembly, it is based on the assembly of small molecules into supramolecular structures of the dimension of several to tens of nanometers.[4] A recent review in the application of coordination chemistry in constructing supramolecular structures is available.[5] Other than small molecules, materials such as polymers[6] and organometallic compounds[7] have been applied in the building of highly ordered self-assembled supramolecular structures.

Block copolymers are another class of materials that have wide application potentials in nanofabrication.[8] Due to the incompatibility between different blocks, nanosized domains will form in solid state as a result of phase separation.[9] In solution state, micelles with various size and shape can also be formed in a specific solvent system. The use of block copolymer for nanopatterning,[10,11] lithography,[12] and construction of supramolecular structures[13,14,15] has been

Frontiers in Transition Metal-Containing Polymers, edited by Alaa S. Abd-El-Aziz and Ian Manners.

addressed extensively. The size and morphology of the domains formed are dependent on the relative size of each block and the interaction parameters between different blocks.[16] These nanodomains may exhibit spherical, cylindrical, lamellar, or gyroid structures. In addition to those block copolymers based on common inorganic elements, it has been shown that block copolymers incorporated with metallic compounds are excellent templates for the fabrication of various metallic or metal-oxides nanostructures. In these examples, the precursor metal-containing block copolymers are functionalized with metal complex/metal ion in one of the blocks.

Metal-containing block copolymers have played a very important role in polymer science and materials engineering due to their great application potentials.[17,18] Recently, the use of metal coordination polymers and organometallic polymers in the fabrication of metallic or metal-oxides nanostructures have received tremendous attention.[19] Metal-containing polymers can also be used as building blocks in the construction of supramolecular structures. One of the most extensively studied examples of organometallic block copolymers is polyferrocenylsilane (PFS) and its derivatives.[20,21] Block copolymers composed of PFS and other polymers such as polystyrene (PS), polysiloxane, poly(ethylene oxide) (PEO), and polyisoprene have been synthesized. These PFS-containing block copolymers were demonstrated to be potential candidates for the fabrication of nanorods/nanotubes,[22–25] water-soluble cylinders[26] and vesicles,[27] magnetic nanoparticles,[28,29,30] inverse opal,[31] semiconducting nanocrystals,[32] organometallic-polypeptide,[33] and in nanolithography[34–38] or catalysis.[39] There are some reviews on the applications of these hybrid organic–inorganic nanocomposites.[40,41]

In this chapter, we focus on the discussion of using metal coordination polymers in nanofabrication. Some examples of nanostructures composed of metallic nanoparticles/semiconducting metal oxides fabricated from polymers coordinated with the precursor compounds are presented. Most of the systems presented involve the coordination between a N or O donor ligand and metal ion/metal complexes. When using metal-containing polymers for nanofabrication, one can make use of the nanodomains formed by phase separation as the templates for the subsequent buildup of nanostructures. The assembly of block polymer molecules is usually based on metal–ligand coordination or aggregation of one of the polymer blocks.[42,43] There are different approaches to fabricate nanostructures from metal coordination polymers, and they are summarized in Figure 5.1. The first approach involves the formation of self-assembled nanostructures from a metal-free block copolymer thin film on a suitable substrate. Metal ions/metal complexes are then introduced on the surface of the copolymer film. The metal ions may adsorb on the film surface or infiltrate into the polymer film due to preferential chemical or physical interactions. This precursor compound can then be treated by physical or chemical processes to convert it into the desired metal-oxide/metallic element. In addition, the organic polymers were also removed by treatment such as oxygen plasma or deep ultraviolet (UV) irradiation. The morphology of the product obtained was the replica of the original diblock copolymer.

In the second approach, metal-ion/complex was first attached to one of the polymer blocks. A thin film of the resulting polymer metal complex was then obtained by spin coating/solution casting. Alternatively, the polymer metal complex may also be dissolved in a suitable solvent system that selectively dissolves one of the blocks. Micelles or nanosized aggregates formed in this case. The micellization of amphiphilic block copolymers and their use in the formation of metal nanoparticles has been discussed previously.[44] A monolayer of micelles was introduced on a substrate surface by dipping or electrostatic attraction. The substrate was then subjected to further chemical or physical treatments as mentioned earlier. The third approach involves the formation of micelles from the metal-free block copolymer in a suitable solvent system. The micelle solution was then added with metal ion, which was selectively coordinated to one of the blocks. These micelle–metal complexes can also be processed by a procedures similar to the second approach.

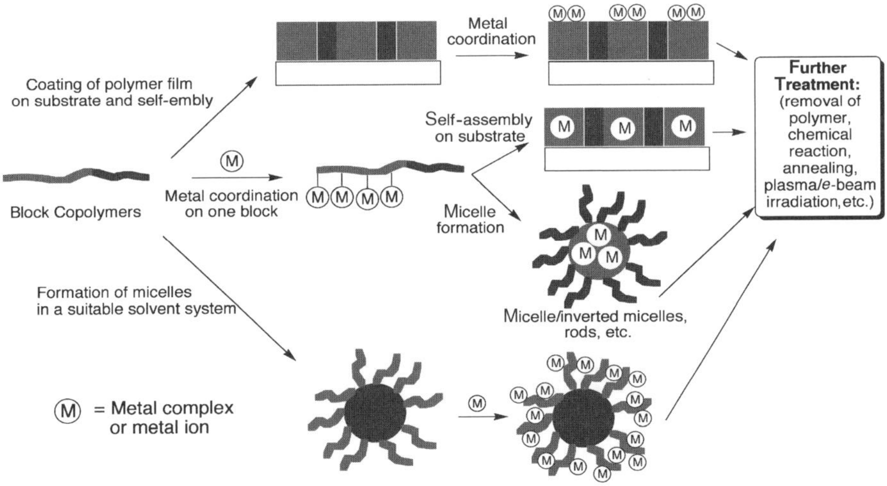

Figure 5.1 Schematic representation of different approaches in fabricating metal-containing block copolymers into self-assembled nanostructures.

II. BLOCK COPOLYMERS FOR NANOFABRICATION

A. Poly(2-vinylpyridine) Derivatives

Pyridine is a very common nitrogen donor ligand that can form a complex with a variety of main group or transition metals. In addition, its basic properties also allow the protonation of the nitrogen atom by acidic metal compounds such as tetrachloroaurucic acid ($HAuCl_4$) or hexachloroplatinic acid (H_2PtCl_6), which are both common precursors for the corresponding metallic nanoparticles. A variety of block copolymers based on 2-vinylpyridine (2VP) and 4-vinylpyridine (4VP) have been synthesized by anionic polymerization[45] or atom-transfer radical polymerization.[46] Block copolymers of PS and poly(2-vinylpyridine) (P2VP)/poly (4-vinylpyridine) (P4VP) are very good candidates for the formation of interesting nanostructures by self-assembly, because the PS and PVP blocks are incompatible in nature. The micellization properties of PS-*b*-P4VP have been studied.[47] As a result of the phase separation, the aggregated polymer metal complex block may exhibited a variety of morphologies. There have been several reports on the fabrication of nanostructures by the formation of hydrogen-bond complexes between PS-*b*-PVP and an organic hydrogen-bond donor.[14,15,48] Some examples of the use of block copolymers that contain 2VP and 4VP for nanofabrication are discussed later in the section.

Metal-free PS-*b*-P2VP with various block sizes were coated on a highly polar ionic substrate (mica).[49] The majority of the mica surface was covered by P2VP. In addition, due to the unfavorable interaction between the PS and P2VP blocks, the PS blocks did not cover the PVP surface, but dewatered the mica substrate, resulting in regularly arranged PS clusters on the substrate surface. The sizes of these clusters were dependent on the relative size of each polymer block. On the surface of these "laterally segregated" diblock copolymer films, a thin layer of titanium was deposited by electron-beam evaporation, followed by exposure to oxygen plasma, resulting in a layer of TiO_2. Due to the preferentially PS/Ti interaction, the height profile of the Ti/TiO_2 film was similar to those of the PS clusters. The results suggest that the aggregated PS may serve as a replica for the deposition of metal oxide. The resulting PS/Ti/TiO_2 also served as the mask for subsequent argon ion sputtering that generated a three-dimensional (3-D) surface pattern for potential lithographic processes.

Other works on the preparation of nanostructures from PS-*b*-P2VP involved the coordination of different metal ions/metal complexes to pyridine ligand, especially those containing gold compounds.[50] In some early examples, micelles of a symmetrical PS-*b*-P2VP were formed from toluene solution.[51,52] The stability of these micelles was increased by the addition of $HAuCl_4$, resulting in a polyanion block (Scheme 5.1). The micelles were then introduced on the surface of a mica plate by dipping the substrate into the micelle solution. After removal of solvent, the morphology of these block copolymer micelles with gold-containing cores was studied by atomic force microscopy (AFM). Figure 5.2 shows the gold-ion-stabilized micelles

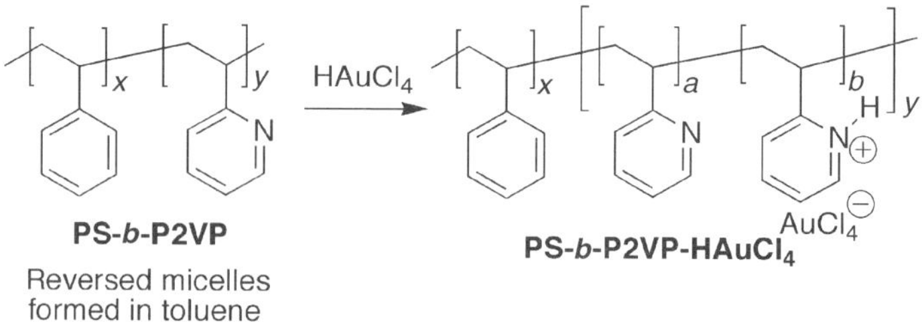

Scheme 5.1

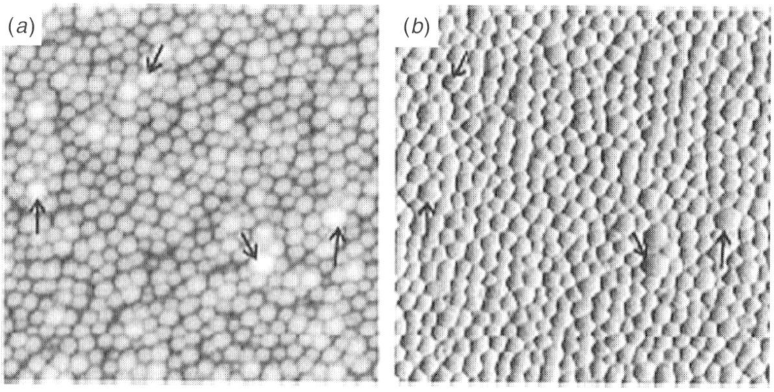

Figure 5.2 Scanning force microscope images of PS-*b*-P2VP-$HAuCl_4$ micelles ($HAuCl_4$/2VP = 0.5): (a) topography image; (b) amplitude image. Scan size = 0.9 × 0.9 μm. (From Spatz et al.[52] Reprinted with permission. Copyright © 1996 American Chemical Society.)

on a mica surface. The micelles are arranged in hexagonal order. When the concentration of the block copolymer solution was decreased from 0.5 to 0.01 mg/mL, no full coverage of the substrate by the micelles was observed. Upon increasing the $HAuCl_4$/2VP ratio from 0.1 to 1.0, the micelle diameter and core diameter were increased from 21 to 33 and 18 to 25 (±2) nm, respectively. The gold ion in the micelles were reduced to gold nanoparticles, which were stabilized by the block copolymer.[53,54] A toluene solution of PS-*b*-P2VP-$HAuCl_4$ micelles was treated with hydrazine such that the Au^{3+} ions were reduced to elemental gold nanoparticle in each micelle. The particle size was controlled by the amount of $AuCl_4^-$ ions originally loaded into the micelle cores. Monolayer of the gold-particle-containing micelles was obtained by dipping a mica substrate in the micelle solution. The gold micelles aggregated into clusters that existed in a hexagonal array. After exposing the film into oxygen plasma, the polymer molecules were removed completely, resulting in an ordered gold nanoparticle array. These gold nanoparticles were anchored with biomolecules such as biotin in a defined pattern.[54]

In later reports, asymmetric PS-*b*-P2VP were used for similar micelles formation and film deposition processes. After exposing the micelles with electron beam (e-beam) on selected areas, the chemical nature of the irradiated polymer was modified. Those nonirradiated areas could be removed selectively by washing with dimethyl formamide (DMF) or toluene, while the irradiated areas remained intact upon exposure to solvent. A schematic diagram of this process is shown in Figure 5.3. When the micelles remaining on the surface were exposed to

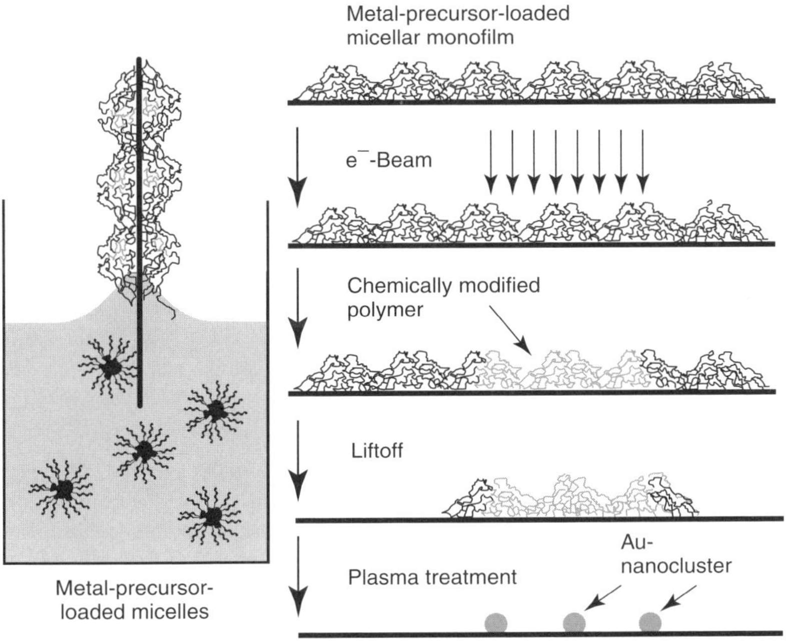

Figure 5.3 Scheme for micronanopatterning of Au nanoclusters on solid surface. The micelles were deposited on the substrate surface by dipping it into the micelle solution (left). The chemical properties of the micelles cast on the substrate surface were modified by exposure to electron beam. Those unexposed areas were removed selectively. After plasma treatment, the gold-containing micelles remaining on the surface were converted to metallic Au nanoclusters. (From Glass et al.[55] Reproduced with permission.)

isotropic hydrogen plasma, gold nanoclusters were formed.[55] The dimension and shape of the gold nanoclusters could be controlled by the e-beam. If the gold-containing PS-*b*-P2VP micelles were deposited on a silicon substrate in an ordered manner, hexagonally ordered gold nanoparticles could also be obtained upon exposure to either hydrogen or oxygen plasma.[56] Examples of some of these nanoparticles after plasma treatment are shown in Figure 5.4. Other substrates based on semiconductors such as GaAs and InP were also used in the deposition of PS-*b*-P2VP-Au micelles. For example, GaAs/InGaAs/GaAs quantum well layers were constructed by making use of the ordered gold nanoparticles on the substrate surface as etch-resistant masks.[57] When PS-*b*-PVP-Au micelles were deposited on GaAs substrate and

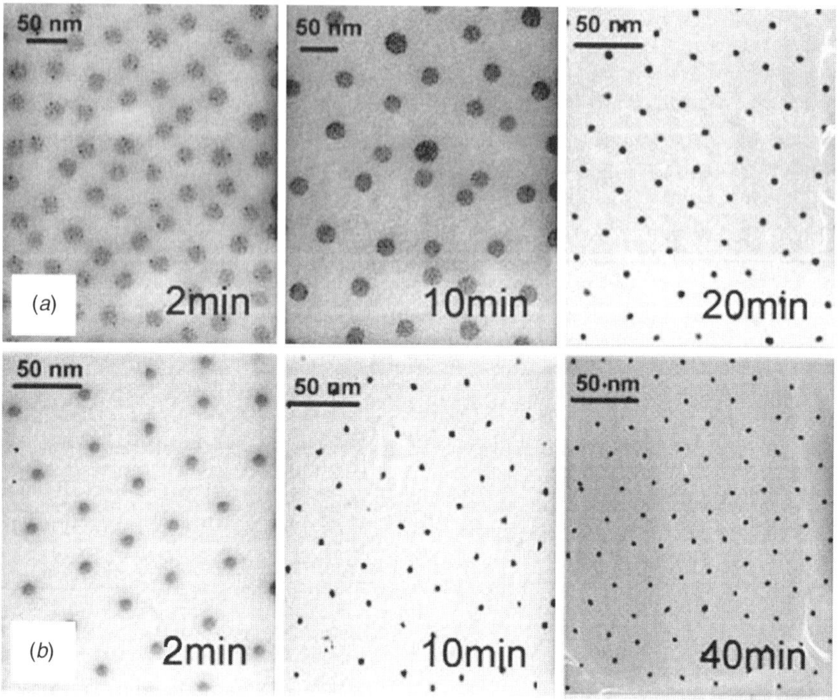

Figure 5.4 TEM images of Au-loaded micelles that were exposed to hydrogen (a) or oxygen (b) plasmas for different times. (From Kastle et al.[56] Reproduced with permission.)

exposed to hydrogen plasma subsequently, the gold nanoparticles on the GaAs surface can protect the surface from being etched by reactive ion etching (RIE). The resulting structured GaAs substrate can be used as a master for the fabrication of nanoporous gold films.[58]

In a report by Selvan et al., PS-*b*-P2VP-$HAuCl_4$ ($[HAuCl_4]/[2VP] = 0.5–1.0$) micelles prepared in toluene solution were treated with different amounts of pyrrole, resulting in gold nanoclusters encapsulated by polypyrrole shell. The oxidative polymerization is associated with the reduction of gold(III) to gold nanoparticles (Scheme 5.2).[59] When the thin film cast from the gold-micelle solution was exposed to pyrrole vapor, the pyrrole molecules adsorbed on the gold micelles initiated the vapor-phase polymerization by which dendritic nanostructures of gold–polypyrrole were obtained (Figure 5.5). It was claimed that these semiconducting gold–polymer nanostructures have potential uses in molecular electronic. The ordered gold nanoparticles were used as catalysts for the oriented growth of zinc-oxide whiskers on sapphire substrates by the vapor–liquid–solid phase transport process.[60] Zinc-oxide wires grew perpendicularly to the substrate surface, and they were more than 500 nm in height and less than 30 nm in diameter. A broad green emission band due to oxygen defects was observed.

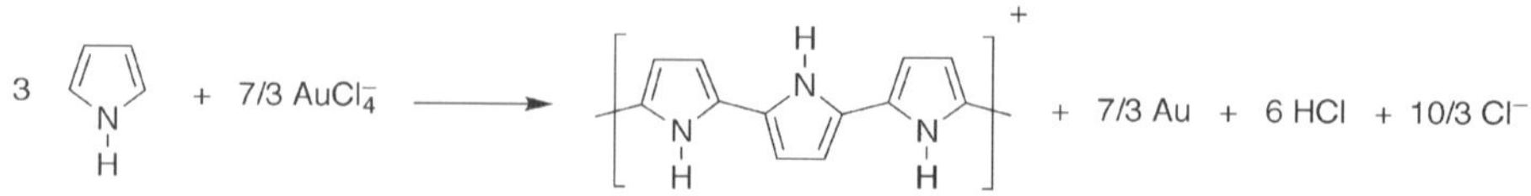

Scheme 5.2

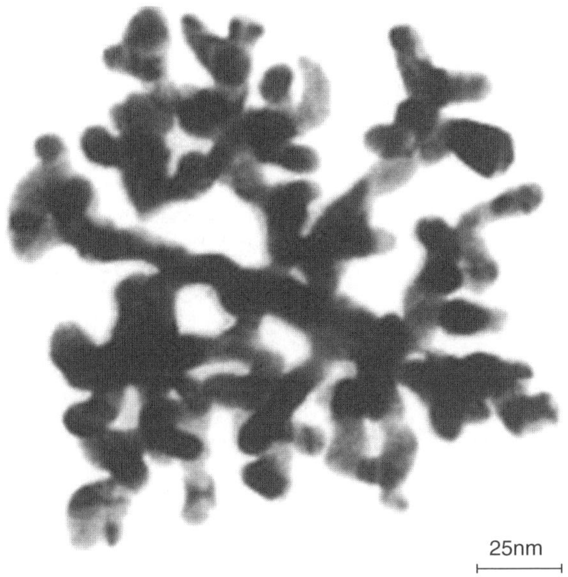

Figure 5.5 TEM picture of dendritic nanostructures of thin films cast from PS-*b*-P2VP-$HAuCl_4$ after vapor-phase polymerization of pyrrole. (Reprinted with permission from Selvan et al., *J. Phys. Chem. B*, **1999**, *103*, 7441. Copyright © 1999 American Chemical Society.)

Besides gold compounds, nanosized micelles formed by PS-*b*-P2VP with other metal salts were also reported. For example, cobalt-containing micelles were prepared by adding $CoCl_2$ into a toluene solution of PS-*b*-P2VP.[61] Similar to the examples just discussed, the reverse micelles obtained were deposited on substrate, resulting in a hexagonally ordered array. Exposure of the polymer film to oxygen resulted in the formation of Co_3O_4, which was confirmed by X-ray photoelectron spectroscopy. The cobalt-oxide particles were reduced to metallic cobalt nanoparticles by hydrogen plasma treatment, and the ordering of the micelle array was preserved. The cobalt nanoparticles showed stepwise oxidation upon exposure to pure oxygen atmosphere, and the thickness of the oxide shell on metal particle surfaces was the function of total exposure to oxygen.

Different block copolymers derived from P2VP and PEO were also used in the formation of various metal-containing micelles. Two copolymers, $P2VP_{135}$-*b*-PEO_{350} and $P2VP_{41}$-*b*-PEO_{205} were used for the preparation of micelles with different metal compounds such as Na_2PdCl_4, H_2PtCl_6, Na_2PtCl_6, $HAuCl_4$, $NaAuCl_4$, $NaPtCl_4$, and $RhCl_3$.[62] Unlike the case of PS-*b*-P2VP, the micelle formation was carried out in aqueous solution. The micelle characteristics were strongly dependent on the nature of the metal compounds. Reduction of the metal ions by $NaBH_4$ yielded the corresponding metal nanoparticles, whose size and stability were dependent on the metal loading and the micelle structures.

B. Poly(4-vinylpyridine) Derivatives

The formation of hydrogen-bond complexes between PS-*b*-P4VP and organic molecules is well known.[63] Both homopolymer and block copolymers based on P4VP have been used in the fabrication of nanostructures. Being an isomer of P2VP, P4VP is able to accommodate metal complexes of larger size because of the less restrictive steric environment. Forster et al. reported the preparation of Au nanoparticles by the chemical reduction of $HAuCl_4$ in the presence of P4VP and $[Os(bpy)_2(P4VP)_{10}Cl]^+$ films.[64] The particle sizes were in the range between 1.6 and 5.2 nm, depending on the mole fraction of osmium complex in the polymer (Figure 5.6).

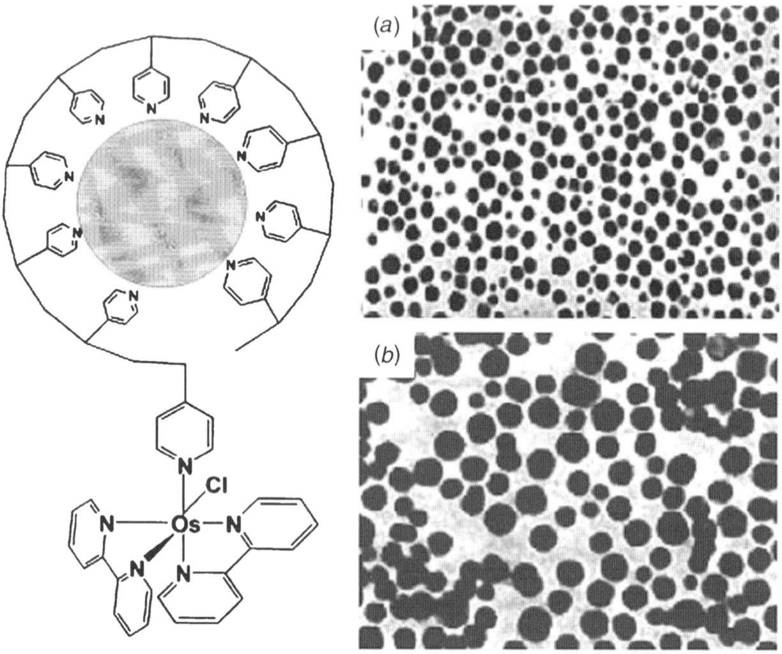

Figure 5.6 TEM micrographs of gold particles in formed by PS-*b*-P4VP-$[Os(bpy)_2(P4VP)_{10}Cl]Cl$, in which the mole fraction of Os complex is (a) 0.84, and (b) 0.02. A schematic diagram of the coordination of 4VP to the Os metal center is also shown. (From Forster and Keane.[64] Reprinted with permission. Copyright © 2003 Elsevier.)

For each gold nanoparticle, it was coordinated to nine pyridine ligands from the P4VP block, and each particle was essentially labeled with an $Os(bpy)_2$(pyridine) complex. Electrochemistry results suggested that the osmium complex enhanced the conductivity of the polymer/nanoparticle composite film.

The binding of P4VP to zinc complex is also known. Ikkala et al. reported the formation of complex between homo P4VP and $Zn(DBS)_2$ (DBS = dodecyl benzene sulfonate) (Scheme 5.3).[65] Small-angle X-ray scanning (SAXS) and birefringence study showed the formation of mesomorphic structures. In another article by the same group, the formation of complex between PS-*b*-P4VP and $Zn(DBS)_2$ was reported.[66] Interesting lamellar structures formed by the PS and P4VP-$Zn(DBS)_2$ blocks were observed. Figure 5.7a and 5.7b shows a schematic diagram for the formation of lamellar structures and the TEM micrograph of PS-*b*-P4VP-$Zn(DBS)_2$, respectively. By treating the zinc complex containing block copolymer film with methanol, $Zn(DBS)_2$ could be washed out, leading to the formation of lamellar porous structures. Interestingly, these structures did not collapse due to the presence of glassy PS block in the film.

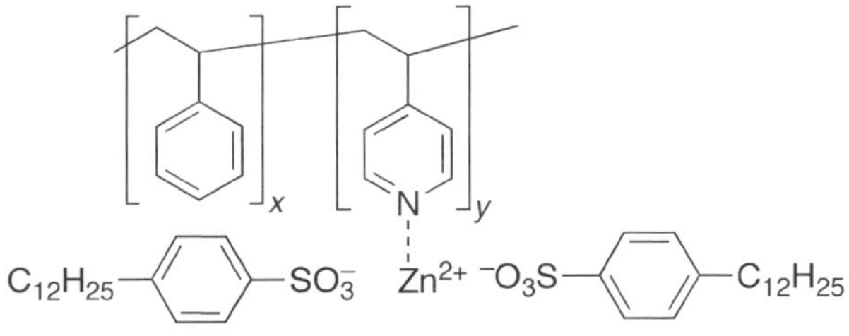

PS-*b*-P4VP[Zn(DBS)$_2$]

Scheme 5.3

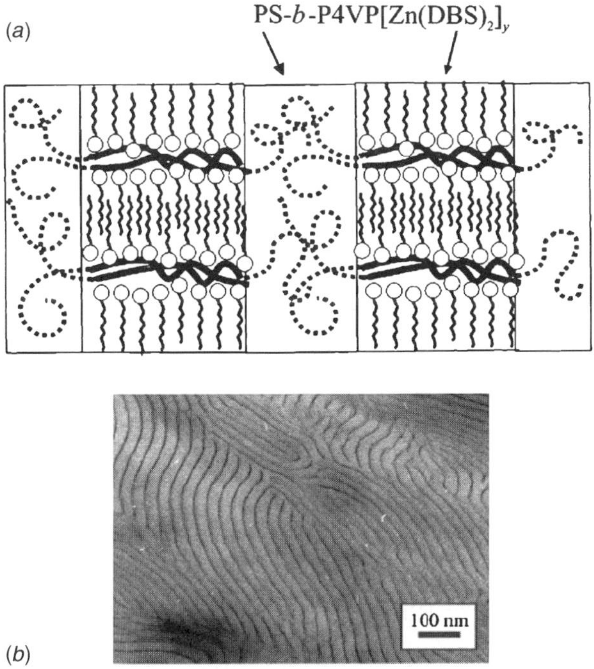

Figure 5.7 (a) Schematic diagram showing the coordination of Zn(DBS)$_2$ to the P4VP block and the formation of lamellar structures between different blocks. (b) TEM micrograph of PS-*b*-P4VP-Zn(DBS)$_2$ (4VP/Zn = 1:0.8) lamellar structures with a long period of approximately 25 nm. (From Valkama et al.[66] Reprinted with permission. Copyright © 2003 American Chemical Society.)

Similar to PS-*b*-P2VP, metallic nanoparticles could also be prepared by using PS-*b*-P4VP as the template. The preparation of palladium nanoclusters with PS-*b*-P4VP was reported by Fahmi et al.[67] The block copolymer first formed hydrogen bond complex with amphiphilic 3-*n*-pentadecylphenol (PDP) using the method developed by Ikkala et al.[13] The self-assembled polymer complex with a lamellar morphology was introduced on a Si wafer substrate. After removal of PDP with ethanol, palladium acetate was infiltrated into the P4VP block, which was followed by reduction with dimethylamine borane. Palladium nanoparticles of size 2–4 nm were obtained as a result.

Metal-containing PS-*b*-P4VP not only exhibited an interesting morphology in solid state, but also formed nanosized micelles in a suitable solvent. PS-*b*-P4VP-Co(DBS)$_2$ micelles were prepared by using copolymers with different block sizes.[68] The micelles were prepared from THF solutions. For the block copolymer with a larger P4VP block size, the shape of the micelles obtained changed from spheres, rods, vesicles, and large compound vesicles, to large compound micelles with a progressive increase in the fraction of cobalt complex [4VP/Co(DBS)$_2$] from 0.05 to 0.45. On the other hand, for the block copolymer with an almost identical block size, the micelle shape changed from rod and bilayer to large compound micelles when the 4VP/Co(DBS)$_2$ ratio was increased.

The formation of micelles with different shapes was also observed in PS-*b*-P4VP coordinated with rhenium complex. Micelles with interesting shapes were prepared from rhenium-containing block copolymer by the reaction between Re(bpy)(CO)$_3$Cl with various PS-*b*-P4VP in the presence of silver perchlorate (Scheme 5.4).[69,70] Instead of converting the metal salt into

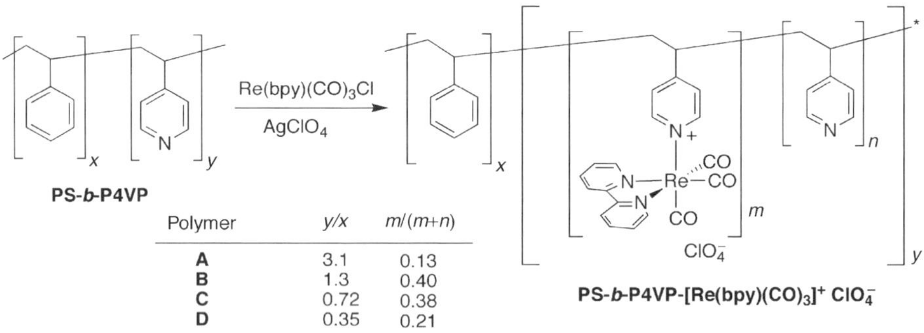

Polymer	y/x	$m/(m+n)$
A	3.1	0.13
B	1.3	0.40
C	0.72	0.38
D	0.35	0.21

Scheme 5.4

metallic nanoparticles, the rhenium complexes in the copolymer acted as "luminescent tags." The resulting copolymers exhibited a very strong emission band ($\lambda_{max} = 530\,\text{nm}$) in the photoluminescence spectra, indicating the attachment of rhenium complex to the block copolymer. Copolymers with different PS/P4VP block-size ratios were used in the study. It was found that the properties of the micelles were dependent on the solvent system used. The micelles were prepared by dissolving the rhenium-containing PS-*b*-P4VP in a "good solvent" (dichloromethane), which was followed by the slow addition of methanol. Due to the insolubility of PS in methanol, the PS block condensed gradually and the P4VP-Re complex block constituted the outer shell of the micelles. Micelles started to form when the added methanol fraction exceeded 30%, which was confirmed by light-scattering results. Figure 5.8 shows the spherical micelles formed when a methanol–dichloromethane mixture was used as the solvent. The diameters of micelles are in the range between 20 and 100 nm, except for the copolymer with a larger PS/PVP block-size ratio, which exhibited a micelle diameter of 300–400 nm. The formation of multilayered micellar structures was suggested. It was also interesting to observe that when other solvent systems were used, the resulting micelles exhibited different morphology. For example, when the micelles were prepared in toluene/dichloromethane and ethyl acetate/dichloromethane solvent pairs, disklike, spherical, rodlike, and branched micelles were observed. The detailed shapes of the micelles were dependent on the block size in the copolymer. Some of the examples are shown in Figure 5.9.

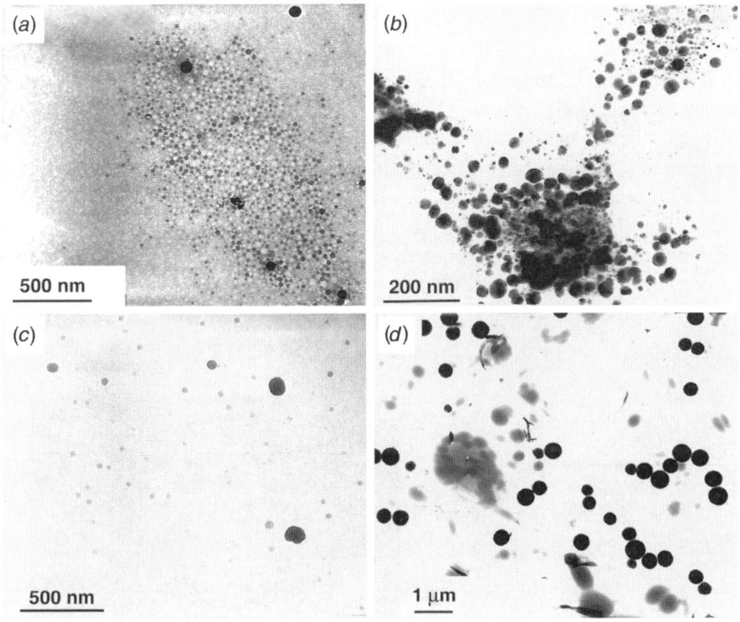

Figure 5.8 TEM micrographs of micelles formed by PS-*b*-P4VP-$[Re(bpy)(CO)_3]^+$ ClO_4^- in methanol–dichloromethane solutions with different block-size ratio: (a) P4VP/PS = 3.1; (b) 1.3; (c) 0.72; (d) 0.35. (From Hou et al.[69] Reprinted with permission. Copyright © 2003 American Chemical Society.)

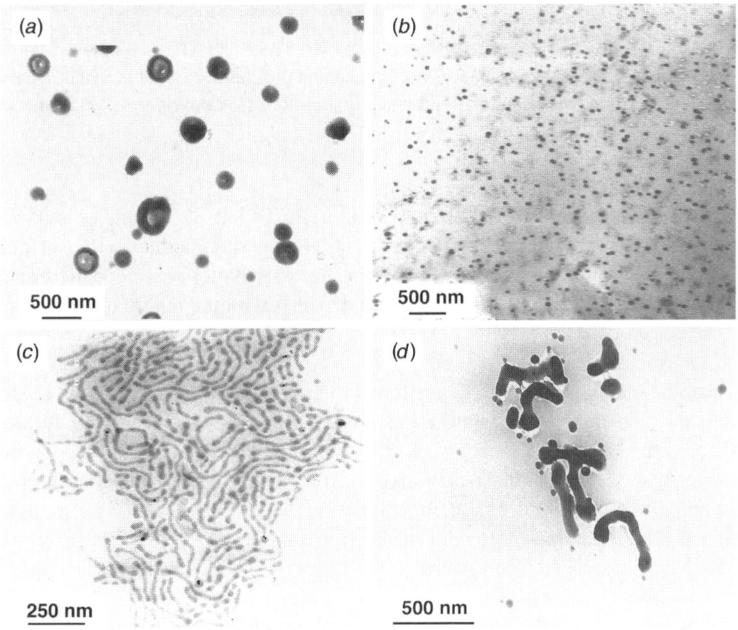

Figure 5.9 TEM micrographs of micelles formed by PS-*b*-P4VP-$[Re(bpy)(CO)_3]^+$ ClO_4^- with different block-size rato in different solvent systems: (a) 4VP/PS = 3.1 in toluene–dichloromethane; (b) 4VP/PS = 0.72 in toluene–dichloromethane; (c) 4VP/PS = 0.35 in toluene–dichloromethane; (d) 4VP/PS = 3.1 in ethyl acetate–dichloromethane. (From Hou et al.[69] Reprinted with permission. Copyright © 2003 American Chemical Society.)

In a more recent article, the use of rhenium-containing PS-*b*-P4VP as template for nanoparticles attachment was demonstrated.[71] PS-*b*-P4VP-[Re(DIAN)(CO)$_3$]$^+$ClO$_4^-$ (DIAN = bis-(phenylimino)acenaphthene) was prepared by a similar method to that discussed earlier. From ^{1}H nuclear magnetic resonance (NMR) spectroscopy, the fraction of P4VP coordinated with rhenium complex was estimated to be 70% (Re/4VP = 2.3). In this copolymer, the rhenium complex was coordinated to a bulky DIAN ligand, which significantly changed the size of the P4VP block and the interaction parameter with the PS block. As a result, the morphological properties of the copolymer were different from those of the metal-free polymer. A TEM micrograph of PS-*b*-P4VP-[Re(DIAN)(CO)$_3$]$^+$ClO$_4^-$ is shown in Figure 5.10. Cylindrical

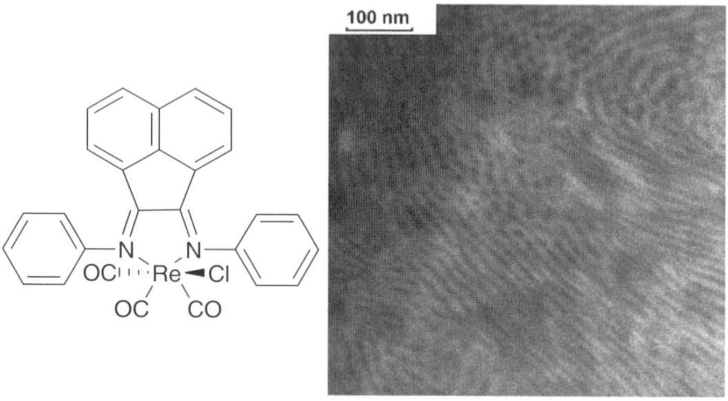

Figure 5.10 Structure of the Re(DIAN)(CO)$_3$Cl and TEM micrograph of PS-*b*-P4VP-[Re(DIAN)(CO)$_3$]$^+$ClO$_4^-$. The sample was not stained and the contrast of the image is due to the presence of rhenium. (From Cheng and Chan.[70] Reprinted with permission. Copyright © 2005 American Chemical Society.)

domains (diameter = 15–20 nm) corresponding to the P4VP-Re complex block can be clearly observed. The cylinders are oriented either parallel or perpendicular to the surface, and some of them are packed in hexagonal arrays. When the copolymer was deposited on a pretreated substrate, the morphology of the thin film was dependent on the nature of the substrate surface. The AFM images of PS-*b*-P4VP-[Re(DIAN)(CO)$_3$]$^+$ClO$_4^-$ coated on various pretreated silicon wafer surfaces are shown in Figure 5.11. Both cylindrical and isolated small island-like domains can be observed. These are attributed to the effect of polarity of the substrate surface.

Due to the presence of the ionic P4VP-Re-complex block on the film surface, nanoparticles decorated with anionic functional groups can be deposited on the copolymer film surface by electrostatic attraction. The copolymer film was immersed into a solution of cadmium sulfide nanoparticles (diameter = 10 nm), functionalized with carboxylate groups on the particle surface. The attachment of cadmium sulfide nanoparticles on the film surface was confirmed by X-ray photoelectron spectroscopy (XPS). In addition, particles deposited on the cylindrical blocks in the copolymer film surface were also observed in AFM image (Figure 5.12). This approach has potential in fabricating nanoparticle–polymer composites on patterned surface.

Diblock PEO-*b*-P4VP and triblock P4VP-*b*-PEO-*b*-P4VP copolymers were also used in the fabrication of metal containing nanostructures.[72] The copolymers were first stirred with $H_2PtCl_6{\cdot}6H_2O$ in water, yielding platinum-containing micelle cores. Coalescence of micelles

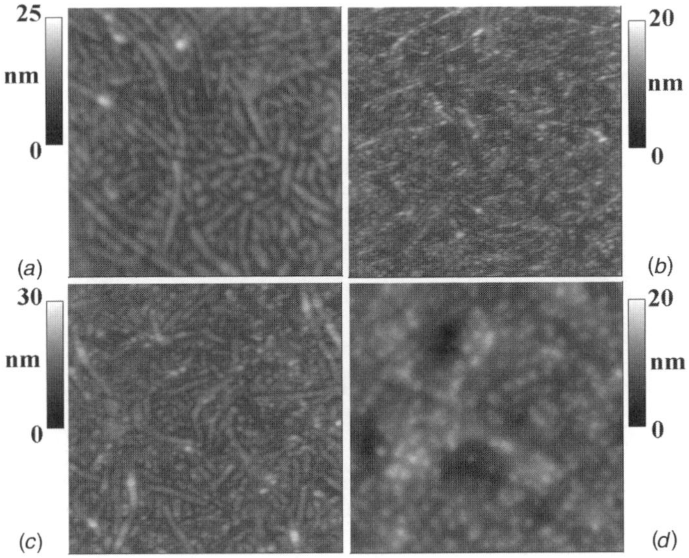

Figure 5.11 AFM images of PS-*b*-P4VP-$[Re(DIAN)(CO)_3]^+ClO_4^-$ spin-coated on different substrate surfaces (film thickness = 40 nm): (a) silicon wafer modified with (3-aminopropyl)trimethoxysilane; (b) silicon wafer modified with *N*-trimethoxysilylpropyl-*N*,*N*,*N*-trimethylammonium chloride; (c) silicon wafer modified with 3-(*p*-methoxyphenyl)propyltrichlorosilane; and (d) silicon wafer with native oxide layer removed. All the films were prepared from toluene solution. Scan size: (a)–(c) 1 × 1 μm; (d) 500 × 500 nm. (From Cheng and Chan.[71] Reprinted with permission. Copyright © 2005 American Chemical Society.)

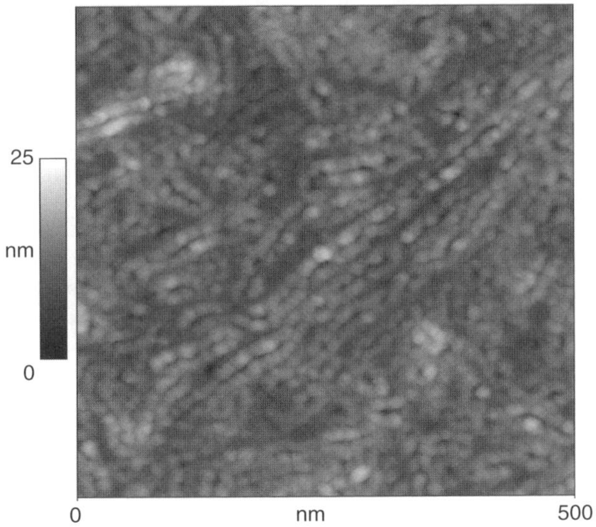

Figure 5.12 AFM image of PS-*b*-P4VP-$[Re(DIAN)(CO)_3]^+ClO_4^-$ deposited with cadmium sulfide nanoparticles. (From Cheng and Chan.[71] Reprinted with permission. Copyright © 2005 American Chemical Society.)

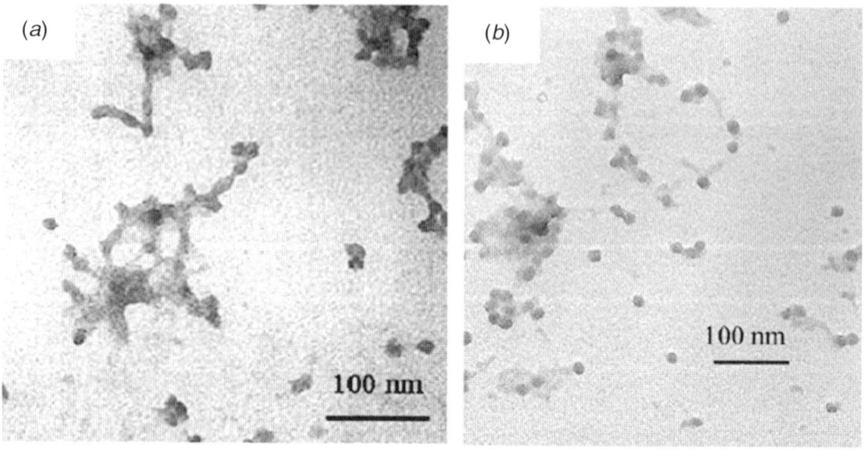

Figure 5.13 TEM images of Pt containing micelles: (a) PEO_{45}-*b*-$P4VP_{28}$-H_2PtCl_6; (b) $P4VP_{29}$-*b*-PEO_{273}-*b*-$P4VP_{29}$-H_2PtCl_6. (From Bronstein et al.[72] Reprinted in part with permission. Copyright © 2005 American Chemical Society.)

in solution was observed (Figure 5.13). It was followed by the reduction by $NaBH_4$ at different pH to obtain platinum nanoparticle clusters surrounded by the block copolymers. The effect of using different reducing agents, solution pH, and polymer structure to the organization of platinum nanoparticles in the micelles was studied by SAXS. It was found that larger platinum particles were formed in triblock copolymer micelles, while smaller cores were formed by the diblock copolymer micelles. This was explained by the participation of a triblock copolymer molecule in the formation of more than one micelle. The pH of the medium in the reduction process mainly affected the organization of metal nanoparticles in the micelle rather than the size of particles.

C. Porphyrin-Containing Polymers

Porphyrin is a macrocyclic structure that is capable of binding to different transition metals. Some examples of using porphyrin-containing polymers have been reported. Zimmerman et al. prepared a dendrimer with a tin(IV) porphyrin complex (Figure 5.14).[73] Oligomerization of the tin complex with succinic acid formed a linkage between the tin centers in the porphyrin moieties. The oligomer obtained was then cross-linked by the Grubbs ruthenium carbene catalyst, by which the vinyl groups at the outer part of the dendrimers were linked together, resulting in organic nanotubes. Subsequent removal of the porphyrin core by transesterification reaction yielded a hollow tube that is soluble in some organic solvents.

Another example of self-assembly of porphyrin-containing polymer was illustrated by Li et al.[73] Polyacetylene functionalized with fullerene and zinc porphyrin pendant groups were synthesized by polymerizing the corresponding fullerene/porphyrin substituted alkyne monomers with rhodium(I) norbornadiene catalyst (Scheme 5.5).[74] Polymers with different ratio of C_{60} and porphyrin were synthesized. The polymers showed photocurrent response when the thin films were irradiated with white light, which was due to the electron transfer from the photo-excited porphyrin to the C_{60} units. In addition, the copolymers aggregated into ellipse-shaped nanorod structures with a diameter of approximately 100 nm and a length of

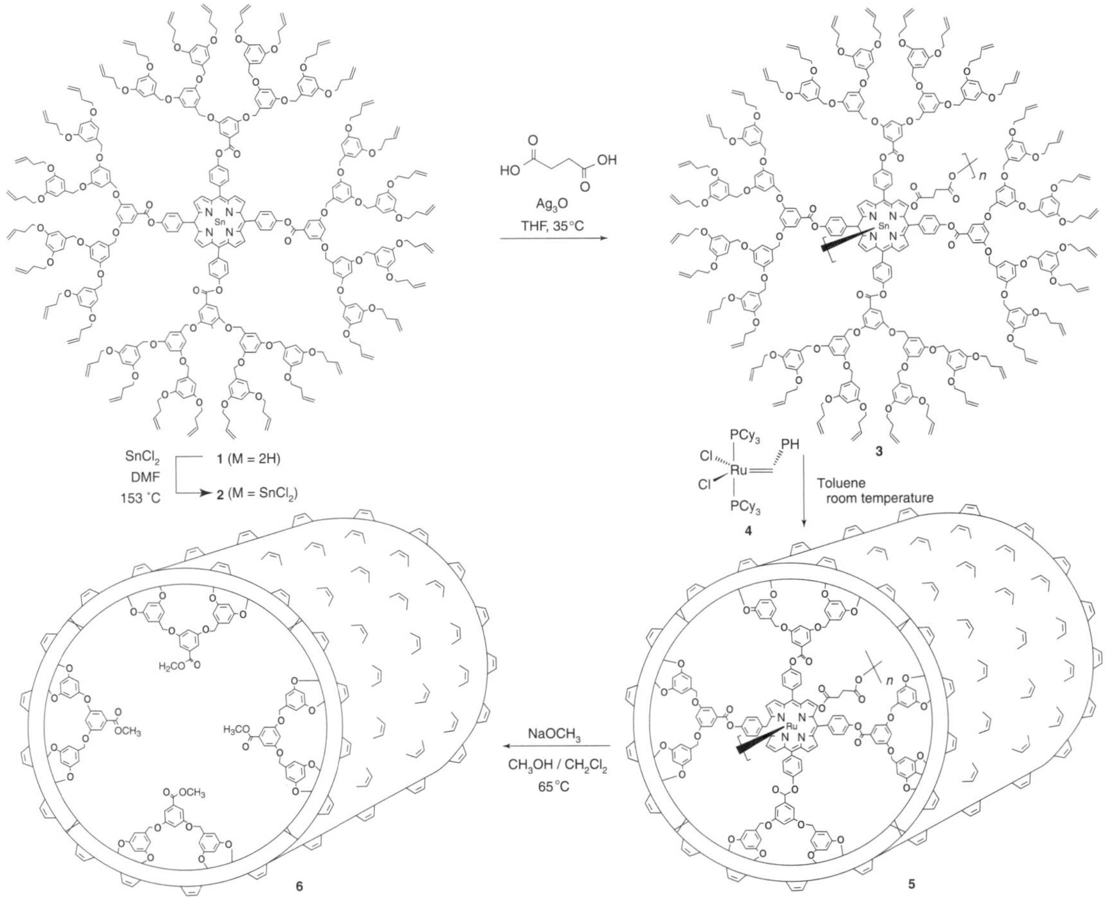

Figure 5.14 Schematic diagram for the self-assembly of tin(IV) porphyrin-containing dendrimer into nanotubes. (From Kim et al.[73])

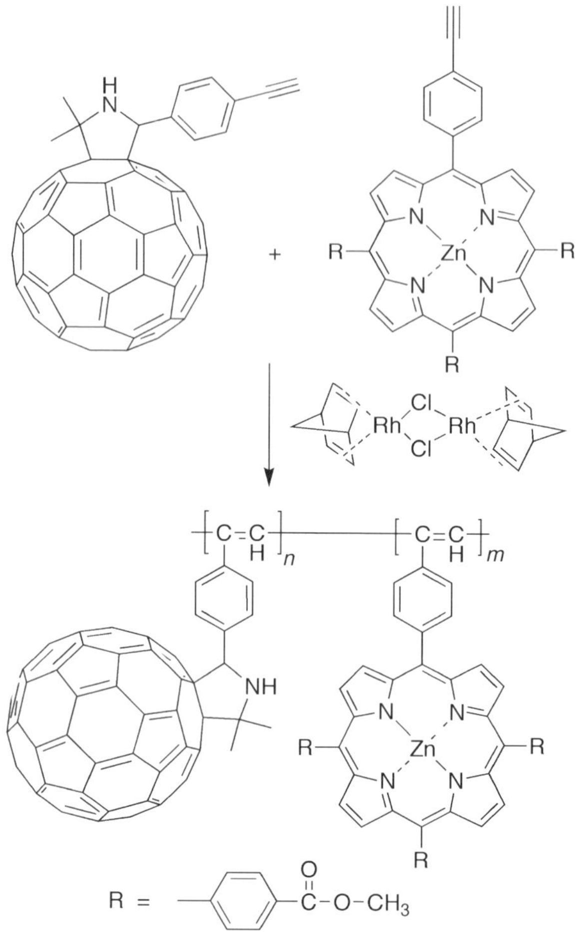

Scheme 5.5

approximately 300 nm. The formation of these nanostructures was suggested to be due to the intermolecular π–π interaction.

D. Polynorborene Derivatives

Besides anionic or controlled radical polymerization, ring-opening metathesis polymerization (ROMP) is another common and versatile method in the synthesis of block copolymers. Many block copolymers synthesized by ROMP are based on polynorborene and its derivatives. For example, the syntheses of various transition metal–containing polynorborene derivatives were reported by Cohen et al.[75,76,77] The polymerization of methyltetracyclododecene (MTD) and the subsequent attachment of the second block with platinum, palladium, and zinc complexes are shown in Scheme 5.6. For the Pt-containing polymer $[Pt]_{50}[MTD]_{113}$,[75] the cast polymer film showed a cylindrical morphology with short-range hexagonal packing of the metal-containing block. Lamellar morphology was observed for the $[Pd]_{50}[MTD]_{113}$ copolymer.

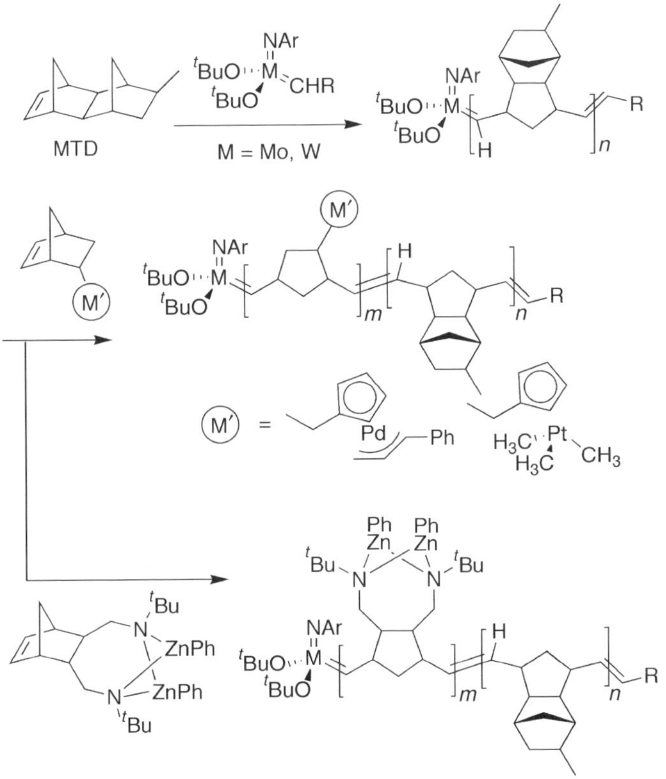

Scheme 5.6

Spherical domains composed of the metal complex were observed for the block copolymer with small Pd complex block($[Pd]_{10}[MTD]_{163}$). These polymer films were then treated with hydrogen in order to reduce the complex to metallic clusters. After reduction, the metal clusters formed in the phase-separated microdomains remained almost the same morphology as their precursor block copolymers, and the size distributions of the metal clusters were quite narrow. In a subsequent report, zinc-containing norborene-based block copolymers were synthesized by using $bTAN(ZnPh)_2$ [bTAN = 2,3-*trans*-bis((*t*-butylamido)methyl)norborn-5-ene] as one of the monomers (Scheme 5.6).[76] Like their platinum- and palladium-containing analogs, the block copolymers also exhibited microphase-separated domains with lamellar or spherical morphologies. Upon treating the polymer film with hydrogen sulfide, zinc sulfide clusters were formed in the polymer matrices. Clusters with diameter up to 30 Å were obtained. It was interesting to observe, from XPS results, that the separation of the 30-Å clusters was measured to be 5.7 eV, which is much larger than that of the bulk zinc sulfide band gap. This was attributed to the quantum size effect.

The growth of zinc-oxide nanoparticles from norborene-based block copolymers was demonstrated by Kofinas et al.[78] The block copolymer consisted of polynorborene and poly(norborene dicarboxylic acid) (block-size ratio = 400:50) was synthesized by ROMP (Scheme 5.7).[78] In a tetrahydrofuran (THF) solution, $ZnCl_2$ was introduced into the dicarboxylic acid units and the resulting Zn^{2+}-containing block film exhibited spherical domains

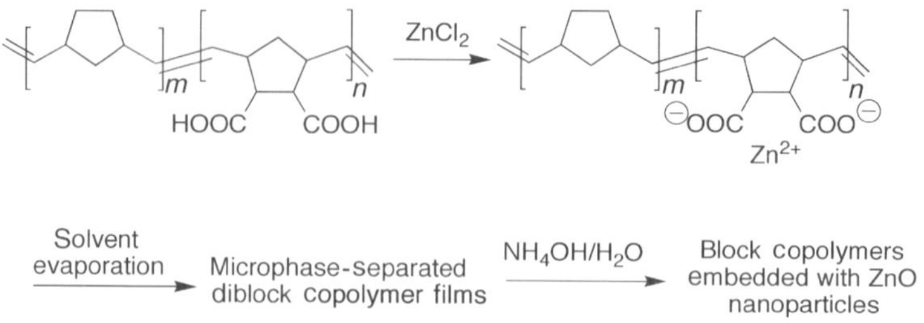

Scheme 5.7

in a polynorborene matrix. Upon treating with NH_4OH, zinc-oxide nanoparticles with diameter of 7–15 nm were formed in the polymer.

E. Block Copolymers With Poly(ethylene oxide) Segments

Poly(ethylene oxide) is an interesting polymer with a hydrophilic main chain, which can also serve as the site for coordination to some metal ions. PEO can be attached to various polar or nonpolar polymers to form block copolymers with amphiphilic properties. The PEO segment can be linked to another polymer molecule by the reaction between the terminal hydroxy end-groups of PEO with the carboxylic acid/acid chloride terminal group of another polymer. In addition, metal complex can also serve as the linker between different blocks. Schubert et al.[79] reported the synthesis of a series of block copolymers in which the two blocks were linked by a bis(2,2′:6′,2″-terpyridine) ruthenium(II) complex (denoted as [Ru]). Some examples of these block copolymers are shown in Figure 5.15. PEB_{70}-[Ru]-PEO_{70} is a block copolymer composed of poly(ethylene-*co*-butylene) (PEB) and PEO that are linked by [Ru].[79] In aqueous solution, the block copolymer formed micelles in which the "soft" PEB block acted as the core and the PEO acted as the corona. Due to the low glass transition temperature exhibited by PEB, the core of the micelles was prone to reorganization upon external stimulation. For example, when the temperature of the micelle solution was increased from 25° to 65°C, a significant change in the size distribution of the micelles was observed. Dynamic light-scattering experimental results revealed that for the micelles solution ($c = 0.5$ g/L) at 25°C, a bimodal size distribution in hydrodynamic diameter, D_h, was observed at 32 and 115 nm. The second peak was attributed to the formation of aggregates by the clustering of several individual micelles. At higher temperature, the breakup of the aggregates into smaller structures or even into individual micelles was observed, which was associated with the monomodal size distribution ($D_h = 32$ nm). This effect is reversible in nature, which can be seen from the reappearance of the original bimodal size distribution upon cooling to 25°C. AFM micrographs showed that when a dilute micelle solution was spin-coated on a silicon wafer, individual micelles were found. The sizes of these micelles/aggregates agreed with those measured by light scattering experiments.

The formation of supramolecular structures by PS_{20}-[Ru]-PEO_{70} were investigated.[80] The micelles formed were studied by ultracentrifugation experiments, which showed that both individual and aggregated micelles were formed. TEM results showed that most of the micelles were in the range between 10 and 25 nm, and some larger particles were also found. Ultracentrifugation also confirmed the formation of large aggregates, and molar mass of the particles formed was around 430,000 g/mol, corresponding to an aggregation number of 85. In the examples just discussed, the counteranions in the ruthenium complexes are hexafluorophosphate.

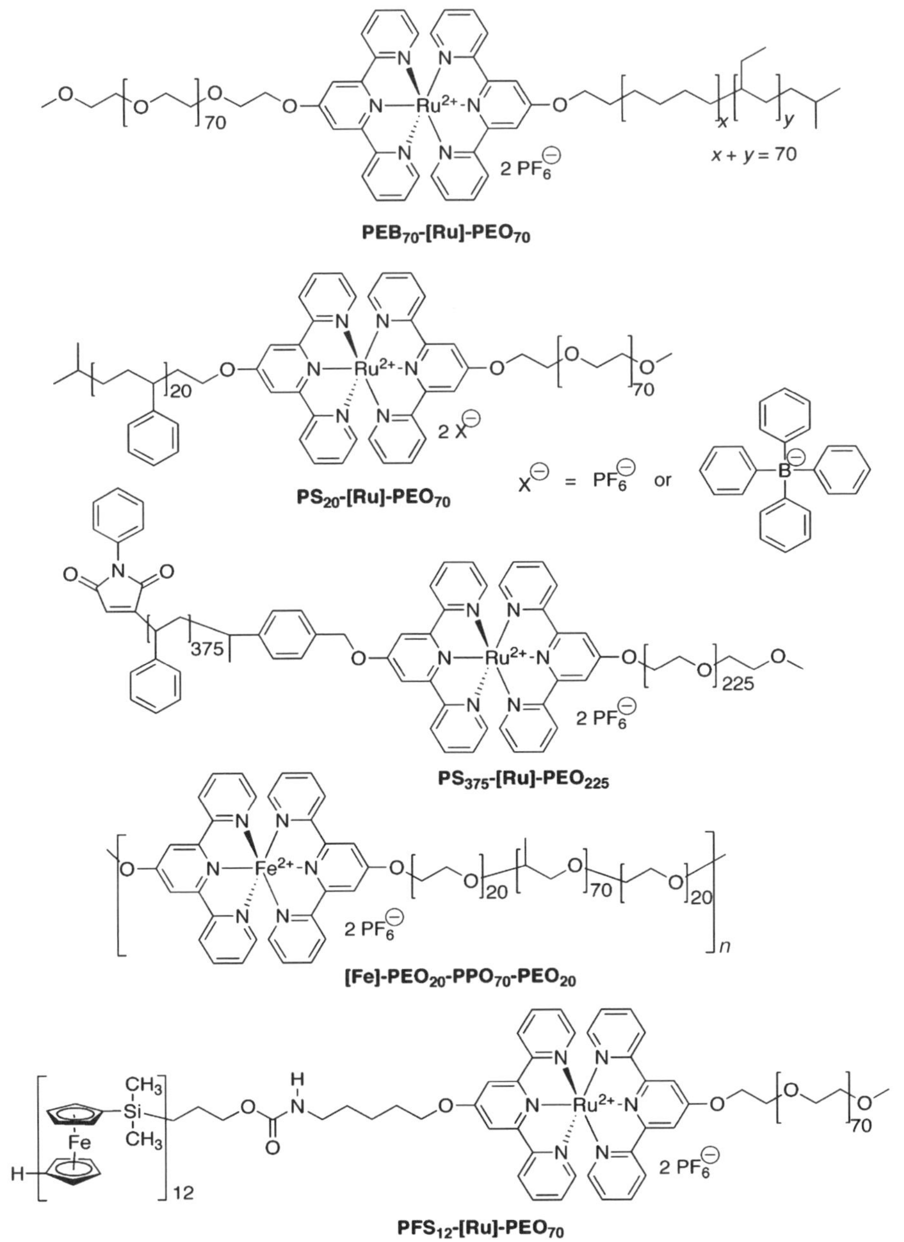

Figure 5.15 Examples of some PEO-based block copolymers with ruthenium or iron bisterpyridine complexes as the link between different blocks.

In a later report, the effect of the presence of bulky tetraphenylborate counteranion on the phase behavior was studied by SAXS.[81] The copolymer with tetraphenylborate anion exhibited a lamellar phase-separated structure compared to the copolymer with hexafluorophosphate counteranion, which showed spherical aggregates. Upon heating to 80°C, disordered melt was observed. Such difference was attributed to the change in interaction energies between the counteranions and the polymer blocks and the volume fraction of polymer metal complex with bulky anion. In a subsequent report by the same research group, the fabrication of nanoporous thin film by a similar block copolymer PS_{375}-[Ru]-PEO_{225} was demonstrated.[82] The nanoporous thin films were prepared by a two-step process. The first step involved the self-assembly of the block copolymer that yielded cylindrical domains oriented perpendicularly to the substrate surface. In the second step, the PS matrix was stabilized by photo cross-linking with deep UV irradiation. The PEO blocks were then removed by cleaving the ruthenium terpyridine complex by oxidizing the ruthenium center with Ce^{4+}. This created an array of perpendicularly oriented channels with a size similar to the previously existing PEO blocks. The AFM images of the polymer film before and after removal of the PEO blocks are shown in Figure 5.16.

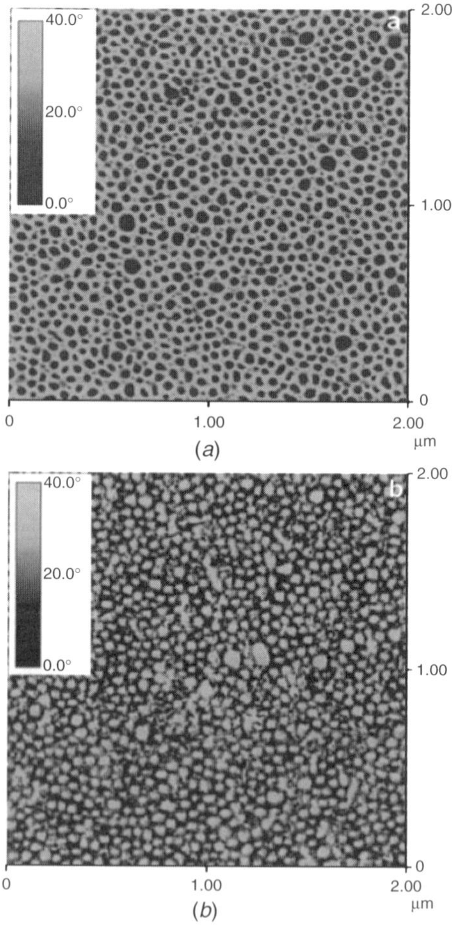

Figure 5.16 AFM phase images of (a) PS_{375}-[Ru]-PEO_{225} film (74 nm in thickness) obtained by spin coating on silicon; and (b) the same polymer film after the removal of PEO blocks by treatment with $Ce(SO_4)_2$. (From Fustin et al.[82])

Shirai et al. reported an amphiphilic PEO_{20}-*b*-PPO_{70}-*b*-PEO_{20} (PPO = poly(propylene oxide)) block copolymer that was modified with terpyridine at both ends.[83] When the ligand-capped polymer was treated with Fe^{2+}, the resulting [Fe]-PEO_{20}-PPO_{70}-PEO_{20} (Figure 5.15) self-assembled into a linear supramolecular polymer with no evidence for the formation of cyclic oligomers. The metal-free polymer may also serve as the template for the sol–gel polymerization of tetraethoxysilane, resulting in well-ordered hexagonal mesoporous silica structures that can accommodate Fe^{2+}.

Since the oxygen atoms in PEO segments can also form coordination with various metal-ion precursors, PEO-based block copolymers were widely used in the fabrication of different metal-oxide/sulfide nanostructures. Of all the metal oxides, titanium dioxide has received the most attention because it is a wide band-gap metal oxide with promising potentials in catalyzing the photochemical splitting of water.[84] It has also been shown to be the key material in the fabrication of efficient organic-based dye-sensitized solar cells (DSSCs).[85,86] TiO_2 nanostructures with different sizes and shapes have been proposed in the fabrication of DSSC. The use of a bicontinuous TiO_2 network fabricated by PEO-based block copolymer has been demonstrated.[87] The PEO segments are capable of coordinating with TiO_2 precursors such as titanium(IV) tetraisopropoxide (TTIP) and $TiCl_4$. The structures of the TiO_2 formed will be dependent on the morphology of the polymer/TiO_2 precursor formed in the block copolymer matrix. Dai et al. reported the synthesis of ordered mixed TiO_2/SiO_2 mesostructures monoliths by the complexation between TTIP/tetraethyl orthosilicate (TEOS) and PEO_{130}-*b*-PPO_{60}-*b*-PEO_{130}.[88] The multivalent Ti(IV) and Si(IV) species preferentially coordinated to the hydrophilic PEO block; a schematic diagram is shown in Figure 5.17a. Because of the phase separation between the PEO and PPO blocks, the mixed oxide would be incorporated into the phase formed by PEO. Amorphous titania inside the silica network was obtained after removal of the organics in the polymer matrix. The TEM image of the mixed Ti/Si oxides is shown in Figure 5.17b. A well-defined cubic mesostructure can be clearly observed. The oxide obtained can be converted into anatase-phase TiO_2 by further calcination. The extent of crystallization was dependent on the calcination temperature. TiO_2 synthesized from PEO-*b*-PPO-*b*-PEO has been applied in the preparation of mesoporous molecular sieve films for photocatalysis.[89] It was demonstrated that the zeolite-like nanocrystalline films exhibited photocatalytic activity in which acetone can be photodecomposed at ambient conditions.

PS-*b*-PEO is another block copolymer that was commonly used in the fabrication of TiO_2 nanostructure. Metal-free PS-*b*-PEO can form various interesting morphologies. For example, PS-*b*-PEO with asymmetric block size can be oriented perpendicularly to the substrate surface, forming nanosized cylindrical domains of PEO embedded in a glassy PS matrix.[90,91] It was used as a scaffold for the growth of nanostructured SiO_2 thin films[92] or gold nanoparticles.[93] By a similar approach, TiO_2 nanostructures could also be prepared. A sol–gel precursor solution (TTIP/EtOH/HCl) was added to a toluene solution of PS-*b*-PEO (volume fraction of PS:PEO = 0.67:0.33).[94] The resulting mixture was spin-coated on a silicon wafer. The TiO_2 sol–gel precursor was selectively incorporated into the ordered cylindrical domains of PEO. After exposing the film to deep UV light that removed the organic components, TiO_2 nanodot arrays were formed. The schematic diagram of the process is shown in Figure 5.18a. The effect of varying the TiO_2 sol–gel precursor was also studied. It was suggested that the concentration of the sol–gel in the PEO-block could affect of volume fraction of the block. The center-to-center distance for each cylinder would decrease as the amount of sol–gel precursor in PEO was increased. The resulting TiO_2 dot arrays are shown in Figures 5.18b and 5.18c, which clearly show that PS-*b*-PEO with 35% sol–gel concentration exhibited higher density arrays of TiO_2 with smaller spacing between different dots when compared to that prepared with a concentration of 15%. It was shown by grazing

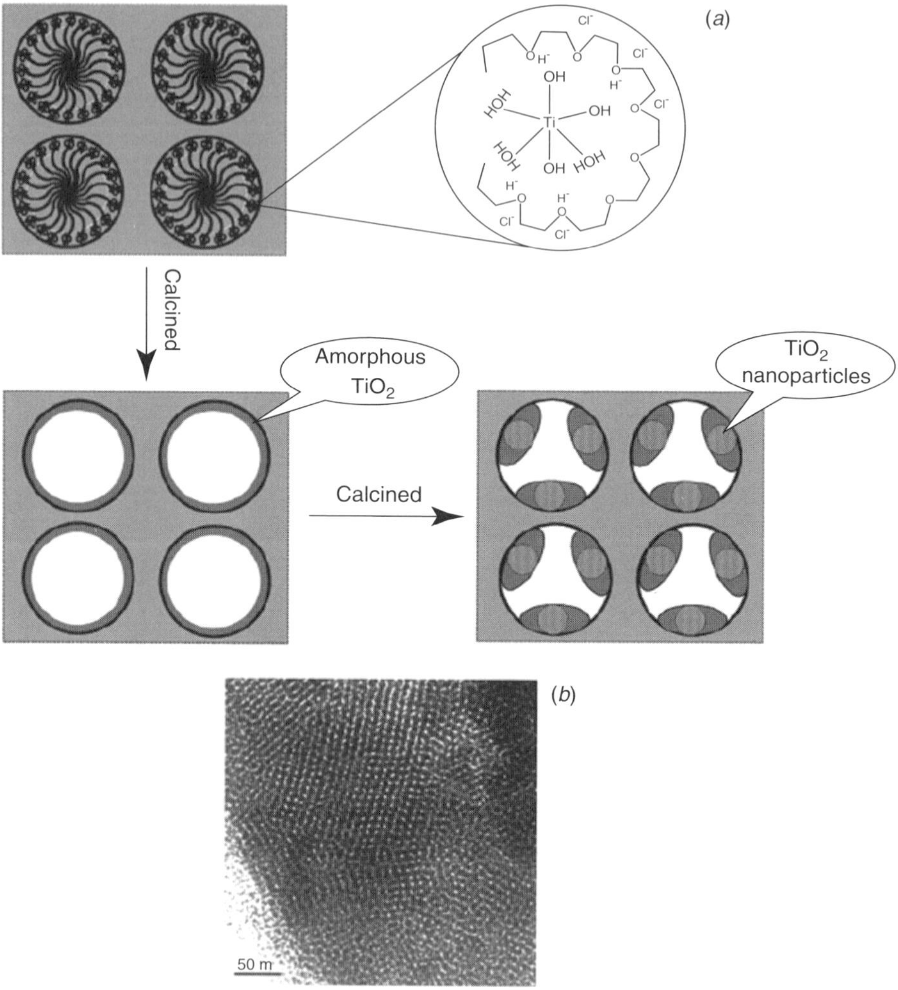

Figure 5.17 (a) Schematic diagram of the formation of mesostructures from TiO_2/SiO_2 in a matrix of PEO-*b*-PPO-*b*-PEO; (b) TEM image of the calcinated sample with a molar ratio Si/Ti = 3.7. (From Zhu et al.[88] Reprinted with permission. Copyright © 2004 American Chemical Society.)

incidence SAXS and SEM that the well-ordered structures were created over large areas of the polymer films.[95]

In another example, an ABC-type triblock copolymer was used for preparing polymer–metal composites for nanofabrication.[96] Micelles of PS-*b*-P2VP-*b*-PEO were prepared by dissolving the polymer in dilute HCl, followed by adjusting the pH by NaOH to a desired level. Different tungsten compounds, such as sodium tungstate (Na_2WO_4), dodecatungstophosphoric acid [$H_3(PW_{12}O_{40})$], and dodecatungstosilicic acid [$H_4(SiW_{12}O_{40})$], were then added to the block copolymer micelle solution. The polymer–metal complexes were formed by the charge interaction between the inorganic polyanions and the P2VP block. Addition of polyanion into the block copolymer resulted in the shrinkage of micelles. It was

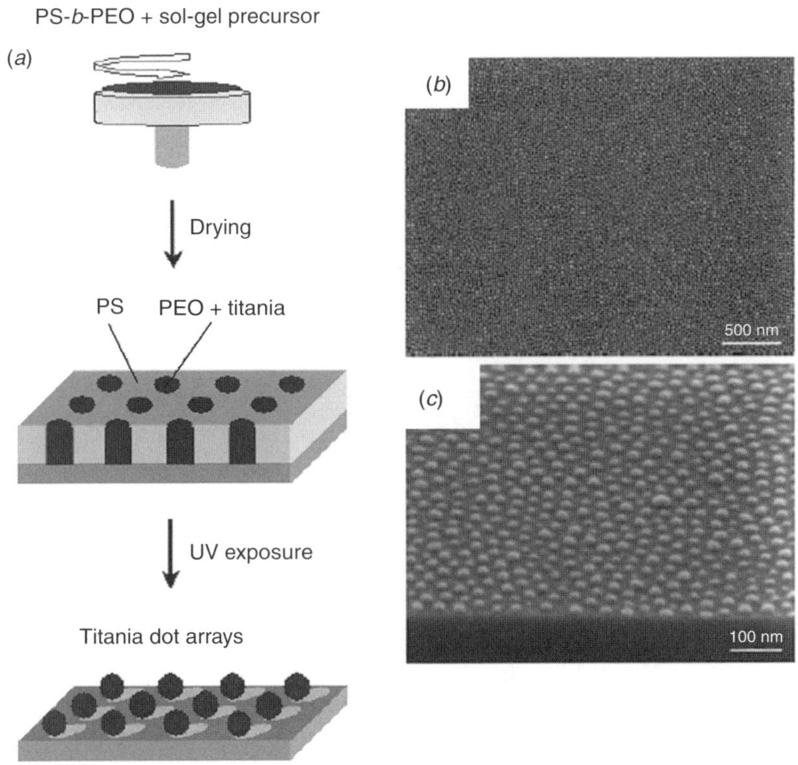

Figure 5.18 (a) Schematic diagram for the preparation of nanopatterned TiO_2 dot array from the PS-*b*-PEO/sol-gel film. (b) Field emissive SEM image of TiO_2 dot arrays prepared from PS-*b*-PEO films with 35% sol–gel precursor after exposure to UV light at room temperature for 6 hours in air. (c) Field-emissive SEM image of TiO_2 prepared from 15% sol–gel precursor. (From Kim et al.[94])

also found that the zeta potentials of the PS-*b*-P2VP-*b*-PEO-polyanion complexes were lower than that of the metal free block copolymer, owing to the charge neutralization.

F. Block Copolymers with Polyacrylates

Some polyacrylate-based block copolymers were also used in the fabrication of metal-containing nanostructures, although the number of examples is less than those from PEO- or PVP-based copolymers. The metal ions are usually coordinated to the carboxylate moieties on the polymer chain. Cadmium sulfide quantum dots were prepared from a blend of polystyrene-*block*-poly(acrylic acid) (PS_{330}-*b*-PAA_{20}) and cadmium acetate.[97] In a benzene/methanol solvent system, reverse micelles with a insoluble PAA-Cd core and soluble PS coronas were formed from the cadmium-containing block copolymer. Micelles in solid form were then obtained by the freeze-drying process. On exposing the micelle powder to H_2S, PS-*b*-PAA-stabilized cadmium sulfide quantum dots were obtained. These quantum dots were easily dispersed into various organic solvents. The time-resolved photoluminescence of the quantum dots were studied in detail, and their hydrodynamic radii and aggregation behaviors were investigated by dynamic light scattering. The cadmium sulfide quantum dots were dispersed evenly into a homo-PS matrix, resulting in optically clear thin films.

Diblock copolymer poly(methyl methacrylate)-*block*-poly(2-hydroxyethyl methacrylate) PMMA-*b*-PHEMA was used as the template for the self-assembly of palladium nanoparticles.[98] Thin films of the metal-free block copolymer were obtained by dip coating from different solvents. The copolymer existed in the ordinary form and inverted micelles when it was dissolved in 1,4-dioxane and methanol, respectively. This was attributed to the hydrophilic properties of the PHEMA block. 2-Methoxyethanol, which is a common solvent to both blocks, was also used in the coating. Films obtained from 1,4-dioxane and methanol solutions showed the formation of spherical domains that were arranged in a 2-D hexagonal lattice, while the film obtained from 2-methoxyethanol showed a stripe pattern. Their AFM topography images are shown in Figure 5.19. The block copolymer thin films were then exposed to

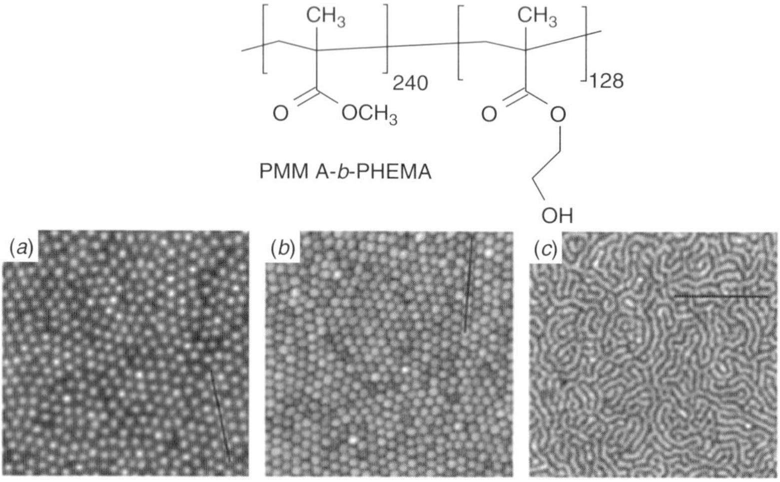

Figure 5.19 AFM topographical images of PMMA-*b*-PHEMA thin films dipped in 0.5 wt % solutions of (a) 1,4-dioxane; (b) methanol; and (c) 2-methoxyethanol. (From Yin et al.[98] Reprinted in part with permission. Copyright © 2005 American Chemical Society.)

vapor of palladium(II) bis(acetylacetonato). It was found that the palladium nanoparticles were selectively located in the PHEMA phase, which was due to its stronger reducing power. The dispersion of palladium nanoparticles can be seen from the TEM images (Figure 5.20).

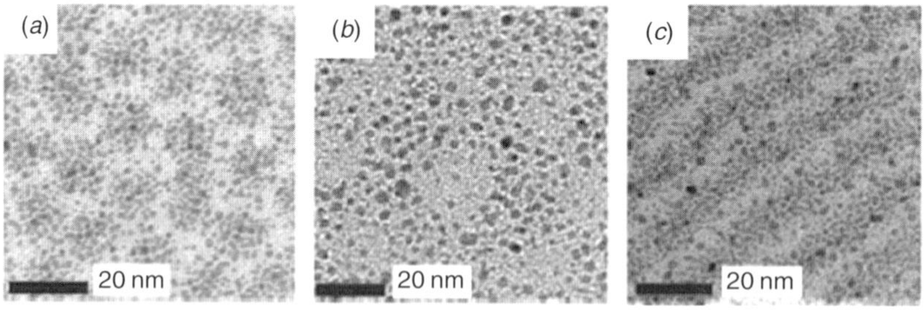

Figure 5.20 TEM micrographs of PMMA-*b*-PHEMA thin films prepared from (a) 1,4-dioxane; (b) methanol; and (c) 2-methoxyethanol after exposure to $Pd(acac)_2$ vapor. Individual palladium particles dispersed in PHEMA phase can be observed in the images. (From Yin et al.[98] Reprinted in part with permission. Copyright © 2005 American Chemical Society.)

The pattern formed by the block copolymer/palladium nanoparticles was transferred onto the silicon wafer substrates by RIE. Oxygen plasma was first applied to the block copolymer/metal composite film, which removed the organic polymer selectively. Subsequently, the oxygen plasma-etched films were treated with CF_4 plasma. Those regions covered with palladium nanoparticles were protected against the plasma etching, and the resulting pattern created on the silicon wafer surface was the replica pattern formed by the palladium nanoparticles. It is possible to modify the pattern created on the silicon wafer surface by simply changing the solvent used in the dip-coating process. This is a simple and versatile approach in preparing masks for nanolithography.

G. Other Block Copolymers

A series of organometallic block copolymers PS-*b*-PPES[$Co_2(CO)_6$] (PPES-poly(4-(phenylethynyl)styrene)) with various block sizes were obtained by the reaction between dicobalt octacarbonyl and PS-*b*-PPES (Scheme 5.8).[99] The metal-free block copolymers were

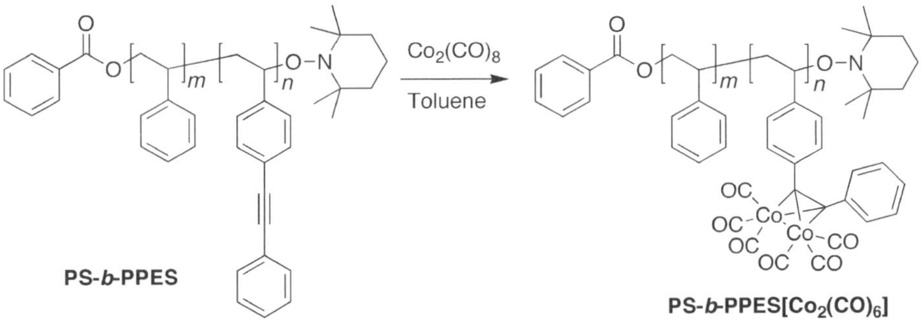

Scheme 5.8

synthesized by the nitroxide-mediated radical polymerization from TEMPO-capped polystyrene (TEMPO = 2,2,6,6-tetramethyl-1-piperidinyloxy free radical). Approximately 80–90% of the alkynyl groups were functionalized with the cobalt cluster. Formation of an ordered segregated phase with cylindrical morphologies was observed. When the polymer was heated at 110°C for over 23 hours, loss of carbon monoxide molecules was observed, as confirmed by Fourier transform infrared (FTIR) spectroscopy. The resulting product became insoluble, which was due to the formation of cross-links by cycloaddition reactions. Further thermolysis at 800°C resulted in the formation of carbonaceous materials that contained carbon nanotubes and cobalt nanoparticles.

III. HOMOPOLYMERS FOR NANOFABRICATION

Compared to block copolymers, there have been relatively fewer examples of using homopolymers for nanofabrication. Nevertheless, some polymers with amphiphilic properties were also used in the fabrication of nanostructures with various metal salts/complexes. For example, polyaniline (PANI) emeraldine base formed self-organized mesomorphic structures when mixed with $Zn(DBS)_2$ by the coordination between Zn^{2+} and the imine nitrogen atoms on the polymer main chain.[100] The resulting supramolecule PANI[$Zn(DBS)_2$]$_{0.5}$ had a comb-shaped structure, and the formation of a lamellar phase was due to the competition between the metal–polymer coordination (attraction) and the repulsion between the dodecyl tails. It was reported later that a mixture of dodecylbenzenesulfonic acid (DBSA)-doped PANI and $Zn(DBS)_2$ also formed comb-shaped supramolecules.[101] Different amounts of zinc complexes were loaded into PANI-DBSA, yielding $PANI(DBSA)_{0.5}[Zn(DBS)_2]_x$, where $x = 0.25, 0.5, 0.75, 1.0$, and 1.5. The formation of the schematic diagram of the complex formation is shown in Figure 5.21.

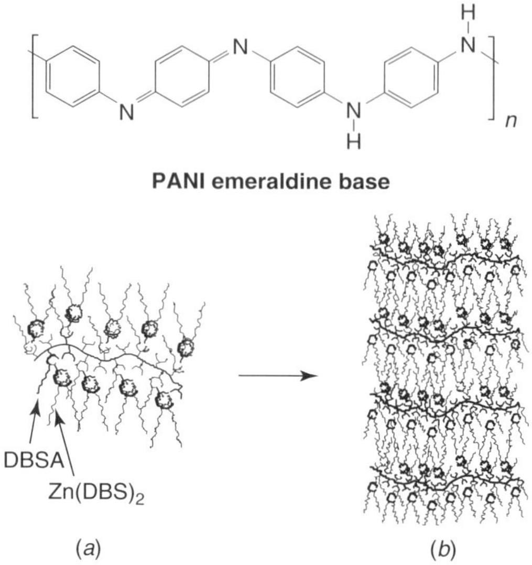

Figure 5.21 Structure of PANI emeraldine base and schematic illustrations of the formation of the lamellar self-organization of $PANI(DBSA)_{0.5}[Zn(DBS)_2]_{0.5}$. (From Hartikainen et al.[101] Reprinted with permission. Copyright © 2001 American Chemical Society.)

The mechanism of the formation of these complexes was studied by the synthesis of single crystalline stoichiometric oligomeric complexes between 4,4′-bipyridine-camphor sulfonic acid [4,4′-Bpy(CSA)] and zinc camphorsulfonate $Zn(CS)_2$. A hydration-induced bonding between $PANI(DBSA)_{0.5}$ and $Zn(DBS)_2$ was proposed.

The growth of gold, palladium, and platinum nanoparticles in the nanostructured matrix of poly(octadecylsiloxane) (PODS) was demonstrated by Bronstein et al.[102] PODS, an amphiphilic polymer with nonpolar alkyl chains and hydrophilic silanol moieties $(\text{—SiO})_x(OH)_y$, forms a bilayer nanostructure composed of these two moieties in its hydrated form. In the presence of an aqueous solution of metal salt (e.g., Na_2PdCl_4, K_2PtCl_4, $AuCl_3$), the siloxy layer absorbed metal anions by hydrogen bonding (Scheme 5.9). After reduction, the metal ions were reduced into nanoparticles, and TEM results revealed that the nanoparticles were distributed in layer structures. The sizes of the particles were measured to be about 1–2 nm, and the narrow size distribution was attributed to the volume availability within the ordered bilayers of PODS.

IV. CONCLUSIONS

The uses of block copolymer and homopolymers for nanofabrication have been presented. A variety of interesting nanostructures composed of metallic or semiconducting materials were

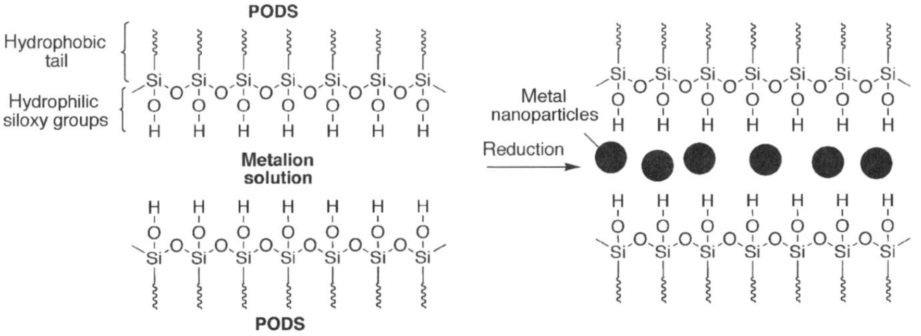

Scheme 5.9

constructed easily by using the block copolymers as the templates. The metal/block copolymer composite precursors were processed by spin coating on substrates, or by the formation of nanosized micelles in a suitable solvent system. By controlling the chemical nature and relative size of different blocks in the copolymer molecules, nanostructures with different shape, size, and morphology were easily obtained.

The use of block copolymers offers an alternative approach in the fabrication of a large area of regular, ordered nanosized patterns. These processes are potentially useful in the fabrication of new generations of molecular devices in various important technological areas.

ACKNOWLEDGMENTS

Financial support from the Research Grants Council of the Hong Kong Special Administrative Region, China (Project No. HKU 7010/05P) is acknowledged. The authors also thank Dr. A. B. Djurisic for her help in the preparation of the manuscript.

REFERENCES

1. J. J. Watkins, D. J. Bishop, *MRS Bull.*, **30**, 937 (2005).
2. Y. Xia, G. M. Whitesides, *Angew. Chem. Engl. Int. Ed.*, **37**, 550 (1998).
3. X. M. Zhao, Y. Xia, G. M. Whitesides, *J. Mater. Chem.*, **7**, 1069 (1997).
4. J. M. Lehn, *Supramolecular Chemistry: Concepts and Perspectives*, Wiley-VCH, Weinheim, Germany, 1995.
5. N. C. Gianneschi, M. S. Masar III, C. A. Mirkin, *Acc. Chem. Res.*, **38**, 825 (2005).
6. J. M. Lehn, *Polym. Int.*, **51**, 825 (2002).
7. I. Haiduc, F. T. Edelmann, *Supramolecular Organometallic Chemistry*, Wiley-VCH, Weinheim, Germany, 1999.
8. I. W. Hamley, *Nanotechnology*, **14**, R39 (2003).
9. F. S. Bates, *Science*, **251**, 4996 (1991).
10. R. A. Segalman, *Mater. Sci. Eng. R*, **48**, 191 (2005).
11. M. Lazzari, M. A. Lopez-Quintela, *Adv. Mater.*, **15**, 1583 (2003).
12. C. J. Hawker, T. P. Russell, *MRS Bull.*, **30**, 952 (2005).
13. J. Ruokolainen, R. Mäkinen, M. Torkkeli, T. Mäkelä, R. Serimaa, G. ten Brinke, O. Ikkala, *Science*, **280**, 557 (1998).

14. J. Ruokolainen, G. ten Brinke, O. Ikkala, *Adv. Mater.*, **11**, 777 (1999).
15. R. Mäki-Ontto, K. de Moel, de Odorico, J. Ruokolainen, M. Stamm, G. ten Brinke, O. Ikkala, *Adv. Mater.*, **13**, 117 (2001).
16. F. S. Bates, G. H. Fredrickson, *Ann. Rev. Phys. Chem.*, **41**, 525 (1990).
17. I. Manners, *Synthetic Metal-Containing Polymers*, Wiley-VCH, Weinheim, Germany, 2004.
18. R. D. Archer, *Inorganic and Organometallic Polymers*, Wiley-VCH, New York, 2001.
19. I. Manners, *Science*, **294**, 1664 (2001).
20. K. Kulbaba, I. Manners, *Macromol. Rapid Commun.*, **22**, 711 (2001).
21. I. Manners, *J. Polym. Sci., Part A: Polym. Chem.*, **40**, 179 (2002).
22. J. Raez, R. Barjovanu, J. A. Massey, M. A. Winnik, I. Manners, *Angew. Chem. Engl. Int. Ed.*, **39**, 3862 (2000).
23. J. Raez, I. Manners, M. A. Winnik, *J. Am. Chem. Soc.*, **124**, 10381 (2002).
24. J. Raez, J. P. Tomba, I. Manners, M. A. Winnik, *J. Am. Chem. Soc.*, **125**, 9546 (2003).
25. X. S. Wang, M. A. Winnik, I. Manners, *Angew. Chem. Int. Ed.*, **43**, 3703 (2004).
26. X. S. Wang, M. A. Winnik, I. Manners, *Macromolecules*, **38**, 1928 (2005).
27. K. N. Power-Billard, R. J. Spontak, I. Manners, *Angew. Chem. Int. Ed.*, **43**, 1260 (2004).
28. M. Ginzburg, M. J. MacLachlan, S. M. Yang, N. Coombs, T. W. Coyle, N. P. Raju, J. E. Greedan, R. H. Herber, G. A. Ozin, I. Manners, *J. Am. Chem. Soc.*, **124**, 2625 (2002).
29. J. Y. Cheng, C. A. Ross, V. Z. H. Chan, E. L. Thomas, R. G. H. Lammertink, G. J. Vancso, *Adv. Mater.*, **13**, 1174 (2001).
30. S. B. Clendenning, S. Fournier-Bidoz, A. Pietrangelo, G. C. Yang, S. J. Han, P. M. Brodersen, C. M. Yip, Z. H. Lu, G. A. Ozin, I. Manners, *J. Mater. Chem.*, **14**, 1686 (2004).
31. J. Galloro, M. Ginzburg, H. Miguez, S. M. Yang, N. Coombs, A. Safa-Sefat, J. E. Greedan, I. Manners, G. A. Ozin, *Adv. Funct. Mater.*, **12**, 382 (2002).
32. P. W. Cyr, M. Tzolov, M. A. Hines, I. Manners, E. H. Sargent, G. D. Scholes, *J. Mater. Chem.*, **13**, 2213 (2003).
33. K. T. Kim, G. W. M. Vandermeulen, M. A. Winnik, I. Manners, *Macromolecules*, **38**, 4958 (2005).
34. R. G. H. Lammertink, M. A. Hempenius, J. E. van den Enk, V. Z. H. Chan, E. L. Thomas, G. J. Vancso, *Adv. Mater.*, **12**, 98 (2000).
35. M. Roerdink, M. A. Hempenius, G. J. Vancso, *Chem. Mater.*, **17**, 1275 (2005).
36. M. A. Hempenius, R. G. H. Lammertink, M. Peter, G. J. Vancso, *Macromol. Symp.*, **196**, 45 (2003).
37. S. B. Clendenning, I. Manners, *Macromol. Symp.*, **196**, 71 (2003).
38. S. B. Clendenning, S. Aouba, M. S. Rayat, D. Grozea, J. B. Sorge, P. M. Brodersen, R. N. S. Sodhi, Z. H. Lu, C. M. Yip, M. R. Freeman, H. E. Ruda, I. Manners, *Adv. Mater.*, **16**, 215 (2004).
39. D. A. Durkee, H. B. Eitouni, E. D. Gomez, M. W. Ellsworth, A. T. Bell, N. R. Balsara, *Adv. Mater.*, **17**, 2003 (2005).
40. C. Sanchez, B. Julian, P. Belleville, M. Popall, *J. Mater. Chem.*, **15**, 3559 (2005).
41. R. B. Grubbs, *J. Polym. Sci., Part A: Polym. Chem.*, **43**, 4323 (2005).
42. J. F. Gohy, B. G. G. Lohmeijer, U. S. Schubert, *Chem. Eur. J.*, **9**, 3472 (2003).
43. B. G. G. Lohmeijer, U. S. Schubert, *J. Polym. Sci., Part A: Polym. Chem.*, **41**, 1413 (2003).
44. M. Vamvakaki, L. Papoutsakis, V. Katsamanis, T. Afchoudia, P. G. Fragouli, H. Iatrou, N. Hadjichristidis, S. P. Armes, S. Sidorov, D. Zhirov, V. Zhirov, M. Kostylev, L. M. Bronstein, S. H. Anastasiadis, *Faraday Discuss.*, **128**, 129 (2005).
45. F. Bossé, H. P. Schreiber, A. Eisenberg, *Macromolecules*, **26**, 6447 (1993).
46. J. Xia, X. Zhang, K. Matyjaszewski, *Macromolecules*, **32**, 3531 (1999).
47. M. Antonietti, S. Heinz, M. Schmidt, C. Rosenauer, *Macromolecules*, **27**, 3276 (1994).
48. A. Sidorenko, I. Tokarev, S. Minko, M. Stamm, *J. Am. Chem. Soc.*, **125**, 12211 (2003).

49. J. P. Spatz, P. Eibeck, S. Mossmer, M. Moller, T. Herzog, P. Ziemann, *Adv. Mater.*, **10**, 849 (1998).

50. R. Glass, M. Moller, J. P. Spatz, *Nanotechnology*, **14**, 1153 (2003).

51. J. P. Spatz, A. Roescher, S. Sheiko, G. Krausch, M. Moller, *Adv. Mater.*, **7**, 731 (1995).

52. J. P. Spatz, S. Sheiko, M. Moller, *Macromolecules*, **29**, 3220 (1996).

53. J. P. Spatz, S. Mossmer, M. Moller, T. Herzog, A. Plettl, P. Ziemann, *J. Lumin.*, **76–7**, 168 (1998).

54. J. P. Spatz, S. Mossmer, C. Hartmann, M. Moller, T. Herzog, M. Krieger, H. G. Boyen, P. Ziemann, B. Kabius, *Langmuir*, **16**, 407 (2000).

55. R. Glass, M. Arnold, J. Blummel, A. Kuller, M. Moller, J. P. Spatz, *Adv. Funct. Mater.*, **13**, 569 (2003).

56. G. Kastle, H. G. Boyen, F. Weigl, G. Lengl, T. Herzog, P. Ziemann, S. Riethmuller, O. Mayer, C. Hartmann, J. P. Spatz, M. Moller, M. Ozawa, F. Banhart, M. G. Garnier, P. Oelhafen, *Adv. Funct. Mater.*, **13**, 853 (2003).

57. J. P. Spatz, T. Herzog, S. Mossmer, P. Ziemann, M. Moller, *Adv. Mater.*, **11**, 149 (1999).

58. M. Haupt, S. Miller, R. Glass, M. Arnold, R. Sauer, K. Thonke, M. Moller, J. P. Spatz, *Adv. Mater.*, **15**, 829 (2003).

59. S. T. Selvan, T. Hayakawa, M. Nogami, M. Moller, *J. Phys. Chem. B*, **103**, 7441 (1999).

60. M. Haupt, A. Ladenburger, R. Sauer, K. Thonke, R. Glass, W. Roos, J. P. Spatz, H. Rauscher, S. Riethmuller, M. Moller, *J. Appl. Phys.*, **93**, 6252 (2003).

61. H. G. Boyen, G. Kastle, K. Zurn, T. Herzog, F. Weigl, P. Ziemann, O. Mayer, C. Jerome, M. Moller, J. P. Spatz, M. G. Garnier, P. Oelhafen, *Adv. Funct. Mater.*, **13**, 359 (2003).

62. L. H. Bronstein, S. N. Sidorov, P. M. Valetsky, J. Hartmann, H. Colfen, M. Antonietti, *Langmuir*, **15**, 6256 (1999).

63. G. O. R. Alberda van Ekenstein, R. Meyboom, G. ten Brinke, *Macromolecules*, **33**, 3752 (2000).

64. R. J. Forster, L. Keane, *J. Electroanal. Chem.*, **554**, 345 (2003).

65. J. Ruokolainen, J. Tanner, G. Tenbrinke, O. Ikkala, M. Torkkeli, R. Serimaa, *Macromolecules*, **28**, 7779 (1995).

66. S. Valkama, T. Ruotsalainen, H. Kosonen, J. Ruokolainen, M. Torkkeli, R. Serimaa, G. ten Brinke, O. Ikkala, *Macromolecules*, **36**, 3986 (2003).

67. A. W. Fahmi, M. Stamm, *Langmuir*, **21**, 1062 (2005).

68. R. P. Zhu, Y. M. Wang, W. D. He, *Eur. Polym. J.*, **41**, 2088 (2005).

69. S. J. Hou, K. Y. K. Man, W. K. Chan, *Langmuir*, **19**, 2485 (2003).

70. S. J. Hou, W. K. Chan, *Macromol. Rapid Commun.*, **20**, 440 (1999).

71. K. W. Cheng, W. K. Chan, *Langmuir*, **21**, 5247 (2005).

72. L. M. Bronstein, S. N. Sidorov, V. Zhirov, D. Zhirov, Y. A. Kabachii, S. Y. Kochev, P. M. Valetsky, B. Stein, O. I. Kiseleva, S. N. Polyakov, E. V. Shtykova, E. V. Nikulina, D. I. Svergun, A. R. Khokhlov, *J. Phys. Chem. B*, **109**, 18786 (2005).

73. Y. Kim, M. F. Mayer, S. C. Zimmerman, *Angew. Chem. Engl. Int. Ed.*, **42**, 1121 (2003).

74. N. Wang, Y. J. Li, F. S. Lu, Y. Liu, X. He, L. Jiang, J. P. Zhuang, X. F. Li, Y. L. Li, S. Wang, H. B. Liu, D. B. Zhu, *J. Polym. Sci., Part A: Polym. Chem.*, **43**, 2851 (2005).

75. Y. N. C. Chan, G. S. W. Craig, R. R. Schrock, R. E. Cohen, *Chem. Mater.*, **4**, 885 (1992).

76. V. Sankaran, J. Yue, R. E. Cohen, R. R. Schrock, R. J. Silbey, *Chem. Mater.*, **5**, 1133 (1993).

77. J. F. Ciebien, R. T. Clay, B. H. Sohn, R. E. Cohen, *New J. Chem.*, **22**, 685 (1998).

78. R. F. Mulligan, A. A. Iliadis, P. Kofinas, *J. Appl. Polym. Sci.*, **89**, 1058 (2003).

79. J. F. Gohy, B. G. G. Lohmeijer, U. S. Schubert, *Macromol. Rapid Commun.*, **23**, 555 (2002).

80. V. Vogel, J. F. Gohy, B. G. G. Lohmeijer, J. A. Van den Broek, W. Haase, U. S. Schubert, D. Schubert, *J. Polym. Sci., Part A: Polym. Chem.*, **41**, 3159 (2003).

81. M. Al-Hussein, W. H. de Jeu, B. G. G. Lohmeijer, U. S. Schubert, *Macromolecules*, **38**, 2832 (2005).

82. C. A. Fustin, B. G. G. Lohmeijer, A. S. Duwez, A. M. Jonas, U. S. Schuhert, J. F. Gohy, *Adv. Mater.*, **17**, 1162 (2005).
83. M. Kimura, Y. Iwashima, K. Ohta, K. Hanabusa, H. Shirai, *Macromolecules*, **38**, 5055 (2005).
84. A. Fujishima, K. Honda, *Nature*, **238**, 37 (1972).
85. B. Oregan, M. Gratzel, *Nature*, **353**, 737 (1991).
86. M. Gratzel, *MRS Bulletin*, **30**, 23 (2005).
87. H. Wang, C. C. Oey, A. B. Djurisic, M. H. Xie, Y. H. Leung, K. K. Y. Man, W. K. Chan, A. Pandey, J. M. Nunzi, P. C. Chui, *Appl. Phys. Lett.*, **87**, 023507 (2005).
88. H. G. Zhu, Z. W. Pan, B. Chen, B. Lee, S. M. Mahurin, S. H. Overbury, S. Dai, *J. Phys. Chem. B*, **108**, 20038 (2004).
89. J. C. Yu, X. C. Wang, X. Z. Fu, *Chem. Mater.*, **16**, 1523 (2004).
90. Z. Lin, D. H. Kim, X. Wu, L. Boosahda, D. Stone, L. LaRose, T. P. Russell, *Adv. Mater.*, **14**, 1373 (2002).
91. S. H. Kim, M. J. Misner, T. Xu, M. Kimura, T. P. Russell, *Adv. Mater.*, **16**, 226 (2004).
92. D. H. Kim, X. Jia, Z. Lin, K. W. Guarini, T. P. Russell, *Adv. Mater.*, **16**, 702 (2004).
93. M. Möller, J. P. Spatz, A. Roescher, *Adv. Mater.*, **8**, 337 (1996).
94. D. H. Kim, Z. C. Sun, T. P. Russell, W. Knoll, J. S. Gutmann, *Adv. Funct. Mater.*, **15**, 1160 (2005).
95. Z. Sun, M. Wolkenhauer, G. G. Bumbu, D. H. Kim, J. S. Gutmann, *Physica B—Condensed Matter*, **357**, 141 (2005).
96. A. Khanal, K. Nakashima, N. Kawasaki, Y. Oishi, M. Uehara, H. Nakamura, Y. Tajima, *Colloid Polym. Sci.*, **283**, 1226 (2005).
97. C. W. Wang, M. G. Moffitt, *Langmuir*, **20**, 11784 (2004).
98. D. Yin, S. Horiuchi, T. Masuoka, *Chem. Mater.*, **17**, 463 (2005).
99. L. A. Miinea, L. B. Sessions, K. D. Ericson, D. S. Glueck, R. B. Grubbs, *Macromolecules*, **37**, 8967 (2004).
100. J. Ruokolainen, H. Eerikainen, M. Torkkeli, R. Serimaa, M. Jussila, O. Ikkala, *Macromolecules*, **33**, 9272 (2000).
101. J. Hartikainen, M. Lahtinen, M. Torkkeli, R. Serimaa, J. Valkonen, K. Rissanen, O. Ikkala, *Macromolecules*, **34**, 7789 (2001).
102. L. M. Bronstein, D. M. Chernyshov, P. M. Valetsky, E. A. Wilder, R. J. Spontak, *Langmuir*, **16**, 8221 (2000).

CHAPTER 6

Rigid-Rod Polymetallaynes

WAI-YEUNG WONG AND CHEUK-LAM HO

Hong Kong Baptist University, Hong Kong, China

I. INTRODUCTION

The incorporation of transition metal elements into a polymeric skeleton was shown to be a good way of affording new polymeric materials with different properties from classic carbon-based polymers.[1–5] Transition metals exhibit a rich diversity of colors, oxidation states, coordination numbers and geometries, and their metal complexes have an array of intriguing catalytic, electrochemical, electronic, magnetic, optical, and photophysical properties. Within the realm of synthetic metal-containing polymers, polymers with direct metal–carbon σ-bonds in the main chain are one of the most important categories of these materials.[1,3] Rigid-rod transition metal acetylide polymers or polymetallaynes have been identified as the most extensively studied subclass of these metallopolymers.[6,7,8]

Since the 1970s, when Hagihara et al. reported the first synthesis of oligomeric and polymeric palladium(II) and platinum(II) σ-acetylide compounds stabilized by phosphine ligands,[9–12] the chemistry and properties of these polyplatinaynes have attracted wide attention, owing to their potential applications in molecular electronics, photonics and optical sensing.[1,2] In the early 1990s, there were several reports by Lewis and co-workers on the preparations of rigid-rod metal-containing polyynes using bis(trimethylstannylacetylide)-functionalized precursors.[13–16] The general structure of such a polymer system has a linear backbone comprising the platinum metal, the spacer group R, and the auxiliary phosphine ligand PR_3 on the metal center. The linear geometry of the alkynyl unit and its unsaturated feature have made metal alkynyls promising and versatile structural building blocks for molecular wires and organometallic polymeric materials that can display interesting functional properties, such as electrical conductivities, rich luminescence properties, nonlinear optical (NLO) properties, liquid crystallinity, and photovoltaic behavior.[6] A wide variety of relevant derivatives have been prepared to date where alkynyl units are bonded to various central spacer groups.[1,6,7] All these materials when prepared are generally soluble in organic solvents and the actual solubility and polymer length can be easily tuned by suitable variation of the ligand and metal groups. This favorable feature accounts for the preference of polymetallaynes over purely

Frontiers in Transition Metal-Containing Polymers, edited by Alaa S. Abd-El-Aziz and Ian Manners.

organic polyyne polymers where solubility is often a major problem when it comes to preparing films for optoelectronic measurements, whereas little can be done to tackle it. The tremendous progress in chemical synthesis, discussed in the literature, has led to an extremely versatile route for producing a large range of application-tailored conjugated polymers of this kind.

Poly(aryleneethynylene)s[17] and their metal-containing derivatives[1–3] have the potential to act as new systems for electronically important materials. Over the past few decades, intensive studies of conjugated polymers allow us to clarify some fundamental issues about the nature of the singlet excited states,[17–20] yet still relatively little is known about the nature of triplet excited states in these materials. Conjugated ethynylated polymers of transition metals have been extensively studied in this context. Much of the recent work has shown that triplet states play an important role in optical and electrical processes within conjugated molecules with direct implications for their technological applications.[21–28] It has been recognized that the ultimate efficiency of light-emitting devices (LEDs) is controlled by the fraction of the generated or harvested triplet states.[29–33] Therefore, a full understanding of triplet photophysics is highly essential if one intends to develop a full picture of basic excitations in conjugated polymers. From both experimental and theoretical considerations, the relative positions of singlet and triplet excited states in conjugated polymers strongly govern the intersystem crossing (ISC) rate into the triplet manifold. For organic systems, this provides a major nonradiative decay mechanism and reduces the luminescence efficiency. In accordance with the energy-gap law, the ISC rate depends exponentially on the energy gap to the nearest triplet excited state.[34] A knowledge of the spatial extent of the singlet and triplet energy levels is thus critical in rational materials design. Transition metal–containing polyynes were widely employed by many research groups throughout the world for directly studying the triplet excited states in conjugated polymers, in which the heavy metals increase the spin-orbit coupling, thus only partially allowing spin-forbidden phosphorescence (triplet emission).

This chapter reviews research work performed in the past 10 years on rigid-rod transition metal polyyne polymers, and a discussion of actual and potential applications of these materials will be highlighted where appropriate. The metallopolymers to be discussed are classified according to different groups of transition metal elements. In the majority of cases, the metal-containing units for the construction of a vast range of polymetallaynes include the organometallic ferrocenyl group, octahedral group-8 ruthenium/osmium units, square-planar group-10 heavy metals, and linear group-11–12 metals. Herein, polymeric metal alkynyl systems of the late transition metals are extensively discussed, and particular attention is focused on the electronic absorption spectroscopy and photoluminescence behavior, thermal stability, and structural aspects of these polymetallaynes.

II. GROUP-8 POLYMETALLAYNES

Efficient synthetic access to an organometallic ruthenium(II) tetrayne complex *trans*-$[Ru(dppe)_2(C{\equiv}C\text{-}p\text{-}C_6H_4\text{-}C{\equiv}CH)_2]$ [dppe = bis(diphenylphosphino)ethane] was developed that can be used to copolymerize with *trans*-$[Pd(PBu_3)_2Cl_2]$ or mesitylborane to form a d^6/d^8 Ru–Pd mixed-metal polymer **P1**[35] and a novel organoboron π-conjugated Ru(II)-containing polymer **P2**,[36] respectively. In the literature, compound **P1** represents the first Ru–Pd polymetallayne to be isolated. For **P2**, the absorption maximum of the metal-to-ligand charge-transfer (MLCT) band experiences a large red shift of 141 nm relative to the starting Ru(II) monomer, since the push–pull effect between the electron-rich Ru group and the electron-deficient organoboron unit greatly facilitates the $d\pi$–$p\pi^*$ transition in this rigid-rod polymer.

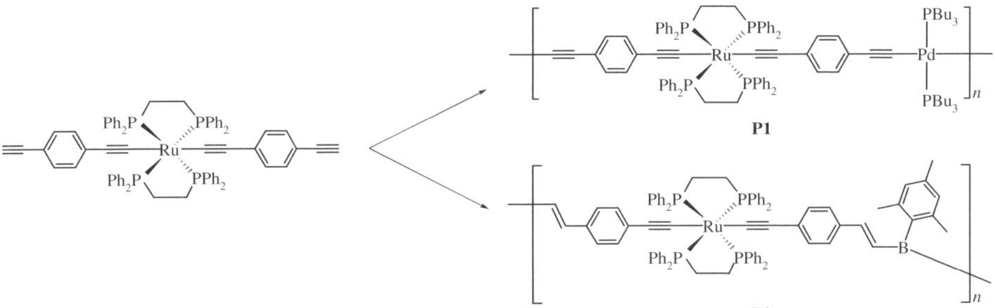

P1

P2

Wolf et al. also reported the generation of new polymetallayne hybrid materials **P3** from the electropolymerization of the appropriate Ru–oligothienylacetylide precursor complex on an electrode surface.[37,38]

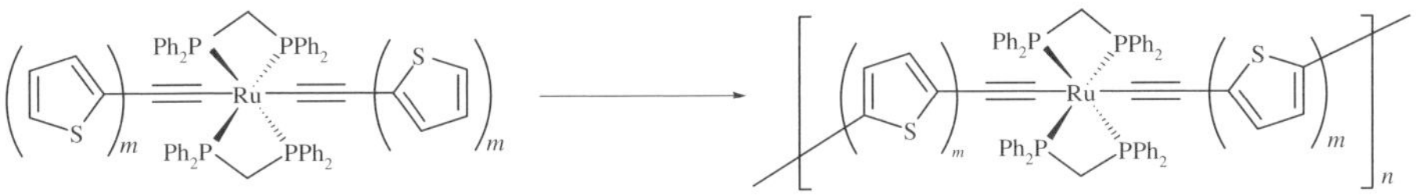

$m = 2, 3$

P3

III. GROUP-10 POLYMETALLAYNES

Since the pioneering work by Hagihara et al. on the prototype polymer of group-10 metals spaced by the phenylene ring **P4**,[9–12] followed by the extension of the method to group-8 metals developed by the Lewis group,[13–16] there has been growing attention to the synthesis and detailed optical and spectroscopic characterization of this class of materials in the scientific community. The following sections summarize the recent research efforts devoted to the developments of group-10 polymetallaynes, in which most of the materials can be prepared by the well-developed polycondensation polymerization routes between a suitable platinum(II) dichloride precursor and the corresponding diethynyl ligands.[1,6,39] First of all, it was shown that vibrational spectroscopy can help delineate structural properties of conjugated acetylenic polymers, which are of pivotal importance to their function as electronic materials. A detailed vibrational spectroscopic investigation of rigid-rod platinum acetylide polymers containing variable acetylenic microstructures **P5**–**P7** was provided.[40] In general, the vibrational data suggest that the backbone is essentially alternating —C≡C— in nature, without any contribution from the allene-type structures (=C=C=) that are typical for purely organic chains. The interruption of the acetylenic linkages with phenylene spacers reduces the π-conjugation. The near-infrared (IR) Fourier transform (FT)–Raman spectra of the polymers show resonance enhancement of selected vibrations that take place along the polymer main chain. Vibrations in these rigid-rod compounds are strongly coupled to the electronic transitions of the delocalized polymeric backbone. These materials can exhibit strong and substituent-sensitive fluorescence in their FT–Raman spectra. Beljonne and co-workers have carried out a thorough experimental and theoretical study of the electronic excitations in **P8** with two different group-10 metals.[41] The effect of metal on the nature of the lowest singlet S_1 and triplet excited states T_1 is characterized. The authors have also analyzed the chain-length dependence of the $S_0 \rightarrow S_1$, $S_0 \rightarrow T_1$, and $T_1 \rightarrow T_n$ transition energies. According to both the experimental and theoretical data, the lowest triplet exciton T_1 is strongly localized on a single C_6H_4 ring, while the S_1 and T_n states extend over a few repeating units. The T_1 excited state was estimated to be ~0.6 eV below the S_1 excited state in the polymer. The Huang–Rhys analysis of the phosphorescence spectrum of these molecules reveals that there is significant and local lattice distortions in the T_1 state, as can be confirmed by a theoretical modeling of the geometric relaxation phenomenon.

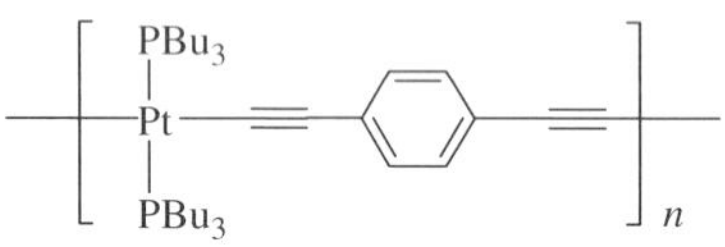

P4

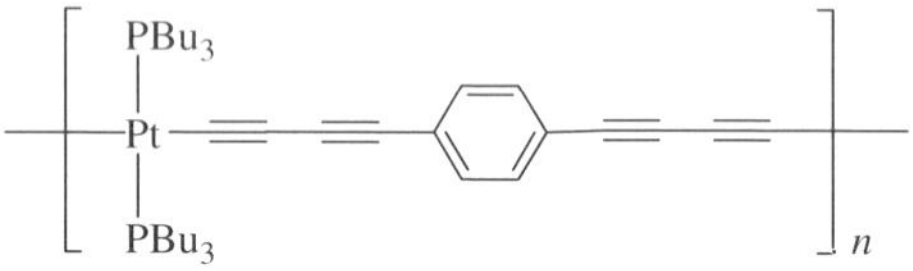

P5

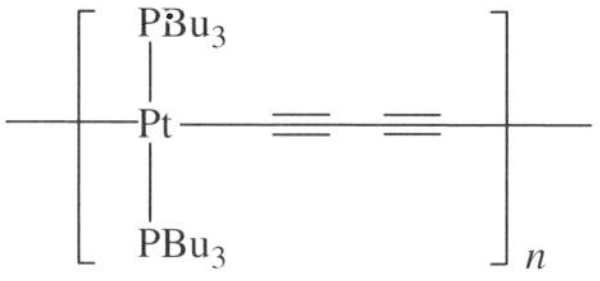

P6

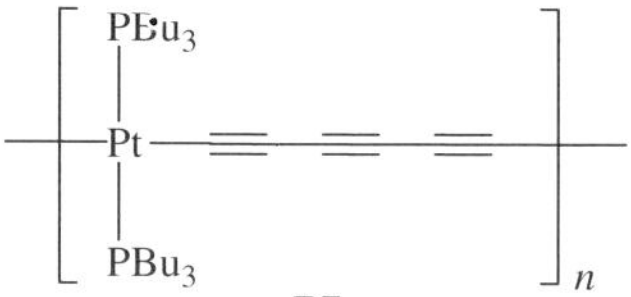

P7

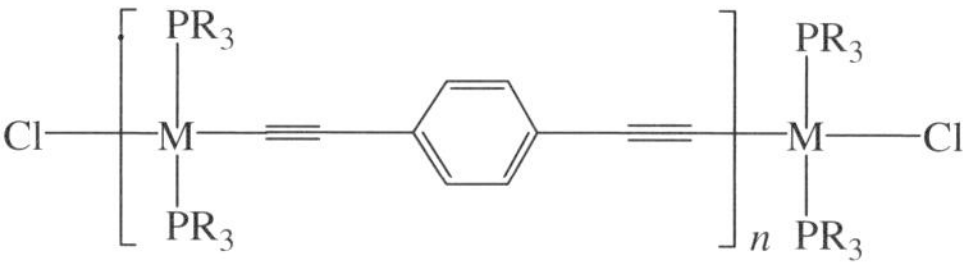

M = Ni, R = Me
M = Pt; R = Bu

P8

More recently, Schanze et al.[42] have extensively reviewed the excited-state properties of a series of Pt–acetylide oligomers, polymers, and copolymers, with the major focus placed on understanding structure–property relationships for the triplet state in such π-conjugated systems. A series of platinum-containing poly(phenyleneethynylene)-type polymers **P9** and **P10** were prepared in which the content of the Pt–acetylide repeat unit was varied.[42] The materials absorb strongly in the near UV region due to optical transitions involving (ππ*) character. The singlet exciton in the Pt–acetylide chain is delocalized over several repeat units; however, the triplet state is spatially confined on a chromophore consisting of two [—Pt-C≡C-C_6H_4-C≡C—] units. Variation of the loading of the Pt–acetylide unit in the copolymers showed that the effect of spin-orbit coupling induced by the Pt-center decreases rapidly as the loading of the metal in the backbone decreases. The mechanisms pertaining to the dynamics of intrachain exciton transfer in these copolymers were discussed and the authors found that the singlet exciton migrates rapidly, whereas the triplet exciton migrates more slowly. They also recently examined the effect of aggregation or interchain interaction on the triplet excited state in Pt–acetylide materials by comparing the optical properties of **P4** with another derivatized polymer, **P11**, incorporating the pentiptycene unit.[43] The interchain interaction is precluded by the sterically demanding pentiptycene moiety in **P11**, and its photophysics are dominated by the intrachain triplet exciton. It was shown that the solid-state phosphorescence spectrum of **P11** is nearly superimposable on the emission profile of a single crystal of the diplatinum model compound. There is also a discussion on the effects of aggregation in **P11** caused by the π–π stacking of the phenylene units and the introduction of low-lying excitations derived from $d\sigma^* \rightarrow p\sigma/\pi^*$ MMLCT (metal–metal-to-ligand charge-transfer) transitions due to interchain Pt–Pt interactions.

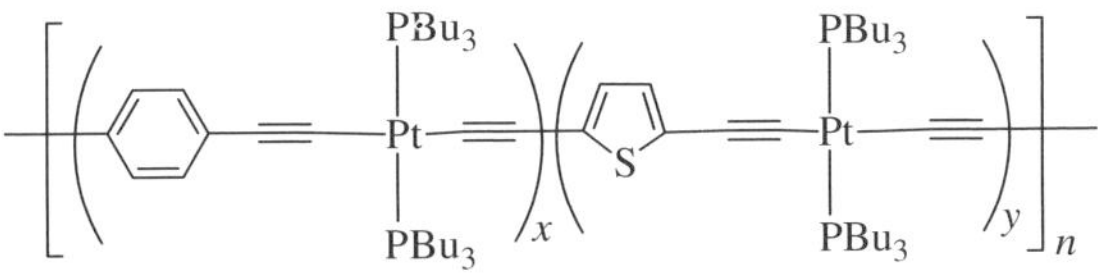

P9

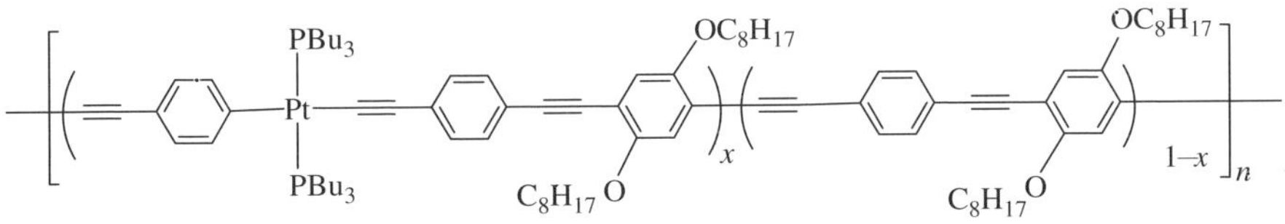

$x = 0.25, 0.5, 1$

P10

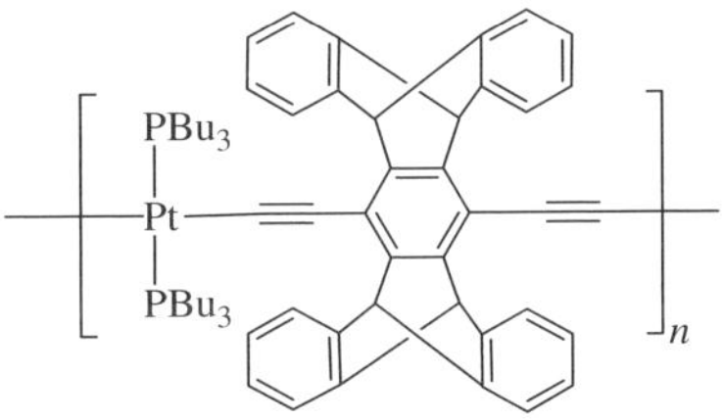

P11

The synthesis, optical, and structural properties of another Pt(II) polyyne-containing biphenyl moiety, **P12**, were reported recently.[44] The system has also been extended to the Au(I) and Hg(II) congeners (see Sections IV and V). The influence of the metal center on the spatial extent of S_1 and T_1 excited states was characterized in detail. The ligand-based phosphorescence emissions can be harvested by the heavy-atom effect of these transition metals, which facilitates efficient intersystem crossing from the S_1 state to the T_1 state.

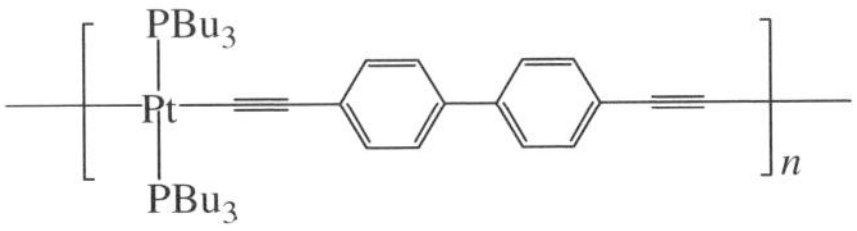

P12

Some organometallic platinum(II) and palladium(II) acetylide polymers spaced by the 2,6-diethynyl-4-nitroaniline bridge, **P13** and **P14**, were isolated in 70–75% yields.[45] Instead of exhibiting the rodlike zigzag structure typical for group-10 polymetallaynes bearing simple phenyl or biphenyl rings, the polymer structures of **P13** and **P14** were shown to consist of a sequence of organic spacers bound to metal centers, arranged in a helical configuration, with the chain length corresponding to ~8 repeat units for **P13** and ~16 repeat units for **P14**. The polymer structure was elucidated using two-dimensional (2-D) correlated spectroscopy (COSY) and nuclear Overhauser effect spectroscopy (NOESY) nuclear magnetic resonance (NMR) spectroscopy.

Optical absorption studies and X-ray photoelectron spectroscopy (XPS) measurements indicated that only a moderate charge delocalization through the metal centers along the chain was apparent.

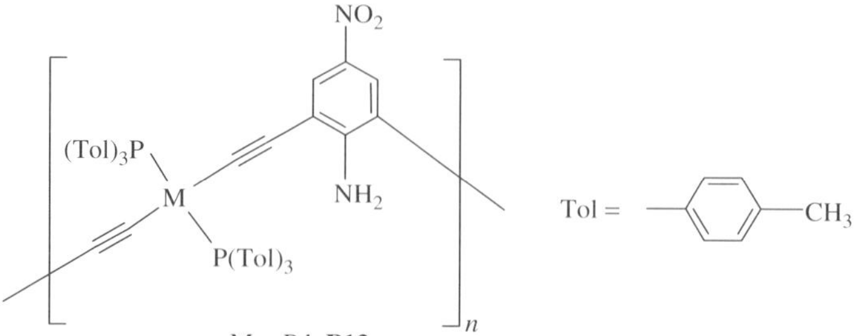

M = Pd **P13**
Pt **P14**

A new Pt(II) polyyne polymer, **P15**, prepared from the reaction of *cis*-$[Pt(PPh_3)_2Cl_2]$ with 1,4-diethynyl-2,5-dihexadecyloxybenzene using the "extended one pot" polymerization route, was tested for its sensing properties and showed fast and reproducible response to relative humidity variations and methanol vapor in surface acoustic-wave (SAW) sensors.[46] A SAW sensor was fabricated from polymer **P15** as a sensitive membrane, and the polymer was deposited as thin film on the surface of SAW delay lines implemented on three different piezoelectric substrates. High sensitivity and reproducibility were recorded for such devices. The acoustic characterization of the polymer film was also studied with the aid of theoretical results obtained by the perturbation theory.

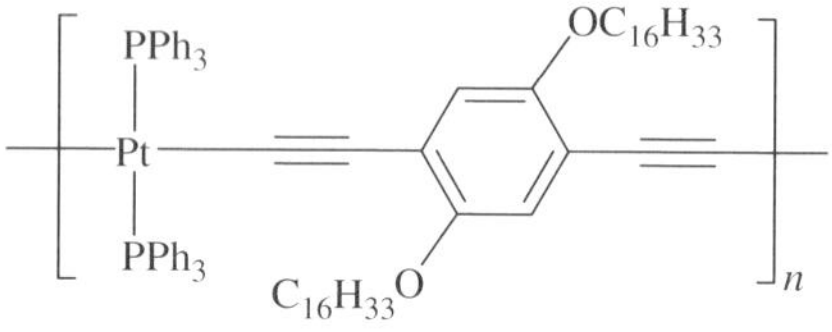

P15

Puddephatt et al. have reported the synthesis of some yellow oligomers **P16** with Pt–Pt bonds in the main chain from the reaction of $[Pt_2Cl_2(\mu\text{-dppm})_2]$ [dppm = bis(diphenylphosphino)methane] with HC≡CRC≡CH using NaOMe/MeOH basic conditions.[47] Since the oligomers are insoluble in organic solvents, no molecular-weight data are available. The XPS spectra of **P16** afforded Pt 4f7/2 binding energies ranging from 72.2 to 72.4 eV, which are very close to that for the model complex $[Pt_2(C{\equiv}CPh)_2(\mu\text{-dppm})_2]$ (72.8 eV).

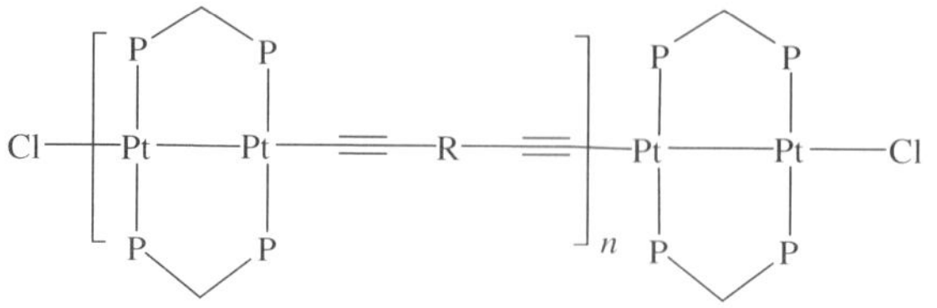

P16

R = p-C_6H_4 $n = 5$
p-$C_6H_2Me_2$ $n = 12$
p-C_6H_4-p-C_6H_4 $n = 3$

The synthesis of several high molecular-weight platinum polyyne polymers carrying chiral (R)-1,1′-bi-2-naphthol bridge, **P17**, was described.[48] These represent the first examples of metal acetylide polymer with an optically active backbone. They display much larger specific negative optical rotations than both the metal-free alkyne precursors and their corresponding dinuclear model compounds. These results support the fact that the main chain of **P17** adopts a one-handed helical conformation and induces the helical chirality of the polymers.

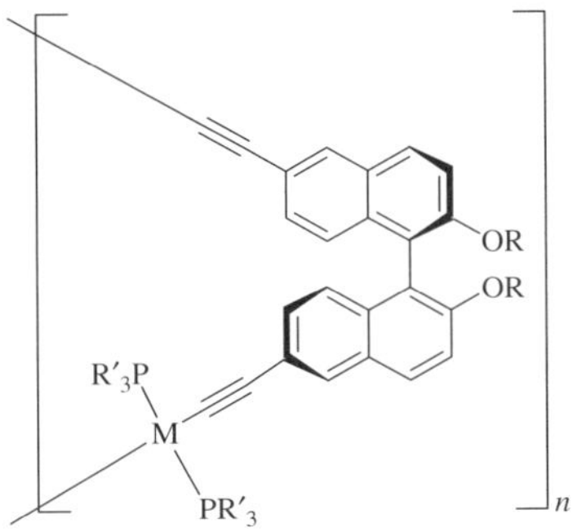

P17

M = Pd, Pt; R′ = Bu, Et; R = Me, Et, ^{i}Pr

A new family of platinum(II) polyyne polymers functionalized with substituted 1,4-diethynylbenzene derivatives **P18–P24** were synthesized and optically characterized by absorption and photoluminescence studies.[49] Thermogravimetry showed that polyynes **P18** and **P19** with the amino and methoxy substituents, respectively, have the lowest thermal stability, while the fluorinated ones, **P22–P24** exhibit increasing thermal stability with increasing extent of fluorination. The high thermal decomposition temperature observed for **P20** is believed to be caused by the presence of the octyloxy chains that can favor stronger interchain van der Waals interactions. In regard to the emission properties, it is interesting to see that the relative intensity of triplet emission to singlet emission increases strongly with the fluorine content in the order **P24** > **P23** > **P22**.

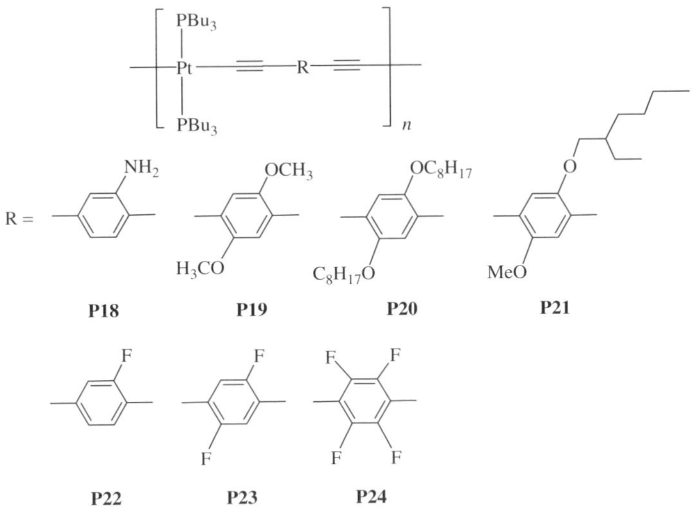

One special current topic is concerned with the use of fluorene spacers in which functionalization at the 9-position of fluorene is possible. Since the first report in 1998,[50] much attention has been paid to the fluorene-based materials of this kind in polymetallaynes and other metallated materials.[39,51] It was demonstrated that substituted oligo- and poly(2,7-fluorene) derivatives hold great promise as active components for organic and polymeric LEDs due to their thermal and chemical stability and their high emission quantum yields.[52–56] The fluorene structural motif offers a rigidly planar biphenyl group in the backbone. In this connection, a series of diethynyl fluorene-based precursors can be used to produce thermally stable polymeric complexes of platinum(II), **P25–P30**.[57,58,59] The corresponding dinuclear compounds were also investigated as model complexes for the electronic and structural properties of the parent polymers in each case. ^{31}P NMR spectroscopy has provided an exceptional tool for the determination of the configuration at the metal center. The 3-D structures of the model compounds for the corresponding polymers **P25**, **P28**, and **P30** have been confirmed by single-crystal X-ray structural analyses. The formation of free-standing films of these materials upon evaporation of solvents suggests that the product is macromolecular in nature.

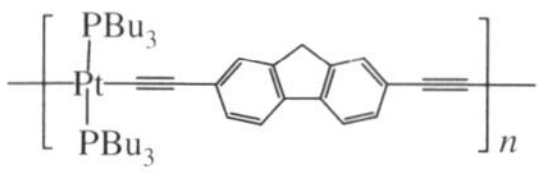

P25

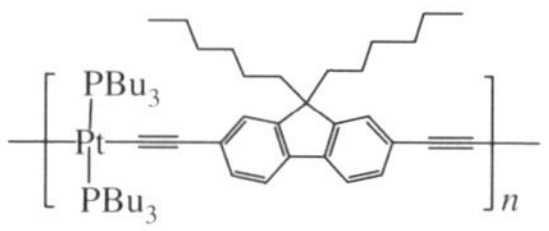

P26

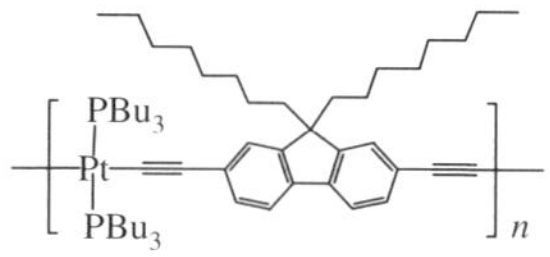

P27

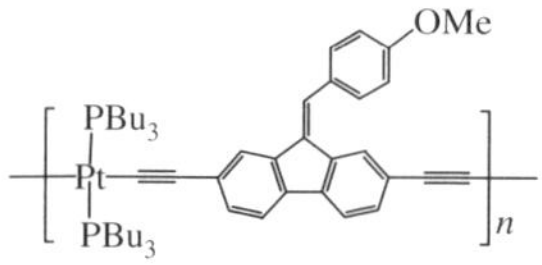

P28

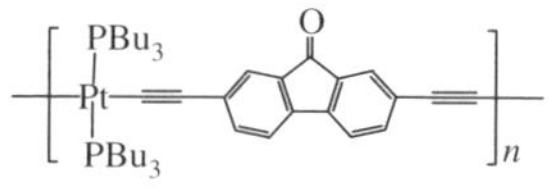

P29

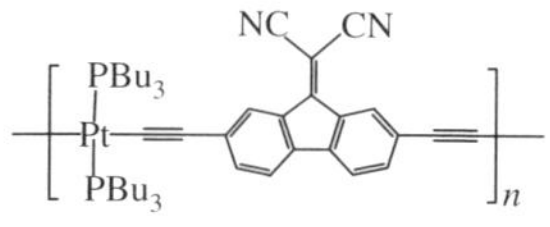

P30

By changing the substituents at the C-9 position of the fluorene ring, the optical band gap (or the onset of absorption) and the absorption and emission properties of these metal-containing polymers can be chemically modified, leading to diversified optoelectronic properties. These materials display a variety of colors and optical band gaps (E_g) in the solid states (Table 6.1). The dependence of the S_1 and T_1 electronic states on the chemical structure of the fluorene spacer groups was examined in detail. In the majority of cases, the versatility of preparing new metal complexes of fluorene derivatives has been demonstrated by insertion of different peripheral groups on the fluorene ring, and variation of the C-9 substituents can readily affect the optical and photophysical properties of these materials.

TABLE 6.1 Colors and Optical Band Gaps (E_g) of Platinum(II) Fluorene-Based Polyynes

Pt(II) Polyyne	Color	E_g(eV)	Reference
P25	Off-white	2.90	50, 59, 95
P26	Off-white	2.92	57
P27	Off-white	2.92	59
P28	Red-orange	2.17	60
P29	Red	2.10	50
P30	Deep blue	1.58	58

The lowest energy absorption peaks in the near UV region can be assigned to ($\pi\pi^*$) transitions of the bridging fluorenyl ligand for **P25–P30** and are red-shifted relative to the free alkynes when the platinum groups are added. This signifies π-conjugation of the ligands through the metal center. In the presence of strong electron-withdrawing cyano side groups in **P30**, the lowest energy absorption peak exhibits an unprecedented large red shift as compared to other fluorene derivatives, demonstrating the importance of donor–acceptor interaction in this class of polymetallaynes.[58] The E_g value of **P30** was estimated to be 1.58 eV, which, to our knowledge, is the lowest among any of the polyplatinaynes known to date and represents a significant step forward in the development of low band-gap materials. This indicates that electron-withdrawing substituents at the periphery exert a more significant electronic effect on narrowing the band gaps of these polymers than electron-donating groups. Hence, the energy of the S_1 state varies significantly with the nature of the fluorene ring and is highest for **P25–P27**.

The photoluminescence (PL) spectra of **P25–P27** feature two emission bands at 11 K, attributable to the fluorescence and the phosphorescence emissions.[57,59] The lower-lying triplet emission was found to be strongly temperature dependent in contrast to the singlet emission. The lowest triplet excited state remains strongly localized, as can be indicated from the small energy difference between triplet emissions in polymers **P25–P27** and in their binuclear model complexes. However, triplet emission was not observed over the measured temperature range for **P28–P30**. Our results suggest that reduced conjugation in **P25–P27** effectively facilitates the ISC rate to render phosphorescence readily measurable using optical spectroscopic techniques.

Raithby and co-workers further compared the photophysical properties of several platinum(II) polyynes **P25, P27**, and **P29** with their organic copolyynes.[59] Since the nonradiative decay rate for the triplet emission, $(k_{nr})_P$, is equal or larger than the corresponding radiative decay rate, $(k_r)_P$, the PL quantum efficiencies of the platinum polyynes are reduced from those for the organic polymers. Optical data reveal that the anchoring of octyl side chains on the fluorenyl spacer reduces interchain interaction in the polyynes, while a fluorenonyl spacer affords a donor–acceptor motif along the rodlike backbone.

The photoconductive properties of **P26**, **P29**, and **P30** were studied by the single-layer photocells in the sandwich-type structure of ITO/Pt polyyne/Al (ITO = indium tin oxide).[57,58] These polymers show moderate photoconductivity. A photocurrent quantum yield of ~0.01% was estimated in most cases, which does not vary much with variation of the central fluorene ring.

The optical-limiting behavior of selected platinum(II) fluorene-based polyynes was investigated by the Z-scan technique. It was demonstrated that the platinum polyyne polymers **P26**, **P28**, and **P30** are excellent optical limiters to nanosecond laser pulse at 532 nm, whose results are comparable to or even better than those of the benchmark fullerene C_{60} and phthalocyanine dyes (Figure 6.1).[60] Examination of their photophysics suggest that these polyplatinaynes can undergo two different optical-limiting mechanisms. One exploits the heavy metal effect to increase the triplet state yield through strong spin-orbit coupling, while the second makes suitable donor–π–acceptor components in the conjugation path of the metal polyynes to facilitate the formation of intramolecular charge-transfer (ICT) states. The optical-limiting thresholds for these polymetallaynes range from 0.06 to 0.13 J/cm^2 at the linear transmittance of 82%. In order to investigate the effect of heavy metal on the optical-limiting action of poly(aryleneethynylene)s, the purely organic poly(2,7-diethynyl-9,9-dihexylfluorene) polymer was prepared and its optical-limiting behavior was also studied. Figure 6.2 compares the input–output curves for the organic and organometallic polyynes, and the former material has its optical-limiting response weaker than that of **P26**. Thus, the optical-limiting capabilities of the organic polyyne can be markedly improved through insertion of heavy metal ion to the polyyne main chain.

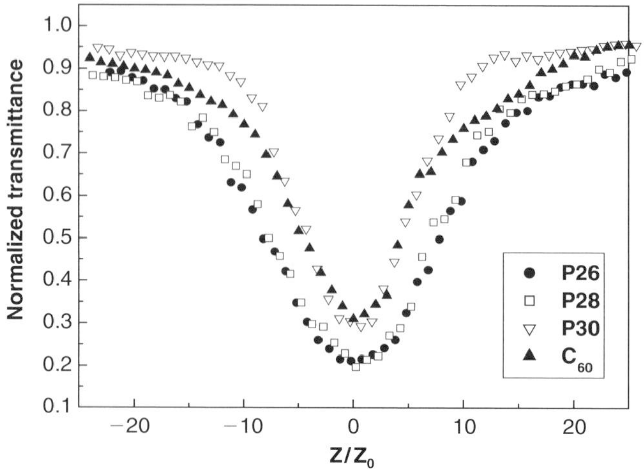

Figure 6.1 Z-Scan results for **P26**, **P28**, and **P30** as compared to C_{60} at the same linear transmittance (T) of 82%.

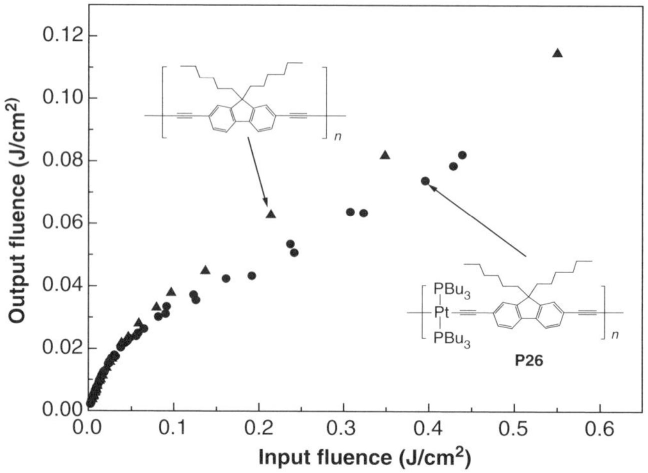

Figure 6.2 Comparison of the optical-limiting response of **P26** and its purely organic polyyne (T = 82%).

Recently, the synthesis and luminescent properties of a metal polyyne containing the conjugation-interrupting diphenylfluorene unit, **P31**, appeared in the literature.[61] Polymer **P31** exhibits good thermal stability with decomposition commencing at 347°C and decomposes without melting. It was shown that introduction of the diphenylfluorene group has the added advantage of increasing the thermal stability of this class of metallopolymers. With reference to the energy-gap law, it is advantageous to work on polymers with high-energy triplets to avoid competition with nonradiative decay. The sp^3-carbon site in diphenylfluroene is an effective conjugation interrupter to limit the conjugation length in metal polyynes, leading to materials with high optical gaps and high-energy triplet states. At 11 K, there is virtually no fluorescence band, but only the phosphorescence band associated with the organic chromophore in **P31**. The hindered conjugation with the use of diphenylfluorene unit in **P31** shifts the phosphorescence to the blue by 0.45 eV as compared to **P12**. It was found that the order of S_1–T_1 crossover efficiency is **P31** > **P12** ≈ **P4**. The $(k_r)_P$ and $(k_{nr})_P$ values for **P31** were computed to be $(9.5 \pm 0.5) \times 10^4\,s^{-1}$ and $(2.5 \pm 0.5) \times 10^4\,s^{-1}$ at 20 K, respectively. The heavy-atom effect of Pt(II) can increase the $(k_r)_P$ value for the triplet emission by four orders of magnitude and incorporation of the diphenylfluorene group in the present system with larger T_1–S_0 gaps can speed up the radiative decay as compared to **P4** ($(k_r)_P = (6 \pm 4) \times 10^3\,s^{-1}$ at 20 K). Therefore, high-energy triplet states can lead to the more efficient phosphorescence in metal-containing aryleneethynylenes.

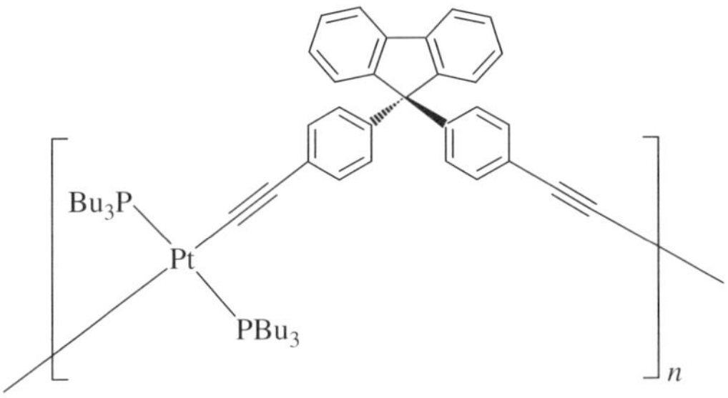

P31

A series of conjugated platinum(II) acetylide polymers containing oligothiophene and bithiazole groups **P32–P35** were synthesized by the classic CuI-catalyzed dehydrohalogenation reactions.[62,63,64] The photophysical, redox, and structural properties of these polymeric metallaynes were extensively studied in terms of the number of oligothienyl rings within the bridging ligand and the nature of the five-membered heterocyclic ring. The solid-state structures of the model complexes for **P32**, **P33**, and **P35** were established by X-ray crystallography in which they provided conclusive proof of the *trans* configuration at the metal center. Figure 6.3 illustrates an example for the model compound of **P32**.

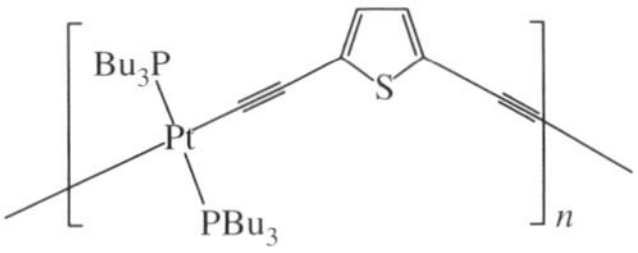

P32

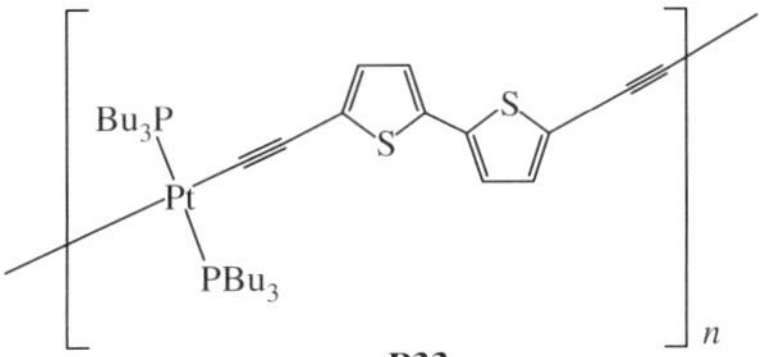

P33

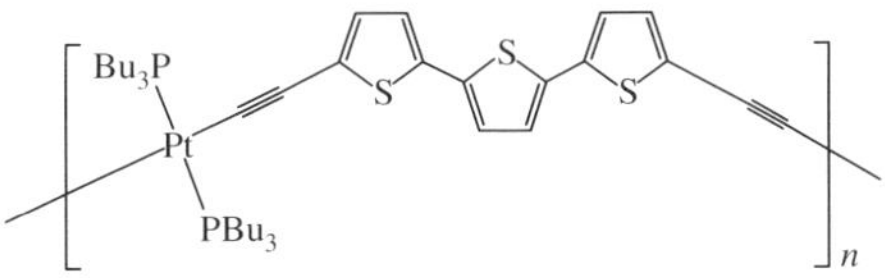

P34

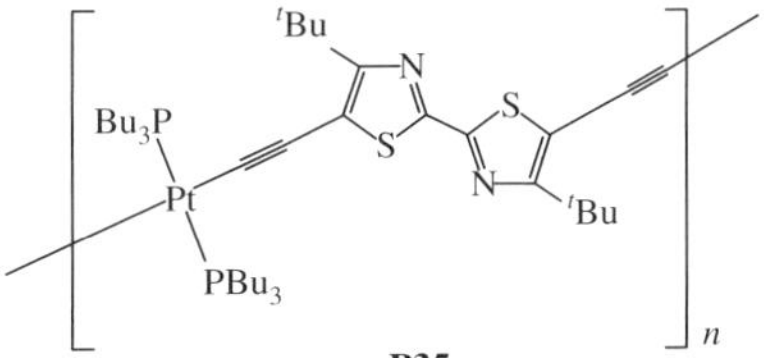

P35

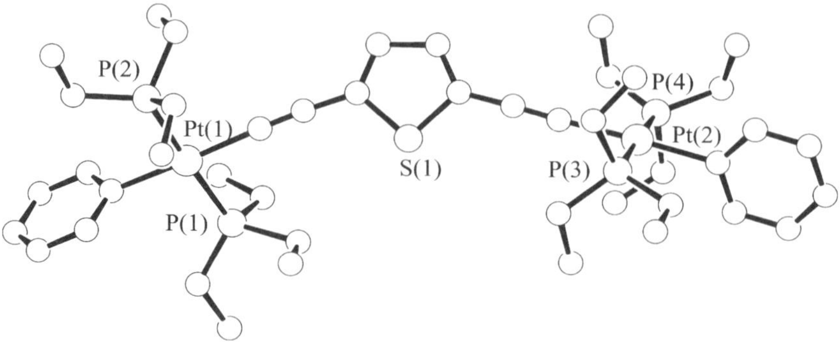

Figure 6.3 X-Ray structure of the model compound for **P32**.

The absorption spectra of **P32–P34** are shown in Figure 6.4. The E_g values of **P32–P34** decrease with the increase in the number of thienyl rings, which is in line with an increased delocalization of π-electrons along the polymer chain. However, the extent of reduction in the E_g value decreases as it moves from monothiophene to terthiophene. The absorption energies of the platinum polyynes are lowered compared to the corresponding model diynes, which is consistent with π-conjugation of the ligands through the metal group. The emission features are shifted to lower energy with an increase in chain length of the thiophene segment. Table 6.2 lists the experimental S_1 and T_1 state energies for **P32–P34**, and the measured S_1–T_1 separations lie essentially constant at 0.75–0.80 eV. The intensity of triplet emission decreases rapidly as more thienyl units are added. We also observe that the energy of the phosphorescence band shifts when adding more thiophene rings in the ligand (from 2.05 to 1.53 eV for **P32–P34**, $\Delta E = 0.52$ eV), suggesting that the triplet excited state should be extended over three or more thiophene rings in the present system. This shift of phosphorescence energy observed in **P32–P34** agrees with the calculations on the evolution of the triplet energy in organic oligothiophene systems.[65,66]

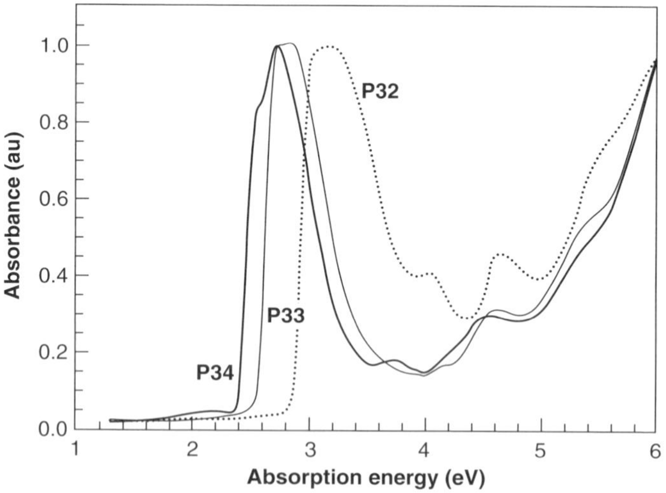

Figure 6.4 Solid-state absorption spectra of **P32–P34**.

TABLE 6.2 Values of S_1 and T_1 State Energies for P32–P34

Pt(II) Polyyne	E_g (eV)	S_1 Energy (eV)	T_1 Energy (eV)	S_1–T_1 Gap (eV)
P32	2.80	2.85	2.05	0.80
P33	2.55	2.44	1.67	0.77
P34	2.40	2.28	1.53	0.75

More interestingly, platinum polyynes **P32–P34** were shown to be good photoconductors.[57] The photocurrent spectra of the Au/**P32**/Al, ITO/**P33**/Al, and ITO/**P34**/Al photocells display two peaks, one at the onset of absorption and one at higher photon energies. Polymers **P32–P34** show a short-circuit quantum efficiency of about 0.04% at the first photocurrent peak, a typical value for single-layer devices. No strong dependence of the quantum efficiency with variation of the thiophene content in these metallopolymers was observed. However, the quantum efficiency of the second peak is different among these three polymers and is very air-sensitive. The effect of air exposure on the photocurrent density in **P32** was studied, and it was shown that air can enhance exciton dissociation easily.[67] The overall photocurrent increases upon exposure to air and is reduced after annealing under vacuum. The enhancement of the photovoltaic response to air was found to be reversible after annealing under vacuum. The *I–V* characteristics at the first peak in the spectral response give open-circuit voltages of 0.47–0.75V and fill factors of 0.30–0.35 for **P32–P34**, and Figure 6.5 shows the typical case for the photovoltaic behavior of **P33** under illumination at 480 nm.

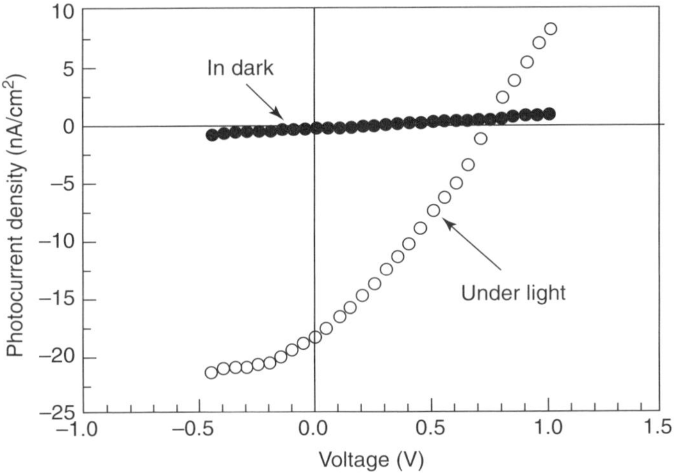

Figure 6.5 *I–V* curve for the ITO/**P33**/Al photocell.

By virtue of the electron-donating and electron-accepting features of the thiazole ring as compared to the thienyl analog, a luminescent bithiazole-bridged polyyne **P35** was also studied.[64] A thiazole unit can be considered as a hybrid of the thiophene and pyridine groups that can be a valuable spacer for controlling the bandgaps of these metallated materials. For **P35**, the absorption peaks show a significant spectral red shift as compared to its bithienyl counterpart **P33** due to the electron-withdrawing effect of the imine nitrogen atoms. No phosphorescence band was detected for **P35** even at low temperatures. The potential of using polymer

P35 as photoconductors was also investigated in a single-layer device.[67] A recent study also showed that polymer **P35** displays a large optical-limiting response to a nanosecond laser pulse with satisfactory threshold value at 0.12 J/cm^2, and its behavior is superior to that for poly(4,4′-diethynyl-5,5′-bithiazole) organic polymer.[60]

Kakkar and co-workers have developed the preparative routes to the Pt-acetylide polymer **P36** with 2,5-diethynylpyridine moiety and its stable quaternized counterparts, **P37** and **P38**.[68] Quaternization of pyridyl nitrogen was achieved by nucleophilic substitution of **P36** with CH_3X, which results in a strong bathochromic shift in the UV-vis absorption spectrum and an enhanced emission quantum yield. There is an improved π-electron delocalization along the backbone upon quaternization. Both polymers are basically insulators in the undoped state. Upon doping with iodine vapor, they become semiconducting and the conductivity of **P37** (3.4×10^{-3} S cm^{-1}) was found to be higher than that for **P36** (2.5×10^{-3} S cm^{-1}). The authors also compared the results for **P36**–**P38** with their organic congeners.[68,69]

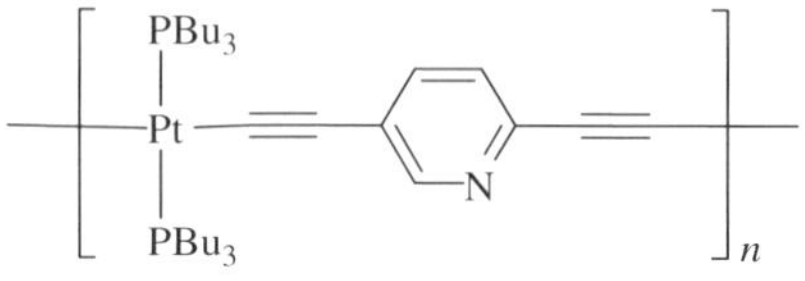

P36

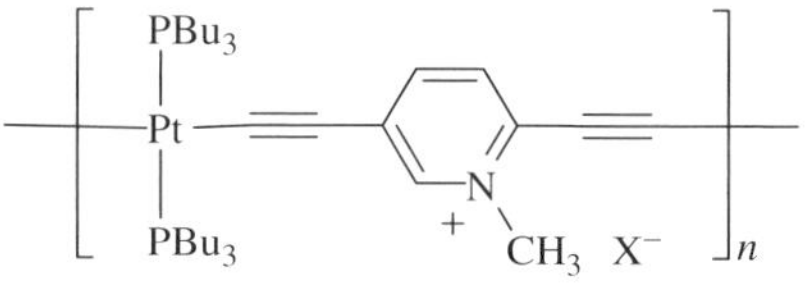

$X^- = I^-$ **P37**

$X^- = CF_3SO_3^-$ **P38**

Researchers in Cambridge have also discussed the dependence of intersystem crossing and the spatial extent of singlet and triplet excited states in **P4**, **P32**, and **P36** as a function of electron delocalization in the organic spacer group. From the optical absorption and photoluminescence data of these materials, conjugation is increased, but the ISC rate is reduced by the electron-rich thiophene ring, whereas the opposite trend prevails for the electron-deficient pyridine moiety as compared to the phenylene unit.[70] With the aid of steady-state photoinduced absorption spectra, an energy scheme for the lower-lying excitations has been constructed and discussed. In all three cases, the T_1 triplet state remains strongly localized and the T_n triplet state remains strongly delocalized even more than the delocalized S_1 singlet state. It was shown that the extent of delocalization is larger for the electron-rich thiophene group. Later on, the same research groups further accomplished the synthesis, structural characterization, and photophysical studies of a series of oligopyridine-linked platinum(II) polyynes **P39**–**P42**, and the results were compared to those for **P36**.[71] The greater the number of pyridine units in the backbone, the less thermally stable the polymer. For **P39**, the inclusion of a second pyridine unit shifts the optical band gap to the red relative to **P36** by 0.1 eV, while for **P40**–**P42**, the band gap is blue-shifted by 0.3 eV compared to **P36**. This is in line with the fact that in **P39**, the alkynyl groups are at the 5,5′-positions and the polymer is fully conjugated, whereas in **P40**–**P42**, the alkynyl units at the 6,6′- or 6,6″-positions can hinder conjugation between the pyridine rings. It appears that the phosphorescence spectra for **P39**–**P41** show some excimer formation at room temperature, but only intrachain emission occurs at 10 K. In particular, the authors observed no fluorescence band, but only the phosphorescence band at 10 K for the kinked bi- and terpyridine-containing compounds **P40** and **P41**. The reduced conjugation shifts the triplet emission peak to the blue by 0.3 eV compared to the linear bipyridine-containing analog **P39**. Polyynes **P40** and **P41** also constitute a class of materials with high-energy triplet states with lowered nonradiative decay rate constants from the T_1 state, which are desirable for applications that harvest the triplet state for light emission.

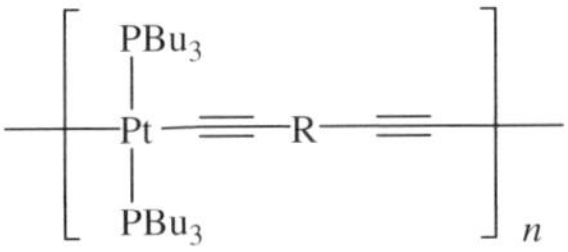

R = 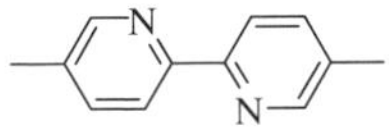 **P39** 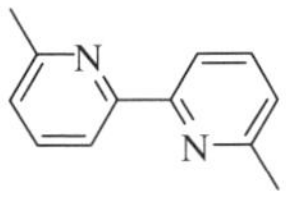 **P40**

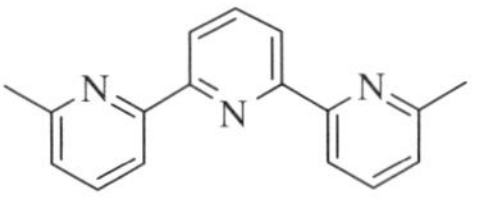 **P41** 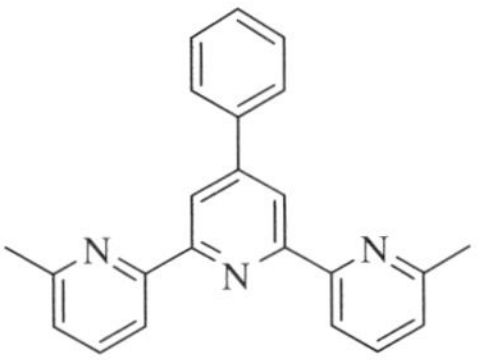 **P42**

The synthesis and optical spectroscopy of two interesting platinum polyynes that possess mixed heterocyclic groups consisting of both thienyl and pyridyl rings **P43** and **P44** were described.[72] They are high molecular-weight polymers with molar masses in the range 85,000–140,000. The optical energy gaps for **P43** and **P44** are 2.67 and 2.55 eV, respectively, which compare to the gaps of 2.55 and 2.40 eV found for **P33** and **P34**, and the blue shifts of 0.12 and 0.15 eV are consistent with a reduction in the donor–acceptor interaction between the metal and the ligand. Again, both polymers exhibit interesting fluorescence and phosphorescence emission features in the emission profile at 10 K.

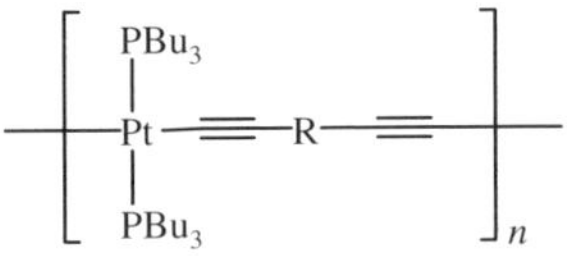

R = 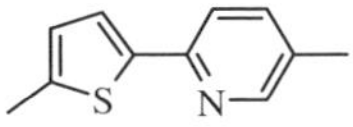 **P43**

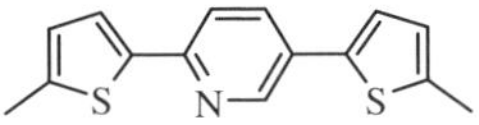 **P44**

One of the most spectacular advances in the area of platinum polyyne polymers is the report of the soluble blue materials containing the thieno[3,4-b]pyrazine spacer **P45**.[73] The polymer was designed by using the concept of alternating donor and acceptor units, as has been successfully exploited for the production of low band-gap organic polymers. Polymer **P45** shows a band gap of 1.77 eV, presumably caused by the strong push–pull interaction between the electron-donating metal acetylene group and the electron-withdrawing thieno[3,4-*b*]pyrazine unit. Polymer **P45** shows a low-energy photoluminescence band at 715 nm. While this polymer is not phosphorescent, it exhibits an unusually high photocurrent quantum efficiency of up to 1% at 400 nm for single-layer sandwich-type photovoltaic cells in air.

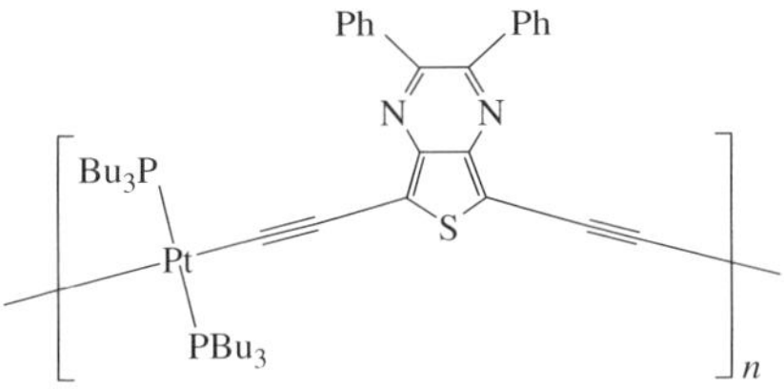

P45

A series of platinum polyynes **P46–P48** functionalized with quinoline, quinoxaline, and benzothiadiazole units have been reported which provide an excellent system to study the evolution of the singlet and triplet excited states with respect to electronegativity of the spacer groups.[74] The absorption spectra of **P46–P48** reveal substantial donor–acceptor interaction between the platinum and the conjugated ligands. Both the singlet and triplet emissions, as well as the absorption bands, decrease in energy with increasing electronegativity of the spacers along the series from **P46** to **P48**. The same authors have also extensively investigated the triplet excited-state properties of a range of polyynes including polymers **P4**, **P21**, and **P46–P48** by means of optical steady state and time-resolved spectroscopy as a function of optical gaps, and they found that the S_1–T_1 energy splitting is independent of the spacer R such that the T_1 state is always 0.7 ± 0.1 eV below the S_1 state.[75] With decreasing optical gap, the lifetimes and intensities of phosphorescence are dramatically reduced. Based on these experimental results, the authors formulated, for the first time, a useful energy-gap law for the triplet states in this class of Pt-containing conjugated polyyne polymers with a tunable triplet energy level ranging from 1.3 to 2.5 eV, while the $(k_{nr})_P$ value increases exponentially with an increasing T_1–S_0 gap.[76] This suggests that future work should be more concentrated on soluble polymers with high-energy triplet states in order to attain the most efficient phosphorescence for practical applications.

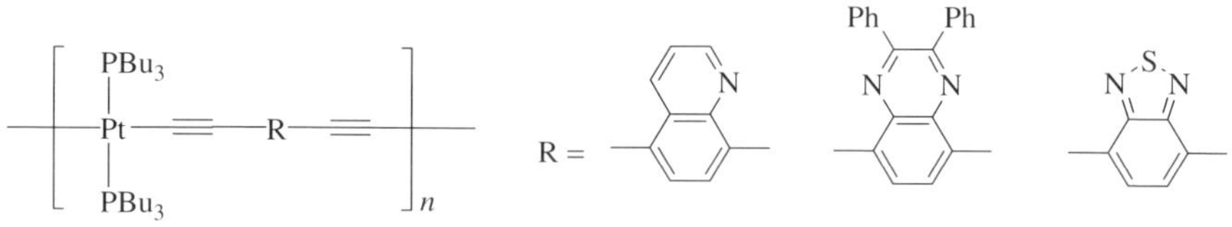

P46 **P47** **P48**

To realize the role of this class of Pt(II) polyyne materials in practical polymer LED devices, Friend and co-workers have reported their seminal work on the spin-dependent exciton formation in the platinum polyyne, **P47**. Direct photoluminescence and electroluminescence emission from the triplet state can be observed and measured for a LED in a structure of ITO/PEDOT-PPS/**P47**/Ca:Al (PEDOT = poly(3,4-ethylenedioxythiophene; PSS = poly (styrenesulphonate)).[77] The average singlet generation fraction of ~0.57 for **P47** was determined, suggesting that a spin-dependent process, favoring singlet formation, was operative in the polymer film. The value is more than double the value expected from simple spin statistics and is above the lower limit for the singlet-generation fraction of 0.35–0.50 measured for poly(*p*-phenylenevinylene) derivatives. The probable processes responsible for this phenomenon have been discussed, and the results demonstrate that singlets are favored over triplets.

To our knowledge, carbazole-containing metal alkynyl compounds are very rare in the literature. The first example is the platinum(II) polyyne **P49** prepared through the CuI-catalyzed dehydrohalogenation between *trans*-[$Pt(PBu_3)_2Cl_2$] and 9-butyl-3,6-diethynylcarbazole.[57] The regiochemical structure of polymer **P49** has been ascertained by single-crystal X-ray analysis on the corresponding diplatinum model complex (Figure 6.6). It was shown that the S_1 state extends over more than a repeat unit, while the T_1 state remains localized to less than one repeat unit. The band gap measured is ca. 3.10 eV. The photoluminescence spectrum of **P49** shows a characteristic singlet emission at 425 nm and a triplet emission peak at 460 nm, both of which are mainly of ($\pi\pi^*$) origin. The phosphorescence is characterized by the strong temperature dependence of the PL spectra, and the triplet emissive band dominates the PL spectrum as the temperature is lowered (Figure 6.7). The study indicates that the use of carbazole unit leads to high-energy triplet states and represents an effective approach in the enhancement of the ISC rate. Polymer **P49** was also shown to be a photoconducting material in a single-layer architecture with a photocurrent quantum efficiency of ~0.01% in the forward-bias mode. Moreover, polymer **P49** is an effective optical limiter for nanosecond pulses of visible light and it is the triplet excited-state absorption that contributes to the optical-limiting action.[60] The polymeric compound **P49** is a better optical limiter than the short-chain diplatinum congener, suggesting that the optical-limiting effect of the polyyne arises not only from the single repeating segment but the fragment that is delocalized over more than one repeat unit. The polyyne **P49** also gives a better optical-limiting performance than the organic counterpart, viz., poly(9-butyl-3,6-diethynylcarbazole).[60]

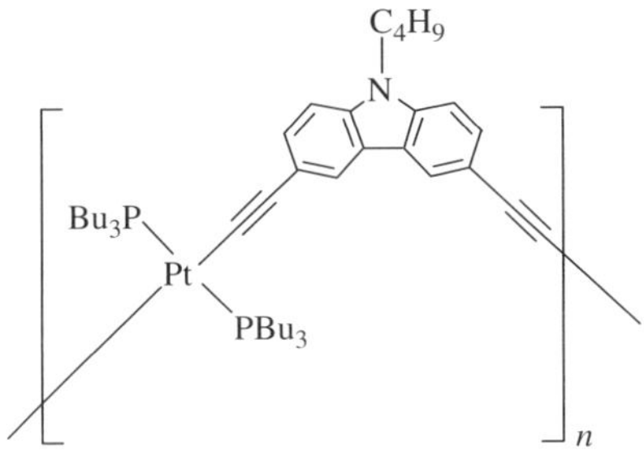

P49

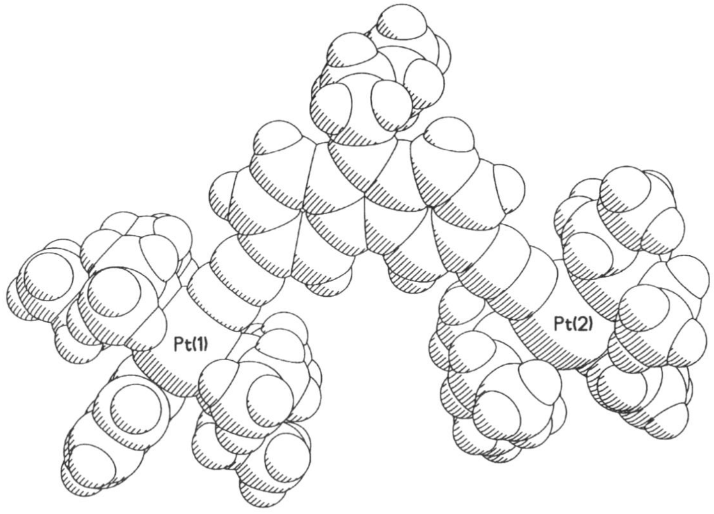

Figure 6.6 Space-filling diagram for the model compound of the platinum polyyne **P49**.

Two poly(cyclodiborazane) polymers containing group-10 transition metal–acetylide groups **P50** and **P51** were prepared by Chujo et al. and represent a new kind of organometallic acetylide polymers functionalized with group-13 boron elements in the main chain.[78] The structures of **P50** and **P51** were confirmed by IR and NMR (^1H, ^{31}P, and ^{11}B) spectra. The optical properties were studied by UV-vis absorption and emission measurements. It was shown that these polymers display extended π-conjugation length via transition metal and boron atom with enhanced air- and moisture stability.

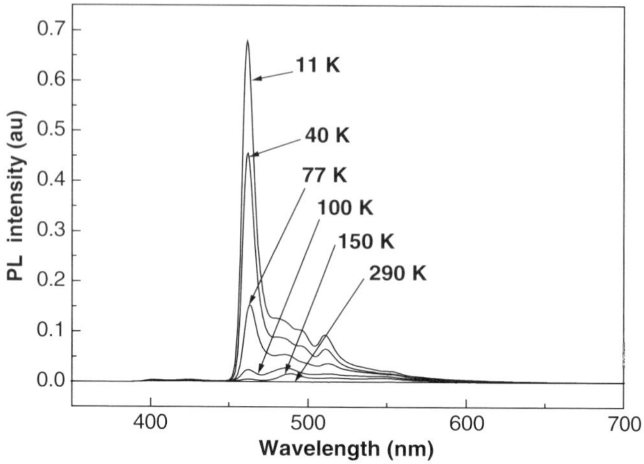

Figure 6.7 Temperature dependence of the PL spectra of **P49**.

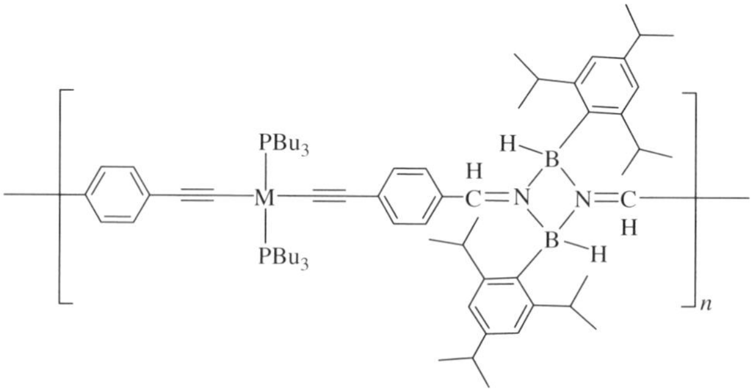

M = Pd **P50**
Pd = Pt **P51**

It was envisaged that the lower ionization energy of the silicon atom (first ionization energy (IE) of Si: 791 kJ mol^{-1}) compared with carbon can enhance the through-bond interaction of ethynyl units along the backbone.[79] π-Conjugated organosilicon systems incorporating aromatic units were widely studied regarding their applications to optoelectronic devices.[80,81,82] Along these lines, the first synthesis, characterization and luminescence behavior of a novel platinum-containing poly(silylacetylenes) **P52** were described.[83] The isolation of **P52** provides us with a unique opportunity to study, for the first time, the effect of main-group inorganic spacer units in polymetallaynes. Its optical and theoretical properties have been discussed in comparison to those polymers possessing purely acetylenic and other (hetero)aromatic groups.

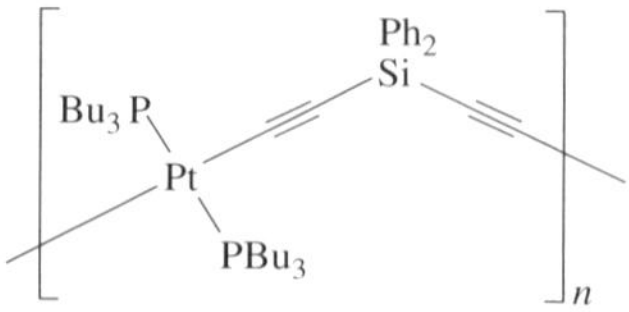

P52

The polymer **P52** is air-stable and has good solubility in common organic solvents. The ^{29}Si NMR chemical shift of **P52** ($\delta = -57.50$) appears at a more downfield position in comparison to $Ph_2Si(C{\equiv}CH)_2$ ($\delta = -48.42$) and does not differ significantly from those for related polymers comprising silylacetylene and π-conjugated organic groups.[84] Polymer **P52** is thermally stable up to ~380°C, with the onset decomposition temperature higher than those in related polymers such as **P4**, **P25**, **P32**, and **P36**, suggesting that the use of the $SiPh_2$ unit can increase the thermal stability of Pt(II) polyynes. The degree of electronic conjugation roughly follows the order $C{\equiv}C \geqslant$ 2,5-thienyl $>$ p-C_6H_4 $>$ 2,5-pyridyl $>$ $SiPh_2$ (Figure 6.8), implying that the energy of the S_1 state is highest for **P52**. At 11 K, the principal emissive peak occurs at ca. 508 nm for **P52**. The triplet emission lifetimes (τ_P) of 11.0 ± 0.1 μs at 290 K and 40.2 ± 0.1 μs at 11 K for **P52** (versus $\tau_P = 30$ μs at 10 K for **P4**) support the $^3(\pi\pi^*)$ character of the phosphorescent state. The hindered conjugation in **P52** shifts the phosphorescence to the blue by 0.06 and 0.38 eV, respectively as compared to **P4** and **P32**.

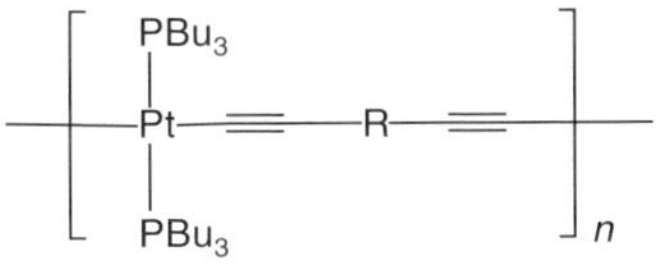

R =	Ethynyl	2,5-Thienyl	1,4-Phenylene	2,5-Pyridyl	Silyl
E_g (eV)	2.82	2.85	2.98	3.05	3.70

Figure 6.8 Optical band gaps (E_g in eV) for platinum polyynes with different spacer groups R.

Synthetic extension to new silylacetylene derivatives fused with additional aromatic chromophore is equally interesting, and a new Pt(II) polyyne **P53** was successfully isolated.[85] Polymer **P54** was prepared as a structural analog to **P53** to evaluate the effect of $SiPh_2$ in such a system. The ^{29}Si NMR chemical shift of **P53** ($\delta = -48.16$) remains relatively unshifted as compared to the free alkyne ($\delta = -48.49$). It is also experimentally evident that addition of $SiPh_2$ units in **P53** can enhance the thermal stability of polyplatinaynes.

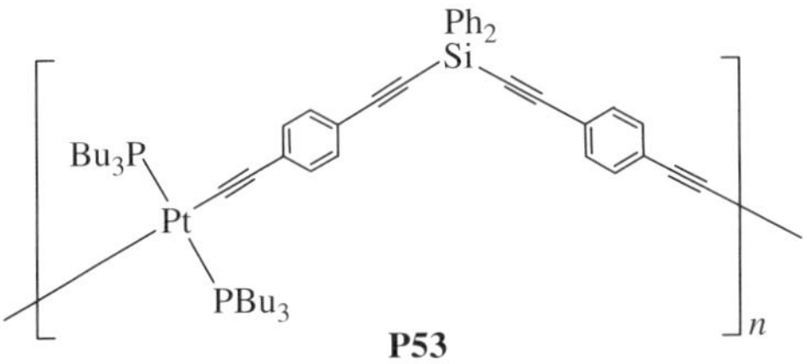

P53

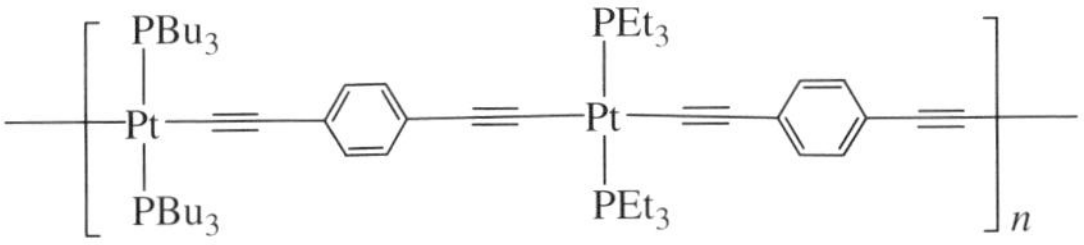

P54

From the absorption data, it was shown that $SiPh_2$ is less effective than the phenylene and $Pt(PEt_3)_2$ units in electronic conjugation. In general, characteristic fluorescence and phosphorescence emission bands were observed at 11 K for **P53** and **P54**. Efficient phosphorescence is easier to obtain for higher S_0–T_1 energy gaps. We note a very high efficiency of triplet emission for **P53**, and the S_1–T_1 crossover efficiency is on the order of **P53** > **P54**. As observed in Figures 6.9 and 6.10, the ratio of integrated intensities of phosphorescence to fluorescence is greater than unity for **P53**, but is less than unity for **P54**. The use of conjugation-interrupting silyl segment in such metal polyynes can break the conjugation and gives rise to efficient crossover between the S_1 and T_1 states. The S_1–T_1 energy gap for **P53** was found to lie within the constant range of 0.7 ± 0.1 eV, regardless of the conjugation interruption by the group-14 element. The photoconductivity of polymer **P53** was also examined, and a photocell in a single-layer structure of ITO/**P53**/Al was fabricated. The quantum efficiency detected was moderate at 0.01%. Alternate substitution of $Pt(PBu_3)_2$ units in **P4** by $SiPh_2$ does not significantly affect the photoconducting properties of this class of platinum polyynes.[85]

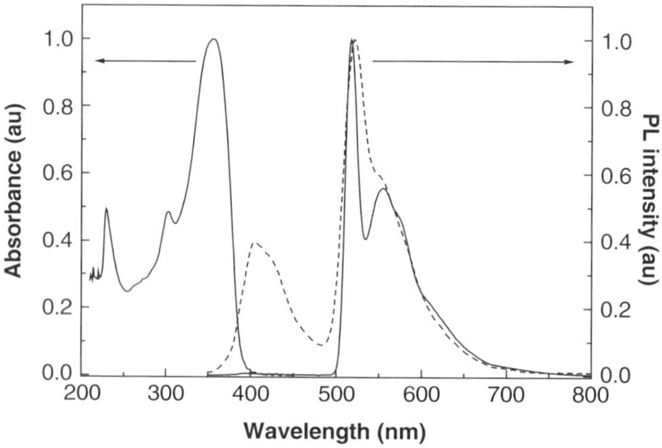

Figure 6.9 Solid-state absorption and PL (11 K) spectra of **P53** (solid line). The solution PL spectrum in CH_2Cl_2 at room temperature is represented by the dashed line.

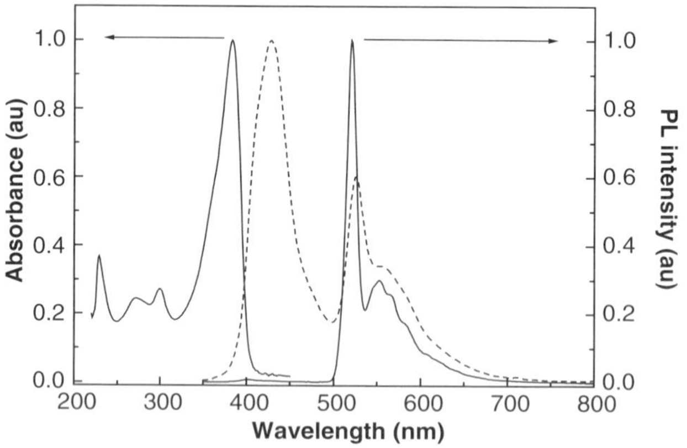

Figure 6.10 Solid-state absorption and PL (11 K) spectra of **P54** (solid line). The solution PL spectrum in CH_2Cl_2 at room temperature is represented by the dashed line.

On the other hand, there is little understood about related metallated macromolecular systems with the heavier group-14 germanium element of even lower IE (first IE of Ge: 762 kJ mol^{-1}). The first examples of thermostable Ge-bridged Pt(II) metallopolymers **P55–P60**, derived from oligo(fluorenyleneethynylenegermylene)s, were reported recently and can display very fast phosphorescence decay rates.[86] To our knowledge, this is the first organogermanium material of this kind to exhibit these properties, and this work represents a significant step forward in the development of triplet emitting materials for organic light-emitting diodes (OLEDs). Inclusion of a conjugation-interrupting sp^3-Ge unit can limit the effective conjugation length and trigger the triplet light emission by taking advantage of the heavy-atom effect of Ge atoms. The onset decomposition temperatures do not depend much on the chain length of the conjugated bridge and the substituent group R. While polymer **P26** commences decomposition at 349°C,[57] addition of GeR_2 unit into the aryl-acetylene segment in **P55–P60** notably increases the thermal stability of these Pt(II) polyynes (ca. 404–418°C). Each of these polymers can readily cast free-standing thin film upon stripping off of the solvent.

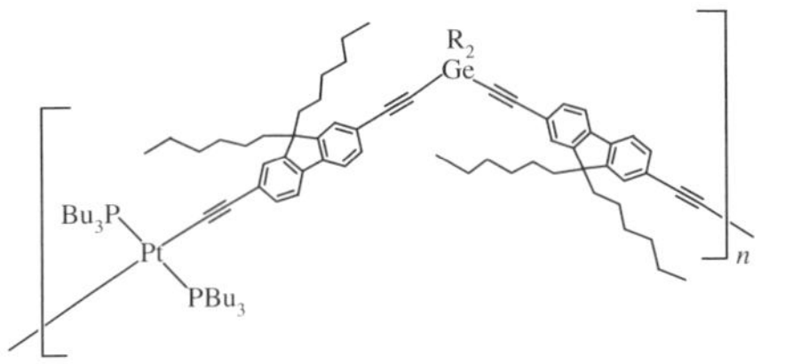

R = Me **P55**
R = Ph **P58**

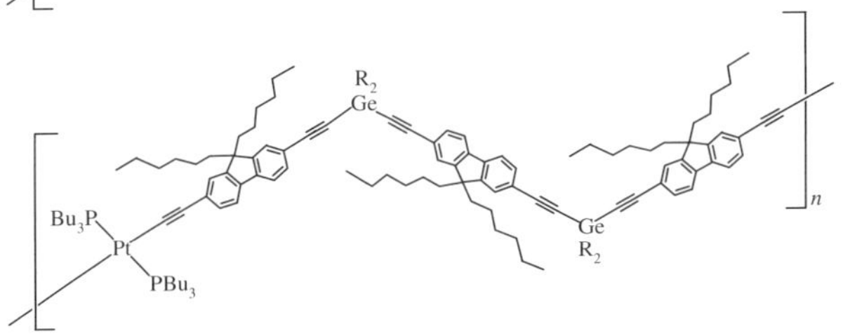

R = Me **P56**
R = Ph **P59**

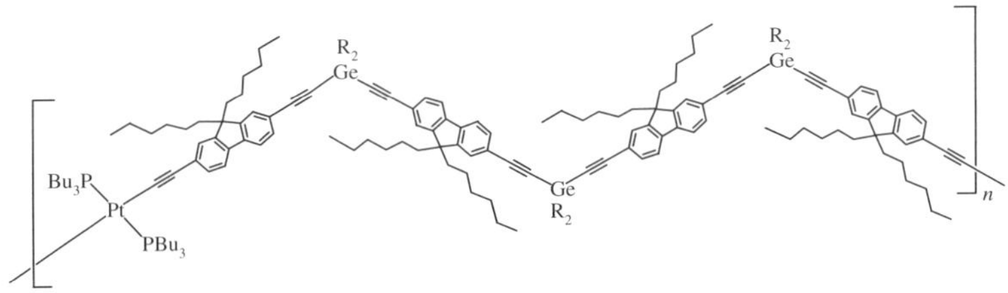

R = Me **P57**
R = Ph **P60**

For **P55–P60**, both fluorescent and phosphorescent emissions arise from ligand-centered $^1(\pi\pi^*)$ transitions. Figure 6.11 depicts the absorption and PL spectra for **P58**, in which the large Stokes shift between the absorption and low-lying emission features ascertains the phosphorescent state of the peak at ca. 548 nm. The strong dependence of the triplet emission on temperature also provides good evidence for its origin (Figure 6.12). In each case, the triplet energy does not vary much with the oligomer chain length, that is, the lowest T_1 state is confined to a single repeating unit. Variation of the R group does not seem to change this strong confinement. Insertion of the conjugation-hindered GeR_2 group in these polymers shifts the phosphorescence bands to the blue relative to **P26**. Values of the energy difference between S_0 and T_1 states were found to be ca. 2.27–2.28 eV for both series, and the S_1–T_1 separations lie within the narrow range of 0.73–0.75 eV.

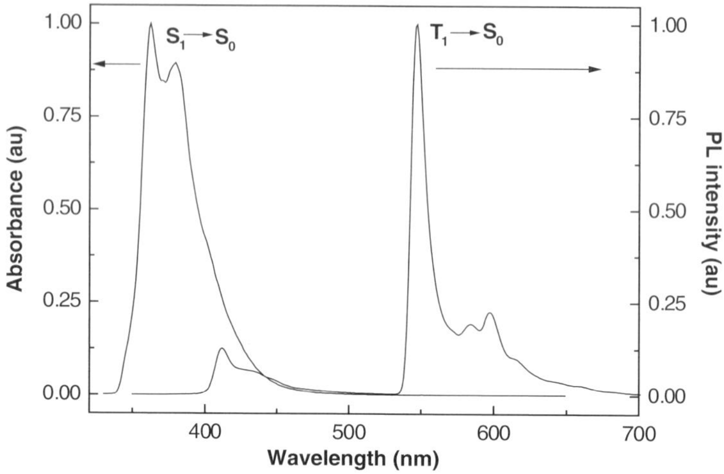

Figure 6.11 Optical absorption and PL spectra (11 K) of **P58**.

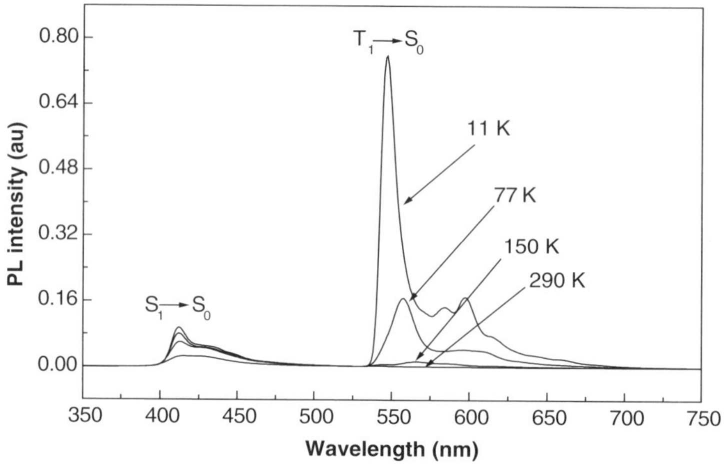

Figure 6.12 Temperature dependence of the PL spectra of **P58**.

Table 6.3 collects the $(k_r)_P$ and $(k_{nr})_P$ values together with the phosphorescence lifetimes, τ_P, and quantum yields, Φ_P, at 20 K for **P55–P60**.[86] While the measured Φ_P values are relatively insensitive to the oligomer chain length, they notably vary with the ER_2 group (E = Si, Ge). The $GeMe_2$ systems give more efficient phosphorescence than the $GePh_2$ congeners by over two times. However, the heavy-atom effect associated with the Ge atoms in **P58–P60** can almost double Φ_P as compared to the $SiPh_2$ congeners (Φ_P ~10–13%). The $(k_r)_P$ values at 20 K are $(2.1–3.5) \times 10^5\,s^{-1}$ for **P55–P57** and $(1.3–1.7) \times 10^5\,s^{-1}$ for **P58–P60**. Relative to **P26** ($(k_r)_P$~$4.4 \times 10^4\,s^{-1}$), insertion of the germylene component can increase $(k_r)_P$ by about one order of magnitude. It is most impressive to get comparable orders of magnitude for $(k_{nr})_P$ and $(k_r)_P$ in **P55–P60**. Hence, heavy-atom derivatization using Pt and Ge atoms in conjunction with conjugation interruption by the latter can greatly boost $(k_r)_P$ values by five orders of magnitude.

TABLE 6.3 Photophysical Data of P55–P60 at 20 K

Pt(II) Polyyne	τ_P (μs)	Φ_P	$(k_{nr})_P$ (s^{-1})	$(k_r)_P$ (s^{-1})
P55	1.27	0.45	4.3×10^5	3.5×10^5
P56	2.08	0.43	2.7×10^5	2.1×10^5
P57	1.41	0.45	3.9×10^5	3.2×10^5
P58	1.32	0.17	6.3×10^5	1.3×10^5
P59	1.21	0.18	6.8×10^5	1.5×10^5
P60	1.16	0.20	6.9×10^5	1.7×10^5

Note: $(k_{nr})_P = (1 - \Phi_P)/\tau_P$ and $(k_r)_P = \Phi_P/\tau_P$.

Very recently, we have also accomplished the synthesis of a soluble silole-containing polyplatinayne, **P61**.[87] The UV-vis absorption spectrum of **P61** in CH_2Cl_2 exhibits intense ($\pi\pi^*$) transitions in the near-UV region and a relatively low-energy shoulder band in the visible range that tails off up to 590 nm. As compared to 2,5-dibromo-1,1-diethylsilole precursor (λ_{max} = 326 nm), the position of the low-lying shoulder bands (λ_{max} = 504 nm) is remarkably red-shifted by ca. 178 nm for **P61** after the inclusion of heavy transition metal chromophores. This is likely due to the intramolecular donor–acceptor interaction between metal ethynyl units and silole rings. The E_g value is 2.10 eV for **P61**, and it is significantly lowered by ca. 0.6–1.0 eV relative to the electron-rich thienyl- (**P62** 2.70 eV)[88] and silyl-bridged (**P52** 3.10 eV) counterparts.[85] The incorporation of electron-accepting silole units in the metallopolymer main chain creates a new narrow band-gap π-conjugated system with unique donor–acceptor characteristics. Polymer **P61** is photoluminescent with a singlet emission band at 537 nm (Figure 6.13). No room temperature emission from the T_1 state was detected over the measured spectral range. The highest occupied molecular orbital (HOMO) and lowest unoccupied molecular orbital (LUMO) levels of **P61** were estimated to be −5.62 and −3.52 eV, respectively, signaling its good electron-transporting and hole-blocking abilities for polymeric LEDs.

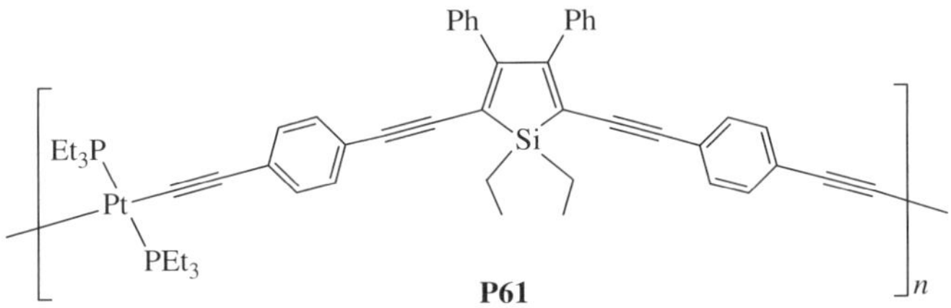

P61

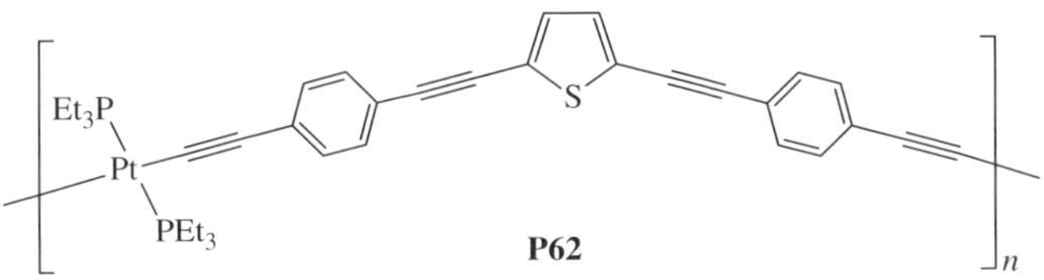

P62

For the first time in the metallopolyyne area, a novel approach based on conjugation interruption was developed for some luminescent group-16 chalcogen-bridged Pt(II) polyyne polymers, **P63–P66**, and these metallopolymers possess high thermal stability with decomposition onsets within 335–363°C.[89] The regiochemical structures of these angular-shaped macromolecules were studied by NMR spectroscopy and single-crystal X-ray structural analyses for the

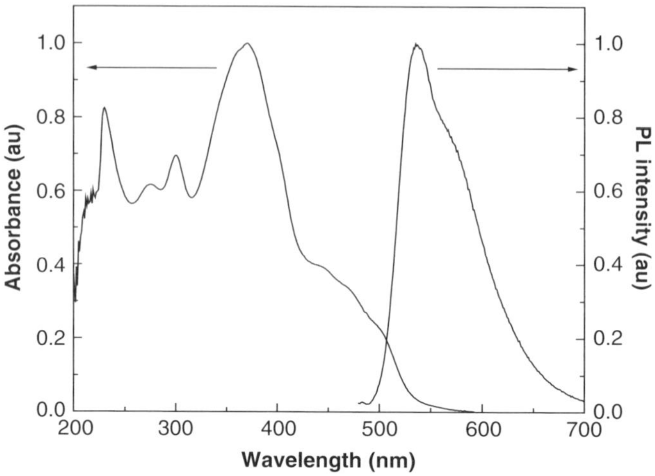

Figure 6.13 Absorption and emission spectra of **P61** in CH_2Cl_2 at 293 K.

diplatinum model complexes. Organic triplet emissions can be harnessed through the strong platinum heavy-metal effects. These metal-containing aryleneethynylenes spaced by chalcogen units were found to have large E_g values and high-energy T_1 states of 2.5 eV or above. A salient feature in such a study is that the use of chalcogen heteroatoms enhances the accessibility of the $^3(\pi\pi^*)$ excited states to significantly boost the phosphorescence decay and renders the room-temperature phosphorescence easily observable for **P63–P66** (Figure 6.14). The results are remarkable in regard to achieving comparable orders of magnitude for $(k_{nr})_P$ and $(k_r)_P$, which is rarely the case for polymetallaynes (Table 6.4). The influence of metal and chalcogen-based conjugation interrupters on the intersystem crossing rate and the spatial extent of the lowest singlet and triplet excitons was elucidated (Figure 6.15). The reduced conjugation in the presence of group-16 elements shifts the optical absorption and the phosphorescence peak to the blue, as compared to **P12** and other heterocycle-spaced metal polyynes.

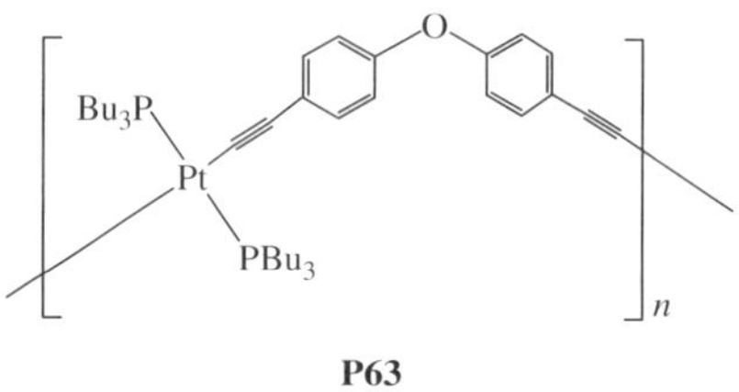

P63

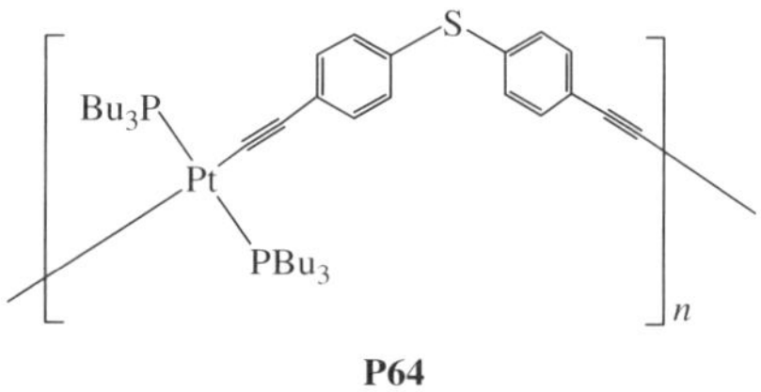

P64

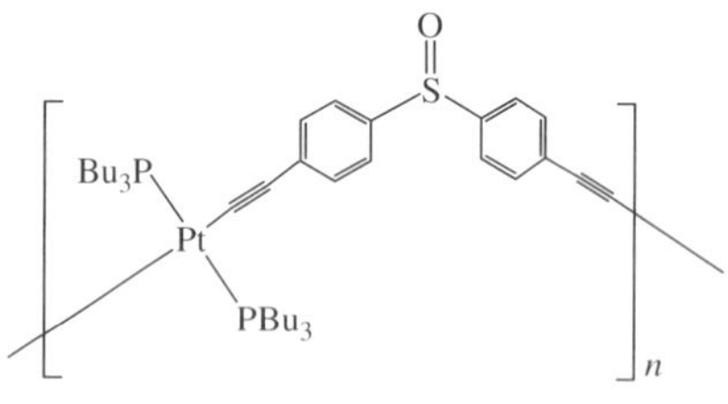

P65

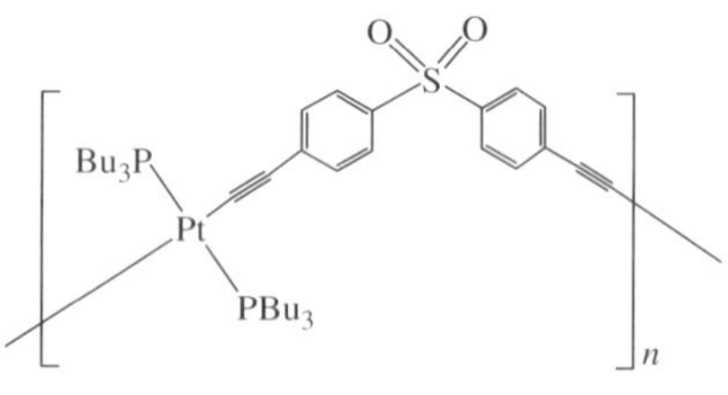

P66

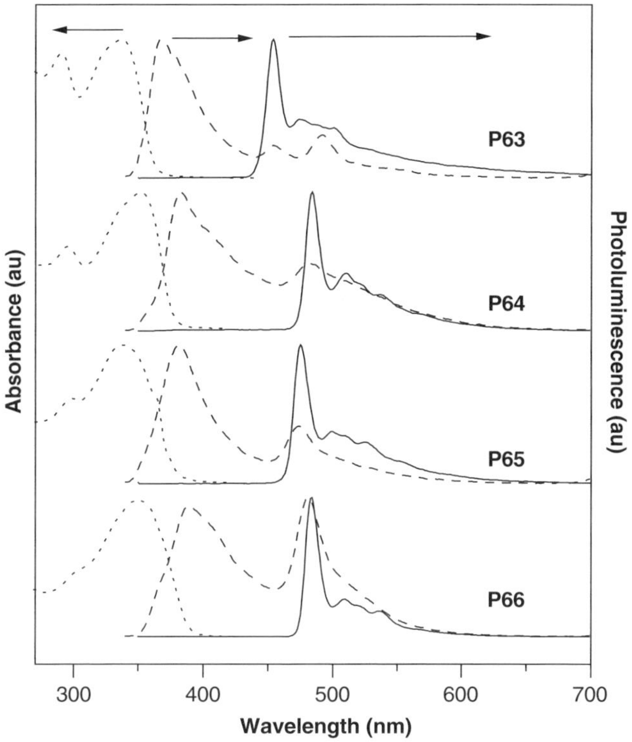

Figure 6.14 The absorption and PL spectra of **P63–P66**. The absorption spectra are the higher-energy dotted lines measured at 290 K. PL spectra were taken at both 290 (dashed lines, in CH_2Cl_2) and 20 K (solid lines, as thin films).

TABLE 6.4 Photophysical Properties for the Triplet Emission at 20 K for P63–P66

Pt(II) Polyyne	Φ_P	τ_P (μs)	$(k_r)_P$ (s^{-1})	$(k_{nr})_P$ (s^{-1})
P63	0.46	6.38	7.18×10^4	8.50×10^4
P64	0.49	6.02	8.08×10^4	8.54×10^4
P65	0.51	7.27	7.06×10^4	6.69×10^4
P66	0.46	8.57	5.41×10^4	6.26×10^4

Since 2002, there have been only a few reports available for the variation of auxiliary ligands on the metal center in the design of polymetallaynes. Group-10 metal complexes containing chelating diphosphine ligand, such as *cis*-[M(dppe)Cl_2] (M = Pd, Pt; dppe = bis(diphenylphosphino)ethane), have been employed, via the reaction with 4,4′-diethynylbiphenyl, to yield zigzag organometallic chains **P67** and **P68** with the metal groups blocked in the *cis* geometry. For **P67**, a polymeric material was obtained, but only an oligomer can be isolated for **P68**.[90] The chemical structures of these materials were also studied using XPS spectroscopy. Very recently, the corresponding monomer model complex *cis*-[Pt(dppe)(C≡C-*p*-C_6H_4-*p*-C_6H_4-C≡CH$)_2$] and its derivatives were synthesized and their optical and structural properties thoroughly examined.[91]

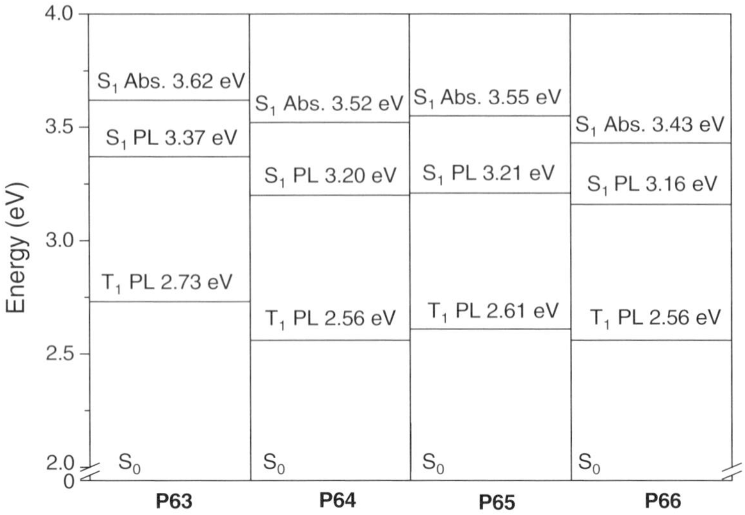

Figure 6.15 Energy-level diagram showing the S_1 and T_1 states for **P63–P66**.

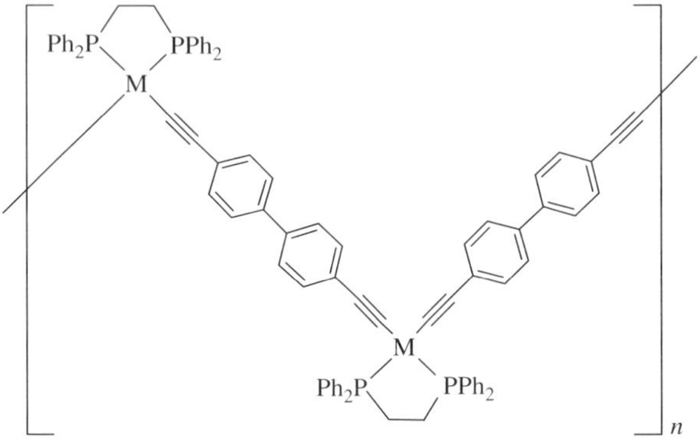

M = Pd **P67**
M = Pt **P68**

Instead of the widely used trialkylphosphine auxiliary ligands on the group-10 metals, the synthesis of a new platinum(II) complex *trans*-[Pt($PPh_2Fc)_2Cl_2$] (PPh_2Fc = ferrocenyldiphenylphosphine) was established, which could then afford a soluble pale yellow platinum polyyne, **P69**, through the Sonogashira-type reaction with 1,4-diethynyl-2,5-dioctyloxybenzene.[92] Gel permeation chromatography (GPC) data on **P69** show a high degree of polymerization with an average molecular weight of ca. 88,000.

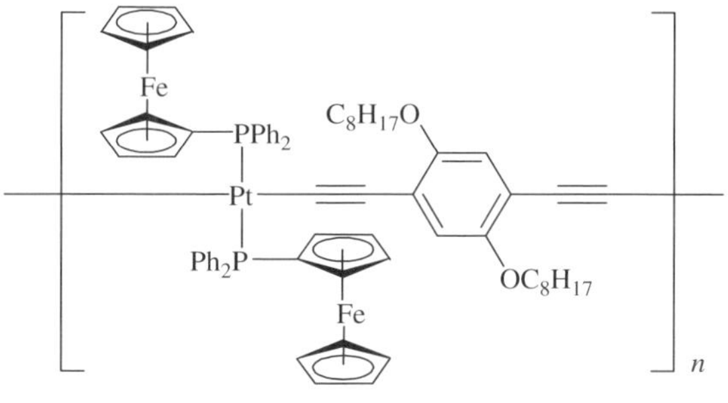

P69

Despite the large body of work on phosphine-stabilized rigid-rod organometallic acetylide polymers of group-10 metals, relatively little was known about these materials bearing nitrogen donor ligands in a *trans* geometry, probably due to the insolubility problem inherent with these materials. For the first time, Adams and others have successfully prepared a novel platinum(II) alkynyl polymer stabilized by 4-(1-butylpentyl)pyridine ligands, **P70**, in which the long alkyl chain can act as the solubilizing group.[93] The authors, however, did not report the molecular weight for the soluble polymer. The crystal structure determined for the mononuclear Pt(II) complex containing two $C{\equiv}C(p\text{-}C_6H_4)NO_2$ ligands ascertains the rodlike skeleton of the as-prepared polymer.

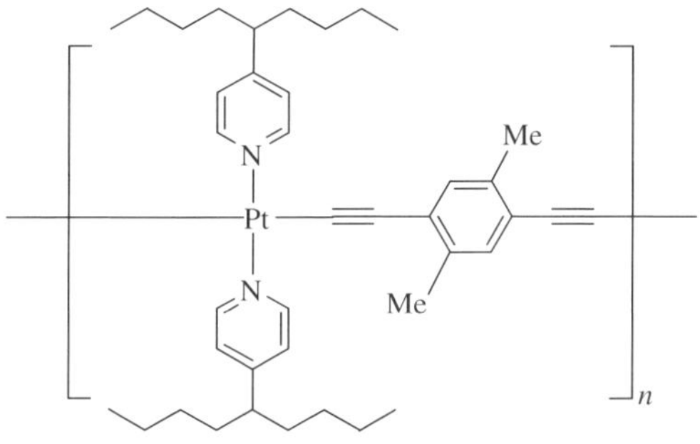

P70

Recently, a water-soluble platinum(II) acetylide polyelectrolyte, **P71**, was prepared and isolated that was shown to be phosphorescent at ambient temperature arising from the $^3(\pi\pi^*)$ state localized on the conjugated backbone.[94] The polymer chains tend to aggregate at a low pH value in which the weakly acidic carboxyl groups can be protonated and the stacked aggregate structure leads to a marked blue shift of its absorption band as well as the strong phosphorescence quenching effect. The triplet emission was shown to be quenched by various viologens and the quenching was being dominated by a dynamic diffusional mechanism. The Stern–Volmer quenching properties of **P71** were studied relative to those of the corresponding fluorescent organic polyelectrolytes, and the results suggest that the amplified quenching effect does not play a key role in **P71**.

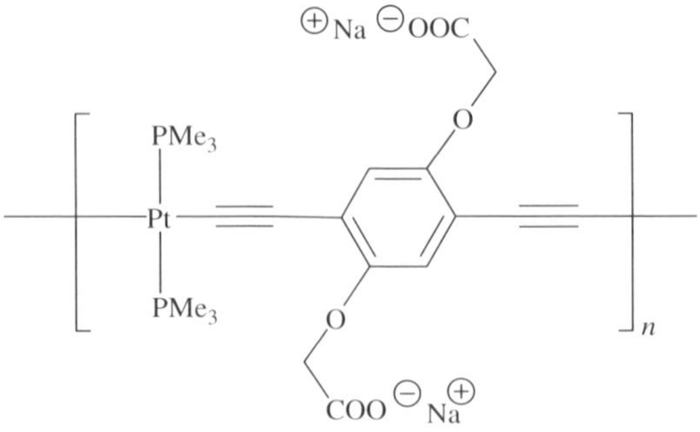

P71

There are several reports on the exploitation of ferrocenyl moiety in the building up of polymetallaynes. A novel heterobimetallic platinum acetylide polymer of ferrocenylfluorene, **P72**, was reported and spectroscopically characterized.[95] The presence of significant donor–acceptor interaction leads to a narrow band-gap value of 2.1 eV for this Pt–Fe mixed material, which is much lower than for the parent polymer, **P25**. The polymer is electroactive, with the half-wave potential of the ferrocene moiety slightly more anodic in **P72** than in the diethynyl ligand precursor, which is in line with the transfer of electron density from the ferrocenyl donor unit to the electron-accepting Pt(II) center through the acetylide linkage.

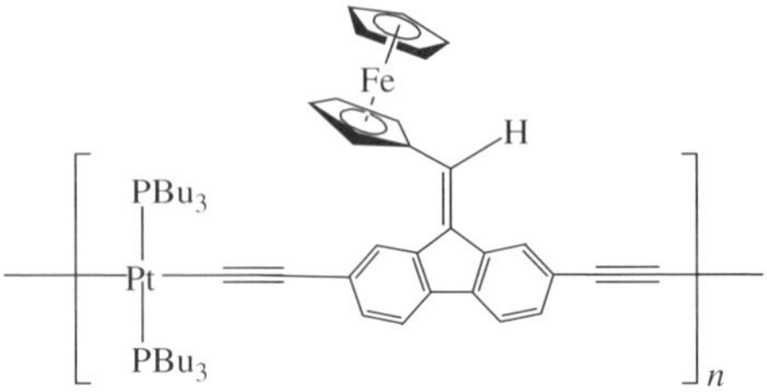

P72

Long and co-workers have incorporated disubstituted alkynylferrocenyl and -biferrocenyl ligands into the main chain of rigid rod Pt-acetylide molecules to furnish polymeric platinum-based alkynyl compounds **P73** and **P74**.[96] Characterization of these molecules was accomplished by optical spectroscopy and, in the cases of their model compounds, by X-ray crystallography. GPC measurements revealed the formation of different-sized oligomers for **P73** (ca. $M_w = 4600$ and $M_n = 2800$). The insolubility of these oligomeric substances and the low-yielding reaction itself both appear to prevent the formation of polymer. Likewise, compound **P74** exists as an oligomeric species of up to 12 repeating units (ca. $M_w = 11{,}800$ and $M_n = 2640$). The greater stability of the biferrocenyl ligands compared to that of the mono-ferrocenyl analog results in a higher reaction yield for **P74** than **P73**. Electrochemical data and extended Hückel theoretical calculations suggest no ferrocene–ferrocene interaction through the Pt-alkynyl bridges. The preparation of the more extended heterobimetallic Fe–Ni and Fe–Pd macromolecules, **P75** and **P76**, has also been described by Dixneuf et al.[35] Both **P75** and **P76** are rigid-rod oligomers arranged in a zigzag geometry owing to the possible rotation of the Fe–C_5H_4 bond.

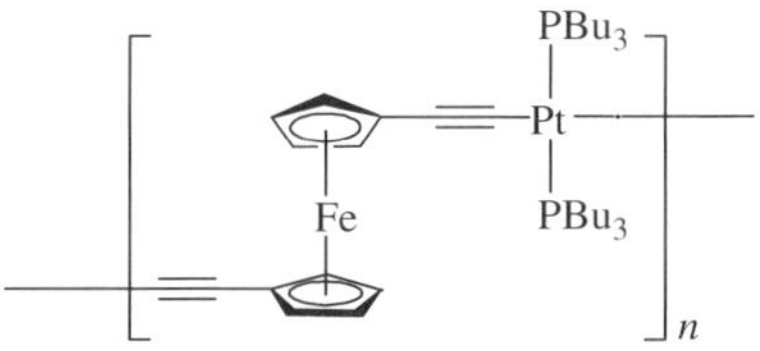

P73

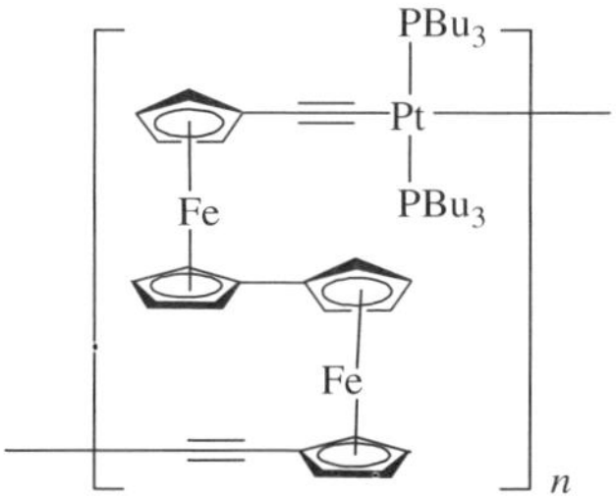

P74

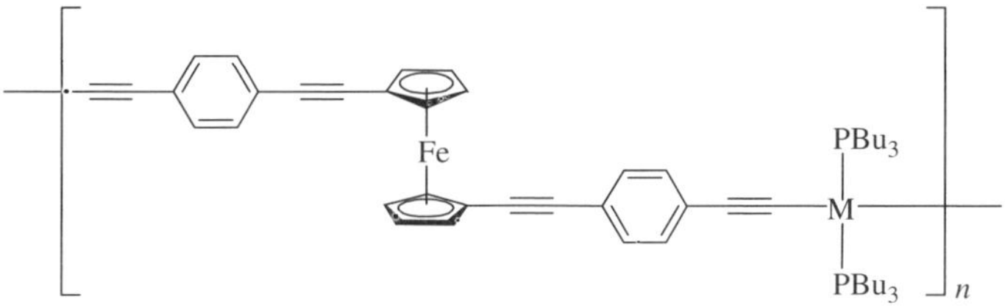

M = Ni **P75**
M = Pd **P76**

The synthesis and spectroscopic characterization of an oligomeric Pt–Zn complex, **P77**, and their derivatives were reported. They can act as useful models for the extended π-conjugated low band-gap organometallic polymers that may be suitable for applications in various optical devices.[97] The oligomeric structure was identified by matrix-assisted laser desorption ionization (MALDI) mass spectrometry. While the use of PBu_3 ligands can afford stable, isolable, and characterizable compounds, it appears that it was not the case when PPh_3 ligands were used.

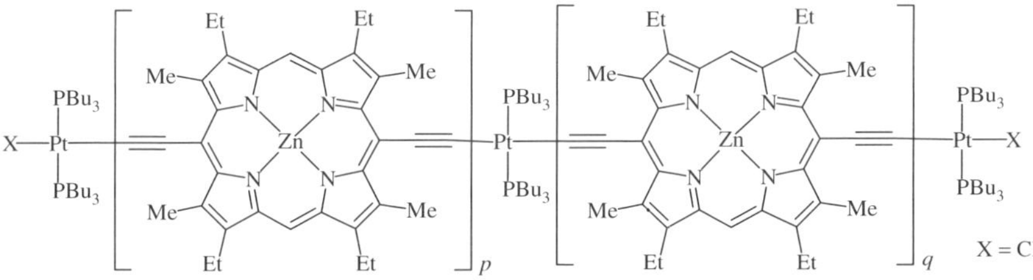

X = Cl or I

P77

IV. GROUP-11 POLYMETALLAYNES

On the other hand, there is a rapidly growing interest in pursuing research on gold(I) acetylide oligomers and polymers.[98,99] Gold(I) ion tends to possess a coordination number of 2 with linear stereochemistry, so it is well suited for forming rigid-rod polymers. Puddephatt et al. have reported on the synthesis and structural as well as luminescent properties of a novel family of rigid-rod alkynylgold(I) polymers bridged by alternating diacetylide and diphosphine ligands, **P78** and **P79**. These polymers are organic-insoluble solids that could only be characterized in the solid state.[100] Their photophysical properties were compared to those for the analogous mononuclear and binuclear alkynylgold(I) model compounds. The role of aurophilicity in the solid-state packing was discussed. As an extension of the work in this area, the same research group explored the use of diisocyanide ligands in the formation of new organometallic acetylide polymers of gold(I), **P80**. Polymers **P80** can be obtained by reacting the linear digold compounds $[AuC{\equiv}CRC{\equiv}CAu]_x$ (R = p-C_6H_4, p-C_6H_4-p-C_6H_4, p-$C_6H_2Me_2$, and CH_2O-p-C_6H_4-p-$C(Me)_2$-p-$C_6H_4OCH_2$) with a range of *para*-substituted diisocyanoarenes, which they were found to be insoluble in common organic solvents. All attempts to introduce organic solubilizing substituents such as *tert*-butyl groups to enhance polymer solubility failed, which was believed to be partly due to the cross-linking effect caused by interchain Au···Au bonding interactions.[101] Characterization of the polymers was achieved using elemental analysis, IR, and XPS methods, and the results conform to the fact that they consist of the same backbone

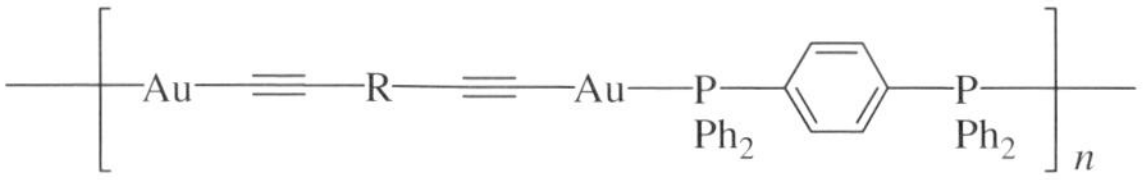

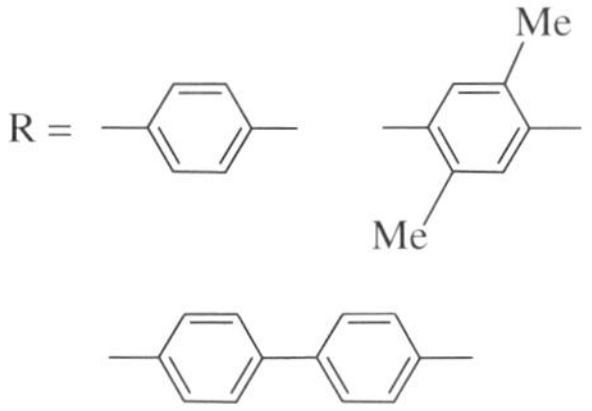

P78

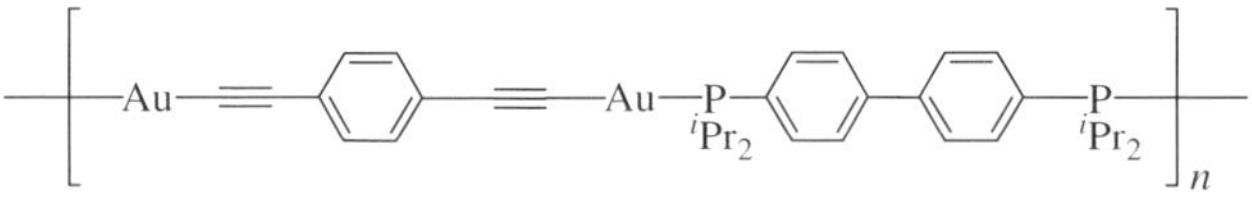

P79

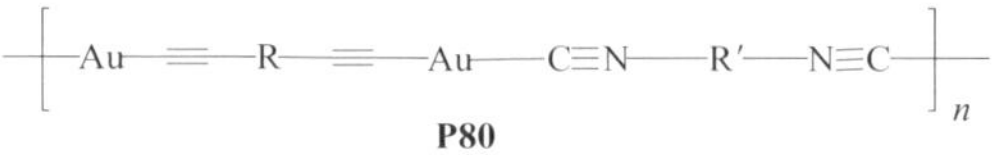

P80

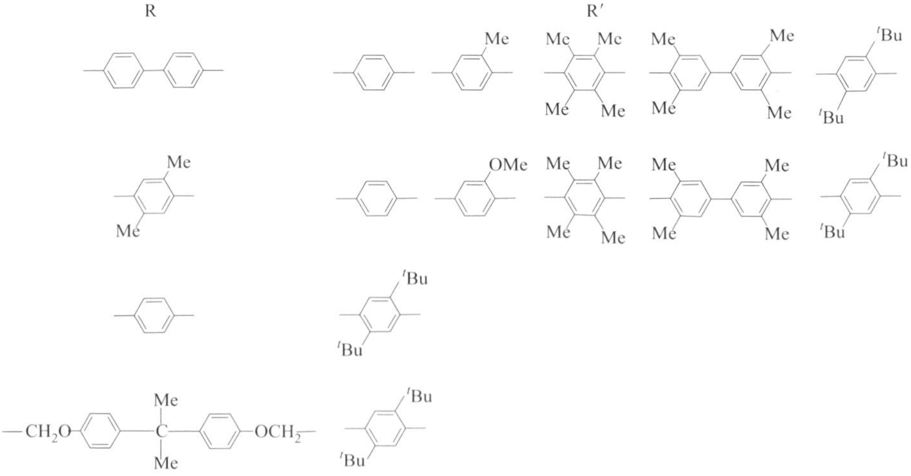

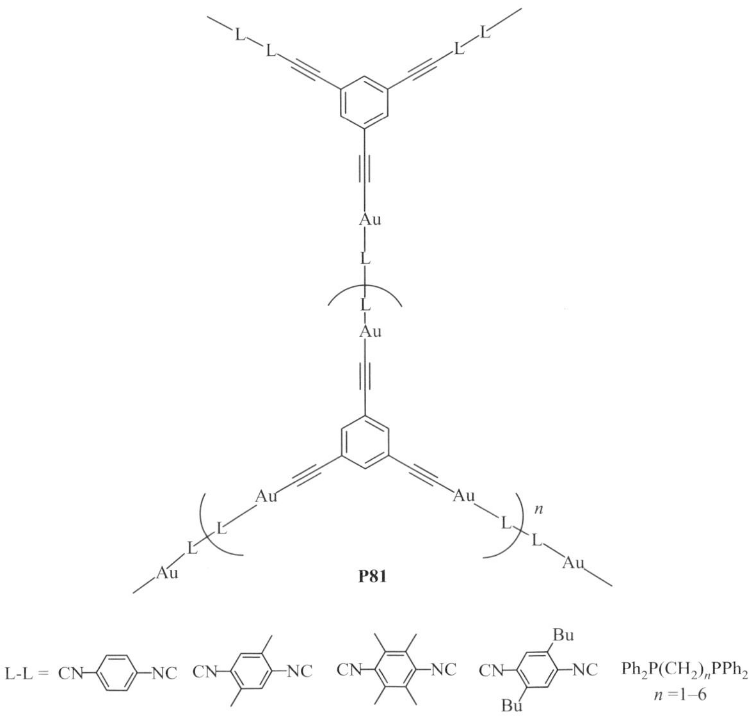

as the digold model compounds. Furthermore, new routes to covalently bridged network polymers of gold(I), **P81**, were developed from [$C_6H_3(C{\equiv}CAu)_3$] and some neutral bidentate ligands.[102] Polymers **P81** appear as pale yellow insoluble solids and their formulations compare very well to the corresponding model compounds based on the IR and XPS data.

In order to gain useful information on the photophysical nature of [-C≡C-Ar-C≡C-]$_n$, two homologous series of gold(I)-derived oligo(aryleneethynylene)s, **P81** and **P82**, were prepared by Che et al., and their detailed steady-state emission spectroscopy, electrochemistry, and photoredox properties were discussed.[103] The advantage of exploiting tricyclohexylphosphine (PCy_3) ligand in the study is that it has no low-lying ligand-localized excited states and its bulkiness precludes metal–metal and π–π oligomerization. The organic triplet emissions of arylacetylide groups can be easily harvested at room temperature by ligation to the $[Au(PCy_3)]^+$ unit. The intensity of phosphorescence relative to fluorescence decreases when the arylacetylide chain is extended in length, in accordance with the established energy-gap law for the triplet excited states. The structure–property relationship between the phosphorescent emission energy and arylacetylide chain length was investigated in detail. A limiting value for the S_0–T_1 energy gap was estimated to be 1.98 eV and 2.04 eV for **P81** and **P82**, respectively. The values compare well with the observed S_0–T_1 gaps of 1.90–2.08 eV for poly(*p*-phenylene) derivatives. The S_0–S_1 separation for $n = \infty$ was deduced to be ~0.8 eV, which corresponds well with that of 0.7 ± 0.1 eV for a series of polymeric Pt(II) polyyne compounds.

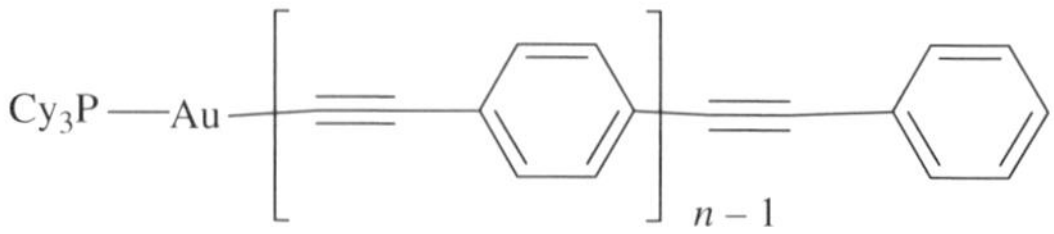

$n = 1–4$

P81

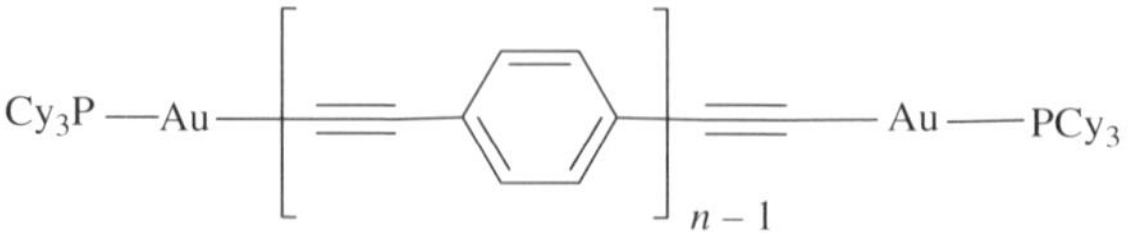

$n = 1–4$

P82

The first anionic polymetallayne of gold(I), **P83**, was prepared in moderate yield (36%) from the reaction of PPN[Au(acac)$_2$] (PPN = bis(triphenylphosphine)iminium) with 1,3-diethynylbenzene in a 1:1 stoichiometric ratio.[104] It was characterized by IR and ^{1}H NMR spectroscopies. More interestingly, efforts were made to make two novel anionic heteronuclear Pt(II)–Au(I) σ-acetylide polymers exhibiting both the linear (**P84**) and branched (**P85**) forms, but these polymers possessed very poor solubility in all common organic solvents.[105]

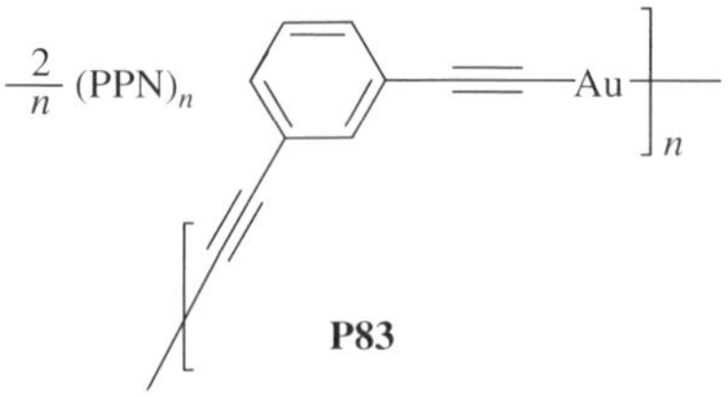

P83

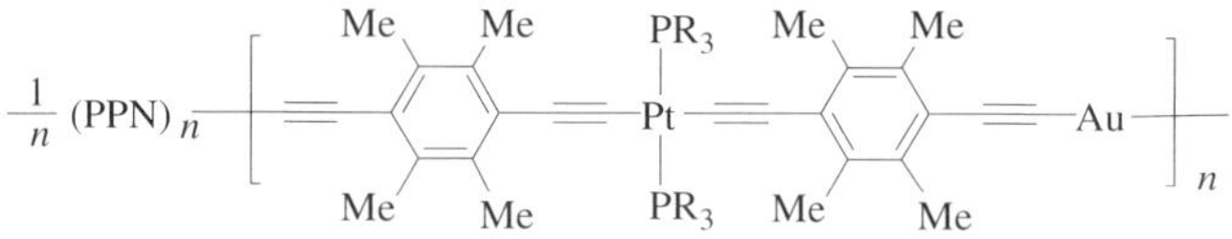

R = Bu, Me, Ph

P84

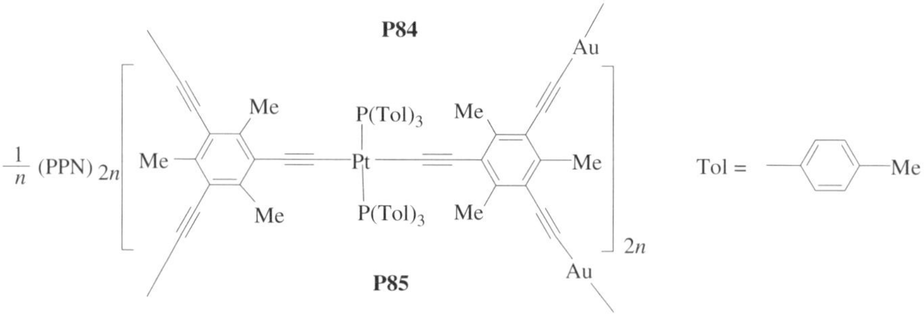

P85

V. GROUP-12 POLYMETALLAYNES

Low solubility of compounds is a major difficulty associated with the preparation of conjugated mercury σ-acetylide oligomers and polymers. Although linear polymeric copper and mercury acetylides of the form [—MC≡C(p-C_6H_4)C≡C—]$_n$ were reported as early as 1960,[106] these materials were often shown to be insoluble and intractable, which hampered their purification and spectroscopic characterization. Wong et al. reported the first examples of a novel series of soluble well-defined high molecular-weight d^{10} mercury(II) polyyne polymers with 9,9-dialkylfluorene spacers, which could render ligand-localized phosphorescence through efficient ISC by ligation to the Hg(II) center.[107] The synthesis of Hg(II) polyyne polymers **P86–P88** was accomplished by the one-pot mercuration reaction of 9,9-dialkyl-2,7-diethynylfluorenes with $HgCl_2$ in the presence of methanolic NaOH at ambient temperature, and the synthetic yields were very high (82–93%). They exhibit excellent film-forming properties and their good solubilities in CH_2Cl_2 and $CHCl_3$ make them readily solution-processable for further optical characterization. Table 6.5 shows that the degree of polymerization (DP) is very high in these polyynes from GPC experiments (DP = 24–47), and the absence of end-group NMR resonances also supports their macromolecular nature. A similar synthetic route using 2,7-diethynylfluorene only gave an intractable off-white solid, suggesting that the long alkyl chains are essential as solubilizing groups to produce these soluble organometallic mercury(II) polyynes. The symmetrical nature and high structural regularity of the polymers were confirmed by NMR studies. These materials have a linear geometry. The X-ray crystal structure of the model complex, in which one coordination site on Hg(II) is capped by a Me group (Figure 6.16), also helped to establish polymer structures in the solid state and to correlate the photophysical results with the structural data.[108,109] The lattice structure reveals the presence of weak intermolecular noncovalent mercuriophilic interactions (ca. 3.738 and 4.183 Å),[110,111] leading to a loose polymeric aggregate in a 3-D network. Polymers **P86–P88** show decomposition temperatures in excess of 200°C. They exhibit increasing decomposition temperatures as the chain length of alkyl groups decreases.

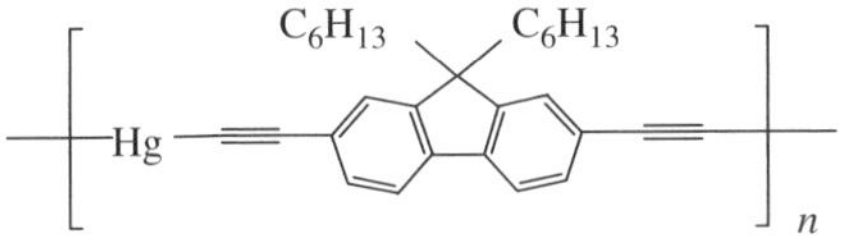

P86

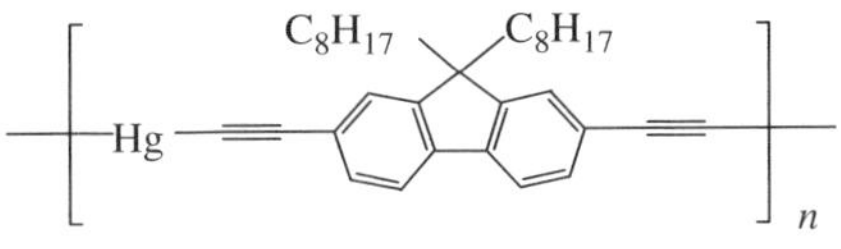

P87

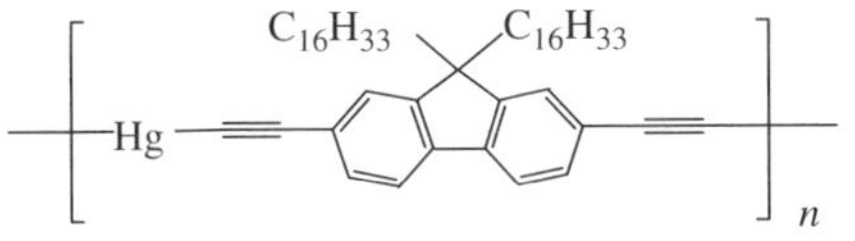

P88

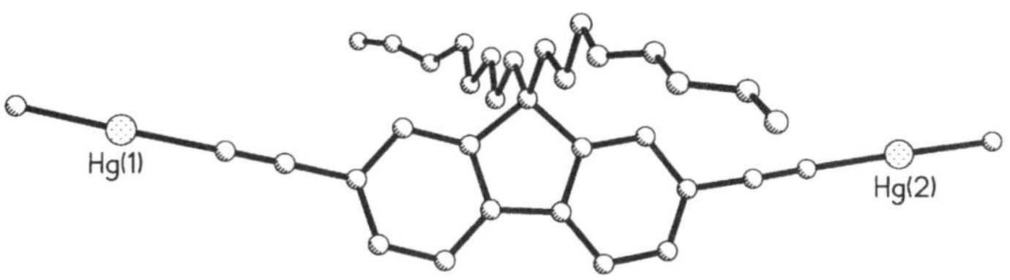

Figure 6.16 X-Ray crystal structure of the model compound for **P87**.

TABLE 6.5 Structural, Thermal, and Photophysical Data of P86–P88

Hg(II) Polyyne	M_w	M_n	T_{decomp} (°C)	λ_{abs} (nm)	λ_{em} (11 K) (nm[a])
P86	28,720	27,100	282	316, 345, 365	427, 570 (28), 610
P87	18,320	15,090	220	316, 345, 364	426, 583 (35), 630
P88	38,650	36,250	200	313, 338, 355	429, 590 (42), 634

[a] Phosphorescence lifetimes (in μs) for the main peaks are shown in parentheses.

All of the Hg(II) polymers show intense absorption bands in the near-UV region, which are mainly due to organic $^1(\pi\pi^*)$ transitions that are possibly mixed with contribution from metal orbitals. We note a red shift of absorption and emission bands in **P86–P88** after the addition of Hg(II) center, which indicates an increase in π-conjugation. The absorption energies and E_g values are independent of the alkyl chain length on the fluorene ring for these polymers. There is an indication of a well-extended singlet excited state in the polymers. Both absorption and X-ray structural studies confirm the presence of solid-state aggregates in polymer thin films.[112] At 11 K, strong triplet emissions appear at around 570–590 nm for **P86–P88**, and the large Stokes shifts of these peaks from the dipole-allowed absorptions (ca. 1.2–1.3 eV), plus the long emission lifetimes at 11 K, suggests the $^3(\pi\pi^*)$ excited states of the parent organic chromophores (see Table 6.5). The ratio for the integrated intensity of the phosphorescence relative to fluorescence increases as the alkyl chain length increases from **P86** to **P88**. Typical of many Pt(II) polyynes, the phosphorescence spectra display a strong temperature dependence (Figure 6.17). The S_1–T_1 crossover efficiency follows the order **P88** > **P87** > **P86**. Values of the energy gap between the S_0 and T_1 levels were found to be 2.10–2.23 eV for **P86–P88**. The S_1–T_1 gaps of 0.73–0.79 eV match well with those of 0.7 ± 0.1 eV for similar π-conjugated Pt(II) and Au(I) polyynes,[76,103] and are close to the gaps estimated for some organic-based polymers.[26,113,114]

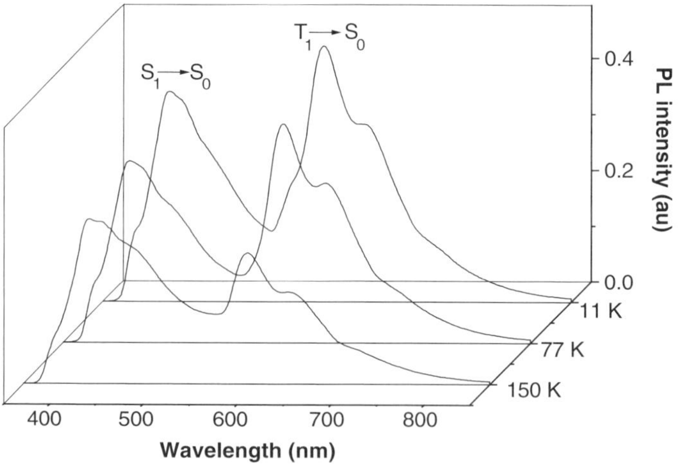

Figure 6.17 Temperature dependence of the PL spectra of **P87**.

Following the first report on the mercury(II)-substituted poly(fluorenyleneethynylene) copolymers, Feng et al. performed detailed theoretical investigations using Hartree–Fock and density functional theory calculations on a series of d^{10} mercury-based diethynylfluorene monomer, oligomer, and polymer in order to establish their structure–property relationships.[115] The results show that there is a weak electronic interaction between the metal-based fragment and the π-conjugated organic groups, and thus the photophysical properties are mainly based on the diethynylfluorene π-conjugated segment with relatively little contribution from the metal center. The role of the metal center can be described as weak delocalization, coupled with strong localization along the organometallic polymer main chain. Both singlet and triplet excited states of the polymer are localized principally on the conjugated ligand. From the chain-length dependence of emission energies, a limiting value for the emission peak at 384.9 nm was estimated for the polymer, which is comparable to the experimental value of 382 nm from the solution-phase PL spectrum of **P86–P88**.[107]

As a continuation of this new area of research, another mercury polyyne, **P89**, was prepared thereafter from the direct base-catalyzed mercuration of $HgCl_2$ with 9,9-bis(4-ethynylphenyl)fluorene.[61] However, only the lower molecular-weight oligomeric fraction of **P89** can dissolve in CH_2Cl_2 and $CHCl_3$. Systematic investigations of the photoluminescence properties of **P89** and its Pt(II), Au(I), and Hg(II) dinuclear model congeners were made. All three metals (viz., Pt, Au, and Hg) can exert heavy-atom effects in the enhancement of the ISC rate, and the effect is most significant for Pt(II) possessing the highest T_1 energy state. The spatial extent of the S_1 and T_1 excitons was evaluated systematically.

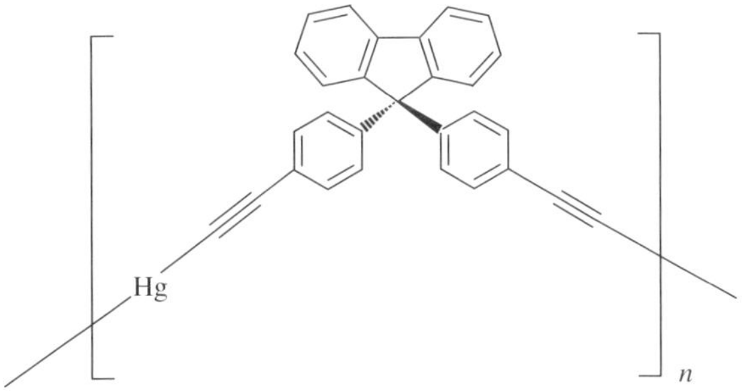

P89

Similarly, evolution of the lowest S_1 and T_1 states with the type of metal centers was examined for biphenyl-bridged polymetallaynes **P12** and **P90**.[44] Since only the low molecular-weight oligomer of **P90** was found to be soluble in chlorinated solvents ($M_w = 2390$, $M_n = 2200$), its photophysical properties were studied instead of the polymer. On the structural aspect, it is interesting to note the observation of a polymeric mercury(II) acetylide system in the solid-state aggregated through Hg···Hg secondary interactions of the model dimercury(II) compound (Figure 6.18), which corroborates the presence of a broad aggregate band in its PL spectrum.

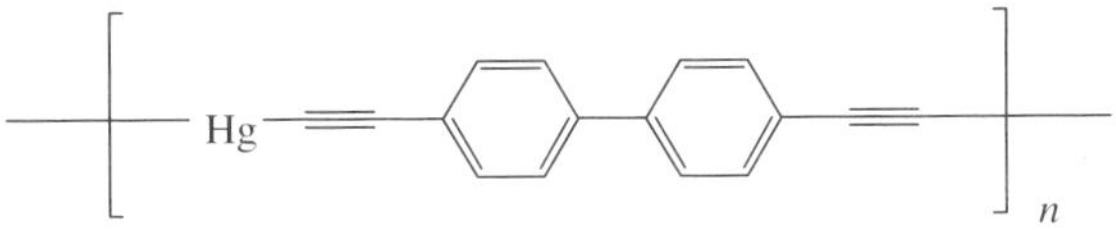

P90

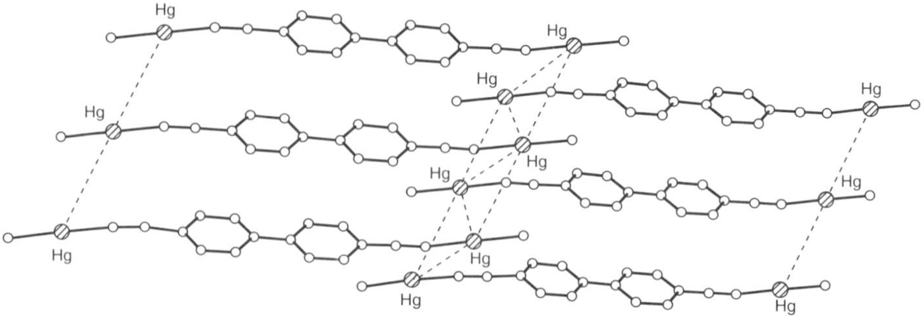

Figure 6.18 Crystal packing diagram for the model compound of **P90**, showing the noncovalent Hg···Hg short contacts.

VI. SUMMARY AND FUTURE DIRECTIONS

A vast body of information from the literature on the chemical syntheses, spectroscopic and structural aspects, material properties, and potential applications of organometallic polyyne polymers of the late transition metals was comprehensively reviewed in this chapter. These polymetallaynes are particularly interesting in light of their remarkable optoelectronic and photonic properties, and their uses both as semiconductors and as triplet emitters are well documented. From a synthetic point of view, polymetallaynes are available in an almost never-ending variety of different chemical structures and topologies. Electronic absorption and photoluminescent characteristics of these metal-based macromolecules can be chemically modified in terms of the identity of metal groups, the spacer units, and the ancillary coligands on metals. Accumulation of the chemical knowledge to understand the factors that control the structure, properties, and function in polymetallaynes would make the synthesis of this class of metallopolymers an essential subject in the development of new advanced functional materials. This would certainly lead to the exciting prospect of the metal polyyne area, in which the functional properties can be engineered through appropriate molecular design and assembly. Current activity in this exciting and emerging area will focus on the development of new narrow band-gap and phosphorescent polymers with high triplet yields. In such a research scenario, metal alkynyl-derived polymeric materials can lead the way to new unique electronic, optical, and photonic materials. The fact that many researchers have circumvented the problem of the triplet exciton being nonemissive by using metal-containing polyyne polymers of the form *trans*-[—MC≡CRC≡C—]$_n$ (M = Pt, Au, Hg) offers much room for further investigation of these materials in optoelectronic applications. The results are significant with regard to a fundamental understanding, as well as providing design concepts for triplet emitting systems. This is desirable for the continual development of LEDs that demand triplet light energy harvesting, and one can benefit from the large triplet emission yield and favorable radiative decay rate of triplet excitons. In this connection, extension of the synthetic methodologies to other main group elements would certainly warrant further examination. While the research interest in polymetallaynes is still growing rapidly, we believe that an in-depth exploration of these functional polymers in optoelectronic work and sensing applications will follow. It is also worthwhile to look at the potential of these types of organometallic polymers as optical limiters and precursors to transition metal–containing ceramics and other nanomaterials.

ACKNOWLEDGMENTS

Our sincere thanks go to all the postgraduate students (especially Dr. Poon Suk-Yue, Dr. Liu Li, Mr. Choi Ka-Ho, and Mr. Wong Chun-Kin) and postdoctoral researchers (Drs. Zhou Gui-Jiang and Lu Guo-Liang) who were involved in the work of my research group. I gratefully acknowledge the financial support from the Research Grants Council of the Hong Kong SAR (HKBU 2022/03P) and the Hong Kong Baptist University (FRG/03-04/II-69). Thanks are also due to Profs. Albert W.-M. Lee, K.-W. Cheah, and P. R. Raithby, and Drs. W.-K. Chan and Z. Lin for various kinds of collaborative efforts.

REFERENCES

1. I. Manners, *Synthetic Metal-Containing Polymers*, Wiley-VCH, Weinheim, Germany, 2004.
2. P. Nguyen, P. Gómez-Elipe, I. Manners, *Chem. Rev.*, **99**, 1515 (1999).
3. I. Manners, *Angew. Chem. Int. Ed. Engl.*, **35**, 1602 (1996).
4. A. S. Abd-El-Aziz, *Macromol. Rapid Commun.*, **23**, 995 (2002).
5. R. P. Kingsborough, T. M. Swager, *Prog. Inorg. Chem.*, **48**, 123 (1999).
6. N. J. Long, C. K. Williams, *Angew. Chem. Int. Ed.*, **42**, 2586 (2003).
7. W.-Y. Wong, *J. Inorg. Organomet. Polym. Mater.*, **15**, 197 (2005).
8. W.-Y. Wong, *Comment Inorg. Chem.*, **26**, 39 (2005).
9. Y. Fujikura, K. Sonogashira, N. Hagihara, *Chem. Lett.*, 1067 (1975).
10. K. Sonogashira, S. Takahashi, N. Hagihara, *Macromolecules*, **10**, 879 (1977).
11. S. Takahashi, M. Kariya, T. Yatake, K. Sonogashira, N. Hagihara, *Macromolecules*, **11,** 1063 (1978).
12. K. Sonogashira, S. Kataoka, S. Takahashi, N. Hagihara, *J. Organomet. Chem.*, **160**, 319 (1978).
13. S. J. Davies, B. F. G. Johnson, M. S. Khan, J. Lewis, *J. Chem. Soc., Chem. Commun.*, 187 (1991).
14. Z. Atherton, C. W. Faulkner, S. L. Ingham, A. K. Kakkar, M. S. Khan, J. Lewis, N. J. Long, P. R. Raithby, *J. Organomet. Chem.*, **462**, 265 (1993).
15. B. F. G. Johnson, A. K. Kakkar, M. S. Khan, J. Lewis, *J. Organomet. Chem.*, **409**, C12 (1991).
16. S. J. Davies, B. F. G. Johnson, J. Lewis, P. R. Raithby, *J. Organomet. Chem.*, **414**, C51 (1991).
17. U. H. F. Bunz, *Chem. Rev.*, **100**, 1605 (2000).
18. J. L. Brédas, J. Cornil, A. J. Heeger, *Adv. Mater.*, **8**, 447 (1996).
19. N. S. Sari•iftci (ed.), *Primary Photoexcitations in Conjugated Polymers: Molecular Exciton versus Semiconductor Band Model*, World Scientific, Singapore, 1997.
20. T. A. Skotheim, J. R. Reynolds, R. L. Elsenbaumer (eds.), *Handbook of Conducting Polymers*, Second Edition, Marcel Dekker, New York, 1998.
21. R. H. Friend, R. W. Gymer, A. B. Holmes, J. H. Burroughes, R. N. Marks, C. Taliani, D. D. C. Bradley, D. A. Dos Santos, J. L. Brédas, M. Lögdlund, W. R. Salaneck, *Nature (London)*, **397**, 121 (1999).
22. J. S. Wilson, A. S. Dhoot, A. J. A. B. Seeley, M. S. Khan, A. Köhler, R. H. Friend, *Nature (London)*, **413**, 828 (2001).
23. A. Köhler, J. S. Wilson, R. H. Friend, *Adv. Mater.*, **14**, 701 (2002).
24. C. Adachi, M. A. Baldo, M. E. Thompson, S. R. Forrest, *J. Appl. Phys.* **90**, 5048 (2001).
25. J. M. Lupton, A. Pogantsch, T. Piok, E. J. W. List, S. Patil, U. Scherf, *Phys. Rev. Lett.*, **89**, 167401 (2002).
26. Y. V. Romanovskii, A. Gerhard, B. Schweitzer, U. Scherf, R. I. Personov, H. Bässler, *Phys. Rev. Lett.*, **84**, 1027 (2000).
27. X. Gong, J. C. Ostrowski, G. C. Bazan, D. Moses, A. J. Heeger, M. S. Liu, A. K.-Y. Jen, *Adv. Mater.*, **15**, 45 (2003).

28. A. S. Dhoot, N. C. Greenham, *Adv. Mater.*, **14**, 1834 (2002).
29. P. K. H. Ho, J. S. Kim, J. H. Burroughes, H. Becker, S. F. Y. Li, T. M. Brown, F. Cacialli, R. H. Friend, *Nature (London)*, **404**, 481 (2000).
30. Y. Cao, I. D. Parker, G. Yu, C. Zhang, A. J. Heeger, *Nature (London)*, **397**, 414 (1999).
31. V. Cleave, G. Yahioglu, P. Le Barny, R. H. Friend, N. Tessler, *Adv. Mater.*, **11**, 285 (1999).
32. M. A. Baldo, D. F. O'Brien, Y. You, A. Shoustikov, S. Sibley, M. E. Thompson, S. R. Forrest, *Nature (London)*, **395**, 151 (1998).
33. M. A. Baldo, M. E. Thompson, S. R. Forrest, *Nature (London)*, **403**, 750 (2000).
34. M. Pope, C. E. Swenberg, *Electronic Processes in Organic Crystals and Polymers*, Second Edition, Oxford Science Publications, Oxford, 1999.
35. O. Lavastre, M. Even, P. H. Dixneuf, A. Pacreau, J. P. Vairon, *Organometallics*, **15**, 1530 (1996).
36. N. Matsumi, Y. Chujo, O. Lavastre, P. H. Dixneuf, *Organometallics*, **20**, 2425 (2001).
37. Y. Zhu, D. B. Millet, M. O. Wolf, S. J. Rettig, *Organometallics*, **18**, 1930 (1999).
38. M. O. Wolf, *Adv, Mater.*, **13**, 545 (2001).
39. W.-Y. Wong, K.-H. Choi, G.-L. Lu, L. Liu, J.-X. Shi, K.-W. Cheah, Chapter 26, in *ACS Symp. Ser. 928 on Metal-Containing and Metallosupramolecular Polymers and Materials*, American Chemical Society, Washington DC, 2006.
40. R. D. Markwell, I. S. Butler, A. K. Kakkar, M. S. Khan, Z. H. Al-Zakwani, J. Lewis, *Organometallics*, **15**, 2331 (1996).
41. D. Beljonne, H. F. Wittmann, A. Köhler, S. Graham, M. Younus, J. Lewis, P. R. Raithby, M. S. Khan, R. H. Friend, J. L. Brédas, *J. Chem. Phys.*, **9**, 105 (1996).
42. E. E. Silverman, T. Cardolaccia, X. Zhao, K. Y. Kim, K. Haskins-Glusac, K. S. Schanze, *Coord. Chem. Rev.*, **249**, 1491 (2005).
43. X. Zhao, T. Cardolaccia, R. T. Farley, K. A. Abboud, K. S. Schanze, *Inorg. Chem.*, **44**, 2619 (2005).
44. L. Liu, S.-Y. Poon, W.-Y. Wong, *J. Organomet. Chem.*, **690**, 5036 (2005).
45. R. D'Amato, I. Fratoddi, A. Cappotto, P. Altamura, M. Delfini, C. Bianchetti, A. Bolasco, G. Polzonetti, M. V. Russo, *Organometallics*, **23**, 2860 (2004).
46. C. Caliendo, I. Fratoddi, M. V. Russo, C. L. Sterzo, *J. Appl. Phys.*, **93**, 10071 (2003).
47. M. J. Irwin, G. Jia, J. J. Vittal, R. J. Puddephatt, *Organometallics*, **15**, 5321 (1996).
48. K. Onitsuka, Y. Harada, F. Takei, S. Takahashi, *Chem. Commun.*, 643 (1998).
49. M. S. Muhammad, M. R. A. Al-Mandhary, M. K. Al-Suti, T. C. Corcoran, Y. Al-Mahrooqi, J. P. Attfield, N. Feeder, W. I. F. David, K. Shankland, R. H. Friend, A. Köhler, E. A. Marseglia, E. Tedesco, C. C. Tang, P. R. Raithby, J. C. Collings, K. P. Roscoe, A. S. Batsanov, L. M. Stimson, T. B. Marder, *New J. Chem.*, **27**, 140 (2003).
50. J. Lewis, P. R. Raithby, W.-Y. Wong, *J. Organomet. Chem.*, **556**, 219 (1998).
51. W.-Y. Wong, *Coord. Chem. Rev.*, **249**, 971 (2005).
52. D. Neher, *Macromol. Rapid Commun.*, **22**, 1365 (2001).
53. B. Tsuie, J. L. Reddinger, G. A. Sotzing, J. Soloducho, A. R. Katritzky, J. R. Reynolds, *J. Mater. Chem.*, **9**, 2189 (1999).
54. K.-T. Wong, Y.-Y. Chien, R.-T. Chen, C.-F. Wang, Y.-T. Lin, H.-H. Chiang, P.-Y. Hsieh, C.-C. Wu, C.-H. Chou, Y. O. Su, *J. Am. Chem. Soc.*, **124**, 11576 (2002).
55. W.-L. Yu, J. Pei, W. Huang, A. J. Heeger, *Adv. Mater.*, **12**, 828 (2000).
56. S. Setayesh, A. C. Grimsdale, T. Weil, V. Enkelmann, K. Müllen, F. Meghdadi, E. J. W. List, G. Leising, *J. Am. Chem. Soc.*, **123**, 946 (2001).
57. W.-Y. Wong, G.-L. Lu, K.-H. Choi, J.-X. Shi, *Macromolecules*, **25**, 3506 (2002).
58. W.-Y. Wong, K.-H. Choi, G.-L. Lu, J.-X. Shi, *Macromol. Rapid Commun.*, **22**, 461 (2001).
59. M. S. Khan, M. R. A. Al-Mandhary, M. K. Al-Suti, B. Ahrens, M. F. Mahon, L. Male, P. R. Raithby, C. E. Boothby, A. Köhler, *Dalton Trans.*, 74 (2003).

60. G.-J. Zhou, W.-Y. Wong, D. Cui, C. Ye, *Chem. Mater.*, **17**, 5209 (2005).
61. W.-Y. Wong, L. Liu, S.-Y. Poon, K.-H. Choi, K.-W. Cheah, J.-X. Shi, *Macromolecules*, **37**, 4496 (2004).
62. N. Chawdhury, A. Köhler, R. H. Friend, W.-Y. Wong, J. Lewis, M. Younus, P. R. Raithby, T. C. Corcoran, M. R. A. Al-Mandhary, M. S. Khan, *J. Chem. Phys.*, **110**, 4963 (1999).
63. J. Lewis, N. J. Long, P. R. Raithby, G. P. Shields, W.-Y. Wong, M. Younus, *J. Chem. Soc., Dalton Trans.*, 4283 (1997).
64. W.-Y. Wong, S.-M. Chan, K.-H. Choi, K.-W. Cheah, W.-K. Chan, *Macromol. Rapid Commun.*, **21**, 453 (2000).
65. J. Cornil, D. Beljonne, D. A. dos Santos, Z. Shuai, J. L. Brédas, *Synth. Met.*, **78**, 209 (1996).
66. D. Beljonne, J. Cornil, J. L. Brédas, R. H. Friend, *Synth. Met.*, **76**, 61 (1996).
67. N. Chawdhury, M. Younus, P. R. Raithby, J. Lewis, R. H. Friend, *Opt. Mater.*, **9**, 498 (1998).
68. K. A. Bunten, A. K. Kakkar, *J. Mater. Chem.*, **5**, 2041 (1995).
69. K. A. Bunten, A. K. Kakkar, *Macromolecules*, **29**, 2885 (1996).
70. N. Chawdhury, A. Köhler, R. H. Friend, M. Younus, N. J. Long, P. R. Raithby, J. Lewis, *Macromolecules*, **31**, 722 (1998).
71. M. S. Khan, M. R. A. Al-Mandhary, M. K. Al-Suti, A. K. Hisahm, P. R. Raithby, B. Ahrens, M. F. Mahon, L. Male, E. A. Marseglia, E. Tedesco, R. H. Friend, A. Köhler, N. Feeder, S. J. Teat, *J. Chem. Soc., Dalton Trans.*, 1358 (2002).
72. M. S. Khan, M. R. A. Al-Mandhary, M. K. Al-Suti, N. Feeder, S. Nahar, A. Köhler, R. H. Friend, P. J. Wilson, P. R. Raithby, *J. Chem. Soc., Dalton Trans.*, 2441 (2002).
73. M. Younus, A. Köhler, S. Cron, N. Chawdhury, M. R. A. Al-Mandhary, M. S. Khan, J. Lewis, N. J. Long, R. H. Friend, P. R. Raithby, *Angew. Chem. Int. Ed.*, **37**, 3036 (1998).
74. M. S. Khan, M. K. Al-Suti, M. R. A. Al-Mandhary, B. Ahrens, J. K. Bjernemose, M. F. Mahon, L. Male, P. R. Raithby, R. H. Friend, A. Köhler, J. S. Wilson, *Dalton Trans.*, 65 (2003).
75. J. S. Wilson, A. Köhler, R. H. Friend, M. K. Al-Suti, M. R. A. Al-Mandhary, M. S. Khan, P. R. Raithby, *J. Chem. Phys.*, **113**, 7627 (2000).
76. J. S. Wilson, N. Chawdhury, M. R. A. Al-Mandhary, M. Younus, M. S. Khan, P. R. Raithby, A. Köhler, R. H. Friend, *J. Am. Chem. Soc.*, **123**, 9412 (2001).
77. J. S. Wilson, A. S. Dhoot, A. J. A. B. Seeley, M. S. Khan, A. Köhler, R. H. Friend, *Nature*, **413**, 828 (2001).
78. F. Matsumoto, N. Matsumi, Y. Chujo, *Polym. Bull.*, **48**, 119 (2002).
79. R. Gleiter, W. Schäfer, H. Sakurai, *J. Am. Chem. Soc.*, **107**, 3046 (1985).
80. K. D. Kim, J. S. Park, H. K. Kim, T. B. Lee, K. T. No, *Macromolecules*, **31**, 7267 (1998).
81. H. K. Kim, M. K. Ryu, S. M. Lee, *Macromolecules*, **30**, 1236 (1997).
82. H. K. Kim, M. K. Ryu, K. D. Kim, S. M. Lee, S. W. Cho, J. W. Park, *Macromolecules*, **31**, 1114 (1998).
83. W.-Y. Wong, C.-K. Wong, G.-L. Lu, K.-W. Cheah, J.-X. Shi, Z. Lin, *J. Chem. Soc., Dalton Trans.*, 4587 (2002).
84. W. E. Douglas, D. M. H. Guy, A. K. Kar, C. Wang, *Chem. Commun.*, 2125 (1998).
85. W.-Y. Wong, C.-K. Wong, G.-L. Lu, A. W.-M. Lee, K.-W. Cheah, J.-X. Shi, *Macromolecules*, **36**, 983 (2003).
86. W.-Y. Wong, S.-Y. Poon, A. W.-M. Lee, J.-X. Shi, K.-W. Cheah, *Chem. Commun.*, 2420 (2004).
87. W.-Y. Wong, C.-K. Wong, S.-Y. Poon, A. W.-M. Lee, T. Mo, X. Wei, *Macromol. Rapid Commun.*, **26**, 376 (2005).
88. A. K. Kakkar, M. S. Khan, N. J. Long, J. Lewis, P. R. Raithby, P. Nguyen, T. B. Marder, F. Wittmann, R. H. Friend, *J. Mater. Chem.*, **4**, 1227 (1994).
89. S.-Y. Poon, W.-Y. Wong, K.-W. Cheah, J.-X. Shi, *Chem. Eur. J.*, **12**, 2550 (2006).
90. G. Iucci, G. Infante, G. Polzonetti, *Polymer*, **43**, 655 (2002).
91. R. Saha, M. A. Qaium, D. Debnath, M. Younus, N. Chawdhury, N. Sultana, G. Kociok-Köhn, L. L. Ooi, P. R. Raithby, M. Kijima, *Dalton Trans.*, 2761 (2005).

92. N. J. Long, A. J. P. White, D. J. Williams, M. Younus, *J. Organomet. Chem.*, **649**, 94 (2002).
93. C. J. Adams, S. L. James, P. R. Raithby, *Chem. Commun.*, 2155 (1997).
94. K. Haskins-Glusac, M. R. Pinto, C. Tan, K. S. Schanze, *J. Am. Chem. Soc.*, **126**, 14964 (2004).
95. W.-Y. Wong, W.-K. Wong, P.-R. Raithby, *J. Chem. Soc., Dalton Trans.*, 2761 (1998).
96. N. J. Long, A. J. Martin, R. Vilar, A. J. P. White, D. J. Williams, M. Younus, *Organometallics*, **18**, 4261 (1999).
97. A. Ferri, G. Polzonetti, S. Licoccia, R. Paolesse, D. Favretto, P. Traldi, M. V. Russo, *J. Chem. Soc., Dalton Trans.*, 4063 (1998).
98. R. J. Puddephatt, *Coord. Chem. Rev.*, **216–217**, 313 (2001).
99. R. J. Puddephatt, *Chem. Commun.*, 1055 (1998).
100. M. J. Irwin, J. J. Vittal, R. J. Puddephatt, *Organometallics*, **16**, 3541 (1997).
101. M. J. Irwin, G. Jia, N. C. Payne, R. J. Puddephatt, *Organometallics*, **15**, 51 (1996).
102. M. J. Irwin, L. Manojlović-Muir, K. W. Muir, R. J. Puddephatt, D. S. Yufit, *Chem. Commun.*, 219 (1997).
103. H. Y. Chao, W. Lu, Y. Li, M. C. W. Chan, C. M. Che, K. K. Cheung, N. Zhu, *J. Am. Chem. Soc.*, **124**, 14696 (2002).
104. J. Vicente, M. T. Chicote, M. D. Abrisqueta, M. M. Alvarez-Falcón, *J. Organomet. Chem.*, **663**, 40 (2002).
105. J. Vicente, M. T. Chicote, M. M. Alvarez-Falcón, *Organometallics*, **24**, 2764 (2005).
106. A. S. Hay, *J. Org. Chem.*, **25**, 1275 (1960).
107. W.-Y. Wong, L. Liu, J.-X. Shi, *Angew. Chem. Int. Ed.*, **42**, 4064 (2003).
108. W.-Y. Wong, K.-H. Choi, G.-L. Lu, J.-X. Shi, P.-Y. Lai, S.-M. Chan, Z. Lin, *Organometallics*, **20**, 5446 (2001).
109. W.-Y. Wong, K.-H. Choi, G.-L. Lu, Z. Lin, *Organometallics*, **21**, 4475 (2002).
110. S. J. Faville, W. Henderson, T. J. Mathieson, B. K. Nicholson, *J. Organomet. Chem.*, **580**, 363 (1999).
111. P. Pyykkö, *Chem. Rev.*, **97**, 597 (1997).
112. N. G. Pschirer, U. H. F. Bunz, *Macromolecules*, **33**, 3961 (2000).
113. D. Hertel, S. Setayesh, H. G. Nothofer, U. Scherf, K. Mullen, H. Bassler, *Adv. Mater.*, **13**, 65 (2001).
114. A. P. Monkman, H. D. Burrows, L. J. Hartwell, L. E. Horsburgh, I. Hamblett, S. Navaratnam, *Phys. Rev. Lett.*, **86**, 1358 (2001).
115. Y. Liao, J. K. Feng, L. Yang, A. M. Ren, H. X. Zhang, *Organometallics*, **24**, 385 (2005).

CHAPTER 7

Polymers with Metal–Metal Bonds Along Their Backbones

DAVID R. TYLER

University of Oregon, Eugene, Oregon

I. INTRODUCTION

This chapter discusses the synthesis, structures, selected properties, and uses of polymers that have metal–metal bonds incorporated into their backbone chains. These materials are of interest for a number of reasons. The primary reason is the promise of new and unusual properties. In particular, as the sections below detail, there is reason to believe that these polymers may have interesting conductivity, nonlinear optical, magnetic, liquid crystalline, and catalytic properties. A second reason for the interest is that these polymers may be useful precursors for other materials. For example, there have been various studies aimed at degrading these materials for the purpose of forming nanoparticles from the metals in the polymer backbone. Other reasons for the interest in these polymers are more specific, but no less important. For example, one of the following sections details how polymers with metal–metal bonds were used as "model" polymers to study the photochemical degradation processes in regular polymers. And finally, many of the polymers discussed here were synthesized because they have unusually interesting and beautiful structures.

Polymers with transition metal–transition metal bonds can be roughly divided into four categories: (1) step-growth polymers made from difunctional (or multifunctional) monomers containing appropriate functional groups, (2) coordination polymers with metal–metal units held together by bidentate ligands, (3) polymers that have chains consisting exclusively of metals, and (4) miscellaneous coordination polymers formed in what are usually serendipitous reactions. Of these four categories, it is noteworthy that the synthetic methods for the step-growth polymer and coordination-polymer categories are well-developed. In contrast, standard synthetic routes to polymers having chains consisting exclusively of metals have not been developed. And, of course, standard synthetic routes to polymers made serendipitously have not been systematically developed.

The following discussion is divided into sections that describe the synthesis and structures of the polymers in the four categories described in the preceding paragraph. Subsections within

Frontiers in Transition Metal-Containing Polymers, edited by Alaa S. Abd-El-Aziz and Ian Manners.

each of these main sections describe the uses, applications, and properties of the polymers. The coverage is generally limited to systems in which the metals are transition metals, but there is some selective coverage of polymers containing transition metals bonded to main-group metals. The coverage does not, in general, include polymers that are (1) one-dimensional (1-D) chains in the solid state, but that are molecular in solution;[1,2] (2) 2-D (layer) or 3-D (network) coordination polymers;[3] or (3) polymers in which the metal–metal unit is appended to the polymer chain rather than incorporated into the chain.

II. STEP-GROWTH POLYMERS

A. Synthesis

Polymers in this category are synthesized by routes similar to the routes used to synthesize regular step-growth polymers. The difference, however, is that the monomer units contain a metal–metal bond. A sample step-growth polymerization reaction is shown in Eq. 7.1, which illustrates the reaction of a metal–metal bonded dialcohol with hexamethylene diisocyanate (HMDI) to form a polyurethane.[4]

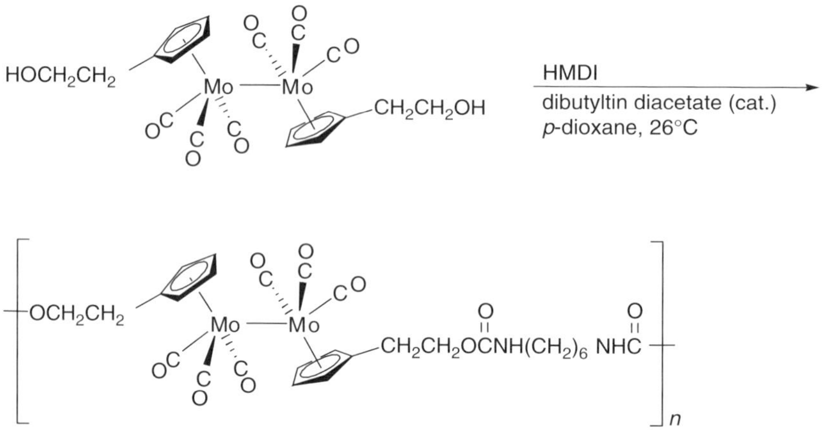

(7.1)

This step-polymerization strategy is quite general, and a number of metal–metal bond-containing polymers have been made from monomers containing functionalized cyclopentadienyl (Cp) ligands.[5–9] In principle, step polymers could be synthesized from monomers containing other derivatized ligands, but in practice not many other ligands are derivatized for this purpose. This probably has to do with the fact that a relatively large number of metal–metal bonded dimers have Cp ligands, whereas dimers with other derivatizable ligands are comparatively rare.

Just as the comparatively weak metal–metal bonds pose problems for the synthesis of the difunctional dimers, they cause similar problems in the synthesis of the polymers. The relative weakness of the metal–metal bonds makes them more reactive than the bonds found in standard organic polymers; thus, under many standard polymerization reaction conditions, metal–metal bond cleavage would result. For example, metal–metal bonds react with acyl halides to form metal–halide complexes. Therefore, the synthesis of polyamides using metal–metal bonded "diamines" and diacyl chlorides would simply lead to metal–metal bond cleavage rather than polymerization. Likewise, metal–metal bonded complexes are incompatible with many Lewis bases because the Lewis bases cleave the metal–metal bonds in disproportionation reactions.[10] This type of reactivity thus rules out many standard condensation polymerization reactions in which bases are used to neutralize any acids produced. Similar reasons prevent the use of acyl chlorides in the synthesis of polyamides. Polymerization strategies must therefore be carefully designed to avoid cleaving the metal–metal bond during the polymerization process.

A sample polymerization reaction showing the synthesis of a polyurethane was shown in Eq. 7.1. Using similar synthetic strategies, various polyurethanes, polyureas (e.g., Eq. 7.2), and polyamides (e.g., Eq. 7.3) were synthesized.[4,5,6,9] Note that the step polymers in the various equations have a metal–metal bond in every repeat unit. Copolymers are straightforwardly synthesized by adding appropriate difunctional organic molecules into the reaction mixture (e.g., Eq. 7.4).[4]

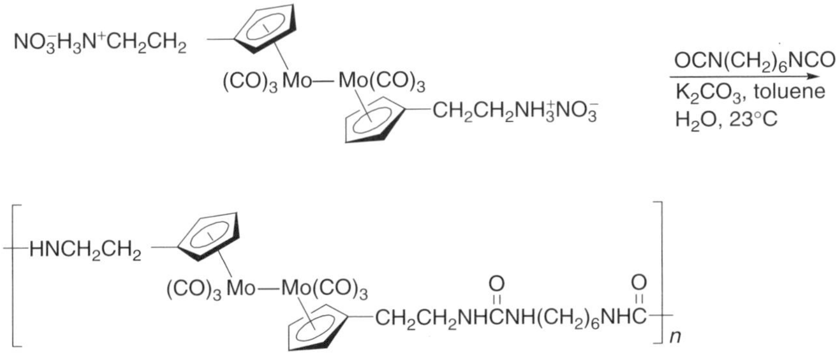

(7.2)

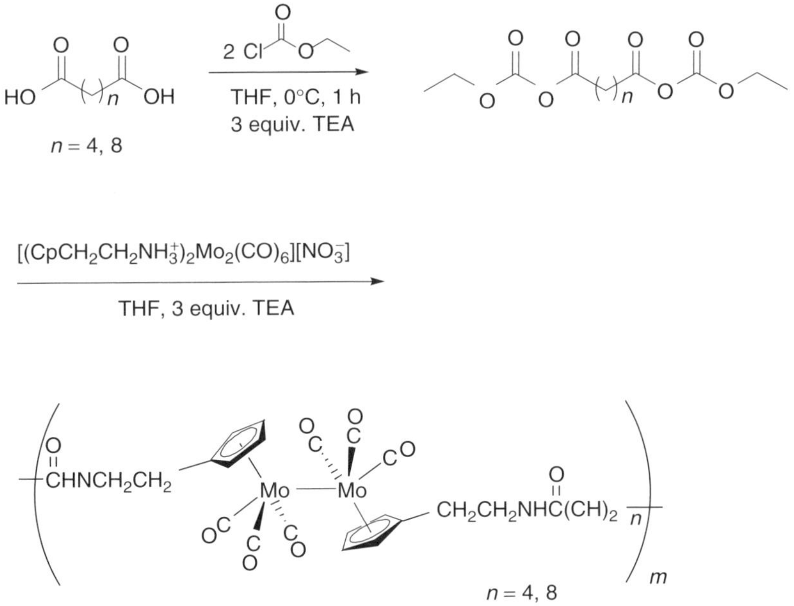

(7.3)

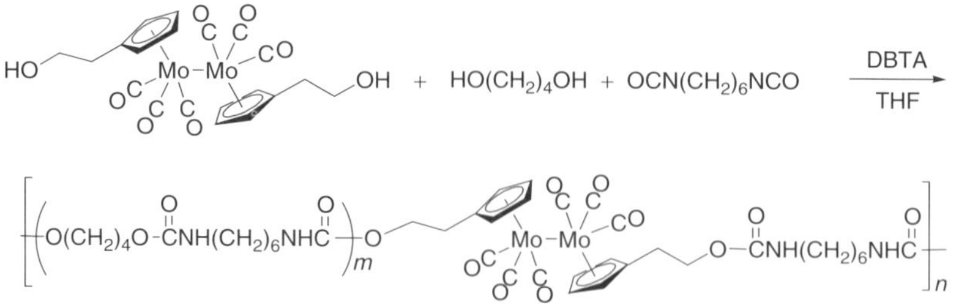

(7.4)

Yet another polymer synthesis strategy is to react the difunctional dimer molecules with prepolymers. Equation 7.5 shows an example of this technique.[5] In this instance, the prepolymer is one of the Hypo polymers sold by W.R. Grace. Analysis of the prepolymer sample showed it contained, on average, three tolyl isocyanate end groups; M_n was about 2000.

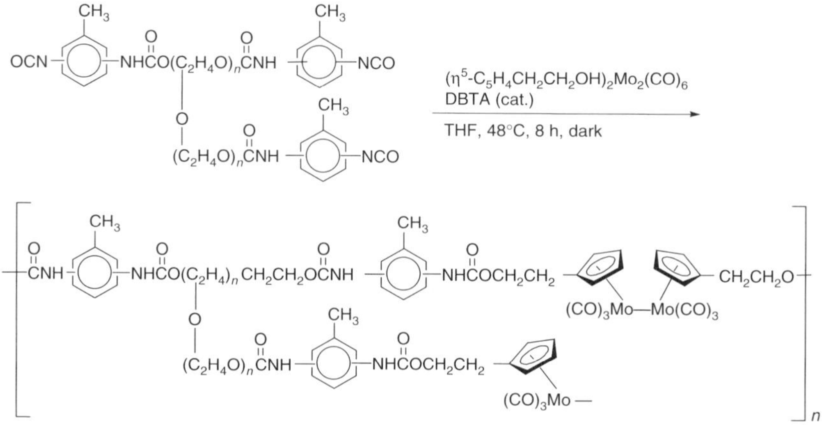

(7.5)

Again, copolymers can likewise be synthesized by using prepolymers and an organic difunctional molecule:[11]

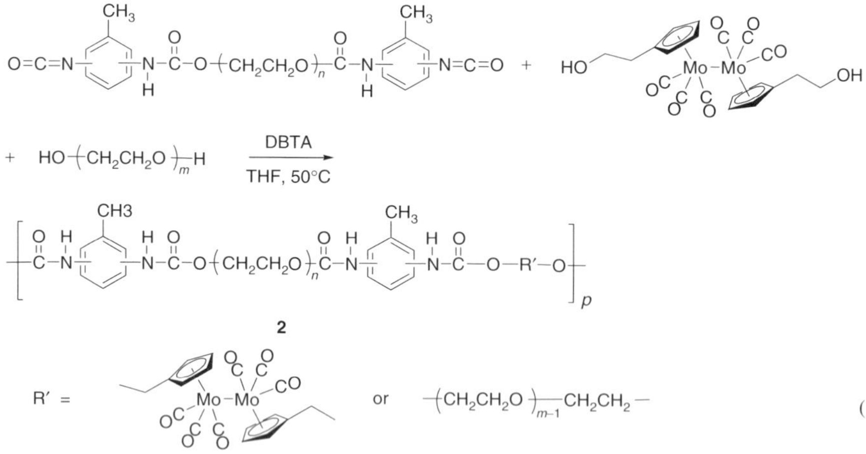

(7.6)

Moran and co-workers reported the synthesis of a number of polysiloxanes that contain Fe–Fe bonds in their backbones. Their syntheses start with the derivatized $Cp_2Fe_2(CO)_4$ dimer follows:[12]

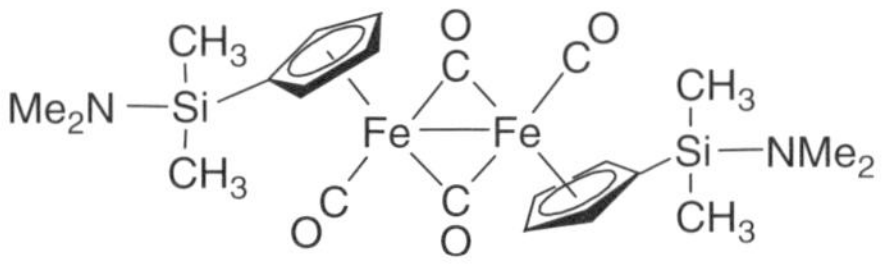

Reaction of this molecule with disilanols gave siloxanes (e.g., Eqs. 7.7 and 7.8).

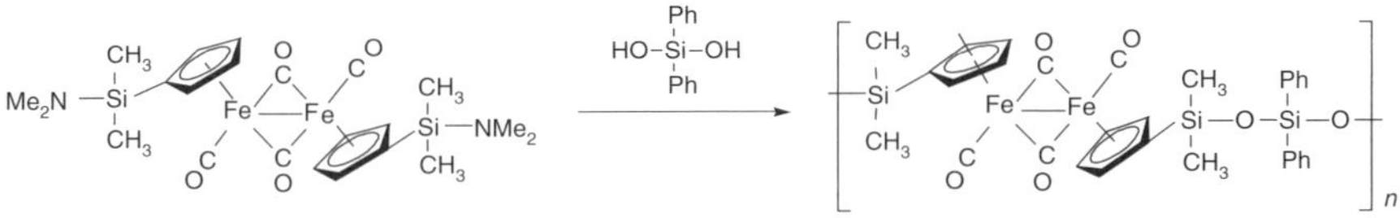

(7.7)

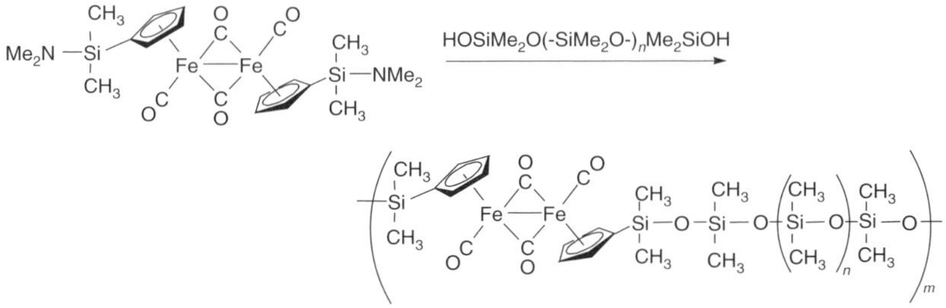

(7.8)

Finally, we note that the step polymerization strategy used to incorporate metal–metal bonded units into polymers can also be used to incorporate metal clusters into polymers. One of only a few examples of this type of reactivity is shown in Eq. 7.9.[13] If polymers containing metal clusters can be shown to have unusual properties or applications, then the synthesis of these polymers will likely burgeon in coming years.

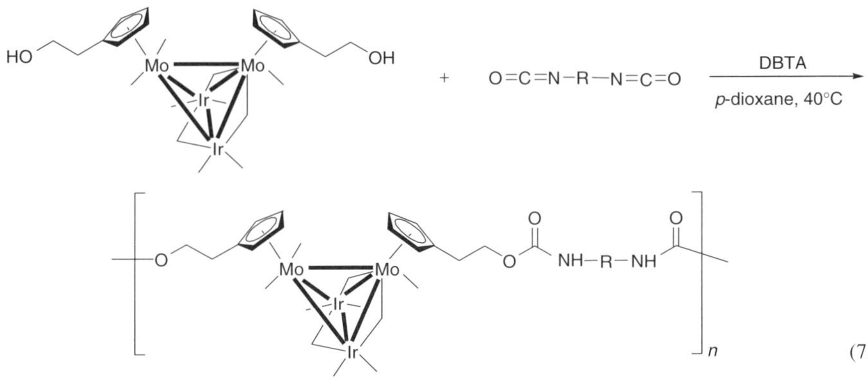

(7.9)

The cluster shown in Eq. 7.9 can also be incorporated into polymers or oligomers using coordination polymerization techniques. This synthetic method is discussed in the section on coordination polymerization.

B. Photochemical Reactions in Solution

Polymers with metal–metal bonds are photochemically degradable because the metal–metal bonds can be cleaved with visible light (Eq. 7.10) and the resulting metal radicals captured with oxygen or other traps (Scheme 7.1).[14,15]

$$L_nM{-}ML_n \xrightarrow{h\nu} L_nM\bullet + \bullet ML_n$$

$$\bullet ML_n = CpMo(CO)_3\ (Cp = \eta^5\text{-}C_5H_5),\ CpW(CO)_3,\ Mn(CO)_5,\ Re(CO)_5,\ CpFe(CO)_2 \qquad (7.10)$$

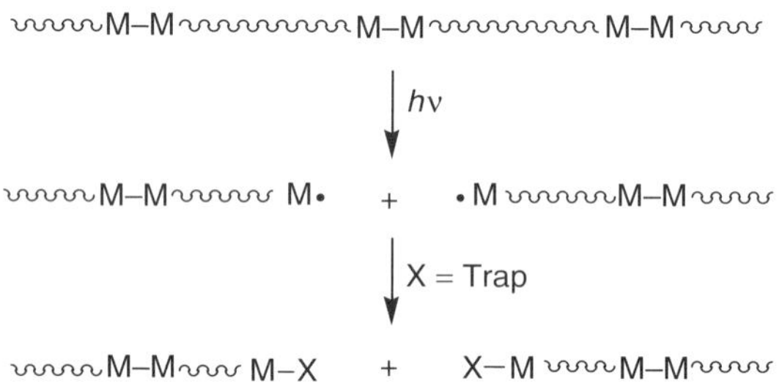

Scheme 7.1 Photochemical degradation of a polymer with metal–metal bonds along its backbone.

Irradiation of metal–metal bonded organometallic complexes into their lowest energy absorption band (≈500 nm) generally leads to one of three fundamental types of reactivity:[6,14,15] (1) the metal radicals produced by photolysis react with radical traps to form monomeric complexes (e.g., Eq. 7.11); (2) the complexes react photochemically with ligands to form ionic disproportionation products (e.g., Eq. 7.12); and (3) the complexes react with oxygen to form metal oxides (Eq. 7.13). (The latter reaction is likely a radical trapping reaction, but may involve excited-state electron transfer.) Note that higher energy excitation leads to M–CO bond dissociation. This type of reactivity is discussed later in this section.

$$Cp_2Mo_2(CO)_6 + 2\ CCl_4 \xrightarrow{h\nu} 2\ CpMo(CO)_3Cl + 2\ [\bullet CCl_3] \qquad (7.11)$$

$$Cp_2Mo_2(CO)_6 + 2\ PR_3 \xrightarrow{h\nu} CpMo(CO)_3^- + CpMo(CO)_2(PR_3)_2^+ + CO \qquad (7.12)$$

$$Cp_2Mo_2(CO)_6 \xrightarrow[O_2]{h\nu} \text{Mo oxides} \tag{7.13}$$

The photochemistry of the polymers in solution is analogous to the reactions of the discrete metal–metal bonded organometallic dimers in solution.[4,5,6,9] As in the photochemical reactions of the discrete dimers, the photochemical reactions of the polymers can be conveniently monitored by electronic absorption spectroscopy. The quantum yields for the reactions are in the ≈0.1 to 0.6 range, depending on the specific polymer and the metal–metal bond.[6] Sample reactions of the polymers showing the three types of reactivity are shown in Eqs. 7.14–7.16.

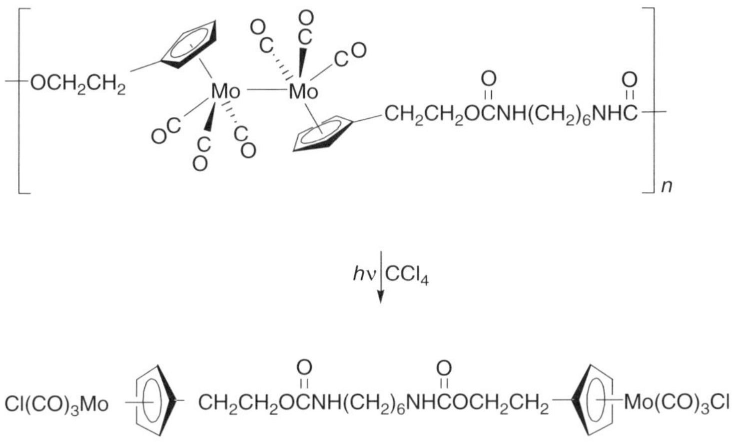

(7.14)

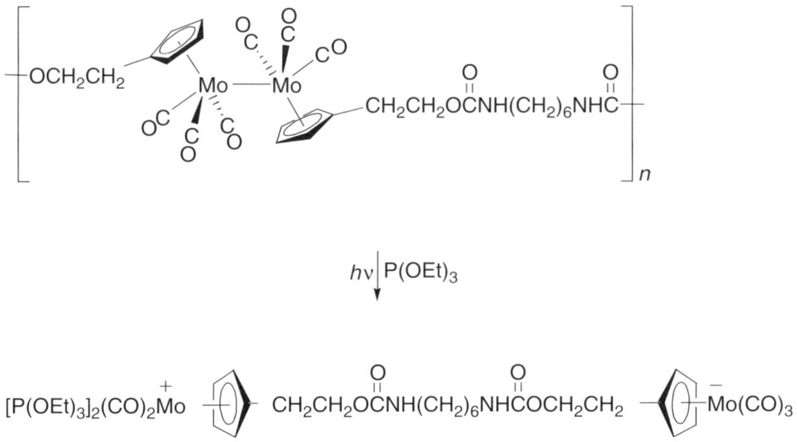

(7.15)

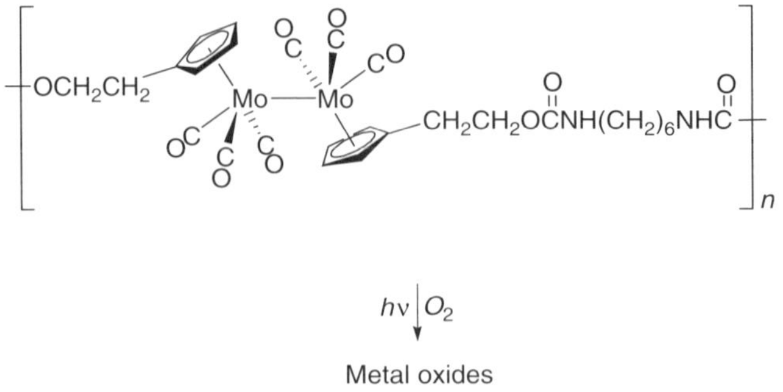

(7.16)

As mentioned, a fourth type of dimer reactivity is dissociation of a CO ligand from the dimer. Generally, this type of reactivity increases in efficiency relative to metal–metal photolysis as the radiation energy increases.[14] In solution, this type of reactivity generally leads to substitution. However, in the case of the $Cp_2Mo_2(CO)_6$ molecule, the reaction in Eq. 7.17 occurs.[6] (Among the dimers, this reaction to form a triply bonded product is unique to the $Cp_2M_2(CO)_6$-type species (M = Cr, Mo, and W; Cp = η^5-C_5R_5).)

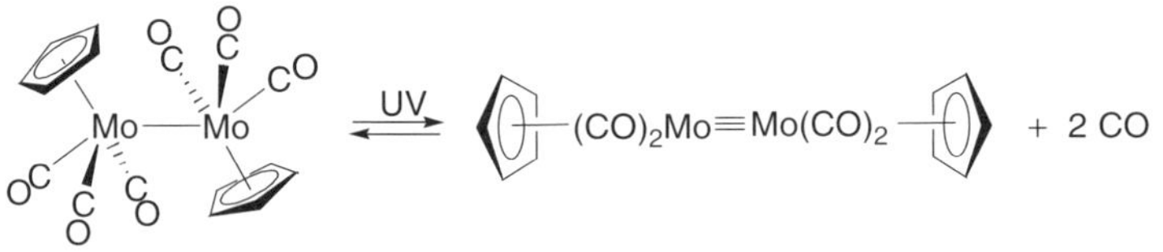

(7.17)

An analogous photoreaction occurs with polymers containing the Mo–Mo unit (Eq. 7.18).[6]

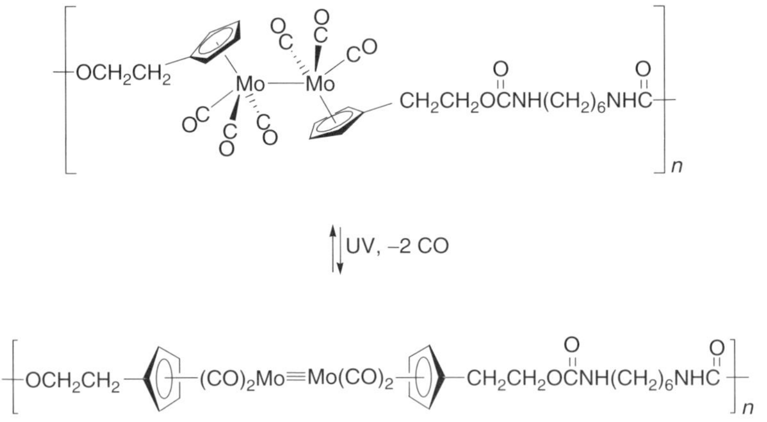

(7.18)

In both Eqs. 7.17 and 7.18, addition of CO to the product solution causes the system to back-react to reform the starting materials. Once again, the main point to be made is that the solution photochemistry of the polymers is analogous to the solution photochemistry of the discrete metal–metal bonded dimers.

Photochemical reactivity in the absence of exogenous radical traps is possible in the case of polymers that have carbon–halogen bonds along their backbones. For example, irradiation of polymers **1**–**3** in solution in the absence of CCl_4 or O_2 led to net metal–metal bond cleavage.[11,16] Spectroscopic monitoring of the reaction showed that metal–metal bond cleavage is accompanied by an increase in the concentration of $CpMo(CO)_3Cl$ units. Photochemical reactions analogous to that in Scheme 7.2 were proposed.

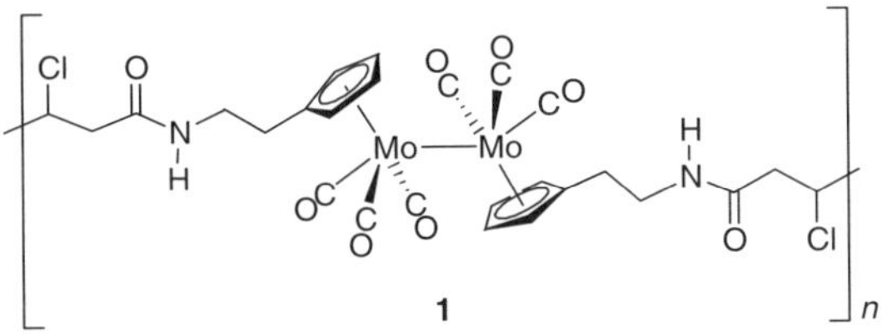

1

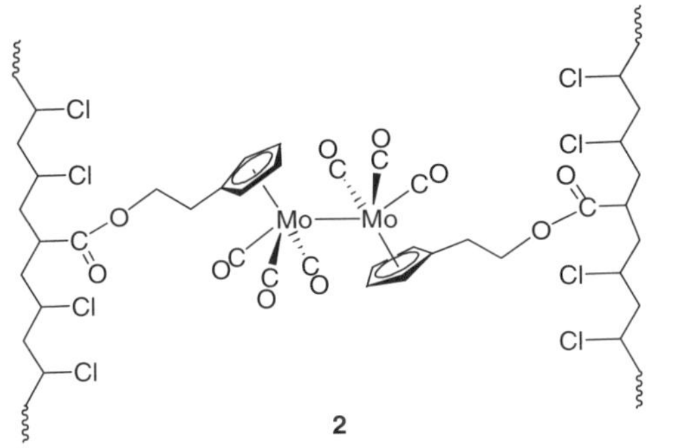

2

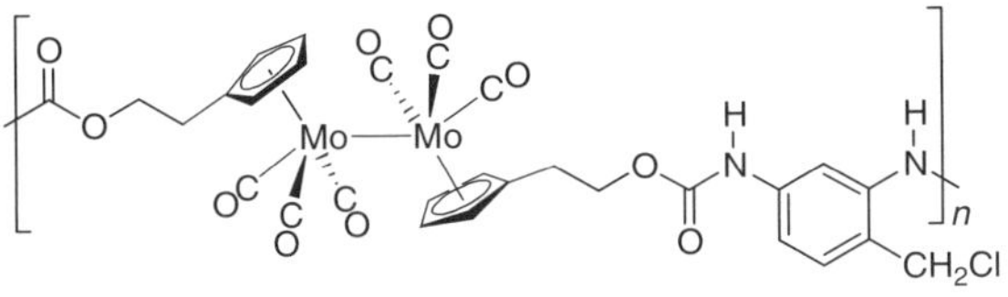

3

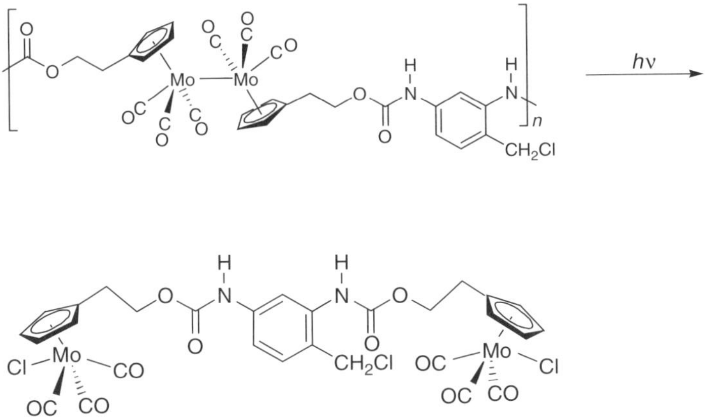

Scheme 7.2 Photochemical reaction of polymer **3** in the absence of an external trapping reagent.

C. Photochemistry in the Solid State

Studies showed that thin films of the polymers containing organometallic metal–metal bonded units (≈0.05 mm in thickness) reacted when they were exposed to *visible* light, whether from the overhead fluorescent lights in the laboratory, from sunlight, or from the filtered output of a high-pressure Hg arc lamp.[6,9,16,17] All of the films were irradiated both in the presence and absence of oxygen. For each film and its dark reaction control, the absorbance of the $d\pi \rightarrow \sigma^*$ transition near 500 nm was monitored periodically over a period of several months. Figure 7.1 is a plot of absorbance at 508 nm versus time for the polymer in Eq. 7.1 under the various experimental conditions. As indicated in the figure, the polymer film that was exposed to sunlight in air completely decomposed in two months. (The $\sigma_{M\text{-}M} \rightarrow \sigma^*_{M\text{-}M}$ electronic absorption band at 390 nm disappeared during this time, confirming that the Mo–Mo bond was not intact.) Thin films stored in the dark in air or irradiated under nitrogen showed only a slight loss of absorbance at 508 nm over a one-year period. From these data, it was concluded that the decomposition of the polymers requires both light and air (oxygen). (The small amount of reaction for those slides stored in the glove box in the light is probably due to reactions with solvent vapors.) Infrared spectra of the decomposition products showed the absence of products with CO ligands, as indicated by the absence of any stretches in the region 1600–2200 cm^{-1}. As mentioned previously, oxide complexes form in the solution phase reactions of $Cp_2Mo_2(CO)_6$ with O_2, and subsequent work showed that the metal-containing decomposition product of the polymer in the solid state is a metal oxide.[11]

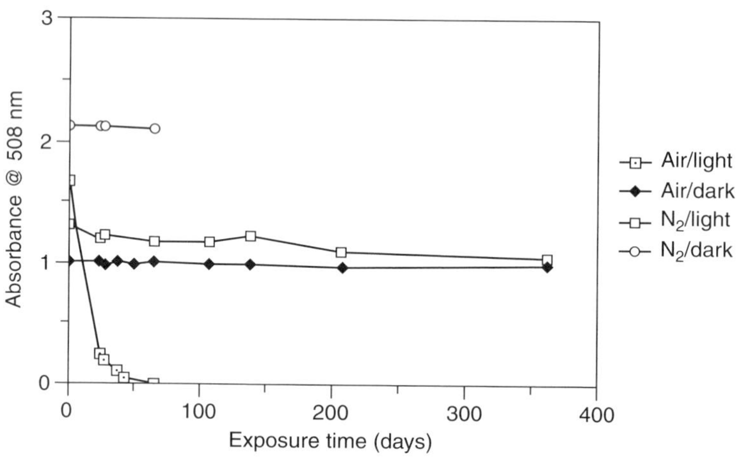

Figure 7.1 A plot of the absorbance at 508 nm vs. time for four thin films of the polymer shown in Eq. 7.1. Only the sample exposed to both light and air degraded. The absorbance at 508 nm is characteristic of the metal–metal bond. (From Tenhaet and Tyler.[6] Reprinted with permission. Copyright © 1991 American Chemical Society.)

These data suggest that oxygen is necessary for the solid-state photochemical reaction to occur. It was proposed that oxygen traps the metal radicals produced in the photolysis of the metal–metal bonds, thereby preventing radical recombination (Eq. 7.19).[11] If oxygen diffusion is rate-limiting, then the relative rates of oligomer photochemical decomposition in the solid state would reflect the oxygen diffusion rate.

$$\sim\!\sim\text{Mo}-\text{Mo}\sim\!\sim \;\underset{}{\overset{h\nu}{\rightleftharpoons}}\; \sim\!\sim\text{Mo}\bullet\;\bullet\text{Mo}\sim\!\sim \;\xrightarrow{O_2}\; \sim\!\sim\text{MoO}_n \quad \text{O}_n\text{Mo}\sim\!\sim \tag{7.19}$$

As described in the previous section, polymers **1–3** were designed to degrade in the absence of exogenous radical traps by building in carbon–chlorine bonds along their backbones. As indicated, all of these polymers did degrade in the absence of oxygen when dissolved in solution, and all three also degraded in the solid state when irradiated in the absence of oxygen.[11]

D. Uses and Applications

i. Mechanistic Applications. An interesting outcome of photodegradation studies on polymers is the finding that tensile and shear stresses can accelerate the rate of photochemical degradation.[18] For example, recent studies of this phenomenon showed that tensile stress will accelerate the photodegradation of numerous polyolefins as well as polycarbonates, nylon, and acrylic-melamine coatings. Conversely, compressive stress will generally retard photodegradation reactions.[19] These observations are of practical importance because most polymers are subjected to light and some form of temporary or permanent stress during their lifetime. In order to control the onset of degradation and the rate of degradation in polymers, it is important to understand the mechanistic origin of the synergism between light and stress in these systems. Furthermore, it is important to understand the photochemical degradation reactions of polymers in order to develop new light stabilizers, to predict the service lifetime of polymers, and to design environmentally friendly degradable materials.[20,21]

Polymers with metal–metal bonds played a key role in the investigation of the stress effects. The reason is that polymers generally photodegrade by a photooxidative pathway involving the autoxidation radical chain mechanism.[22] The photooxidative mechanism, while well understood, is intricate, involving multiple steps, cross-linking, and side reactions. These features make pinpointing the effects of stress difficult. For example, one formidable complication is that O_2 diffusion can be the rate-limiting step in many photooxidative degradations. This adds to the intricacy of the kinetics analysis because O_2 diffusion rates are frequently time-dependent. To circumvent these experimental and mechanistic complexities and therefore make it less difficult to interpret data and obtain fundamental insights, polymer **2** was studied because this polymer photochemically degrades in the absence of oxygen. By eliminating the need for external O_2 to act as a radical trap, the complicating kinetics features of diffusion-controlled oxidation reactions were eliminated.

Infrared spectroscopic analysis demonstrated that the chlorine atoms along the polymer backbone did act as built-in traps for Mo-centered radicals formed by photolysis of the Mo–Mo bonds. The effect of stress on the degradation quantum yield of polymer **2** is shown in Figure 7.2. Note that stress initially increased the quantum yields for degradation, but the quantum yields reached a maximum value and then decreased with higher stress. These results support the "decreased radical recombination efficiency" (DRRE) hypothesis, one of several hypotheses that have been proposed in the literature to explain the effect of stress on polymer photodegradation rates and efficiencies. Specifically, the DRRE hypothesis proposes that the function of stress is to increase the initial separation of the photochemically generated radical pair, which has the effect of decreasing their recombination efficiency and thus increasing the degradation efficiency. The hypothesis predicts an eventual downturn in degradation efficiency because of polymer chain ordering; the increased order hinders diffusion apart of the radicals and thus increases their probability of recombination. Wide-angle X-ray diffraction and infrared spectroscopy confirmed that chain orientation increased with increasing stress on polymers **2** and **2′**.

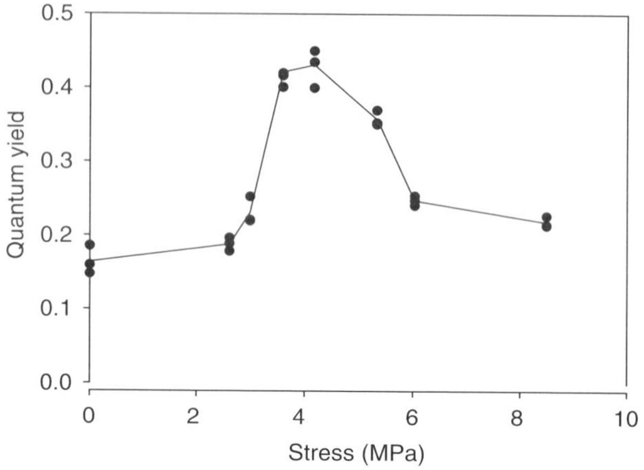

Figure 7.2 Quantum yields for the degradation of polymer **2** vs. applied tensile stress. The results of three independent measurements at each stress are shown. (From Chen and Tyler.[17] Reprinted with permission. Copyright © 2004 American Chemical Society.)

ii. Electrochemistry. The electrochemistry of polymers containing metal–metal bonds has not been extensively explored. One of the few electrochemical studies that has been done dealt with the polymers in Eqs. 7.7 and 7.8. The electrochemistry of these polymers is unusual in comparison to related nonpolymer species, but probably will be found to be typical for metal–metal bond-containing polymers once further studies are done on these materials. In general, irreversible Fe–Fe bond cleavage occurs as a result of electrochemical oxidation or reduction of the $Cp_2Fe_2(CO)_4$ unit in nonpolymer molecules. At slow scan rates, similar behavior was observed for the polymers containing $Cp_2Fe_2(CO)_4$ units. However, at high scan rates (>50 mV/s), the polymers reformed after oxidation and cleavage of the Fe–Fe bond. This result was attributed to the presence of the siloxane chains, which presumably prevent diffusive separation of the two oxidized halves of the Fe–Fe unit. At present, this observation of reversible electrochemical metal–metal bond cleavage is only a curiosity, as this reactivity has not been exploited in any applications. This reactivity would seem to be ready for development, however. In particular, the color change associated with disappearance of the metal–metal bond chromophore might be exploited in an electrochemical lithographic processes. In fact, metal–metal bonds have been used in a related reversible electrochemical lithographic process, as described in a later section.

III. CHAIN-GROWTH POLYMERS

A. Synthesis

Few chain-growth polymers with metal–metal bonds have been reported. The general synthetic route to these materials is to substitute a ligand on a metal–metal bonded dimer with a polymerizable olefin. In the case of $Cp_2M_2(CO)_x$-type molecules (M = Mo, W, Fe), it is very difficult to synthesize a dimer that has only one substituted Cp ring, and hence both Cp rings are substituted with polymerizable olefins. This leads to cross-linked polymers with metal–metal bonds in the chain. Examples of this reactivity are shown in Scheme 7.3.

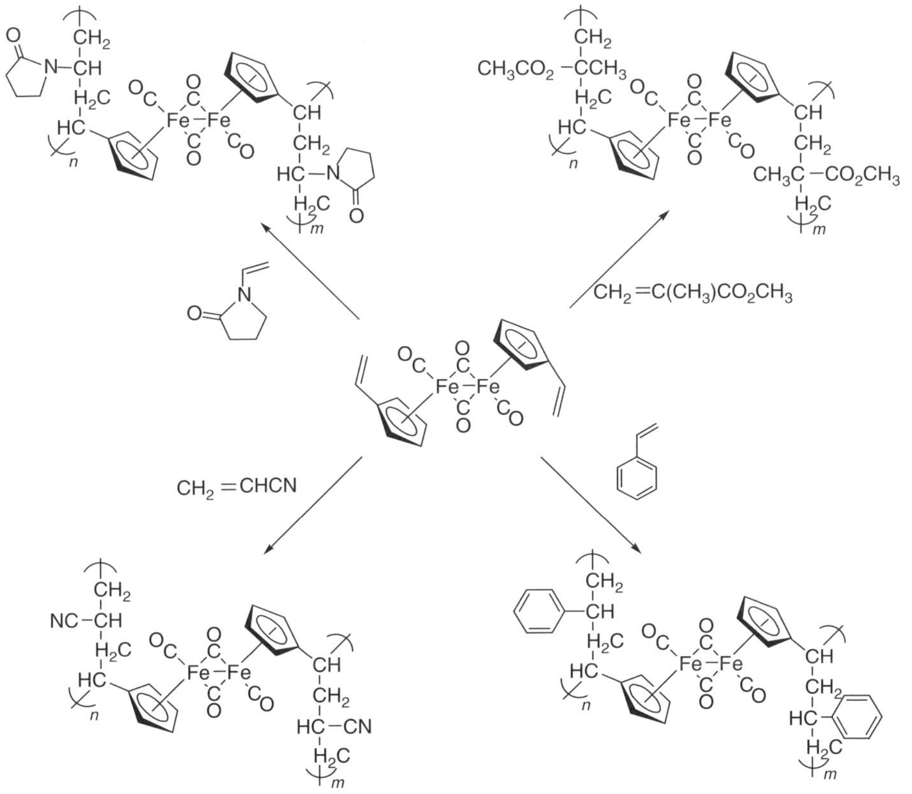

Scheme 7.3 Synthesis of chain-growth polymers containing metal–metal bonds along their backbones.

An example of a chain-growth polymer that uses an olefin-substituted ligand other than cyclopentadienyl is shown in Eq. 7.20.[23]

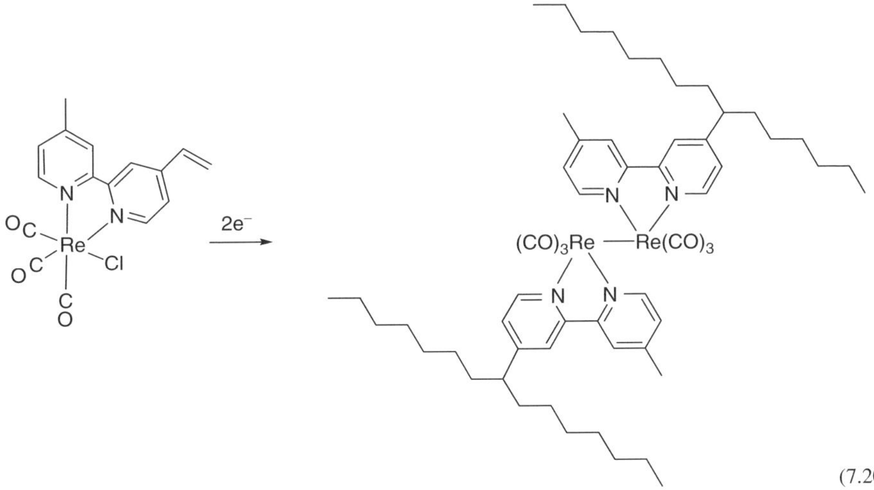

(7.20)

This polymer is particularly interesting because the metal–metal bond can be reformed by electrochemical reduction following photochemical cleavage in the presence of a CCl_4 radical trap (Scheme 7.4).

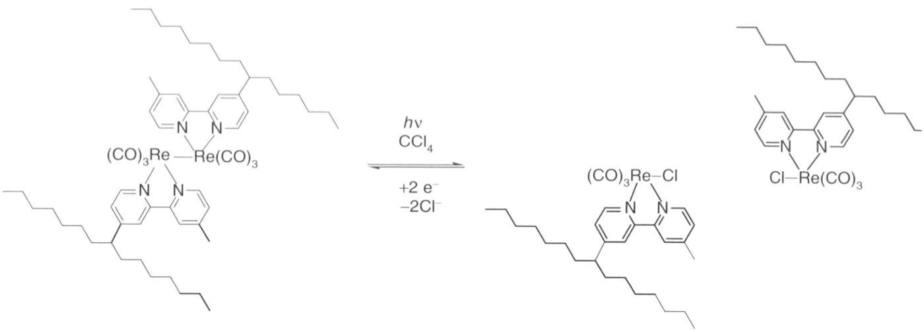

Scheme 7.4 Photochemical cleavage of the Re–Re bond and electrochemical formation of the Re–Re bond in poly[(vbpy)Re(CO)$_3$]$_2$.

It was demonstrated that the reversibility of the Re–Re bond cleavage (and similar reactivity in other systems) may be useful in applications involving reversible imaging. Visible photolysis of the green polymer films in the presence of CCl_4 led to the formation of yellow poly[(vbpy)Re(CO)$_3$Cl]. The yellow-on-green image could be erased in a subsequent electrochemical reduction step (with reformation of the Re–Re bond). The image could be made permanent by reacting the irradiated film with an organic oxidant or with Ag^+. This led to the deposition of metallic silver in the nonirradiated, dark areas of a light-generated pattern. It was found that further deposition of silver could be carried out selectively in the unexposed regions by electrochemical reduction of Ag^+. The process is outlined schematically in Figure 7.3.

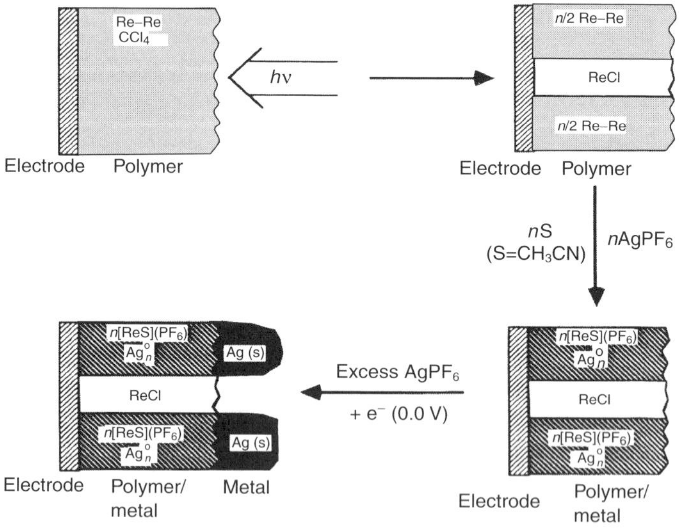

Figure 7.3 Photochemical–chemical–electrochemical cycle for spatially selective silver deposition in poly[(vbpy)Re(CO)$_3$]$_2$. (From O'Toole et al.[23] Reprinted with permission. Copyright © 1989 American Chemical Society.)

Although the resolution of this lithographic method by current standards is somewhat low, this method nevertheless demonstrates that metal–metal bond-containing polymers may be useful in applications where reversible imaging is needed and where the image can be made permanent if need be. Because this example is the only example of reversible lithography using polymers of this type, further study would seem to be called for.

IV. COORDINATION POLYMERS

A. Synthesis

The generic synthesis scheme for polymers of this type is shown in Eq. 7.21.

$$L_nM{-}ML_n + L'{-}L' \longrightarrow \left(L_{n-1}M{-}ML_{n-1}L'{-}L' \right)_n + 2\,L \qquad (7.21)$$

In this equation, L′–L′ is a bidentate ligand that substitutes a metal–metal dimer (or forms an adduct with the dimer) to give a polymer consisting of metal–metal units linked by the L′–L′ chains. The metal–metal unit in this reaction was a variety of species, including $Pt_2Cl_2(\mu\text{-dppm})_2$, $Pd_2(dmb)_2Cl_2$, $Pd_2(dba)_3 \cdot S$ (dba = dibenzylideneacetone; S = benzene or $CHCl_3$), $Pt_2(dba)_3 \cdot CHCl_3$, $M_2(O_2CR)_4$, $[Ru(CO)_2(\mu\text{-}O_2CR)(NCMe)]_2$, and various types of metal cluster complexes. The bidentate ligand is typically either a diphosphine, a diisocyanide, diamine, or a diacetylide ligand. Specific examples of coordination polymers containing metal–metal bonds are discussed later in the section.

B. Polymers Containing Pt–Pt Bonds

Puddephat and co-workers studied the reactivity of the $Pt_2Cl_2(\mu\text{-dppm})_2$ species with diacetylides, diisocyanides, and diphosphines. The reactions resulted in low molecular-weight oligomers and are summarized in Eqs. 7.22–7.25.[24]

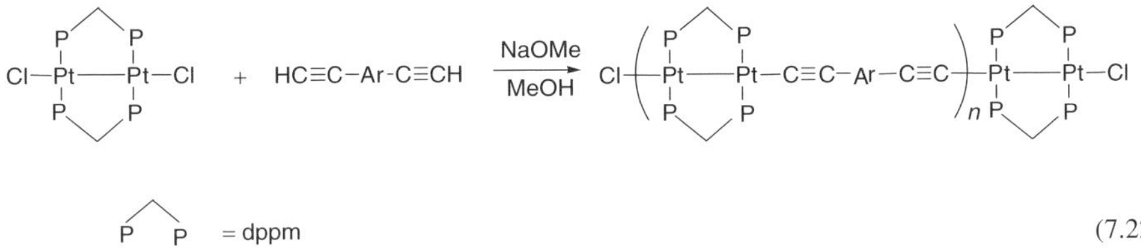

(7.22)

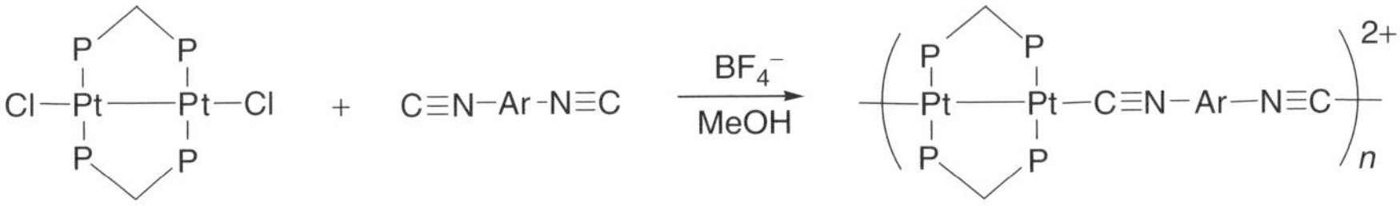

(7.23)

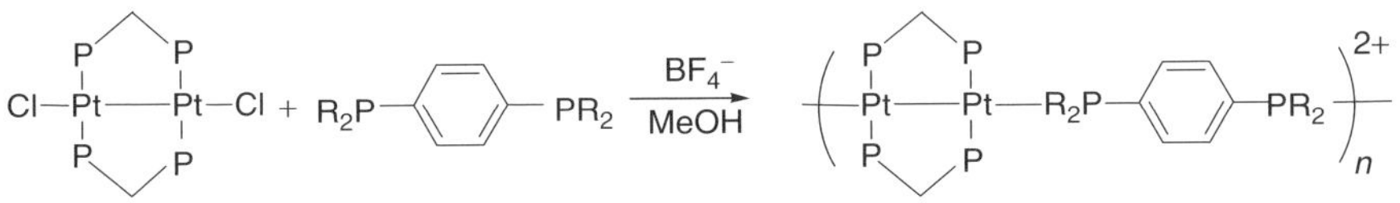

(7.24)

Note that the polymers are neutral if the bridging ligands are diacetylides, but cationic if the bridging ligands are diphosphines or diisocyanides. In general, the oligomers were insoluble so it was not possible to obtain molecular weights. However, elemental analytical data indicated number-average degrees of polymerization of about 3. It was suggested that the molecular weights are low because the oligomers are insoluble and they precipitate from solution before extensive chain growth has occurred. The thermal properties of the polymers were studied, and it was found that although each polymer underwent major weight loss in the temperature range 330–480°C, the residue obtained did not correspond to pure Pt metal.

The work by Puddephatt and co-workers[24] is notable because it reemphasizes and demonstrates an important technique for characterizing polymers with metal–metal bonds. As is the case with many polymers, including many metal–metal bonded polymers, the polymers described in the preceding paragraph are insoluble in most solvents. Consequently, their characterization is challenging. A useful technique for characterizing such materials is to synthesize a model complex that consists of the repeat unit of the supposed polymer. The spectroscopic properties of the polymer and the model complex should in general be very similar, if not identical. This technique was successfully used by Puddephatt and co-workers to characterize the $Pt_2Cl_2(\mu\text{-dppm})_2$-containing polymers (e.g., Eq. 7.25). They also used their syntheses of the model complexes to fine-tune the synthetic conditions used in the polymerization reactions.

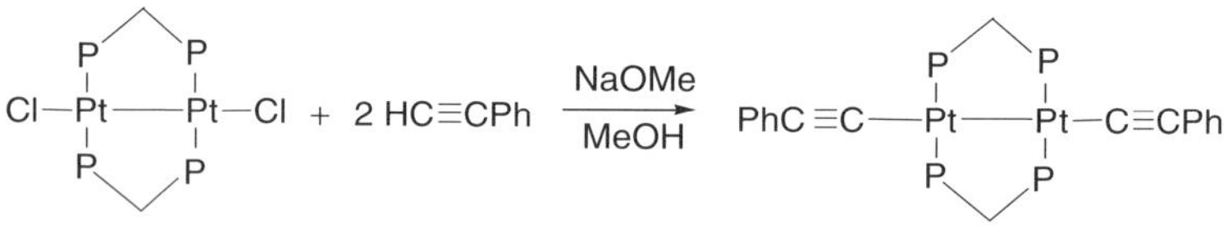

(7.25)

C. Polymers with the dmb Ligand

An emerging synthetic theme in the chemistry of coordination polymers that contain metal–metal bonds is to use diphosphines, diisocyanides, and diacetylides to link units together that contain metal–metal bonds. Harvey and co-workers have had notable success in this area using metal–metal units intramolecularly bridged by the bidentate dmb (1,8-diisocyano-p-menthane) ligand or intermolecularly connected by the dmb ligand.[25,26,27]

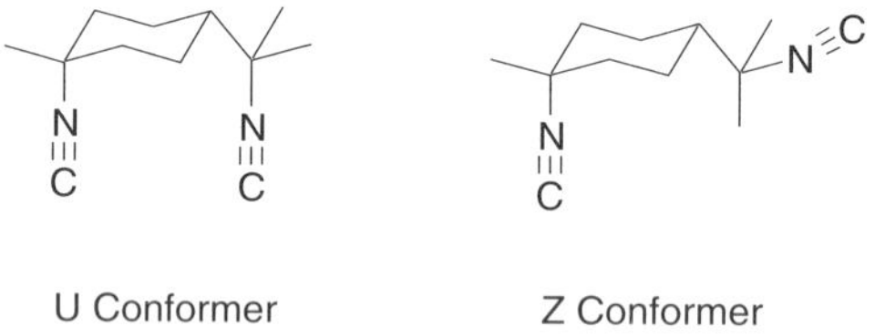

Using this ligand, Harvey and co-workers synthesized and studied a number of polymers containing Pd_4 and Pt_4 metal–metal bonded units (Fig. 7.4):[25,26,28,29]

$$Pd_2(dba)_3{\cdot}S + xs\ dmb + Pd(O_2CCH_3)_2 \longrightarrow [(Pd_4(dmb)_4(dmb)^{2+})(CH_3CO_2^-)_2]_n$$

dba = dibenzylideneacetone; S = benzene or $CHCl_3$ (7.26)

$$Pt_2(dba)_3\cdot CHCl_3 + 2\ dmb + 1/2\ diphos \longrightarrow (Pt_4(dmb)_4(diphos)^{2+})_n$$

dba = dibenzylideneacetone; diphos = $Ph_2P(CH_2)_mPPh_2$, $m = 4, 5, 6$ (7.27)

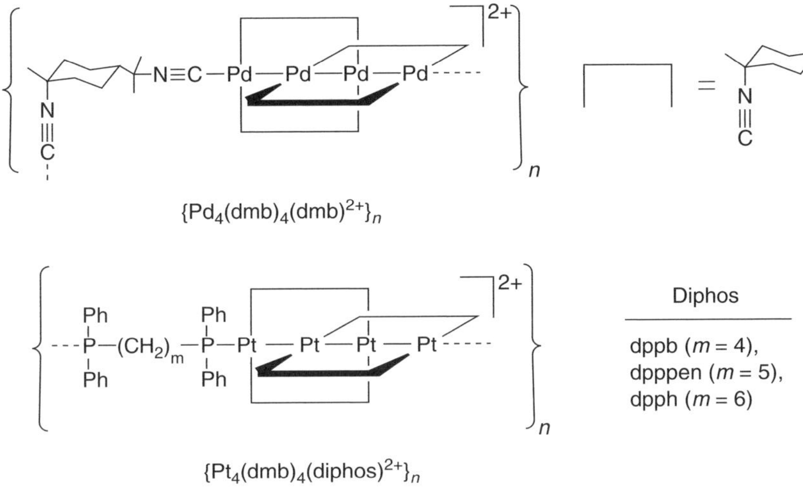

Figure 7.4 Structures of the $\{Pd_4(dmb)_4(dmb)^{2+}\}_n$ and $\{Pt_4(dmb)_4(diphos)^{2+}\}_n$ polymers synthesized in Eqs. 7.26 and 7.27. (From Sicard et al.[25] Reprinted with permission. Copyright © 2004 American Chemical Society.)

The synthetic method of linking axially substituted linear polynuclear fragments together with bidentate phosphines has been applied to other species, such as $Pd_2(dmb)_2Cl_2$. Reaction of this and related clusters with diphosphines [dppb = bis(diphenylphosphino)butane; dpppen = bis(diphenylphosphino)pentane; and dpph = bis(diphenylphosphino)hexane] gave polymers of the type $\{[Pd_2(dmb_2(diphos)](ClO_4)_2\}_n$:[30]

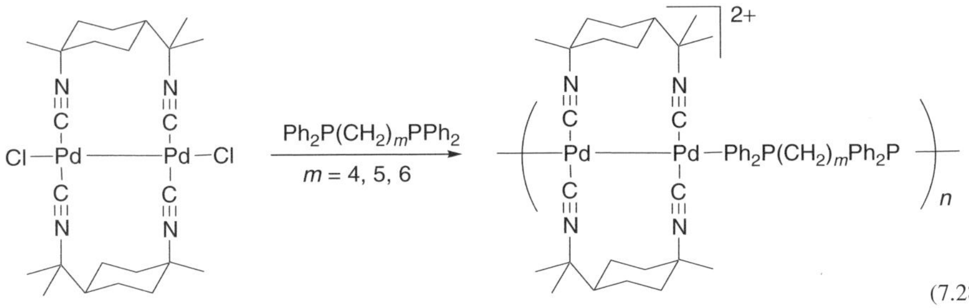

(7.28)

A number of experimental techniques confirmed the structure of these materials. Particularly noteworthy was the presence of the $d\sigma \rightarrow d\sigma^*$ band in the UV-vis spectrum, indicative of the presence of the intact $Pd_2(dmb)_2(P)_2^{2+}$ unit. Molecular-weight measurements indicated these species are oligomers in solution containing about 12–16 repeat units.

Different reactivity was found when $Pd_2(dmb)_2Cl_2$ was reacted with the dppe [bis(diphenylphosphino)ethane] or dppp [bis(diphenylphosphino)propane] phosphines. In these reactions, the diphosphine acts as a chelate and the polymer contains a "$Pd_2(diphos)_2(CNR)_2{}^{2+}$" unit (Eqs. 7.29 and 7.30).[25]

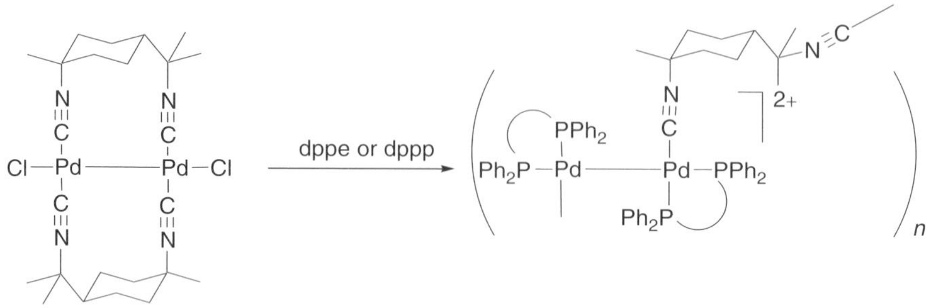

(7.29)

$$Pd_2(dmb)_2Cl_2 + 2\ \text{diphos} \longrightarrow (1/n)\ Pd_2(\text{diphos})_2(dmb)^{2+}{}_n + dmb + 2\ Cl^- \quad (7.30)$$

Molecular-weight measurements indicated these species are oligomers in solution containing about eight repeat units. Note that, unlike the $\{[Pd_2(dmb_2(\text{diphos})](ClO_4)_2\}_n$ polymers, the $\{Pd_2(\text{diphos})_2(dmb)^{2+}\}$ polymers have unsupported metal–metal bonds in the backbone.

The structure of the $Pd_2(dmb)_2Cl_2$ species is itself likely polymeric with Pd–Pd bonds. Because the material is amorphous, the X-ray structure could not be obtained, but modeling suggested a polymer structure, perhaps the structure shown below:[25]

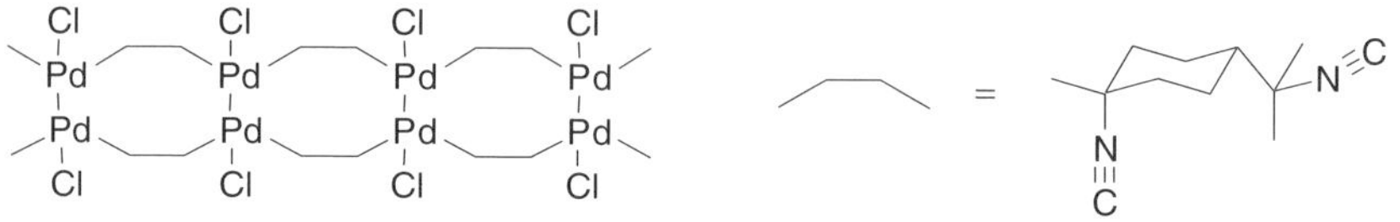

D. Polymers With Metal Clusters Along the Backbone Chain

The synthetic strategy presented in the preceding section using bidentate ligands to incorporate metal–metal bonded units into polymers can also be used to incorporate metal clusters into polymers. A sample reaction is shown in Eq. 7.31.[31] For this specific reaction, a polymer with n = 1000–1020 was obtained, as estimated by electron microscopy. Conducting wires were obtained from this polymer by exposing a polymer sample supported on silica to an electron beam.[31] Initial exposure to the beam caused the sheath of nonconducting carbonyl ligands to be lost (as inferred from the contraction in volume and the weight loss, as measured by thermogravimetric analysis (TGA) with the formation of nanoparticle chains. Further irradiation then fuses the nanoparticles into larger clusters, forming a conducting chain. (The authors note, however, that a continuous metal wire is not likely formed by this process. The observed conductivity was attributed to electron hopping between nanoparticle groups.) Although still in the early stages of development, the synthesis of conducting nanowires by this technique would seem to hold great promise.

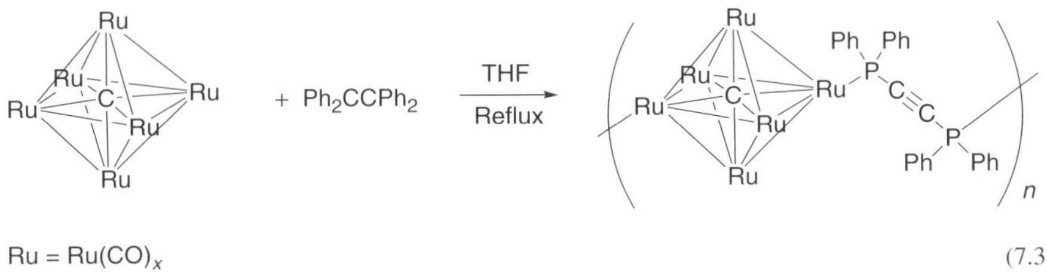

(7.31)

Other polymers (actually oligomers because of their relatively low molecular weights) have been synthesized by reacting bidentate phosphines or isocyanides with metal clusters. Thus, mixing $Co_4(CO)_{12}$ and dppe in varying ratios gave oligomeric structures containing Co_4 tetrahedra, the largest being $Co_4(CO)_{11}\{(\mu\text{-dppe})Co_4(CO)_{10}\}_3(\mu\text{-dppe})Co_4(CO)_{11}$.[32] Using a similar synthetic strategy, the $[(\mu\text{-}N,N'\text{-hexamethylenetetraamine})Cu_2\{Co(CO)_4\}_2]_n$ polymer was reported (but n was not measured).[33] The $\{[\{Pt_3(\mu\text{-dppm})_3\}(\mu\text{-1,4-CNC}_6R_4NC)][PF_6]\}_n$ polymer was also reported.[34] With the exception of the Ru_6-containing polymer in Eq. 7.31, none of these polymers and oligomers dissolves to give solutions that contain polymers or oligomers; the cluster–bidentate ligands bonds are too readily cleaved in solution. Consequently, the point has been made that the coordination polymerization approach to making polymers with clusters along their backbones is probably not the best synthetic approach to making these materials.[13] Humphrey and Lucas have shown that a step-growth polymerization approach to the synthesis of these types of polymers leads to more robust and soluble materials (see Eq. 7.9).[13]

To conclude this section, it is noted that coordination polymers consisting of hexanuclear rhenium–chalcogenide clusters and transition metal ions have been synthesized.[35,36] In a typical example, *trans*-$[Re_6(\mu_3\text{-Se})_8(PEt_3)_4(4,4'\text{-dipyridyl})_2]^{2+}$ (molecule **4**) was reacted with $M(NO_3)$ (M = Cd, Co, Zn) to yield polymers containing $[Re_6(\mu_3\text{-Se})_8]^{2+}$ cluster units along the backbone. The resulting polymers have intriguing differences in their crystal structures, but in all of the polymers the key linkage feature is that two of the *trans*-$[Re_6(\mu_3\text{-Se})_8(PEt_3)_4(4,4'\text{-dipyridyl})_2]^{2+}$ molecules act as "expanded ligands" and coordinate to the transition metal ions through two $N_{pyridyl}$–M dative bonds (Figure 7.5). It is frequently suggested that supramolecular materials such as these will be useful in a host of applications because they may possess unusual magnetic, electronic, optical, and catalytic properties.

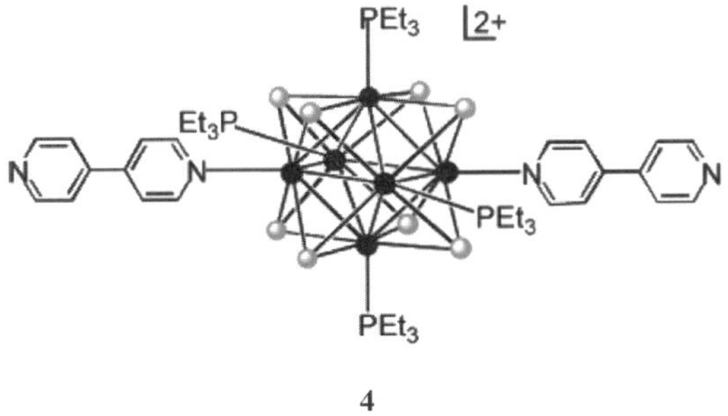

4

The $[Re_6(\mu_3\text{-Se})_8]^{2+}$ clusters have apparently not been used in the synthesis of coordination polymers prepared by linking reactions analogous to that shown in Eq. 7.31, that is, by a linking reaction with a bidentate ligand.[36] However, supramolecular arrays containing $[Re_6(\mu_3\text{-Se})_8]^{2+}$ clusters have been prepared by this route. Specifically, molecular squares, molecular stars, and molecular trees have been synthesized by using the rigid clusters to

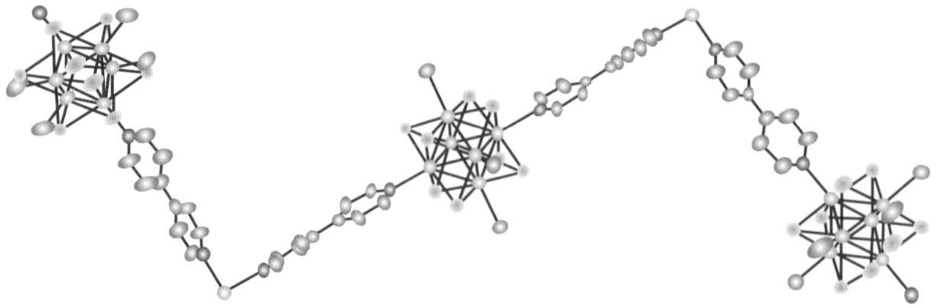

Figure 7.5 A section of a single chain of the polymer formed in the reaction of $Zn(NO_3)_2$ with molecule **4**. (From Selby et al.[36] Reprinted with permission. Copyright © 2003 American Chemical Society.)

impart the desired directionality in the assembly of the products. A sample reaction showing the synthesis of a molecular square is shown in Figure 7.6.[37] The claim is that materials with these architectures may have desirable magnetic, catalytic, and optical properties, but these properties have yet to be thoroughly investigated.[36]

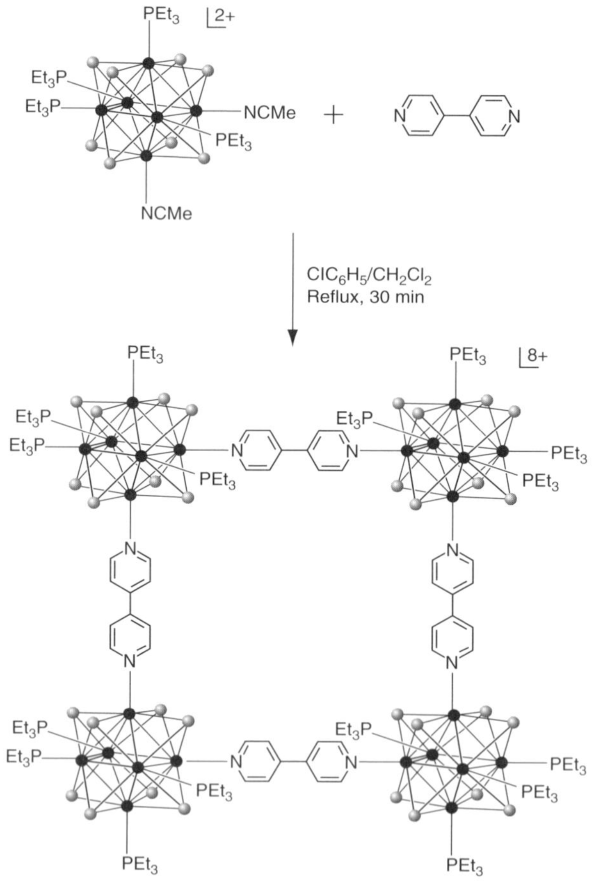

Figure 7.6 Synthesis of a molecular square containing $[Re_6(\mu_3\text{-}Se)_8]^{2+}$ clusters. (From Selby et al.[36] Reprinted with permission. Copyright © 2003 American Chemical Society.)

E. Another Method for Synthesizing Polymers and Oligomers Containing Metal Clusters

A step-growth synthetic method for incorporating metal clusters into polymers was shown in Eq. 7.9. And, in the preceding section, metal clusters were incorporated into polymers and oligomers using bidentate ligands and the techniques of coordination polymerization. Yet a third synthetic route to these materials is to react selected metal clusters with acetylenes.[13,38,39,40] In many such reactions, the two carbons of the acetylene unit are incorporated into the metal cluster. An example is shown in Eq. 7.32.

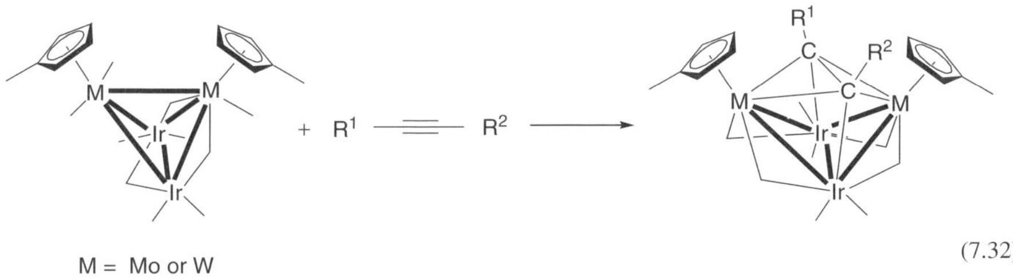

(7.32)

If molecules containing more than one C≡C unit are used in the synthesis, then one metal cluster is generally incorporated at each C≡C site. Conceptually, polymers of considerable molecular weight could be synthesized by this route, but in practice the method is limited by the size of the available acetylene molecule. Humphrey and co-workers have used this technique to make a variety of oligomers containing metal clusters, and an example from their work that demonstrates this method is shown in Eq. 7.33.[38]

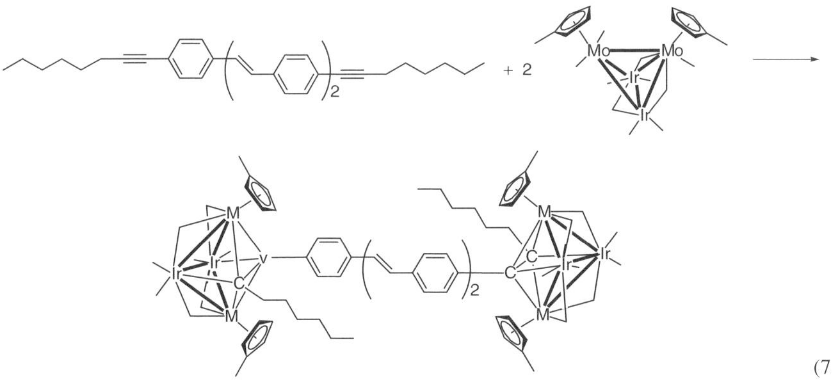

(7.33)

It has been suggested that metal clusters may be useful as optical limiters.[41] (These are materials that display decreasing transmittance as a function of incident light intensity. Optical limiters are needed for a range of applications in optical-device protection, for example, in protecting the eyes and sensors from intense light pulses. They are also useful in laser mode locking and optical pulse shaping.) Humphrey and co-workers incorporated metal clusters into oligomers and polymers because the applications for optical limiters require that they be in film-processable form.[41] Generally, metal clusters are obtained as crystalline materials, in which form they are not suitable as optical limiters, and the oligomers and polymers containing these materials would more likely be film-processable. Studies on oligomers synthesized by the route in Eqs. 7.32 and 7.33 showed that the optical limiting response originated from the ligated metal core and was not attributable to an intercluster electronic interaction. Although this is the only reported *polymer* application for these materials, it is noted that metal clusters linked by π-delocalizable bridging units (like those in Eq. 7.33) are also of interest for their electrochemical properties. Studies have shown such materials can act as electron reservoirs with tunable redox properties.[38]

F. Polymers with Weak Au···Au Bonding

Au(I) generally has a coordination number of 2 with a linear geometry, so it is well suited to form linear rigid-rod polymers. Puddephatt noted that to obtain a neutral polymer with Au(I), one must use one neutral ligand and one anionic ligand. Suitable ligand combinations therefore include neutral isocyanides and anionic acetylides. Puddephatt's strategy to obtain these polymers was to react isocyanides with gold acetylides. The latter were obtained as shown in Eq. 7.34.[42]

$$2\,[Me_2SAuCl] + HC{\equiv}C{-}Ar{-}C{\equiv}CH + 2\,B \longrightarrow \{Au{-}C{\equiv}C{-}Ar{-}C{\equiv}C{-}Au\}_n + 2\,Me_2S + 2\,BH^+Cl^- \tag{7.34}$$

With these species in hand, coordination polymers were prepared by reaction with isocyanides according to Eq. 7.35:[42]

$$1/n\,\{Au{-}C{\equiv}C{-}Ar{-}C{\equiv}C{-}Au\}_n + 2\,CNR \longrightarrow [R{-}N{\equiv}C{-}Au{-}C{\equiv}C{-}Ar{-}C{\equiv}C{-}Au{-}C{\equiv}N{-}R] \tag{7.35}$$

The products formed in Eq. 7.35 are polymeric in the solid state because of weak Au···Au interactions and π–π stacking (Figure 7.7). The Au···Au interactions are on the order of

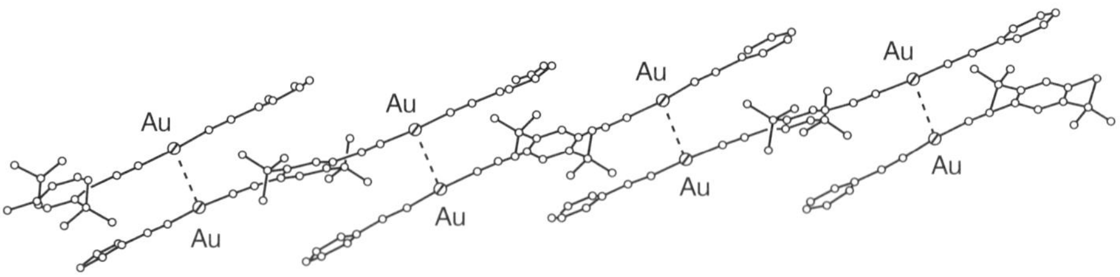

Figure 7.7 The intermolecular association present in the solid-state structure of [PhC≡CAuC≡NBu$^t_2$C$_6$H$_2$N≡CAuC≡CPh] due to (a) π-stacking and (b) Au···Au bonding. The Au···Au distance is 3.174(1) Å. (From Puddephatt et al.[42] Reprinted with permission. Copyright © 1998 Royal Society of Chemistry.)

5–10 kcal/mole and are proposed to arise from relativistic London forces. Reactions with diisocyanides formed polymers (Eq. 7.36):

$$1/n\,\{Au{-}C{\equiv}C{-}Ar{-}C{\equiv}C{-}Au\}_n + C{\equiv}N{-}Ar'{-}N{\equiv}C \longrightarrow$$

$$\left(Au{-}C{\equiv}C{-}Ar{-}C{\equiv}C{-}Au{-}C{\equiv}N{-}Ar'{-}N{\equiv}C\right)_n \qquad (7.36)$$

These materials are effectively cross-linked because of interchain Au···Au interactions. Finally, it is noted that a variety of structurally interesting branched structures with Au···Au interactions formed when acetylides such as species **5** were used in the reactions with diisocyanides or diphosphines.

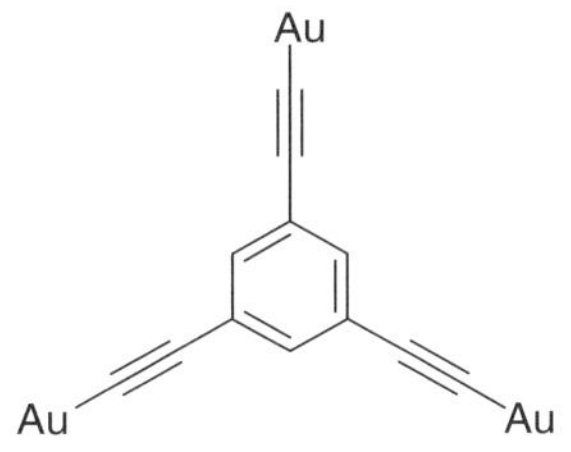

5

A number of polymers with weak Au···Au interactions have been synthesized and these materials have been reviewed.[42]

G. Coordination Polymers with $M_2(\mu\text{-}L_2)_4$ Units

The axial coordination sites of $M_2(\mu\text{-}L_2)_4$-type complexes (where L_2 is a bridging bidentate ligand) are generally available for bonding to ligands, and polymers will result if bidentate ligands are bonded at these sites:

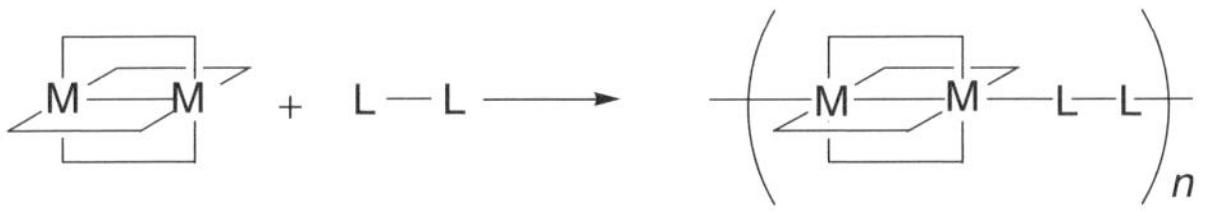

(7.37)

A number of coordination polymers have been synthesized by reaction of bimetal carboxylates, $M_2(O_2CR)_4$, with bidentate ligands. Thus, the addition of pyrazine to aqueous solutions of dicopper acetate gave a polymer with $Cu_2(O_2CCH_3)_4(pyz)$ units (pyz = pyrazine).[43] (The Cu–Cu interaction in the copper acetate dimer is attributed to an antiferromagnetic exchange interaction and not an ordinary metal–metal bond. Electron paramagnetic resonance (EPR) experiments showed that, in addition to the intradimer Cu-Cu exchange interaction, there is also a weak interdimer interaction in the polymer, mediated by the pyrazine bridge.) In a similar reaction, an isomorphous polymer formed in the reaction of pyrazine with dichromium acetate.[44] A linear polymer formed in the reaction of phenazine with dirhodium tetracarboxylate, and a zigzag polymer formed in the reaction of durenediamine with dirhodium tetracarboxylate.[45] An illustrative example showing the polymer that forms in the reaction of $Mo_2(O_2CCH_3)_4$ with the bidentate dmpc [bis(dimethylphosphino)ethane] ligand is shown in Eq. 7.38.[46] As is often the case with these coordination polymers, the molecular weights of the polymer materials were not reported. Surprisingly perhaps, some of the polymers are soluble in selected solvents, but in general the study of their properties was focused on their solid-state structures.

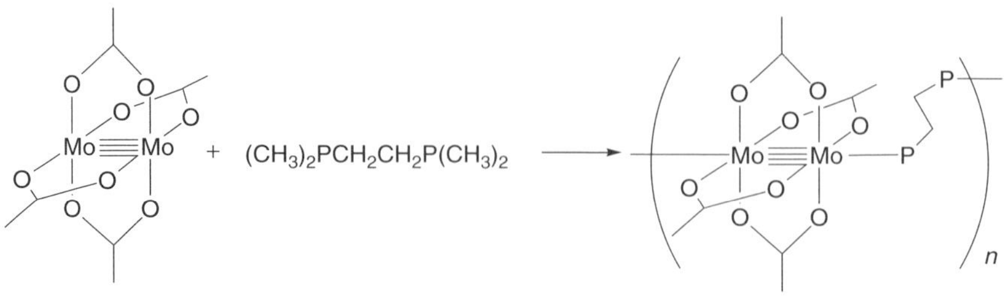

(7.38)

In related work, reaction of the carboxylate-bridged $[Ru(CO)_2(\mu\text{-}O_2CR)(NCMe)]_2$ dimers or $[Ru(CO)_2(\mu\text{-}O_2CR)(NCMe)]_n$ polymers ($R = CH_3$ or CH_2CH_3) with bidentate ligands ($L_2 = Ph_2PCH_2PPh_2$; $Me_2PCH_2PMe_2$; $PhSCH_2SPh$; or $CH_3SCH_2SCH_3$) gave coordination polymers of formula $[Ru_2(CO)_4(\mu\text{-}O_2CR)_2(L_2)]_n$, in which the binuclear units are linked by diphosphine or dithioether ligands (Eq. 7.39).[47] As was the case with the polymers in the preceding paragraph, neither the molecular weights of these polymer materials nor their physical properties were reported.

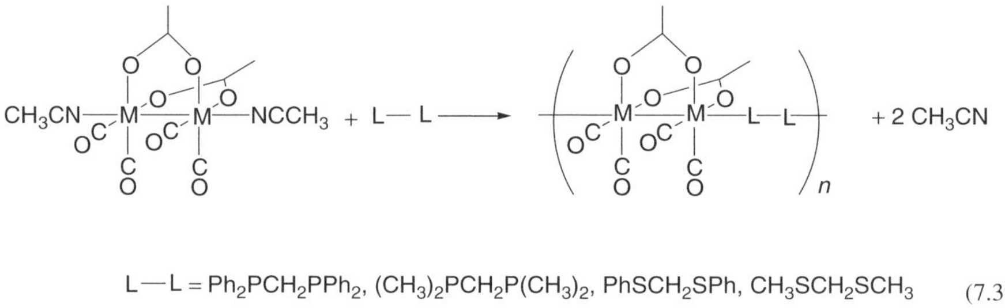

(7.39)

It is noted that most recent attention in this area has focused not on the synthesis of polymers prepared according to Eq. 7.37 but rather on the synthesis of supramolecular arrays based on dimetal building blocks. Synthetic strategies have been worked out for linking the dimetal units at the axial sites (as in Eq. 7.37) or at the equatorial sites using dicarboxylate ligands such as oxalate (Scheme 7.5).[48] Using these synthetic methods, a variety of interesting structures containing metal–metal bonds have been synthesized, including molecular loops, molecular

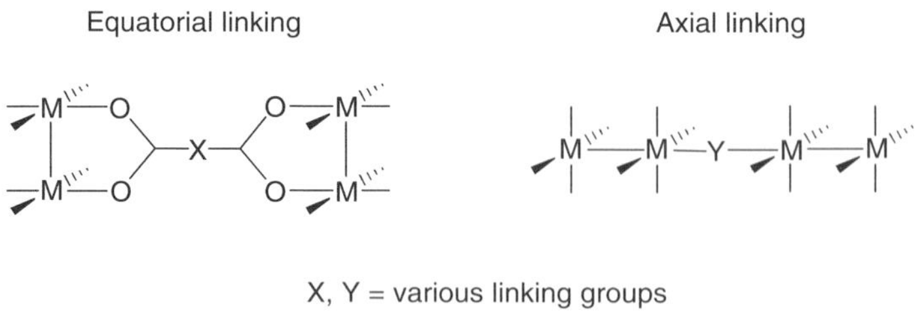

Scheme 7.5 Bonding motifs showing the equatorial and axial linking of dimetal units.

triangles, molecular squares, molecular polyhedra, and various 1-D, 2-D, and 3-D extended structures. A generic example of a 1-D extended structure with metal–metal bonds is shown in Scheme 7.6; specific examples of these structures can be found in the references.[48–51]

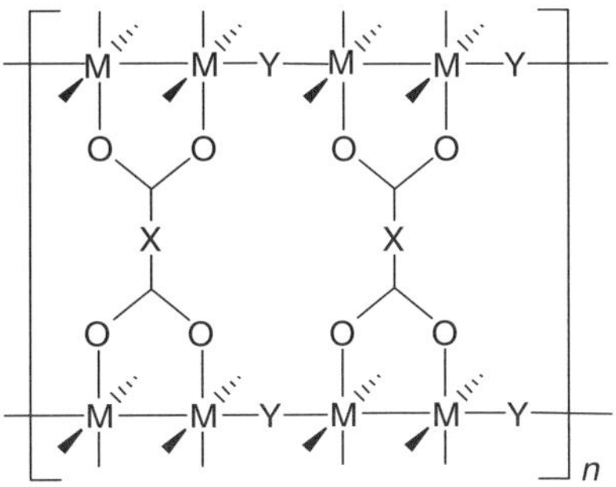

Scheme 7.6 A generic 1-D structure with both axial and equatorial linkers.

H. Dendrimers

Dendrimers that have metal–metal bonds along their interior branches are not known. However, a small number of dendrimers have been synthesized that contain a metal cluster at the interior core or that have metal clusters attached to the surface of the dendrimer. In one example, Zheng reported that the labile acetonitrile ligands in $[Re_6Se_8(MeCN)_6]^{2+}$ were replaced by dendritic ligands (see Figure 7.8) to give the cluster-supported dendrimer complexes shown in Figure 7.8.[52]

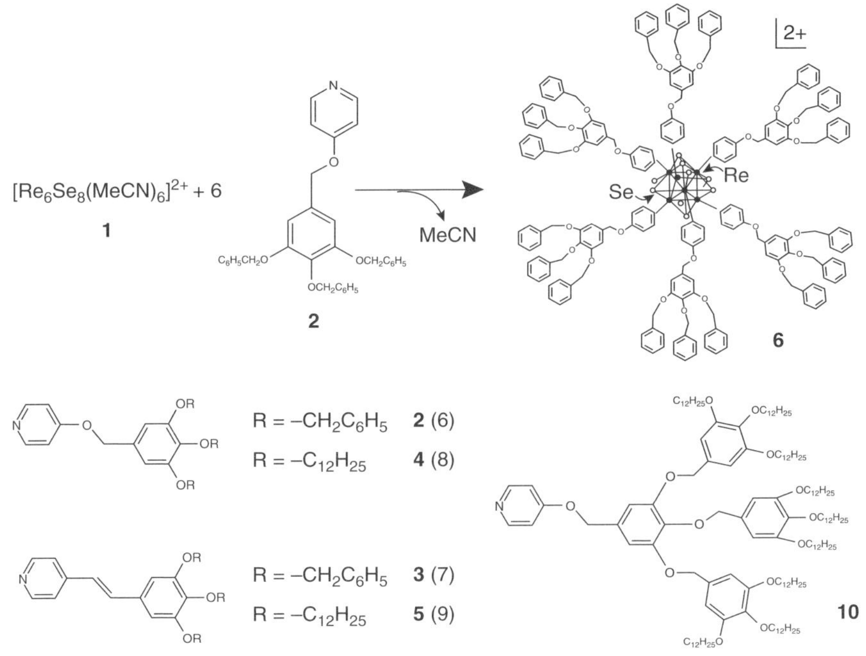

Figure 7.8 Synthesis of a dendrimer with a $[Re_6Se)_8]^{2+}$ core. (From Wang and Zheng.[52] Reprinted with permission. Copyright © 1999 American Chemical Society.)

The point was made that polynuclear clusters, for example, the $[Re_6Se_8(MeCN)_6]^{2+}$ cluster, make superb scaffolds for dendrimer constructions. One reason is the rich variety of magnetic, electrochemical, and photophysical properties that clusters possess and that may be imparted to the dendrimer materials. A second reason is that the octahedral disposition of the six metal sites (in the case of the $[Re_6Se_8(MeCN)_6]^{2+}$ cluster) reduces the number of structural defects due to steric congestion at higher dendritic generations. Other reasons include the ease of synthesis by coordination polymerization techniques and the existence of multiple coordination sites, which presents the opportunity for various degrees of ligand substitution, and hence, property tunability.

In another example of a dendrimer with a metal cluster core, Gorman and co-workers reported that dendritic ligands with phenoxide groups substituted for triflate or methoxide ligands around an Mo_6Cl_8 core to form molecules of the type $Mo_6(\mu_3\text{-}Cl)_8(OR)_6$, where R = dendrons with zero to two hyperbranches.[53]

It is noted that a growing number of dendrimers and hyperbranched polymers have been synthesized that have metal–metal bonded dimers or metal clusters attached to the surface of the dendrimer. These materials are analogous to polymers that have metal–metal units appended to the polymer backbone rather than incorporated into the backbone chain, and as such they are not covered in this chapter. The field was recently reviewed by Rossell and co-workers.[54] The interest in these materials stems from the catalytic properties of the metal complexes attached to the dendrimer surface. In particular, these dendrimer species offer the opportunity to combine the advantages of homogeneous and heterogeneous catalysis, because the materials can be recycled by nanofiltration.

V. POLYMERS WITH INFINITE METAL CHAINS

There is considerable interest[55] in 1-D solids consisting of linear chains of transition metal atoms because it is thought that these materials will be useful for electronics applications, for example, as conducting molecular wires. In particular, systems with unsupported metal–metal interactions are suggested to be particularly promising. Examples of infinite metal–metal-based chains are very rare, and this has encouraged considerable new research into these materials. Examples of infinite metal–metal-based chains can be divided into two categories. One category consists of infinite 1-D mixed-valence rhodium chains prepared by reduction of binuclear complexes. Examples include the mixed-valence compounds $\{[Rh(MeCN)_4](BF_4)_{1.5}\}_n$ and $\{[Rh_2(\mu\text{-}O_2CMe)_2(L)_2](BX_4)dotH_2O\}_n$ (L = bipyridine or phenanthroline derivatives; X = F, Ph) (Eq. 7.40).[56,57]

$$[Rh_2(MeCN)_{10}][BF_4]_4 \xrightarrow[CH_3CN]{e^- \text{ (Pt electrode)}} [Rh(MeCN)_4(BF_4)_{1.5}]_x \qquad (7.40)$$

The second category consists of materials with the general formula $[M^0(L)(CO)_2]_n$, where M = Ru or Os.[55,58] In a typical example with M = Ru and L = bipyridine, the Ru–Ru bond distance is 2.95 Å.

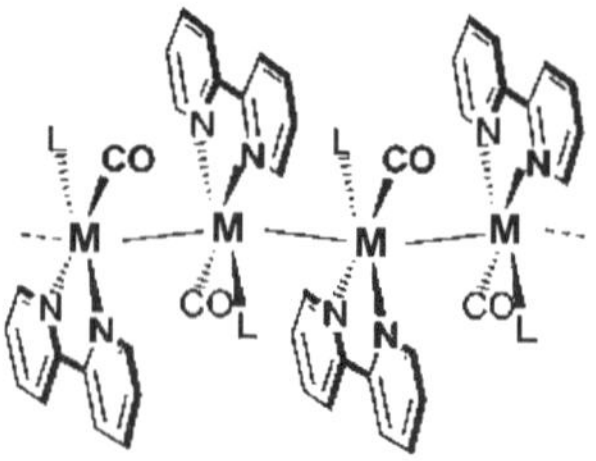

Polymers in the second category are formed by electrochemical reduction of appropriate precursor molecules:[55]

$$n[\mathrm{Ru^{II}(L)(CO)_2Cl_2}] + 2ne^- \longrightarrow [\mathrm{Ru^0(L)(CO)_2}]_n + 2n\mathrm{Cl}^- \quad (7.41)$$

The mechanism of polymer formation is suggested to follow the route shown in Scheme 7.7.[55] In the case of the analogous Os complexes, several of the intermediates in this scheme were long-lived enough to be spectroscopically characterized.

$$[\mathrm{Ru^{II}(L)(CO)_2Cl_2}] + e^- \longrightarrow [\mathrm{Ru^{II}(L^{\cdot -})(CO)_2Cl_2}]^-$$

$$[\mathrm{Ru^{II}(L^{\cdot -})(CO)_2Cl_2}]^- \longrightarrow [\mathrm{Ru^{II}(L^{\cdot -})(CO)_2Cl}] + \mathrm{Cl}^-$$

$$[\mathrm{Ru^{II}(L^{\cdot -})(CO)_2Cl}] \longleftrightarrow [\mathrm{Ru^{I}(L)(CO)_2Cl}]$$

$$2[\mathrm{Ru^{I}(L)(CO)_2Cl}] \longrightarrow \text{dimer} \longrightarrow \longrightarrow$$

$$\text{tetramer} \longrightarrow \longrightarrow [\mathrm{Ru^0(L)(CO)_2}]_n$$

Scheme 7.7 Proposed mechanism of $[\mathrm{Ru^0(L)(CO)_2}]_n$ formation by electrochemical reduction of $[\mathrm{Ru(L)(CO)_2Cl_2}]$ (L = bipyridine).

These polymers show particular promise as molecular wires because they are soluble, or at least soluble enough, in selected solvents so that there is some possibility they could be processed into appropriate shapes and form.

As of this writing, applications of these chains as conductive wires have not been reported, but these materials were shown to be potentially useful as catalysts in the electroreduction of CO_2 to CO in aqueous media[59] and in the water–gas shift reaction.[60] In the case of the water–gas shift reaction, the $[\mathrm{Ru(bpy)(CO)_2}]_n$ polymer was dispersed onto a SiO_2 support. The catalytic activity was low, however, and it was suggested that this was not necessarily due to poor catalytic activity of the polymer, but simply to poor dispersion of the polymer on the SiO_2, which kept the surface area of the catalyst low. For the electroreduction of CO_2, films of the polymer were generated on carbon electrodes. For films with a regular bpy ligand or having donor substituents, CO is the main reduction product. With films based on a bpy ligand disubstituted with electron-withdrawing groups like esters, the selectivity is changed and $HCOO^-$ is formed quantitatively. The proposed mechanism for CO_2 electroreduction is shown in Figure 7.9.

A problem in the practical application of these electrodes is their sensitivity to oxidation in air: molecular oxygen was found to induce oxidation of the polymer into monomeric Ru(II) species and disintegration of the cathode. To circumvent this problem, the polymer material was electrochemically generated within preformed films of polypyrole containing functionalized bipyridine ligands. The polypyrole apparently acts as a barrier to oxygen and the stability of

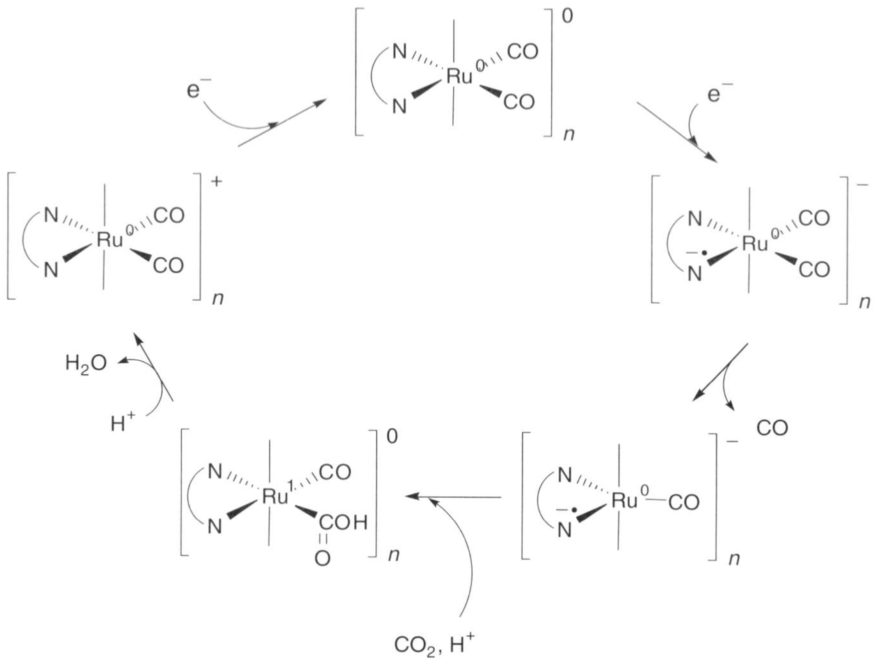

Figure 7.9 Proposed mechanism for CO_2 electroreduction using $[Ru^0(L)(CO)_2]_n$.

the metal–metal units in the polymer is increased accordingly. The proposed structure of the stabilized polymer is shown here.

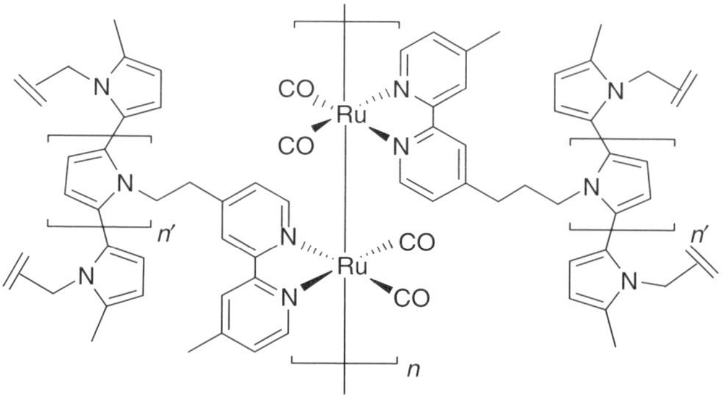

Finally, it is noted that an intriguing report suggests that the $[Ru(L)(CO)_2]_n$ polymers [L = bpy (2,2′-bipyridine) or dmbpy (4,4′-dimethyl-2,2′-bipyridine)] are reactive to light.[61] Irradiation of an electrode on which one of these polymers has been deposited results in a marked increase in the current. When the light is switched off, the current returns to its original value. The implication is that the polymer can act as a conductive wire, but this phenomenon remains to be thoroughly investigated.

VI. MISCELLANEOUS POLYMERS WITH METAL–METAL BONDS ALONG THE BACKBONE CHAIN

1. McArdle and co-workers proposed a general synthetic strategy for incorporating metal–metal bonds into polymer backbones.[62] The route is outlined in Scheme 7.8. The key strategy in this scheme is to start with a metal–dimer system that has linked cyclopentadienyl rings. In the first step, the metal–metal bond is cleaved, either by reaction with iodine to produce the diiodide species or by reduction to form the dianion. In the second step, metal–metal bonds are reformed by reaction of the anion with, say, $SnPh_2Cl_2$, or in a metal–metal bond–forming reaction with the tethered metal–iodide unit (with concomitant formation of I^-). Using this strategy, the polymer shown in Scheme 7.9 was formed. The polymer formed in this reaction had molecular weights around 8000 to 8500. The synthetic route in Scheme 7.8 is conceptually elegant, but in practice the polymers formed using this route contain impurities. No subsequent reports on polymers synthesized by this route have appeared, and this may be an indication that this route to metal–metal bond–containing polymers is not as promising as it once appeared.

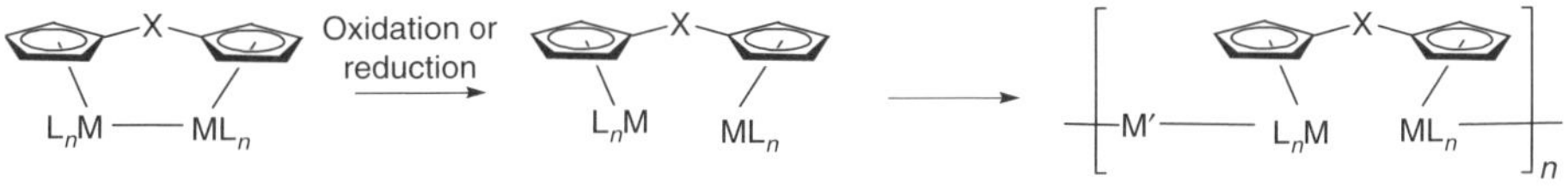

Scheme 7.8 General synthetic strategy for incorporating metal–metal bonds into polymer backbones using oxidation or reduction and salt elimination reactivity. See text for details about the oxidation and reduction reactions and about the salt elimination reactions.

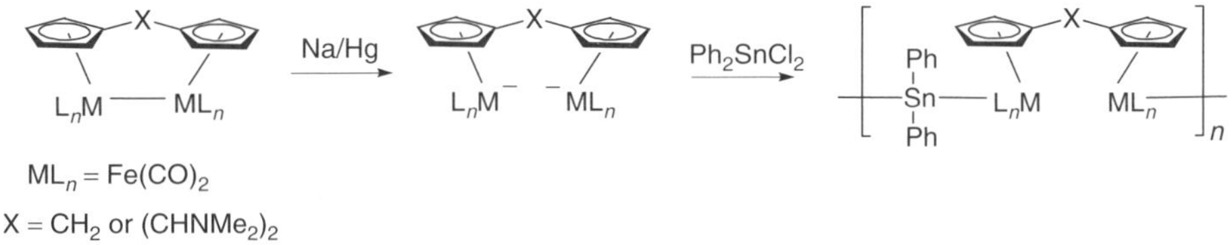

Scheme 7.9 Specific example of the general polymerization scheme depicted in Scheme 7.8.

2. Modec and co-workers used the oxalate ligand as a tetradentate ligand to link together $\{Mo_2O_2(\mu\text{-}O)_2\}^{2+}$ units to form polymers.[63] (Note the $\{Mo_2O_2(\mu\text{-}O)_2\}^{2+}$ unit has a Mo–Mo bond.) The hydrothermal synthetic route used $(pyH)_2[MoOCl_5]$ as a starting material in the presence of oxalic acid and pyridines. The anionic polymer products contained alternating pairs of $\{Mo_2O_2(\mu\text{-}O)_2\}^{2+}$ units containing edge-sharing octahedra and planar oxalates. The bisbidentate oxalate ligands act as bridges between two of the dinuclear units. A section of the polymer chain is illustrated in Figure 7.10. For completeness, it is noted that extended solid-state structures containing $\{Mo_2O_2(\mu\text{-}O)_2\}^{2+}$ are known for a variety of molybdenum phosphate materials.[64]

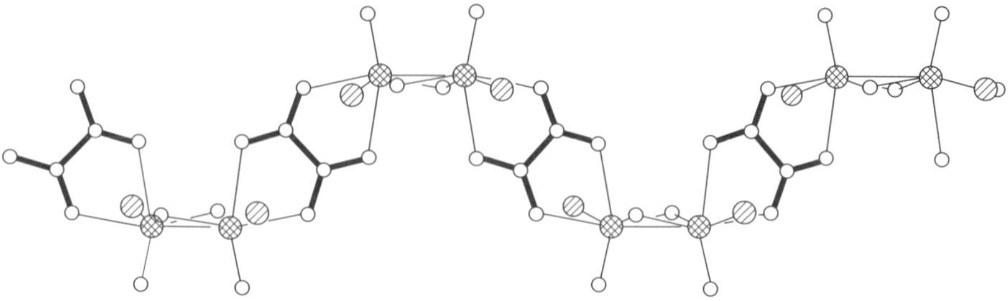

Figure 7.10 A section of a single $[Mo_2O_4(C_2O_4)Cl_2]_n^{2n-}$ chain. Mo sites are crosshatched; chlorides are lined bottom left to top right; oxygen atoms are unshaded and carbon atoms shaded, small-sized spheres, respectively. (From Modec et al.[63] Reprinted with permission. Copyright © 2002 Royal Society of Chemistry.)

3. Connelly and co-workers reported that thermolysis of organometallic complex, **6**, resulted in the formation of polymer **7**.[65] The polymer is insoluble, but it reacted with iodine to cleave the Fe–Fe bonds and form Fe–I bonds.

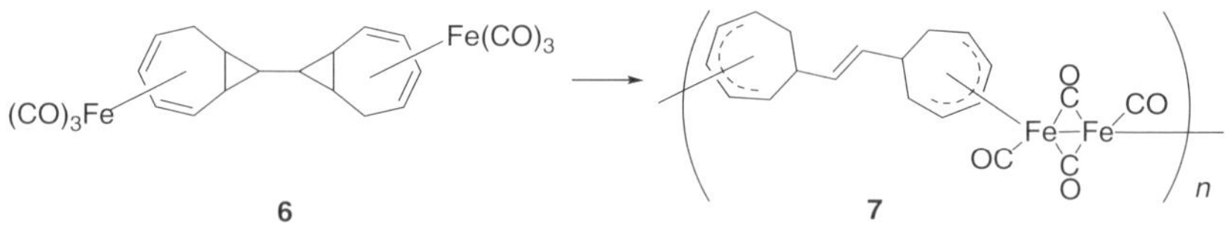

6 **7** (7.42)

VII. OUTLOOK

A. Synthetic Methods

As documented in the preceding paragraphs, several general strategies have been developed for the synthesis of polymers with metal–metal bonds. Certainly, the synthesis of these materials by step-growth polymerization techniques is now well developed, as is the technique of linking metal–metal bonded units with bidentate phosphines, diisocyanides, diacetylides, and other bidentate ligands. Less developed and generalized are: (1) most organometallic routes (e.g., Scheme 7.8); (2) the reduction routes used to synthesize 1-D infinite metal–metal bond–containing chains; and (3) chain-growth polymerization techniques. Research to explore the scope and limitations in these three areas is certain to continue.

A number of interesting synthetic routes have been reported in the literature to yield short-chain oligomers; extension of these routes to true polymer materials may yet be possible. An example is the reaction of $Os(CO)_4^{2-}$ with $SnPh_2Cl_2$.[66] In this case, a 12-membered ring with the formula $[Os(CO)_4SnPh_2]_6$ formed as the major product rather than a polymer, but further study may find a method for obtaining a polymer. (Note that the analogous $[Fe(CO)_4Bi(^nBu)]_n$ polymer exists,[67] but it was not prepared by a reaction analogous to the $Os(CO)_4{}^{2-}/SnPh_2Cl_2$ ring-forming reaction.)

The synthetic routes discussed in this chapter generally involved metal units that contained two metal centers. In principle, many of these routes can also be used with metal clusters. Although a few reactions involving metal clusters were cited previously, the incorporation of metal clusters into polymers has not been extensively studied. The discovery of intriguing or useful properties in these materials would almost certainly intensify research activity in this area.

Two final points regarding synthetic methods are the following: (1) synthetic routes making use of ring-opening polymerization (ROP) or ring-opening metathesis polymerization (ROMP) methods have not been applied to polymers containing metal–metal bonds; and (2) dendrimers that contain metal–metal bonds within their branches are rare. Both areas should be explored. (As discussed in a previous section, a small number of dendrimers have been synthesized that contain a metal cluster at the interior core or that have metal clusters attached to the surface of the dendrimer. However, these materials are not common.)

B. Applications

Numerous papers cite potential uses for metal–metal bond–containing polymers in the areas of nonlinear optics, magnetic materials, conducting polymers, liquid crystals, and catalysis. However, given the small number of applications actually reported, it is fair to say that applications of metal–metal-bonded polymers are still in the early stages of development.[68,69] While applications will undoubtedly be found for these materials, the real question is whether the metal–metal bonds will impart any special qualities to these applications beyond those found in polymers containing mononuclear metals or polymers containing no metals. For these materials to have maximum impact, it is suggested that the search for applications focus on areas where the properties of the metal–metal bonds can be exploited in a polymer setting. To illustrate, consider the following. If it is known, for example, that a particular metal–metal-bonded complex has certain nonlinear optical properties, then it will likely be worthwhile to investigate the nonlinear optical properties of polymers containing that metal–metal-bonded species. In the absence of such data, it is unlikely that simply incorporating metal–metal bonds into a polymer will endow the polymer with new and unusual properties. The point is that research in the applications area is best focused on exploiting the properties that metal–metal bonds are known to have. An example discussed in this chapter (and admittedly from the author's laboratory) is the ability of metal–metal bond–containing polymers to act as model polymers in a mechanistic study of polymer photodegradation. This study took advantage of the known electronic absorption properties of metal–metal bonds to allow straightforward measurements of photodegradation quantum yields by UV-vis (vis = visible) monitoring rather than by the more laborious methods using molecular-weight measurements that had been used in the past. In summary, the suggestion is that polymers with metal–metal bonds will initially find the most applications in areas where the properties of the metal–metal bonds can be exploited in a polymer setting.

ACKNOWLEDGMENT

The author thanks the National Science Foundation for support of the author's research mentioned herein.

REFERENCES

1. O. Crespo, A. Laguna, E. J. Fernandez, J. M. Lopez-de-Luzuriaga, A. Mendia, M. Monge, E. Olmos, P. G. Jones, *J. Chem. Soc., Chem. Commun.*, 2233 (1998).
2. H. Miyasaka, C. Kachi-Terajima, T. Ishii, M. Yamashita, *J. Chem. Soc., Dalton Trans.*, 1929 (2001).

3. C. Song, Y. Z. Li, Y. Q. Chen, S. J. Xiao, X. Z. You, *Acta Crystallogr., Sect. E: Struct. Rep. Online, E60*, m1741–m1743 (2004).
4. S. C. Tenhaeff, D. R. Tyler, *Organometallics*, **10**, 473 (1991).
5. S. C. Tenhaeff, D. R. Tyler, *Organometallics*, **11**, 1466 (1992).
6. S. C. Tenhaeff, D. R. Tyler, *Organometallics*, **10**, 1116 (1991).
7. D. R. Tyler, *Coord. Chem. Rev.*, **246**, 291 (2003).
8. G. F. Nieckarz, J. J. Litty, D. R. Tyler, *J. Organomet. Chem.*, **554**, 19 (1998).
9. G. F. Nieckarz, D. R. Tyler, *Inorg. Chim. Acta*, **242**, 303 (1996).
10. A. E. Stiegman, D. R. Tyler, *Coord. Chem. Rev.*, **63**, 217 (1985).
11. R. Chen, J. Meloy, B. C. Daglen, D. R. Tyler, *Organometallics*, **24**, 1495 (2005).
12. M. Moran, M. C. Pascual, I. Cuadrado, J. Losada, *Organometallics*, **12**, 811 (1993).
13. N. T. Lucas, M. G. Humphrey, A. D. Rae, *Macromolecules*, **34**, 6188 (2001).
14. T. J. Meyer, J. V. Caspar, *Chem. Rev. (Washington, DC)*, **85**, 187 (1985).
15. G. L. Geoffroy, M. S. Wrighton, *Organometallic Photochemistry*; Academic Press, New York, 1979.
16. R. Chen, M. Yoon, A. Smalley, D. C. Johnson, D. R. Tyler, *J. Am. Chem. Soc.*, **126**, 3054 (2004).
17. R. Chen, D. R. Tyler, *Macromolecules*, **37**, 5430 (2004).
18. D. R. Tyler, *J. Macromol. Sci. Polym. Rev.*, **44**, 351 (2004).
19. L. Tong, J. R. White, *Polym. Degrad. Stab.*, **53**, 381 (1996).
20. N. Grassie, G. Scott, *Polymer Degradation and Stabilization*; Cambridge University Press, New York, 1985.
21. J. Guillet, *Polymer Photophysics and Photochemistry: An Introduction to the Study of Photoprocesses in Macromolecules*, Cambridge University Press, New York, 1985.
22. G. Geuskens, *Compr. Chem. Kinet.*, **14**, 333 (1975).
23. T. R. O'Toole, B. P. Sullivan, T. J. Meyer, *J. Am. Chem. Soc.*, **111**, 5699 (1989).
24. M. J. Irwin, G. Jia, J. J. Vittal, R. J. Puddephatt, *Organometallics*, **15**, 5321 (1996).
25. S. Sicard, J.-F. Berube, D. Samar, A. Messaoudi, D. Fortin, F. Lebrun, J.-F. Fortin, A. Decken, P. D. Harvey, *Inorg. Chem.*, **43**, 5321 (2004).
26. E. Fournier, S. Sicard, A. Decken, P. D. Harvey, *Inorg. Chem.*, **43**, 1491 (2004).
27. P. D. Harvey, *Macromol. Symp.*, **196**, 173 (2003).
28. T. Zhang, M. Drouin, P. D. Harvey, *Inorg. Chem.*, **38**, 1305 (1999).
29. T. Zhang, M. Drouin, P. D. Harvey, *Inorg. Chem.*, **38**, 957 (1999).
30. P. D. Harvey, *Coord. Chem. Rev.*, **219–221**, 17 (2001).
31. B. F. G. Johnson, K. M. Sanderson, D. S. Shephard, D. Ozkaya, W. Zhou, H. Ahmed, M. D. R. Thomas, L. Gladden, M. Mantle, *J. Chem. Soc., Chem. Commun.*, 1317 (2000).
32. S. Aime, R. Gobetto, G. Jannon, D. Osella, *J. Organomet. Chem.*, **309**, C51 (1986).
33. R. Fuchs, P. Kluefers, *J. Organomet. Chem.*, **424**, 353 (1992).
34. A. M. Bradford, E. Kristof, M. Rashidi, D.-S. Yang, N. C. Payne, R. J. Puddephatt, *Inorg. Chem.*, **33**, 2355 (1994).
35. H. D. Selby, P. Orto, Z. Zheng, *Polyhedron*, **22**, 2999 (2003).
36. H. D. Selby, B. K. Roland, Z. Zheng, *Acc. Chem. Res.*, **36**, 933 (2003).
37. H. D. Selby, Z. Zheng, T. G. Gray, R. H. Holm, *Inorg. Chim. Acta*, **312**, 205 (2001).
38. N. T. Lucas, E. G. A. Notaras, M. P. Cifuentes, M. G. Humphrey, *Organometallics*, **22**, 284 (2003).
39. E. G. A. Notaras, N. T. Lucas, M. G. Humphrey, A. C. Willis, A. D. Rae, *Organometallics*, **22**, 3659 (2003).
40. N. T. Lucas, E. G. A. Notaras, S. Petrie, R. Stranger, M. G. Humphrey, *Organometallics*, **22**, 708 (2003).

41. N. T. Lucas, E. G. A. Notaras, M. G. Humphrey, M. Samoc, B. Luther-Davies, *Proc. SPIE—The International Society for Optical Engineering*, **5212**, 318 (2003).
42. R. J. Puddephatt, *J. Chem. Soc., Chem. Commun.*, 1055 (1998).
43. J. S. Valentine, A. J. Silverstein, Z. G. Soos, *J. Am. Chem. Soc.*, **96**, 97 (1974).
44. F. A. Cotton, T. R. Felthouse, *Inorg. Chem.*, **19**, 328 (1980).
45. F. A. Cotton, T. R. Felthouse, *Inorg. Chem.*, **20**, 600 (1981).
46. M. C. Kerby, B. W. Eichhorn, J. A. Creighton, K. P. C. Vollhardt, *Inorg. Chem.*, **29**, 1319 (1990).
47. S. J. Sherlock, M. Cowie, E. Singleton, M. M. d. V. Steyn, *Organometallics*, **7**, 1663 (1988).
48. F. A. Cotton, C. Lin, C. A. Murillo, *Acc. Chem. Res.*, **34**, 759 (2001).
49. F. A. Cotton, C. Lin, C. A. Murillo, *J. Chem. Soc., Chem. Commun.*, 11 (2001).
50. F. A. Cotton, C. Lin, C. A. Murillo, *Inorg. Chem.*, **40**, 5886 (2001).
51. F. A. Cotton, C. Lin, C. A. Murillo, *J. Chem. Soc., Dalton Trans.*, 499 (2001).
52. R. Wang, Z. Zheng, *J. Am. Chem. Soc.*, **121**, 3549 (1999).
53. C. B. Gorman, W. Y. Su, H. Jiang, C. M. Watson, P. Boyle, *J. Chem. Soc., Chem. Commun.*, 877 (1999).
54. O. Rossell, M. Seco, I. Angurell, *C.R. Acad. Sci., Ser. IIc: Chim.*, **6**, 803 (2003).
55. F. Hartl, T. Mahabiersing, S. Chardon-Noblat, P. D. Costa, A. Deronzier, *Inorg. Chem.*, **43**, 7250 (2004).
56. M. E. Prater, L. E. Pence, R. Clerac, G. M. Finniss, C. Campana, P. Auban-Senzier, D. Jerome, E. Canadell, K. R. Dunbar, *J. Am. Chem. Soc.*, **121**, 8005 (1999).
57. J. K. Bera, K. R. Dunbar, *Angew. Chem., Int. Ed. Engl.*, **41**, 4453 (2002).
58. S. Chardon-Noblat, G. H. Cripps, A. Deronzier, J. S. Field, S. Gouws, R. J. Haines, F. Southway, *Organometallics*, **20**, 1668 (2001).
59. S. Chardon-Noblat, A. Deronzier, R. Ziessel, D. Zsoldos, *J. Electroanal. Chem.*, **444**, 253 (1998).
60. S. Luukkanen, P. Homanen, M. Haukka, T. A. Pakkanen, A. Deronzier, S. Chardon-Noblat, D. Zsoldos, R. Ziessel, *Appl. Catal., A: Gen.*, **185**, 157 (1999).
61. E. Eskelinen, M. Haukka, T. Venaelaeinen, T. A. Pakkanen, M. Wasberg, S. Chardon-Noblat, A. Deronzier, *Organometallics*, **19**, 163 (2000).
62. P. McArdle, L. O'Neill, D. Cunningham, A. R. Manning, *J. Organomet. Chem.*, **524**, 289 (1996).
63. B. Modec, J. V. Brencic, D. Dolenc, J. Zubieta, *J. Chem. Soc., Dalton Trans.*, 4582 (2002).
64. R. C. Haushalter, L. A. Mundi, *Chem. Mater.*, **4**, 31 (1992).
65. N. G. Connelly, A. R. Lucy, R. M. Mills, J. B. Sheridan, P. Woodward, *J. Chem. Soc., Dalton Trans.*, 699 (1985).
66. W. K. Leong, R. K. Pomeroy, R. J. Batchelor, F. W. B. Einstein, C. F. Campana, *Organometallics*, **16**, 1079 (1997).
67. M. Shieh, Y. Liou, M.-H. Hsu, R.-T. Chen, S.-J. Yeh, S.-M. Peng, G.-H. Lee, *Angew. Chem. Int. Ed. Engl.*, **41**, 2384 (2002).
68. D. Woehrle, A. D. Pomogailo, (eds.), *Metal Complexes and Metals in Macromolecules*, Wiley-VCH, Weinheim, Germany 2003.
69. I. Manners, *Synthetic Metal-Containing Polymers*, Wiley, Hoboken, NJ, 2002.

CHAPTER 8

Structures and Properties of One-Dimensional Transition Metal-Containing Coordination/ Organometallic Polymers and Oligomers Built Upon Assembling Diphosphine and Diisocyanide Ligands

PIERRE D. HARVEY

Université de Sherbrooke, Sherbrooke, Québec, Canada

I. INTRODUCTION

The field of metal-containing polymers or coordination and organometallic polymers is experiencing an astonishing growth. Important applications are reported on a regular basis, including conducting, semi- and photoconducting and optical materials, catalysis, and nanomaterials. A myriad of reviews on these topics have been published in the past 3–4 years or so.[1–33] The most remarkable feature is that the construction of such metal–ligand-based materials are primarily centered on M–N and M–O coordinations.[1–33] On the other hand, the M–P and M–CN (isocyanide) bonds were exploited to a much lesser extent in this respect, and only short reviews or short sections discussing them in larger reviews have recently appeared.[34–44] This chapter is dedicated to such materials, focusing on recent works, but earlier works are also mentioned for the benefit of the reader, thus providing a better review of the field. The goal of this chapter is to provide the first exhaustive account of the recent development of one-dimensional (1-D) transition metal–containing coordination and organometallic polymers and oligomers (excluding dendrimers) built upon bridging dihosphine and diisocyanide ligands. Reports on polydentate isocyanides and phosphines. Two- and 3-D transition metal–containing coordination and

Frontiers in Transition Metal-Containing Polymers, edited by Alaa S. Abd-El-Aziz and Ian Manners.

organometallic polymers and oligomers and polymers built upon M···M interactions are not included in this chapter.

II. LIST OF BRIDGING LIGANDS

A list of common bridging ligands is provided for the convenience of the reader. This list is separated into four categories (alkyl diisocyanides, aryl diisocyanides, alkyl diphosphines, and unsaturated diphosphines (aryl, alkynyl, and alkenyl)).

Common alkyl diisocyanides:

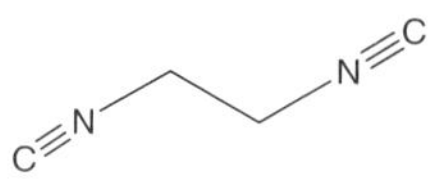

1,2-diisocyanoethane (1,2-die)

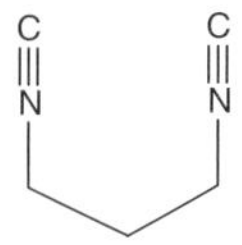

1,3-diisocyanopropane (1,3-dip)

1,4-diisocyanocyclohexane (1,4-dichx)

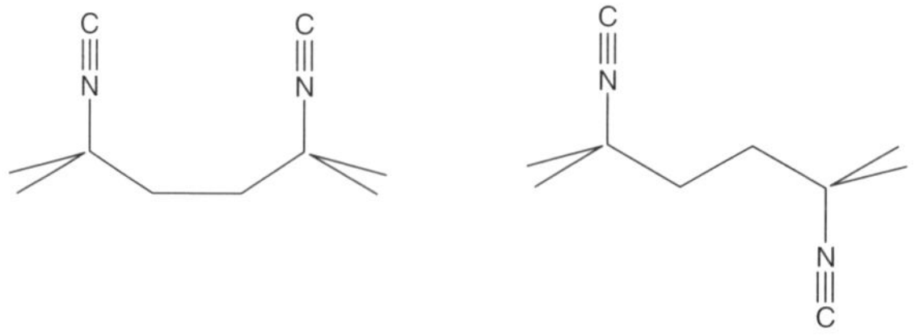

2,5-diisocyano-2′,5′-methylhexane (tmb); *syn* and *anti*

$C{\equiv}N{-}(CH_2)_6{-}N{\equiv}C$

1,6-diisocyanododecane (1,6-dinx)

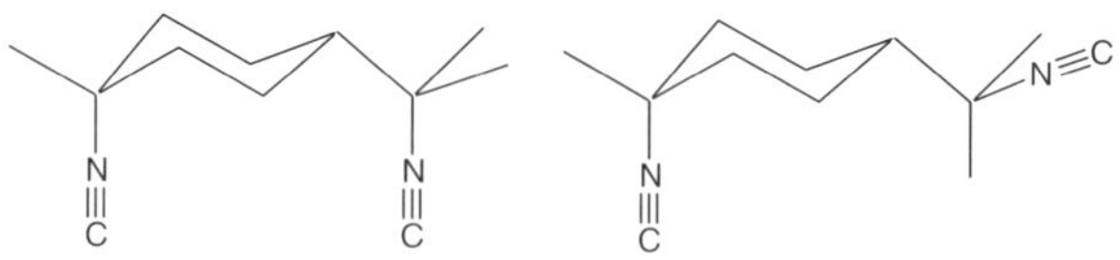

1,8-diisocyano-*p*-menthane (dmb); U- and Z-conformation

$C{\equiv}N{-}(CH_2)_{12}{-}N{\equiv}C$

1,12-diisocyanododecane (1,2-didod)

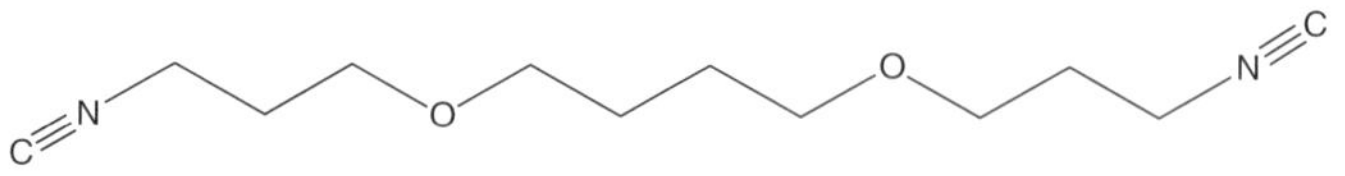

1,12-diisocyano-4,9-dioxadodecane (1,12-dioxdod)

Common aryl diisocyanides:

1,4-diisocyanobenzene (dib)

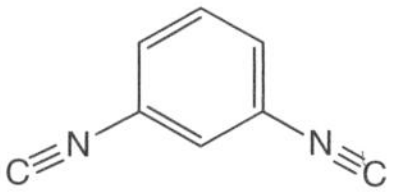

1,3-diisocyanobenzene (1,3-dib)

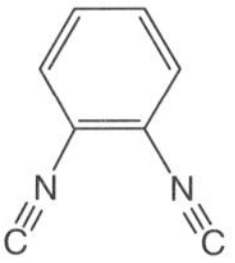

1,2-diisocyanobenzene (1,2-dib)

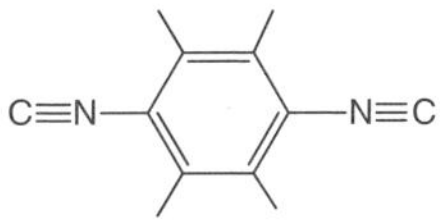

1,4-diisocyanotetramethylbenzene (Me_4dib)

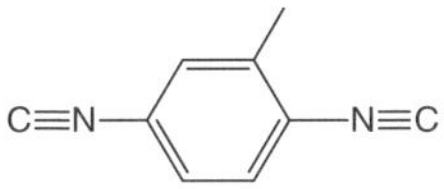

1,4-diisocyanomethylbenzene (Medib)

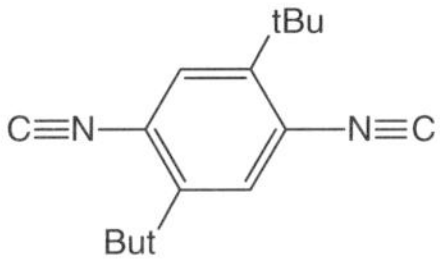

1,4-diisocyanodi-*t*-butylbenzene (*t*-Bu_2dib)

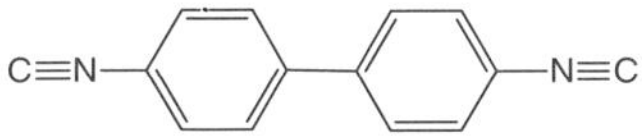

4,4′-diisocyanodiphenylene (Phdib)

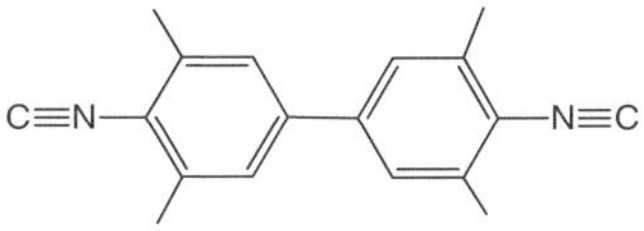

4,4′-diisocyano-3,3′,5, 5′-diphenylene(Me_4Phdib)

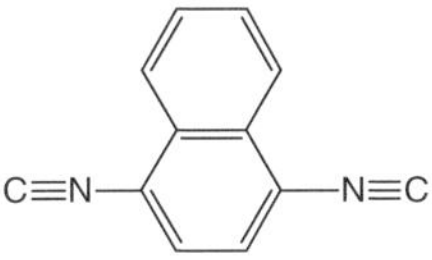

1,4-diisocyanonaphthalene (1,4-din)

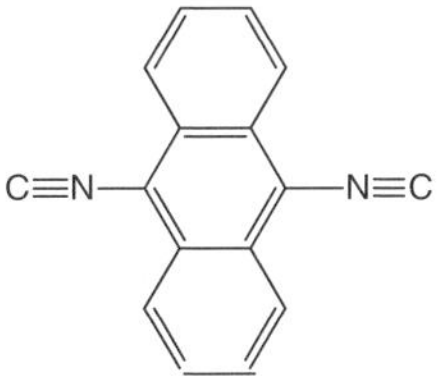

9,10-diisocyanoanthracene (9,10-din)

Common unsaturated diphosphines:

bis(diphenylphosphino)acetylene (dpa)

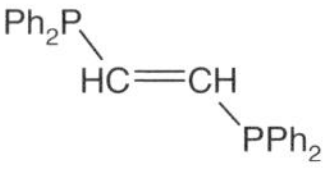

trans-bis(diphenylphosphino)ethylene (dppet)

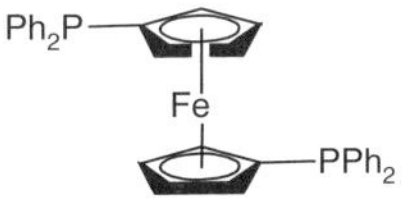

1,1′-bis(diphenylphosphanyl)ferrocene (dppf)

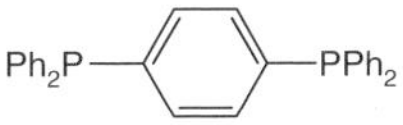

1,4-bis(diphenylphosphino) benzene

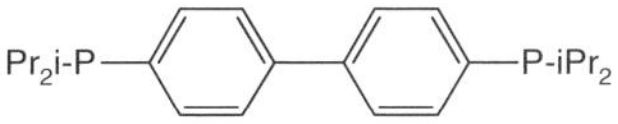

4,4′-bis(di-isopropylphosphino) diphenyl

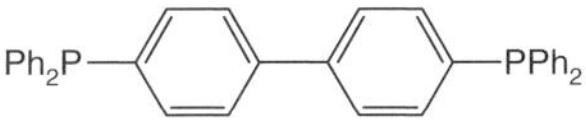

4,4′-bis(diphenylphosphino) diphenyl (dppbp)

Common alkyl diphosphines:

bis(diphenylphosphino)methane (dppm)

bis(dimethylphosphino)methane (dmpm)

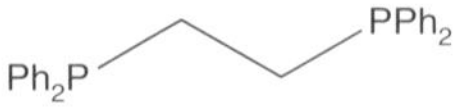

1,2-bis(diphenylphosphino)ethane (dppe)

1,3-bis(diphenylphosphino)propane (dppp)

1,4-bis(diphenylphosphino)butane (dppb)

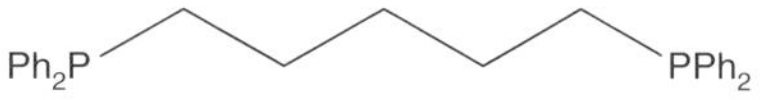

1,5-bis(diphenylphosphino)pentane (dpppen)

1,5-bis(diphenylphosphino)hexane (dpph)

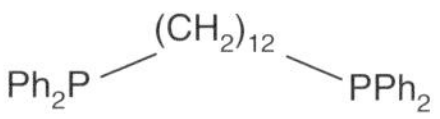

1,12-bis(diphenylphosphino)dodecane

III. SINGLY BRIDGED 1-D POLYMERS

A. Using Bridging Diisocyanides

Some well-characterized cyclic and acyclic oligomers of transition metal complexes were recently reported. For example, the addition of dmb to the dimeric $d^{10} - d^{10}$ species $Cu_2(dmpm)_3^{2+}$ and $Ag_2(dmpm)_2^{2+}$ in acetonitrile leads to oligomers **1** and **2** formulated as $\{Cu_2(dmpm)_3(dmb)_{1.33}^{2+}\}_n$ and $\{Ag_2(dmpm)_2(dmb)_{1.33}^{2+}\}_n$, respectively.[45] These oligomers were fully characterized by elemental analysis, infrared (IR), ^{1}H, and ^{31}P nuclear magnetic resonance (NMR) spectroscopy, thermo-gravimetric analysis (TGA), and X-ray powder diffraction (XRD). In both cases, the solid-state IR spectra exhibit the presence of uncoordinated C≡N— groups indicating short chains. The model compound $Cu_2(dmpm)_3(CN\text{–}t\text{-}Bu)_2^{2+}$ exhibits a Cu•••Cu distance of 4.011 Å (X-ray structure), indicating an absence of M•••M interactions.

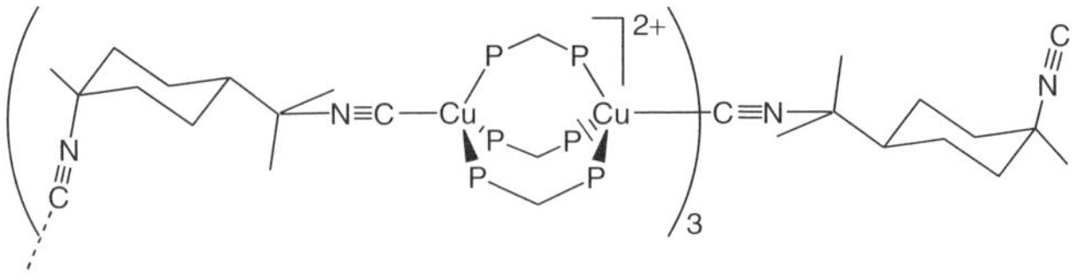

1 (P = PMe$_2$)

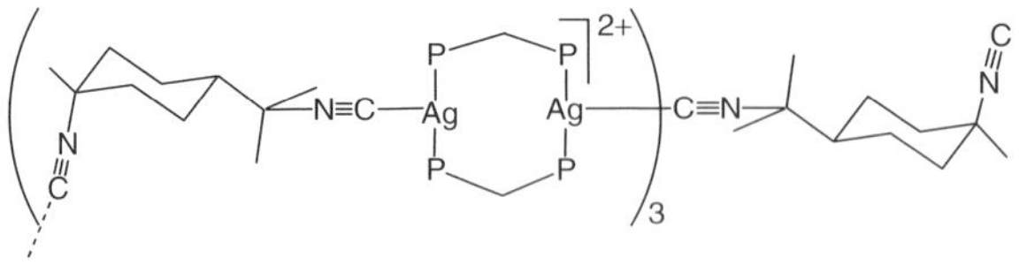

2 (P = PMe$_2$)

Oligomers **1** and **2** are luminescent and exhibit a strong luminescence centered at 482 and 447 nm, with emission lifetimes (τ_e) of 257 and 31 μs, respectively. These data compare favorably to that of the model compound $Cu_2(dmpm)_3(CN\text{–}t\text{-}Bu)_2^{2+}$ (for **1**) and the binuclear complex

$Ag_2(dmpm)_2^{2+}$ (also a model compound for **2**), for which the emission maximums and lifetimes are 476 and 445 nm, and 291 and 41 μs, respectively.

Other diisocyanide-containing oligomers/polymers were recently reported.[46] These new materials are based on the 5,11- and 5,17-diisocyano-25,26,27,28-tetra-*n*-propoxycalix[4]arene ligands and react with simple silver(I) salts to form oligomers/polymers **3** and **4**. Although no IR absorption associated with uncoordinated —N≡C groups were observed, the true nature of these amorphous materials (cyclic oligomer versus long polymer) was not elucidated due the extensive light sensitivity in the solution. Based on molecular modeling, dimeric species are not possible due to important ring stress. The materials proved to be very thermally unstable where weight losses were observed at temperatures as low as 90°C. In addition, no T_g is observed between −20 and 90°C (differential scanning calorimetry (DSC)). Both **3** and **4** were not luminescent in the solid state at room temperature.

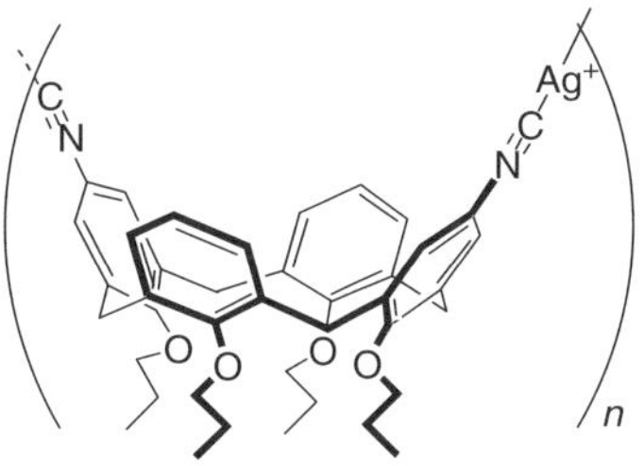

3

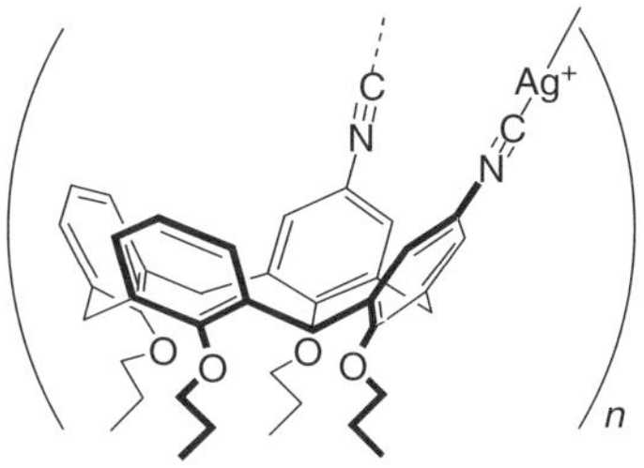

4

Recent examples of cyclic oligomers include 1,2-dib and platinum(II) centers as in **5** and **6**.[47] Oligomer **5** was characterized by X-ray crystallography, and the nonbonding Pt•••Pt distances are 5.65 Å. The C_6F_5 groups are more or less coplanar within the $Pt(C_6H_5)_2$ fragments, and the orientation of these fragments indicated that the molecule belongs to the C_3 point group.

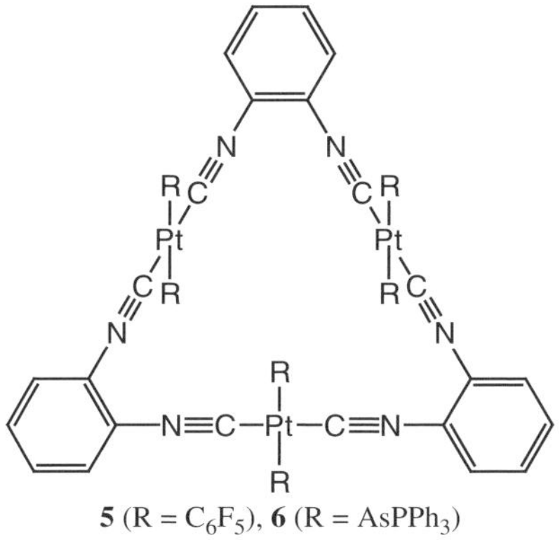

5 ($R = C_6F_5$), **6** ($R = AsPPh_3$)

Another triangular oligomer (**7**) built with diisocyanides involves the use of two dmb ligands.[48] This oligomer was characterized by X-ray crystallography, ultraviolet-visible (UV-vis), and Raman spectroscopy. The Ag•••Ag distances are 3.633(4), 3.760(4), and 2.805(4) Å for the two dmb-bridged fragments, and the μ_3-I_3Ag_2 residues, respectively. The solid-state Raman spectra exhibit a ν(Ag•••Ag) scattering at 122 cm^{-1}, which is associated with a force constant of ~0.47 mdyn Å^{-1}. Such a force constant corresponds to about one-half of a single metal–metal bond. The presence of strong non-bonding interactions is consistent the short Ag•••Ag separation.

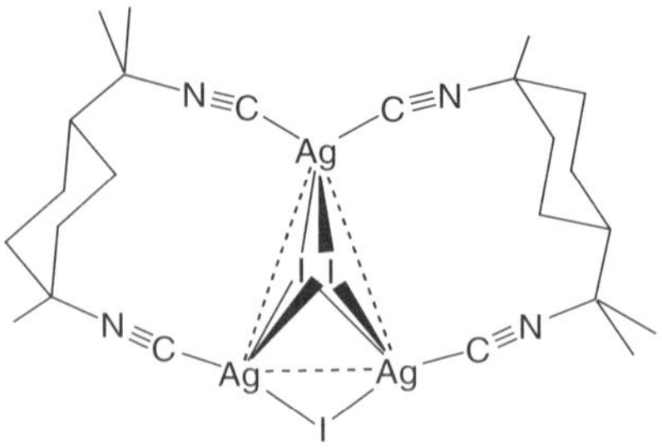

7

Other examples of cyclic oligomers are the mixed-ligand compounds **8–16**.[49,50] These oligomers are best described as “dimers of dimers” forming squares and rectangles. Their syntheses stem from the dimeric materials $[Cp^*MCl_2]_2$ (M = Rh, Ir; $Cp^* = C_5Me_5$) to which is

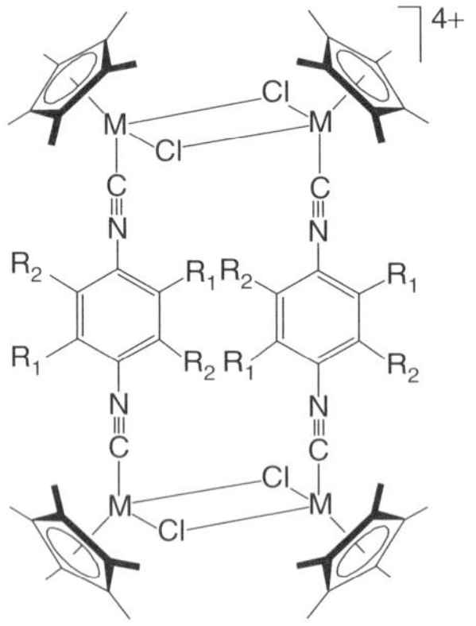

8 (M = Ir; R_1 = Me, R_2 = H)
9 (M = Ir; $R_1 = R_2$ = Me)
10 (M = Rh; $R_1 = R_2$ = Me)

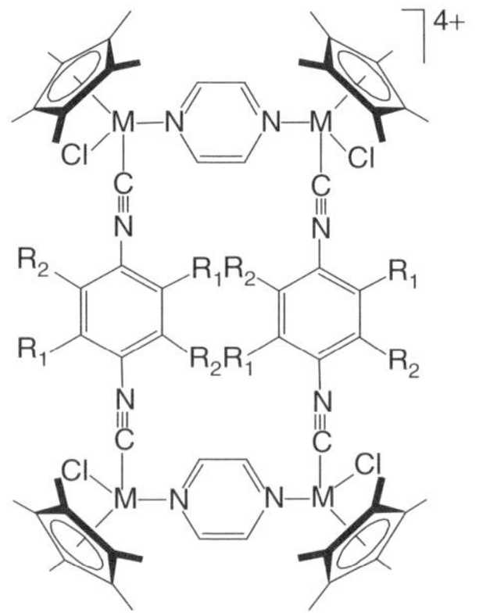

11 (M=Ir; R_1 = Me, R_2 = H)
12 (M = Ir; $R_1 = R_2$ = Me)
13 (M = Rh; $R_1 = R_2$ = Me)

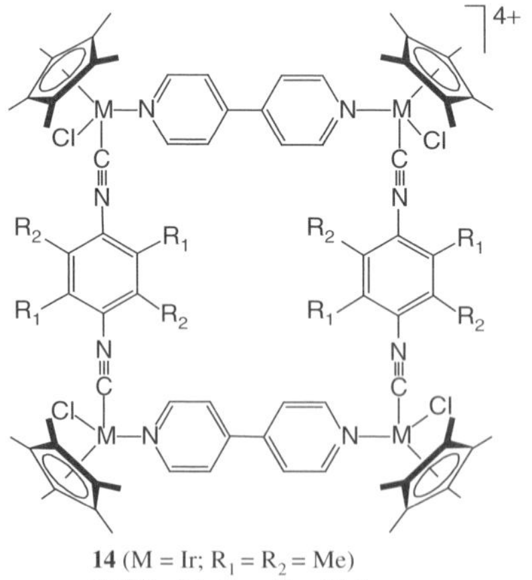

14 (M = Ir; $R_1 = R_2$ = Me)
15 (M = Rh; $R_1 = R_2$ = Me)

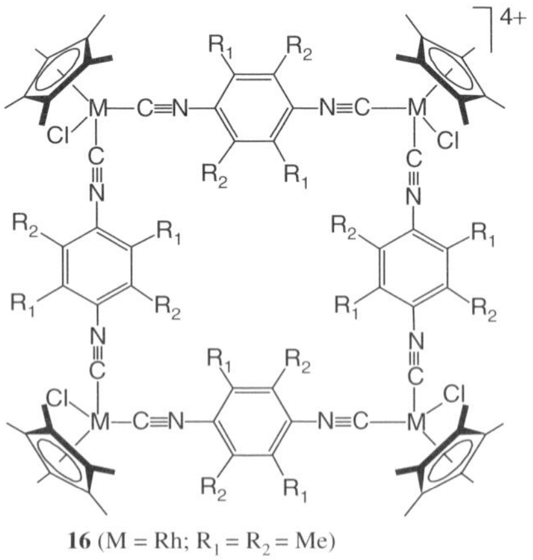

16 (M = Rh; $R_1 = R_2$ = Me)

added various bridging ligands such as pyrazine, 4,4′-pyrazine or a 4,4′-diisocyanodimethyl- or tetramethylbenzene, in appropriate stoichiometric amounts.

Oligomers **11** and **15** were characterized by XRD. These rectangles form large cavities where Ir•••Ir distances in **11** are 11.6 (diisocyanide) and 7.00 Å (pyrazine) with a diagonal distance of 13.6 Å. Ring-opening polymerization (ROP) was not investigated, but these cyclic oligomers provide a good opportunity to do so. Zigzag structures are anticipated due to the presence of M–L single bonds, where free rotation can take place.

Several singly bridged polymers using diisocyanides were also reported. Earlier examples are polymers **17** and **18**, which are prepared from the reaction of diruthenium precursors **19** and **20** with 4,4′-diisocyanotetramethylbenzene.[51] These insoluble polymers were characterized by IR in the $\nu(C{\equiv}O)$ region in comparison with the model compounds. There was no evidence of the presence of uncoordinated C≡N— groups, so the polymers are probably very long.

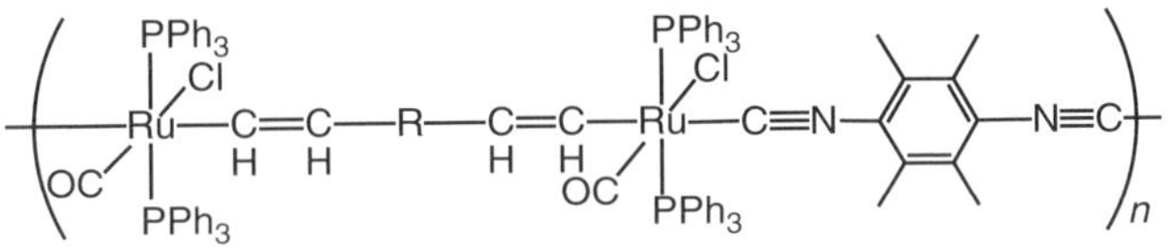

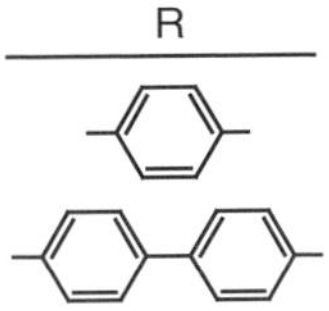

17

18

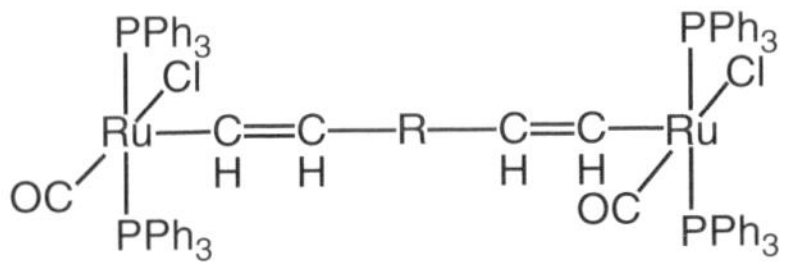

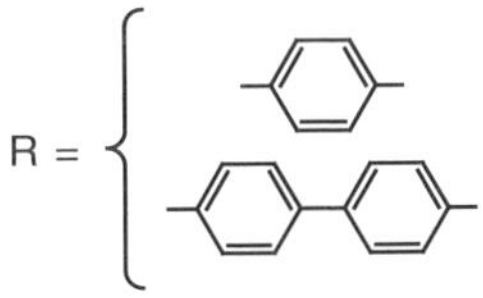

19

20

More recent examples of diisocyanide-containing polymers are **21–24**.[52] These materials are synthesized by reacting dmb with the corresponding $[M(dppe)(BF_4)]$ and $[M_2(dppp)_2](BF_4)_2$ (M = Cu, Ag) in acetonitrile. These amorphous materials were characterized by NMR, IR, TGA, XRD, and DSC. The absence of a $\nu(C{\equiv}N)$ signal associated with uncoordinated C≡N— groups in the solid state IR spectra suggests long polymers, but the delays of 1–3 seconds of the polymer solutions against the pure solvent (acetonitrile) from the measurements of the intrinsic viscosity, provides evidence of oligomers in solution. The number of units (~8–9 for M = Cu (**21** and **23**), were too small to be measured for M = Ag) and were approximated using PMMA standards (polymethylmethacrylate). The method was checked against a known polymer $(Ag(dmb)_2^+)_n$ (M_w ~ 4000 (7–9 units), which is discussed below). The fact that these materials behave as long polymers in the solid state and short oligomers in solution is a common property for diisocyanide- and diphosphine-containing polymers, and is noted several times in this chapter. Polymer **21** also exhibits a T_g at 82°C (ΔC_p = 0.43 J/(g·dep)).

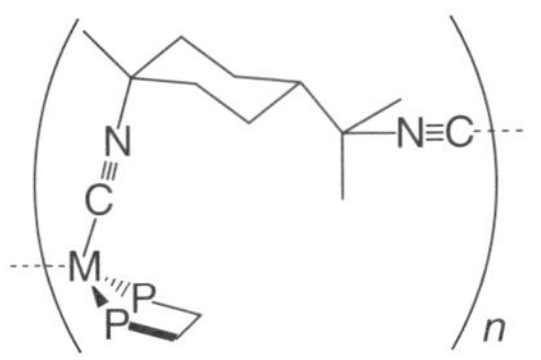

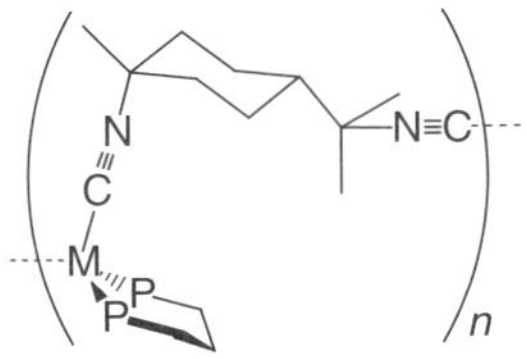

P = PPh$_2$; **20** (M = Cu), **21** (M = Ag) **22** (M = Cu), 23 (M = Ag)

These polymers are luminescent in the solid state, exhibiting emission maximums between 480 and 548 nm, and τ_e between 22 and 48 μs. These emissions are assigned to ^{3}MLCT (metal-to-ligand charge transfer).

Similarly, the reaction between the silver tetrafluoroborate and silver nitrate with the assembling ligand tmb leads to corresponding polymers **25** and **26**.[53] Weak anion–silver interactions were depicted based on shorter interatomic distances measured in the X-ray structures.

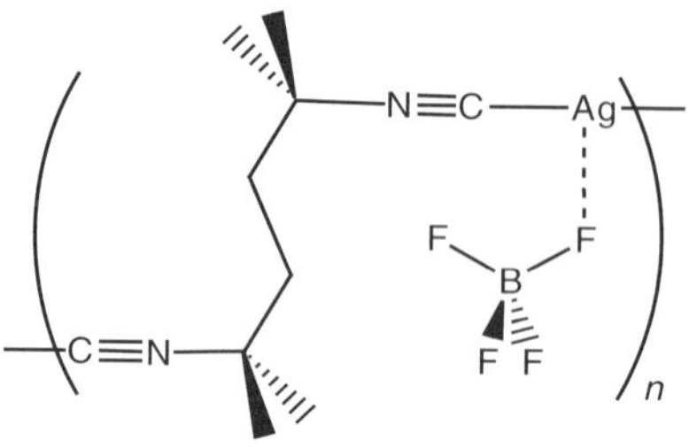

25

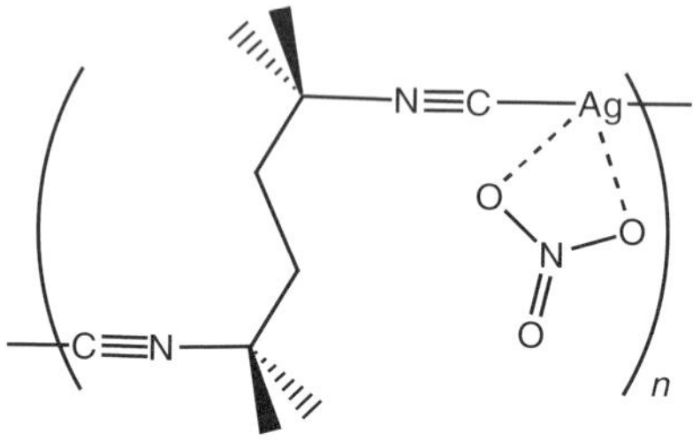

26

Another important class of diisocyanide-containing polymers is the "shish-kebab" structured polymers of porphyrin and porphryin-like materials. M. Hanack (University of Tübingen) explored this area in detail from 1983 to 2000. The most recent review is found in the *Porphyrn Handbook* published in 2003,[44] so only a summary is presented in this chapter and no numbering for the oligomers/polymers is assigned. The bulk of the references are still provided for the convenience of the readers.[54–70] Some other works from other groups also appeared.[71,72]

The general structure of the polymers is shown below (the square represents the macrocycle and L is the diisocyanide bridging ligand), and their syntheses involve reacting a given metallo-macrocycle with a desired bridging ligand. The polymer precipitates due to relatively weak solubility, and characterization is often limited to the solid state.

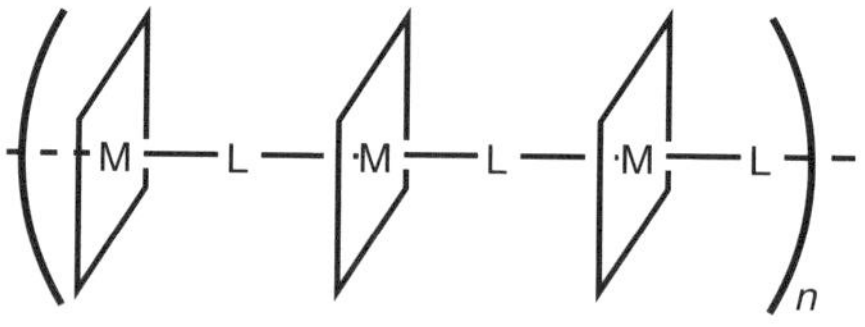

The list of many relevant oligomers and polymers defining the metal, macrocycle, substituents on the macrocycle and bridging ligand is provided in Table 8.1. The drawings of the various macrocycles are shown below.

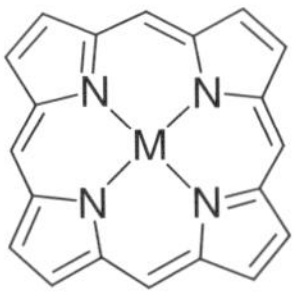

Metal–pophyrine

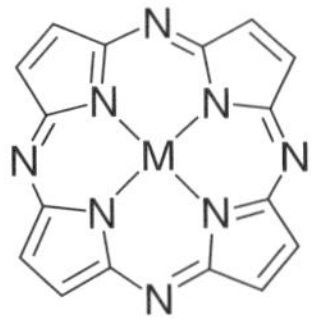

Metal–phthalocyanine

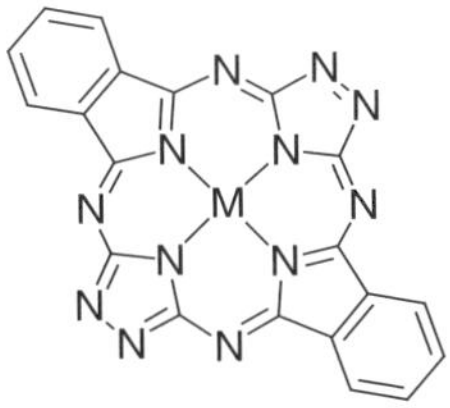

Metal–triazolehemiporphyrazine

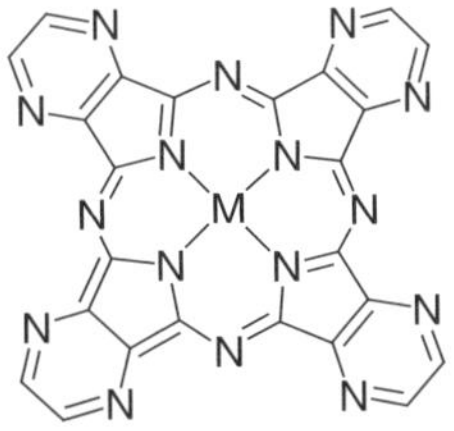

Metal–tetraazaporphyrin

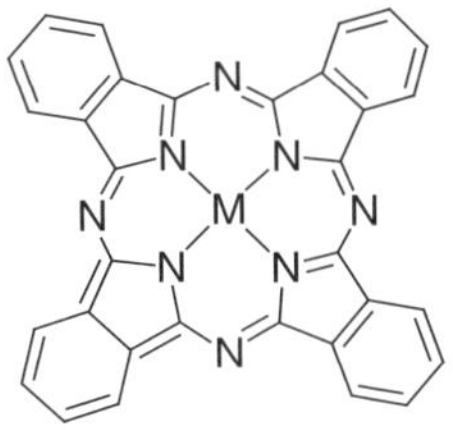

Metal–phthalocyanine

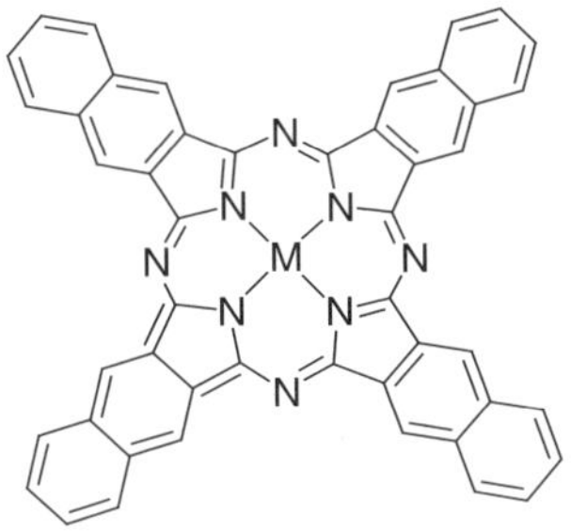

Metal–naphthalocyanine

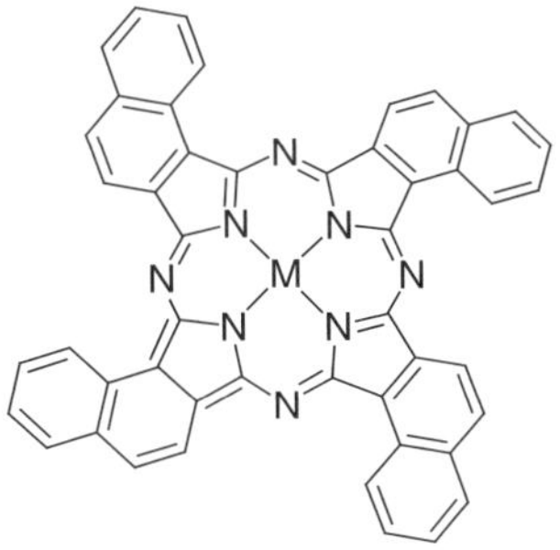

Metal–2,3-naphthalocyanine

TABLE 8.1 List of Some Rigid-Rod Porphyrine- and Prophyrine-Like–Containing Oligomers and Polymers

Metal	Macrocycle	R Groups	Bridging Ligand	Reference
Co	Phthalocyanine	H	dib	54
Fe, Ru, Co	Phthalocyanine	H	dib	55
	Naphthalocyanine	—	—	
	Porphyrine	meso-Ph	—	
Fe	Phthalocyanine	*t*-Bu, 2-Et-C_6H_{12}	Me_4dib	56
Fe	1,2-Naphthalocyanine	Me, *t*-Bu	dib	57
Fe, Ru	Phthalocyanine	*t*-Bu	dib, Me_4dib	58
—	Tetrapyraazaporphyrazine	—	— —	
Fe, Pd, Pt	Phthalocyanine	O-*n*-C_5H_{11},	Me_4dib	59
—	—	O-*n*-C_8H_{17},	—	
		O-2-Et-C_6H_{12}		
Fe	Phthalocyanine	H	1,2-die, 1,6-dinx,	60
—	—	—	1,2-didod,	
—	—	—	1,2-dioxdod,	
			1,4-dichx	
Fe	Phthalocyanine	*n*-C_7H_{15}, *n*-C_5H_{11},	dib	61
		n-C_6H_{13}		

(Continued)

TABLE 8.1 (*Continued*)

Metal	Macrocycle	R Groups	Bridging Ligand	Reference
—	2,3-Naphthalocyanine	—	—	
Ru	2,3-Naphthalocyanine	H	dib, Me_4dib	62
Fe, Ru	Triazolehemiporphyrazine	$C_{12}H_{25}$ and $OC_{12}H_{25}$	dib	63
Fe	2,3-Naphthalocyanine	H	1,2-die,	64
—	—	—	1,6-dinx, 1,2-didod,	
—	—	—	1,2-dioxdod,	
—	—	—	1,4-dichx, 1,4-din, 9,10-dia	
Ru	Phthalocyanine	O-C_5H_{11},	dib	65
—	—	O-C_5H_{11}, O-2-Et-hex	Me_4dib	
Ru	Tetrapyraazaporphyrazine	Ph	Me_4dib	66
Fe	Phthalocyanine	Ph	Me_4dib	67
Ru	Phthalocyanine	H	Me_4dib, dib,	68
—	—	—	Me_2dib, t-Bu_2dib	
Fe	Phthalocyanine	4-C_6H_4-n-hex	9,10-dia, dib	71
Fe	Phthalocyanine	CN	dib	72

These materials are characterized by common techniques such as ^{1}H NMR, IR, elemental analysis, DSC, and TGA.[54,58,59,60,64,68] Other techniques such as electron spin resonance (ESR),[54] magnetic susceptibilities[54] and Mössbauer spectroscopy (when they apply),[56,58,60,64,71] UV-visible spectroscopy,[58,59,60,61,64–67] and solid-state electric conductivity measurements[54,56,59–63,66,67,71] were also employed. These materials were carefully compared to model bis(mono-isocyanide) adducts for better understanding of the physical properties. Important solubility problems are often observed when no alkyl side chain is used. So, these "more soluble" substituents are incorporated either on the macrocycles or the bridging ligands for better characterization.

In fact, the substituents play a key role in the conductivity properties and the degree of polymerization. The conductivity is believed to be due to π–π overlaps between the macrocycles where the shish-kebab oligomers or polymers stack in a intercalated zipper form. For example, the dib-polymeric complexes of (octaphenyl-porphyrazinato)iron (porphyrazine = tetraazaporphyrine) is an insulator ($\sigma_{RT} < 10^{-11}$ S/cm), while the corresponding dib-phthalocyanine–containing polymers exhibit semiconducting properties.[67] Also, the incorporation of 4-hexylphenyl substituents at the meso positions of porphyrin allowed an end-group analysis (^{1}H NMR) and provided an estimate of the average number of units in the oligomers for $\{Fe(porph)(9,10\text{-dia})\}_n$ ($n = 10$) and $\{Fe(porph)(9,10\text{-dia})\}_n$ ($n = 5$).[71] Many of these materials are semiconducting, but this observed conductivity (in absence of doping agent) increases after doping with I_2.[55,57,63] However, the conductivity is a function of two parameters. When soluble side chains are incorporated in the macrocycle rendering the material more soluble, the nature of the polymer is best described as an oligomer. Short oligomers are poorer conductors and sometimes insulators. The second parameter is steric hindrance. Bulky substituents give poor results, as the oligomer or polymer chains do not stack properly in order to favor good π-contacts in the solid.

Interestingly, these dib-containing materials turn out to be relatively stable with temperature, as the first ligand loss (i.e., decomposition) occurs between 210 and 290°C for the phthalocyanine derivatives.[44] The presence of chlorides, either on the macrocycle or the bridging ligand, increases the temperature of decomposition with respect to the nonsubstituted analogs. The existence of intermacrocycle Cl•••Cl interactions and the considerable electronic effects of the Cl atoms may explain this gain in stability.[44] On the other hand, because the temperature of decomposition is the same for Cl-substituted dib versus Me_4dib, one concludes that the improved thermal stability with respect to the unsubstituted dib is due to a size effect, not to an electronic effect.

The last series of examples of diisocyanide-containing polymers is the mixed-ligand polymers where two different bridging ligands are used to hold the polymer backbone. The polymers are of the general formula $(Au—C\equiv C—R—C\equiv C—Au—C\equiv N—R'—N\equiv C)_n$ where the list for R and R′ is given in Table 8.2.[73,74]

TABLE 8.2 List of Polymers of the Type $(Au—C\equiv C—R—C\equiv C—Au—C\equiv N—R'—N\equiv C)_n$

R	$C\equiv N—R'—N\equiv C$	Refs.
4,4′-C_6H_4–1,1′-C_6H_4	dib (**27**), Medib (**28**), Me_4dib (**29**), Me_4Phdib (**30**), *t*-Bu_2dib (**31**)	73, 74
4,4′-C_6H_4–3,6-Me_2	dib (**32**), Medib (**33**), Me_4dib (**34**), Me_4Phdib (**35**), *t*-Bu_2dib (**36**)	73, 74
1,4-C_6H_4	*t*-Bu_2dib (**37**)	73, 74
$(CH_2O)_2$–1,4-$C_6H_4CMe_2$	*t*-Bu_2dib (**38**)	73

Unfortunately, these polymers are insoluble, and they were characterized by IR, DSC, X-ray photoelectron spectroscopy (XPS), and luminescence spectroscopy. The insolubility is argued to be due to the presence of Au•••Au interactions. To overcome the problem of insolubility for characterization, model compounds were investigated in parallel. The IR signature in the ν(CN) region is the same for the model compounds and polymers ($\sim$2200 cm^{-1}), and the ν(CN) values for these coordinated isocyanide groups is 90 cm^{-1} higher than that of the uncoordinated ones. The polymers exhibit an endothermic transition over a wide temperature range of 30–170°C and decompose at about 200°C before melting. The binding energy of the Au 4f measured by XPS are almost identical to each other, but vary between 85.1 and 85.5 eV. The polymers are found to be luminescent at room temperature and display a weak and broad emission band at 585 nm. The presence of Au•••Au interactions is also claimed on the basis of a 13-nm red-shift with respect to the corresponding model compounds. However, the presence of such interactions is also possible for the same model compounds.

B. Using Bridging Diphosphines

There are also some well-characterized cyclic and acyclic examples of oligomers of transition metal complexes using diphosphine assembling ligands. Earlier examples of acyclic diphosphine-containing oligomers were reported by Hor and collaborators.[75] Their rational syntheses are based on the use of the electroactive (dppf)-M-(dppf) building blocks to which outer metallic fragments are added to construct homo- and hetero-tri- or polymetallic oligomers (excluding the presence of Fe in the dppf bridging ligand). These species are well characterized from IR, ^{1}H, and ^{31}P NMR techniques and elemental analysis. Below P = PPh_2 for clarity.

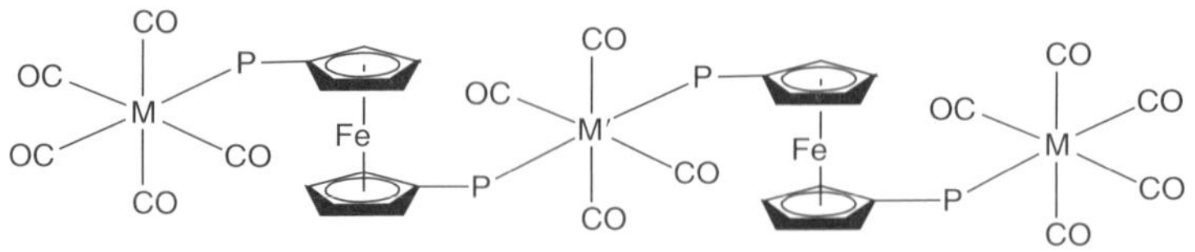

	39	40	41	42	43	44	45
M	Cr	Mo	W	Mo	Cr	Cr	W
M′	Cr	Cr	Cr	Mo	Mo	W	W
M	Cr	Mo	W	Mo	Cr	Cr	W

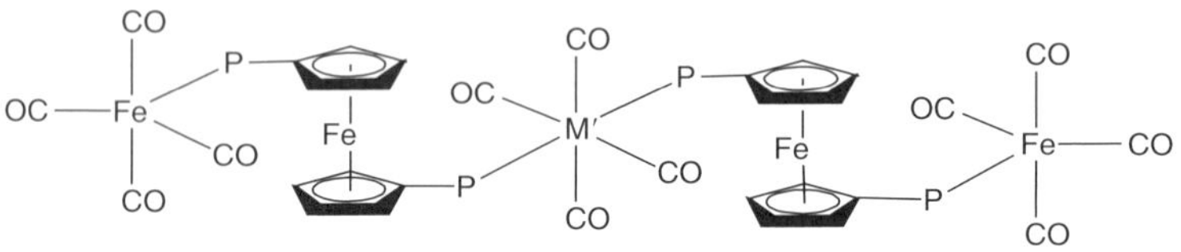

46 (M′= Cr), **47** (M′= Mo)

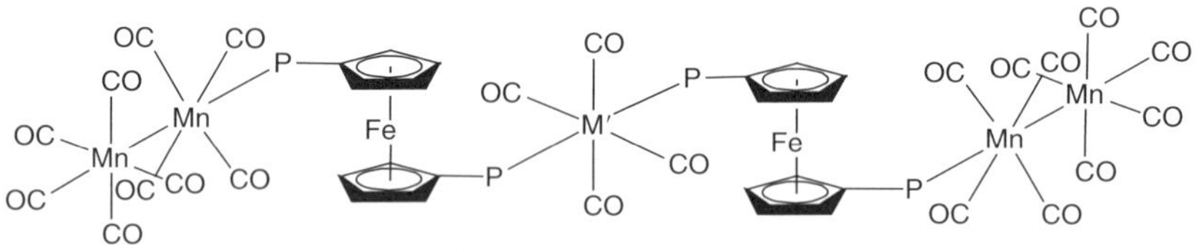

48 (M′= Cr), **49** (M′= Mo)

The ferrocenyl fragment is known for its 1-electron reversible oxidation ($Fe^{2+} - 1e^- = Fe^{3+}$), and species **39** to **49** are ideal candidates to investigate the electronic communication between the outer metallic fragments using electrochemical and spectro-(IR)-electrochemical methods. The same group also prepared dppe (**50**) and dppp analogs (**51**) to compound **40** ($P = PPh_2$).

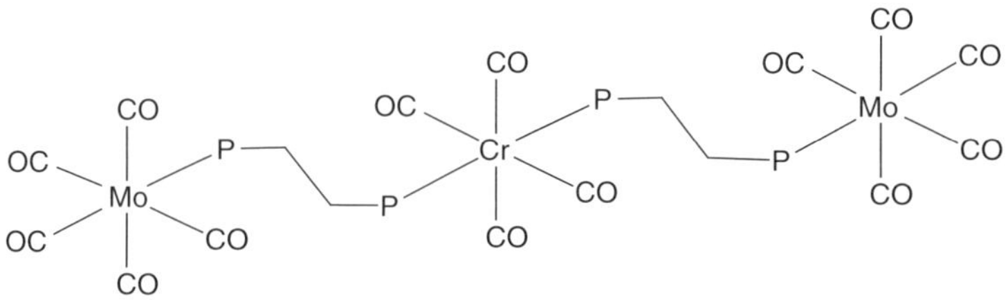

50

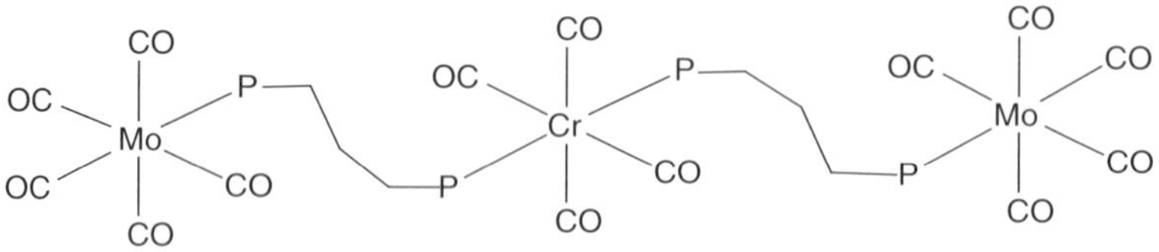

51

These oligomers represent good starting materials for the syntheses of larger oligomers or polymers, either by using similar metallic fragments employing the same strategy, or by reacting them with a different bridging ligand prior to using another metallic fragment. These options are not without the obvious problem of site selectivity for reactivity. This may explain why no further work on this topic was seen in the literature.

Another example of electronic communication through the oligomer is found in the luminescent oligomers **52** and **53** in which three bipyridine-rhenium(I) or phenantroline-rhenium(I) centers are held together by two dppet ligands.[76] The Re(I)-bpy center (bpy = 2,2′-bispyridine) is known for its MLCT luminescent excited state. Using model complexes for the interior chain and capping groups, it was possible to establish that the end-groups exhibited higher energy ^{3}MLCT states (λ_e = 534 nm), while the interior fragment displayed the lower energy ^{3}MLCT state (λ_e = 637 nm). In the trimers, the λ_e are 615 and 602 nm for **52** and **53**, respectively, and a large decrease in emission quantum yields is observed. The photophysical behavior is associated with an intramolecular triplet–triplet energy transfer, but at this time it is not possible to determine whether this transfer occurs through bond or through space.

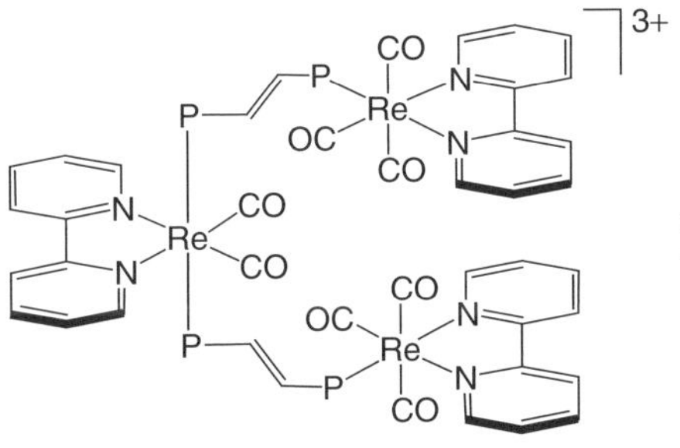

P = PPh$_2$; **52**

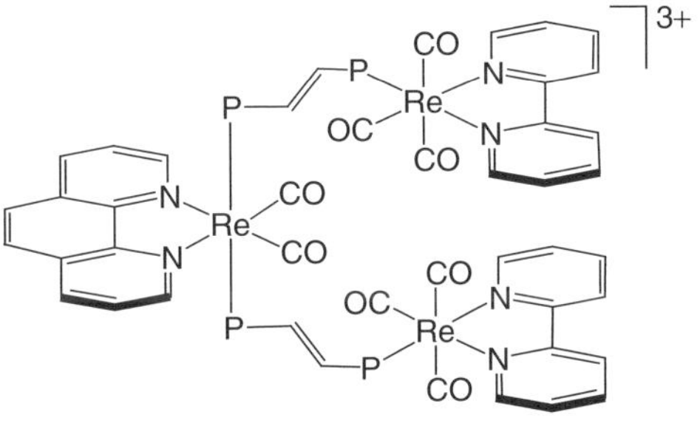

P = PPh$_2$; **53**

An earlier example of a cyclic oligomer includes the one reported by Matt and Dorsselär.[77] Indeed, the reaction of Pd(acac)$_2$ with the 1,3-bis(diphenylphosphinoacetyl)benzene bridging ligand leads to oligomer **54**. Its identification was made on the basis of mass spectroscopic measurements. The absence of end-group–containing fragments in the mass and ^{1}H NMR spectra suggests that this oligomer is probably cyclic. The proposed drawing indicates a *cis*-isomer for the Pd centers, based on the ^{31}P NMR chemical shifts in comparison with other known complexes, leading to a crown structure, probably driven by the steric hindrance imposed by the phenyl groups.

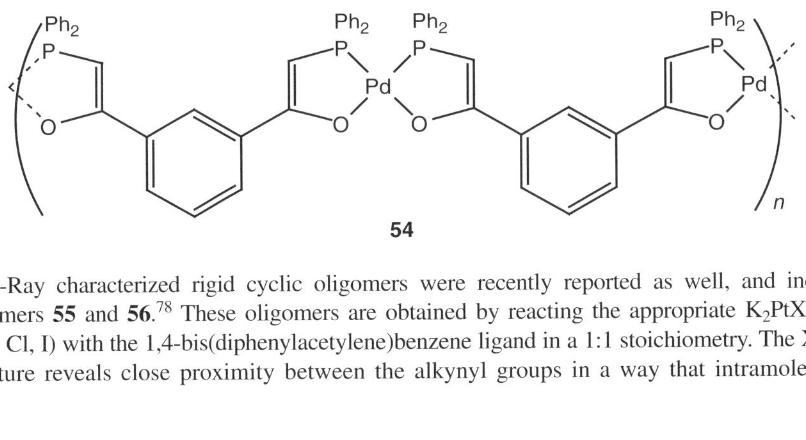

54

X-Ray characterized rigid cyclic oligomers were recently reported as well, and include oligomers **55** and **56**.[78] These oligomers are obtained by reacting the appropriate K_2PtX_4 salt (X = Cl, I) with the 1,4-bis(diphenylacetylene)benzene ligand in a 1:1 stoichiometry. The X-ray structure reveals close proximity between the alkynyl groups in a way that intramolecular cyclization is possible. In fact, oligomer **57** results from the reactivity of **56** in a $CHCL_3$/CH_2Cl_2/CH_3CN (1:1:1) mixture at room temperature for four days. This reaction is an example of taking advantage of the supramolecular autoassembling for the generation of new materials.

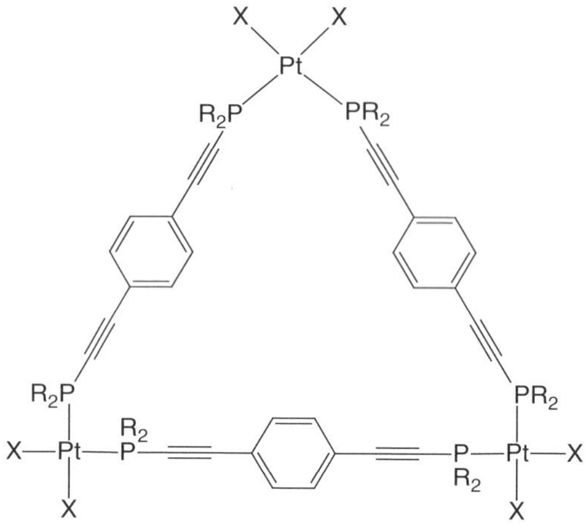

R = Ph; **55** (X = Cl), **56** (X = I)

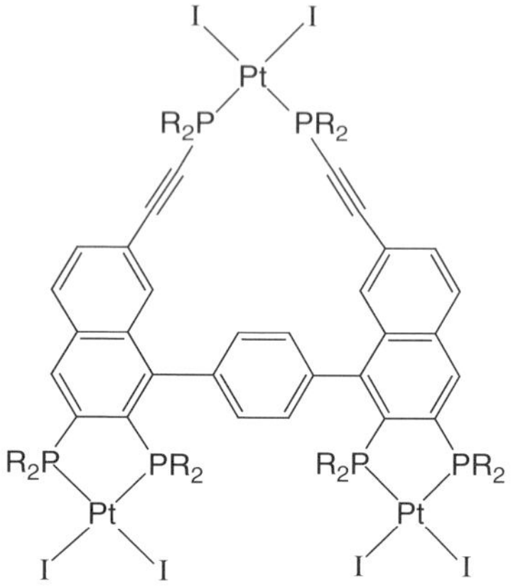

R = Ph; **57**

A recent work was reported on palladium-containing oligomers of the type *trans*- and *cis*-$[PdCl_2(L–L)]_m$, where *m* is 1 (monomer), 2 (dimer), or of larger value, and L–L is the diphosphines $Ph_2P(CH_2CH_2O)_nCH_2CH_2PPh_2$ and $Ph_2P(CH_2)_{12}PPh_2$, forming the metallo-crown ethers **58–61**.[79] Using variable-temperature $^{31}P\{^1H\}$ NMR, it was demonstrated that these species exist in equilibrium with each other (*cis-trans* and monomer–oligomer) when in $CDCl_3$. The NMR data were adequately modeled to show that the polymerization occurs in two steps: dimerization, then oligomerization. This work is important because it illustrates the possibility of ROP from spectral evidence for coordination polymers and oligomers in solution, but also the *cis-trans* isomerization in "long CH_2-chain" diphosphines.

trans-$[PdCl_2(L–L)]_m$ ⇌ *trans*-$[PdCl_2(L–L)]$

Oligomer Monomer

⇅ ⇅

cis-$[PdCl_2(L–L)]_m$ ⇌ *cis*-$[PdCl_2(L–L)]$

Oligomer Monomer

L–L = $Ph_2P(CH_2CH_2O)_nCH_2CH_2PPh_2$, $Ph_2P(CH_2)_{12}PPh_2$

58 (*n* = 3), **59** (*n* = 4), **60** (*n* = 5), **61** (L–L = $PPh_2(CH_2)_{12}PPh_2$)

Another example of a small cyclic diphosphine-containing oligomer is the triangular $Cu_3I_3(dppe)_3(L)$ (**62**) where L is the pyridine-2-thione ($2\text{-}SC_5H_4NH$).[80] It is produced from the reaction between the polymer $\{Cu_6(\mu_3\text{-}L)_4(\mu_2\text{-}L)_2(I)_4(\mu\text{-}I)_2\}_n$ (L = $2\text{-}SC_5H_4NH$) and dppe.

The Cu•••Cu separations are 3.199(3) for the Cu-(μ-I)$_2$-Cu fragment, and 6.503(3) and 6.440(3) Å for the other two bridges.

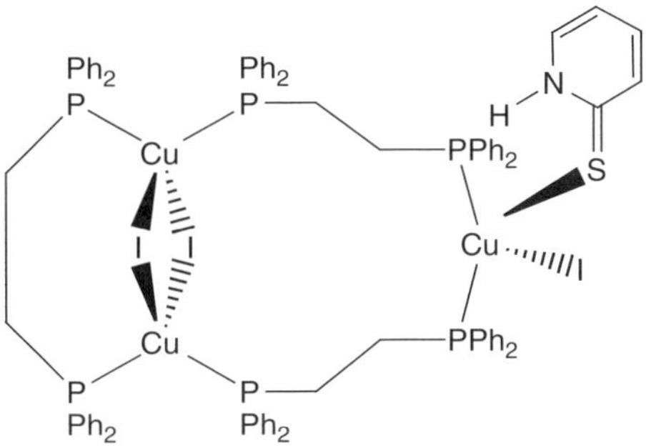

59

Other larger oligomers were also recently characterized by X-ray crystallography by Fenske and his collaborators.[81] The reactions between with $(AuX)_2$(diphos) (diphos = dpppen or dppbp; X = Cl, Br) and $Se(SiMe_3)_2$ lead to oligomers **63** and **64**. If the diphosphine is the tetradentate 1,1,1-tris(diphenylphosphinoethyl)phosphine (tpep is similar to triphos presented earlier), the oligomer exhibits the formula $[Au_{34}Se_{14}(tpep)_6(tpep^{Se})_2]Cl_6$, where $tpep^{Se}$ is 1,1-bis(diphenylphosphinoethyl)-1-(diphenylselenophosphinoethylphosphine). These species are mixed-ligand oligomers in the sense that two types of ligands bridge the different metal atoms (as for **7** to **14**).

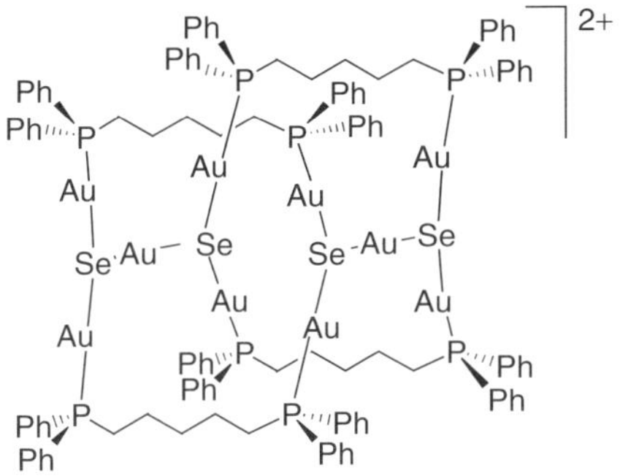

63

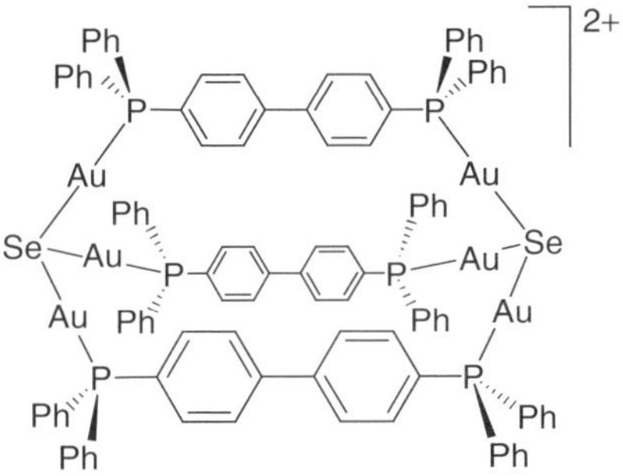

64

Mixed-ligand oligomers where one of the bridging ligands is a diphosphosphine are common in the literature. Puddephatt and collaborators have investigated the chemistry of various diphosphines (dppm, dppe, dppp, dppb, dpppen, dpph), rigid bispyidines, notably with 4,4′-bispyridine and *trans*-1,2-bis(4-pyridyl)ethylene, and the diisocyanide ligand dib in some detail.[82–86] Large macrocycles or cyclic oligomers containing four gold or silver atoms were characterized by X-ray crystallography and are often formed when the number of CH_2-groups in the chain is 1, 3, and 5, favoring the *syn*-conformation, as shown for oligomers **65–72**. But this rule is not rigid, as a macrocycle formed with dppe was also observed for oligomer **69**. Similarly, with dppp and *trans*-1,2-bis(4-pyridyl)ethylene, a polymer is obtained (as discussed later in the section). When dppm is used, Au•••Au and Ag•••Ag contacts are observed (3.106(1) and 3.084(1) Å for **62**, and 3.0479(8) Å for **68**).[85] Polymer–oligomer equilibrium is suspected to occur in solution. The bipyrine derivatives in the following figure are found to be luminescent in solution and in the solid state with λ_{max} in the range of 390–490 and 416–440 nm, respectively. Two types of emissions were observed; broad and vibronically structured. The latter resolved emission is unquestionably related to a π–π^* state arising from the bipyridine ligand, while for the broad ones, experimental evidence provided could not address the nature of the excited state with certainty. Whether Au···Au interaction, exciplex or two versus three coordinated excited states were involved, still remain to be seen.

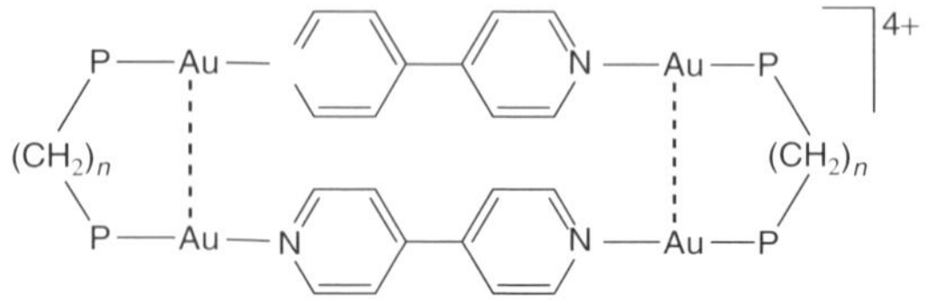

P = PPh_2; **65** (n = 1), **66** (n = 3), **67** (n = 5)

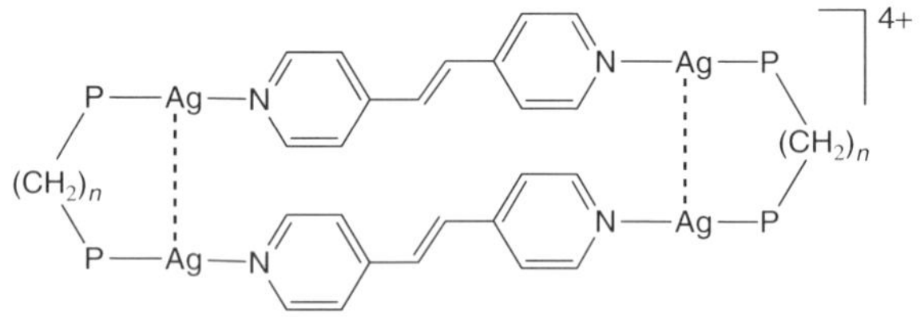

P = PPh_2; **68** (n = 1), **69** (n = 2), **70** (n = 5)

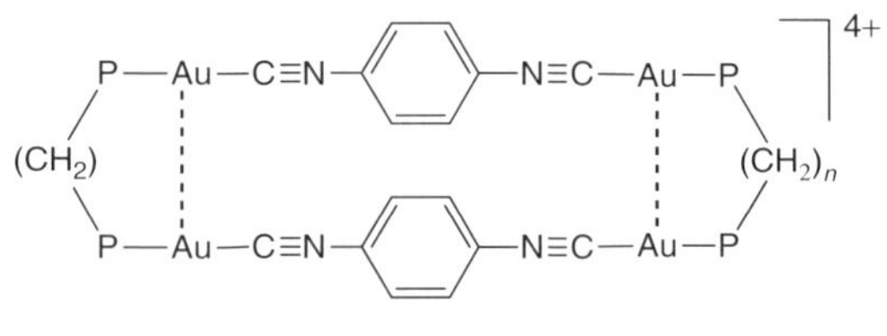

P = PCy_2; **71**

The preparation of materials suspected of being coordination polymers built on the use of diphosphines goes back as early 1981.[87] Among the examples, one finds a material described as $\{Ru(OEP)(dppb)\}_n$, where OEP is the common macrocycle octaethylporphyrin. Since there was no further characterization, no formal assignment was provided.

The organometallic polymers **72** and **73** were obtained earlier from the reaction between $[Cr(\eta^6\text{-}1,4\text{-}(Ph_2P)_2C_6H_4)(CO)_2L]$, where L = CO or PPh_3, with the dimeric $[Rh(CO)_2Cl]_2$

complex.[88] The true identity of the materials, oligomer versus polymer, was not addressed. The coordination of both PPh_2 groups on the aryl group by Rh centers leads to a hindered rotation about the Cr-(η^6-arene) axis with $\Delta G^{\ddagger}$ of 11.7 ± 0.1 kcal mol^{-1} for **73**.

These species represent potential precursors for catalysts for the hydroformylation reaction of alkenes, bearing in mind that the lower solubility of the materials may be useful in order to recuperate, by filtration, the expensive rhodium(I) materials. In addition, if robust enough, the chromium fragment can be functionalized with various monophosphine ligands, such as PR_3, where R can be an alkyl or aryl group. The introduction of chiral ligands on the chromium metals, such as Binaph-type ligands (2,2′-bis(diphenylphosphino)-1,1′-dinaphthyl), leads to regio- and enantio-selectivity for the hydroformylation reaction.

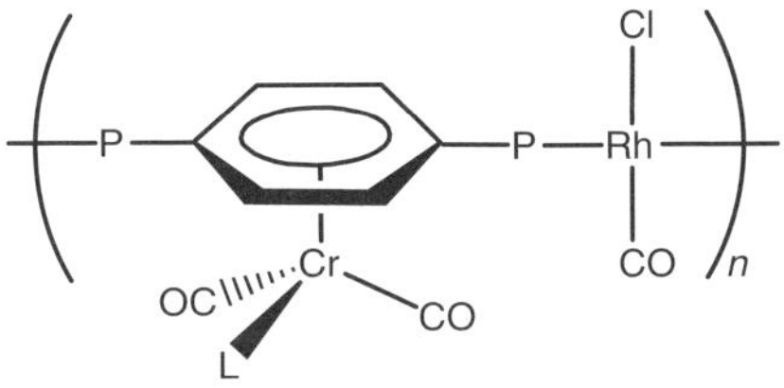

P = PPh_2; **72** (L = CO), **73** (L = PPh_3)

Conducting organometallic polymer **74** was recently investigated by Gray and collaborators.[89] It uses the bridging ligand 5,5′-bis(diphenylphosphino)-2,2′-bithiophene, $Ph_2P(C_4H_4S)_2PPh_2$, with the $Mo(CO)_4$ metallic fragment. In a 1:1 stoichiometric amount, a large cyclic oligomer is formed, for which the IR spectra indicate the *cis*-conformation of the two phosphorous atoms about the metal, and the $^{31}P\{^1H\}$ spectra display no evidence for a P-containing end-group. Size exclusion chromatography in THF determines that M_w and M_n are 6000 and 9700, respectively. Electrochemical polymerization should prove very interesting if the $Mo(CO)_4$ fragment remains intact, taking advantage of the ν(CO) IR probe.

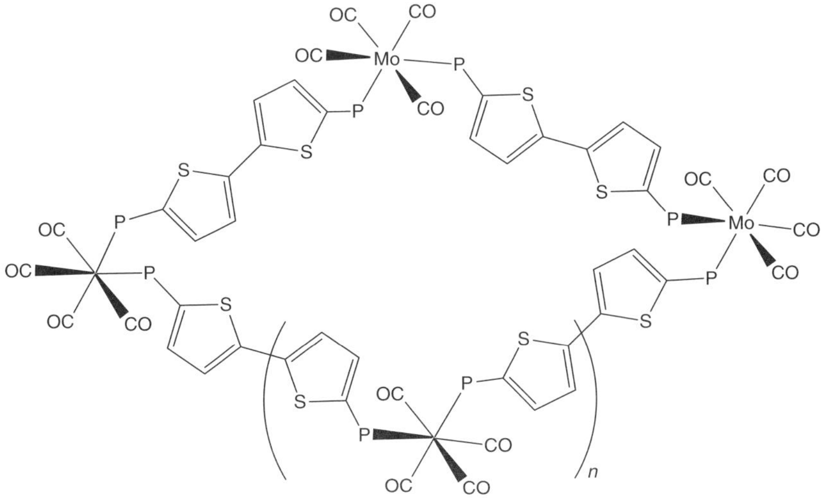

P = PPh_2; **74**

Polymer (or oligomer) **75** is part of a mixture resulting from a reaction between Pd(COD)(Me)(Cl) (COD-chemical oxygen demand) and dppb to which is added $AgSO_3CF_3$ as a Cl-abstractor in acetonitrile.[90] The other species present in solution are the mononuclear $Pd(dppb)(Me)(NCCH_3)^+$ and the dimeric complex $Pd_2(dppb)_2(Me)_2(NCCH_3)^{2+}$. The presence of many species in solution is completely consistent with the ^{31}P NMR work reported by Gray and collaborators, as previously discussed.[79] Upon a reaction with CO, the corresponding acetyl derivatives, including the oligomer/polymer **76**, are obtained. Dekker et al. rightfully argue that the flexibility of the alkyl chain is responsible for the presence of macrocycles (dimers) and oligomers.

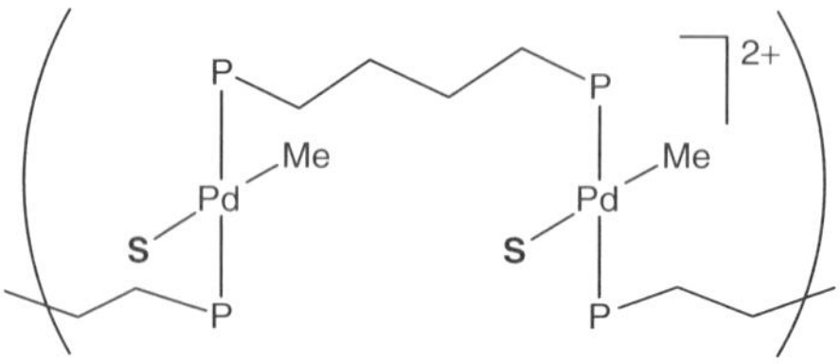

75 (**S** = CH_3CN, P = PPh_2)

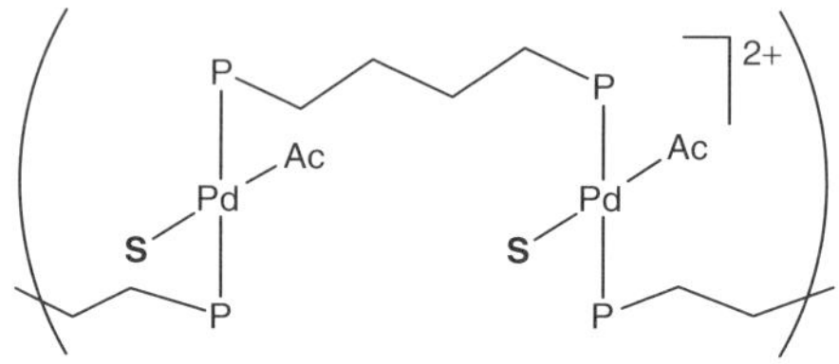

76 (**S** = CH_3CN, P = PPh_2)

Similar to this work, the recent use of 5,17-bis(diphenylphosphinomethylene)-25,26,27,28-tetra-*n*-propoxycalix[4]arene in the presence of the square planar Pd(II) and Pt(II) fragments leads to polymers **77** and **78**.[91] The assignment for these polymeric species was based on NMR, where large peaks were observed. No further characterization was provided, so it is not possible to say whether cyclic polymers are present in the solution.

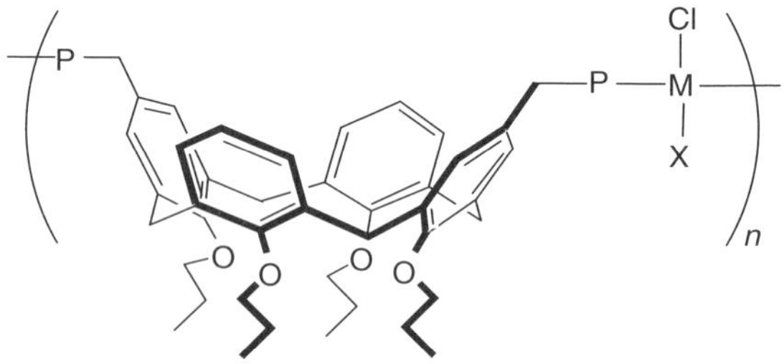

P = PPh_2; **77** (M = Pd, X = Me), **78** (M = Pt, X = Cl)

On the other hand, the reaction of 1,12-bis(diphenylphosphino)dodecane with palladium dichloride generates a mixture of 34% macrocycle (**79**) and 61% polymer (**80**) that are in equilibrium with each other.[92] This equilibrium represents a good evidence for ROP. The materials were characterized from ^{31}P NMR in CD_2Cl_2 solution and size exclusion chromatography. Samples that were analyzed immediately after a slow evaporation followed by a redissolution contained polymers consisting of up to 80 units. Much higher M_n were observed in the secondary electron conduction (SEC) traces for polymers obtained by melt polymerization at 185°C for 15 min. Polymers consisting of up to 500 units were reported.

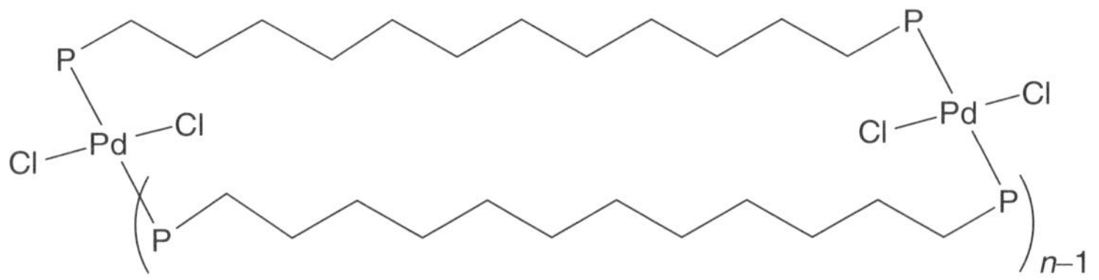

P = PPh_2; **79**

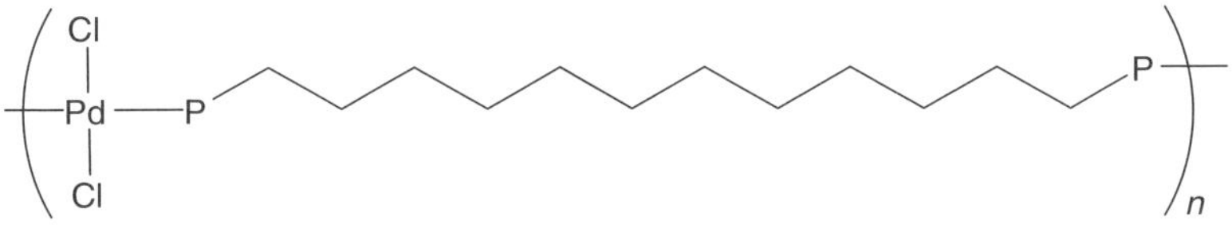

P = PPh_2; **80**

The two "doubly bridged" oligomers **81** and **82** and the very insoluble polymer **83** were obtained using the bridging ligand dpa.[93] These materials were characterized by ^{1}H, ^{31}P, and ^{19}F NMR. The *trans*-conformation for **83** was established on the basis of IR findings (argument based on the number of bands).

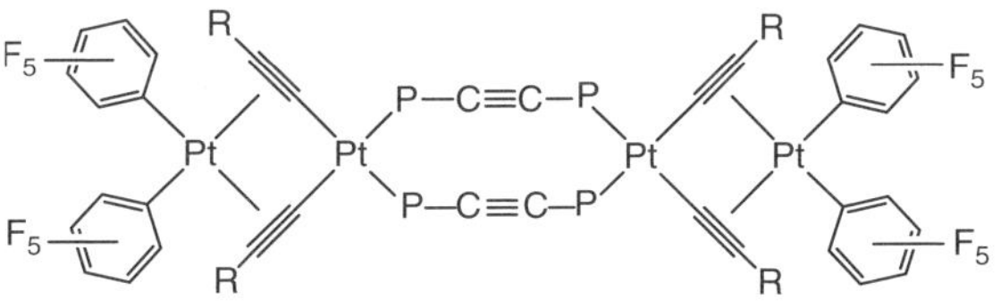

P = PPh_2; **81** (R = ^{t}Bu), **82** (R = Ph)

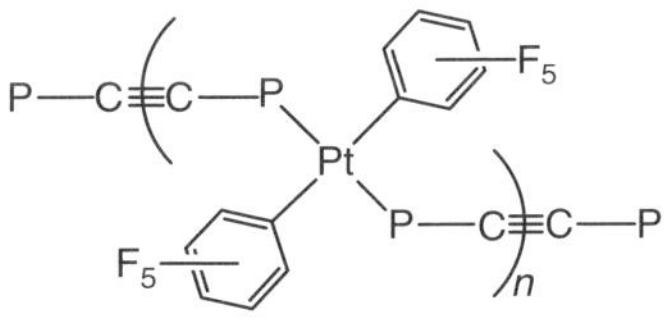

P = PPh_2; **83**

Complex $W_2(SPh)_2(CO)_8$ reacts with dppm in toluene at 298 K, to form a neutral dark-green polymer $\{W_2(SPh)_2(CO)_8(dppm)\}_n$ (**84**).[94] This polymer is almost insoluble in CCl_4, CH_3CN, benzene, toluene, acetone, and methanol, and is moderately soluble in CH_2Cl_2, $(CH_3)_2SO$, cyclohexane and hexane. The assignment for the polymer is based on the comparison with a model compound $(W_2(SPh)_2(CO)_8(PPh_3)_2$ and its spectroscopic IR signature, its low solubility, and the difficulty in getting crystals of suitable quality for X-ray crystallography. The *trans*-conformation is also assigned on the basis of the IR resemblance with the model compound.

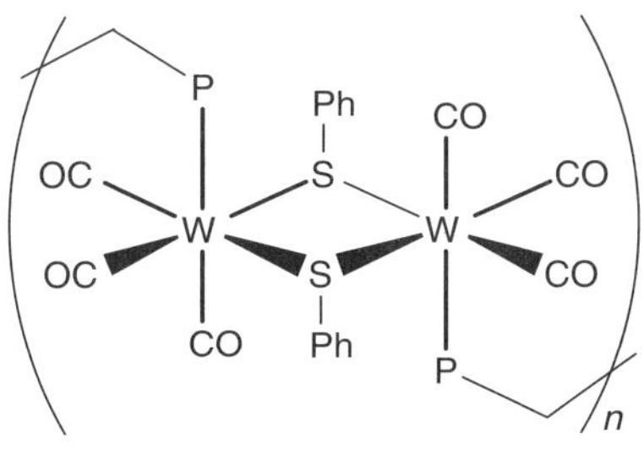

P = PPh_2; **84**

Although this work goes back to 1977, and its preparation method is different from traditional coordination chemistry, the reaction between Me_2PCl with liquid Hg leads to the interesting diphosphine-containing polymer **85**.[95] Liquid mercury reduces the chloro-monophosphine to presumably generate the free radical Me_2P•, which in turn dimerizes as $(Me_2P)_2$. The oxidized Hg (Hg(II)) coordinates the Cl^- anions and the new P–P bond–containing ligand.

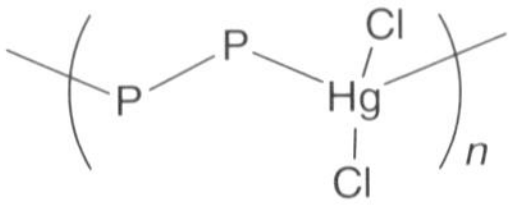

P = PMe_2; **85**

Linear 1-D polymers of the type $\{M(diphos)(CN\text{-}t\text{-}Bu)_2^+\}_n$ (M = Cu, Ag; diphos = dppb, dpppen, dpph) and $\{Ag(dpppen)(CN\text{–}t\text{-}Bu)^+\}_n$ (**86–92**) were recently prepared and fully characterized by ^1H, ^{13}C, ^{31}P NMR, IR, elemental analysis, X-ray structures (for **90** and **92**), DSC, TGA, XRD, and electronic spectroscopy.[96] The polymers are weakly soluble in CH_3CN, CH_2Cl_2, methanol, and $CHCl_3$, allowing for better characterization. The materials are stable to at least 150°C (TGA) and photochemically stable to sunlight, at least for several months. The TGA traces are characterized by two thermal events between 150 and 350°C. The first weight loss is due to CN–*t*-Bu, and the second, to the diphosphine ligands.

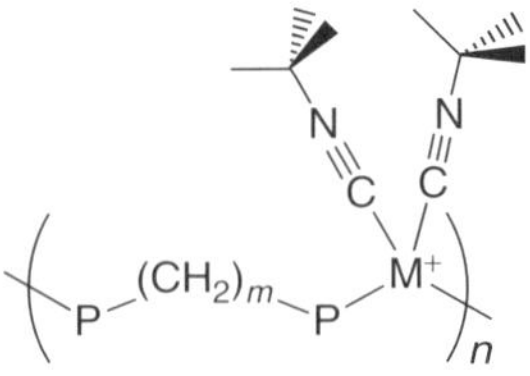

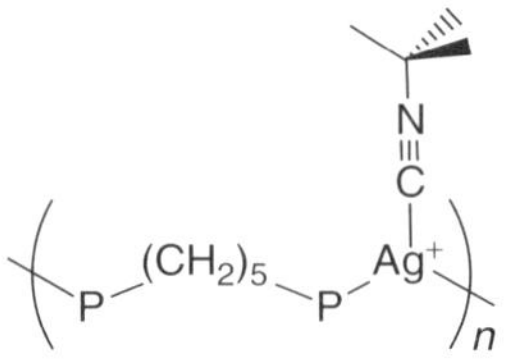

diphos: dppb (m = 4), dpppen (m = 5), dpph (m = 6)				
Cu	**86**	**87**	**88**	
Ag	**89**	**90**	**91**	**92**

The Ag-containing materials are generally more soluble and form brittle films upon slow evaporation. Conversely, the Cu-polymers form stand-alone film with the same treatment or spin coating. The XRD patterns display traces indicating crystalline or semicrystalline materials, but in general, the Ag materials are more crystalline than the Cu ones. These observations suggest that the Ag materials must be of shorter chains (i.e., smaller number of repetitive units). The solid-state luminescence data at 298 K are as follow: **86** (475, 22 ± 5), **87** (483, 31 ± 5), **88** (485, 12 ± 3), **89** (481, 14 ± 3), **90** (491, 55 ± 4), and **91** (486 nm, 56 ± 4 μs). The excited state responsible for the luminescence is an MLCT arising from the d^{10} metal to the both π-systems of the C≡NR and PPh_2 groups.

The dppf diphosphine ligand is also a popular assembling reactant. A recent example includes the 1-D polymer $\{Ag(dppf)(\mu^2\text{-}O_2CCF_3)$ **93**.[97] The latter is prepared from a binuclear complex $Ag_2(O_2CCF_3)_2(dppf)$, which reacts with an excess of dppf. The polymer was characterized by X-ray crystallography, and ^1H, ^{31}P, and ^{19}F NMR, as well as IR. Evidence for H-bonds of the type CF_3•••HC(Ph) was provided.

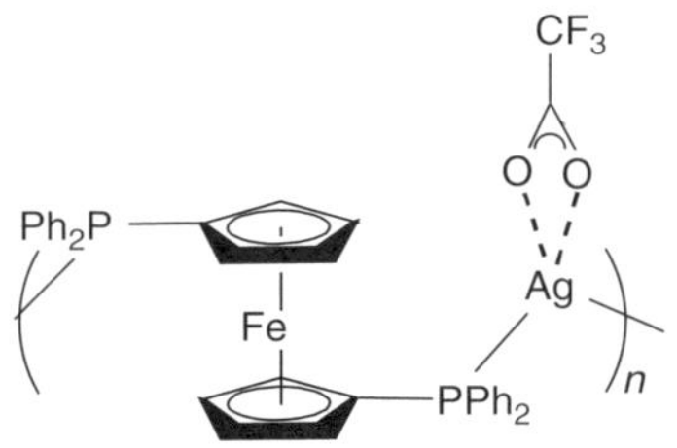

93

This type of repetitive motif $(Ag(L)_2(diphos)^+)$ is common. Other examples are polymers **94–100**,[98] which are obtained by reacting the $AgNO_3$ salt with dppe in the presence of commercial thiones.

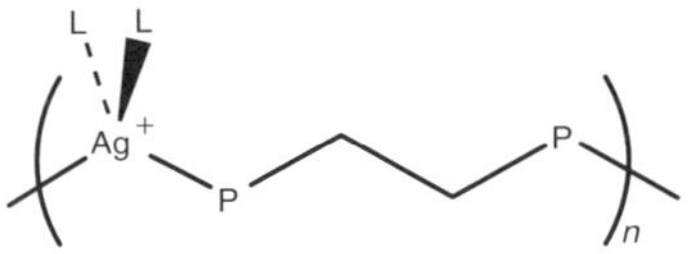

P = PPh_2; L =

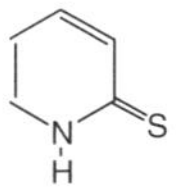 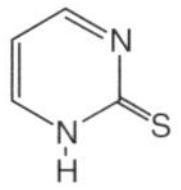 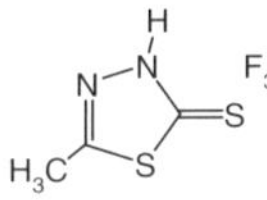 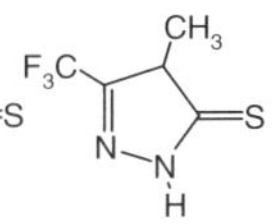 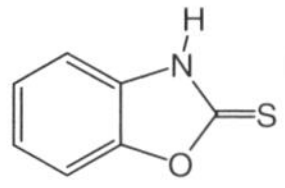 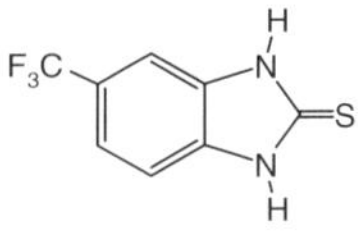 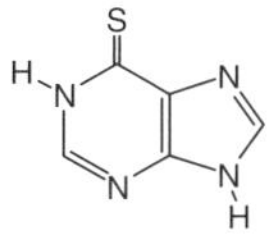

94 **95** **96** **97** **98** **99** **100**

The polymers were characterized from elemental analyses, TGA, IR, UV-vis spectroscopy, and ionic conductivity measurements in solution. Polymer **95** was also characterized from X-ray crystallography.

Fenske and his collaborators reported the reaction between $PhP(S)(SSiMe_3)_2$ and Ag(OAc) in the presence of dppe, which produces the polymer $\{[Ag_2(PhS_2P\text{-}PS_2Ph)(dppe)]\bullet(\mu\text{-dppe})\}_n$ (**101**).[99] This reaction bears some similarities with that of polymer **85**, in the sense that a redox process inducing the P–P bond formation is observed. This bond formation is rationalized by a disproportion reaction of $PhP(S)(SSiMe_3)_2$. The polymer was characterized by X-ray diffraction and ^{31}P NMR. In solution, two species, presumably rotamers, are depicted.

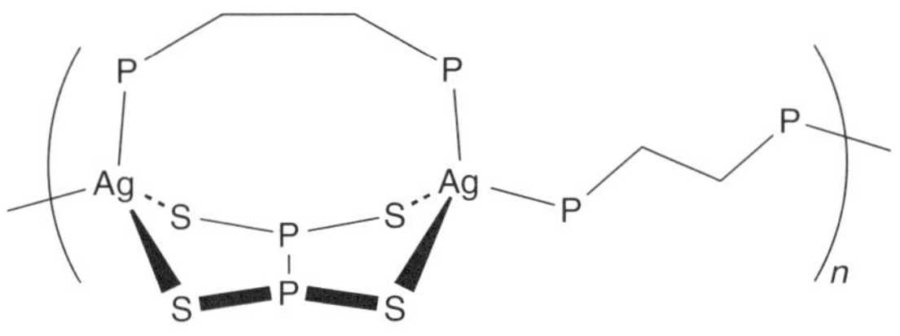

P = PPh_2; **101**

This polymer bears structural similarity with the diarsine-containing polymers **102** and **103**,[100] which were obtained from the reaction of the AgBr and Ag(NCO) salts with 1,2-bis(diphenylarsino)ethane (dpam).

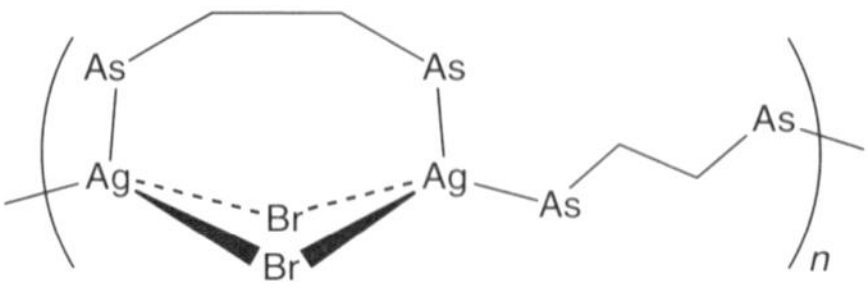

P = $AsPh_2$; **102**

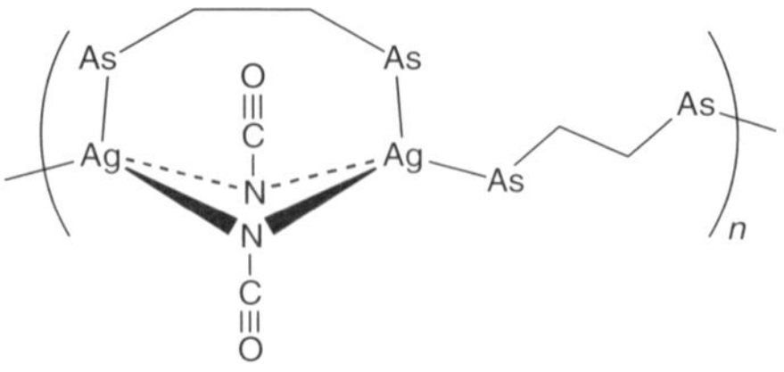

103

Polymers **101–103**, exhibit a common structure of the type $\{[M_2L_2](diphos)\}_n$, where L is a bridging ligand (not necessarily a disphosphine) keeping together two metal atoms. Examples of this type where all three ligands are the same (notably dpa, dppt, and dppbp) or are different, such as trifluoroacetate, and coordinate Ag atoms include polymers **104** to **108**.[101–104]

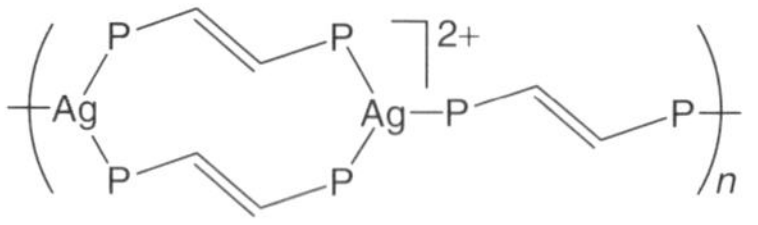

P = PPh_2; **104**

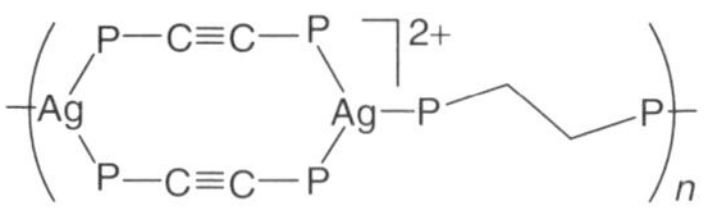

P = PPh_2; **105**

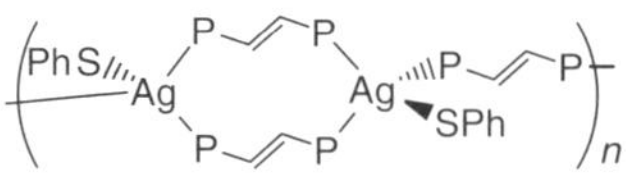

P = PPh_2; **106**

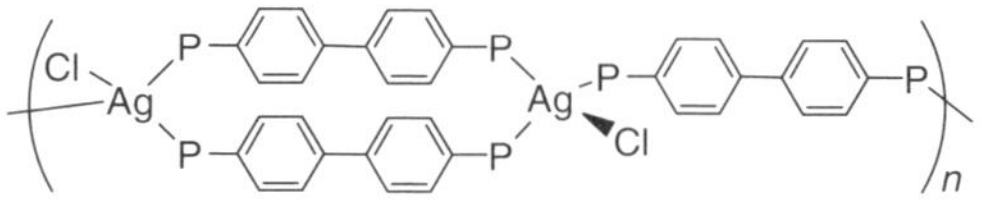

P = PPh_2; **107**

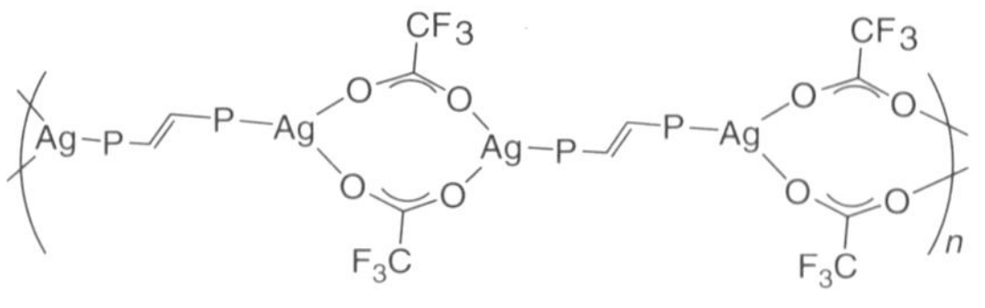

P =PPh_2; **108**

Another example for a mixed-bridging ligand-containing polymer is **109**.[86] This 1-D polymer was also characterized by X-ray crystallography. There is no evidence for a second diphosphine ligand bridging the silver metals. Instead, the trifluoroacetate ligand acts a monodentate ligand. Clearly, the large difference in bite distance and the stronger pyridine and phosphine ligands prevent the trifluoroacetate from acting as a bridging ligand.

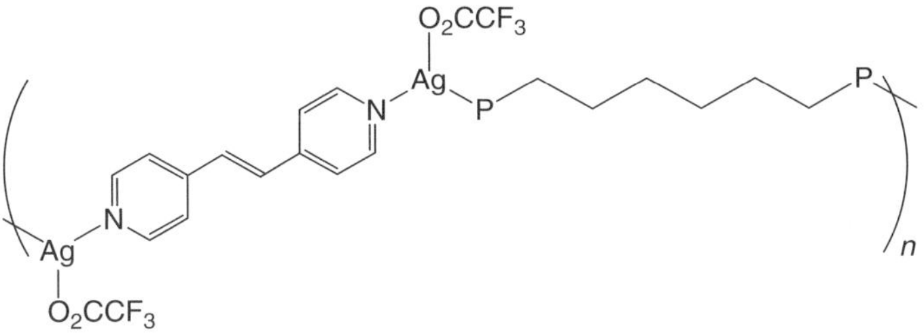

P = PPh_2; **109**

Polymer **110** was prepared from the reaction between potassium (1, 2, 4-triazolyl)borate, dppm, and silver nitrate.[105] It was characterized by X-ray crystallography, IR, and ^1H and ^{31}P NMR. Two Ag···Ag distances are reported (3.5214(9) and 3.9143(8) Å) and are too long for any significant interactions.[106,107]

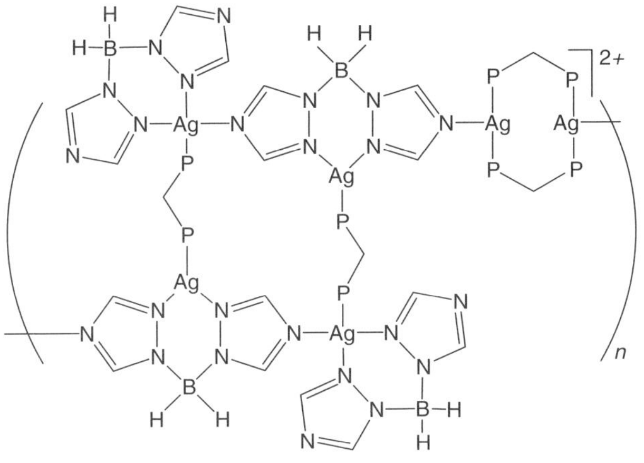

P = PPh_2; **110**

Polymer **111** exhibits an unusual coordination pattern and is built upon the tridentate ligand *N,N′*-(bis(diphenylaminomethyl)pyridine.[108] The repetitive unit consists of two macrocycles using the diphosphine–pyridine ligand given earlier, and a monophosphine–pyridine ligand *N*-(diphenylphosphino)aminomethyl)pyridine. One macrocycle is constructed with a head-to-tail geometry (*N*-(diphenylphosphino)-aminomethyl)pyridine), and the other, head-to-head. The latter fragment uses the diphosphine end to bridge two silver(I) atoms in a η^2-μ^2 fashion that are also coordinated by another diphosphine linker, hence inducing polymerization. The two-bridged metals are separated by 3.0363 Å, and are each coordinated by an acetonitrile molecule, which is placed nearly perpendicular to the P_4Ag_2 plane.

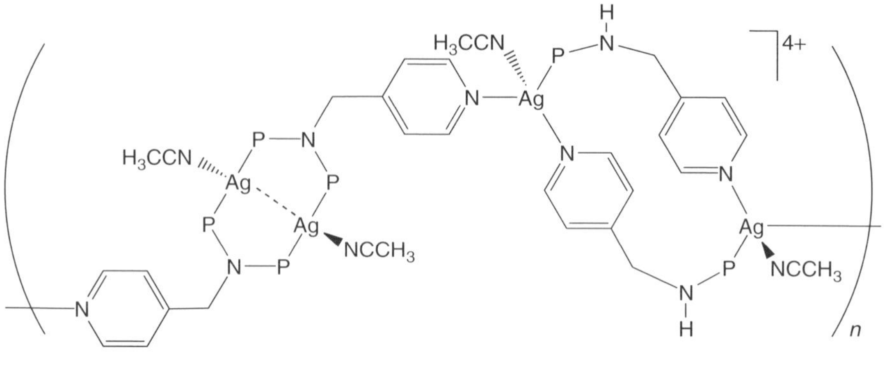

P = PPh_2; **111**

Among the transition elements, gold is the metal that is the most exploited or most successful at making coordination and organometallic polymers. The similarities with silver are often obvious, but differences are also observed. The last part of the section on singly bridged 1-D oligomers and polymers is devoted to this element.

Polymer **112** and oligomer **113** built upon the dppf ligand were reported by Mingos and collaborators and by Hor and collaborators.[109,110] Polymer **112** is a yellow nonelectrolyte material that can be obtained in two conformational forms. The two different crystals can be separated manually by their difference in shapes (round diamond-shape versus rhombic prismatic). The differences in conformations arise through a combination of torsions about the Au–P and P–Cp bonds. The Cl/NO_3 metathesis leads to $\{Au(dppf)(NO_3)\}_n$, for which we do not have an X-ray structure, but the addition of sodium formate provides the dication oligomer **113** in which two dppf ligand act as chelates and a third one, as a bridge. No significant Au···Au interaction is noted. The proposed mechanism of formation of the polymer **112** is that it passes through a monomer Au(dppf-P,P′)Cl, then a dimer $[Au(\mu\text{–}dppf)(Cl)]_2$, which reversibly polymerizes to **112** in $CDCl_3$.

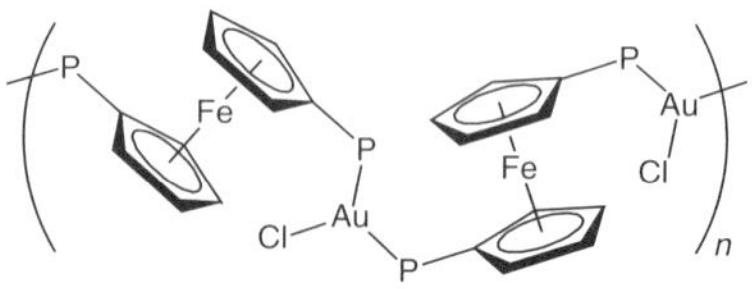

P = PPh_2; **112**

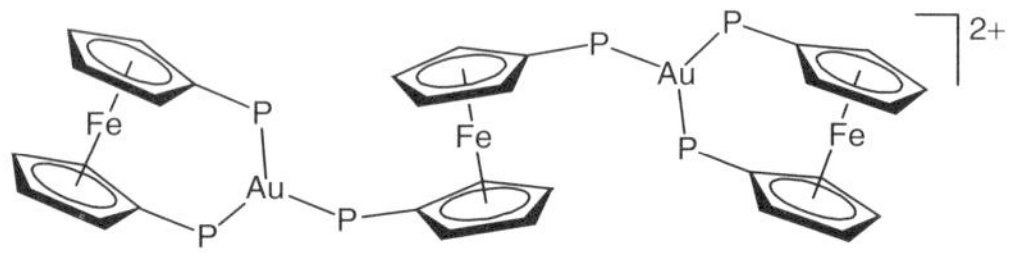

P = PPh_2; **113**

Dinuclear Complexes of the type $[Au_2(diphos)_2^{2+}]$ are reported for diphos ligands dppm, dppe, dppp, and dpppen, but for dppb, a polymer (**114**) is discovered and characterized by X-ray crystallography.[111] These species are prepared from the Cl-abstraction reaction of the corresponding complexes $Au_2(diphos)Cl_2$ in the presence of the diphos (diphos = dppm, dppe, dppp, dppb, dpppen). It is suspected that equilibrium occurs between the dimers and polymers.

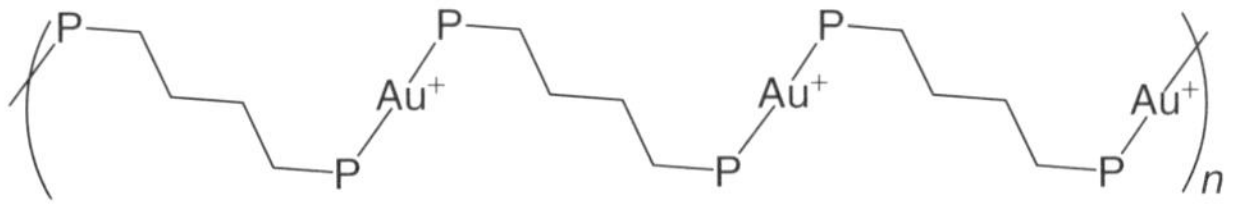

P = PPh_2; **114**

Similarly, polymer **115** is also synthesized from the addition of silver trifluoroacetate to Au_2(dppet)Cl_2 in the presence of dppet.[112] The difference with polymer **113** is that two uncoordinated dppet's reacted with the starting material.

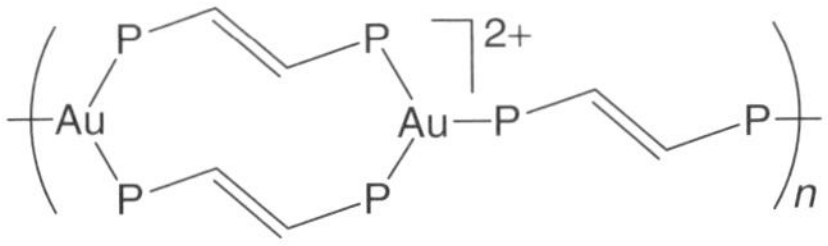

P =PPh_2; **115**

Other examples of mixed-ligand Au-containing polymers exist, such as **116–123**, which were investigated by Puddephatt and collaborators.[82–86] These colorless air-stable 1-D polymers are in equilibrium with the corresponding oligomers **65–70**. They have good thermal stability and melt sharply with immediate decomposition. They are electric insulators with poor conductivities of (2–10) $\times$ $10^{-9}\,\Omega^{-1}\,cm^{-1}$. Polymers **120** and **121** were characterized by X-ray crystallography, which reveals a folded *syn*-conformation for **120** where *m* is an odd number favoring stacking and perhaps Au···Au interactions leading to folded structure somewhat similar to an accordion. A zigzag *anti*-conformation for **121** is depicted.[82] It is assumed that when *m* is odd or even, an accordion or a zigzig structure is always obtained, respectively.

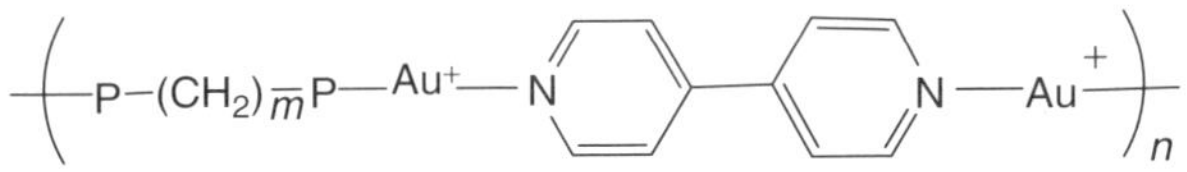

P = PPh_2; **116** ($m = 3$), **117** ($m = 4$), **118** ($m = 5$), **119** ($m = 6$)

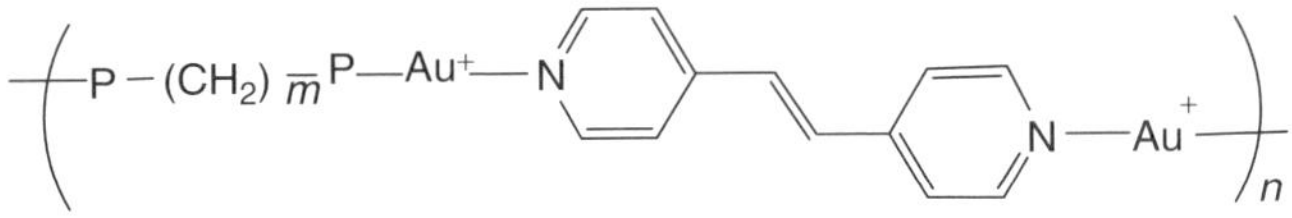

P = PPh_2; **120** ($m = 3$), **121** ($m = 4$), **122** ($m = 5$), **123** ($m = 6$)

Other examples include the rigid diphosphine–diacetylide-containing polymers **124–125**.[113] They were prepared from the reaction between the polymer {CC-R-CC-Au}$_n$

with the diphosphine 4,4′-bis(di-isopropylphosphinophosphine)diphenylene. The pale-yellow products were air-stable and insoluble in most organic solvents, except for halogenated solvents such as dichloromethane and chloroform. They were characterized by elemental analysis, NMR, and IR spectroscopy, and compared to model bis-monoethynyl digold compounds. The end-groups were coordinated chlorides on gold.

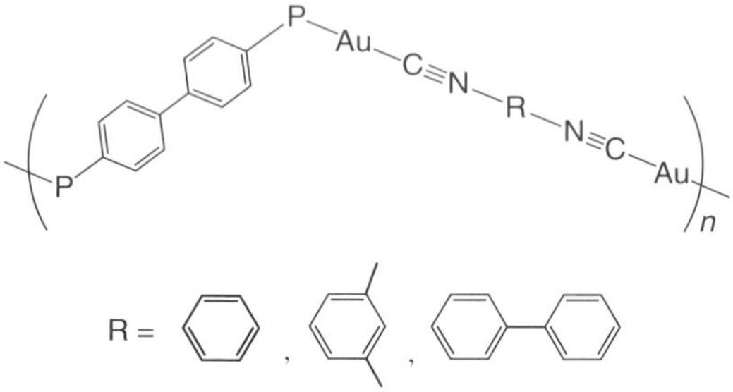

P = P-^{i}Pr$_2$; **124** **125** **126**

More flexible versions of mixed-ligand gold-containing polymers (**127**, **128**) were also reported.[114,115] The weakly soluble polymer **127** was prepared from the reaction between *N,N′*-bis(pyridin-3-yl)-1,3-benzenedicarboxamide with complex [Au_2(dppe)(O_2CCF_3)$_2$].[114] The polymer was characterized from X-ray crystallography, which revealed two different alternating conformations for the dppe ligand: *syn* and *anti*. The *syn*-conformation promoted Au···Au interactions (2.99 Å) curving the chain, while the *anti*-one induced a linear section of the chain with a long Au···Au separation (6.97 Å). Two nitrate counteranions are found to be H-bonded with the N–H groups of the amide residues.

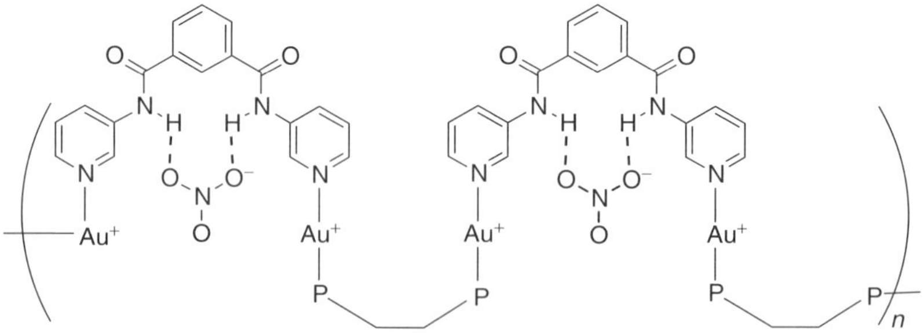

P = PPh$_2$; **127**

Polymer **128** is built upon dppb and forms a zigzag polymer as demonstrated by X-ray studies, and supports the trend where an even number of methylenes in the chain promotes such a structure.[115] The use of dppp leads to a cyclic oligomer (**129**). Dynamic ^{31}P NMR investigations and mass-spectrometry analyses demonstrate the presence of equilibrium between cyclic oligomers and polymers.

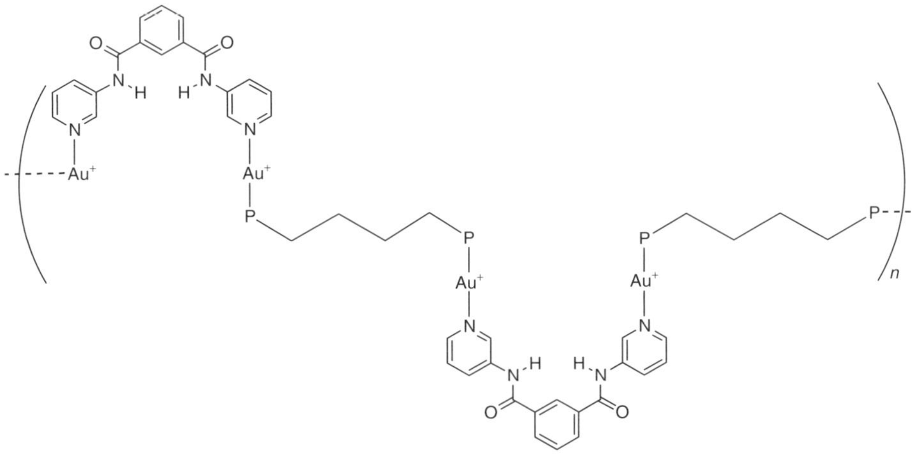

P = PPh_2; **128**

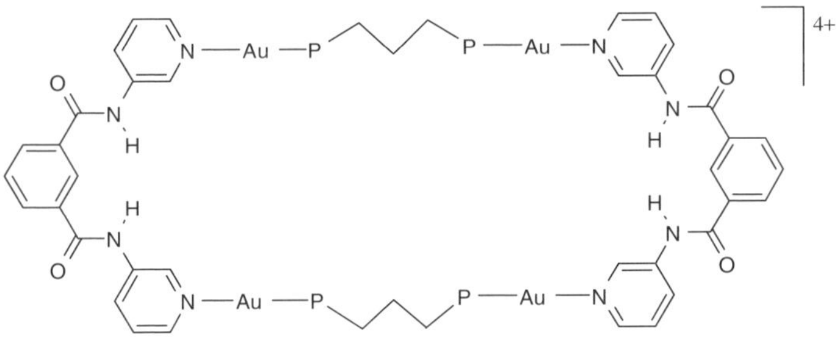

P = PPh_2; **129**

Both polymers **130** and **131** were also characterized by the X-ray method.[116] The 1-D polymer adopts a zigzag structure where the P–Au–S vectors are perfectly parallel to each other, and so adopts a perfect *anti*-conformation. The conformation of the dithiolate ligands places the phenyl and naphthyl groups at a center of inversion, so no chirality is observed within this polymer.

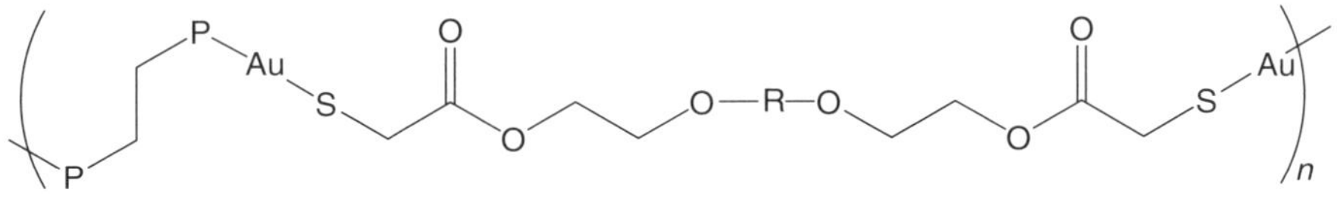

P = PPh_2; **130** (R = 1,4-C_6H_4), **131** (R = 1,5-$C_{10}H_6$)

Polymer **132** is also an example of mixed-ligand gold-containing polymer.[81] It is synthesized from the reaction between $[(AuBr)_2(dppbe)]$ (dppbe = 1,4-bis(diphenylphosphino) benzene and $Se(SiMe_3)_2$. The polymer forms a helix where dppbe adopts an anticonformation and the An–Se–Au angle is small (78.3°), allowing for the folding. The projection along the polymer 1-D axis shows that the dppbe ligands are placed perpendicular to each other. The Au···Au separation is 3.018 Å.

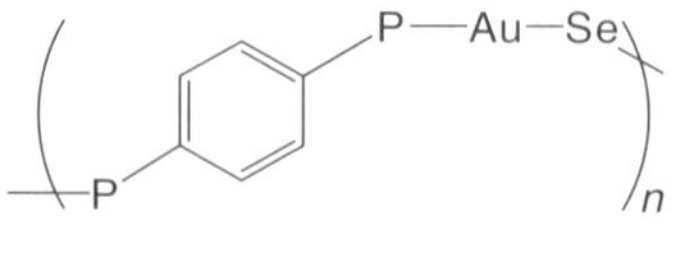

P = PPh_2; **132**

An interesting case of a polydentate ligand is the 2,6-bis(diphenylphosphino)pyridine.[117] It reacts with Ag(I) salts of the type $[Ag(NCCH_3)_4](ClO_4)$ to form polymer 133. The N-atom remains uncoordinated.

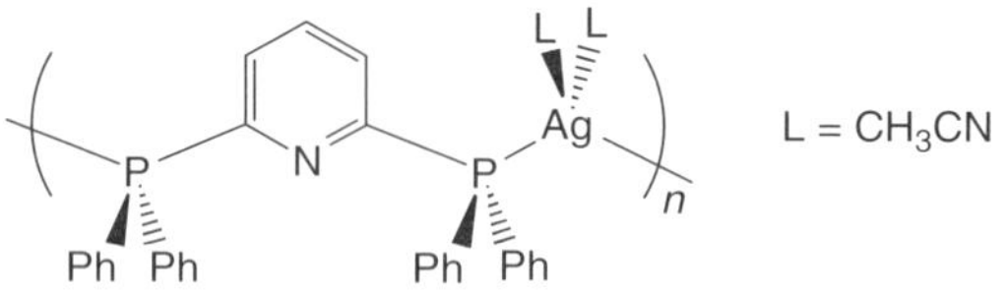

133

IV. DOUBLY BRIDGED 1-D POLYMERS

The doubly bridged 1-D polymers are obviously a subgroup of the 1-D polymers, but at the same time they are constructed with enough bridging ligands to be structurally related to the 2-D and 3-D materials. Some doubly bridged oligomers are also known. Dpa-containing oligomers **81** and **82**, which contain 4 Pt atoms placed in line, are examples.

Oligomer **134** is composed of two $Ag_2(dmb)_2^{2+}$ macrocycles held together by three tetracyanoquinodimethane radical anions ($TCNQ^-$) with long —CN···Ag^+ coordination.[118] The Ag···Ag separation is 4.113(1) Å. These TCNQs are placed face-to-face with a separation of 3.333(1) Å, a distance that is within the sum of the van der Waals radii of C (1.77 Å). The $[Ag_4(dmb)_4(TCNQ)_3]^+$ units form a 1-D chain using π-contacts of the external TCNQs separated by 3.372(1) Å. The other counteranion is also a $TCNQ^-$. Because of the radical-type nature of the TCNQs, magnetic properties were investigated. The magnetic data are best explained by extended chains of antiferromagnetically coupled centers S = 3/2 and S = ½ for the $[Ag_4(dmb)_4(TCNQ)_3]^+$ and $TCNQ^-$ moieties. No hyperfine structure has been observed in the EPR spectra between 106 and 290 K, indicating the presence of a rapid exchange in the paramagnetic system. In solution, the polymer dissociates completely and the TCNQs are all separated from each other.

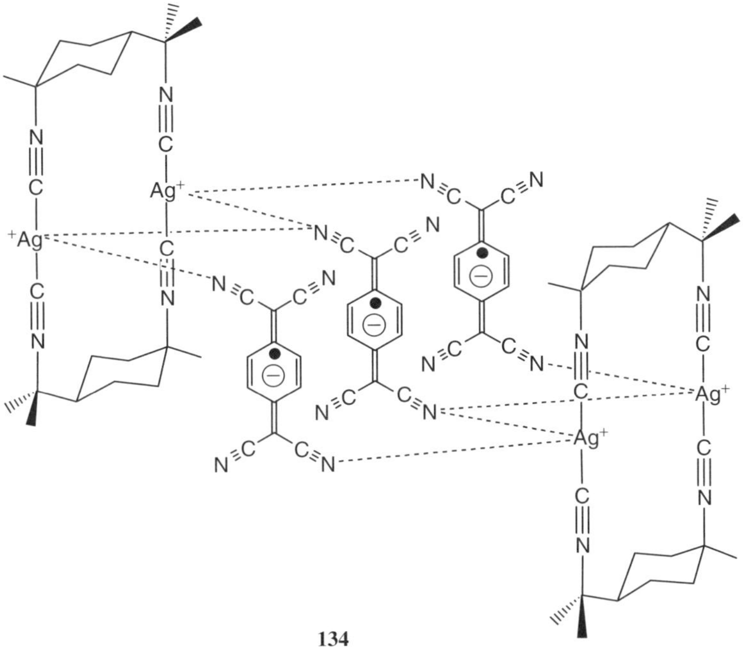

134

The oligomer and polymer dissociation in solution is a very important aspect of this field. Polymer **135**, which was first reported in 1992,[119] was fully characterized by X-ray crystallography as a BF_4^- salt. However, in solution, the polymer dissociates to form oligomers with 7–9 units, depending on the counteranion (BF_4^-, PF_6^-, ClO_4^-). The evaluation of the number of units was performed using spin-lattice relaxation time (T_1) and nuclear Overhauser enhancement constant (NOE) measurements.[120]

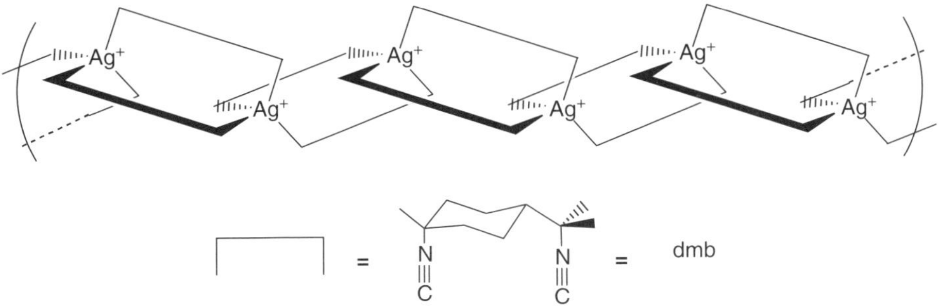

135

Doubly bridged 1-D polymers are scarce in comparison with singly bridged materials, as presented in the preceding section. The first full description of the polymers of the type $\{[M(dmb)_2]Y\}_n$, where M = Ag (**135**) and Cu (**136**), and $Y^- = BF_4^-$, PF_6^-, NO_3^-, ClO_4^-, and $CH_3CO_2^-$ was reported in 1997.[121] They are prepared from the corresponding Ag salts in the presence of an excess of dmb. For Cu, the reactants $[Cu(NCCH_3)_4](BF_4)$ or $Cu(BF_4)_2$ can also be used directly with dmb in excess. In the latter case, a redox process occurs to change the oxidation state II into state I. The dmb itself acts as a redox sacrificial agent.

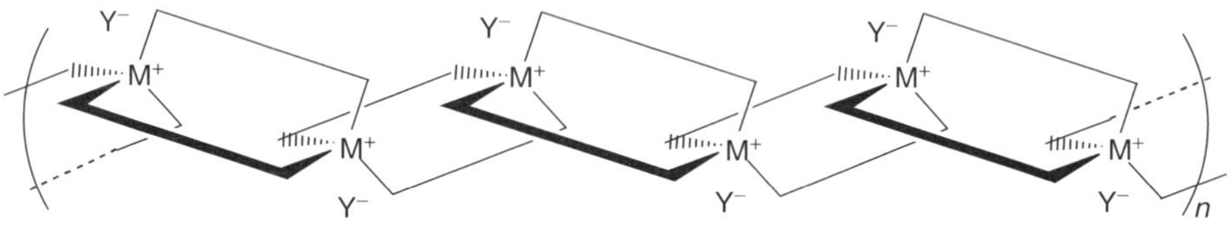

136; M = Cu

The various 1-D polymers were characterized by X-ray crystallography (M = Ag; $Y^- = BF_4^-$, NO_3^-, ClO_4^-). The Ag···Ag separation and the Ag_3 angle were about 5°C and 140°C in all cases, showing that the counteranion acts as a spectator. The Cu-analogs were characterized by light scattering and M_w is about 160,000 (300 units). While the Ag polymers were crystalline according to XRD analysis, the corresponding Cu polymers were amorphous, or at best semicrystalline. Other characterization techniques were used and include solid-state ^{13}C NMR (magic-angle spinning), DSC, and IR spectroscopy. No evidence of uncoordinated CNR groups was observed, corroborating the long dimension of the polymers. Weak second- order phase transitions resembling glass transitions were depicted in the DSC traces. The change in heat capacity ranged from 0.1 to 1.3 (J/g)deg, and the temperature for transition occurred between 38 and 96°C. The white (Ag) and beige (Cu) materials absorbed 202 and 245 nm, respectively. They emitted broad emission bands centered around 500 nm with emission lifetimes found in the μs regime. Based on DFT calculations (Density Functional Theory), an MLCT transition was assigned for the absorption and emission. The emission was also assigned from an intrachain exciton photoprocess occurring in the triplet states.

When the counteranions were exchanged by $TCNQ^-$, polymers of the type $\{[M(dmb)_2]TCNQ\ solvent\}_n$ (M = Cu, Ag) were obtained and were structurally characterized by X-ray as well.[122,123] The anion exchange does not perturb the skeleton of the polymer. Subsequently, when neutral TCNQ was added to these polymers, new materials of the type $\{[M(dmb)_2]TCNQ \cdot x\ TCNQ^0 \cdot y\ solvent\}_n$ (M = Cu, Ag; $x = 0$, 0.5, 1.0, 1.5; solvent = none, THF, and toluene) were obtained and were also fully characterized (X-ray, DSC, XRD, ...).[122] When $x = 0$ and 0.5, the materials were electric insulators. When $x = 1.0$ and 1.5, semiconducting materials were discovered. Performances (resistivities) were found to be directly related to the morphology of the materials. The Ag polymers were more crystalline and exhibit less resistivity. One X-ray structure was resolved ($x = 1.5$ and M = Ag) and revealed 1-D polymers stacking side-by-side, forming a layer, and above it, another layer of perpendicular mixed-valence TCNQs forming a "carpet" structure where π–π contacts were promoted, explaining the presence of conductivity. This bilayer structure repeated itself to form a multilayer edifice. For M = Cu and $x = 1.0$ and 1.5, the two materials were photoconducting as well. The mechanism proceeded by an initial photoinduced electron transfer from the Cu(I) center, the "$(TCNQ)_5^{2-}$" layer rendering the bulk more conductive. Photovoltaic cells built upon glass/SnO_2/polymer

$\{[Cu(dmb)_2]TCNQ \cdot 1.5\ TCNQ^0\}_n$/Al were designed and investigated. By replacing the neutral TCNQ by other derivatives or C_{60} (as doping agent), performance could be changed (photocurrent versus steric and redox properties of the agent).

Belonging to this same family of 1-D polymers, the $\{M(dppm)(dmb)^+\}_n$ polymers **137** and **138** (M = Cu, Ag) were also recently reported.[96] They were prepared from the corresponding dimers $M_2(dppm)_2^{2+}$ (M = Cu, Ag) reacting with dmb in slight excess. The X-ray structure for M = Ag was solved and the Ag···Ag distances were 4.028(1) and 9.609(1) Å for the dppm- and dmb-bridged fragments, respectively. The dmb adopts the Z-conformation because of the phenyl–phenyl steric hindrance induced by the side-by-side $M_2(dppm)_2$-$M_2(dppm)_2$ unit interactions. This polymer was the unique case of mixed-ligand polymer (diisocyanide and diphosphine) in the doubly bridge series. In fact, the 1-D polymer exhibited a staircase structure. The metathesis of the counterion by $TCNQ^-$ and $TCNQ^0$ as doping agent did not lead to conducting or semiconducting materials, indicating that the formation of a p-stacked mixed-valent TCNQ 1-D chain did not occur. This behavior was consistent with the staircase structure of the polycationic chain.

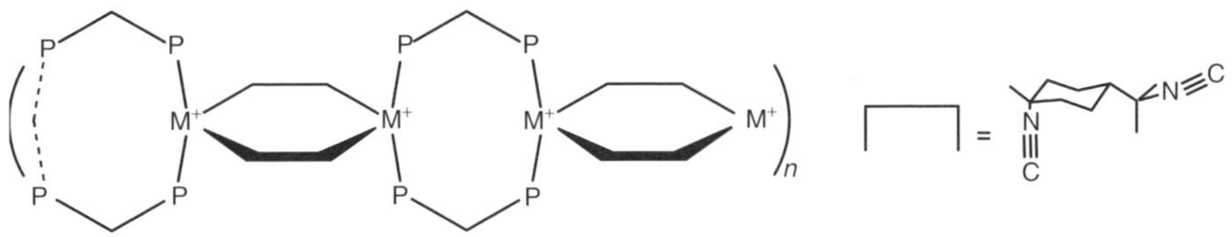

P = PPh_2; **137** (M = Cu), **138** (M = Ag)

Polymer **139** is another example of a mixed-ligand 1-D polymer. The difference with **137** and **138** is that bromide ions replace the diisocyanides.[124] The X-ray structure reveals two Ag–Br distances (2.7431(13) and 2.9453(14) Å), and the dimension indicates ionic interactions. The structure also exhibits a staircase geometry as previously and the nonbonding Ag···Ag separations are 3.605(2) and 3.916(2) Å for the dmpm- and Br-bridged fragments, respectively. Raman spectra exhibit ν(Ag···Ag) of 48 cm^{-1}, giving a weak approximate force constant of 0.03 mdyn $Å^{-1}$.

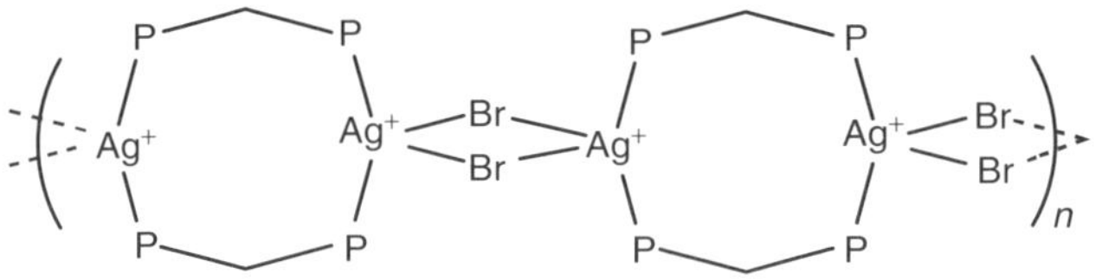

P = PMe_2; **139**

This polymer is also very similar to those recently reported by Pettinari and collaborators[125] and Brandys and Puddephatt.[104] Indeed, polymers **140–144** exhibit the same basic structure as **139**, as revealed by X-ray crystallography.

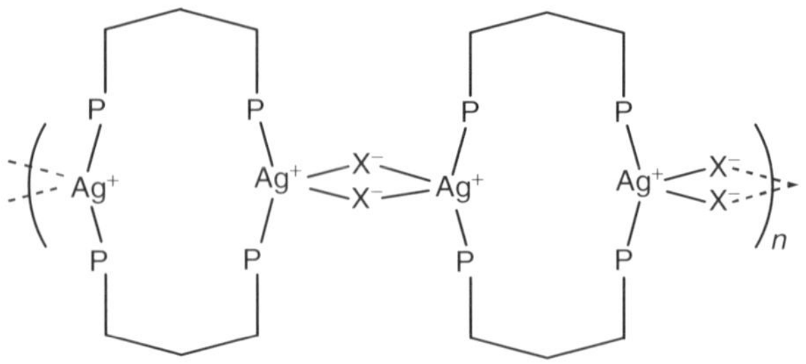

P = PPh_2; **140** (X = Cl), **141** (X = SCN); **142** (X = NCO); **143** (X = O_2CCH_3)

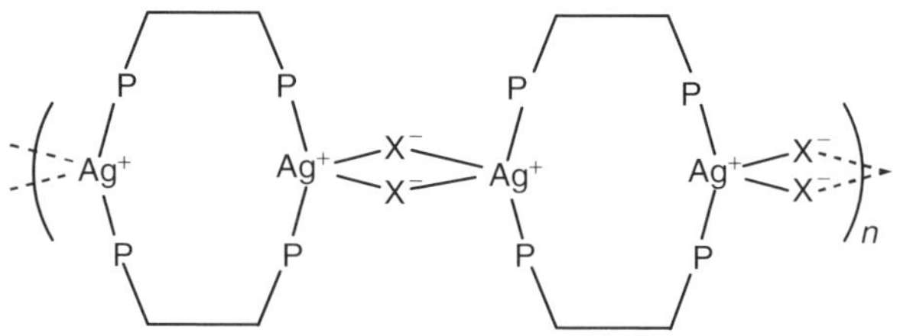

P = PPh_2; **144** (X = O_2CCH_3)

Another interesting 1-D doubly bridged polymer is **145**.[117] It is prepared from the direct reaction from 3,6-bis(diphenylphosphino)pyridazine with the $Ag(ClO_4)$ salt in acetonitrile. The Ag···Ag separations for the two crystallographically independent N–P units are 3.005(2) and 3.185(2) Å, while the distance for the interligand P–P segment, is 3.535 Å.

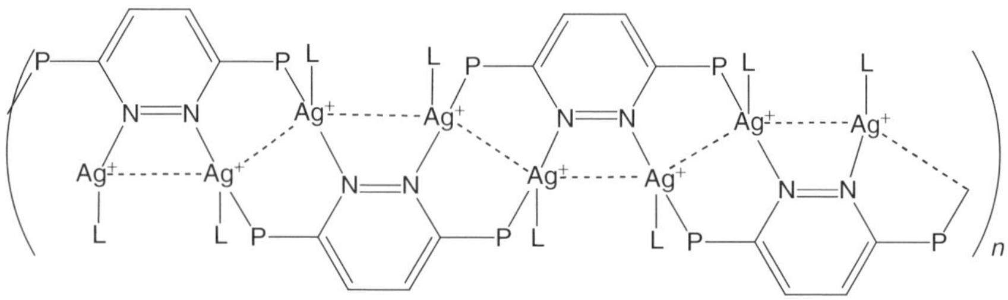

P = PPh_2; L = CH_3CN; **145**

Scheer and collaborators reported three papers on very interesting polymers (**146–150**) using unusual metalla-P_2- and metalla-P_5-containing fragments, such as $Cp_2(CO)_4Mo_2P_2$ and Cp^*FeP_5, which were directly reacted with the corresponding CuX or AgX salts.[126,127,128] These crystallographically characterized materials were mixed-ligand doubly bridged polymers and were also characterized by ^{31}P MAS solid-state NMR (**146–148**) and discrete

Fourier transform (DFT) calculations to interpret the spectra, notably with respect to P–Cu coupling constants.[127]

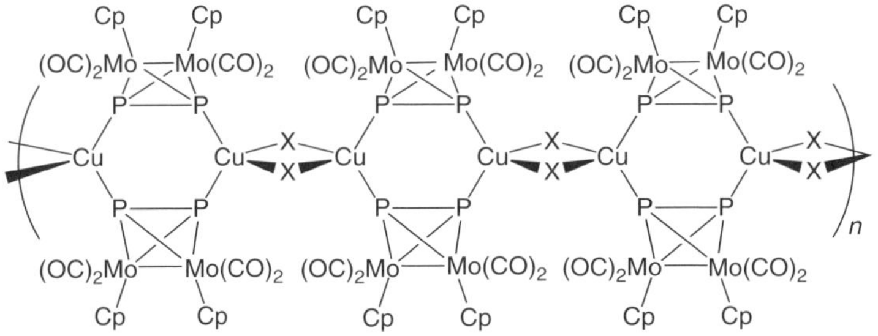

Cp = η^5-C_5H_5; **146** (X = Cl), **147** (X = Br), **148** (X = I)

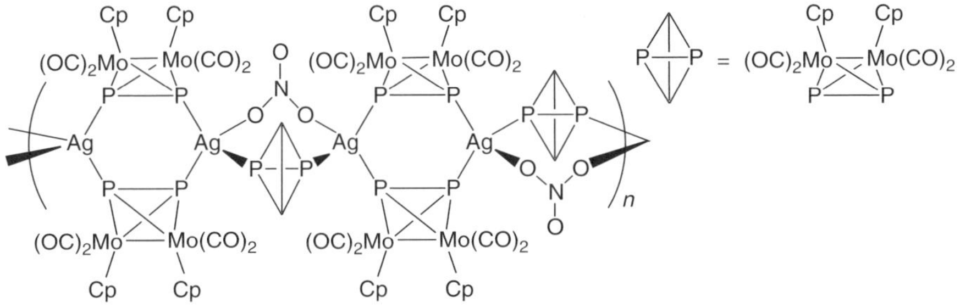

149

Polymer **150** exhibits the versatile multidentate ligand P_5, which provides the possibility of acting as a η^1, η^2, η^3, η^4, and η^5 ligand in theory. In this example, this fragment binds FeCp* (Cp* = C_5Me_5) in a η^5 fashion, and bridges two Cu atoms in a μ:η^1:η^1 manner.

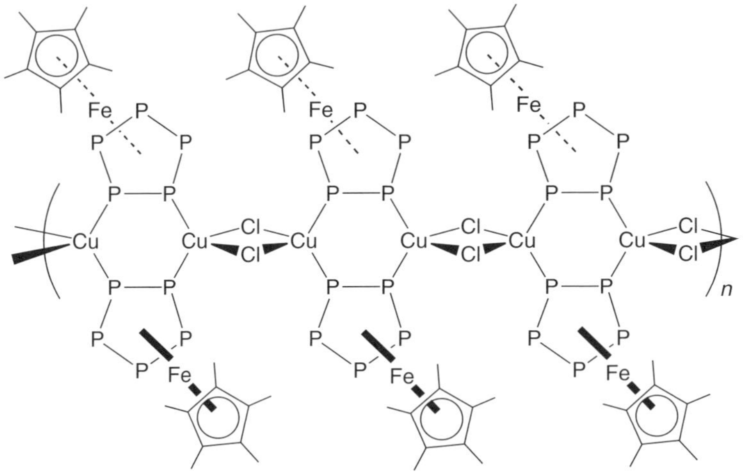

150

Similar to the preceding examples, the ligand P_4S_3 formed the doubly bridged polymer **151** from the direct reaction with the AgAl(hfip) salt (hfip = $OC(CH_3)(CF_3)_2$).[129] The polymer exhibited two types of binding modes for the polyphosphinepolysulfur ligand. First, it used a P–P η^1:η^1 bridging mode to form $Ag_2(P_4S_3)_2^{2+}$ units exhibiting a center of inversion. Then, two other ligands bridged these units in a S–P^1:η^1 mode to form a rigid-rod polymer, similar to mixed-ligand materials discussed in this chapter. This polymer was characterized by X-ray, Raman, and solution NMR spectroscopy. Optimized structures of the ligand, and Ag-containing fragments were also computed with the aid of DFT, focusing on the total energies of these fragments and the charge delocalization within the ligand. The low-frequency Raman peaks associated with ν(P–S), ν(P3), δ(P–S) were assigned with the help of DFT calculations as well.

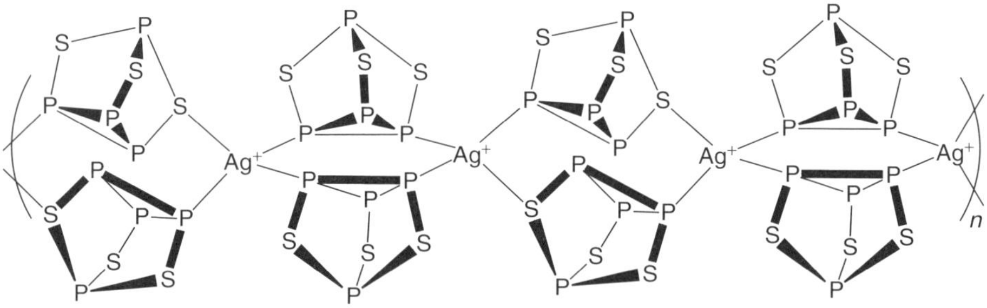

151

V. OLIGOMERS AND POLYMERS WITH M–M BONDS IN THE BACKBONE (INCLUDING ACYCLIC LINEAR CLUSTERS)

Such species are rare in comparison with non-M–M bonded polymers simply because reliable characterization often requires the presence of X-ray structures. On the other hand, oligomers have the advantage of being more soluble most of the time, hence increasing the chance for X-ray characterization. An earlier example of an M–M bond containing oligomer supported by diisocyanides was reported by Mann and collaborators.[130] The tetranuclear dicationic species $[Rh_2Mn_2(tmb)_4(CO)_{10}]^{2+}$, **152**, was prepared as PF_6^- salt and characterized by X-ray diffraction. The supported Rh–Rh bond is surprisingly long (2.922(2) Å), as are the unsupported axial Rh–Mn bonds (2.905(5) and 2.883(4) Å). The oxidation state of the Rh metal atom is +1 (assuming that Mn is 0) leading to the first assumption that the electronic configuration is d^8. In binuclear species, d^8–d^8 electronic configuration leads to non-M–M bonding interactions. The long Rh–Rh distance is not surprising, and so the tmb ligands truly support the formation of this oligomer. However, if this nonbonding interaction really exists, then the presence of axial $Mn(CO)_5$ fragments must be explained. The UV-vis spectrum reveals a single strong and narrow absorption at 605 nm attributable to a $d\sigma$–$d\sigma^*$ electronic transition. The rest of the spectra is free of any other absorption (i.e., no other $d\sigma$-$d\sigma^*$ electronic band for the Rh–Mn bonding) meaning that the Rh_2Mn_2 frame is entirely delocalized. One must reassess the electronic configuration to d^7 for the Rh atom (i.e., Rh(II)) in agreement with the way the species was prepared ($Rh_2(tmb)_4^{2+}$ + $Mn_2(CO)_{10}$ + $h\nu$).

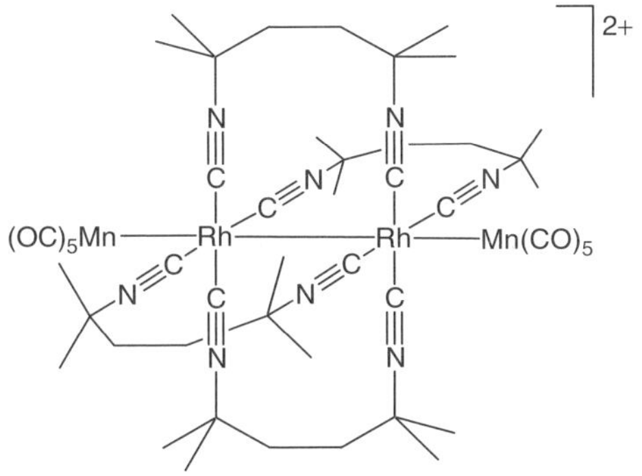

152

Other examples of diisocyanide-containing oligomers were also reported by Mann and collaborators and by Gray and collaborators. These species were described in a recent review and are not described here.[41]

A more recent diisocyanide-containing oligomer was reported by Kadish and collaborators.[131] Oligomer **153**, $(dpf)_4Rh_2(CNC_6H_4NC)Rh_2(dpf)_4$ (dpf = *N,N′*-diphenylformamidine) is prepared by reacting the known $Rh_2(dpf)_4$ with the corresponding bridging diisocyanide. This compound exhibits two overlapping "one"-electron reversible oxidations in CH_2Cl_2 at $E_{1/2}$ = 0.22 and 1.20 V vs. SCE. There is no evidence of electronic interactions between the two dinuclear centers, either during oxidation or reduction.

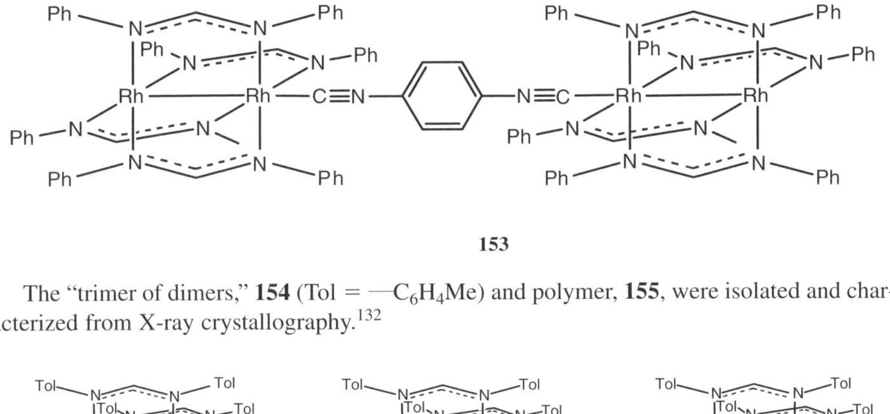

153

The "trimer of dimers," **154** (Tol = —C_6H_4Me) and polymer, **155**, were isolated and characterized from X-ray crystallography.[132]

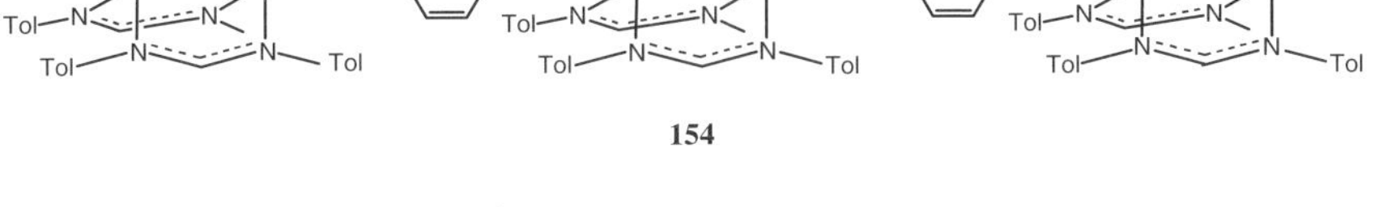

154

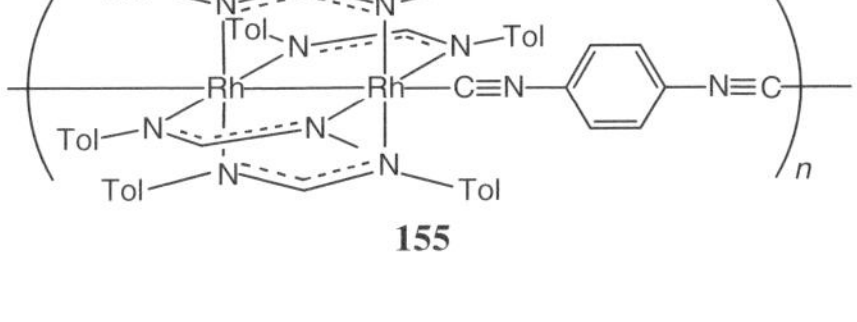

155

The Rh–Rh bond distance is 2.5697(7) Å in **155**, and 2.520(2) (inner) and 2.4862(13) Å (outer) in **154**.

An earlier work on diphosphine-containing oligomers was reported by Deeming and collaborators.[133] Oligomers **156** and **157** were prepared by reacting the Os–Os bonded precursor $(PPh_3)Os_2(MeCO_2)_2(CO)_5$ with diphos (dppm or dppe), where the axial COs are substituted by the diphosphines, and not the equatorial COs, presumably because of lesser steric problems at this position. The complexes where characterized by IR among other techniques allowing one to monitor reactions. The presence of PPh_3 as the end-group prevents further coordination at the axial position and so oligomerization stops at two units in a controlled fashion.

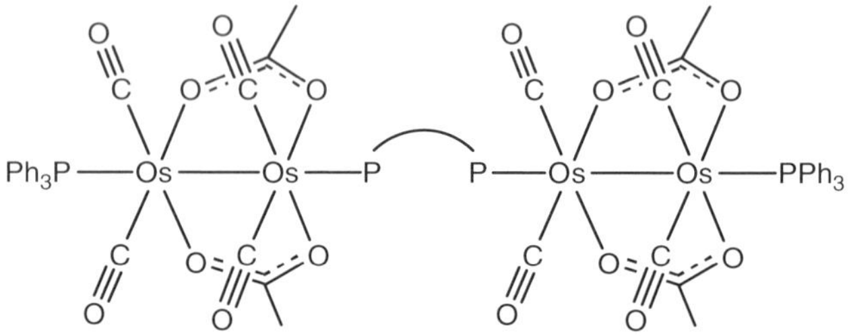

156 (P–P = dppm)
157 (P–P = dppe)

On the other hand, a 2:1 stoichiometric addition of $Os_2(MeCO_2)_2(CO)_6$ with dppm or dppe leads to **158** and **159**, respectively.

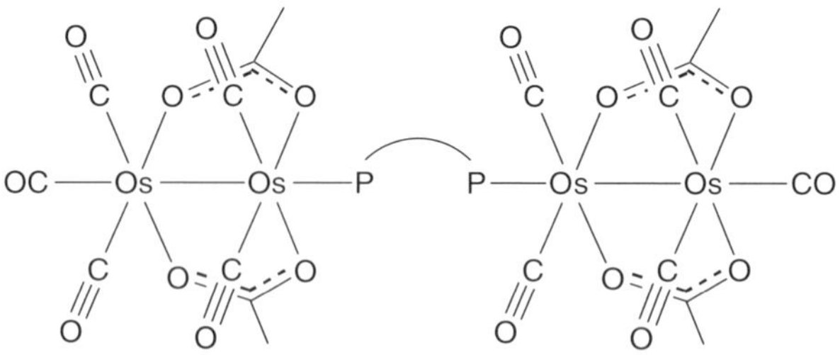

158 (P–P = dppm)
159 (P–P = dppe)

The controlled formation of a three-unit species (**160**) can be achieved by reacting two equivalents of $Os_2(MeCO_2)_2(CO)_6$ with a difunctionalised dinuclear complex $(dppm)Os_2(MeCO_2)_2(CO)_4(dppm)$. Again, IR and ^{31}P NMR are particularly useful techniques for confirming formation of such oligomers. Diarsine-containing materials (diarsine = $Ph_2AsCH_2AsPh_2$) were prepared and characterized in a similar way.

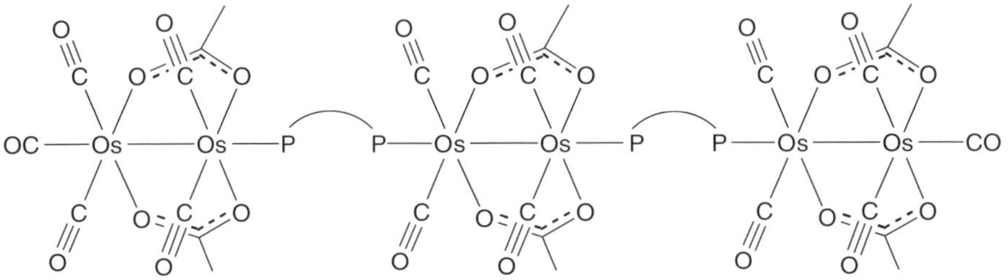

160

Cowie and collaborators synthesized and characterized the structurally related polymers **161** and **162**, $\{Ru_2(MeCO_2)_2(CO)_4(diphos)\}_n$, (diphos = dppm, dmpm).[134]

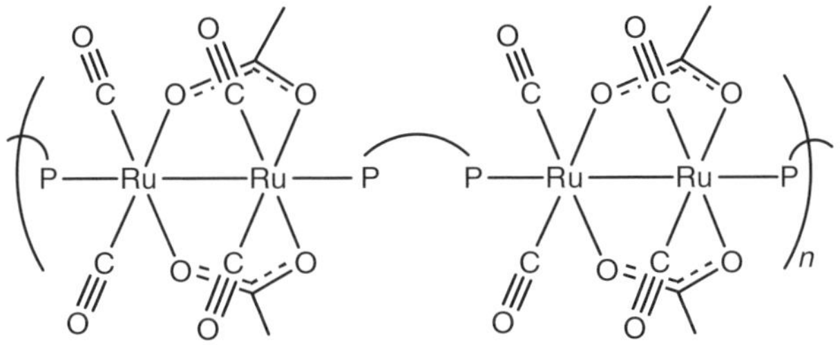

161 (P–P = dppm)
162 (P–P = dmpm)

Their syntheses consist of reacting either the dinuclear $[Ru(CO)_2(O_2CMe)(NCMe)]_2$ or the polymer $[Ru_2(CO)_4(O_2CMe)_2]_2$ with dppm or dmpm in a 1:1 ratio. An excess of dppm leads to the formation of the dinuclear complex $[Ru(CO)_2(O_2CMe)(dppm)]_2$.

Kerby and collaborators reported the synthesis and characterization of a multiply bonded M–M containing a polymer built upon a diphosphine ligand.[135,136] Polymer **163** is prepared from the quadruply bonded precursor $Mo_2(O_2CMe)_4$ in the presence of one equivalent of bis(dimethylphosphino)ethane (dmpe). The polymer was characterized by X-ray crystallography and the Mo_2 distance is 2.105(3) Å, typical for a quadruple bond.

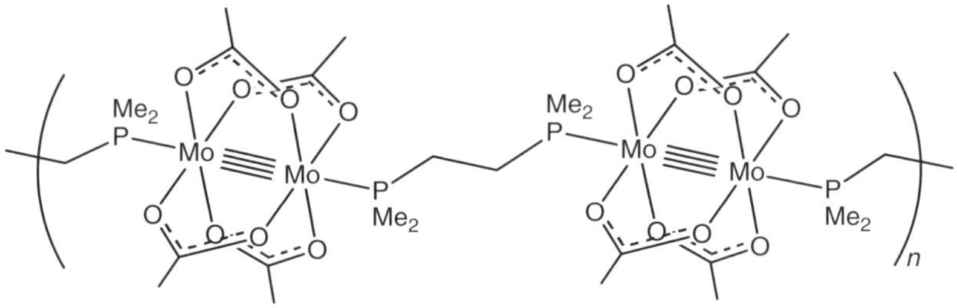

163

Other examples of M–M bond-containing polymers were reported by Puddephatt and collaborators.[137] Examples of these include the rigid-rod polymers **164** and **165** and the zigzag materials **166** and **167**. They were prepared from the d^9-d^9 M–M bonded dimer, $Pt_2(dppm)_2Cl_2$, and the corresponding diphosphine or diisocyanide. The polycationic materials of the type **164** and **165** exhibit no IR absorption associated with uncoordinated CN—ligand, which may act as the end-group, suggesting that the polymer chain must be long in the solid state. For **167**, the presence of cyclic oligomers was detected in solution, but in all cases, no molecular weight was obtained. The oligomer turned out to be a dimer based on X-ray data. The dppm-containing polymer, **166**, does not form an oligomer, at least not a detectable one, and this result was explained by computer modeling where steric hindrance plays a major role in the form of the polymer.

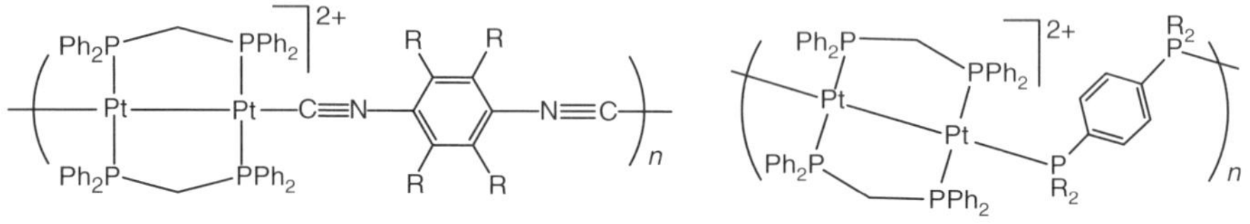

164 (R = H), **165** (R = Me) **166** (R = Ph), **167** (R = *i*-Pr)

In a closely related work, Harvey and collaborators revisited the diisocyanide-containing polymer **165**, and investigated related polymers, **168–170**, in relation with electronic communication along the backbone.[138] These amorphous polymers were prepared in a similar way, where the starting materials, $M_2(dppm)_2Cl_2$ (M = Pd, Pt) were reacted with the corresponding diisocyanide in the presence of a chloride scavenger. The new materials were exhaustively characterized by spectroscopic techniques in order to address the size of the oligomers in solution, their thermal stability and behavior, and their optical and photophysical properties. Based on light scattering, T_1/NOE, and viscosity measurements, the size of the oligomers were found to be around 4 to 15 units in acetonitrile. These amorphous solids exhibited clear signs of polymers with long chains. First, stand-alone polymer films were obtained upon spin-coating using acetonitrile as solvent. Second, important swelling of the solids was observed when these materials are dissolved in acetonitrile. Luminescence in the 640–750-nm range was detected in butyronitrile solution at 77 K, luminescence that was assigned to a triplet $d\sigma d\sigma^*$ excited state. The excited-state lifetimes ranged from 6 to 21 ns for M = Pd and from 3 to 4 μs for M = Pt at this temperature. The overall conclusion was that some electronic communication was observed, but these were not exhaustive.

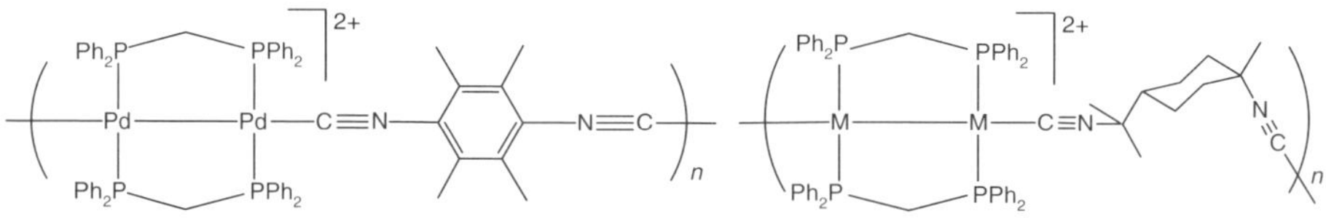

168 **169** (M = Pd), **170** (M = Pt)

Other Pd–Pd bond-containing polymers and oligomers were exhaustively investigated by this same group.[52,139] Polymers **171–173** are prepared from reacting the starting material $Pd_2(dmb)_2Cl_2$ with the chelating ligands dppe and dppp—for instance, in a 1:2 stoichiometry—in the presence of $LiClO_4$ as a chloride scavenger.[52] The chelating preference for these ligands forces the axial and equatorial positions to be coordinated so that the bridging ligand dmb no longer supports the Pd–Pd bond. In this case, some rotation about the Pd–Pd bond becomes possible. The polymer length in solution is about 2 (T_1/NOE) or 8 (intrinsic viscosity) units, but molecular modeling predicts that the formation of rings appears to be difficult in these cases. The addition of side chains (**173**) does not change the number of units.

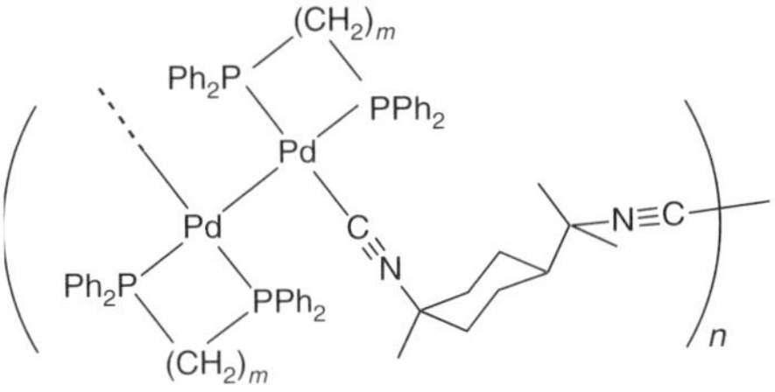

171 ($m = 2$), **172** ($m = 3$)

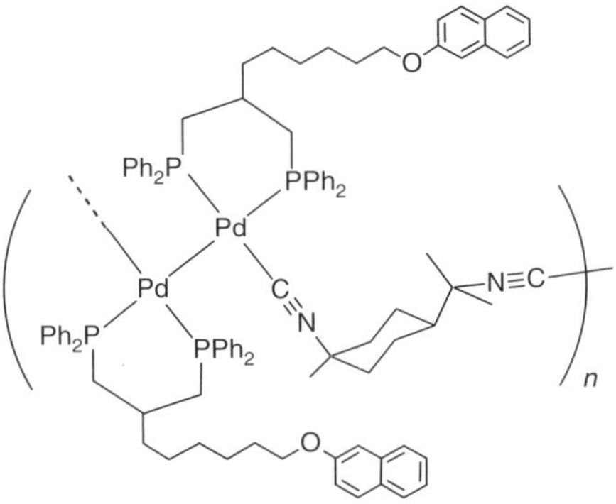

173

When m is larger ($m = 4$–6) or when dpa is employed, the diphosphine ligands act as bridging unit only and for a 1:1 $Pd_2(dmb)_2Cl_2$/diphos ratio, the polymers **174–177** are obtained.[139] These species exhibit longer chain lengths in solution (four units according to T_1/NOE, and 12–16 units according to intrinsic viscosity). Again, no evidence of ring formation was obtained. The discrepancy between the two types of measurements for **171–177** stems from the lack of an appropriate standard in the intrinsic viscosity measurements, degrees of freedom of rotation about the single bond (giving the wrong impression that the tumbling material is smaller than real (T_1/NOE), and the unavoidably approximate assumption that the oligomers are spherical in the T_1/NOE analyses).

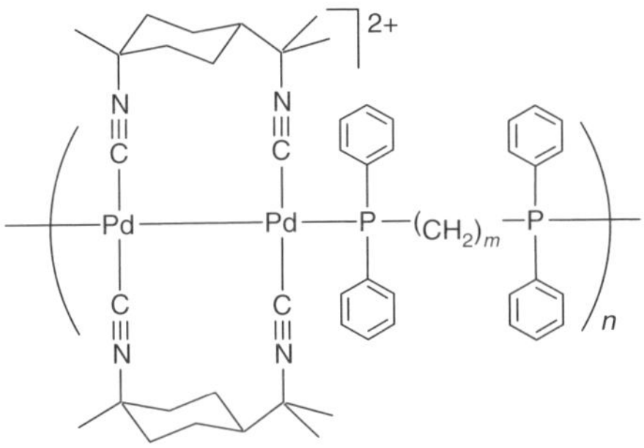

174 ($m = 4$), **175** ($m = 5$), **176** ($m = 6$)

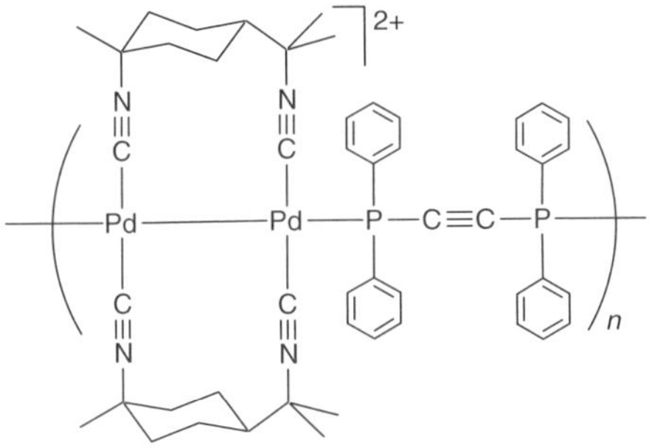

177

These materials, **171–177**, are also amorphous in the solid state and form stand-alone film during drying or spin-coating. They are also strongly luminescent at 77 K (in PrCN glasses), again arising from the triplet $d\sigma d\sigma^*$ excited state. The emission quantum yields a range from 0.03 to 0.17, and lifetimes from 1.50 to 2.75 ns. The short lifetimes, in comparison with Pt–Pt bond-containing polymers discussed earlier (**167** and **170**), is due to the relatively facile photoinduced Pd–Pd and Pd–L bond dissociation in the excited state.

The starting material, **178**, is also interesting.[139] Structurally related to the d^9–d^9, $Pd_2(dppm)_2Cl_2$, $Pd_2(dmb)_2Cl_2$ is most definitely a dinuclear complex in solution based on

T_1/NOE experiments. Molecular modeling strongly indicates the presence of a ring stress due to the large difference between the ligand bite distance (4.4 Å) and the Pd–Pd bond length (2.7 Å). However, matrix-assisted laser desorption ionization–time of flight (MALDI–TOF) measurements indicate the presence of larger molecular-weight species and the amorphous material (based on the X-ray powder diffraction patterns) exhibits a T_g at 6°C ($\Delta Cp = 1.55$ J/°C) and form stand-alone films upon drying acetonitrile solutions. Compound **178** is a polymer in the solid state, but dissociates in solution to become soluble. The behavior appears to be common in this series (**167–178**).

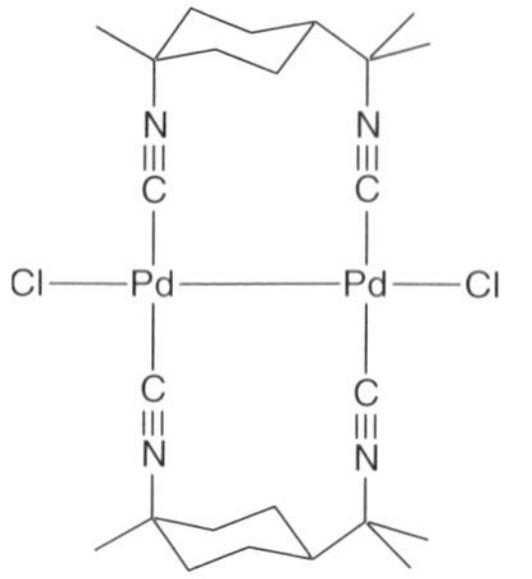

178 (solution)

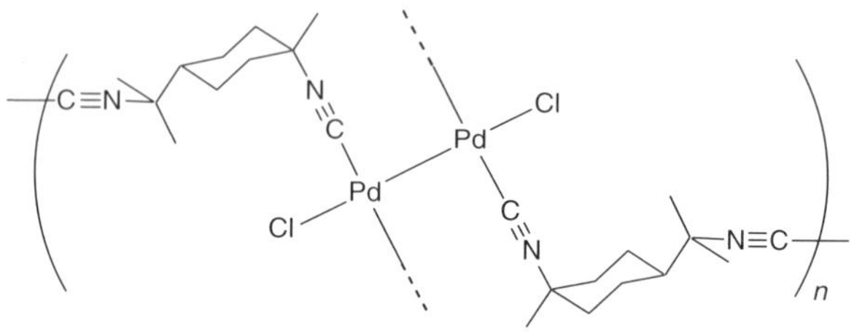

178 (solid state)

Tanase and collaborators reported the syntheses and characterization of two Pt_3/diisocyanide-containing polymers (**179**, **180** as PF_6^- and I^- salts; PPP (dpmp) = $Ph_2PCH_2PPhCH_2PPh_2$).[140] The Pt_3 core represents a linear acyclic cluster (the Pt_3 angle ranges from 176° to 179°) where electronic communication between the axial aryl diisocyanides is possible. The polymers were characterized by extended X-ray absorption fine structure (EXAFS), allowing extraction of structural parameters. The Pt–Pt bond lengths range from 2.70 to 2.72 Å, typical for a Pt_3^{2+} core as verified for model compounds in this same work. The Pt–Pt bonds are very reactive as they add H^+, NO, and tetracyanoethylene, which could be useful for sensor design. The presence of electronic communication along the Pt–Pt bonds and the aryl diisocyanides could not be addressed by electrochemical methods. Both the oxidation and reduction waves measured by cyclic voltametry exhibit irreversible waves, indicating the instability of these species upon the addition or subtraction of electrons. DFT computations predict that the highest occupied molecular orbital (HOMO) is derived from a σ-bonding interaction between the hybridized *dσ* orbitals of the terminal Pt atoms and the *pσ* orbital of the central Pt atom. The lowest unoccupied molecular orbital (LUMO) includes a delocalized *pπ* orbital of the central Pt atom, the hybridized *pπ*/*dπ* systems, and the *pp* orbital of the isocyanides moieties. The replacement of alkyl and aryl groups in the computations suggests that electronic communication is possible from a computation standpoint.

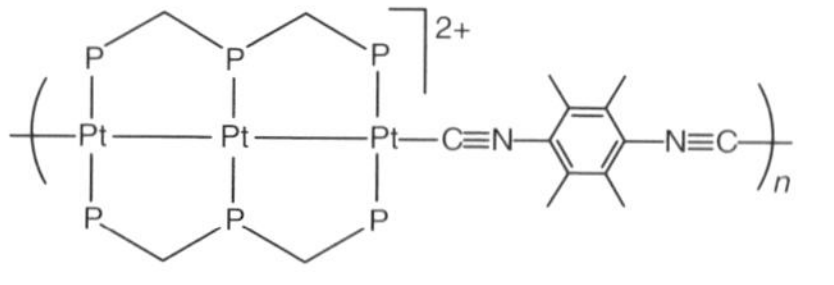

179a (as PF_6^- salt), **179b** (as I^- salt)

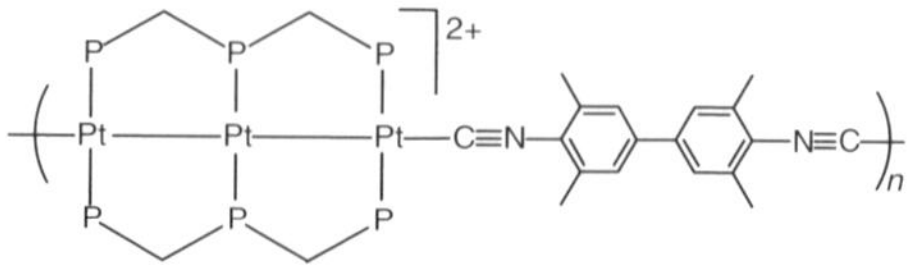

180 (as PF_6^- salt)

Harvey and collaborators also reported the syntheses and characterization of polymers **181–184**.[141,142] They are unique, as these polynuclear acyclic clusters of Pd_4^{2+} and Pt_4^{2+} were, and still are the only examples of linear 58-electron cluster species. The dmb frame corresponds to a catenane structure. The polymers were prepared in a redox manner, where the d^{10}-d^{10} precursor, $M_2(dba)_3{\bullet}CHCl_3$ species (M = Pd, Pt; dba = dibenzylideneacetone), reacted in the presence of dmb and $Pd(O_2CMe)_2$ (for M = Pd only). For M = Pt, the presence of chloroform contributed to the oxidation of the zero-valent Pt species in solution, so the formation of the Pt_4 clusters with the oxidation state of 0.5 was obtained as well. While **181** was characterized by X-ray crystallography, **182–184** were amorphous (based upon X-ray diffraction patterns), and were characterized by spectroscopic methods and the measurements of the intrinsic viscosity. M_n values in the order of 84,000 to 203,000 were found using $\{Cu(dmb)_2^+\}_n$ as a comparative standard. The polymers were also characterized by UV-vis and luminescence spectroscopy. The UV-vis spectrum of **181** exhibited a single and strong absorption at 490 nm, which became very narrow upon cooling the solution, a behavior that was consistent with a $d\sigma d\sigma^*$ transition. The (EHMO) calculations confirmed this assignment. Polymer **181** emitted at approximately at 700 nm with τ_e of about 1 ns in PrCN at 77 K, while polymers **182–184** emitted at about 736–755 nm, with emission lifetimes of 4.8 to 5.2 ns. For all four polymers, the emission arose from the $d\sigma d\sigma^*$ triplet excited state.

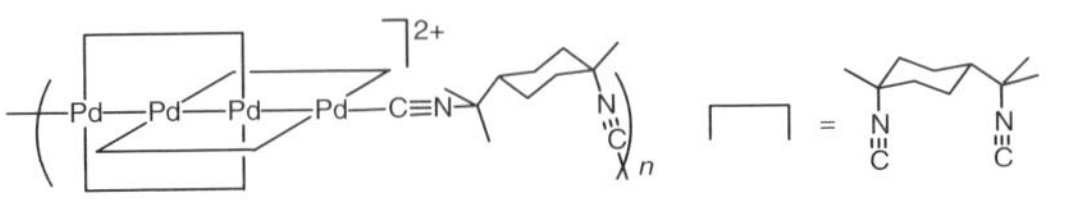

181

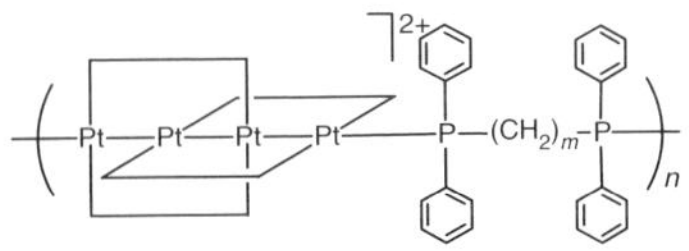

182 (m = 4), **183** (m = 5), **184** (m = 6)

The polymers turned out to be difficult to make, as the choice of the oxidizing agent or coreactants was important. For example, neutral tetracyanoquinodimethane, which is reduced to the corresponding monoanion, was capable of oxidizing the Pd atoms to 2+. In this case, the sole product was the d^8–d^8 dimer $Pd_2(dmb)_4^{4+}$.[142] Similarly, when $Au(PPh_3)_2^+$ was used in the presence of the d^{10}–d^{10} $Pt_2(dba)_3$ and d^9–d^9 $Pt_2(dmb)_2Cl_2$ (with the aim of incorporating a gold atom in the polymer chain), a redox reaction occurred where the mixed-metal cyclic cluster $Pt_2Au_2(dmb)_2(PPh_3)_4^{2+}$ was formed as a side product. The other side products, $Au(PPh_3)Cl$ (acting as a halide scavenger) and the model cluster, $Pt_4(dmb)_4(PPh_3)_2^{2+}$ where no gold atom was introduced, were also observed at the end of the reaction.[143]

VI. POLYMERS AND OLIGOMERS OF CYCLIC CLUSTERS

Oligomers of clusters are obviously precursors and models for their corresponding polymers, and some have been well characterized. For example, oligomer **185** was recently synthesized and characterized by spectroscopic methods.[144] Based on an X-ray structure of a single-cluster model, the structure of oligomer **185** could be established with certainty. It consisted of one central $Pt_6(CO)_4(t\text{-}Bu_2P)_4$ and two peripheral $Pt_3(CO)_2(t\text{-}Bu_2P)_3$ clusters rigidly linked together by 1,4-diethynylbenzene assembling ligands. This oligomer was characterized by $^{31}P\{^1H\}$ NMR, MALDI-TOF and elemental analysis. Using a well-characterized model compound ($Pt_6(CO)_4(t\text{-}Bu_2P)_4(C{\equiv}CC_6H_5)_2$), the spectroscopic signature was crucial in the identification of **185**.

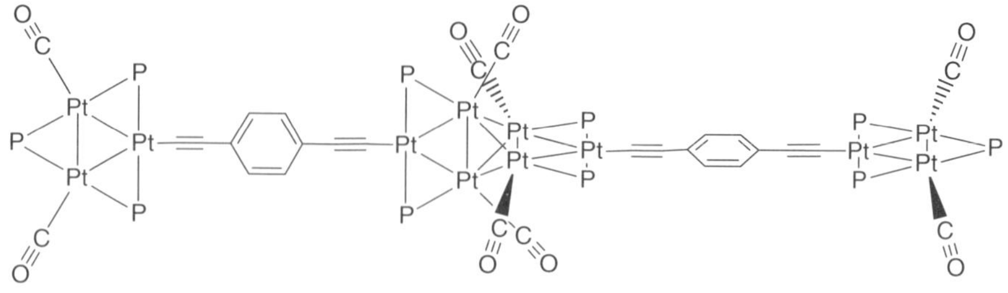

185(P = $^{t}Bu_2P^-$)

Similarly, the unsaturated $Ni_3(dppm)_3(I)^+$ cluster reacted with 1,6-diisocyanohexane to form the "dimer of clusters" **186**.[145] The oligomer was unambiguously identified by plasma desorption and mass fast-atom bombardment (FAB) spectrometry. Although the iodide represents a potential leaving group, its extraction may be very difficult under mild conditions or without a strong Lewis acid.[146]

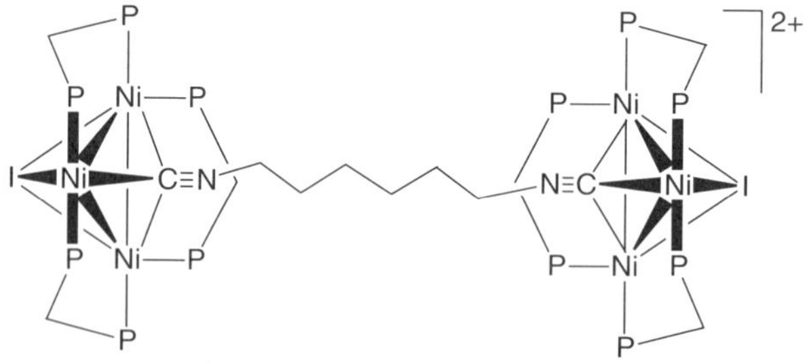

186 (P^P = dppm)

Other oligomers of clusters were reported using the bridging ligand dppf and considering the $Co_2(C_2)$ core as a cluster.[147] Several examples were prepared, including **187–190**, and were analyzed by cyclic voltammetry. Elemental analyses and ^{1}H NMR established the relative stoichiometry of dppf versus cluster. No evidence of communication between the redox centers was observed.

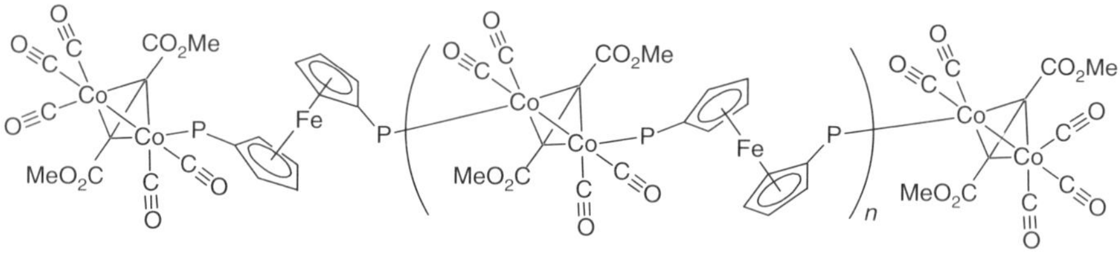

187 (n = 0), **188** (n = 1), **189** (n = 2), **190** (n = 3)

Puddephatt and collaborators also reported a series of oligomers (**191**, **192**) and polymers of clusters (**193**, **194**).[148,149] The syntheses consist of reacting the known precursors $M_3(dppm)_3(CO)^{2+}$ (M = Pd, Pt) with Me$_4$dib in the appropriate ratios in CH_2Cl_2. Their identification and characterization were performed using model clusters and monoisocyanides, and spectroscopy. The polymers exhibit extensive dissociation and fluxion motions of the isocyanide ligand.

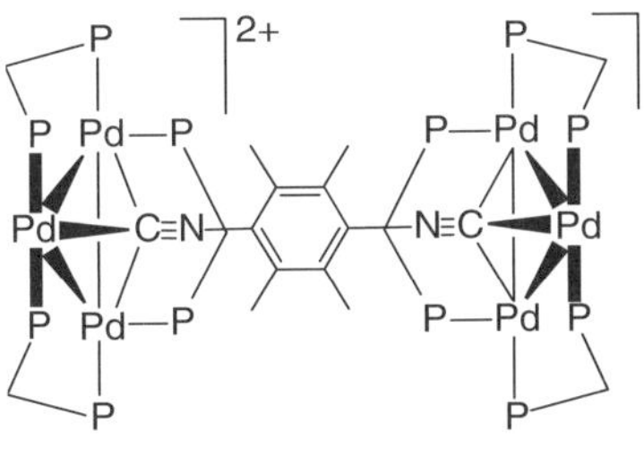

191 (P^P = dppm)

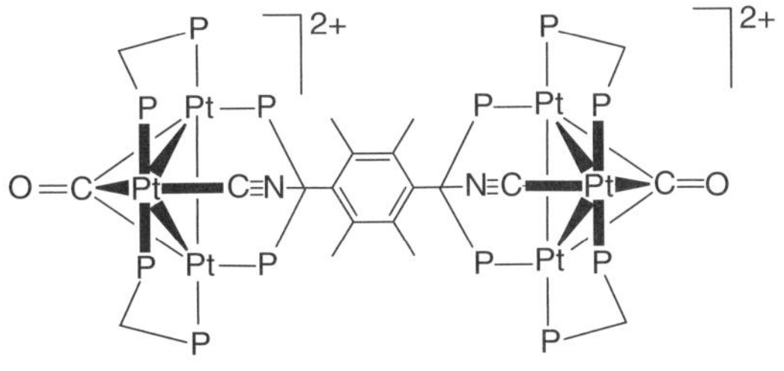

192 (P^P = dppm)

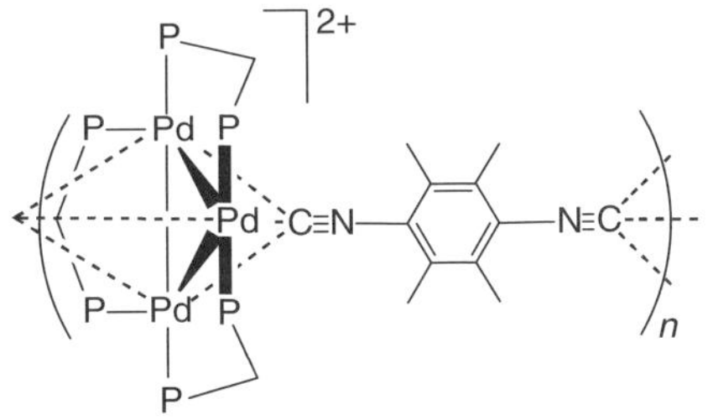

193 (P^P = dppm)

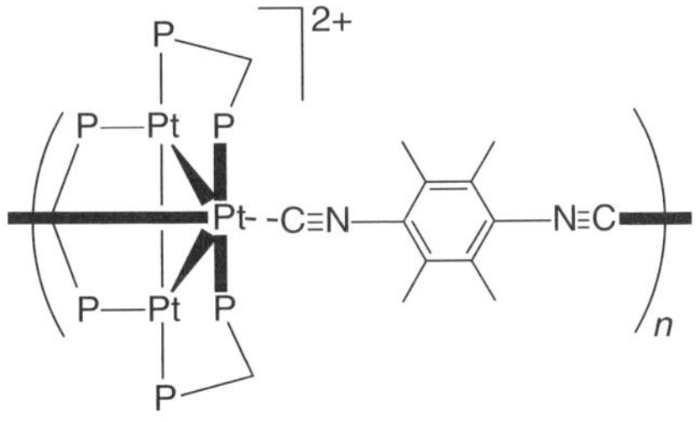

194 (P^P = dppm)

Johnson and collaborators recently reported the synthesis and characterization of an original polymer of cluster **195** (the CO groups are not shown for clarity) and the corresponding formation of nanoparticles and nanowires induced by an electron beam.[150,151,152] The synthesis consists of refluxing the known hexanuclear cluster $Ru_6C(CO)_{17}$ with dpa in THF. The proposed formula was verified using elemental analysis. The estimated molecular weight was obtained by electron microscopy (1000–1020) using several samples. Upon irradiation of the samples on a silicon wafer, the polymer lost its "insulating" carbonyl groups, and conducting nanoparticles and nanowires were obtained. It was proposed that the conductivity operates through an electron hopping.

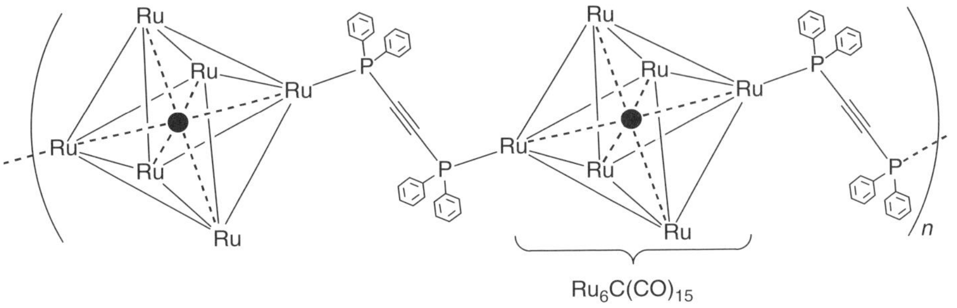

195

Another example of a coordination polymer of clusters is **196**.[153] The reaction consisted of mixing CuI with pyridine-2-thione. The crystal structure revealed that the Cu_6 core was held together by six pyridine-2-thione ligands via Cu–S coordination according to the formulation $\{Cu_6(\mu_3\text{-}SC_5H_4NH)_4(\mu_2\text{-}SC_5H_4NH)(I)_4(\mu\text{-}I)_2\}_n$. Two μ-I bridging ligands secured the polymer structure as a doubly bridged 1-D polymer.

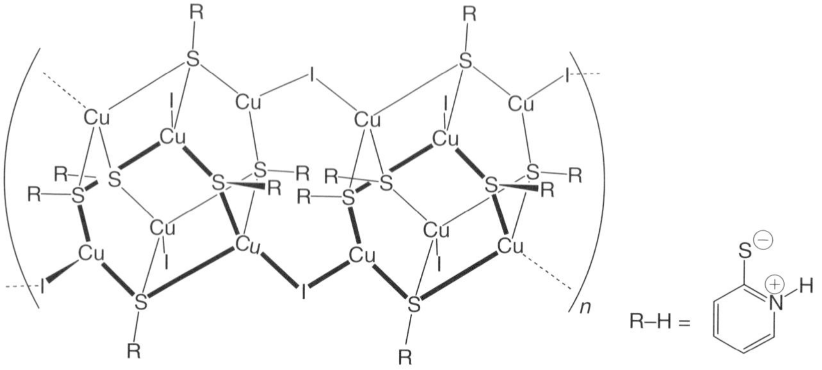

196

VII. CONCLUSION

Despite the fact that numerous examples have been described in this chapter, the field of 1-D transition metal-containing coordination/organometallic polymers and oligomers built upon assembling diphosphine and diisocyanide ligands is still in its infancy in comparison with those issued from N-containing bridging ligands. In addition, applications for these types of materials are very scarce, as most works focus on the preparation and characterization. There are clearly opportunities for future investigations in areas such as nanomaterials, liquid crystals, catalysis, and photonics. Many of the oligomers and polymers described in this chapter exhibit luminescence. Another concept that has yet to be introduced is chirality in the polymer backbone. Rapid developments are anticipated in the future.

ACKNOWLEDGMENT

The Natural Sciences and Engineering Research Council (NSERC) is acknowledged for funding over the years. The author thanks the undergraduate and graduate students and postdoctoral fellow that did the work. Their names are listed with the references.

REFERENCES

1. L. Han, M. Hong, *Inorg. Chem. Commun.*, **8**, 406 (2005).
2. Q. Ye, X.-S. Wang, H. Zhao, R.-G.Xiong, *Chem. Soc. Rev.*, **34**, 208 (2005).
3. T. Itaya, K. Inoue, *Recent Res. Develop. Macromol.*, **6**, 259 (2002).
4. T. Uemura, S. Kitagawa, *Chem. Lett.*, **34**, 132 (2005).

5. B.-H. Ye, M.-L. Tong, X.-M. Chen, *Coord. Chem. Rev.*, **249**, 545 (2005).
6. S. Kitagawa, K. Uemura, *Chem. Soc. Rev.*, **34**, 109 (2005).
7. P. D. Harvey, *J. Inorg. Organomet. Polym.*, **14**, 211 (2004).
8. A.S. Abd-El-Aziz, I. Manners, *J. Inorg. Organomet. Polym. Mater.*, **15**, 157 (2005).
9. J.-C. Dai, Z.-Y. Fu, X.-T. Wu, *in Encyclopedia of Nanoscience and Nanotechnology*, H. S. Nalwa, Ed., Vol. 10, p. 247, American Scientific Publishers, Stevenson Ranch, CA, 2004.
10. X. Chen, C. M. Drain, in *Encyclopedia of Nanoscience and Nanotechnology*, H. S. Nalwa, Ed., Vol. 9, p. 593, American Scientific Publishers, Stevenson Ranch, CA, 2004.
11. R. Knapp, S. Kelch, O. Schmelz, M. Rehahn, *Macromol. Symp.*, **204**, 267 (2003).
12. B. Kesanli, W. Lin, *Coord. Chem. Rev.*, **246**, 305 (2003).
13. L. Carlucci, G. Ciani, D. M. Proserpio, *Coord. Chem. Rev.*, **246**, 247 (2003).
14. A. Erxleben, *Coord. Chem. Rev.*, **246**, 203 (2003).
15. S. L. Zheng, M.-L. Tong, X.-M. Chen, *Coord. Chem. Rev.*, **246**, 185 (2003).
16. S. A. Barnett, N. R. Champness, *Coord. Chem. Rev.*, **246**, 145 (2003).
17. A. S. Abd-El-Aziz, E. K. Todd, *Coord. Chem. Rev.*, **246**, 3 (2003).
18. T. Chivers, *Topics Current Chem.*, **229**, 143 (2003).
19. S. R. Batten, K. S. Murray, *Coord. Chem. Rev.*, **246**, 103 (2003).
20. D. Woehrle, "Binding of Metal Ions and Metal Complexes to Macromolecular Carriers," in *Metal Complexes and Metals in Macromolecules*, D. Woehrle, A. D. Pomogailo, Eds., p. 279, Wiley-VCH, Weinheim, 2003.
21. A. S. Abd-El-Aziz, E. K. Todd, *Polymer News*, **26**, 5 (2001).
22. B. Moulton, M. J. Zaworotko, *Current Opinion Solid State Mat. Sci.*, **6**, 117 (2002).
23. Y. Liu, Y. Li, K. S. Schanze, *J. Photochem. Photobio.*, part *C: Photochem. Rev.*, **3**, 1 (2002).
24. P. R. Andres, U. S. Schubert, *Advanced. Mat.*, **16**, 1043 (2004).
25. D. R. Tyler, R. Chen, *Macromol. Symp.*, **209**, 231 (2004).
26. I. Tomita, M. Ueda, *Macromol. Symp.*, **209**, 217 (2004).
27. C. Moorlag, O. Clot, Y. Zhu, M. O. Wolf, *Macromol. Symp.*, **209**, 133 (2004).
28. D. R. Tyler, *Coord. Chem. Rev.*, **246**, 291 (2003).
29. K. Onitsuka, S. Takahashi, *Topics Current Chem.*, **228**, 39 (2003).
30. L. Han, M. Hong, *Inorg. Chem. Commun.*, **8**, 406 (2005).
31. H. Nishihara, M. Kurashina, M. Murata, *Macromol. Symp.*, **196**, 27 (2003).
32. U. S. Schubert, C. Eschbaumer, *Angew. Chem. Engl. Int. Ed.*, **41**, 2892 (2002).
33. D. P. Gates, *Annual Reports Prog. Chem., Sec. A: Inorg. Chem.*, **98**, 479 (2002).
34. R. J. Puddephatt, *Macromol. Symp.*, **196**, 137 (2003).
35. S. L. James, X. Xu, R. V. Law, *Macromol. Symp.*, **196,** 187 (2003).
36. S. L. James, *Macromol. Symp.*, **209**, 119 (2004).
37. S. L. James, *Chem. Chem. Rev.*, **276,** 32 (2003).
38. P. D. Harvey, *Macromol. Symp.*, **209**, 81 (2004).
39. P. D. Harvey, *Macromol. Symp.*, **196**, 173 (2003).
40. P. D. Harvey, Chapter 4 in *Macromolecule Containing Metal and Metal-like Elements*, A. S. Abd-El-Aziz, C. E. Carraher, Jr., C. U. Pittman, Jr., M. Zeldin, Eds., Vol. 5, p. 83, Wiley, New York, 2005.
41. P. D. Harvey, *Coord. Chem. Rev.*, **219–221**, 17 (2001).
42. T. Tanase, *Bull. Chem. Soc. Jap.*, **75**, 1407 (2002).
43. S.-W. A. Fong, T. S. A. Hor, *J. Cluster Sci.*, **9**, 351 (1998).
44. M. Hanack, D. Dini, in *Porphyrin Handbook*, K. M. Kadish, K. M. Smith, R. Guilard, Eds., Vol. 18, p. 251, Elsevier, San Diego, CA, 2003.

45. É. Fournier, A. Decken, P. D. Harvey, *Eur. J. Inorg. Chem.*, 4420 (2004).
46. P. Mongrain, P. D. Harvey, *Can. J. Chem.*, **22**, 2862 (2003).
47. P. Espinet, K. Soulantia, J. P. H. Charmant, A. G. Orpen, *Chem. Commun.*, 915 (2000).
48. P. D. Harvey, M. Drouin, A. Michel, D. Perreault, *J. Chem. Soc., Dalton Trans.*, 1365 (1993).
49. Y. Yamamoto, H. Suzuki, N. Tajima, K. Tatsumi, *Chem. Eur. J.*, **8**, 372 (2002).
50. H. Suzuki, N. Tajima, K. Tatsumi, Y. Yamamoto, *Chem. Comm.*, 1801 (2000).
51. G. Jia, W. F. Wu, R. C. Y. Yeung, H. P. Xia, *J. Organomet. Chem.*, **53**, 539 (1997).
52. É. Fournier, S. Sicard, A. Decken, P. D. Harvey, *Inorg. Chem.*, **43**, 1491 (2004).
53. M. Dartiguenave, Y. Dartiguenave, A. Mari, A. Guitard, M. J. Olivier, A. L. Beauchamp, *Can. J. Chem.*, **66**, 2386 (1988).
54. J. Metz, M. Hanack, *Chem. Ber.*, **120**, 1307 (1987).
55. M. Hanack, *Mol. Cryst. Liq. Cryst.*, **133**, 160 (1988).
56. M. Hanack, A. Hirsch, H. Lehmann, *Angew. Chem. Int. Ed. Engl.*, **29**, 1467 (1990).
57. M. Hanack, G. Renz, J. Strähle, S. Schmid, *J. Org. Chem.*, **56**, 3501 (1991).
58. M. Hanack, A. Hirsch, S. Kamezin, R. Thies, P. Vermehen, *Synth. Metals* **41–43**, 2633 (1991).
59. M. Hanack, S. Knecht, E. Witke, P. Haisch, *Synth. Metals* **55–57**, 873 (1993).
60. H. Ryu, S. Knecht, L. R. Subramanian, M. Hanack, *Synth. Metals*, **72**, 289 (1995).
61. M. Hanack, S. Knecht, R. Polley, L. R. Subramanian, *Synth. Metals*, **80**, 183 (1996).
62. S. Knecht, R. Polley, M. Hanack, *App. Organomet. Chem.*, **10**, 649 (1996).
63. O. Fernandez-Rodriguez, F. Fernandez-Lazaro, B. Cabezon, M. Hanack, T. Tores, *Synth. Metals*, **84**, 369 (1997).
64. H. Ryu, Y.-G. Kang, S. Knecht, L. R. Subramanian, M. Hanack, *Synth, Metals*, **87**, 69 (1997).
65. M. Hanack, M. Hees, E. Witke, *New. J. Chem.*, 169 (1998).
66. P. A. Stuzhin, S. I. Vagin, M. Hanack, *Inorg. Chem.*, **37**, 2655 (1998).
67. S. I. Vagin, P. A. Stuzhin, M. Hanack, *Russ. J. Gen. Chem.*, **69**, 319 (1999).
68. M. Hanack, S. Kamenzin, C. Kamezine, L. R. Subramanian, *Synth. Metals* **110**, 93, (2000).
69. W. Kobel, M. Hanack, *Inorg.Chem.*, **25**, 103 (1986).
70. M. Hanack, L. R. Subramanian, *Handbook of Organic Conductive Molecules and Polymers*, Vol. 1, p. 687, Wiley, Chichester, 1997.
71. J. Pawlik, C. Kautz, M. Baumgarten, *J. Inorg. Organomet. Polym.*, **4**, 237 (1994).
72. Y. Asai, K. Onishi, S. Miyata, S.-J. Kim, M. Matsumoto, K. Shigehara, *J. Electrochem. Soc.*, **148**, A305 (2001).
73. M. J. Irwin, J. J. Vittal, R. J. Puddephatt, *Organometallics*, **16**, 3541 (1997).
74. M. J. Irwin, G. Jia, N. C. Payne, R. J. Puddephatt, *Organometallics*, **15**, 51 (1996).
75. L.-T Phang, K.-S Gan, H. K. Lee, T. S. A. Hor, *J. Chem. Soc., Dalton Trans.*, 2697 (1993).
76. S. M. Wössner, J. B. Helms, K. M. Lantzky, B. P. Sullivan, *Inorg. Chem.* **38**, 4378 (1999).
77. D. Matt, A. van Dorsselär, *Polyhedron*, **10**, 1521 (1991).
78. T. Baumgartner, K. Huynh, S. Schleidt, A. J. Lough, I. Manners, *Chem. Eur. J.*, **8**, 4622 (2002).
79. D. C. Smith, Jr., G. M. Gray, *J. Chem. Soc., Dalton Trans.*, 677 (2000).
80. T. S. Lobana, R. Sharma, E. Bermejo, A. Castineiras, *Inorg. Chem.*, **12**, 7728 (2003).
81. P. Sevillano, T. Langetepe, D. Fenske, *Z. Anorg. Allg. Chem.*, **629**, 207 (2003).
82. M. J. Irwin, J. J. Vittal, G. P. A. Yap, R. J. Puddephatt, *J. Am. Chem. Soc.*, **118,** 13101 (1996).
83. R. J. Puddephatt, *Chem. Commun.*, 1055 (1998).
84. M. J. Irwin, L. M. Rendina, J. J. Vittal, R. J. Puddephatt, *Chem. Commun.*, 1281 (1996).
85. M.-C. Brandys, M. C. Jennings, R. J. Puddephatt, *J. Chem. Soc., Dalton Trans.*, 4601 (2000).
86. M.-C. Brandys, R. J. Puddephatt, *Chem. Commun.*, 1508 (2001).

87. R. G. Ball, G. Domazetis, D. Dolphin, B. R. James, J. Trotter, *Inorg. Chem.*, **20**, 1556 (1981).
88. M. E. Wright, L. Lawson, R. T. Baker, D. C. Roe, *Polyhedron*, **11**, 323 (1992).
89. R. D. Myrex, C. S. Colbert, G. M. Gray, *Organometallics*, **23**, 409 (2004).
90. G. P. C. M. Dekker, C. J. Elsevier, K. Vrieze, P. W. N. M. van Leeuwen, *Organometallics*, **11**, 1598 (1992).
91. X. Fan, B. L. Scott, J. G. Watkin, C. A. G. Carter, G. J. Kubas, *Inorg. Chim. Acta*, **317**, 276 (2001).
92. J. M. J. Paulusse, R. P. Sijbesma, *Chem. Commun.*, 1494 (2003).
93. L. R. Falvello, J. Forniés, J. Gomez, E. Lalinde, A. Martin, F. Marinez, M. T. Moreno, *J. Chem. Soc., Dalton Trans.*, 2132 (2001).
94. B. Zhuang, H. Sun, G. Pan, L. He, Q. Wei, Z. Zhou, S. Peng, K. Wu, *J. Organometal. Chem.*, **640**, 127 (2001).
95. R. Hoge, R. Lehnert, K. F. Fisher, *Cryst. Struct. Comm.*, **6**, 359 (1977).
96. É. Fournier, F. Lebrun, M. Drouin, A. Decken, P. D. Harvey, *Inorg. Chem.*, **43**, 3127 (2004).
97. X. L. Lu, W. K. Leong, L. Y. Goh, A. T. S. Hor, *Eur. J. Inorg. Chem.*, 2504 (2004).
98. P. Aslanis, P. J. Cox, S. Divanidis, P. Karagiannidis, *Inorg. Chim. Acta*, **357**, 2677 (2004).
99. D. Fenske, A. Rothenberger, M. S. Fallah, *Eur. J. Inorg. Chem.*, 59 (2005).
100. A. Cingolani, C. D. Nicola, Effendy, C. Pettinari, B. W. Skeleton, N. Somers, A. H. Whaite, *Inorg. Chim. Acta*, **358**, 748 (2005).
101. E. Lozano, M. Niewenhuyzen, S. L. James, *Chem. Eur. J.*, **12**, 2644 (2001).
102. E. Lozano, M. Niewenhuyzen, S. L. James, *Chem. Commun.*, 617 (2000).
103. X.-J. Wang, T. Langetepe, D. Fenske, B.-S. Kang, *Z. Anorg. Allg. Chem.*, **628**, 1158 (2002).
104. M.-C. Brandys, R. J. Puddephatt, *J. Am. Chem. Soc.*, **124**, 3946 (2002).
105. G. G. Lobbia, M. Pellei, C. Pettinari, C. Santini, B. W. Skelton, A. H. White, *Polyhedron*, **24**, 181 (2005).
106. D. Perreault, M. Drouin, A. Michel, P. D. Harvey, *Inorg. Chem.*, **32**, 1903 (1993).
107. P. D. Harvey, *Coord. Chem. Rev.*, **153**, 175 (1996).
108. R. P. Feazell, C. E. Carson, K. K. Klausmeyer, *Acta Cryst.*, **C60**, M598 (2004).
109. A. Houlton, D. M. P. Mingos, D. M. Murphy, D. J. Williams, L.-T. Phang, *J. Chem. Soc., Dalton Trans.*, 3629 (1993).
110. L.-T. Phang, T. S. A. Hor, Z.-Y. Zhou, T. C. W. Mak, *J. Organomet. Chem.*, **469**, 253 (1994).
111. M.-C. Brandys, R. J. Puddephatt, *Chem. Commun.*, 1280 (2001).
112. M.-C. Brandys, R. J. Puddephatt, *J. Am. Chem. Soc.*, **123**, 4839 (2001).
113. G. Jia, R. J. Puddephatt, J. D. Scott, J. J. Vittal, *Organometallics*, **12**, 4771 (1993).
114. Z. Qin, M. C. Jennings, R. J. Puddephatt, *Chem. Eur. J.*, **8**, 735 (2002).
115. W. J. Hunks, M. C. Jennings, R. J. Puddephatt, *Chem. Commun.*, 1834 (2002).
116. T. J. Burchell, D. J. Eisler, M. C. Jennings, R. J. Puddephatt *Chem. Commun.* 2228 (2003).
117. S.-M. Kuang, Z.-Z. Zhang, Q.-G. Wang, T. C. W. Mak, *Chem. Commun.*, 581 (1998).
118. D. Fortin, M. Drouin, P. D. Harvey, F. G. Herring, D. A. Summers, R. C. Thompson, *Inorg. Chem.*, **38**, 1253 (1999).
119. D. Perreault, M. Drouin, A. Michel, P. D. Harvey, *Inorg. Chem.*, **31**, 3688 (1992).
120. M. Turcotte, P. D. Harvey, *Inorg. Chem.*, **41**, 1739 (2002).
121. D. Fortin, M. Drouin, M. Turcotte, P. D. Harvey, *J. Am. Chem. Soc.*, **119**, 531 (1997).
122. D. Fortin, M. Drouin, P. D. Harvey, *Inorg. Chem.*, **39**, 2758 (2000).
123. D. Fortin, P. D. Harvey, *Coord. Chem. Rev.*, **171**, 351 (1998).
124. D. Perreault, M. Drouin, A. Michel, V. M. Miskowski, W. P. Schaefer, P. D. Harvey, *Inorg. Chem.*, **31**, 695 (1992).
125. Effendy, C. D. Nicola, M. Fianchini, C. Pettinari, B. W. Skelton, N. Summers, A. H. White, *Inorg. Chim. Acta*, **358**, 763 (2005).

126. J. Bai, A. V. Virovets, M. Scheer, *Angew. Chem. Engl. Int. Ed.*, **41**, 1737 (2002).
127. M. Scheer, L. Gregoriades, J. Bai, M. Sierka, G. Brunklaus, H. Eckert, *Chem. Eur. J.*, **11**, 2163 (2005).
128. J. Bai, E. Leiner, M. Scheer, *Angew. Chem. Engl. Int. Ed.*, **41**, 783 (2002).
129. A. Adolf, M. Gonsior, I. Krossing, *J. Am. Chem. Soc.*, **124**, 7111 (2002).
130. D. A. Bohling, T. P. Gill, K. R. Mann, *Inorg. Chem.*, **20**, 194 (1981).
131. J. L. Bear, B. Han, Z. Wu, E. Van Caemelbecke, K. M. Kadish, *Inorg. Chem.*, **40**, 2275 (2001).
132. M. Handa, M. Yasuda, Y. Muraki, D. Yoshioka, M. Mikuyira, K. Kasuga, *Chem. Lett.*, **32**, 946 (2003).
133. A. J. Deeming, D. Nuel, N. P. Randle, C. Whittaker, *Polyhedron*, **8**, 1537 (1989).
134. S. J. Sherlock, M. Cowie, E. Singleton, M. M. de Steyn, *Organometallics*, **7**, 1663 (1988).
135. M. C. Kerby, B. W. Eichorn, J. A. Creighton, K. P. C. Vollhardt, *Inorg. Chem.*, **29**, 1319 (1990).
136. M. C. Kerby, B. W. Eichhorn, J. A. Creighton, K. P. C. Vollhardt, *Macromol. Chem.*, **1**, 288 (1990).
137. M. J. Irwin, G. Jia, J. J. Vittal, R. J. Puddephatt, *Organometallics*, **15**, 5321 (1996).
138. J.-F. Bérubé, K. Gagnon, D. Fortin, A. Decken, P. D. Harvey, *Inorg. Chem.*, **45**, 2817 (2006).
139. S. Sicard, J.-F. Bérubé, D. Samar, A. Messaoudi, D. Fortin, F. Lebrun, J,-F. Fortin, A. Decken, P. D. Harvey, *Inorg. Chem.*, **43**, 5321 (2004).
140. T. Tanase, E. Goto, R. A. Begum, M. Hamaguchi, S. Zhan, M. Iida, K. Sakai, *Organometallics*, **23,** 5975 (2004).
141. T. Zhang, M. Drouin, P. D. Harvey, *Inorg. Chem.*, **38**, 1305 (1999).
142. T. Zhang, M. Drouin, P. D. Harvey, *Inorg. Chem.*, **38**, 957 (1999).
143. T. Zhang, M. Drouin, P. D. Harvey, *Inorg. Chem.*, **38**, 4928 (1999).
144. P. Leoni, F. Marchetti, L. Marchetti, M. Pasquali, *Chem. Commun.*, 2372 (2003).
145. D. A. Morgenstern, C. C. Bonham, A. P. Rothwell, K. V. Wood, C. P. Kubiak, *Polyhedron*, **14**, 1129 (1995).
146. F. Lemaître, D. Lucas, K. Groison, P. Richard, Y. Mugnier, P. D. Harvey, *J. Am. Chem. Soc.*, **125**, 5511 (2003).
147. C. J. McAdam, N. W. Duffy, B. H. Robinson, J. Simpson, *J. Organomet. Chem.*, **179**, 527 (1997).
148. M. Rashidi, E. Kristof, J. J. Vittal, R. J. Puddephatt, *Organometallics*, **33**, 1497 (1994).
149. A. M. Bradford, E. Kristof, M. Rashidi, D.-S. Yang, N. C. Payne, R. J. Puddephatt, *Inorg. Chem.*, **33**, 2355 (1994).
150. B. F. G. Johnson, K. M. Sanderson, D. S. Shepard, D. Ozkaya, W. Zhou, H. Ahmed, M. D. R. Thomas, L. Gladden, M. Mantle, *Chem. Commun.*, 1317 (2000).
151. M. D. R. Thomas, H. Ahmed, K. M. Sanderson, D. S. Shepard, B. J. G. Johnson, W. Zhou, *Applied Phys. Lett.*, **76**, 1773 (2000).
152. M. D. R. Thomas, H. Ahmed, K. M. Sanderson, D. S. Shepard, B. J. G. Johnson, D. Ozkaya, N. Sharma, C. Humphreys, *Applied Phys. Lett.*, **90**, 947 (2001).
153. T.S. Lobana, R. Sharma, E. Bermajo, A. Castinerous, *Inorg. chem.*, **42**, 7728 (2003).

CHAPTER 9

Redox-Based Functionalities of Multinuclear Metal Complex Systems

HIROSHI NISHIHARA
The University of Tokyo, Tokyo, Japan

I. INTRODUCTION

The reversible redox nature of transition metal complexes plays a key role in natural systems such as metalloproteins[1], and artificial functional systems such as electrocatalysis[2]. In some cases, multistep and/or multielectron redox reactions are crucial in the systems to specifically activate the target molecules. These redox reactions are often derived from combinations of multiple redox components, because electronic interaction between redox sites causes the stabilization of mixed valence states with low-energy electronic transition. Also, the multiredox systems exhibit the change in magnetic and optical properties by changing their oxidation states, and thus they are candidates of molecular switches and memories applicable to the molecular devices.[3,4,5] The present chapter presents redox properties and redox-based functionalities of multinuclear metal complex systems in several different categories. They are π-conjugated redox conjugated linear polymers of oligo- and polyferrocenylene derivatives,[6–9] two- and three-dimensional multinuclear complexes of metalladichalcogenolenes, organometallic π-conjugated conducting polymers made of metallacycles,[9,10,11] and surface immobilized metal complex polymers prepared by stepwise coordination reactions.

II. REDOX, OPTICAL, AND PHOTOCONDUCTING PROPERTIES OF CONJUGATED FERROCENE OLIGOMERS AND POLYMERS

Ferrocene is often incorporated in polymers because of its inherent organic as well as inorganic nature, including its high thermal stability, good solubility in organic media, versatility in the synthesis of derivatives using a reactivity similar to that of benzene, and reversible redox ability, leading to easy electrochemical and chemical handling of the oxidation state due to the iron center.[12] Ferrocene oligomers in which the ferrocene units electronically interact with each other are part of a novel group exhibiting a formation of mixed-valence states where Fe^{II} (ferrocene) and Fe^{III} (ferrocenium ion) coexist within a molecule, and provide fundamental

Frontiers in Transition Metal-Containing Polymers, edited by Alaa S. Abd-El-Aziz and Ian Manners.

information on the redox-based phenomena of the multiredox nuclei system.[13,14] In this section, π-conjugated ferrocene oligomers and polymers were focused on two aspects: (1) the recognition of the physical property sequence that occurs in the change from dimer to polymer, and (2) the combination of unique physical properties of conjugated ferrocene oligomers with other photo- and electrofunctional molecules or materials.[6,7,8]

A. Oligo- and Polyferrocenylenes

In contrast to many studies on the mixed valence state of redox binuclear complexes, only a few systematic studies on the mixed-valence state of the complexes with more than two redox nuclei are available.[12] The first experimental recognition of the dependence on the redox nuclei in the linearly combined multiredox system was carried out using oligo(1,1′-dihexylferrocenylene)s up to a heptamer, **1_1–1_7**.[15,16] For an analysis of their redox behavior, the neighboring-site interaction model was employed based on the interaction energies between neighboring redox sites, u_{OR}, u_{OO}, and u_{RR}, where OR, OO, and RR denote combinations of an oxidized site (Ox) and a reduced site (Red), Ox and Ox, and Red and Red, respectively.[15,16,17] In the case of oligoferrocenylenes, it can be assumed that u_{OR} and u_{OO} consist primarily of resonance exchange energy and electrostatic repulsion energy, thus negative and positive, respectively.

Simulation of the redox potentials of the oligo(1,1′-dihexylferrocene)s by the theory based on the interaction only between nearest neighboring centers[18] showed a discrepancy from the experimental results for the oligomers higher than pentamers (Figure 9.1). This discrepancy was reasonably interpreted by the introduction of an additional parameter, u_{OXR}, denoting the donor–acceptor interaction over three redox centers. The simulation indicates that $u_{OO} = 9\,\text{kJ mol}^{-1}$, $u_{OR} = -10.5\,\text{kJ mol}^{-1}$, and $u_{OXR} = -3.8\,\text{kJ mol}^{-1}$, assuming $u_{RR} = 0$ (because there is no electrostatic interaction or electron delocalization between Red and Red). The fact that the magnitude of u_{OXR} is only one-third that of u_{OR} suggests that the positive charge in the molecule is considerably localized on the "Ox" sites.

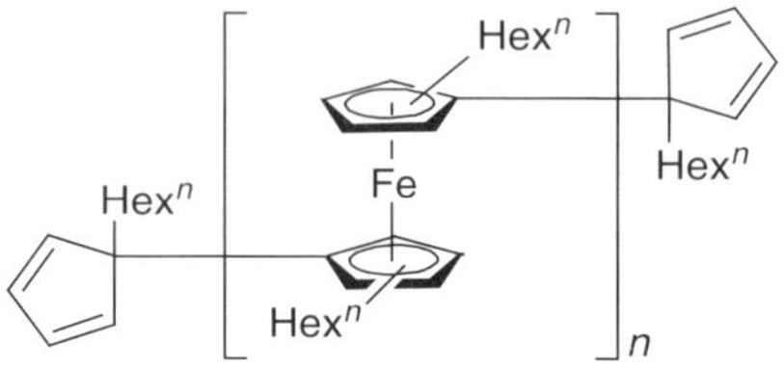

1_1–1_7: $n = 1$–7

Infrared spectroscopy of (η^4-cyclopentadiene)Fe(CO)$_3$ moiety-attached ferrocene and its oligomers, **2**, gave information on the electronic structure of the mixed-valence complexes;[19] the ν(CO) bands of the Fe(CO)$_3$ moiety[20,21] could sense the oxidation state of the nearest terminal ferrocenyl group.

In Table 9.1 are summarized the wavenumbers of the ν(CO) peaks of **2_1–2_3** in each oxidation state generated electrochemically with $\Delta\nu_{CO}$, which is the shift of ν_{CO} from the neutral form to a given oxidized form, and $\theta(\Delta\nu_{CO})$, which denotes the ratio of $\Delta\nu_{CO}$ for a given oxidation state to $\Delta\nu_{CO}$ for a fully oxidized state. A higher wavenumber shift upon oxidation of the monomer **2_1** in the magnitude of 12 to 15 cm^{-1} took place (see Table 9.1) around the formal potential, and the direction of this wavenumber shift is reasonable because the positively charged ferrocenium

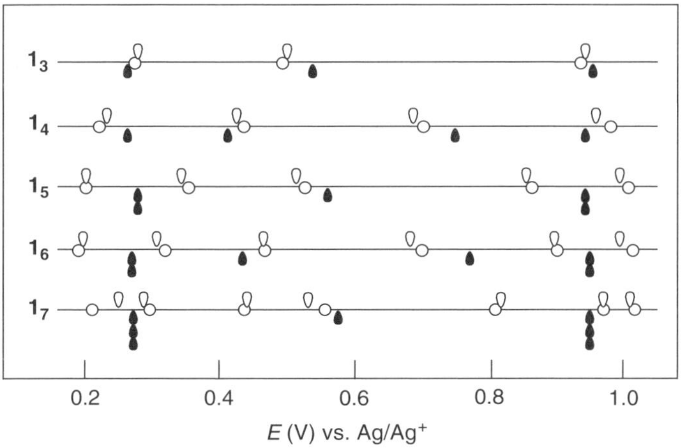

Figure 9.1 Formal potentials of oligo(1,1′-dihexylferrocenylene)s, **1**, obtained by Osteryoung square-wave voltammetry (○), those calculated from the first neighboring site interaction energies (●) with $u_1 = 15\,\text{kJ mol}^{-1}$ and $u_2 = 4.5\,\text{kJ mol}^{-1}$, and those calculated from both the first and second neighboring site interaction energies (0) with $u_1 = 15\,\text{kJ mol}^{-1}$, $u_2 = 4.5\,\text{kJ mol}^{-1}$, and $u_{OXR} = -3.8\,\text{kJ mol}^{-1}$.

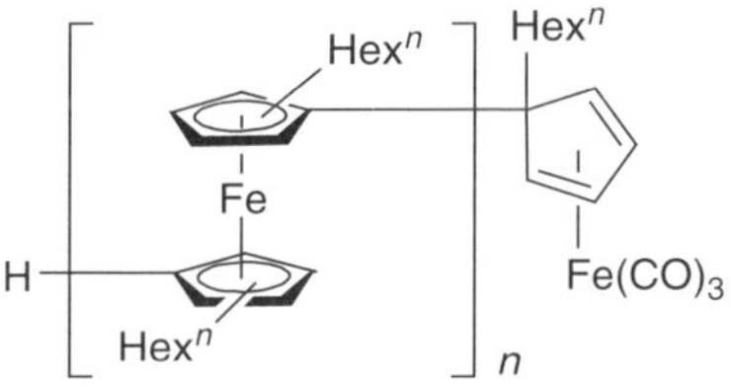

$\mathbf{2}_1$–$\mathbf{1}_3$: $n = 1$–3

TABLE 9.1 ν_{CO} at Different Oxidation States of 3_1–3_3

Complex	Wavenumber/cm^{-1}	$\Delta\nu_{CO}$/cm^{-1}		$\theta(\Delta\nu_{CO})$
$\mathbf{2}_1^0$	2039, 1965	—	—	0
$\mathbf{2}_1^+$	2051, 1980	12	15	1
$\mathbf{2}_2^0$	2038, 1965	—	—	0
$\mathbf{2}_2^{1+}$	2047, 1976	9	11	0.6 ~ 0.7
$\mathbf{2}_2^{2+}$	2051, 1982	4	6	1
$\mathbf{2}_3^0$	2038, 1964	—	—	0
$\mathbf{2}_3^{1+}$	2041, 1970	3	6	0.2 ~ 0.3
$\mathbf{2}_3^{2+}$	2050, 1979	9	9	0.8 ~ 0.9
$\mathbf{2}_3^{3+}$	2052, 1982	2	3	1

site withdraws electrons from Fe(II) in the (cyclopentadiene)Fe(CO)$_3$ moiety, and consequently, the back-donation from Fe to CO must be reduced.

The wavenumber shift from the fully reduced form to the fully oxidized form for the dimer $\mathbf{2}_2$, was 13–17 cm^{-1}, which is similar to the shift of the monomeric complex (see Table 9.1). The monocationic form gave the peaks at wavenumbers between those for the neutral form and the dicationic form, and the peak width was broadened. This is attributed to the rate of electron exchange between two electronic isomers, Red–Ox–Fe(CO)$_3$ and Ox–Red–Fe(CO)$_3$, where Fe(CO)$_3$ refers to a [(η^4-cyclopentadiene)Fe(CO)$_3$] moiety. The rate is slower than or comparable to the time scale of infrared (IR) (10^{11}~10^{13} s^{-1}).[22] This suggests that the (cyclopentadiene)Fe(CO)$_3$-attached ferrocene and the nonattached ferrocene are electronically similar, and thus the attached ferrocene unit can be either Red or Ox.

IR spectra of timer **2$_3$** showed three-step ν(CO) changes corresponding to the oxidation process, and the most significant wavenumber shift, 9 cm^{-1}, was observed at the second step. This is consistent with the neighboring-site interaction model that proposes the three-step oxidation pathway of the trimer as Red–Red–Red–Fe(CO)$_3$ → Red–Ox–Red–Fe(CO)$_3$ → Ox–Red–Ox–Fe(CO)$_3$ → Ox–Ox–Ox–Fe(CO)$_3$, where the (cyclopentadiene)Fe(CO)$_3$-attached ferrocene site changes from Red to Ox at the second oxidation step. Small wavenumber shifts were observed even at the first and the third oxidation steps, and a simple evaluation is that the shifts at the first and the third steps correspond to about 20% and 80% oxidation of the terminal ferrocene site. These values of partial charge density can be regarded as the degree of electron delocalization.

Appearance of intervalence charge transfer (IVCT) bands in the electronic spectra is a characteristic feature of the mixed-valence complexes, and their analysis is efficient to elucidate the magnitude of the internuclear electronic interaction. A systematic study on the effects of IVCT bands on the number of nuclei and the oxidation number was carried out for oligo(1,1′-dihexylferrocenlene)s, **1$_2$**–**1$_6$**.[23,24] Figure 9.2 displays the dependence of ν_{max} values of the IVCT band on the oxidation state and the number of ferrocene units. Characteristic features in the figure are a higher energy shift in ν_{max} as the oxidation number for each oligomer increases and a lower energy shift in ν_{max} for the monocationic form of the oligomers as the number of ferrocene units increases.

The shift in ν_{max} of IVCT bands could be analyzed based on a model, assuming that the mixed-valence states are a linear combination of Red and Ox and that, for example, in terferrocene, the photochemical intervalence charge transfer of Red–Ox–Red (or Ox–Red–Ox) requires more energy compared with the case of biferrocene, not only due to the energetic difference between unequal ground states before and after photoelectron transfer but also due to a

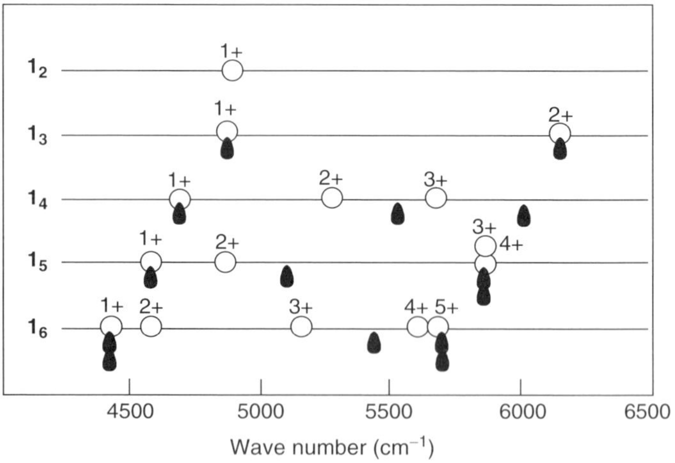

Figure 9.2 The ν_{max} values of IVCT bands for the mixed-valence states of **1$_2$**–**1$_6$**; measured in CH$_2$Cl$_2$–acetone (1:1 in volume) (○), and calculated with u_{ex} = 13 kJ mol^{-1} (●). The numbers in the figure refer to oxidation numbers.

strain derived from the difference in internuclear distance between Red–Ox and Red–Red (or Ox–Ox) combinations. As can be deduced, the extra energy for the change from Red–Ox to Ox–Ox is the largest. This energy, u_{ex}, was employed as the factor for rationalization of the ν_{max} shift due to the change in oxidation number of the higher oligomers. The dependency of the ν_{max} on the number of ferrocene units and the oxidation states was qualitatively rationalized by taking into account that $u_{ex} = 13\,kJ\,mol^{-1}$, as shown in Figure 9.2.

Photoconductivity has been found for the complexes of oligo(1,1′-dihexylferroceneylene)s, $\mathbf{1}_2$–$\mathbf{1}_6$, and poly(1,1′-dihexylferrocenylene) (**3**) with tetracyanoethylene (TCNE).[25] The complexes exhibited both ferrocene-TCNE charge-transfer (CT) and Fe^{II}–Fe^{III} IVCT bands in the near-IR region. The ratio of Fe^{III} to total Fe estimated from the peak area and intensity of the ^{57}Fe Mössbauer spectra of $\mathbf{1}_3$-TCNE and $\mathbf{1}_4$-TCNE was 0.4–0.5 in the temperature range of 50–293 K. Photoconductivity of the **3**-TCNE charge-transfer complex under near-IR light irradiation was higher than that under visible light irradiation. A positive dependence was observed between the near-IR photoconductivity and the number of ferrocene units in the oligomers, indicating that both IVCT- and CT-band excitation cause photoconductivity. The CT-band excitation would alter the charge distribution on the poly(1,1′-ferrocenylene) chain, and IVCT-band excitation would accelerate the intrachain charge-transport rate.

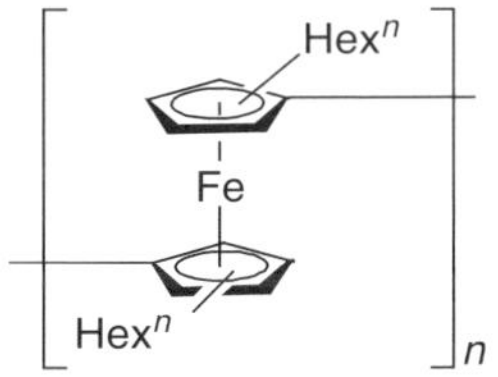

3

B. Azo-Bridged Ferrocene Oligomers and Polymers

Redox behavior and optical properties in the mixed-valence states were examined for azo-bridged ferrocene (Fc) oligomers such as azoferrocene, **4**, and trimers, Fc–N═N–Fc′–N═N–Fc (**5**) and Fc–Fc′–N = N–Fc (**6**), and a polymer composed of [-(Fc′–N = N–Fc′)$_{0.6}$— (Fc′–Fc′)$_{0.4}$—]$_n$ (**7**).[26,27] The π-conjugated *trans*-azo bridge acted as a spacer efficient for assisting the electron exchange between ferrocene moieties and resulted in a formation of thermally stable mixed-valence states for **4**–**6**. The trimer **5** exhibited reversible 2e$^-$ and 1e$^-$ oxidation waves in a cyclic voltammogram in CH_2Cl_2 or THF, contrary to the behavior of terferrocene (three 1e$^-$ waves), as noted earlier. Analysis based on the neighboring site interaction model indicates that the behavior of **5** can be rationalized by the assumption that the positive charge in the monocation is localized mostly on the terminal ferrocene unit (correspondingly, Fc^+–N_2–Fc′–N_2–Fc) due to a strong electron-withdrawing effect of the azo group.[26] This charge distribution in the mixed-valence states of **5** was supported by the characteristics of IVCT bands.[26] An asymmetrical complex, **6**, underwent a three-step 1e$^-$ oxidation, and the two mixed-valence forms can be roughly expressed as Fc^+–Fc′–N_2–Fc and Fc^+–Fc′–N_2–Fc^+.

Azo-bridged ferrocene oligomers, **4**, **5**, and **6** showed a marked dependence on the redox potentials and IVCT band characteristics of the solvent (Figure 9.3).[26] The absorption of the metal-to-ligand charge transfer (MLCT) band at 533–542 nm diminished and a new band assignable to the ligand-to-metal charge transfer (LMCT) band appeared at 630–678 nm with the oxidation of **4**, **5**, and **6**. More donating solvent afforded higher IVCT and LMCT energy of **4**, **5**, and **6** in the mixed-valence states, as shown in Figure 9.3, indicating the hole-transfer mechanism.[28]

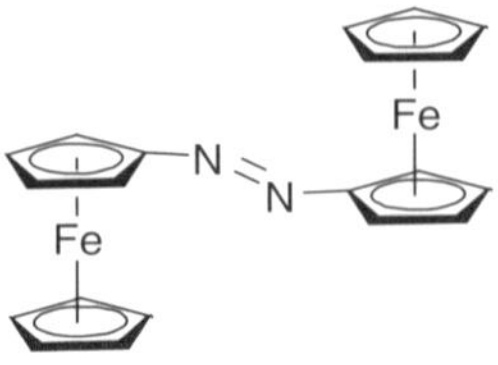

4

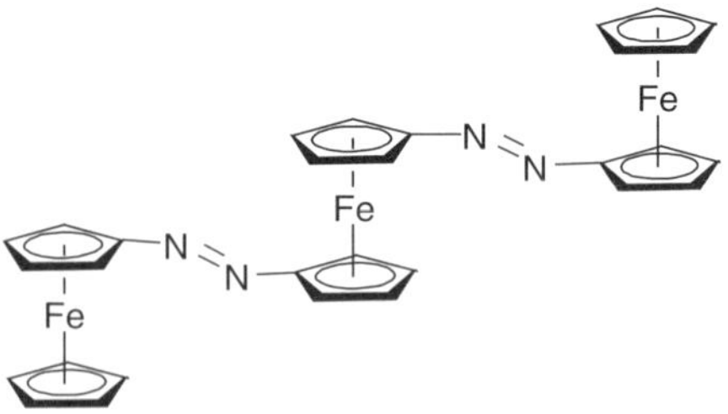

5

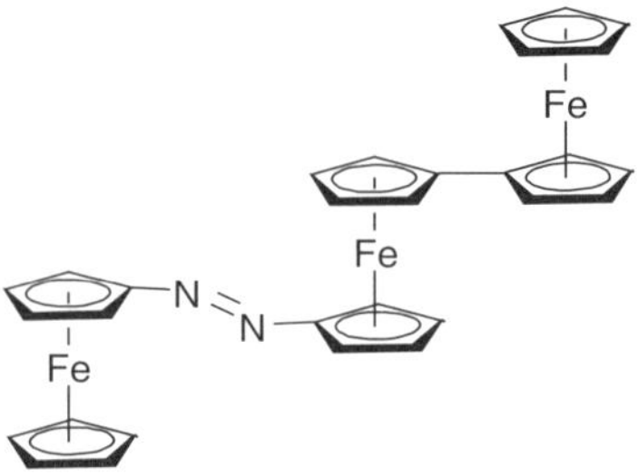

6

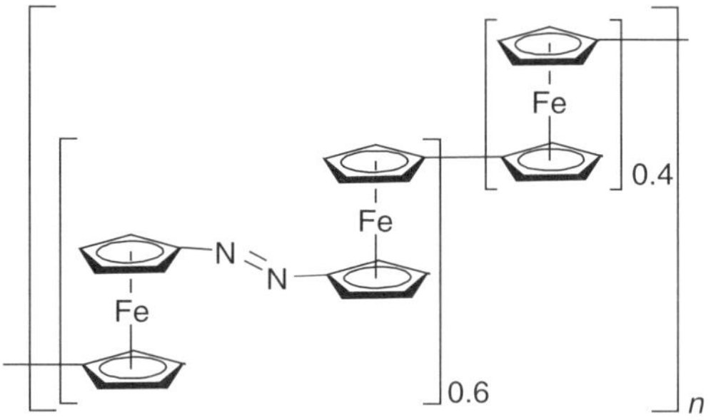

7

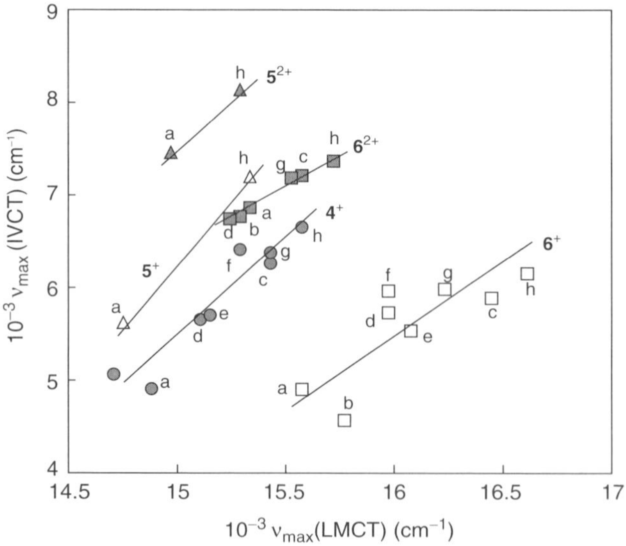

Figure 9.3 Plots of ν_{max}(IVCT) vs. ν_{max}(LMCT) for mixed-valence states of **4**, **5**, and **6** in (a) CH_2Cl_2, (b) 1,2-dichloroethane, (c) nitromethane, (d) nitrobenzene, (e) PhCN, (f) MeCN, (g) acetone, and (h) THF.

The ultraviolet-visible (UV-vis) absorption spectrum of the polymer **7** showed an MLCT band at 535 nm, and the π–π^* band edge shifted to a longer wavelength, probably due to the elongation of the π-conjugation. Cyclic voltammetry of **7** in Bu_4NClO_4–CH_2Cl_2 showed a chemically quasi-reversible broad redox wave at $E^{0'} = 0.32$ V vs. Ag/Ag$^+$. A spin-coated film of **7** on indium-tin oxide (ITO) also gave a broad redox wave with $E^{0'} = 0.26$ V vs. Ag/Ag$^+$ in Bu_4NClO_4–MeCN. Spectral changes of the film with a potential shift in the positive direction indicated a decrease in intensity of the MLCT band and the appearance of a broad IVCT band around 1000 nm at 0.25 V, generating the mixed-valence state Bu_4NClO_4–MeCN.

Photoreaction of azoferrocene and its related molecules undergo unique photochemical reaction due to the absorption of green light (517 nm) exciting the MLCT band. In the case of azoferrocene, the photoirradiated species accepts a rapid attack of proton, causing chemical degradation.[29,30,31] On the other hand, ferrocenylazobenzenes cause reversible isomerization between *trans* and *cis* forms of the azobenzene moiety by the combination of the 517-nm light irradiation and the Fe^{III}–Fe^{II} redox change.[32,33,34]

C. π-Conjugated Cyclobutadienecobalt Polymer Containing Ferrocenyl Groups

A mononuclear complex $\{\eta^4\text{-}C_4Fc_2(o\text{-}FcC_6H_4)_2\}CoCp$ (**8**) and a polymer $[p\text{-}C_6H_4\{(\eta^4\text{-}C_4Fc_2)CoCp\}]_n$ (**9**) were obtained by the reaction of $CpCo(PPh_3)_2$ with two geometric isomers of bis(ferrocenylethynyl)benzene[35] (Scheme 9.1).

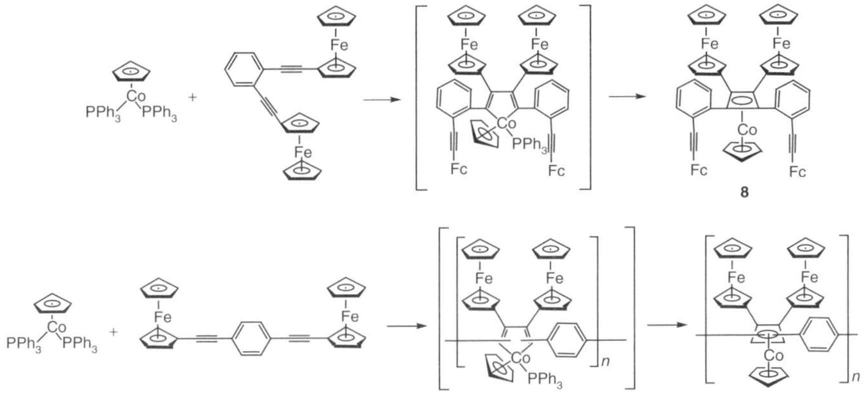

Scheme 9.1

The monomeric complex **8** underwent reversible two-step 1e$^-$ and 3e$^-$ oxidation of four ferrocenyl groups in Bu_4NPF_6–CH_2Cl_2 at $E_1^{0'} = 0.12$ V and $E_2^{0'} = 0.35$ V vs. Ag/Ag$^+$, respectively, besides an irreversible oxidation of the cobalt site at 0.944 V. It was proposed that the ferrocenyl groups directly attached to the cyclobutadiene ring interacted electronically, leading to separation of redox potentials due to the formation of a thermodynamically stable mixed-valence state. In contrast, the ferrocenylethynyl groups at the *ortho* position of the phenylene groups were almost entirely electronically independent, giving one-step two-electron oxidation. Therefore, the 1e$^-$ + 3e$^-$ oxidation waves were interpreted as the overlapping of two 1e$^-$ waves due to the ferrocenyl groups directly bound to cyclobutadiene and one 2e$^-$ wave due to the o-ferrocenylphenyl groups. This consideration was supported by the spectroscopic data for the oxidation of **8**, in which a new broad band appeared at 1340 nm, which could be assigned to the intervalence-transfer band in the mixed-valence state. Its intensity reached a maximum at 1e$^-$ oxidation, and then decreased with further oxidation.

In a cyclic voltammogram of the polymer complex, **9**, in Bu_4NPF_6-CH_2Cl_2, two waves attributed to oxidation of ferrocenyl groups appeared at $E^{0'} = 0.14$ and 0.22 V vs. Ag/Ag^+, with the second wave being broad and both waves being chemically reversible, besides an irreversible wave due to the oxidation of the cyclobutadienecobalt moiety at 1.02 V (Figure 9.4). The separation into two waves indicates the existence of strong electronic coupling between the ferrocenyl groups, and it is thought to be caused by the formation of the $Fe^{II}Fe^{III}$ mixed-valence state for two ferrocenyl moieties at the 1,2 position of a cyclobutadiene ring. The voltammetric waves of **9** were as broad as those of other conjugated ferrocene polymers.[36,37] It thus can be deduced that the broadness of the waves is caused by the existence of electronic communication between ferrocene units not only on the same cyclobutadiene ring but also on different cyclobutadiene rings, even though the latter interaction is weaker than the former, and/or by the existence of different structures within the polymer such as the coexistence of both 1,2- and 1,3-diferrocenylcyclobutadiene structures.

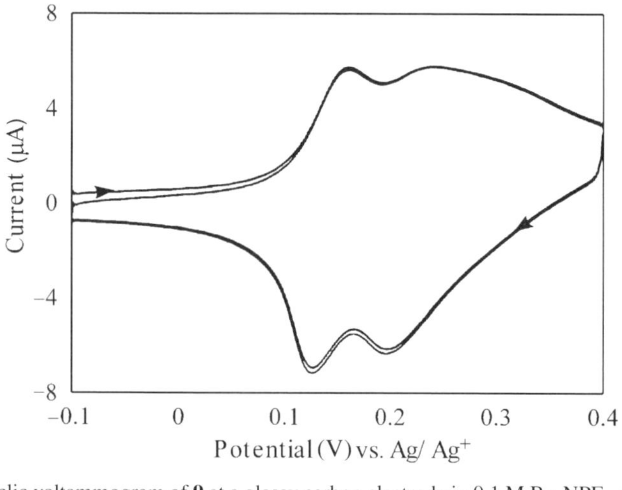

Figure 9.4 Cyclic voltammogram of **9** at a glassy carbon electrode in 0.1 M Bu_4NPF_6–CH_2Cl_2 at a scan rate of 0.1 Vs^{-1} at 25°C.

III. REDOX, OPTICAL, AND MAGNETIC PROPERTIES OF METALLACHALCOGENOLENE MULTINUCLEAR COMPLEXES

Transition metal complexes involving a chalcogen atom, S or Se, are of importance as functional materials exhibiting peculiar electromagnetic properties[38–41] and an active site of a metal enzyme,[42,43] which participates in vital functions. A metalladichalcogenolene ring is a unique metal chelate ring, which exhibits interesting physical and chemical properties such as reversible redox activity, deep colors, and various substitution and addition reactions due to the quasi-aromaticity and electronic unsaturation of the metalladithiolene ring.[44–50] In this section, physical properties of triangular metalladitiolene trimers and metal–metal bond-containing cluster complexes are described as the examples of two- and three-dimensionally expanded systems of metalladichalcogenolenes.

A. π-Conjugated Cyclic Cobaltadithiolene Trimers

Dinuclear complexes, **11a** and **11b,** and trinuclear complexes, **12a** and **12b**, were synthesized by the reaction of monomers, **10a** and **10b**, with HBr and dioxygen.[51,52,53] Their UV-vis spectra showed a strong absorption peak attributed to an LMCT band from S to Co as usually seen in metalladithiolene complexes.[54] The maximum wavelength of **10a**, **10b**, **11a**, **11b**, **12a**, and **12b** are 551, 588, 640, 647, 688, and 696 nm, respectively, indicating the shift to low energy with the increase in the number of nuclei from **10** to **11** and **12** according to the expansion in conjugation length.

10a: R = H
10b: R = Me

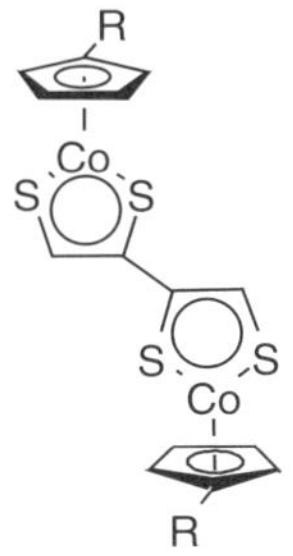

11a: R = H
11b: R = Me

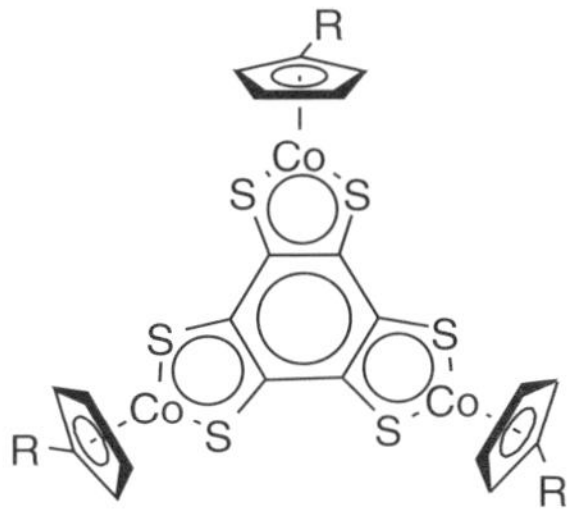

12a: R = H
12b: R = Me

Triangular trinuclear complexes, **12a** and **12b**, are good examples showing large matrix (solvent and counterion) effects on the electronic and magnetic interaction between the cobaltadithiolene nuclei. They showed redox properties of three 1e$^-$ reduction waves in various electrolyte solutions, indicating formation of two mixed-valence states, [Co^{II}Co^{III}Co^{III}] and [Co^{II}Co^{II}Co^{III}], with strong dependencies of their redox potentials on the type of solvent and electrolyte. In this case, comproportionation constants,[55] K_{c1} and K_{c2}, in Eq. 9.1 and Eq. 9.2 were calculated from the redox potential differences between $\mathbf{12}^{0/1-}$ and $\mathbf{12}^{1-/2-}$ and between $\mathbf{12}^{1-/2-}$ and $\mathbf{12}^{2-/3-}$ according to equations (1) and (2), respectively.

$$K_{c1} = [\mathbf{12}^-]^2/[\mathbf{12}][\mathbf{12}^{2-}] = \exp(|E_1^{0\prime} - E_2^{0\prime}|F/RT) \qquad (9.1)$$

$$K_{c2} = [\mathbf{12}^{2-}]^2/[\mathbf{12}^-][\mathbf{12}^{3-}] = \exp(|E_2^{0\prime} - E_3^{0\prime}|F/RT) \qquad (9.2)$$

The redox potentials depended on the cation size; the larger the size, the higher the K_{c1} and K_{c2} values, as is seen in a series of tetraalkylammonium ions in MeCN as shown in Table 9.2. Alkali metal ions give smaller K_c values compared to tetraalkylammonium ion. Addition of crown ethers to the solution of alkali metal ions increased K_c values, and this behavior could be simulated using the complexation constants of crown ether with cation obtained by conductivity analysis and several equations concerning association, ion-pairing equilibrium, comproportionation and charge balance (Figure 9.5).[52] There also existed a strong dependence of comproportionation constants on solvent; solvent with higher polarity gave higher K_c values. Dependence of K_c on both the cation size and the solvent polarity indicates that the solvated cation size dominated the K_c values.

TABLE 9.2 Redox Properties of 12a and 12b

	$E^{0'}$ for Reduction vs. Fc^+/Fc					
Compound	Electrolyte Solution[a]	1st	2nd	3rd	log K_{c1}	log K_{c2}
12a	NBu_4ClO_4–THF	−1.06	−1.30	−1.74	4.1	7.4
12a	$LiClO_4$–THF	−1.18	−1.39	−1.57	3.6	3.0
12a	$LiClO_4$–THF[b]	−1.01	−1.22	−1.61	3.6	6.9
12a	$NaBPh_4$–THF	−1.11	−1.25	−1.36	2.4	1.9
12a	$NaBPh_4$–THF[c]	−1.06	−1.32	−1.67	4.4	5.9
12a	$NaBPh_4$–THF[d]	−1.11	−1.43	−1.80	5.4	6.3
12a	NBu_4ClO_4–MeCN	−0.96	−1.22	−1.61	4.4	6.6
12a	NEt_4ClO_4–MeCN	−1.01	−1.21	−1.54	3.4	5.6
12a	$NaBPh_4$–MeCN	−0.96	−1.19	−1.33	3.9	3.4
12a	$NaBPh_4$–DMF	−1.00	−1.29	−1.63	4.9	5.7
12a	$NaBPh_4$–DMSO	−0.95	−1.22	−1.60	4.6	6.4
12b	NBu_4ClO_4–THF	−1.13	−1.36	−1.80	3.9	7.4
12b	$NaBPh_4$–THF	−1.16	−1.27	−1.39	1.9	2.1
12b	$NaBPh_4$–THF[d]	−1.16	−1.38	−1.50	3.7	2.1
12b	NBu_4ClO_4–MeCN	−1.03	−1.29	−1.66	4.4	6.3

[a] Electrolyte concentration was 0.1 M unless otherwise stated.
[b] Added 0.25 M 12-crown-4.
[c] Added 0.2 M 15-crown-5.
[d] Added 0.2 M 18-crown-6.

UV-vis-NIR (NIR-near infrared) spectra of the reduced forms of **12** formed by stepwise reduction by Na in THF and by $[Co(\eta^5\text{-}C_5Me_5)_2]$ in propylene carbonate (PC) showed a significant difference in the IVCT bands at the mixed-valence states. The mixing coefficient, α, and H_{AB}, which exhibited a degree of internuclear interaction according to Hush's theory Eq. (9.3) and Eq. (9.4)[56,57]

$$\alpha^2 = (4.24 \times 10^{-4}/m_d m_a)(\varepsilon_{max}\Delta\nu_{1/2}\nu_{max}{}^{-1})r^{-2} \tag{9.3}$$

$$H_{AB} = \alpha\nu_{max} \tag{9.4}$$

where ν_{max}, ε_{max}, $\Delta\nu_{1/2}$, m_d, m_a, and r are the wave numbers of the peak, the molar extinction coefficient, the half-height wave-number width, the number of donor sites, the number of acceptor sites, and the donor–acceptor distance, respectively, were estimated from the analysis of the IVCT bands. There appears to be a dependence of both the α and H_{AB} values on the nature of solvent and counterion and a significant correlation between α and K_c values. The α value was in a range between 0.02 and 0.04, indicating that $\mathbf{12}^-$ and $\mathbf{12}^{2-}$ can be classified as Robin and Day class II mixed-valence complexes.[58]

At 77 K electron paramagnetic resonance (EPR) spectra of $\mathbf{12a}^-$ and $\mathbf{12a}^{3-}$ generated by chemical reduction of **12a** with Na in THF, a representative system of low α and K_c values, showed different behaviors from those by $[Co(\eta^5\text{-}C_5Me_5)_2]$ in PC, a representative system of high α and K_c values. Both the spectra of monoanion $\mathbf{12a}^-$ in the Na–THF and in $[Co(\eta^5\text{-}C_5Me_5)_2]$-PC systems were similar to each other and exhibited hyperfine splitting due to Co nucleus with $I = 7/2$. The A_z value evaluated from the splitting of the $g_{//}$ signals for $Na^+\mathbf{12a}^-$ in THF is 9.2 (Figure 9.6a), larger than that for $[Co(\eta^5\text{-}C_5Me_5)_2]^+\mathbf{12a}^-$ in PC, which is 8.8, denoting that the interaction of the electronic spin with the Co nuclear spin is stronger in the

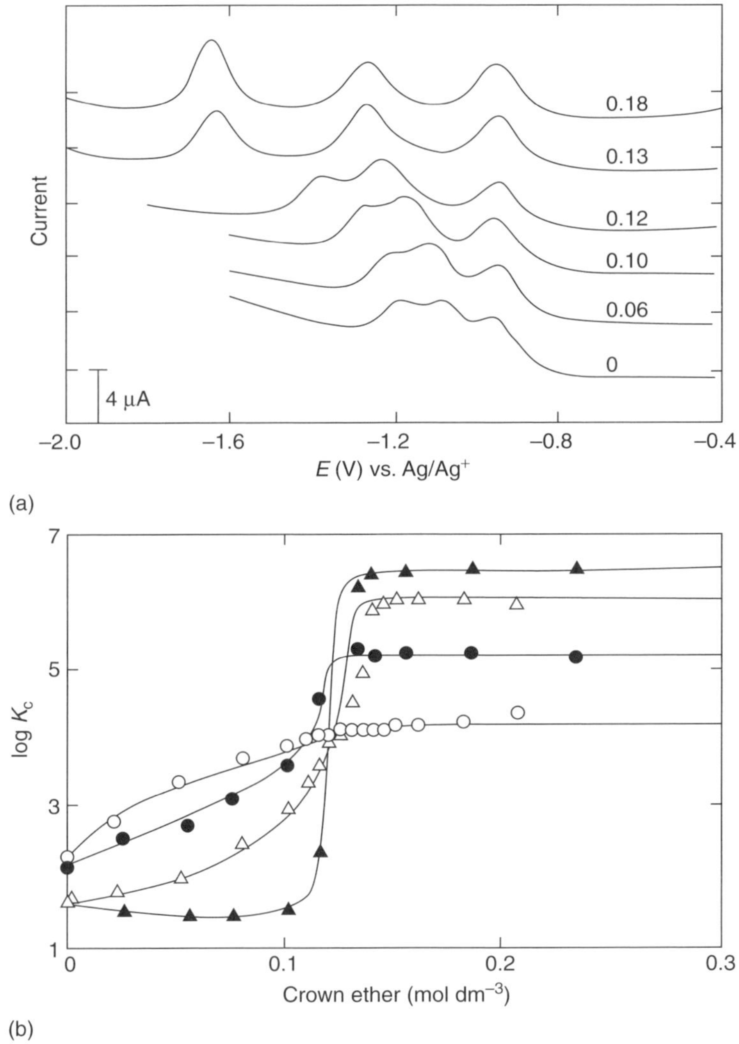

Figure 9.5 (a) Osteryoung square-wave voltammogram of **12a** in 0.1 M $NaBPh_4$–THF with and without 18-crown-6 (18C6) at a glassy carbon electrode with a frequency of 15 Hz and a potential step of 4 mV. Numbers in the figure refer to the concentration of crown ether (M). (b) Experimental (●, K_{c1} for 18C6; ▲, K_{c2} for 18C6; ○, K_{c1} for 15-crown-5 (15C5); △, K_{c2} for 15C5) and simulated (lines) data for changes of log K_c of **12a** with concentration of crown ether in 0.1 M $NaBPh_4$–THF.

former than in the latter. The $\mathbf{12a}^{2-}$ ion was EPR-silent in both media, indicating diamagnetism or antiferromagnetic interaction. The most significant difference appeared between the spectrum of $Na^+{}_3\mathbf{12a}^{3-}$ in THF and that of $[Co(\eta^5\text{-}C_5Me_5)_2]^+{}_3\mathbf{12a}^{3-}$ in PC (Figure 9.6d). The spectrum for the latter was similar to the spectra of **13a**, indicating that $S = 1/2$, with the A_z value of 9.1 being larger than that for $[Co(\eta^5\text{-}C_5Me_5)_2]^+\mathbf{12a}^-$, indicating a stronger electron–Co nucleus interaction. Contrarily, the spectrum of $Na^+{}_3\mathbf{12a}^{3-}$ in THF was the typical

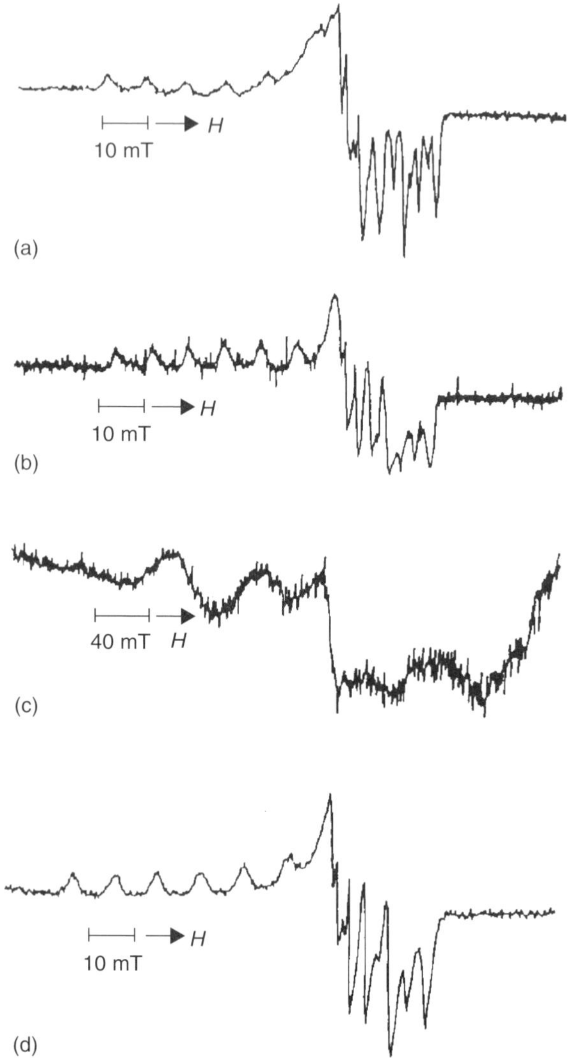

Figure 9.6 (a) EPR spectra of 2×10^{-5} M Na^+**12a**$^-$ and (b) 2×10^{-5} M $[Co(\eta^5\text{-}C_5Me_5)_2]^+$**12a**$^-$, and (c) $Na^+{}_3$**12a**$^{3-}$ in THF, and (d) $[Co(\eta^5\text{-}C_5Me_5)_2]^+{}_3$**12a**$^{3-}$ in PC at 77 K. Parameters obtained are $g_{\parallel} = 2.225$, $A_z = 9.2$, and $g_{\perp} =$ ca. 2.00 for (a), $g_{\parallel} = 2.238$, $A_z = 8.8$, and $g_{\perp} =$ ca. 2.00 for (b), $g = 2.057$ and $D = 0.102\,\text{cm}^{-1}$ for (c), and $g_{\parallel} = 2.235$, $A_z = 9.1$ and $g_{\perp} =$ ca. 2.00 for (d).

pattern of $S = 3/2$ in the nearly axial field with $g = 2.057$ and $D = 0.102\,\text{cm}^{-1}$.[59] This result implied that the spin-spin interaction was changed drastically by the matrix effects.

The results on both the redox properties, IVCT bands, and EPR spectra of the trimers described earlier demonstrated a stronger electrostatic interaction of the complex anion with a smaller counterion that could locate more closely to the complex ion than the larger one, resulting in perturbation of the delocalization more significantly than larger anions.

B. Metalladichalcogenolene Dinuclear and Trinuclear Complexes Containing Metal–Metal Bonds

As noted earlier in this section, metalladichalcogenolenes of late transition metals exhibited unique electronic properties due to their quasi-aromaticity caused by strong $d\pi$–$p\pi$ interaction, in which the metal center could be stabilized in unsaturated $16e^-$ forms by the contribution of lone-pair electrons in chalcogen (S, Se) atoms.[44–50] This electron deficiency of the metal center suggested the possibility of utilizing mononuclear metalladichalcogenolenes as building blocks of metal cluster complexes. Actually, metal–metal bond formation reactions of metalladithiolene complexes shown in Scheme 9.2 have been found to give dinuclear and trinucler complexes with frameworks of $CoFeS_2$ (**13**), $RuCoS_2$ (**14**), $CoWSe_2$ (**15**), Co_2MoS_4 (**16**), Co_2WS_4 (**17**), Co_2MoSe_4 (**18**), $RhCo_2S_2$ (**19**), $IrCo_2S_2$ (**20**), and $RuCo_2S_2$ (**21**).[60,61,62] These compounds can be divided into

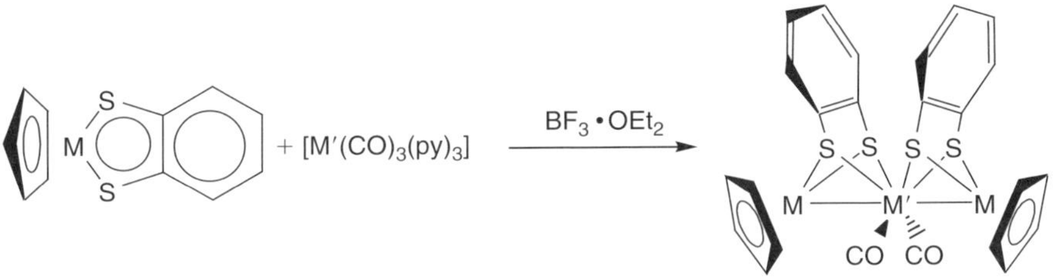

Scheme 9.2

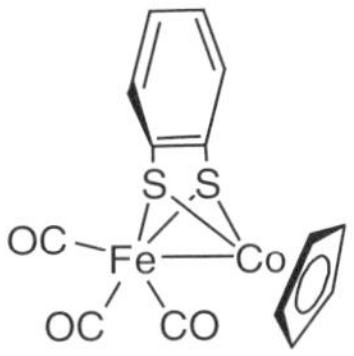

13

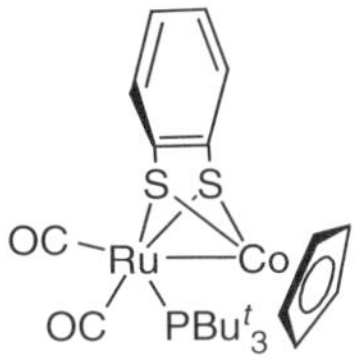

14

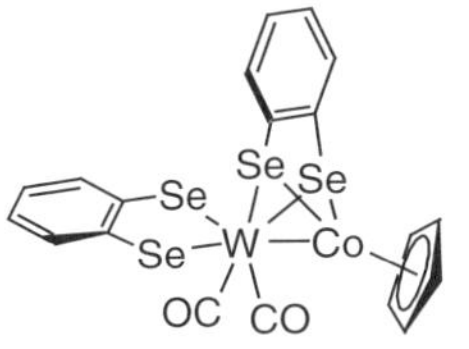

15

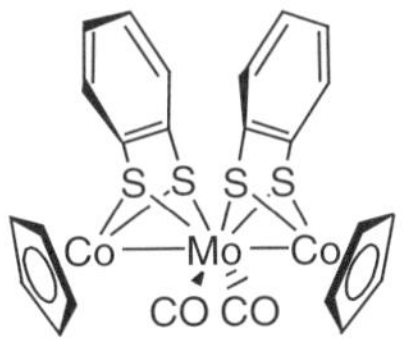

16

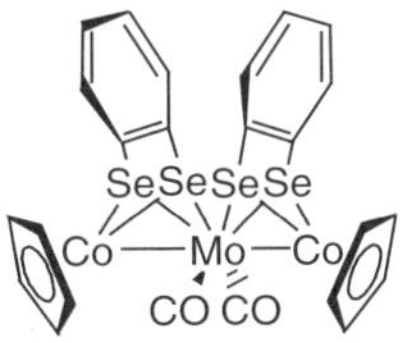

17

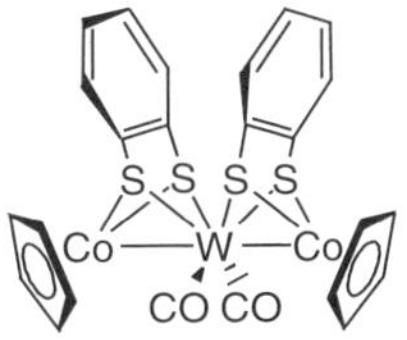

18

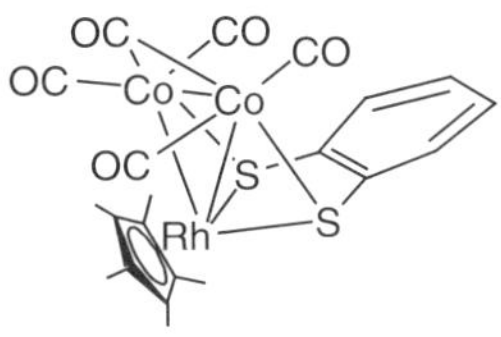

19

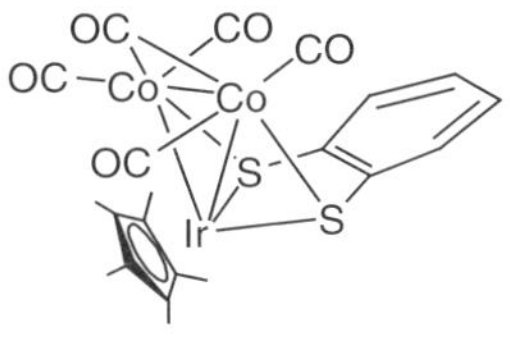

20

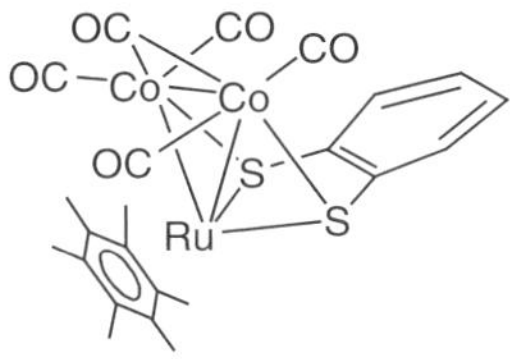

21

two groups based on the planarity of the metalladithiolene ring in the molecular structure. In compounds **13–18**, the metalladithiolene ring was bent because of the donation of a lone pair of electrons on S, thus leading to the loss of the aromaticity of the ring.[60,62] On the other hand, the planarity of the metalladithiolene ring in the triangular compounds **19–21** was maintained after the formation of the metal–metal bonds.[61] Theoretical calculation indicated that two Co atoms were located just above the S atoms of the metalladithiolene rings, and the $p\pi$ orbitals of the S atoms could interact effectively with those of the Co atoms without perturbing the sp^2 configuration of the S atoms. An interesting property of thermochromism was found in the case of **20**. The color of a solution of **20** in EtOH–MeOH–dimethylformamide (DMF) was dark brown at 300 K, but it turned to green at 77 K. Similar color changes were observed when **20** was dissolved in other solvents such as CH_2Cl_2 and when **20** was in the form of a KBr pellet. The UV-vis–NIR and IR spectra have indicated that the origin of the thermochromism was not a drastic change in the chemical structure, but rather the increased rigidity of the framework.

TABLE 9.3 Redox Properties of Metalladithiolene Cluster Complexes

	$E^{0\prime}$ for Reduction vs. Fc^+/Fc		
Compound	1st	2nd	$\Delta E^{0\prime}$
13	−1.48	−1.78	0.30
14	−1.72(ir)[a]		
15	−1.37[b]		
16	−1.41	−1.74	0.33
17	−1.49	−1.73	0.24
18	−1.13	−1.55	0.42
19	−1.04		
20	−1.14		
21	−1.14		

[a] Irreversible; others are reversible.
[b] In CH_2Cl_2; others are in MeCN.

Most of the metallachalcogenolene dinuclear and trinuclear complexes underwent reversible reductions based on the redox activity of the metalladichalcogenolene moieties.[60,61,62] Redox properties of the metalladithiolene dinuclear and trinuclear complexes are summarized in Table 9.3. Analysis of the cyclic voltammograms of Co_2MoS_4 (**16**) (see Figure 9.7), Co_2WS_4 (**17**), Co_2MoSe_4 (**18**) has revealed the redox process as the EEC mechanism given in equations (9.5)–(9.7),

$$\mathbf{16} + e^- = \mathbf{16}^- \tag{9.5}$$

$$\mathbf{16}^- + e^- = \mathbf{16}^{2-} \tag{9.6}$$

$$16^{2-} \xrightarrow{k_1} 2[(\eta^5\text{-}C_5H_5)Co(S_2C_6H_4)]^- + \text{'}Mo(CO)_2\text{'} \tag{9.7}$$

The rate constant of the decomposition reaction, k_1, of $\mathbf{16}^{2-}$, $\mathbf{17}^{2-}$, and $\mathbf{18}^{2-}$ was estimated by a computer simulation of the voltammograms to be $0.19\,s^{-1}$, $0.02\,s^{-1}$, and $0.04\,s^{-1}$, respectively.[60,61,62] This denotes that the Se bridge stabilized the complex much more than did the S

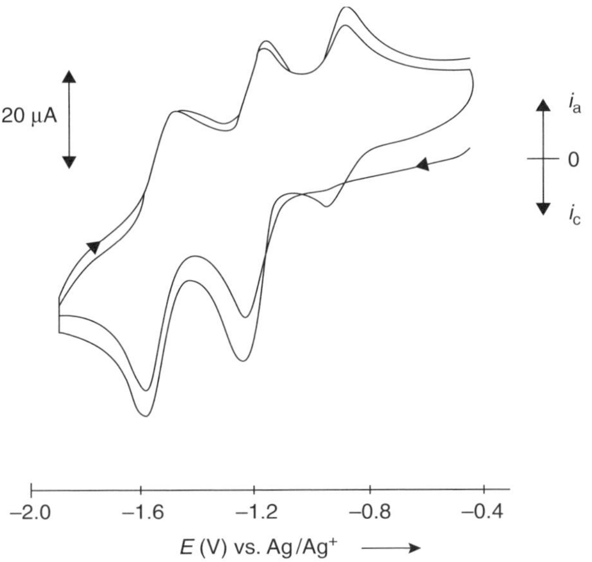

Figure 9.7 Cyclic voltammogram of **16** at a glassy carbon electrode in 0.1 M Bu_4NClO_4–MeCN at a scan rate of $0.1\ Vs^{-1}$.

bridge, and the W–Co bond was stronger than the Mo–Co bond in the 2-valence state. In comparing the redox potentials among the three complexes, it was reasonable to assume that the two Co centers, but not the Mo or W atoms, underwent reduction. The formal oxidation states in the neutral form could be written as $Co^{III}Mo^{0}Co^{III}$ (**16**), $Co^{III}W^{0}Co^{III}$ (**17**), and $Co^{III}Mo^{0}Co^{III}$ (**18**). This implied that the valency balance of monoanions of **16**, **17**, and **18** could be expressed as $Co^{III}M^{0}Co^{II}$ (M = Mo, W), suggesting that the Mo or W bridge assisted the electronic interaction between Co sites to form a thermodynamically favorable mixed-valence state.[51,52,53] This was supported by the electronic spectrum of $\mathbf{16}^{-}$ generated by the reduction of **16** with Na in THF, where a broad band attributable to an intervalence transfer[36,37] appeared at 1160 nm ($\varepsilon = 60\ M^{-1}cm^{-1}$). Judging from the $\Delta E^{0'}$ values, the thermodynamic stability of the mixed-valence state increased in the order **17** < **16** < **18**. The unoccupied *d*-orbital of W, which was at a higher level than that of Mo, might have assisted less effectively in the electronic communication between the two Co sites, and that one introduced electron was more liable to delocalize via the Se bridge than via the S bridge, because Se was softer and easier to polarize than S.

The CoFe complex **13** in THF underwent two $1e^-$ reversible reductions at $E^{0'} = -1.48$ and -1.78 V vs. Fc^+/Fc. The CoRu complex **14**, in contrast, showed only a broad irreversible reduction at $E_{p,c} = -1.72$ V vs. Fc^+/Fc, suggesting that the skeleton of **14** is easily decomposed upon reduction.

In the cyclic voltammograms of **19**, **20**, and **21** in Bu_4NClO_4–MeCN, one quasi-reversible $1e^-$ reduction wave was observed in the region between -1.04 and -1.14 V vs. Fc^+/Fc. These reduction potentials were located between those of the mononuclear metalladithiolenes, $[(\eta^5\text{-}C_5H_5)Co(S_2C_6H_4)]$, $[(\eta^5\text{-}C_5Me_5)Rh(S_2C_6H_4)]$, and $[(\eta^5\text{-}C_5Me_5)Ir(S_2C_6H_4)]$ (-1.58 to -2.00 V in dichloromethane; $1e^-$, reversible) and the reduction potential of a related compound ($Co_2(CO)_8$, -0.42 V in THF; $1e^-$, irreversible). The lowest unoccupied molecular orbitals (LUMOs) of **20** and **21** were for the most part localized around the Co_2 moiety, and were less localized around the metalladithiolene; this environment was consistent with the position of the reduction potentials. The difference of the redox potentials around **19**, **20**, and **21** was only 0.1 V; however, the difference of the corresponding mononuclear complexes exceeded 0.4 V. The difference in redox properties between the multinuclear complexes with nonplanar metalladichalcogenolene rings and those with planar metalladichalcogenolene rings was important for the design of electrofunctional molecules.

IV. REDOX, OPTICAL AND MAGNETIC PROPERTIES OF π-CONJUGATED METALLACYCLE POLYMERS

One category involved in "organometallic conducting polymers"[63] is poly(arylene metallacyclopentadienylene), wherein the framework of the polymer is π-conjugated and in part composed of metallacyclopentadienes.[9,10,11] In this polymer, the metallacyclopentadiene ring with a *d*-block heteroatom is structurally analogous to the rings found in other representative π-conjugated polymers, such as poly(pyrrole) and poly(thiophene), which contain *p*-block heteroatoms. One efficient synthetic method of metallacyclopentadine polymers is metallacycling polymerization (MCP), which is based on successive metallacyclization of conjugated diacetylenes to metal centers (Scheme 9.3).[64,65,66] In this section, physical properties of π-conjugated cobaltacyclopentadiene and ruthenacyclopentatriene polymers mainly prepared by MCP are described.

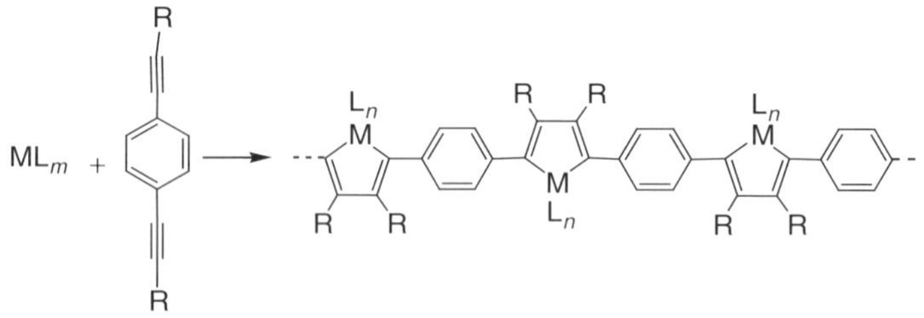

Scheme 9.3

A. Poly(Arylene Cobaltacylopentadienylene)s

Poly(arylene cobaltacylopentadienylene)s were the first series of π-conjugated metallacycle polymers prepared by MCP, and various kinds of compounds have been reported in this category.[9,10,11] There was, however, a problem in the synthesis of poly(arylene cobaltacylopentadienylene)s by MCP, which was the imperfect regioselectivity of the metallacyclization (Scheme 9.4a).[10,11] A mixture of 2,4-diarylcobaltacyclopentadiene moieties, which was an unfavorable structure for π-conjugation, at more than 15% and the favorable 2,5-diarylcobaltacyclopentadiene moieties have been produced. Perfectly π-conjugated poly (arylene cobaltacylopentadienylene) **22** was prepared by reductive coupling of the dibromocobaltacyclopentadiene derivatives (Scheme 9.4b),[67] and its physical properties were compared with the corresponding polymer prepared by the MCP method.[68]

The color of poly(arylene cobaltacylopentadienylene)s was dark brown, and the UV-vis absorption spectra of the films of insoluble polymers **23–26** coated as grown on quartz glass show strong bands that have an edge at 500 to 600 nm.[68] As for the soluble polymer, the band edge shifted to the higher wavelength according to an increase in polymerization degree for **27** in CH_2Cl_2, indicating the formation of π-conjugated structure.[68] The perfectly π-conjugated polymer **22** prepared by polycondensation on dihalogenated complex exhibited a shift of the

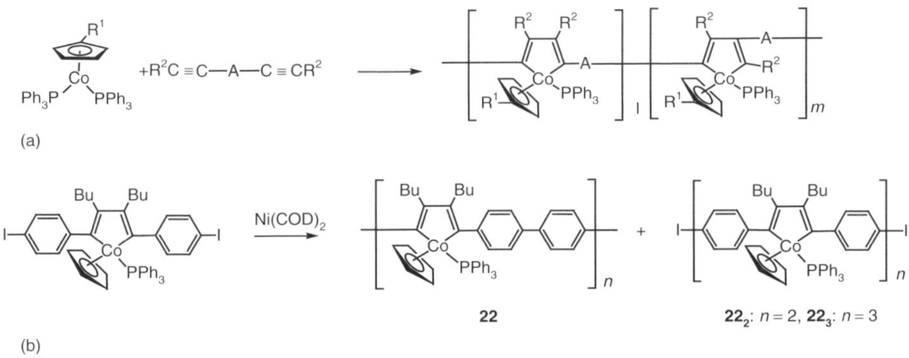

Scheme 9.4

peak edge to the longer wavelength compared with the corresponding polymer prepared by metallacycling polymerization.[67] Evaluated band-gap energies, E_g, were 2.1–2.3 eV,[68] which corresponded roughly to the value observed in poly(thiophene) (2.0 eV), and were relatively small when compared to the band gaps among previously known π-conjugated organic polymers.[69]

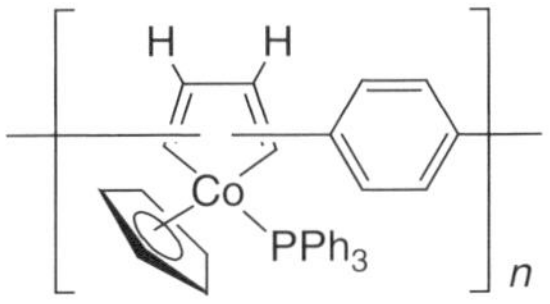

23

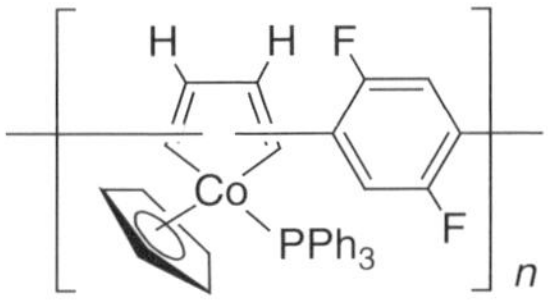

24

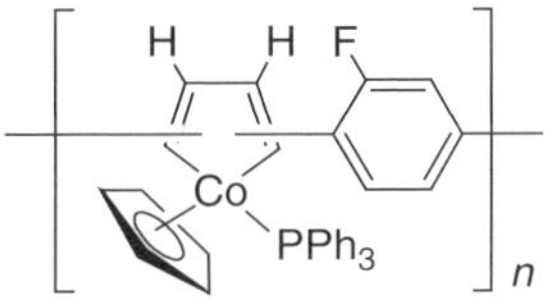

25

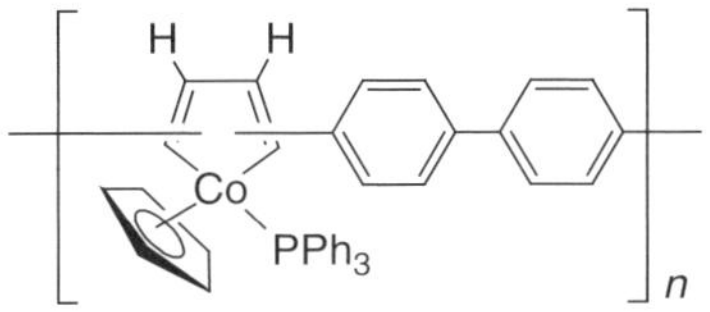

26

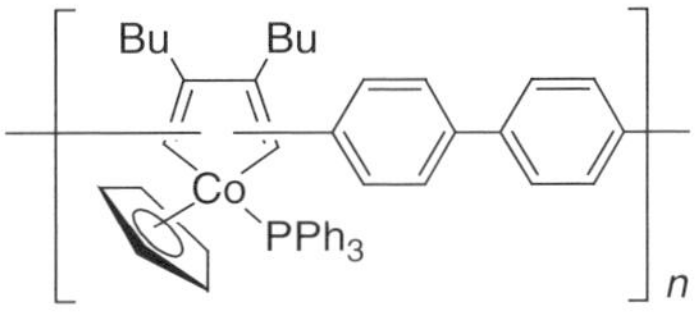

27

Cobaltacylopentadiene complexes undergo 1-electron oxidation, and their potential and chemical reversibility depend strongly on the substituents on the metallacycle.[70–73] This directly reflects the redox properties of the cobaltacylopentadiene polymers. The oxidation potential of the polymers became more positive when the electron-withdrawing substituents, such as —COMe, were bound to the metallacycle,[74] and the chemical reversibility increased in the order of the substituents, —COMe (**28**) < —H (**26**) < alkyl (—Me (**29**), Bu (**27**)).[74] These results support our consideration that the highest occupied molecular orbital (HOMO) based on *d*-orbital of the metal atoms in the polymer exists between the valence band (VB) and the conduction band (CB) derived from π-conjugation, and the oxidation occurs at metal sites. In the cyclic voltammograms of a dimer $\mathbf{22}_2$ and a trimer $\mathbf{22}_3$, the waves were broader compared with that of the monomer, although only a single oxidation peak was observed for the dimer and the trimer (Figure 9.8),[67] suggesting that the Co^{III} and Co^{IV} sites were weakly interacted and the mixed-valence states were generated within a narrow potential range. The oxidation potentials estimated by the simulation of the voltammogram were $E_1^{0'} = -0.285\,\text{V}$ and $E_2^{0'} = -0.212\,\text{V}$ vs. Fc^+/Fc for the dimer, and $E_1^{0'} = -0.291\,\text{V}$, $E_2^{0'} = -0.248\,\text{V}$, and $E_3^{0'} = -0.189\,\text{V}$ vs. Fc^+/Fc for the trimer. The $\Delta E^{0'}$ values indicated that the interaction energy between ferrocene and ferrocenium sites, u_{OR} is estimated to be $-2\,\text{kJ mol}^{-1}$ based on the neighboring site interaction model,[18] and the value was about one-fifth compared with that for oligo(1,1′-dihexylferrocenylene)s.[15,16]

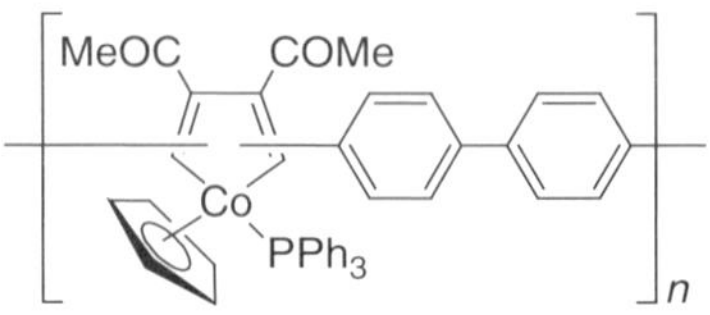

28

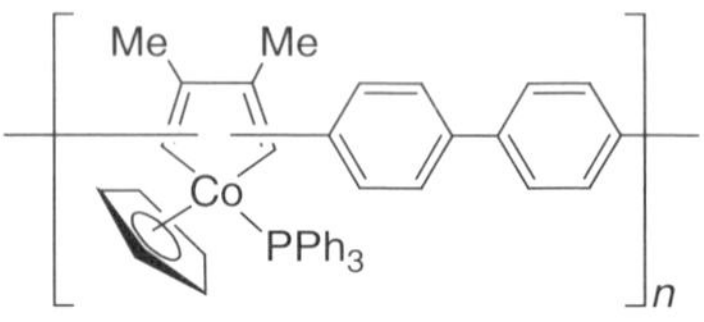

29

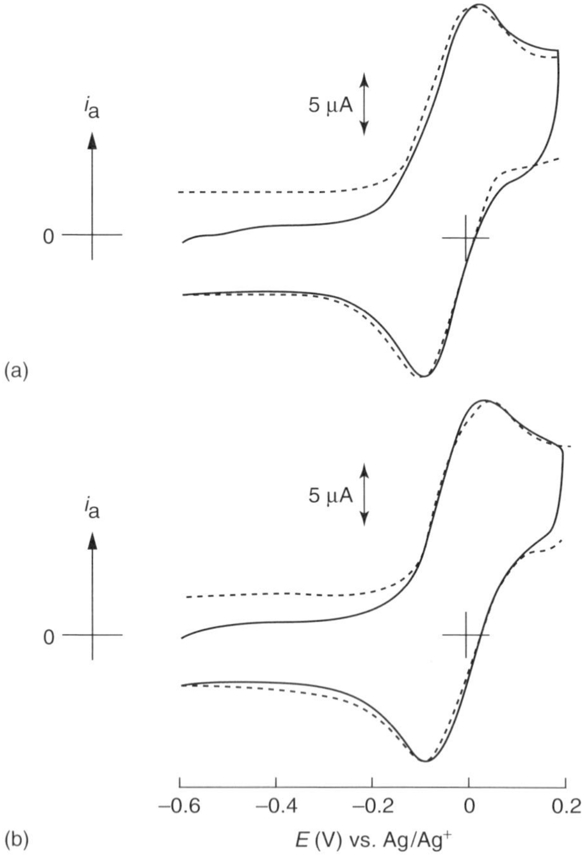

Figure 9.8 (a) Cyclic voltammograms of compounds $\mathbf{22}_2$ and (b) $\mathbf{22}_3$ at a glassy carbon electrode in 0.1 M NBu_4ClO_4–CH_2Cl_2 at a scan rate of $0.1\,\text{Vs}^{-1}$ (full lines) and their simulation based on the open boundary finite-diffusion model (broken lines).

The oxidation wave in the cyclic voltammograms of thenylene-bridged cobaltacyclopentadiene polymer, **30**, is broader than that of the phenylene-bridged one, **31** (Figure 9.9).[75] This is because the energy level for the highest occupied π-orbital of thiophene is closer than that of phenylene to the *d*-orbital level of the cobalt site, so that the internuclear electronic interaction through the thiophene ring is considered to be stronger. In the oxidation process, therefore, more than one oxidation wave due to the formation of mixed-valence states overlap, resulting in a broad wave in the cyclic voltammogram.

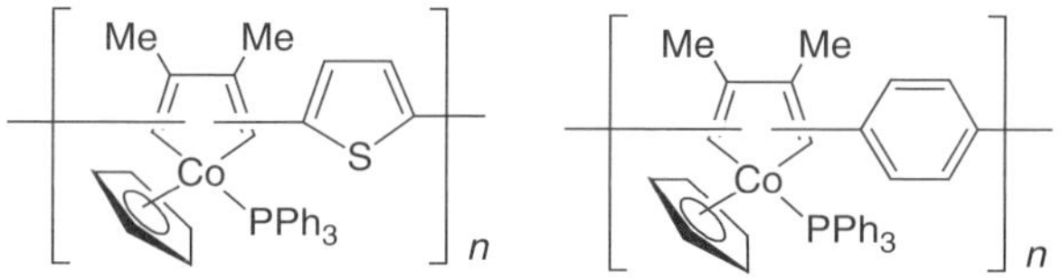

30 **31**

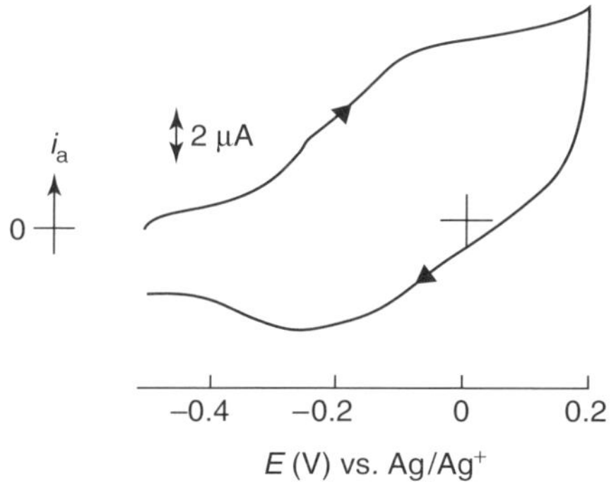

Figure 9.9 Cyclic voltammogram of a cobaltacyclopentadiene polymer complex, **30**, at a glassy carbon electrode in 0.1 M NBu_4ClO_4–CH_2Cl_2 at a scan rate of 0.1 Vs^{-1}.

The cobaltacyclopentadiene polymers **26** and **29** show an electrical conductivity of 10^{-12}–10^{-6} Scm^{-1} in the neutral form at room temperature.[74,76] When **29** was treated with I_2, the conductivity increased up to 10^{-4} Scm^{-1}. This result could be interpreted by the consideration that I_2-doping generates Co^{III}/Co^{IV} mixed-valence states in the polymer chain; this state is stable to some extent, as suggested by the electrochemical properties; and consequently, Co^{III} and Co^{IV} sites are interacted through a π-conjugated chain, causing the mixed-valence conductivity.

The intrinsic photoconductive property was found for a cobaltacyclopentadine polymer, **32**.[68] Photoresponse of current–voltage (I–V) characteristics for ITO/**32**/ITO indicated that the polymer had a low conductivity in the dark and the photocurrent was four times larger than the dark current. It was proposed that the metal *d*-character orbitals localized at cobalt sites, plus their energy level lying between valence and conduction bands acted as the trapping sites of holes generated by photoactivation of electrons from the valence band to the conduction band.

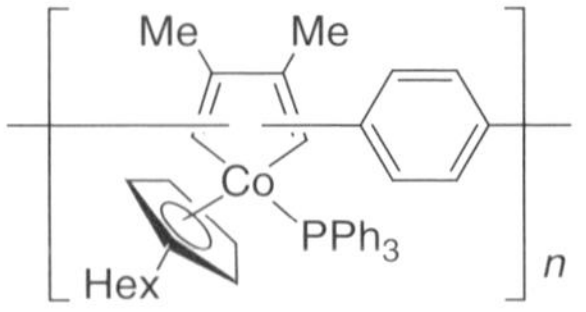

32

B. Poly(arylene ruthenacylopentatrienylene)s

An important advantage of the synthesis of poly(arylene ruthenacylopentatrienylene)s is the perfect regioselectivity of the metallacyclizaion to give 2,5-diaryl ruthenacycles (Scheme 9.5).[77,78,79] UV-vis–NIR absorption spectra of the ruthenacyclopentatiene polymer **33** in dichloromethane displayed the bands at 296, 400, and 525 nm, which were shifted to longer wavelength compared with the bands of the corresponding monomer, **34**, at 256, 378, and 505 nm, respectively, due to the extension of the conjugated system for the polymer, while the band at 694 nm remained unchanged between **33** and **34**,[80] This band can be ascribed to LMCT (carbene $\pi \rightarrow$ Ru d) at the ruthenacyclopentatriene unit.

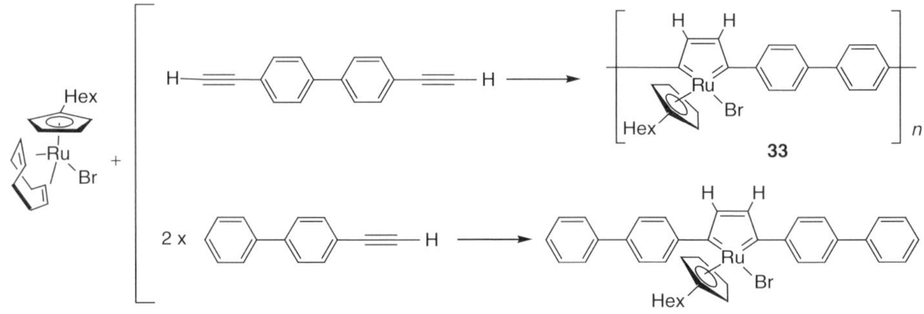

Scheme 9.5

Ruthenacyclopentadiene complexes undergo reversible 1e$^-$ reduction, as shown in the cyclic voltammogram of **34**, indicating the role of ruthenacyclopentatriene as a good electron acceptor. This redox property is quite different from those of cobaltacyclopentadiene, for which the oxidation potential is 0–0.5 V vs. Fc^+/Fc, indicating that it functions as a fine electron donor.[68] The cyclic voltammogram of the polymer **33** in Figure 9.10 indicates that it also undergoes reversible 1e$^-$ reduction ascribed to the ruthenacyclopentatriene unit at $E^{0'} = -1.01$ V for **33** and -1.03 V for **34** vs. Fc^+/Fc. The peak of **33** is broader than that of **34**, indicating an existence of electronic interaction between ruthenacycle units in **33**. The location of the modestly localized Ru-centered orbital between the π and π^* orbitals of the conjugated chain is consistent with the electronic spectra, as noted earlier in this section.

The ability of the ruthenacycle to assist the interaction between the ferrocenyl moieties was evaluated by the stability of the mixed-valence state of the 2,5-diferrocenyl derivative, **35**.[79] A reversible reduction wave of the ruthenacycle was observed at -1.40 V vs. Fc^+/Fc, and two quasi-reversible waves of the ferrocene moieties appeared at -0.01 and 0.23 V vs. Fc^+/Fc in the cyclic voltammogram (Figure 9.11). The separation of the oxidation waves by the ferrocenyl moieties in **35** indicated the existence of an electronic interaction between the two ferrocene moieties. The difference in potential of the two waves, 0.24 V, was smaller than

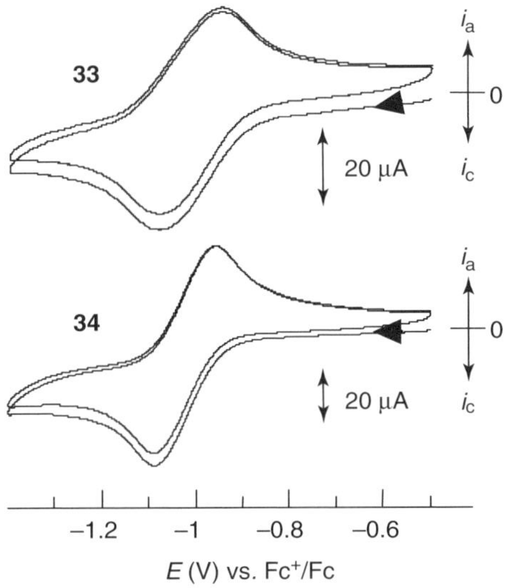

Figure 9.10 Cyclic voltammograms of **33** (top) and **34** (bottom) on a glassy carbon electrode in 0.1 M Bu_4NClO_4–CH_2Cl_2 at a scan rate of 0.1 Vs^{-1}.

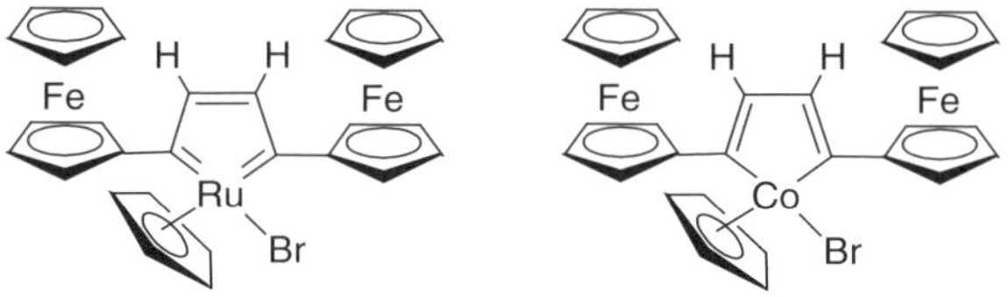

35 **36**

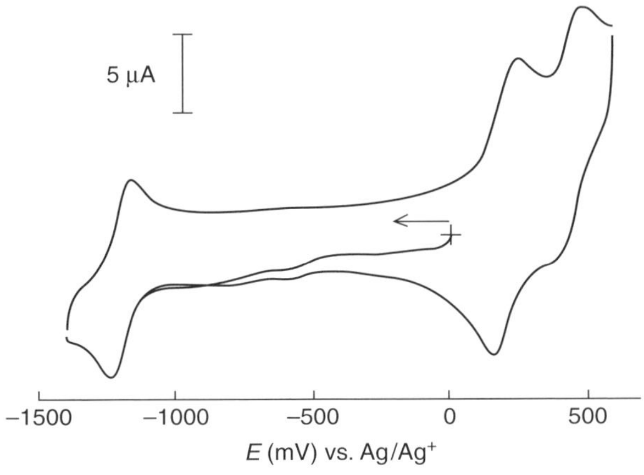

Figure 9.11 Cyclic voltammogram of **35** on a glassy carbon electrode in 0.1 M Bu_4NClO_4–CH_2Cl_2 at a scan rate of 0.1 Vs^{-1}.

those for simple biferrocene (0.42 V)[81] and cobaltacyclopentadienyl-bridged complex **36** (0.47 V),[82] whereas the value was larger than those for most bis-ferrocenyl compounds with π-conjugated organic bridges, such as 1,2-diferrocenylethylene (0.17 V)[83] and 1,2-diferrocenylacetylene (0.13 V).[84] Analysis of the (IVCT) band of the 1e$^-$ oxidized form of **35** in dichloromethane indicated that λ_{max} = 1180 nm (8484 cm^{-1}), with the half-height width = 3840 cm^{-1}. Based on Marcus–Hush theory,[56,57] the mixing coefficient was calculated to be 0.06, which was smaller than or comparable to those for biferrocene (0.09),[84] 1,2-diferrocenylethylene (0.09),[83] and 1,2-diferrocenylacetylene (0.07).[84] From these measurements, the ruthenacyclopentatriene moiety of **35** was found to conduct significant electronic interaction between its 2- and 5-substituents, such as π-conjugated organic bridges, although the magnitude of this interaction appeared to be smaller than that of the cobaltacyclopentadiene analog, **36**. The larger internuclear interaction in **36** than in **35** should be due to the stronger donor ability of the cobaltacycle than the ruthenacycle. This enhances the hole transfer in the electron exchange between the ferrocenyl moieties.

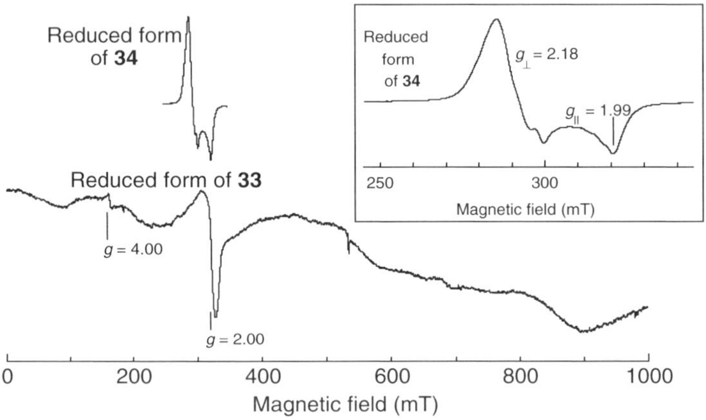

Figure 9.12 EPR spectra of reduced forms of **34** (top) and **33** (bottom) at 4 K. Inset: Enlarged view of EPR spectrum of reduced form of **34**.

Comparison of the EPR spectra between the reduced forms of **33** and **34** yielded information on the magnetic interaction between Ru sites via π-conjugated linker in the reduced form of **33** (Figure 9.12).[80] The reduced form of **34** in frozen THF afforded axial symmetry with weak rhombic spectra ($g_\perp$ = 2.18, $g_\parallel$ = 1.99), which were similar to that of an S = 1/2 spin in a low-spin d^5 configuration of Ru^{III}.[85,86,87] Hyperfine interaction with ^{99}Ru and ^{101}Ru isotopes (I = 5/2, 12.7% and 17%, respectively) was resolved in the spectra, especially for g_2 (a = 3 mT), indicating that the unpaired electron of the reduced form of **34** localized in the ruthenacycle moiety and did not interact with other molecules. On the other hand, the EPR spectrum of the reduced form of **33** showed several peaks in a wide magnetic field, and was completely different from that of **34**. This difference indicated the existence of ferromagnetic interaction between Ru sites in the polymer chain. The signal at g = 2 was caused mainly by isolated spin in the ruthenacycle, but it should be the sum of this state and the triplet state, indicated by a signal at g = 4. Broad signals at 50, 205, 553, and 855 mT indicated interactions between more than two ruthenacyclopentatriene units. The instability of the reduced form of **34** as well as that of **33** made the measurement of magnetic susceptibility to confirm

S difficult. These results manifested the possibility of developing magnetically functional materials using organometallic conducting polymers.

V. STEPWISE PREPARATION OF LINEAR π-CONJUGATED BIS(TERPYRIDINE)METAL POLYMER CHAINS ON GOLD SURFACE

One of the interesting things about the redox polymers is their use in the creation of the molecular electronic devices.[3–5] Redox polymer films on electrodes have been fabricated using chemical modification, electrochemical polymerization, polymer coating, and so on.[88] Recently, stepwise complexation methods have been employed to fabricate multiple complex layers.[89,90,91] In this section, the stepwise preparation of bis(tpy)metal polymer chains by combining terpyridine (tpy) ligand self-assembled monolayer (SAM) formation and metal–tpy coordination reactions is described as an example. This method realized the formation of a desired number of polymer units and a desired sequence of Co–Fe heterometal structures in the polymer chain.[92]

A typical method for fabricating multiple complex layers is as follows. First, Au–S–AB–tpy SAM was prepared by immersing an Au/mica or Au/ITO plate in a chloroform solution of tpy–AB–SS–AB–tpy (**37**). In the case of connecting Fe^{II} ion, the tpy-terminated surface was immersed in $Fe(BF_4)_2$ (aq) or $(NH_4)_2Fe(SO_4)_2$ (aq). The bis(tpy)iron structure was prepared by immersing the film with the metal-terminated surface in a chloroform solution of tpy–AB–tpy (**38**). The latter two processes were repeated for the preparation of multilayered (namely, polymeric) bis(tpy)iron(II) complex films (Figure 9.13).

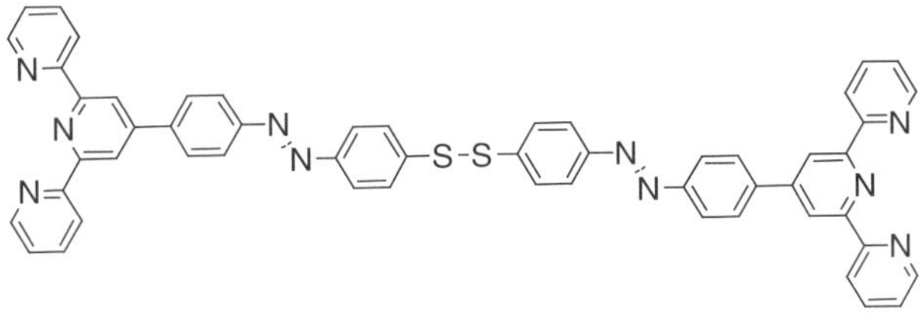

37 (tpy–AB–SS–AB–tpy)

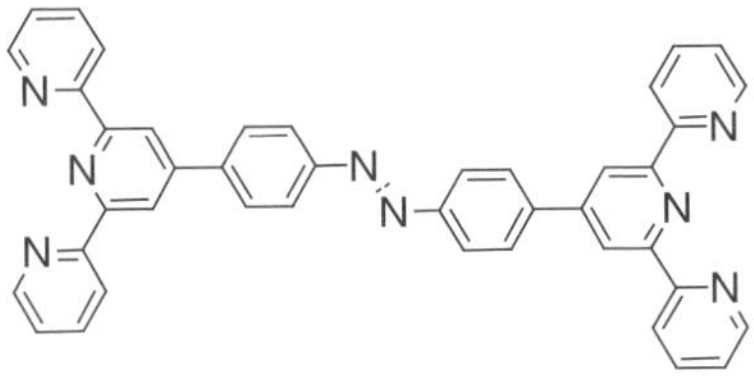

38 (tpy–AB–tpy)

In contrast, a combination of immersion in $CoCl_2$ (aq) and immersion in a chloroform solution of **38** with the Au–S–AB–tpy SAM was not sufficient for accumulating bis(tpy)cobalt complex units. But the addition of a process of electrochemical oxidation from Co^{II} to Co^{III} resulted in the successful construction of multilayered bis(tpy)cobalt complex films.

Figure 9.13 Stepwise coordination method to prepare metal complex molecular wire on Au.

UV-vis spectra and cyclic voltammetry confirmed the stepwise quantitative film formation. In the UV-vis spectrum of the bis(tpy)iron complex film, absorbance of the peak at 592 nm attributed to the MLCT transition increased linearly with the number of stepwise complexations, n, and the peak current density, j_p, of the reversible peak of the Fe^{III}/Fe^{II} couple at 0.67 V vs. Fc^+/Fc in Bu_4NClO_4–CH_2Cl_2 in the cyclic voltammogram increases proportionally with *n*, indicating a linear increase of the coverage of electroactive species (Figure 9.14).

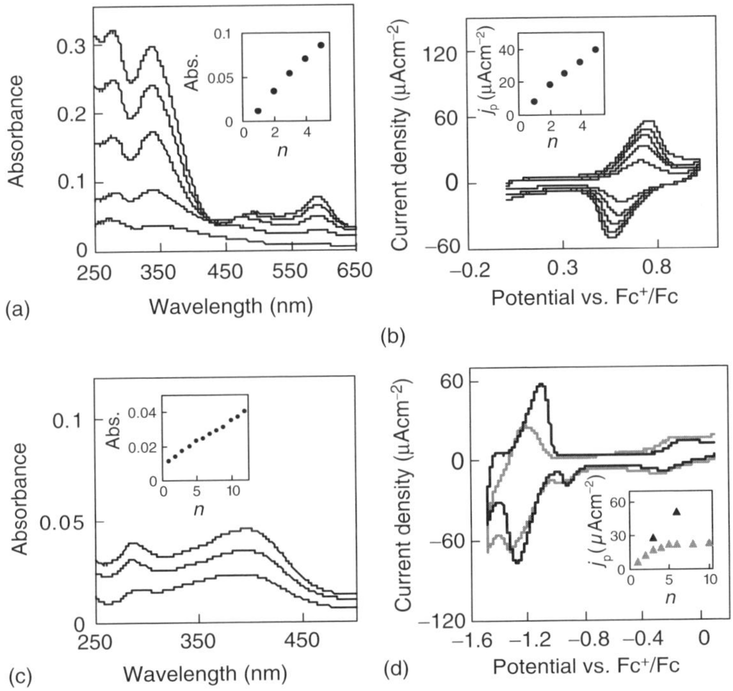

Figure 9.14 (a) Absorption spectra of [***n*****Fe**] ($n = 1$–5), and the plots of absorbance at 592 nm vs. *n* (inset). (b) Cyclic voltammograms of [***n*****Fe**] ($n = 2, 4, 6, 8$, and 10) at 0.1 Vs^{-1}, and the plots of j_p vs. *n* (inset). (c) Absorption spectra of [***n*****Co**] ($n = 5$, 10, and 15), and the plots of absorbance at 394 nm vs. *n* (inset). (d) Cyclic voltammograms of [**3Co**] (gray) and [**6Co**] (black) at 0.1 vs^{-1}, and the plots of j_p vs. *n* (inset).

A nearly closed packing of 6-nm-o.d. circular domains, indicating the stacking of molecular chains, and the formation of a fairly smooth surface (apparent roughness of the film is within 1 nm less than the length of one complex unit, 2 nm) were observed in the scanning transmission electron microscope (STM) image of the bis(tpy)iron complex film with two iron complex layers, abbreviated as [**2Fe**], as shown in Figure 9.15a. The scanning electron microscope (SEM) photograph of [**47Co**], shown in Figure 9.15b, indicated the growth of the film to a thickness of 100 nm on Au in the side-view image; the thickness was nicely consistent with the number of product layers, 47, times the molecular unit length, 2 nm.

One merit of this polymer preparation method is that it allows creation of heterometal polymer chains with the intended sequence.[93] The stepwise formation of a heterometal double-layer film [**1Co1Fe**] was monitored by cyclic voltammetry during film construction (Figure 9.16a). When the bis(tpy)iron complex units were connected to the already prepared bis(tpy)cobalt complex layer, the redox activity of the Fe^{III}/Fe^{II} couple appeared without

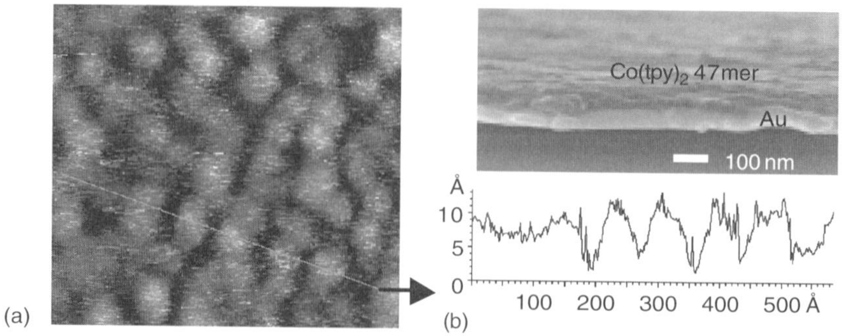

Figure 9.15 (a) STM image of [**2Fe**] and its topological profile. (b) Side view of the SEM image of [**47Co**].

changing the redox activity of the Co^{III}/Co^{II} couple. The result was that the surface coverage of the bis(tpy)iron complex units, estimated from the cyclic voltammetry, $\Gamma_{Fe,CV} = 1.4 \times 10^{-10}\,mol\,cm^{-2}$, was equal to that of bis(tpy)cobalt complex units ($\Gamma_{Co,CV} = 1.4 \times 10^{-10}\,mol\,cm^{-2}$), indicating that each bis(tpy)cobalt unit was connected to one bis(tpy)iron complex unit, and that there was no other sequences of the units in the film. In the cyclic voltammogram of heterostructured polymer films made of polymer chains composed of [**10Co5Fe**] (Figure 9.16b), the peak current for Co^{II}/Co^{I} was larger than that of Co^{III}/Co^{II}, due to the difference in electron self-exchange rate constant values, as noted earlier in this section. Also, the redox wave of Fe^{III}/Fe^{II} was less than two-thirds the size of the Co^{II}/Co^{I} redox wave, due to the existence of a cobalt complex sequence between the electrodes, which acted as a barrier to the electron transfer of the Fe complex, even though Fe^{III}/Fe^{II} was a fast electron-exchange couple. However, the retardation was very small, given that the barrier layer thickness was large (21 nm). This should be due to the high electron-transport ability of the inner π-conjugated $Co(tpy)_2$ polymer chain. These results indicated that the quantitative formation of heterostructured film can be achieved at the surface by selecting the conditions of stepwise complexation reactions. This surface bottom-up method will give a new strategy for the molecular design of electronically functional molecular chains suitable for the development of "molecular electronic devices."

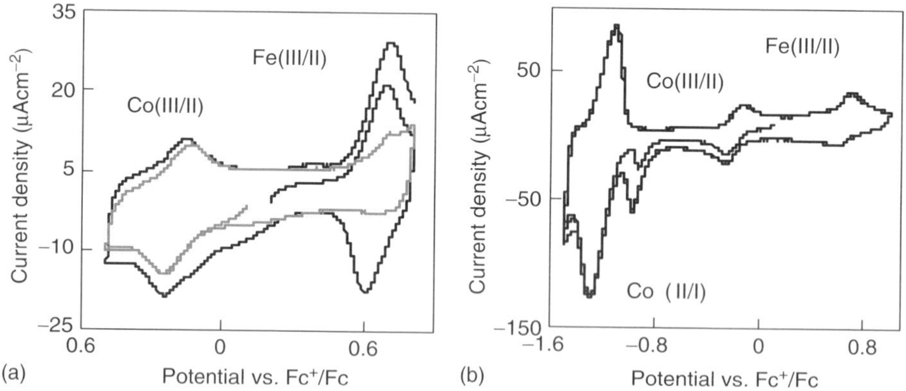

Figure 9.16 (a) Cyclic voltammograms of [**1Co**] (gray) and [**1Co1Fe**] (black), and (b) [**10Co5Fe**] on gold at $0.1\,Vs^{-1}$.

VI. CONCLUDING REMARKS

In the present chapter, several redox active multinuclear metal complex systems were presented. They involve more than two metal centers that have electronic communication through π-conjugated bridging ligands or by direct metal–metal bonds. They exhibited many functionalities, such as multistep redox reaction, low-energy light absorption, photoconductivity, ferromagnetic spin-spin interaction, electrochromism, photochromism, thermochromism, and facile electron transport due to the formation of mixed-valence states, and magnetic interaction of electron spins on the metal centers. These original functionalities can be combined with other molecular functions to construct a molecular electric and photonic circuit, which is a key target of nanoscience and nanotechnology in this century.

REFERENCES

1. H. B. Gray, W. R. Wllis, Jr., Chapter 6 in *Bioinorganic Chemistry*, I. Bertini, H. B. Gray, S. J. Lippard, J. S. Valentine, Eds., pp. 315–363, University Science Books, Sausalito, CA, 1994.
2. A. Deronzier, J.-C. Moutet, Chapter 9 in *Comprehensive Coordination Chemistry*, Vol. II, J. A. McCleverty, T. J. Meyer, Eds., pp. 471–507, Elsevier, Oxford, 2004.
3. C. Joachim, J. K. Gimzewski, A. Aviram, *Nature*, **408**, 541 (2000).
4. J. Park, A. N. Pasupathy, J. I. Goldsmith, C. Chang, Y. Yalsh, J. R. Petta, M. Rinkoski, J. P. Sethna, H. D. Abruña, P. L. McEuen, D. C. Ralph, *Nature*, **417**, 722 (2002).
5. C. E. D. Chidsey, R. W. Murray, *Science*, **231**, 25 (1986).
6. H. Nishihara, M. Murata, *J. Inorg. Organomet. Polym. Mater.*, **15** 147 (2005).
7. H. Nishihara, *Bull. Chem. Soc. Japan*, **74**, 19 (2001).
8. M. Kurihara, K. Kubo, T. Horikoshi, M. Kurosawa, T. Nankawa, T. Matsuda, H. Nishihara, *Macromol. Symp.*, **156**, 21 (2000).
9. H. Nishihara, Chapter 19 in *Handbook of Organic Conductive Molecules and Polymers*, H. S. Nalwa, Ed., Vol. 2, pp. 799–832, Wiley, Weinheim, 1997.
10. H. Nishihara, M. Kurashina, M. Murata, *Macromol. Symp.*, **196**, 27 (2003).
11. M. Kurashina, M. Murata, H. Nishihara, *Macromol. Symp.*, **209**, 141 (2004).
12. H. Nishihara, *Adv. Inorg. Chem.*, **53**, 41 (2002).
13. P. Nguyen, P. Gómez-Elipe, I. Manners, *Chem. Rev.*, **99**, 1515 (1999).
14. Y. Okamoto, M. C. Wang, *J. Polym. Sci., Polym. Lett. Ed.*, **18**, 249 (1980).
15. T. Hirao, M. Kurashina, K. Aramaki, H. Nishihara, *J. Chem. Soc., Dalton Trans.*, 2929 (1996).
16. H. Nishihara, T. Hirao, K. Aramaki, K. Aoki, *Synth. Metals*, **84**, 935 (1997).
17. K. Aoki, J. Chen, H. Nishihara, T. Hirao, *J. Electroanal. Chem.*, **416**, 151 (1996).
18. K. Aoki, J. Chen, *J. Electroanal. Chem.*, **380**, 35 (1995).
19. T. Hirao, K. Aramaki, H Nishihara, *Bull. Chem. Soc. Japan*, **71**, 1817 (1998).
20. G. Davidson, *Inorg. Chim. Acta*, **3**, 596 (1969).
21. M. A. Busch, R. J. Clark, *Inorg. Chem.*, **14**, 219 (1975).
22. T. Ito, T. Hamaguchi, H. Nagino, T. Yamaguchi, J. Washington, C. P. Kubiak, *Science*, **277**, 660 (1997).
23. T. Horikoshi, K. Kubo, H. Nishihara, *J. Chem. Soc., Dalton Trans.*, 3355 (1999).
24. H. Nishihara, T. Horikoshi, *Synth. Metals*, **102**, 1523 (1999).
25. H. Nishihara, M. Kurashina, K.Aramaki, K. Kubo, *Synth. Metals*, **101**, 457 (1999).
26. M. Kurosawa, T. Nankawa, T. Matsuda, K. Kubo, M. Kurihara, H. Nishihara, *Inorg. Chem.*, **38**, 5113 (1999).

27. M. Kurihara, M. Kurosawa, T. Matsuda, H. Nishihara, *Synth. Metals*, **102**, 1517 (1999).
28. C. Creutz, M. D. Newton, N. Sutin, *J. Photochem. Photobiol. Part A: Chem.*, **82**, 47 (1994).
29. M. Kurihara, T. Matsuda, A. Hirooka, T. Yutaka, H. Nishihara, *J. Am. Chem. Soc.*, **122**, 12373 (2000).
30. M. Kurihara, T. Matsuda, A. Hirooka, T. Yutaka, H. Nishihara, *J. Am. Chem. Soc.*, **126**, 4740 (2004).
31. Y. Men, S. R. Korupoju, M. Kurihara, J. Mizutani, H. Nishihara, *Chem. Eur. J.*, **11**, 7322 (2005).
32. M. Kurihara, A. Hirooka, S. Kume, M. Sugimoto, H. Nishihara, *J. Am. Chem. Soc.*, **124**, 8800 (2002).
33. A. Sakamoto, A. Hirooka, K. Namiki, M. Kurihara, M. Murata, M. Sugimoto, H, Nishihara, *Inorg. Chem.*, **44**, 7547 (2005).
34. M.-C. Daniel, A. Sakamoto, J. Ruiz, D. Astruc, H. Nishihara, *Chem. Lett.*, **35**, 38 (2006).
35. M. Murata, T. Hoshi, I. Matsuoka, T. Nankawa, M. Kurihara, H. Nishihara, *J. Inorg. Organomet. Polym.*, **10**, 209 (2000).
36. H. Atzkern, P. Bergerat, H. Beruda, M. Fritz, J. Hiermeier, P. Hudeczek, O. Kahn, F. H. Köhler, M. Paul, B. Weber, *J. Am. Chem. Soc.*, **117**, 997 (1995).
37. H. Plenio, J. Hermann, J. Leukel, *Eur. J. Inorg. Chem.*, **2063** (1998).
38. P. T. Manoharan, J. H. Noordik, E. de Boer, C. P. Keijzers, *J. Chem. Phys.*, **74**, 1980 (1981).
39. P. Kuppusamy, P. T. Manoharan, *Chem. Phys. Lett.*, **118,** 159 (1985).
40. P. Cassoux, L. Valade, H. Kobayashi, A. Kobayashi, R. A. Clark, A. E. Underhill, *Coord. Chem. Rev.*, **110**, 115 (1991).
41. R. M. Olk, B. Olk, W. Dietzch, R. Kirmse, E. Hoyer, *Coord. Chem. Rev.*, **117**, 99 (1992).
42. R. Hille, *Chem. Rev.*, **96**, 2757 (1996).
43. J. K. Hsu, C. J. Bonangelino, S. P. Kaiwer, C. M. Boggs, J. C. Fettinger, R. S. Pilato, *Inorg. Chem.*, **35**, 4743 (1996).
44. M. Fourmigué, *Coord. Chem. Rev.*, **178–180**, 823 (1998).
45. R. P. Burns, C. A. Mcauliffe, *Adv. Inorg. Chem. Radiochem.*, **22**, 303 (1979).
46. A. Vogler, H. Kunkely, *Inorg. Chem.*, **21**, 1172 (1982).
47. M. Kajitani, G. Hagino, M. Tamada, T. Fujita, M. Sakurada, T. Akiyama, A. Sugimori, *J. Am. Chem. Soc.*, **118**, 489 (1996).
48. J. A. McCleverty, *Prog. Inorg. Chem.*, **2**, 72 (1969).
49. R. Eisenberg, *Prog. Inorg. Chem.*, **12**, 295 (1970)
50. A. Sugimori, T. Akiyama, M. Kajitani, T. Sugiyama, *Bull. Chem. Soc. Japan*, **72**, 879 (1999).
51. M. Okuno, K. Aramaki, S. Nakajima, T. Watanabe, H. Nishihara, *Chem. Lett.*, 585 (1995).
52. M. Okuno, K. Aramaki, H. Nishihara, *J. Electroanal. Chem.*, **438**, 79 (1997).
53. H. Nishihara, M. Okuno, N. Akimoto, N. Kogawa, K. Aramaki, *J. Chem. Soc., Dalton Trans.*, 2651 (1998).
54. G. N. Schrauzer, *Acc. Chem. Res.*, **2**, 72 (1969).
55. C. Creutz, *Prog. Inorg. Chem.*, **30**, 1 (1983).
56. N. S. Hush, *Prog. Inorg. Chem.*, **8**, 391 (1967).
57. N. S. Hush, *Coord. Chem. Rev.*, **64**, 135 (1985).
58. M. B. Robin, P. Day, *Adv. Inorg. Chem. Radiochem.*, **10**, 247 (1967).
59. W. Weitner, Jr., *Magnetic Atoms and Molecules*, Dover, New York, 1989, pp. 236–243.
60. M. Nihei, T. Nankawa, M. Kurihara, H. Nishihara, *Angew. Chem. Int. Ed.*, **38**, 1098 (1999).
61. N. Nakagawa, T. Yamada, M. Murata, M. Sugimoto, H. Nishihara, *Inorg. Chem.,* **45**, 14 (2006).
62. M. Murata, S. Habe, S. Araki, K. Namiki, T. Yamada, N. Nakagawa, T. Nankawa, M. Nihei, J. Mizutani, M. Kurihara, H. Nishihara, *Inorg. Chem.*, **45**, 1108 (2006).
63. H. Nishihara, "Organometallic Conductive Polymers," in *Handbook of Organic Conductive Molecules and Polymers*, H. S. Nalwa, Ed., Vol. 2, pp. 799–832, Wiley, New York, 1997.
64. A. Ohkubo, K. Aramaki, H. Nishihara, *Chem. Lett.*, 271 (1993).

65. J. R. Nitschke, S. Zurcher, T. D. Tilley, *J. Am. Chem. Soc.*, **122**, 10345 (2000).
66. I.Tomita, J. –C. Lee, T. Endo, *J. Organomet. Chem.*, **611**, 570 (2000).
67. I. Matsuoka, K. Aramaki, H. Nishihara, *J. Chem. Soc., Dalton Trans.*, 147 (1998).
68. T. Shimura, A. Ohkubo, N. Matsuda, I. Matsuoka, K. Aramaki, H. Nishihara, *Chem. Mater.*, **8**, 1307 (1996).
69. H. S. Nalwa (ed.), *Handbook of Organic Conductive Molecules and Polymers*, Wiley-VCH, Weinheim, Germany, 1997.
70. R. S. Kelly, W. E. Geiger, *Organometallics*, **6**, 1432 (1987).
71. B. T. Donovan, W. E. Geiger, *J. Am. Chem. Soc.*, **110**, 2335 (1988).
72. B. T. Donovan, W. E. Geiger, *Organometallics*, **9**, 865 (1990).
73. A. Ohkubo, T. Fujita, S. Ohba, K. Aramaki, H. Nishihara, *J. Chem. Soc., Chem. Commun.*, 1553 (1992).
74. I. Matsuoka, K. Aramaki, H. Nishihara, *Mol. Cryst. Liq. Cryst.*, **285**, 199 (1996).
75. I. Matsuoka, H. Yoshikawa, M. Kurihara, H. Nishihara, *Synth. Metals*, **102**, 1519 (1999).
76. H. Nishihara, A. Ohkubo, K. Aramaki, *Synth. Metals*, **55**, 821 (1993).
77. M. O. Albers, D. J. A. deWaal, D. C. Lies, D. J. Robinson, E. Singleton, M. B. Wiege, *J. Chem. Soc., Chem. Commun.*, 1681 (1986).
78. C. Ernst, O. Walter, E. Dinjus, S. Arzberger, H. Görls, *J. Pract. Chem.*, **8**, 341 (1999).
79. Y. Yamada, J. Mizutani, M. Kurihara, H. Nishihara, *J. Organomet. Chem.*, **637–639**, 80 (2001).
80. M. Kurashina, M. Murata, T. Watanabe, H. Nishihara, *J. Am. Chem. Soc.*, **125**, 12420 (2003).
81. C. Lavenda, K. Bechgaard, D. O. Cowan, *J. Org. Chem.*, **41**, 2700 (1970).
82. A. Ohkubo, T. Fujita, S. Ohba, K. Aramaki, H. Nishihara, *J. Chem. Soc., Chem. Commun.*, 1553 (1992).
83. A. C. Ribou, J. P. Launay, M. L. Sachtleben, H. Li, C. W. Spangler, *Inorg. Chem.*, **35**, 3735 (1996).
84. C. Lavenda, K. Bechgaard, D. O. Cowan, *J. Org. Chem.*, **41**, 2700 (1970).
85. A. Pramanik, N. Bag, G. K. Lahiri, A. Chakravorty, *J. Chem. Soc., Dalton Trans.*, 3823 (1990).
86. A. Ceccanti, P. Diversi, G. Ingrosso, F. Laschi, A. Lucherini, S. Magagna, P. Zanello, *J. Organomet. Chem.*, **526**, 251 (1996).
87. D. Bhattacharyya, S. Chkraborty, P. Munshi, G. K. Lahiri, *Polyhedron*, **18**, 2951 (1999).
88. P. G. Pickup, W. Kutner, C. R. Leidner, R. W. Murray, *J. Am. Chem. Soc.*, **106**, 1991 (1984).
89. M. Abe, T. Michi, A. Sato, T. Kondo, W. Zhou, S. Ye, K. Uosaki, Y. Sasaki, *Angew. Chem. Int. Ed.*, **42**, 2912 (2003).
90. M. Maskus, H. D. Abruña, *Langmuir*, **12**, 4455 (1996).
91. M.-A. Haga. T. Takasugi, A. Tomie, M. Ishizuya, Y. Yamada, M. D. Hossain, M. Inoue, *J. Chem. Soc., Dalton Trans.*, 2069 (2003).
92. K. Kanaizuka, M. Murata, Y. Nishimori, I. Mori, K. Nishio, H. Masuda, H. Nishihara, *Chem. Lett.*, **34**, 534 (2005).
93. H. Nishihara, M. Noguchi, K. Aramaki, *Chem. Commun.*, 628 (1987).

CHAPTER 10

Metallodendrimers and Their Potential Utilitarian Applications

SEOK-HO HWANG AND GEORGE R. NEWKOME

The University of Akron, Akron, Ohio

I. INTRODUCTION

Since the mid-1980s, two of the most intriguing and rapidly expanding areas in chemistry have been highly branched macromolecular architectures[1,2] and self-assembly processes, both of which have become key fundamental research efforts for scientists worldwide. These seemingly disparate fields have, in fact, merged beautifully in recent years to create one of the pillars of nanoscience with a wide variety of potential applications. Numerous reasons for the rapid emergence of these molecular curiosities can be cited; for instance, unprecedented control over structural unit positioning, a general ease-of-construction, well-behaved and tunable solubility features that facilitate characterization and further modification, and the overwhelming abundance of documented proof that such species can, in fact, be constructed for utilitarian as well as aesthetic purposes. However, it is important to note that with the advent of more recent mass-spectroscopy techniques, absolute purity and perfect branching are, in many cases, difficult to obtain. But perhaps one of the most important attributes of dendritic chemistry, which tends to propagate this new polymeric regime, is its vast potential to be easily integrated into, and meld with, more mature chemistries.

The merger of supramolecular chemistry, defined as "chemistry beyond the molecule" by Lehn,[3] and the chemistry of dendrimers led directly to "supramacromolecular" chemistry[4] and the creation of the metallodendritic regime. Incorporation of metal ions into the dendritic infrastructure was initiated by the Balzani[5,6,7] and Newkome[8,9] research groups in the early 1990s, either by the use of metal branching centers or by internal metal complexation and encapsulation at a specific binding site(s), respectively. From a structural viewpoint, metallodendrimers can be classified into categories based on structural positioning (Figure 10.1), for example, metal centers are located at the core, branching points, connectors between branching centers, terminal groups, and incorporated as structural auxiliaries, in which metals are introduced to the framework after dendritic construction. In general, the metal loci possess positional homogeneity within the structure, but metals at different framework positions have also been created. Although

Frontiers in Transition Metal-Containing Polymers, edited by Alaa S. Abd-El-Aziz and Ian Manners.

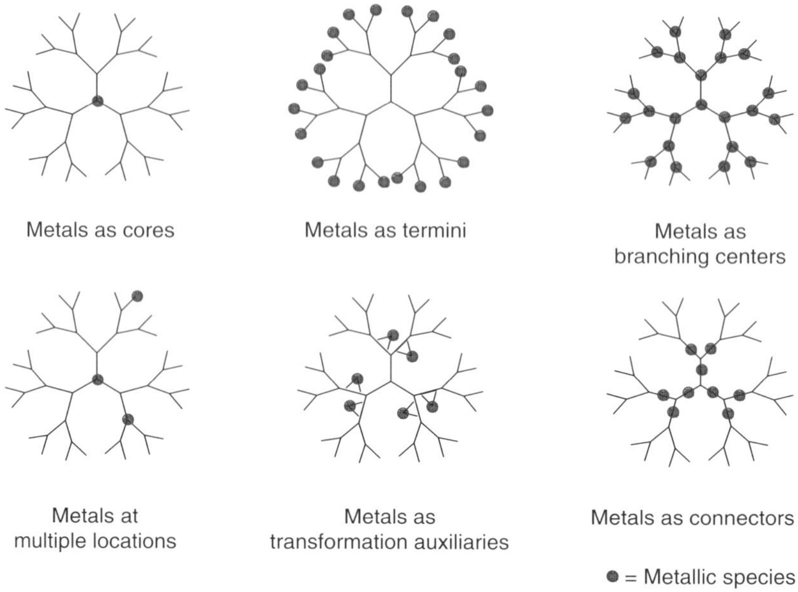

Figure 10.1 Metals serving in different roles in metallodendrimers.

metallodendrimers have been discussed to some degree in recent literature,[10–15] this chapter addresses recent developments in the chemistry and potential applications; as well, selected multibranched, metal-based constructs that are not strictly within the dendritic arena (i.e., construction with only 2–3 repetitive reaction sequences) are also considered.

II. GENERAL STRUCTURE AND THEIR POTENTIAL APPLICATIONS

A. Catalysts

Among the potential applications of metallodendrimers, catalysis appears to be one of the most promising areas,[10,16–25] since the dendritic infrastructure can be controlled to ensure the appropriate size, shape, and constitution. The ability to preprogram an appropriate surface, so necessary to instill the desired solubility characteristics, is readily possible due to one's synthetic ability to tailor the specific (macro)molecular infrastructures; advantages of homogeneous and heterogeneous catalysis can also be probed without the loss of well-defined molecular features.[26] In 1994, the first publication appeared in which organometallic and inorganic catalysts were prepared using either organic macromolecules, such as lightly cross-linked polystyrene, or inorganic polymers, such as silica.[27–31] Since dendritic materials can be created with controlled shape as well as internal and external catalytic sites, their diverse microenvironments afford new opportunities to explore specific utilitarian applications.

i. Ni-Based Dendritic Catalysts. One of the first reported examples of a metallodendrimer catalyst[32] resulted from the attachment of monoanionic chelating "N-C-N"-type pincer moieties to the periphery of a carbosilane dendrimer, followed by the complexation of nickel(II) ions.[22,23,33,34] These pincer-based carbosilane metallodendrimers (Figure 10.2) have been used

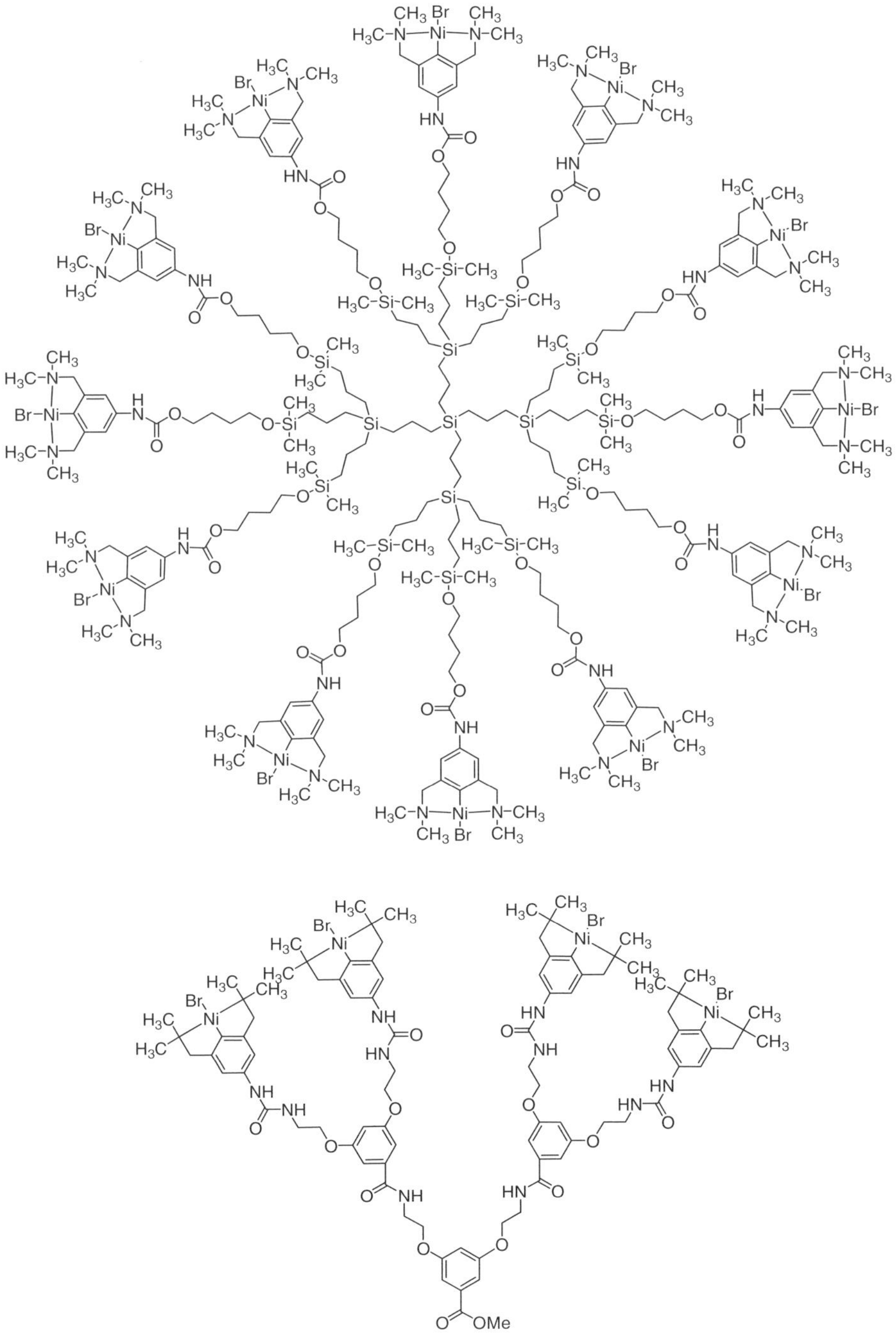

Figure 10.2 Van Koten's Ni-containing metallodendrimer and urea-linked metallodendron catalysts for the Kharasch addition reaction.

as a Kharasch addition[35] catalyst, whereby an atom-transfer radical-addition reaction between an olefin and a polyhalogenated alkane was effected. These dendritic catalysts are based on the unique feature of the [(N-C-N)NiX] moiety possessing a low Ni(II)/Ni(III) redox potential [$E_{1/2} = 0.14$ V vs. standard calomel electrode (SCE)];[36] however, the catalytic activity of these metallodendrimers was shown to decrease with increasing generation during the Kharasch-type addition of CCl_4 to methyl methacrylate (MMA) under standard reaction conditions. It was suggested that the Ni(II) sites were partly irreversibly oxidized to Ni(III), which is indicative of a back-reaction of the Ni(III) sites with intermediate radical (Cl_3C–MMA·) or initial radical (CCl_3·) that becomes less efficient with time. Based on molecular modeling,[37] the Ni–Ni inter-site distance in these metallodendrimers was relatively small, causing Cl-bridging in mixed-valance intermediates, affording a rationale for the observed inhibition in these congested metallodendrimers. Preliminary investigation of these dendritic systems within a membrane reactor has shown another possible application for these metallodendrimers.[36] Van Koten's group also described the use of amino acid–based dendrons, as molecular scaffolds, for the urea-mediated attachment of catalytically active organometallic Ni "pincer" complexes.[38]

Although nickel complexes, such as *o*-diphenylphosphinophenols,[39] have been used to catalyze the oligomerization of ethylene monomers by the Shell Higher Olefins Process,[40] recently, van Leeuwen's group has constructed a nickel catalyst embedded at the core of a carbosilane dendrimer.[41] Since these dendron-based diphenylphosphinophenol ligands can form more stable bis(P,O)-Ni complexes than those of the corresponding parent ligand in toluene, the former out performed the latter for the oligomerization of ethylene. This metallodendrimers (Figure 10.3) is one of the rare examples in which the core-functionalized dendritic catalyst is far more active than its parent complex, due inpart to site-isolation of the catalytic nucleus.

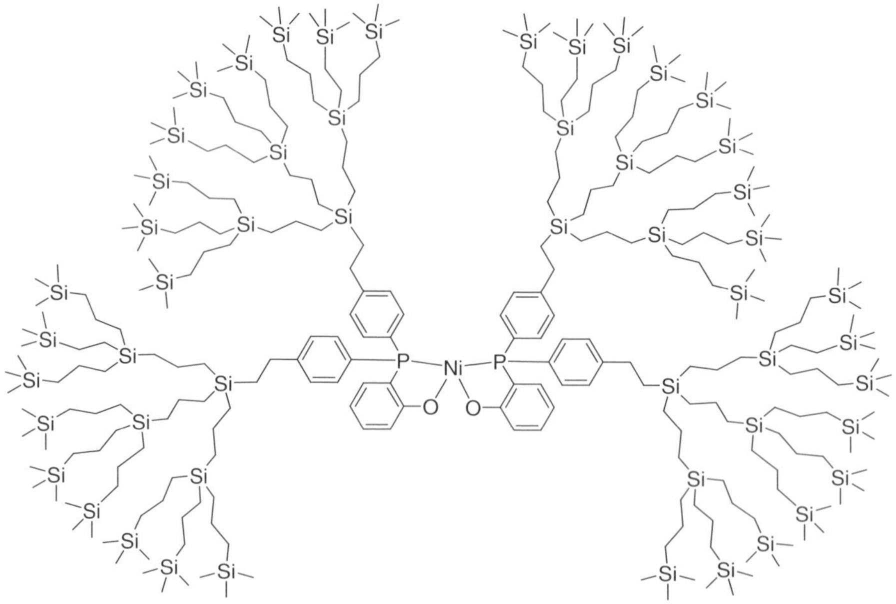

Figure 10.3 Van Leeuwen's Ni-containing metallodendrimer.

ii. Cu-Based Dendritic Catalysts. Chow et al. designed a Cu(II)-bis(oxazoline) dendritic catalyst (Figure 10.4) to accelerate the Diels–Alder reaction between cyclopentadiene and

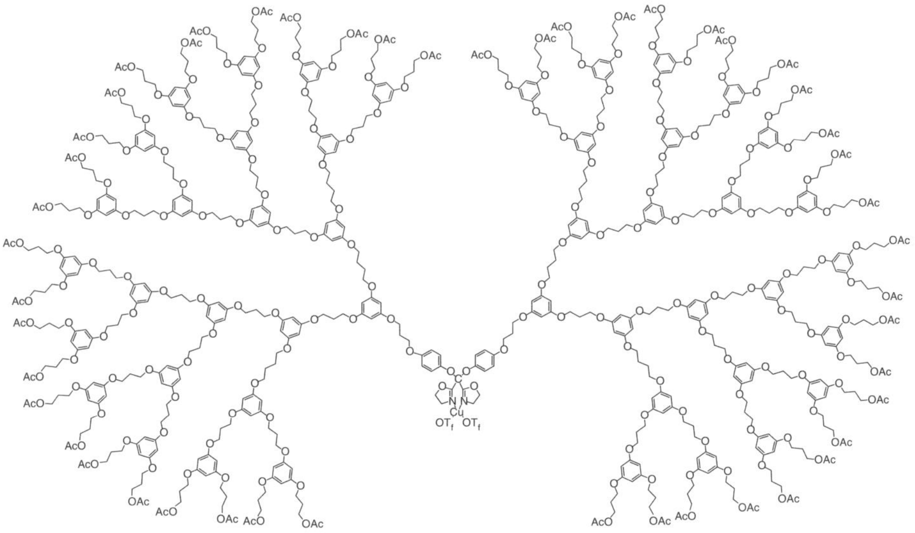

Figure 10.4 Chow's dendritic copper(II)–bis(oxazoline) metallodendrimer developed for the Diels–Alder reaction.

crotonyl imide. A kinetic study supporting a two-step mechanism for this Diels–Alder reaction was demonstrated, in which a reversible binding of the dienophile to the copper complex was followed by the rate-determining step between the resulting dienophile–catalysts and the diene. In addition, the initial complexation constant for this dienophile–catalyst complex gradually decreased with increasing dendrimer generation, while the Diels–Alder reaction rate constant remained similar for zero to second-generation catalysts; it notably decreased at the third generation.[42] These observations were rationalized as a consequence of back-folding of the dendritic arms at the third generation, thereby inhibiting access to the catalytic center. Thus, the increased steric size impeded both the reactivity and binding profile of the catalyst. These results were described as an *exo-* to *endo*-active site transition that hindered the approach of the diene toward the dienophile–catalyst complex.[43,44]

Buckminsterfullerene [C_{60}] can be used as a compressed spherical core in the facile construction of globular dendrimers[45,46,47] and lipofullerenes with an octahedral pattern.[48,49,50] The use of C_{60} as the core allowed not only the synthesis of adducts with T_h symmetry,[51] but also those with C_{2v}, C_s, and C_3 symmetries. Hirsch et al.[52] synthesized enantiomerically pure C_3-symmetrical fullerene dendrimers, such as *all-R-ᶠA*, *all-S-ᶠC*, *all-R-ᶠC*, and *all-S-ᶠA*, as well as *all-S-ᶠA* and *all-S-ᶠC* lipofullerenes involving an octahedral [3:3] addition pattern (Figure 10.5). The novel near-core incorporation of the bis(oxazoline) moieties that subsequently form internal Cu(II) complexes was capable of serving as catalytic sites for the cyclopropanation of styrene with ethyl diazoacetate.[53] The stereoselectivity caused by these prototypes (termed dendrizymes) was shown to be very low when compared to those obtained with other bis(oxazoline) catalysts. The ee-*trans* values for the cyclopropanation of styrene using the *all-S-ᶠA* and *all-S-ᶠC* in the presence of Cu(OT_f), as CO-catalyst, were determined to be 2 and 1, respectively; whereas, the ee-*cis* values were found to be 9 and 7, respectively.

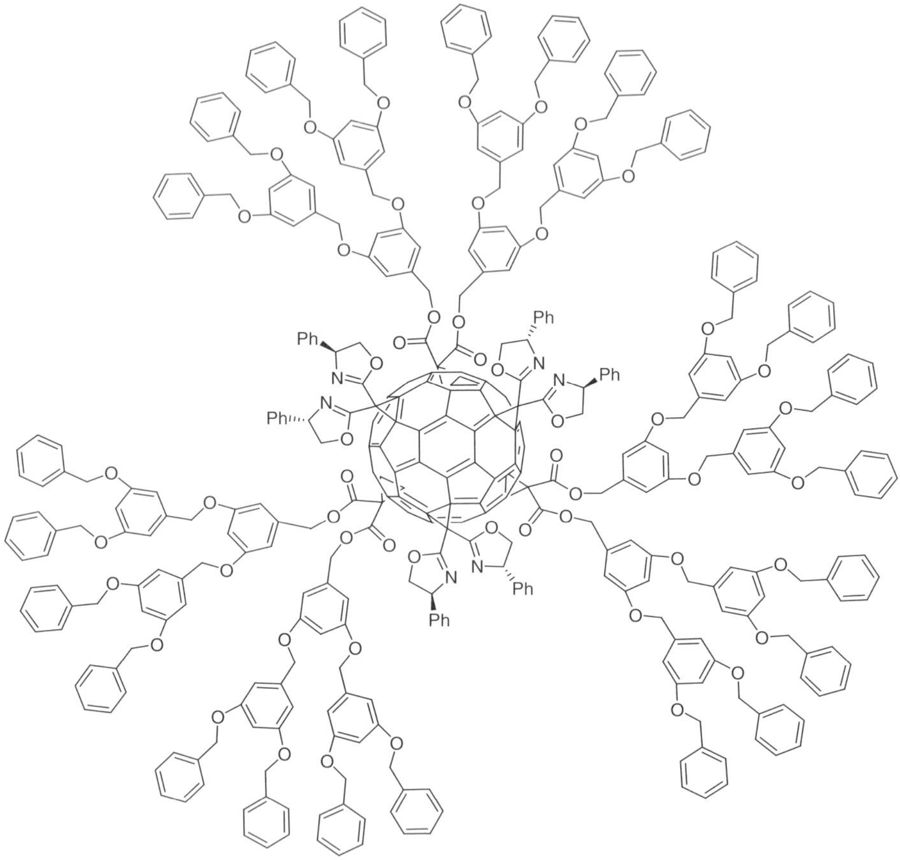

Figure 10.5 A fullerene dendritic ligand for Cu(OT$_f$) complexation.

iii. Rh-Based Dendritic Catalysts. Hydroformylation has been widely used for homogeneous catalytic processes to produce aldehydes and alcohols.[54] Reek et al.[55] have reported the synthesis of dendrimers possessing both mono- and bidentate end-groups through the hydrosilylation of different generations of carbosilane dendrimers[56] with $HSiMe_2Cl$ or $HSiMeCl_2$, followed by reaction with $LiCH_2PPh_2$-(*N,N,N',N'*-tetramethylethylenediamine (TMEDA).[57] These Ph_2P-functionalized carbosilane dendrimers were used as ligands for the Rh-catalyzed hydroformylation of 1-octene. The rhodium metallodendrimers were shown to possess the same selectivity when compared to the related monomeric analogs; the catalytic activity was shown to be dependent on the generation and flexibility of the metallodendrimers. These results are in agreement with the observations of Alper et al. employing rhodium complexes coordinated to the bidentate phosphine-functionalized polyamidoamine dendrimers on silica.[58]

Reetz et al.[59] have introduced polypropylenimine (PPI) dendrimers as the core for building phosphine-coated constructs that can complex with Rh(COD) · BF_4, where COD = 1,5-cyclooctadiene, to instill the desired catalytic character. Hydroformylation of 1-octene with these metallodendrimers was shown to have turnover numbers that were comparable to those of monomeric analogs. It was pointed out that these catalysts could be easily recovered by means of membrane separation technology.[60] Gong et al. have used water-soluble, phosphonated dendritic ligands based on the poly(amidoamine) (PAMAM) frame for a similar transformation.[61] These metallodendrimers were tested as the catalyst in a two-phase hydroformylation using either styrene or 1-octene under mild (40°C, 20 atm) reaction conditions. In the case of styrene, Rh(I) complexes exhibited a higher selectivity to form 2-phenylpropionaldehyde (branched-type) over hydrocinnamaldehyde (linear-type) products. In contrast to styrene, the hydroformylation of 1-octene was shown to favor 1-nonanal (linear-type) over 2-methyloctanal.

Jeffrès and Morris[62] have developed dendrimers that utilized a polyhedral oligosilsesquioxane (POSS) core with up to 72 terminals. Using a similar strategy, Cole-Hamilton et al.[63,64,65] introduced the polyhedral oligosilsesquioxane core to construct alkylphosphine-coated dendrimers capable of reacting with either [Rh(acac)(CO)$_2$] or [Rh$_2$(OAc)$_4$]. The hydrocarbonylation of propen-1-ol using the first generation dendritic ligand shown in Figure 10.6 led to the formation of identical products (butane-1,4-diol and 2-methylpropan-1-ol) to those reported elsewhere;[66] however, the formation of the linear alcohol (butane-1,4-diol) clearly occurred via a two-step reaction, that is, hydroformylation to form 4-hydroxybutan-1-al and subsequent reduction. The construction of 2-methylpropane-1-ol probably occurred via a hydroxycarbene intermediate with intermolecular protonation of the acyl moiety.[63]

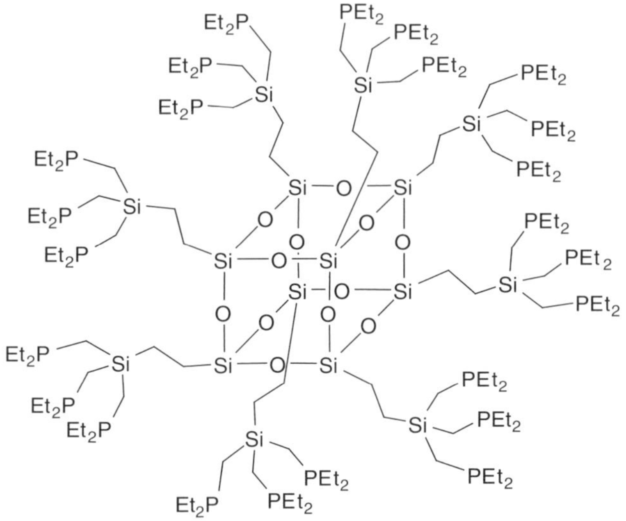

Figure 10.6 Cole-Hamilton's dendrimer based on a polyhedral oligomeric silsesquioxanes core.

Another dendrimer possessing 16 PPh$_2$ terminals was shown to exhibit much higher linear selectivity (14:1) than those of the monomeric analogs (3 ~ 4:1) in the hydroformylation of oct-1-ene catalyzed by the Rh(I) complex.[64,65] In this metallodendrimer, the phosphorus atoms were separated by 4~7 Å within one arm, while its distance was in the 5~10 Å range between arms (from molecular modeling). Indeed, analogous metallodendrimers containing two CH$_2$ units between the Si and P atoms showed no special selectivity enhancement over that of the monometallic catalysts. This positive dendritic effect was explained by steric crowding coupled with diminished arm length inducing an eight-membered bidentate coordination that enhanced the linear selectivity.

Kakkar et al.[67,68] reported the divergent synthesis of a novel series of organophosphine dendrimers possessing both a phosphorus core and branching centers with [—$OSiMe_2O$—] connectivity as well as terminal hydroxy groups. The treatment of these multitiered phosphorus centers with $[Rh(\mu\text{-}Cl)(\eta^4\text{-}1,5\text{-}COD)]_2$ generated metallodendrimers containing both internal and external $[RhCl(\eta^4\text{-}COD)PR_3]$ organometallic centers; such complexes were shown to catalyze the hydrogenation [25°C, 20 bar H_2, 30 min, tetrahydrofuran (THF)] of 1-decene in a 1:200 metal-to-substrate ratio. There was a slight decrease in turnover frequencies with the larger $Rh(I)_{46}$ metallodendrimer. After one hydrogenation cycle using this $Rh(I)_{46}$ metallodendrimer, the organic products were extracted into pentane and the dendritic catalyst was isolated, recrystallized from THF/hexane, and reused with only a 5% decrease in the catalyst's activity [resulting in a turnover number of ca. 200 $mol_{prod}(mol_{cat})^{-1}$ and turnover frequency of ca. 400 $mol_{prod}(mol_{cat})^{-1}h^{-1}$], which was found to be similar to that of the corresponding monometallic complex. Indeed, other metallodendrimers[68] were derived from these organophosphine dendrimers by treatment with $[Rh(COD)Cl]_2$ by means of a bridge-splitting procedure. Catalytic hydrogenation of 1-decene with these 3,5-dihydroxybenzyl alcohol-based organometallic dendritic materials[69] under standard conditions (25°C, 20 bar H_2) was found to be dependent on two distinct factors: reaction time and generation number; the maximum conversion was accomplished in a 5-hour time frame.

Also, the use of dendrimers as templates to prepare network carriers containing cavities of predetermined size and disposition has been reported.[70] Recently, Rossell et al.[71] have constructed cationic Rh-containing carbosilane metallodendrimers, whose catalytic activity was tested for the hydrogenation of 1-hexene under mild conditions (25°C, 10 bar H_2, 1h, Me_2CO) in a 1:500 metal-to-substrate ratio. There was a slight decrease in turnover frequency with the larger Rh(I) metallodendrimers, but they obtained comparable results with Kakkar's group.[67] Similar ruthenodendrimers were also evaluated as catalysts in the hydrogen transfer of cyclohexanone with 2-propanol.[72]

Reek et al.[73] developed a new bicarbazolediol-[74] (BICOL)-based, chiral monodentate phosphoramidite ligand, in which the *N*-sites in the bicarbazole skeleton permitted the easy introduction of metal centers. As a model reaction, the Rh-catalyzed asymmetric hydrogenation of methyl 2-acetamidocinnamate was evaluated. Using a ligand to rhodium ratio of 2.2, the enantioselectivity induced by the rhodium complex (Figure 10.7) was 93% at full conversion,

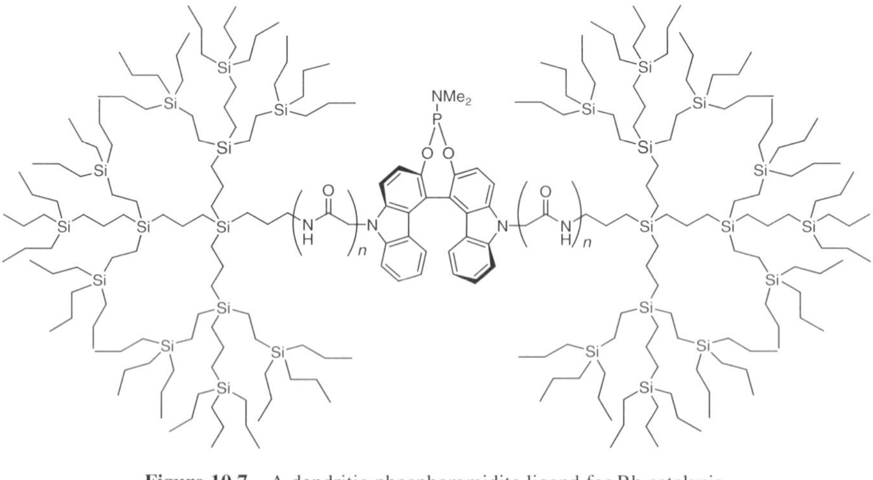

Figure 10.7 A dendritic phosphoramidite ligand for Rh catalysis.

which is comparable to the results obtained by Feringa et al.[75,76] using 1,1′-binaphthol-(BINOL)-derived monodentate phosphoramidite monophos. These results demonstrated the ability of the bicarbazole bidentate to induce a high degree of enantioselectivity.

Arya et al. initiated a new strategy to explore the application of immobilized dendritic ligands anchored onto silica gel[58,77] and polystyrene-based beads[78,79] for use in hydroformylation reactions in order to minimize the recycling cost of the Rh catalyst. To examine the recycling behavior of the polystyrene-supported Rh catalyst in a hydroformylation reaction, *p*-methoxystyrene was selected as a model compound. The first-generation catalyst was found to be very reactive up to the fourth cycle (99%), but the conversion to the product decreased to 56% for the fifth cycle. While the second-generation catalyst was found to be equally reactive (>99% for the fourth cycle), there also was a continued reactive behavior observed for the fifth cycle (>85%). It was generally believed that the polymer-supported catalysts are less reactive compared to homogeneous catalysts; however, these results emphasize the enhanced recycling potential, since this approach facilitated the recovery of the catalyst using various-size exclusion techniques.

Busseto et al.[80–83] have reported the synthesis and catalytic properties for various rhodium complexes of the type [Rh<$C_5H_4CO_2(CHR)_2OH$>(L,L)], [R = H, Me, Ph; L,L = 2CO, 2,5-norbornadiene (NBD)] prepared from the readily available sodium β-hydroxyalkoxycarbonyl-cyclopentadienides, Na[$C_5H_4CO_2(CHR)_2OH$], and [Rh(L,L)Cl]$_2$. The rhodium complexes, [Rh<(1*R*,2*S*)-$CpCO_2(CHMe)_2OH$>(NBD)] and [Rh<(1*S*,2*S*)-$CpCO_2(CHMe)_2OH$>(NBD)], where Cp = η^5-C_5H_5, are active catalysts for the hydroformylation of hex-1-ene. They synthesized a new family of DAB-*dendr*-[NH(O)$COCH_2CH_2OC(O)C_5H_4Rh(NBD)_n$] ($n$ = 4, 8, 16, 32, 64)[84] metallodendrimers by treatment of the poly(propylenimine) (PPI) dendrimers with alkoxycarbonylcyclopentadienyl complexes of Rh(I).

Chiral diphosphines are the most widely used ligands in the asymmetric, catalytic hydrogenation of C=C and C=O bonds.[85] 3,4-bis(Diphenylphosphino)pyrrolidine (pyrphos), which was easily synthesized from natural tartaric acid, contains an amino group to which organic or inorganic supports may be directly attached. Rh-pyrphos complexes and its derivatives have been studied for the asymmetric hydrogenation of dehydroamino acid and found to exhibit 99% ee.[86,87] Chan et al.[88] synthesized the novel chiral dendritic diphosphine ligand for Rh(I)-catalyzed asymmetric hydrogenation of α-acetamidocinnamic acid, as a model reaction. In the case of dendritic pyrphos-Rh catalysts (from the first–fourth generation), a dramatic change in catalytic activity was observed in the transition from the third to fourth generation, which might correlate with a change in the shape of the metallodendrimer from a flattened or pancake-like motif to a more globular structure.

N-Heterocyclic carbene (NHC) ligands have attracted considerable attention in both homogeneous catalysts and organometallic chemistry because their strong coordination to metals compared to conventional ligands, such as phosphines, prevented their dissociation from the metal center.[89,90] Tsuji et al.[91] synthesized imidazolium salts bearing Fréchet-type polybenzyl ether dendrons and investigated their catalytic performance as well as the dendritic effect on the hydrosilylation of ketones (Figure 10.8). A positive dendritic effect was found in solvents, such as benzene and CH_2Cl_2; the rationale was that the dendron's aromatic character interacted with the rhodium metal. Indeed, the dendritic effect was also influenced by the concentration of the reaction. When the hydrosilylation of acetophenone was conducted under dilute conditions, the yield of 1-phenylethanol decreased considerably to 44% after 46 hours with the first generation catalyst, while with the higher generation (second–fourth) catalysts, the yields were unchanged.

iv. Pd-Based Dendritic Catalysts. There has been great interest in olefin oxidation since the successful application of the Wacker process[92] for the industrial production of acetaldehyde.

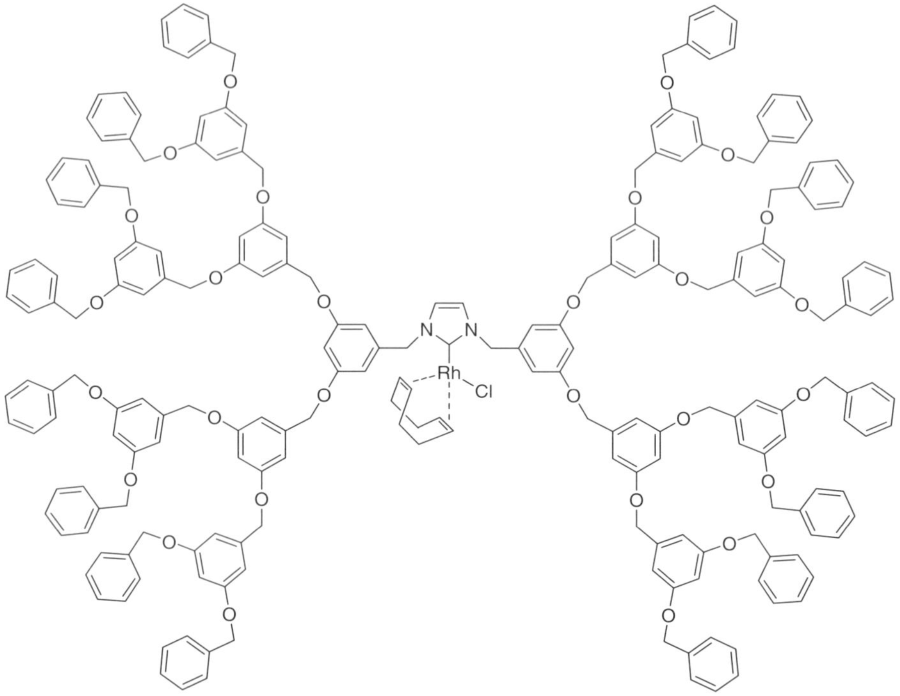

Figure 10.8 An *N*-heterocyclic carbene complex that coordinates Rh(I).

This catalytic system consisted of an aqueous acidic solution of a palladium salt with a copper(I) salt and oxygen or air as an oxidant; however, application of the Wacker process to longer chain alkenes has been a challenge due to their low solubility in aqueous media. This has led to research in biphasic systems using tetraalkylammonium salts,[93] polyethylene glycols,[94] as well as cyclodextrins.[95,96]

Alper et al.[97,98] immobilized palladium-modified PAMAM-like dendron complexes onto silica for carbonylation of iodobenzene in MeOH to generate methyl benzoate under low pressures of CO at 100°C. Product yields were high overall and the catalyst could be recycled 4∼5 times without significant loss of activity. This type of PAMAM-type dendron–Pd complex was also used for the oxidation of terminal alkenes to generate methyl ketones under mild conditions;[99] oxidative selectivity toward the terminal versus internal double bonds was also determined.

Kragl et al.[100] described the retention of diaminopropyl-type metallodendrimers bearing palladium phosphine complexes on ultra- or nanofiltration membranes and their use as catalysts for allylic substitution in a continuously operating chemical membrane reactor. Their results demonstrated a viable procedure for catalyst recovery, because these metallodendrimers acting as catalyst supports offered an advantage in that the intrinsic viscosity of the solution is smaller, facilitating filtration.

Screttas et al.[101] synthesized a PPI dendrimer with an iminophosphine surface functionality, DAB-*dendr*-$[1,2\text{-}(N{=}CHC_6H_4PPh_2)]_{32}$ ("DAB-32-imiphos"), and the corresponding reduced aminophosphine-based constructs, DAB-*dendr*-$[1,2\text{-}(NHCH_2C_6H_4PPh_2)]_{32}$ ("DAB-32-amiphos") as well as evaluated these *P,N*-dendritic ligands, as a catalyst for the Heck reaction by coordination with Pd.[102] Palladium acetate (1%) in the presence of 3 equivalents of

DAB-32-imiphos, lithium acetate, and a solvent mixture of tri-*n*-butylamine and acetic acid at 130°C for 15.5 hours converted bromobenzene completely to a mixture of ca. 90% *trans*-stilbene and ca. 10% isomers. The reaction seemed to proceed better in a novel equimolar solvent mixture of a tertiary amine and acetic acid. A marked dependence of the conversion on the [equiv. of dendritic ligand]/[Pd] ratio has been observed; the higher this ratio, the lower the conversion, which became zero for ratios greater than 15. This result implied that there is a rather strong interarm interaction within the dendrimer to which the metal is coordinated, preventing the substrate from competing for a coordination site on the metal.[103] However, due to the diminished thermal stability of the ligands, it probably underwent degradation under the reaction conditions and the extensive formation of Pd-black. Thus, these catalytic systems were not recoverable.

Reek et al.[57,104] functionalized carbosilane dendrimers with diphenylphosphane-termini, then transformed these moieties into the corresponding palladium complexes. Similar to the work discussed earlier,[100] these Pd-functionalized dendrimers were used to explore their catalytic behavior in continuously operated membrane reactors. The second-generation carbosilane dendrimer, serving as the starting point, was a white solid whose X-ray crystal structure was determined and whose molecular volume of 2414 $Å^3$ was anticipated to be large enough for catalyst separation from the reaction mixture by nanofiltration. These phosphine-functionalized dendrimers were synthesized by repeated hydrosilylation of double bonds with $HSiMe_2Cl$ or $HSiMeCl_2$, followed by termination with Ph_2PCH_2Li/TMEDA. The phosphine-terminated dendrimers were metalated with $[PdCl(\eta^3\text{-}C_3H_7)]_2$ to yield either bidentate palladium phosphine metallodendrimers or mixtures, when monodentate dendritic phosphines were used. All these metallodendrimers were used as catalysts in allylic alkylation of allyl trifluoroacetate and diethyl methylsodiomalonate, yielding diethyl allylmethylmalonate. The reaction was first conducted in a batch process; all metallodendrimers showed a very high catalytic activity. Using 0.2% of the catalyst, the yield was greater than 80% after 30 min and only small differences of reaction rates were observed for the different catalysts. The metallodendrimer containing 12 chelated palladium atoms with a calculated volume of ca. 7600 $Å^3$ was used as a catalyst in a continuous process. Retention of this catalyst in the membrane reactor was determined to be 98.1%, which corresponds to a calculated value of only 25% decreased activity after flushing the reactor (15 times). Samples taken from the flow stream were not catalytically active, which confirmed that the observed decrease of activity was due to decomposition of the palladium complex, and not due to the loss of dendritic catalyst.

The Sonogashira reaction[105,106] couples aryl or vinyl halides to terminal alkynes and has widespread synthetic appeal due to an efficient conversion without requiring the preparation of an organometallic intermediate. Astruc et al.[107,108] synthesized a series of bis(*tert*-butylphosphine)- and bis(cyclohexylphosphine)-functionalized Pd(II) monomers and PPI-coated catalysts, and investigated the Sonogashira carbon–carbon coupling reaction under efficient copper-free procedures. The *tert*-butyl metallodendrimer series was shown to catalyze the Sonogashira reaction of iodoarenes with phenylacetylene at 25°C, although longer reaction times than with monomeric catalysts were required; whereas, the cyclohexyl counterparts were active at 80°C, they maintained the same catalytic efficiency after five recycling operations, and were demonstrated to be easier to recover than the former *tert*-butyl series due to solubility considerations.

Van Koten's group[109] investigated the aldol condensation of benzaldehyde and methylisocyanate to afford oxazolines that were catalyzed by cationic cyclopalladated carbosilane dendrimers serving as Lewis acids. Their catalytic performance was compared to that of the parent mononuclear derivative $[Pd(C_6H_3\{CH_2NMe_2\}\text{-}2\text{-}(SiMe_3)\text{-}5)(pyr)(H_2O)]BF_4$ (pyr = pyridine). Results indicated that up to the second-generation, the selectivity of the reaction was not affected; however, the rate of reaction decreased with increasing steric congestion at the dendrimer's periphery.

Using divergent procedures, Moberg et al.[110] prepared the first- to fourth-generation dendritic substituents based on 2,2-bis(hydroxymethyl)propionic acid and (1*R*,2*S*,5*R*)-methoxyacetic acid. The resultant dendrons were attached to 2-(hydroxymethyl)pyridinooxazoline and bis[4-(hydroxymethyl)oxazoline]; the resulting ligands were assessed in Pd-catalyzed allylic alkylation. Introduction of a chiral dendritic substituent on the pyridinooxazoline ligand has less effect on the enantioselectivity, two diastereomers (*R*)- and (*S*)-chiral dendron ligands each resulted in 79% ee. In contrast, substitution of the bis(oxazoline) ligand with the same chiral dendron yielded the (*S*,*S*)-chiral dendron ligand, which exhibited higher stereoselectivity (94% ee) than the bis(oxazoline) ligand carrying achiral dendrons. The lesser enantioselectivity in these reactions using the pyridinooxazoline ligands was explained by the ring distance between the dendritic substituent and the catalytic center and/or the high flexibility of the system. In the case of the bis(oxazoline) series, the dendritic wedges seemed to be situated close enough to the catalytic center to have an effect on both stereoselectivity and reactivity, thereby leading to increased stereoselectivity, but decreased catalytic activity.

Recently, van Koten et al.[111] designed a specific metallodendrimer possessing encapsulated (within a tailored macrocycle) catalytic sites to influence regio- as well as stereocontrol (Figure 10.9). The *N*,*C*,*N*-pincer moiety was linked to the core and the other was used to coordinate the active metal atom, as well as to provide the next branching point at a remote —$Si(Ph_2)$— site within a macrocycle for further extension of the dendritic structure. Following this approach, they synthesized a new macrocyclic carbodiazasilane ligand and a *para*-OH functionalized ligand as cages, then their attachment to a central core and the subsequent transformation to the corresponding palladium(II) complexes by oxidative addition. The aldol condensation of benzaldehyde and methyl isocyanoacetate was higher using the tricyclopalladated catalyst than with other mononuclear models tested; however, the steric chiral environment around the Pd center did not have an influence on the diastereoselectivity of the resultant reaction.

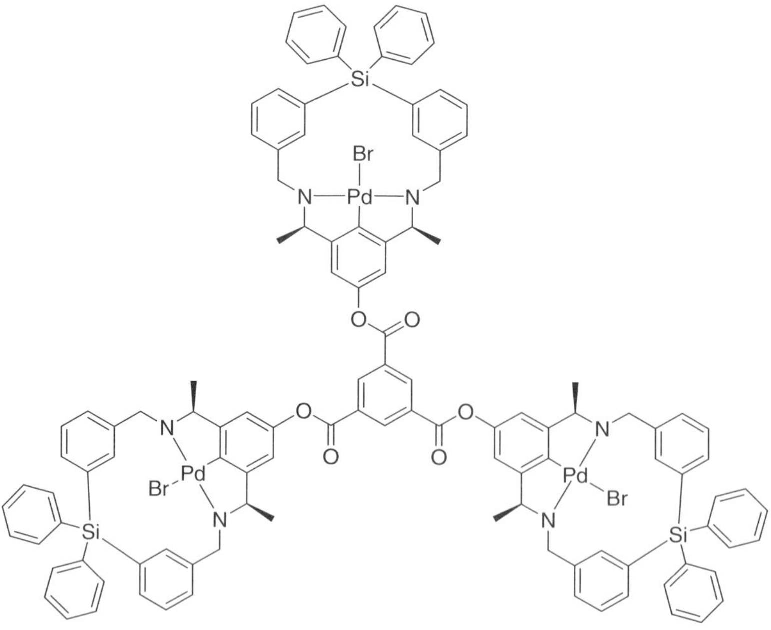

Figure 10.9 A tricyclopalladated metallodendrimer.

The Pd^{II} allyl and Pd(0) fumaronitrile complexes bearing pyridyl-dithioether–based metallodendrimers as ancillary ligands, have been constructed by Canovese and Chessa et al.[112] The reactivity (based on the second-order rate constant and equilibrium constant) of these complexes was tested with respect to allyl amination using piperidine. No macroscopic effects were observed on going from a nondendritic model to the second-generation dendritic substrates, although the third-generation complex exhibited notable variations in the observed rate and equilibrium constant. This phenomenon was rationalized as a rearrangement that induced an increase in steric hindrance at the allyl fragment and a concomitant distortion of the ligand environment at the metal core. This suggested a justification for the decreased rate-constant and increased equilibrium-constant values.

Heck olefin arylation,[113] one of the most widely used reactions in synthetic organic chemistry, has been successfully conducted in solution with a variety of aryl iodides, bromides, and chlorides, using numerous dendritic catalytic systems. Recently, Portnoy et al.[114] constructed the first poly(aryl benzyl ether) dendronized polystyrene resin and then decorated it with $Pd(dba)_2$ (dba = dibenzylideneacetone) in THF at 25°C. Treatment of bromobenzene with methyl acrylate under typical Heck conditions was conducted, from which the second- and third-generation catalysts proved superior in all parameters to the monodendron analog. Thus, the selectivity of the catalysis was significantly improved with higher dendron generation. To examine whether these observed dendritic effects were limited only to electron-deficient olefinic substrates, the performance of the metallodendrimers with butyl vinyl ether, as a substrate, was tested. According to the regio- and stereoselectivity results, the proportion of the α-arylated product increased as the generation number of the polymer template was increased. The decrease in β-/α-arylation ratio for butyl vinyl ether, was attributed to a higher portion of the olefin insertion occurring through cationic intermediates (enforcing α-arylation) versus neutral intermediates (yielding mixtures of α- and β-arylated enol ether).[115] This change in the distribution between the two alternative catalytic pathways may result from the more polar environment of the catalytic units in the case of the higher-generation catalysts. Alternatively, or additionally, the reason for this change may be associated with the increased local loading of phosphines in higher generation of dendrons.

Kaneda et al.[116] constructed a dendrimer-bound Pd(II) complex, which was prepared by complexation of the diphenylphosphinomethyl-capped PPI dendrimer, DAB-*dendr*-$[N(CH_2PPh_2)_2]_{16}$, with $PdCl_2(PhCN)_2$. This catalyst showed selective activity in the hydrogenation of conjugated dienes to monoenes under an atmospheric pressure of H_2. The dendritic catalyst was easily recovered from reaction mixtures and could be reused without any loss in catalytic activity. Allylic amination using dendrimer-bound Pd(0) and Pd(II) complexes was also investigated,[117,118,119] in which these catalysts showed high stereoselectivity ascribed to the surface congestion. Employing a thermomorphic system made it possible to efficiently recycle the dendritic catalysts.

Gade et al.[120] reported a strong positive dendritic effect in the asymmetric catalysis in the allylic amination of 1,3-diphenyl-1-acetoxypropene with morpholine when using pyrphos-palladium-functionalized PPI and PAMAM dendrimers. A remarkable and unprecedented increase in catalyst selectivity was observed as a function of the dendrimer's generation. This steady increase of ee values for these allylic aminations was less pronounced for the PPI-derived Pd-catalysts than for the corresponding PAMAM-catalysts, for which an increase in selectivity from 9% ee for a mononuclear reference system to 69% ee for the Pd_{64}-dendrimer was realized.

The synthesis of artificial active sites[121–125] for hosting transition metal catalysts with unique reactivity profiles is an important issue, but it can be technically difficult. Gagné et al.[126] first synthesized a polymerizable amino acid containing a diphosphine ligand possessing eight styrenyl units; this ligand was utilized in the synthesis of a series of P_2PdX_2 complexes [X_2 = (*R*)-BINOL, (*S*)-BINOL, Cl_2, and 1,3-Ph_2-allyl$^+$]. Their metallomonomers were used as comonomers in the synthesis of permanently porous ethylene dimethacrylate (EDMA)-base polymers, which in turn served as catalysts for the alkylation of allylic acetates. Removal of the large ligands from the metallodendrimer postpolymerization ensured that the catalyst had sufficient room at the active site to accommodate the reactants necessary to accomplish catalysis.

v. Ru-Based Dendritic Catalysts. Hoveyda's group[127] synthesized two efficient and recyclable dendritic Ru-based metathesis catalysts (Figure 10.10). The structure of the monomeric catalyst differs from the Grubbs' catalyst, $[Ru(PPh_3)_2Cl_2({=}CHPh)]$,[128,129] in that a chelating carbene (2-isopropoxystyrene) takes the place of the benzylidene and one phosphine. In the 4-directional Si-centered dendrimer, the $[{-}(CH_2)_3SiMe_2(CH_2)_3OC(O)(CH_2)_2{-}]$ branches are units connected to the styrenyl ether ligand. The yield of ring-closing metathesis of $TsN(CH_2CH{=}CH_2)_2$ using only 5 mol % of ruthenium-based dendritic catalyst was shown to proceed at 99% conversion. The catalyst was recovered with 13% vacant styrenyl ligand sites (i.e., 13% Ru loss). Repetitive use of the dendritic catalyst led to complete conversion of $TsN(CH_2CH{=}CH_2)_2$ in 91% yield. The dendritic catalyst remained active after six cycles (although 59% of the Ru content was lost), affording yields of 87% in the ring-closing metathesis process. The high level of activity was suggested to arise, at least partially, to the release of a highly active monophosphine Ru complex into the solution. It suggested that the catalytically active Ru species was released from the dendrimer into the reaction mixture and could be trapped again by a styrenyl ether ligand arm of the dendrimer. Another analogous 4-directional metallodendritic catalyst (Figure 10.10) was prepared in which the last phosphine ligand was replaced by a diamino–carbene ligand used in the Grubbs catalyst for the design of an extremely efficient next-generation metathesis catalyst.[130] This dendritic catalyst exhibited greater activity than the former since it promoted formation of trisubstituted allylic alcohol and was recovered with only an 8% Ru loss. It also catalyzed tandem ring-opening or ring-closing metathesis, and was easily separable from reaction mixtures because of its polarity and high molecular weight. Ring-closing metatheses of diethyl diallylmalonate to the favored five-membered cyclopentene ring using chelating dendritic ligands coordinated to Ru have also been achieved by the van Koten group.[131]

Using an efficient electron-transfer chain reaction with $[Ru_3(CO)_{12}]$ catalyzed by $[FeCp(\eta^6\text{-}C_6Me_6)]$, Astruc et al.[132] introduced 32 or 64 $[Ru_3(CO)_{11}]$ clusters into a 32-branched dendritic phosphine synthesized by Reetz et al.[59] using the double phosphine-methylation of a third-generation PPI dendrimer[133] with PPh_2CH_2OH.[134] The bis(phosphine)s end units possessing two cyclohexyl groups on each phosphorus were reacted with monoruthenium carbine complexes, which were made by replacement of PCy_3 to PPh_3 in the Hoveyda's metathesis catalyst.[127,135] These constructs provided four generations of new, stable metallodendrimers containing ruthenium–benzylidene units at the periphery.[136] The fourth-generation metallodendrimer containing 32 ruthenium–benzylidene units, however, was found to have rather low solubility in common organic solvents, unlike the first–third-generation complexes that contained 4-, 8-, and 16-ruthenium–benzylidene moieties, respectively. This poor solubility of the 32-Ru metallodendrimer was presumably due to peripheral steric congestion. The X-ray crystal structure of the model mononuclear complex in which the dendritic branch was replaced by a benzyl group revealed distorted square pyramidal geometry and the classic geometric features of a Ru=C double bond. These first–third generations of metallodendrimers and the model complex were efficient catalysts for the ring-opening metathesis polymerization of norbornene under ambient conditions giving dendrimer-cored star-shape polymers. Analyses of the molecular weights by secondary electron conduction (SEC) gave data that were close to theoretical, which indicated that all the branches were efficiently polymerized. It was determined that dendrimer-cored star-shape polymers with ca. 100 norbornene units on each dendritic branch were synthesized with these ruthenium carbene-based dendrimers.

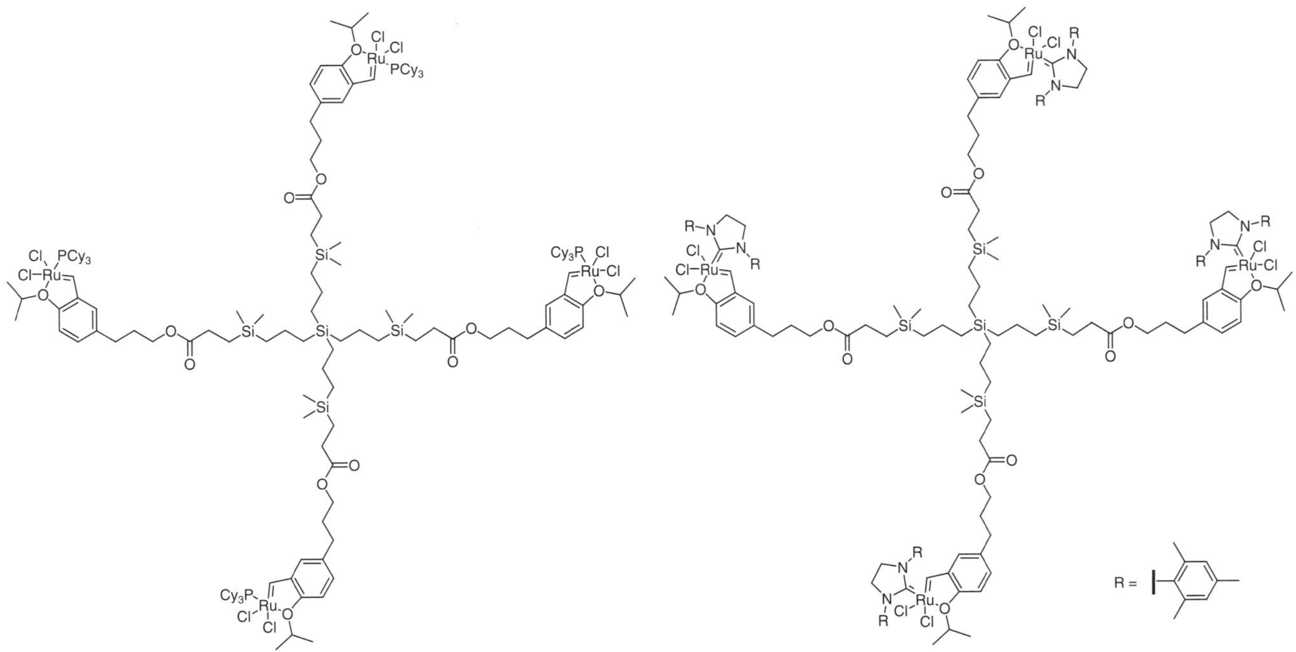

Figure 10.10 Hoveyda's dendritic Ru-based metathesis catalysts.

Moss et al.[137] investigated the Fischer–Tropsch synthesis[138] using the hexa-branched ruthenium metallodendrimer supported on silica. The dendritic branches each possessed 18-electron [RuCp(CO)$_2$alkyl] terminals. Under Fischer–Tropsch conditions, this metallodendrimer was metastable. The "time-on-stream" behavior of this catalyst in the CO hydrogenation was compared to an impregnated Ru/SiO_2 catalyst and showed no Fischer–Tropsch-specific product pattern, which was substantiated by the rate-of-formation of some specific hydrocarbons. This study showed that a single ruthenium site was not sufficient for this particular reaction.

Majoral's group[139] used a third-generation metallodendrimer containing 24 terminal ruthenium diphosphine complexes to catalyze the Knoevenagel condensation[140] between malononitrile and cyclohexanone, leading to the corresponding unsaturated nitrile and Michael addition between ethyl cyanoacetate and diethyl ethylidenemalonate. The activity was compared with that of the simple monomeric complexes, but was often higher; recycling was also possible without significant loss of catalytic activity.

Catalytic hydrogenation of organic substrates is a widely studied chemical transformation.[141] An alternative method, which avoids the use of gaseous hydrogen and uses standard reflux techniques, is provided by hydrogen-transfer processes where the hydrogen is supplied by a hydride donor (such as cyclohexadiene or formic acid), which itself undergoes oxidation (dehydrogenation) during the course of the reaction. These catalytic systems are relatively stable, easy to handle, and environmentally friendly.[142,143] Rossell et al.[72] synthesized a series of metallodendrimers by treatment of phosphanyl-terminated carbosilane dendrimers displaying only one phosphorus ligand per arm with [$RuCl_2$(*p*-cymene)]$_2$ (Figure 10.11). Activity of the multi-(η^6-arene)-ruthenium(II) species was compared with the corresponding ruthenium(II) mononuclear complexes in the reduction of cyclohexanone by a hydrogen transfer reaction. Dendritic systems showed lower activities than that found for related mononuclear complexes. Among first-generation metallodendrimers, the neutral species was more active than the corresponding cationic analog, following the same trend found for mononuclear species. Under standard conditions, the cationic species resulted in only 62% conversion, showing possible catalyst decomposition. The third-generation metallodendrimer was less active than that of the first generation. Nevertheless, these systems showed higher activities than those reported in the literature for other dendritic systems.[144]

vi. Ti-Based Dendritic Catalysts. Tilley et al.[145–148] synthesized a series of dendritic Ti-containing catalysts for the epoxidation of cyclohexene; these catalysts were shown to be very selective and more active in terms of yields and initial rates then the Shell catalyst[149] prepared by the treatment of silica with [$Ti(O^iPr_4)_4$]. The second- and third-generation alkoxysilyl-terminated carbosilane dendrimers have been used as building blocks for the synthesis of high-surface-area xerogels. The acid-catalyzed hydrolysis of $Si[CH_2CH_2CH_2Si(CH_2CH_2Si(OEt)_3)_3]_4$ and $Si<CH_2CH_2CH_2Si[CH_2CH_2CH_2Si(CH_2CH_2Si(OEt)_3)_3]_3>_4$ in benzene-generated monolithic gels; the resulting xerogels possessed surface areas of 600 and 800 m^2/g, respectively, in which the surface area of these xerogels increased with increasing dendritic radii. These xerogel-like gels were treated with $Ti(O^iPr)_4$ to yield the desired catalyst system. All three gels were subsequently used as catalysts in the epoxidation of cyclohexene and shown to be very selective and significantly more active (in terms of yield and initial rate) then the traditional Shell catalyst.

Seebach et al.[150,151] synthesized a hexa-armed dendrimer attached the C_2 symmetry ligand, TADDOL ($\alpha,\alpha,\alpha',\alpha'$-tetraaryl-1,3-dioxolane-4,5-dimethanol), complexing with $Ti(OCHMe_2)_4$ at the periphery. Using this chiral metallodendrimer, the enantiomeric addition[152] of diethylzinc to benzaldehyde proceeded with the same enantioselectivity (ee, 97%) as that of the monomeric chiral catalyst. Since this metallodendrimer had a molecular weight of only 3833 Da, its recovery

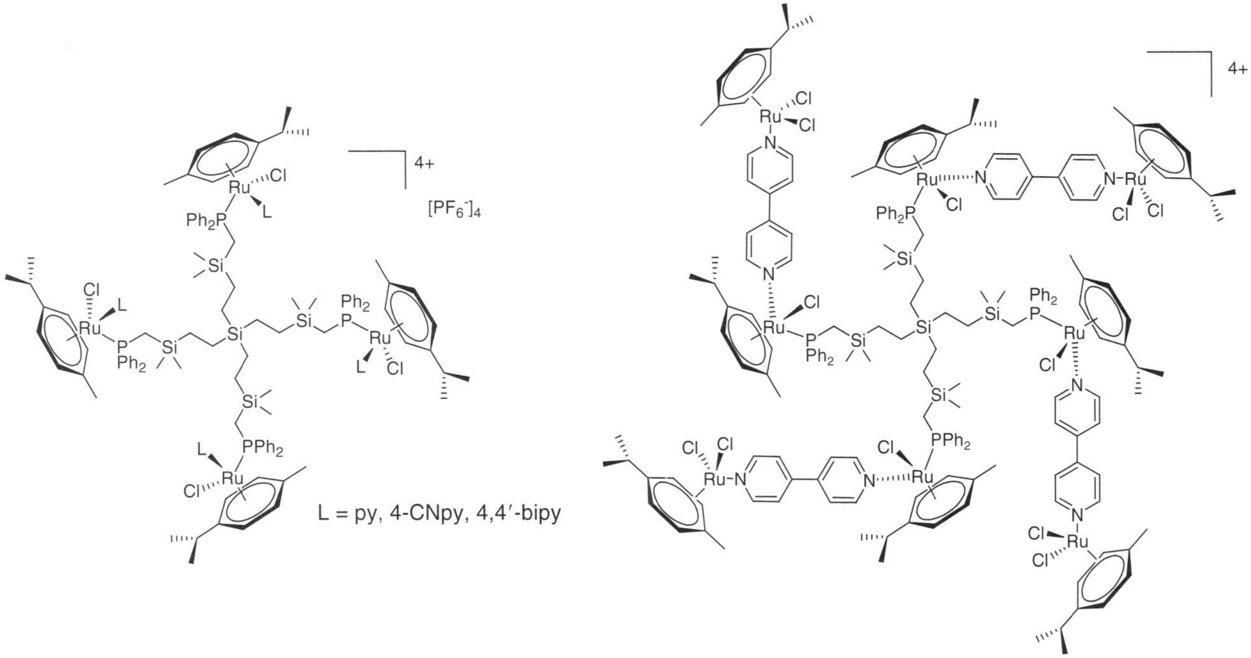

Figure 10.11 Single and double metallic layer–containing ruthenium metalloconstructs.

had to be conducted by column chromatography rather than by an ultrafiltration technique. To simplify the recovery process, Fréchet-type dendrons with styrenyl end-groups to cross-link the catalyst into a polystyrene support were used, whereby the TADDOL core was subsequently coordinated with Ti(IV).[150,153–156] This polymeric catalyst showed (1) an enantioselectivity above 9:1 for all polymers of low loading (i.e., 0.1 mmol TADDOL per gram of polymer), only the dendritic polymer gave rise to a constant selectivity of 98:2 in 20 sequential applications; (2) the catalytic performance dropped when the chain length of the spacers between the TADDOL core and polymer backbone was increased; (3) the low-loaded dendritic catalyst beads with the shortest spacer kept their swelling properties high even after 20 runs, while all others did not swell even after multiple reuse; and (4) the rate-of-reaction was the same with or without stirring using the beads of dendritic catalyst that had the shortest spacer, thus filling the whole reaction volume under standard conditions. Thus, free diffusion of reactants and products to and from the active center was obtained. Also, Seebach et al.[157] used dendritic Ti-TADDOLates with the first–fourth-generation Fréchet-type dendrons in the enantioselective addition of Et_2Zn to benzaldehyde. There was no detectable decrease of selectivity (98:2) up to the second generation, and the rates marginally decreased up to the third generation. Enantiomeric branches caused no change for stereoselectivity within experimental error. It was denoted that there might be applications for special properties such as high molecular weight, good solubility, and spacing of central sites from cross-linked polymer matrices.

Pu et al.[158,159] have used 4,4′,6,6′-tetrabromo-1,1′-bi-2-naphthol to construct a novel series of rigid and cross-conjugated optically active dendrimers. Following complexation with $Ti(O^iPr)_4$, the chiral dendrimer exhibited high enantioselectivity in the catalysis of the reaction of 1-naphthaldehyde with diethylzinc; notably, the system catalyzed this reaction with 100% conversion and 90% ee in 5 hours without side products. The advantage of this metallodendrimer over BINOL was that it could be easily removed from the reaction mixture by simple precipitation with MeOH.

Using (*R*)-6,6′-dibromo-1,1′-binaphthol, Yoshida et al.[160] synthesized optically active Ti-1,1′-binaphthol metallodendrimers containing poly(benzyl ether) wedges at the 6,6′-positions. The Ti-binaphthol-catalyzed allylation of the aldehyde and allyl stanane as the model reaction,[161] was demonstrated. Enantioselectivity remained constant with increased generation 0–3 (90 ± 2% ee; nondendritic parent, 87% ee).

Renewed interest in β-diketiminato transition metal complexes is partially due to this ligand's ability to stabilize coordinatively unsaturated complexes,[162] which can function as catalysts for nonmetallocene olefin polymerizations.[163] Gómes et al.[164,165] reported the synthesis and ethylene polymerization using mixed cyclopentadienyl(β-diketiminato) complexes possessing either titanium or zirconium catalytic sites engulfed with the first generation carbosilane dendritic wedges linked to the β-diketiminato ligand (Figure 10.12). Mixed cyclopentadienyl (β-diketiminato) complexes of titanium were obtained by direct reaction of $(\eta^5\text{-}C_5H_5)TiCl_3$ with dendritic β-diketimine in the presence of triethylamine acting as a Lewis base promoting protonolysis of the metal–chloride bond. The catalysts' efficacy was better than the performance of the metallocene complexes $(\eta^5\text{-}C_5H_5)MCl_3$ (M = Ti, Zr); however, they displayed slightly higher activities than their nondendritic counterparts.

vii. Fe-Based Dendritic Catalysts. Catalytic reduction of nitrate has been known for some time, but use of the iron-based redox catalyst $[Fe(II)(\eta^5\text{-}C_5H_4CO_2^-)(\eta^6\text{-}C_6Me_6)][PF_6]$ was the first example of an organometallic catalyst to effect this reaction.[166,167] Astruc et al.[168] synthesized an analogous water-soluble hexametallic redox catalyst employing the $CpFe^+$ group to induce hexa-allylation of hexamethylbenzene. The kinetics of an $[FeCp(arene)]^+$-centered star and dendritic-based star bearing the $[FeCp(arene)]^+$ group at the periphery were compared.

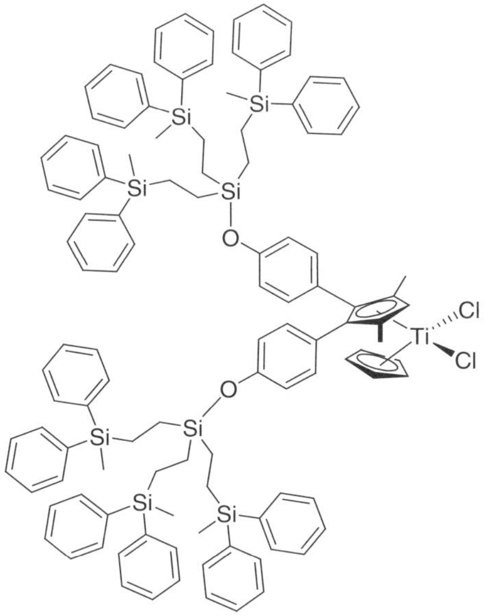

Figure 10.12 A dendritic β-diketiminato titanium complex.

Remarkably, the kinetics of catalysts bearing the $[FeCp(arene)]^+$ moiety attached to the dendritic core at the center of a star was determined to be one order-of-magnitude lower than that of such a star bearing the catalyst at the periphery.

Togni et al.[169] synthesized dendrimers using a cyclophosphazene core and 16 peripheral chiral ferrocenyl ligands (Figure 10.13). The periphery of these ferrocenyldiphosphine metallodendrimers was decorated with $[Rh(COD)_2][BF_4]$ to instill the desired catalytic properties necessary to reduce dimethyl itaconate. These metalloconstructs showed 98% ee, which compared well with the 99% ee obtained with monomeric Rh-based josiphos catalyst.

viii. Co-Based Dendritic Catalysts. Jacobsen et al.[170] reported the synthesis of dendrimer-bound [Co(III)-(salen)] complexes, which demonstrated significantly enhanced catalytic activity in the hydrolytic kinetic resolution of terminal epoxides (Figure 10.14). This metal–salen [H_2salen = bis(salicylidene)ethylenediamine] catalyst has led to a proposed mechanism for asymmetric ring-opening (ARO) reactions involving the simultaneous activation of both epoxide and nucleophile by different metal–salen units.[171] These metallodendrimers generated from the G0–G2 PAMAM cores were compared, and it was found that the best results were obtained with a simple core (4-branched) construct over that derived from either G1 or G2 cores. This catalytic system was shown to possess a positive dendritic effect because of cooperativity in the key mechanistic step, whereby a coordinated nucleophile attacks an epoxide coordinated to adjacent cobalt atom. Thus, it was the proximity of the two different cobalt complexes that favored the reaction.

Phthalocyanines and their metal complexes have interesting catalytic, electronic, and optical properties.[172] Kimura et al.[173] synthesized a phthalocyanine-centered dendrimer possessing Newkome-type dendrons,[174] and subsequently used the second-generation cobalt-metalated phthalocyanine dendrimer as the catalyst for the oxidation of mercaptoethanol by dioxygen

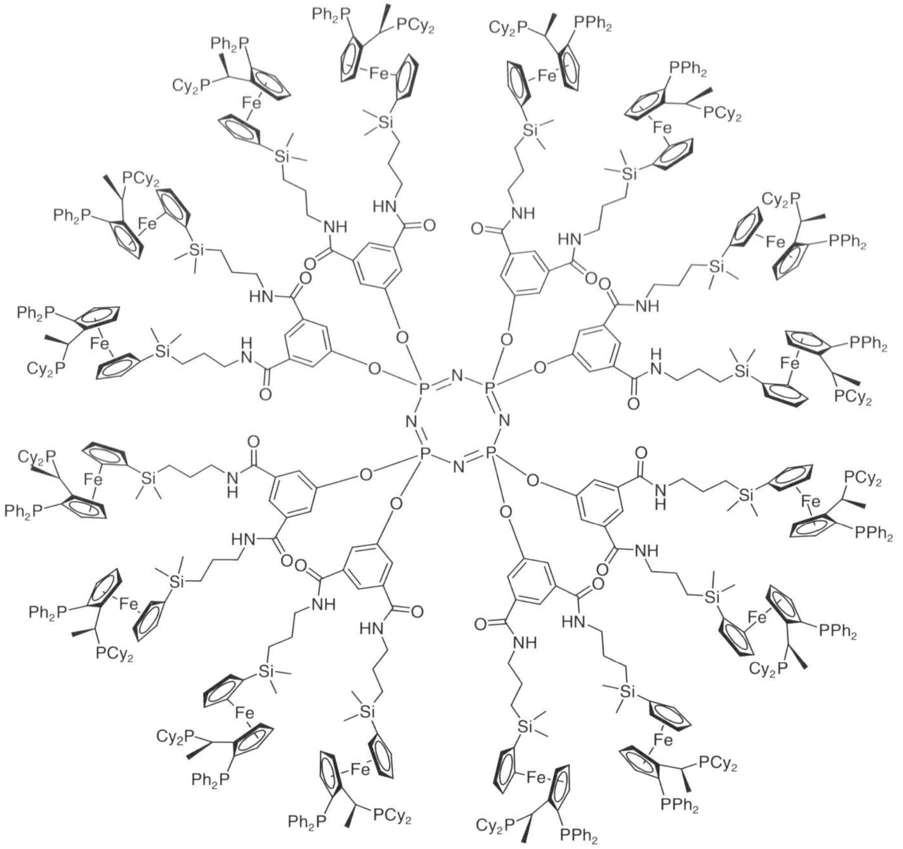

Figure 10.13 Togni's ferrocenyldiphosphine dendritic ligand for rhodium complexation.

(Figure 10.15). Catalytic stability was enhanced by the encapsulation within the dendritic structure due to the catalytic activity of metallophthalocyanines being influenced by phthalocyanine aggregation resulting from strong intermolecular stacking.

For the Pauson–Khand [2+2+1] cycloaddition[175] of alkyne, alkene, and carbon monoxide, as well as being promoted by $Co_2(CO)_8$, Portnoy et al.[176] synthesized a dendronized support modified with 2- and 4-(diphenylphosphino)benzoic acid groups; $Co_2(CO)_8$ was then incorporated into the support. A notable increase in catalytic activity and selectivity for the intramolecular Pauson–Khand reaction was found for the Co complexes immobilized on the second- and third-generation dendron-functionalized polystyrene, when compared with the analogous nondendronized support.

ix. Mn-Based Dendritic Catalysts. Suslick et al.[177,178,179] designed dendritic chloromanganese(III) porphyrins for the catalysis of alkene epoxidation with iodosylbenzene, as an oxygen donor (Figure 10.16). The dendritic wedges were derived from the first- and second-generation aromatic polyesters that provided a confined environment, thus instilling better intra- and intermolecular regioselectivities than those obtained with the unsubstituted 5,10,15,20-tetraphenylporphyrinatomanganese(III) cation. The least hindered double bond of unconjugated

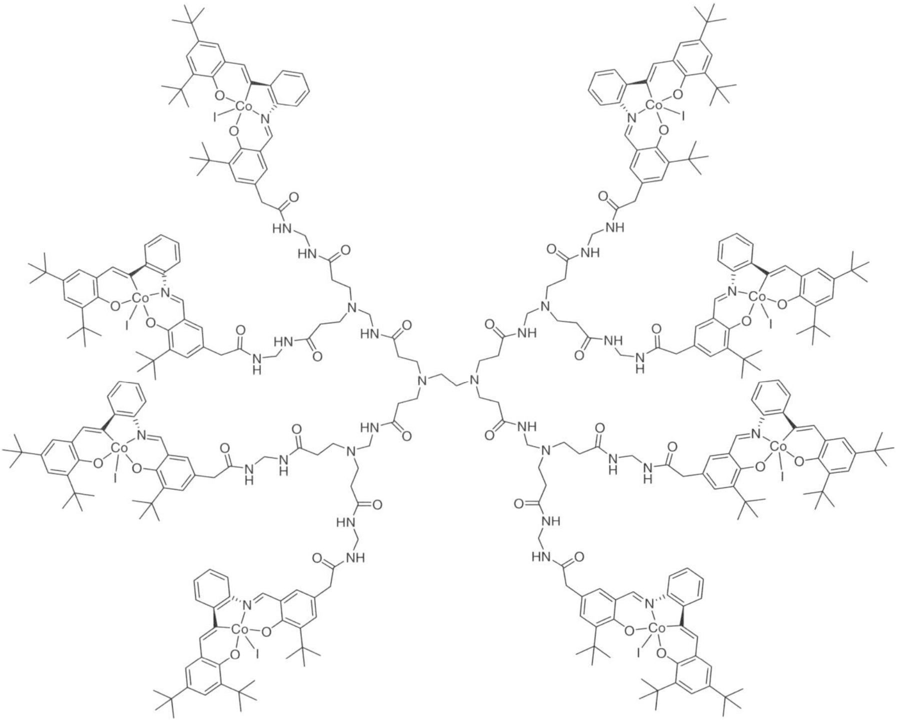

Figure 10.14 Breinbauer and Jacobsen's dendritic PAMAM–Co(salen) complex.

dienes, such as are found in 1,4-heptadiene and limonene, was epoxidized preferentially. Similarly, epoxidation of a terminal monoalkene and cyclooctene mixture using a second-generation dendritic metalloporphyrins showed a 2- to 3-fold higher selectivity toward the 1-alkenes relative to a nondendritic catalyst. This regioselectivity was significantly reduced when compared to that of the classic picket-fence porphyrin, 5,10,15,20-tetrakis(2′,4′,6′-triphenylphenylporphyrin).[180,181]

Kawi et al.[182] reported the anchoring of different PAMAMs to ultrafine silica, then a Mn(II)–salen complex was immobilization onto this functionalized silica surface to afford a bound dendritic catalyst for subsequent olefin epoxidation. Activity was found to increase with increasing generation, indicating the importance of amino group concentration on the periphery as well as the Mn loading of the Mn–selen complex. However, the length of the dendritic backbone chain also played an important role in improving the accessibility between the active sites and the reactant species to enhance the catalytic activity.

B. Luminescence

Luminescence can be defined as the emission of light (as in the broad sense of ultraviolet, visible, or near-infrared radiation) by electronic excited states of atoms or molecules. Luminescence is an important phenomenon that is useful for monitoring excited-state behavior,[183] as well as for utilitarian applications (e.g., laser, display, sensors).[184] Since dendrimers are complex multifunctional constructs possessing well-defined chemical tree-like structures, a high degree-of-order, and capable of containing selected chemical units within predetermined sites in their structure, their incorporation into luminescence science can lead to systems capable of performing very

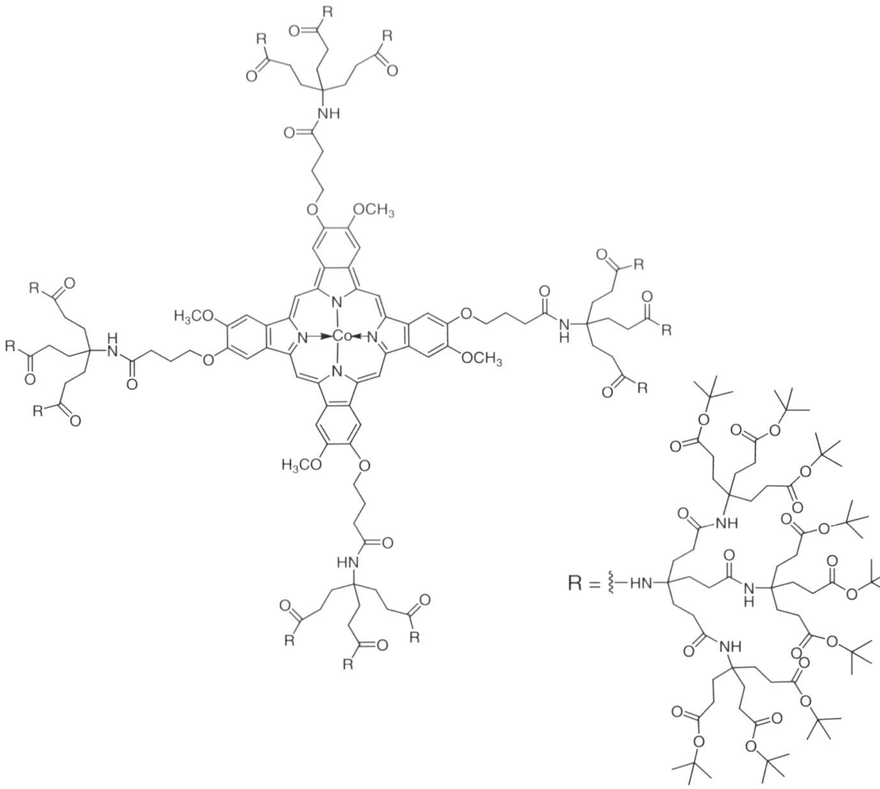

Figure 10.15 Kimura's dendritic metallophthalocyanines.

interesting functions. In this section, we review recent advances in the field of luminescent metallodendrimers.

Balzani's group has paid a great deal of attention to the synthesis of mono- and polynuclear transition metal complexes and the study of their photophysical, photochemical, and electrophysical properties.[17,185–191] Balzani et al.[192] designed a polynuclear metal complex of nanometric size and dendritic shape, constructed on an osmium(II)-based core and containing 21 ruthenium(II)-based units in the branches. Divergent and convergent synthetic approaches were both possible.[187] The light harvesting capability increased with increasing nuclearity and these metallodendrimers were shown to exhibit luminescence in solution at room temperature. Campagna et al.[193] reported analogous new trinuclear dendrons: $[Cl_2Os<(\mu\text{-}2,3\text{-dpp})Ru(bpy)_2>_2]^{4+}$, $[Cl_2Os<(\mu\text{-}2,3\text{-dpp})Os(bpy)_2>_2]^{4+}$, and $[(bpy)Ru<(\mu\text{-}2,3\text{-dpp})Os(bpy)_2>_2]^{6+}$ [bpy = 2,2′-bipyridine and 2,3-dpp = 2,3-bis(2-pyridinyl)pyrazine]. The redox behavior, absorption, and luminescence properties have been measured. The luminescence properties were dominated by the photophysical properties of the subunit(s) in which the lowest-lying excited state of their structure is localized suggesting fast intercomponent energy transfer in their new trinuclear dendrons. However, the properties of these metallodendrimers were not fully satisfactory, because the lowest-lying excited state involving the intermediate chromophore(s) was at higher energy than the lowest-lying excited states of both the central and peripheral chromophores. As a consequence, these intermediate chromophore(s) constitute a barrier to the

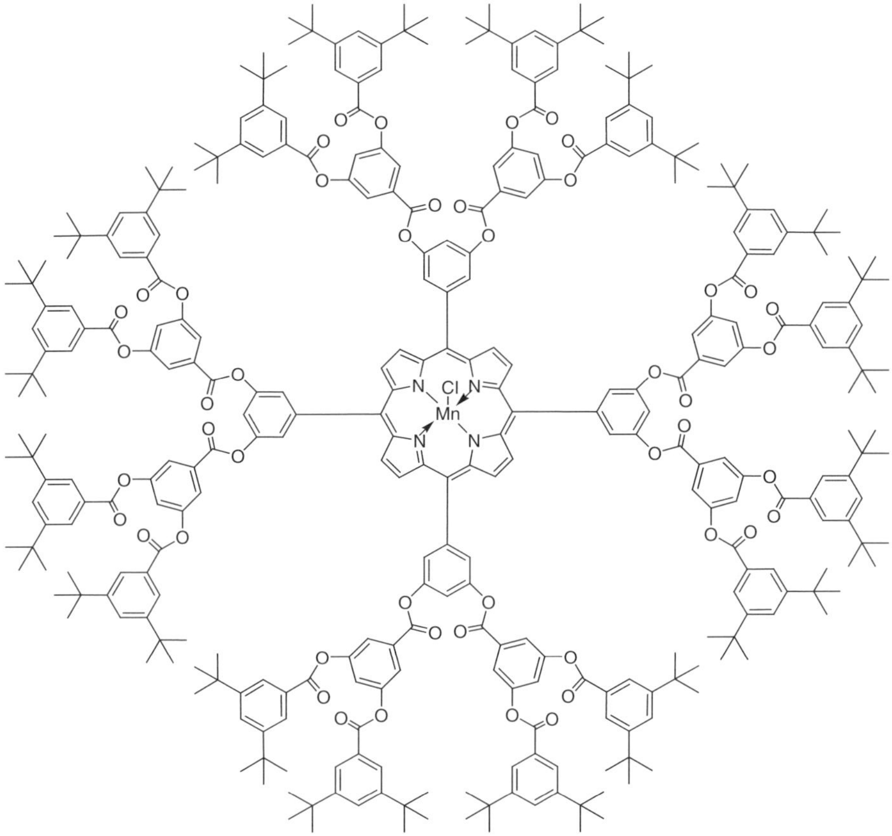

Figure 10.16 Suslick's dendritic chloromanganese(III) porphyrin.

periphery-to-center energy transfer. To overcome these problems, suitable alternative metals that facilitate efficient periphery-to-center energy transfer in larger species were introduced.[194] They prepared [Os<(μ-2,3-dpp)Ru[(μ-2,3-dpp)$PtCl_2$]$_2$>]$(PF_6)_8$ or $OsRu_3Pt_6$, in which the external layer of metal subunits was coated with $PtCl_2$ units. They also studied[195] the racemic species [$(phen)_2$Ru(phen-5,6-dione)]$(PF_6)_2$ (phen = 1,10-phenanthroline), [$(phen)_2$ Ru(phen-5,6-diamine)]$(PF_6)_2$, [Ru(phen-5,6-dione)$_3$]$(PF_6)_2$, [$(phen)_2$Ru(tpphz)]$(PF_6)_2$ (tpphz = tetrapyrido[3,2-*a*:2′,3′-*c*:3″,2″-*h*:2″,3″-*j*]phenazine), [$(phen)_2$Ru(μ-tpphz)Ru$(phen)_2$]$(PF_6)_4$, and [<$(phen)_2$Ru(μ-tpphz)>$_3$Ru]$(PF_6)_8$, the stereochemically pure dinuclear species ΔΔ-[$(phen)_2$Ru (μ-tpphz)Ru$(phen)_2$]$(PF_6)_4$, ΛΛ-[$(phen)_2$Ru(μ-tpphz)Ru$(phen)_2$]$(PF_6)_4$, and ΔΛ-[$(phen)_2$Ru (μ-tpphz)Ru$(phen)_2$]$(PF_6)_4$, and the stereochemically pure dendritic tetranuclear complexes [(Δ-$(phen)_2$Ru(μ-tpphz))$_3$-Δ-Ru]$(PF_6)_8$, [(Δ-$(phen)_2$Ru(μ-tpphz))$_3$-Λ-Ru]$(PF_6)_8$, and [(Λ-$(phen)_2$ Ru(μ-tpphz))$_3$-Λ-Ru]$(PF_6)_8$ (Figure 10.17). In all cases, luminescence decay was shown to be nonexponential, with lifetimes in the range from 10^{-5} to 10^{-8} s. Also, energy transfer occurred in the dendritic tetranuclear complexes from the central chromophore to those on the periphery.

Turro's group[196] has investigated and conducted a statistical analysis of luminescence quenching of the photoexcited sensitizer, $[Ru(phen)_3]^{2+}$, using $[Co(phen)_3]^{3+}$ on the surface

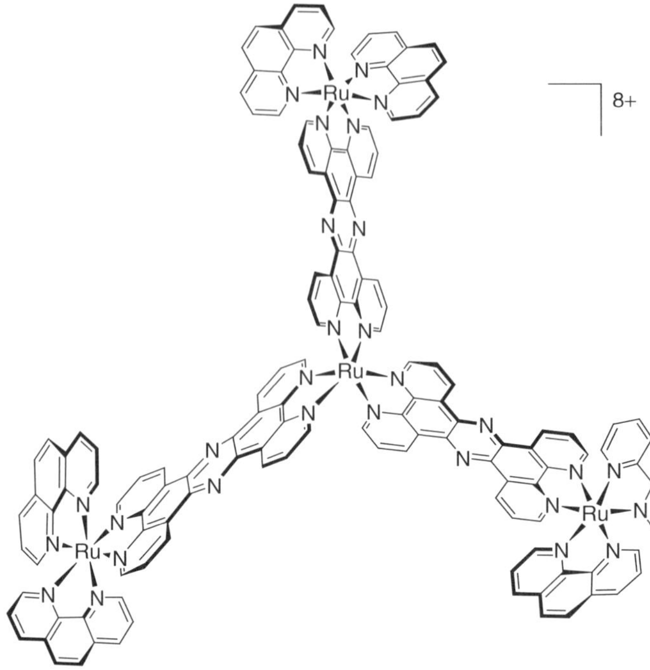

Figure 10.17 Balzani's dendritic tetranuclear Ru(II) complex.

of a PAMAM polyanion. An extension of the random-deposition model,[197] previously developed for the "one-dimensional" case of deoxyribonucleic acid (DNA) to analyze the luminescence quenching, showed that the electron-donor and -acceptor metal complexes do not bind randomly on the dendrimer surfaces.

Fréchet-type dendrons bearing carboxylate-focal moieties were self-assembled around Er^{3+}, Tb^{3+} or Eu^{3+} ions, as the resultant core, leading to the formation of dendrimers, as shown in Figure 10.18.[198–202] Experiments conducted in toluene revealed that the UV excitation of the chromophoric groups, contained in the branches, caused the sensitized emission of the lanthanide ion, presumably by energy transfer based on the Förster mechanism.[203] The lower sensitization effect found for the coordinated Eu^{3+}, when compared to that of the Tb^{3+}, was ascribed to a weaker spectral overlap, but it could also be related to the fact that Eu^{3+} can quench the donor excited state by electron transfer. As the shell size increased, these metallodendrimers exhibited enhanced luminescence activity, both in solution and the bulk. This enhancement was attributed to both an "antenna effect," the energy transfer from the ligand to the core, and a "shell effect"; the site isolation of the steric exclusion phenomenon, which keeps the Ln^{3+} cores apart from one another, leads to a decrease in their rate of self-quenching.

The Burn group[204] synthesized a metallodendrimer consisting of a *fac*-tris(2-phenylpyridine) iridium core, phenylene dendrons, and 2-ethylhexyloxy surface groups. This device was prepared using a neat solution-processed method consisting of a iridium metallodendrimer emissive layer with an evaporated 2,9-dimethyl-4,7-diphenyl-1,10-phenanthroline-based (BCP) electron-transport/hole-blocking layer. This unique metallodendrimer device exhibited bright green luminescence with a turn-on voltage of 4.5 V and peak brightness of 1680 cd/m^2 at 12 V, but the maximum efficiency of the device was low, only reaching 0.14 lm/W (0.47 cd/A) at 9.5 V.

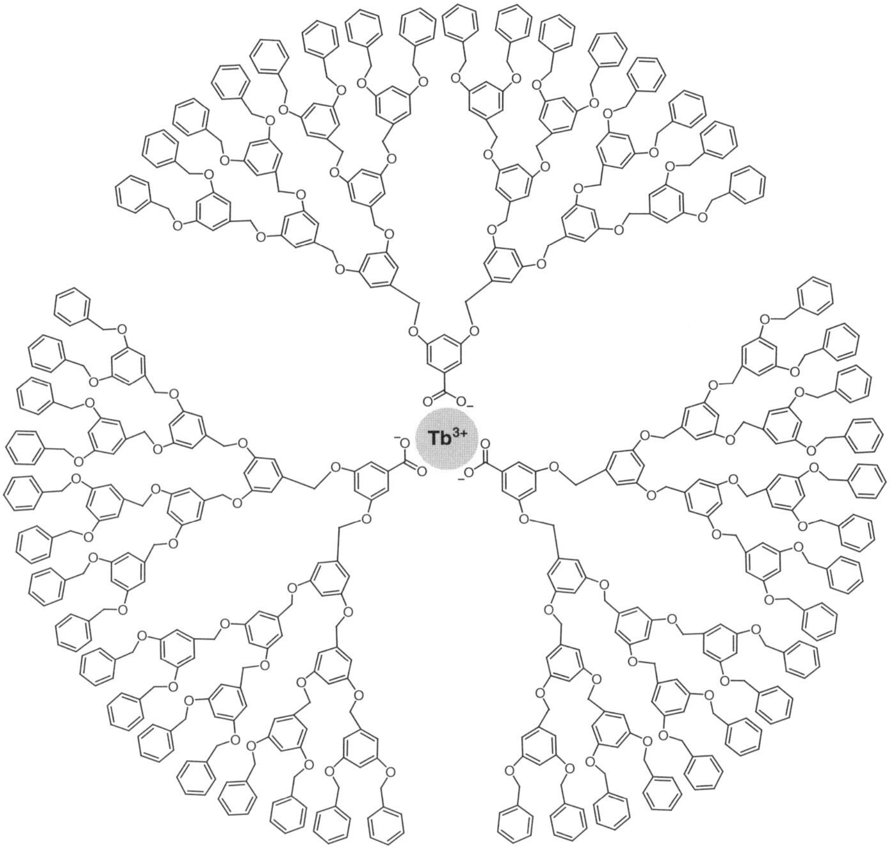

Figure 10.18 Kawa and Fréchet's Tb-cored metallodendrimer.

Blend devices such as ITO/IrppyD:CBP:BCP:LiF/Al, ITO/IrppyD/TCTA/BCP/LiF/Al, and ITO/IrppyD:TCTA/TPBI/LiF/Al have also been examined, where IrppyD is the metallodendrimer, CBP is 4,4′-bis(*N*-carbazolyl)biphenyl, TCTA is 4,4′,4″-tris(*N*-carbazolyl)triphenylamine, and TPBI is 1,3,5-tris(2-*N*-phenylbenzimidazolyl)benzene. It was observed that the efficiency of these dendritic devices remained relatively constant over a range of brightness. This resulted from a balanced charge injection and uniform distribution of iridium metallodendrimer in the host layer.

Lanthanide ions are known to display a very long-lived luminescence, which is quite a useful property for several applications, such as in sensors[205,206,207] and fluoroimmunoassay.[208] Because of the forbidden nature of their electronic transitions, lanthanide ions exhibit very weak absorption bands, which can be a severe drawback for applications based on luminescence. Thus, there have been several efforts to circumvent this difficulty, such as attachment of a dendrimer or dendron, which has strong chromophores. These examples[209–212] in solution have shown that (1) the absorption spectrum of the dendrimer was unchanged; (2) the fluorescence of the dansyl units was quenched; (3) the quenching effect was very large for Nd^{3+} and Eu^{3+}, moderate for Er^{3+} and Yb^{3+}, small for Tb^{3+}, and very small for Gd^{3+}; and (4) in the case of Nd^{3+}, Er^{3+}, and Yb^{3+}, the quenching of the dansyl fluorescence was accompanied by a sensitized, near-infrared emission of the lanthanide ion. These observations were explained using the energy levels and redox potentials of both the dendrimer (Figure 10.19) and metal ions.

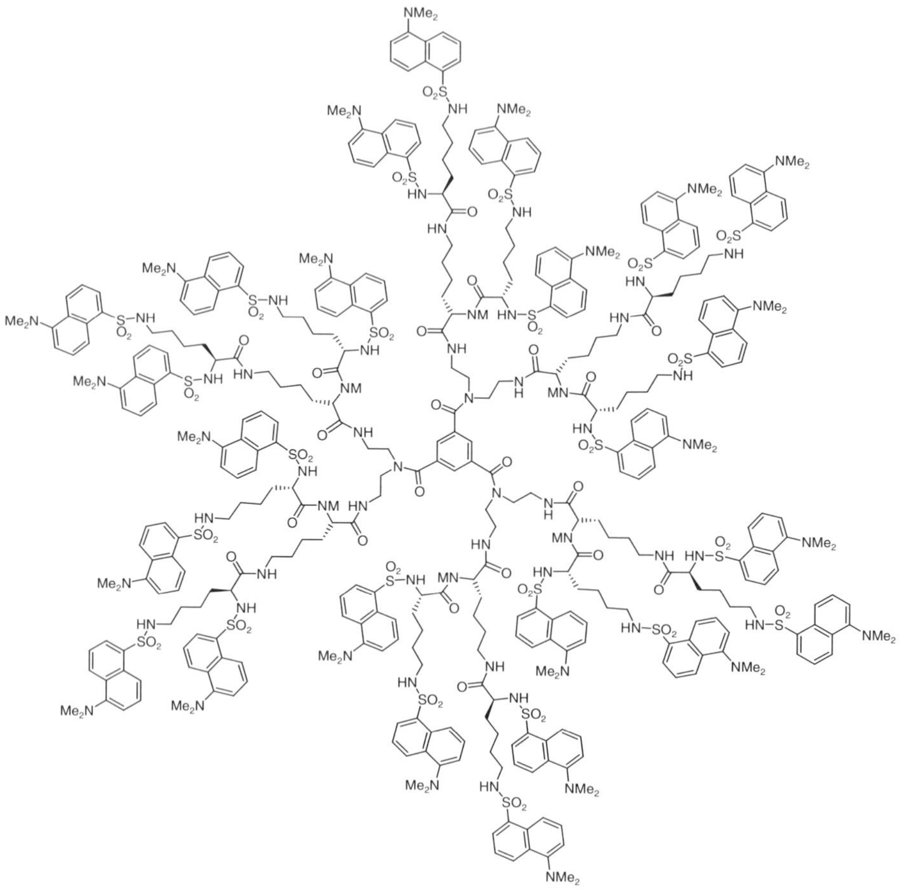

Figure 10.19 Polylysine dendritic ligand for lanthanide ion complexation.

1,4,8,11-Tetraazacyclotetradecane (cyclam) is one of the more extensively investigated ligands in coordination chemistry.[213,214] Both cyclam and its 1,4,8,11-tetramethyl derivative in aqueous solution can be mono- and di-protonated and can coordinate numerous metal ions, such as, Co^{II}, Ni^{II}, Cu^{II}, Zn^{II}, Cd^{II}, and Hg^{II} that display large stability constants. Extensive investigations have been performed on metal-ion interactions of dendrimers, which have a cyclam as a core unit, and Fréchet-type dendrons as peripheral units.[215–218] Complexation with Zn^{2+} engages the *N*-lone pairs, and thereby prevents exciplex formation, with a resulting intense naphthyl fluorescence.[217] This strong fluorescent signal was quite suitable for monitoring the formation of complexes in dendrimer–metal titration experiments. Another complexation study of dendritic ligands with lanthanide ions (M = Nd^{3+}, Eu^{3+}, Gd^{3+}, Tb^{3+}, Dy^{3+}) has led qualitatively to similar results: an increase of the naphthyl units emission band at 337 nm and the complete disappearance of the exciplex band at 480 nm have been observed.[215]

In another effort to obtain strong fluorescence, Halet et al.[219] designed heterometallic dendritic molecules using branched Pd^{II} and Pt^{II} alkynyl complexes. A series of Re^{I} alkynyl precursors, $[1,3\text{-}(HC{\equiv}C)_2\text{-}5\text{-}<(N\text{^}N)(CO)_3ReC{\equiv}C>C_6H_3]$ (N^N = ^{t}Bu$_2$bpy, Me$_2$bpy or bpy), and their trinuclear branched Re^I–Pd^{II} mixed-metal alkynyl complexes, $[1,3<Cl(PEt_3)_2PdC{\equiv}C>2\text{-}5\text{-}<(N\text{^}N)(CO)_3ReC{\equiv}C>C_6H_3]$ has been prepared. All complexes exhibited an intense low-energy absorption at ca. 410~430 nm; this was tentatively assigned to the $[d\pi(Re) \rightarrow \pi^*(\text{diimine})]$ metal-to-ligand charge transfer (MLCT) transition, probably with some mixing of alkynyl-to-diimine $[\pi(C{\equiv}C) \rightarrow \pi^*(\text{diimine})]$ ligand-to-ligand charge transfer (LLCT) character.

C. Sensors

Metallodendrimers have attracted widespread interest for various sensing applications, such as ion detection, gas sensing, and redox-active switches, because of their polyvalent characteristics.[220] Several recent examples of successful sensors using the metallodendrimers have appeared.

Castellano et al.[221] reported the formation of a luminescence lifetime-based sensor for cyanide and other counterions using Ru^{II} diimines possessing MLCT excited states with the anion recognition capabilities of 2,3-di(1*H*-2-pyrrolyl)quinoxaline (DPQ). Using time-resolved photoluminescence decay, its viability as a lifetime-based sensor for anions has been tested. There were significant changes to the UV-vis and steady-state emission properties after the addition of several ions (e.g., fluoride, cyanide, and phosphate).

Ferrocene can be used to detect carbon monoxide (CO) by coordination, which increases the electron acceptability by the formation of a stable Fe–CO bond. Kim et al.[222] have synthesized and investigated a ferrocenyl-based dendrimer as a CO gas sensor. The metallodendrimer sensor showed a linear increase by 10 times up to 40% volume concentration, above which it was saturated. The transient responses exhibited a delay time of 50 seconds, a rising time of 150 seconds, and a falling time of 420 seconds, which were fair for a gas sensor when the diffusion time of gas into the chamber and the continued reaction with the remaining gas in the chamber after the introduction of the gas has been terminated are considered.

Since phosphorescence quenching is very sensitive to and selective for oxygen, it is a well-know method for the quantification of dissolved molecular oxygen (O_2).[220,223,224,225] Vinogradov et al.[226,227,228] have designed a new metallodendrimer consisting of an encapsulated Pd or Pt porphyrins. These porphyrins are widely used as basic phosphors. The mass of the outer shell of the dendrimer, which surrounds luminescent chromophores, has been used to increase diffusion barriers for oxygen and other quenchers in solution. A phosphorescent oxygen nanosensor using the dendritic structure and two-photon absorbing antenna (Figure 10.20) was recently reported.[229] Two-photon light harvesting and intramolecular energy transfer were accompanied by intersystem crossing within the acceptor chromophore, thereby inducing its phosphorescence.

Van Koten et al.[230–233] designed a series of square-planer platinum(II) complexes using *N,C,N′*-tridentate-coordination monoanionic "pincer" ligand, $[PtX(4\text{-}E\text{-}2,6\text{-}<CH_2NRR'>_2\text{-}C_6H_2)]$ (X = Cl, Br, I, tolyl; R, R′ = Et, Me; E = H, OH, OSiMe$_2$-^{t}Bu). These complexes spontaneously adsorb SO_2 gas to form penta-coordinated adducts. Adduct formation in the solid state or in solution was fast as well as reversible, and was shown by a characteristic color change of the material from colorless to bright orange. Since facile methods have been developed to remove SO_2 from the adducts and to regenerate the square-planer starting complexes, these complexes fulfill several essential prerequisites of sensor materials for repeated diagnostic SO_2 detection. A platinum sensor has been found to be highly selective for SO_2 and to be particularly sensitive for submillimolar to molar amounts. Their response capacity is tunable by electronic and steric modifications of the ligand by introduction of, for example, different *N* substituents. The periphery of these metallodendrimers was shown to be an appropriate

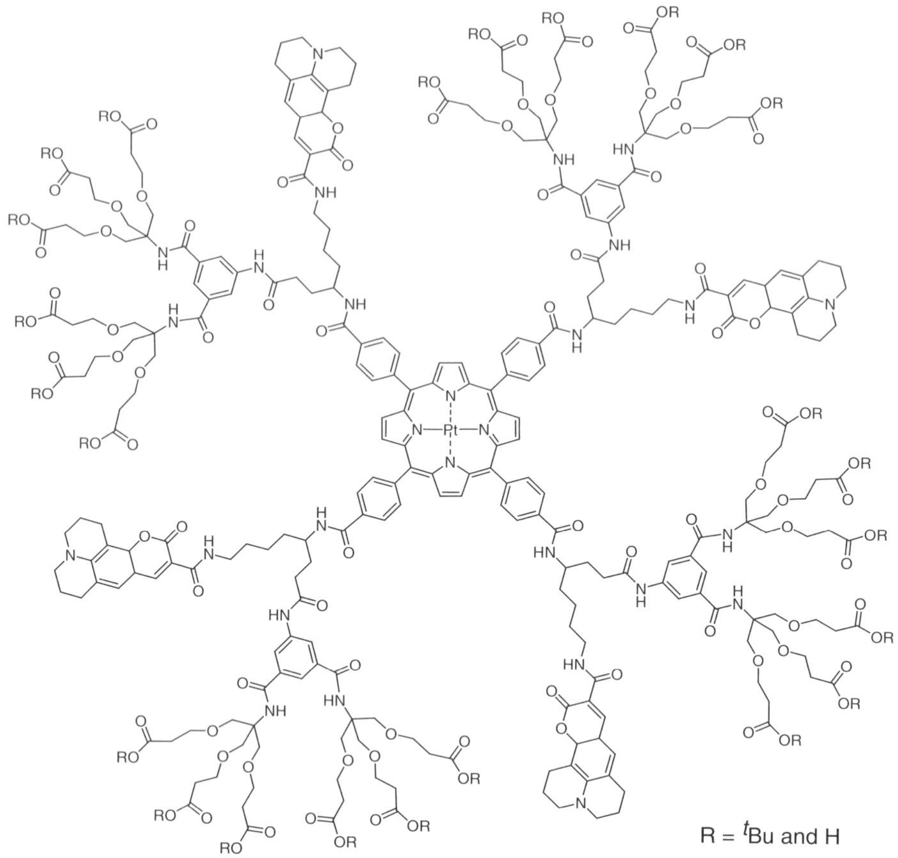

R = ^{t}Bu and H

Figure 10.20 A dendritic Pt-containing porphyrin.

macromolecular support for anchoring the detection-active sites, thus allowing full recovery of the sensor materials for repetitive use. Recently, a technique for signal detection using quartz crystal microbalance (QMB) has been reported.[230] In such a sensor device, substrate binding is accomplished by changing the net mass of the surface (host-guest complex versus host only). QMBs are highly sensitive balances, which translate small mass changes (typically on the order of nanograms) of the QMB disk into an inverse modification of its resonance frequency that can readily be recorded.[234,235]

Nanoparticle-cored dendrimers can be useful for sensing. A new mode of construction involved the assembly of dendrons onto gold nanoparticles giving stable nanoparticle-cored dendrimers (Figure 10.21).[236–239] These metallodendrimers can selectively recognize oxo-anions and ATP^{2-} (ATP-ademosine triphosphate) using cyclic voltammetry (redox recognition by variation of the redox potential of the ferrocenyl group). The selective recognition and titration of these oxo-anions can be effected in the presence of other less interacting anions, such as, hydrogenosulfate, chloride or the anion of the supporting electrolyte (BF_4^- or PF_6^- in large quantity) by the following: (1) supramolecular interaction between the oxygen atom of the anion and the amido or silyl group conjugated with the ferrocenyl group; (2) electrostatic

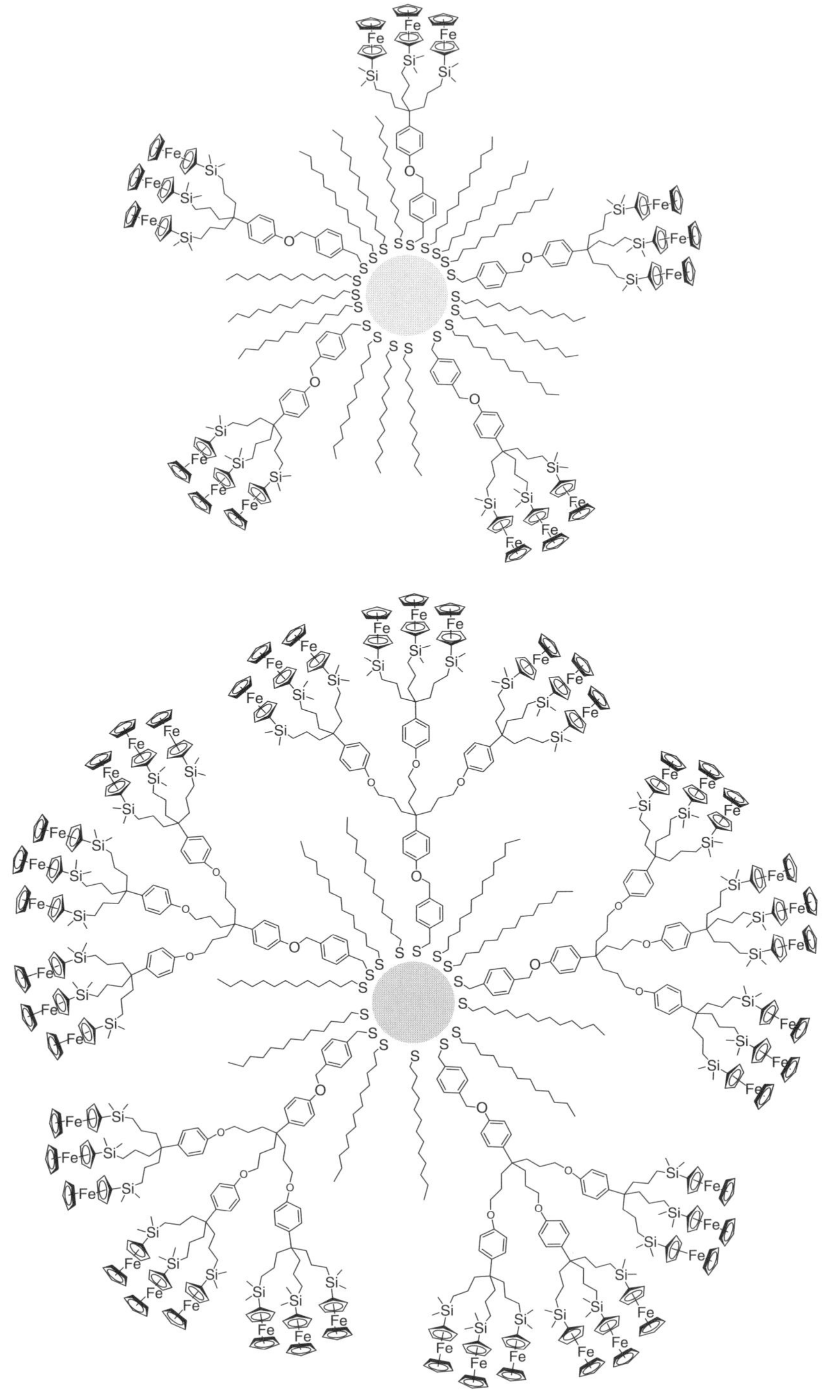

Figure 10.21 Dendronized gold colloids.

interaction between the oxo-anion and the cationic ferrocenium generated upon anodic oxidation; and (3) topological effects at the dendrimer periphery, whereby the small microcavities offer a narrow, selective channel for the supramolecular interaction. In the absence of this topology effect, the change in potential was very weak.[240] A platinum electrode can be modified with the dendrimer by scanning the ferrocenyl region to give a very stable modified electrode that can be efficiently used as the sensor. Upon washing this electrode with dichloromethane (DCM), the ATP^{2-} was removed and the original cyclic voltammetry wave of the dendrimer was recovered for repeated use. This operation can be conducted many times, because the supramolecular forces involved in the recognition are weak.

D. Others

Dendritic magnetic resonance imaging (MRI) contrast agents,[241] consisting of Gd^{III} complexes of the chelator 2-(4-isothiocyanatobenzyl)-6-methyldiethylenetriaminepentaacetic acid anchored to amino-terminated PAMAMs, have been developed.[242] These complexes enhanced MRIs and were found to be more effective contrast agents than other commercially available macromolecule-chelated complexes, such as those formed using albumin, polylysine, and dextrin. Recently, Tóth et al.[243,244,245] synthesized a series of multivalent lanthanide(III)-glycoconjugates, based on 1,4,7,10-tetrakis(carboxymethyl)-1,4,7,10-tetraazacyclododecane-(DOTA) monoamide functionalized chelators (Figure 10.22). These DOTA-like chelators are

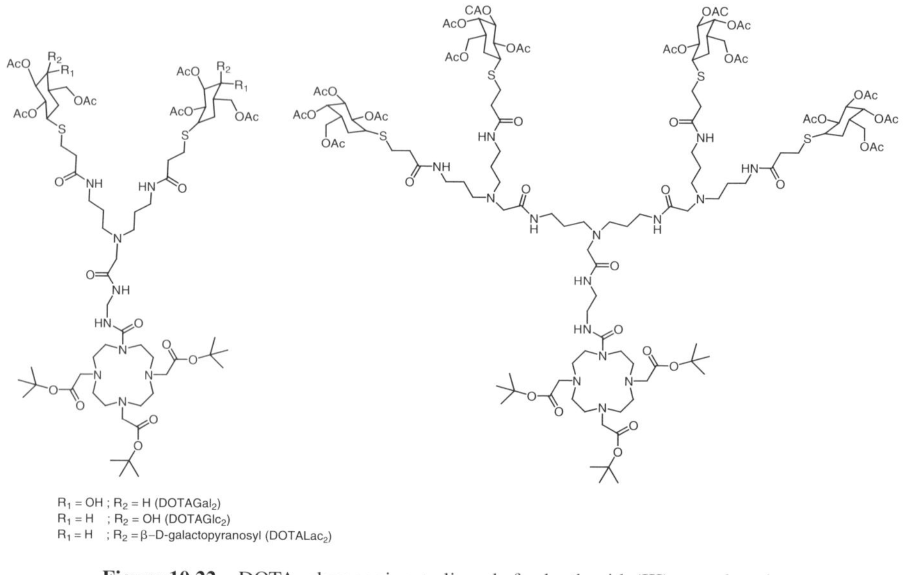

Figure 10.22 DOTA–glycoconjugate ligands for lanthanide(III) complexation.

well known to form Ln^{III} complexes possessing high thermodynamic and kinetic stability, which is of crucial importance for in vivo applications.[246] The in vitro relaxivity of Gd^{III}-glycoconjugates using the lectin-mediated molecular imaging agents has been measured; the feasibility of in vivo MRI applications based on receptor binding, in general, depends on the receptor concentration, as well as on the relaxivity of the receptor-bound species. However, the flexibility of the glycodendrimer moiety in solution limited the relaxivity of their Gd^{III} complexes to values lower than expected from their molecular weight. In spite of the limits, the lectin-glycoconjugated interaction can slow down the tumbling rate considerably, and therefore increase the relaxivity of the Gd^{III} chelates.

Metallodendrimers containing numerous metal elements that can provide robust redox systems are electron reservoirs that can be used as molecular batteries[247,248,249] or in nanodevices for molecular electronics.[250–256] Astruc et al.[257] have synthesized metallodendrimers containing 54 ferrocenyl units using a precise convergent method as well as other metallodendrimers resulting from divergent construction and theoretically containing 81 and 243 ferrocenyl groups[248] at the dendrimer periphery. The ferrocene–ferrocenium redox couple for these ferrocenyl metallodendrimers, however, was not suitable as an electron reservoir, due to a low redox potential of the silylferrocenyl group. Later, an excellent electron-reservoir system $[FeCp(\eta^6\text{-}C_6Me_6)]^{+/0}$ was crafted onto the commercial fifth-generation PPI dendrimer (theoretically 64 amine terminals) using the chlorocarbonyl complex $[Fe^{II}(\eta^5\text{-}C_5H_4COCl)(\eta^6\text{-}C_6Me_6)][PF_6]$ to give the corresponding amido-based dendrimer *dendr*-$<[(NHCOC_5H_4)Fe^{II}(\eta^6\text{-}C_6Me_6)][PF_6]>_{64}$. This metallodendrimer has been reduced to its deep-purple 19-electron Fe^I form using the prototype 19-electron, electron-reservoir complex $[Fe^ICp(\eta^6\text{-}C_6Me_6)]$. The latter reacted with 64 equivalents of C_{60} by single exergonic electron transfer from each Fe^I site to C_{60} to give the *dendr*-$<[(NHCOC_6H_4)Fe^{II}(\eta^6\text{-}C_6Me_6)][C_{60}^{-}]>_{64}$ hybrid that was characterized *inter alia* by its Mössbauer and electron paramagnetic resonance (EPR) spectra.[249]

Photodynamic therapy (PDT) is a promising technique for the localized treatment of tumors; an increasing number of the photosensitizers has been recently explored in preclinical and clinical study.[258–263] Dendritic porphyrins can transport absorbed energy over relatively large distances using dendritic architectures, thereby mimicking the antenna complex and bacteriochlorophyl photosystem.[264,265] On the other hand, numerous porphyrins have been known to effectively produce highly toxic singlet oxygen through excitation by light at a characteristic wavelength; some are being used as photosensitizers for PDT (Figure 10.23). Consequently, dendritic porphyrins may have the potential to act as novel photosensitizers used in PDT. Kataoka et al.[266] synthesized third-generation aryl ether dendritic porphyrin with either 32 quaternary ammonium groups (32[+]DPZn) or 32 carboxylic groups (32[−]DPZn) and evaluated them as a novel, metallodendrimer class of photosensitizers for PDT. Notably, 32[+]DPZn achieved remarkably higher singlet oxygen-induced cytotoxicity against Lewis lung carcinoma (LLC) cells than protoporphyrin IX, demonstrating their highly selective photosensitizing effect in combination with a reduced systemic toxicity.

III. OUTLOOK

The unique features of dendritic architecture and the rich chemistry of organo-transition metal complexes have been combined in metallodendrimers to create the potential for a wide range of utilitarian applications. Because dendrimers allow scientists to probe the twilight zone between homogeneous and heterogeneous catalysis as well as to apply the techniques associated with combinatorial-type chemistry, diverse new areas of the nanoworld have became accessible. Since many new avenues in supramolecular chemistry have been opened by organometallic complexes, metallodendrimers will continue to play an important role in not only organometallic chemistry and polymer science, but also in material science. These new interfaces will be rich areas for future science to pursue.

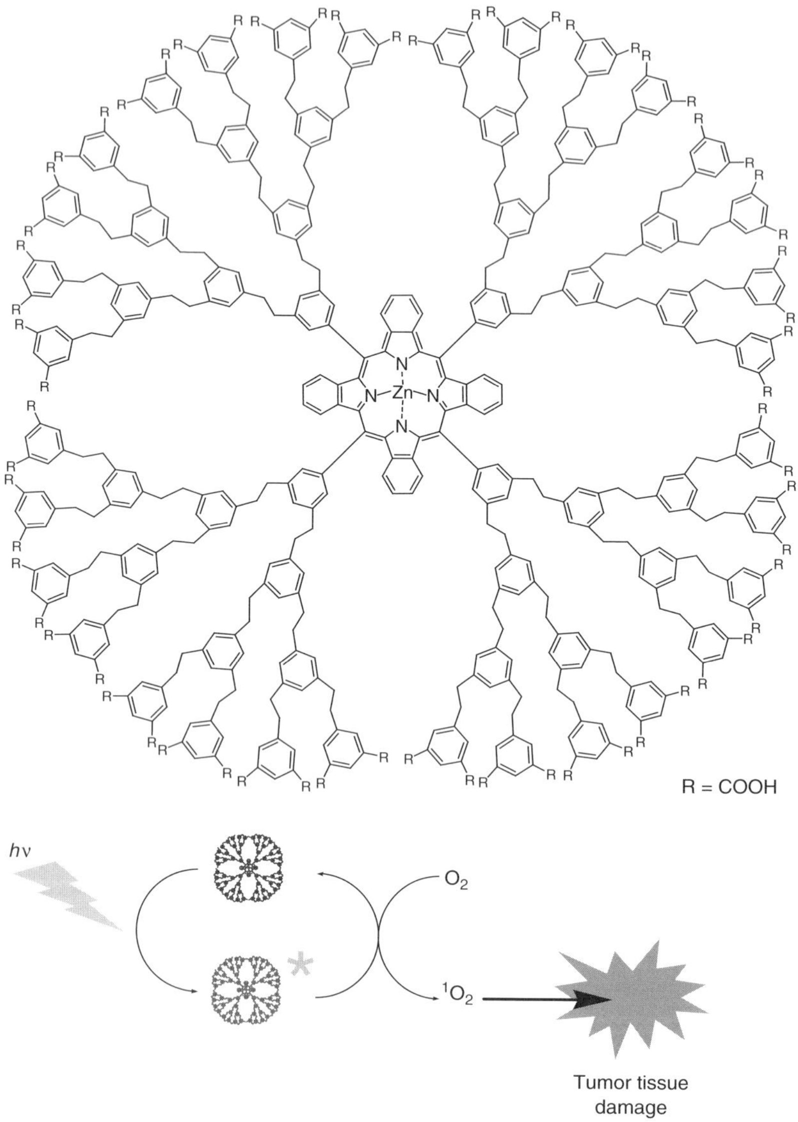

Figure 10.23 Schematic depiction of photodynamic therapy (PDT) using a dendrimer with a protoporphyrin photosensitizer core, which upon irradiation with light and subsequent reaction with oxygen creates tissue-damaging singlet oxygen.

ACKNOWLEDGMENTS

The authors thank the National Science Foundation and Air Force Research Office for financial assistance as well as Dr. Charles Moorefield and members of our research group for technical support.

REFERENCES

1. G. R. Newkome, Z. Yao, G. R. Baker, V. K. Gupta, *J. Org. Chem.*, **50**, 2003 (1985).
2. D. A. Tomalia, H. Baker, J. Dewald, M. Hall, G. Kallos, S. Martin, J. Roeck, J. Ryder, P. Smith, *Polym. J. (Tokyo)*, **17**, 117 (1985).
3. J.-M. Lehn, *Supramolecular Chemistry: Concepts and Perspectives*, VCH, Weinheim, Germany, 1995.
4. J.-M. Lehn, *Polym. Int.*, **51**, 825 (2002).
5. G. Denti, S. Campagna, S. Serroni, M. Ciano, V. Balzani, *J. Am. Chem. Soc.*, **114**, 2944 (1992).
6. G. Denti, S. Serroni, S. Campagna, V. Ricevuto, V. Balzani, *Inorg. Chim. Acta*, **182**, 127 (1991).
7. S. Campagna, G. Denti, S. Serroni, M. Ciano, A. Juris, V. Balzani, *Inorg. Chem.*, **31**, 2982 (1992).
8. G. R. Newkome, C. N. Moorefield, G. R. Baker, A. L. Johnson, R. K. Behera, *Angew. Chem., Int. Ed. Engl.*, **30**, 1176 (1991).
9. G. R. Newkome, F. Cardullo, E. C. Constable, C. N. Moorefield, A. M. W. C. Thompson, *J. Chem. Soc., Chem. Commun.*, 925 (1993).
10. D. Astruc, J.-C. Blais, E. Cloutet, L. Djakovitch, S. Rigaut, J. Ruiz, V. Sartor, C. Valério, *Topics Curr. Chem.*, **210**, 229 (2000).
11. C. Gorman, *Adv. Mater.*, **10**, 295 (1998).
12. M. A. Hearshaw, J. R. Moss, *Chem. Commun.*, 1 (1999).
13. G. R. Newkome, E. He, C. N. Moorefield, *Chem. Rev.*, **99**, 1689 (1999).
14. K. Onitsuka, S. Takahashi, *Topics Curr. Chem.*, **228**, 39 (2003).
15. F. J. Stoddart, T. Welton, *Polyhedron*, **18**, 3575 (1999).
16. D. Astruc, F. Chardac, *Chem. Rev.*, **101**, 2991 (2001).
17. V. Balzani, P. Ceroni, A. Juris, M. Venturi, S. Campagna, F. Puntoriero, S. Serroni, *Coord. Chem. Rev.*, **219–221**, 545 (2001).
18. R. M. Crooks, M. Zhao, L. Sun, V. Chechik, L. K. Yeung, *Acc. Chem. Res.*, **34**, 181 (2001).
19. G. Denti, S. Campagna, L. Sabatino, S. Serroni, M. Ciano, V. Balzani, "Towards an Artificial Photosynthesis. Di-, Tri-, Tetra-, and Hepta-Nuclear Luminescent and Redox-Reactive Metal Complexes," in *Photochemistry, Conversion and Storage of Solar Energy*, E. Pelizzetti, M. Schiavello, Eds., pp. 27–45, Kluwer Academic Publishers, Dordrecht, The Netherlands, 1991.
20. C. B. Gorman, *C. R. Chimie*, **6**, 911 (2003).
21. R. H. E. Hudson, *J. Organomet. Chem.*, **637–639**, 47 (2001).
22. A. W. Kleij, A. Ford, J. T. B. H. Jastrzebski, G. van Koten, "Dendritic Polymer Applications Catalysis," in *Dendrimers and Other Dendritic Polymers*, J. M. J. Fréchet, D. A. Tomalia, Eds., pp. 485–514, Wiley, West Sussex, UK 2001.
23. G. E. Oosterom, J. N. H. Reek, P. C. J. Kamer, *Angew. Chem. Int. Ed.*, **40**, 1828 (2001).
24. S. Serroni, S. Campagna, F. Puntoriero, F. Loiseau, V. Ricevuto, R. Passalacqua, M. Galletta, *C. R. Chimie*, **6**, 883 (2003).
25. M. A. Hearshaw, A. T. Hutton, J. R. Moss, K. J. Naidoo, "Organometallic Dendrimers: Synthesis, Structural Aspects, and Applications in Catalysis," in *Advances in Dendritic Macromolecules*, G. R. Newkome, Ed., pp. 1–60. JAI Press, Stamford, CT, 1999.
26. G. R. Newkome, C. N. Moorefield, F. Vögtle, *Dendrimers and Dendrons: Concepts, Syntheses, Applications*, Wiley-VCH, Weinheim, Germany, 2001.
27. A. Corma, H. Garcia, *Dalton Trans.*, 1381 (2005).
28. A. Corma, *Chem. Rev.*, **97**, 2372 (1997).
29. C. Halm, M. Kurth, *Angew. Chem. Int. Ed.,* **37**, 510 (1998).
30. S. Kobayashi, S. Nogayama, *J. Am. Chem. Soc.*, **120**, 2985 (1998).
31. C. P. Mehnert, D. W. Weaver, J. Y. Ying, *J. Am. Chem. Soc.*, **120**, 12289 (1998).

32. J. W. J. Knapen, A. W. van der Made, J. C. de Wilde, P. W. N. M. van Leeuwen, P. Wijkens, D. M. Grove, G. van Koten, *Nature*, **372**, 659 (1994).

33. A. W. Kleij, R. A. Gossage, R. J. M. K. Gebbink, N. Brinkmann, E. J. Reijerse, U. Kragl, M. Lutz, A. L. Spek, G. van Koten, *J. Am. Chem. Soc.*, **122**, 12112 (2000).

34. N. J. Hovestad, A. Ford, J. T. B. H. Jastrzebski, G. van Koten, *J. Org. Chem.*, **65**, 6338 (2000).

35. J. March, *Advanced Organic Chemistry*, Fourth Edition, Wiley, London, 1992, pp. 5–33.

36. G. van Koten, J. T. B. H. Jastrzebski, *J. Mol. Catal. Part A: Chem.*, **146**, 317 (1999).

37. A. W. Kleij, R. A. Gossage, J. T. B. H. Jastrzebski, J. Boersma, G. van Koten, *Angew. Chem. Int. Ed.*, **39**, 176 (2000).

38. R. A. Gossage, J. T. B. H. Jastrzebski, J. van Ameijde, S. J. E. Mulders, A. J. Brouwer, R. M. J. Liskamp, G. van Koten, *Tetrahedron Lett.*, **40**, 1413 (1999).

39. T. B. Rauchfuss, *Inorg. Chem.*, **16**, 2966 (1977).

40. D. Vogt, "Nonaqueous Organic Organic Separation (SHOP Process)," in *Aqueous-Phase Organometallic Catalysts*, B. Cornits, W. A. Herrmann, Eds., pp. 639–645, Wiley-VCH, Weinheim, Germany, 2004.

41. C. Müller, L. J. Ackerman, J. N. H. Reek, P. C. J. Kamer, P. W. N. M. van Leeuwen, *J. Am. Chem. Soc.*, **126**, 14960 (2004).

42. H.-F. Chow, Z.-Y. Wang, Y.-F. Lau, *Tetrahedron*, **54**, 13813 (1998).

43. C. C. Mak, H.-F. Chow, *Macromolecules*, **30**, 1228 (1997).

44. H.-F. Chow, C. C. Mak, *J. Org. Chem.*, **62**, 5116 (1997).

45. X. Camps, H. Schönberger, A. Hirsch, *Chem. Eur. J.*, **3**, 561 (1997).

46. X. Camps, E. Dietel, A. Hirsch, S. Pyo, L. Echegoyen, S. Hackbarth, B. Röder, *Chem. Eur. J.*, **5**, 2362 (1999).

47. A. Herzog, A. Hirsch, O. Vostrowsky, *Eur. J. Org. Chem.*, 171 (2000).

48. M. Hetzer, H. Clausen-Schaumann, S. Bayerl, T. M. Bayerl, X. Camps, O. Vostrowsky, A. Hirsch, *Angew. Chem. Int. Ed.*, **38**, 1962 (1999).

49. M. Hetzer, S. Bayerl, X. Camps, O. Vostrowsky, A. Hirsch, T. M. Bayerl, *Adv. Mater.*, **9**, 913 (1997).

50. M. Hetzer, T. Gutberlet, M. F. Brown, X. Camps, O. Vostrowsky, H. Schönberger, A. Hirsch, T. M. Bayerl, *J. Phys. Chem. A*, **103**, 637 (1999).

51. I. Lamparth, A. Maichle-Mössmer, A. Hirsch, *Angew. Chem., Int. Ed. Engl.*, **36**, 1607 (1995).

52. F. Djojo, E. Ravanelli, O. Vostrowsky, A. Hirsch, *Eur. J. Org. Chem.*, 1051 (2000).

53. H. Fritschi, U. Leutenegger, *Helv. Chim. Acta*, **71**, 1553 (1998).

54. A. M. Trzeciak, J. J. Ziólkowski, *Coord. Chem. Rev.*, **190–192**, 883 (1999).

55. D. de Groot, P. G. Emmerink, C. Coucke, J. N. H. Reek, P. C. J. Kamer, P. W. N. M. van Leeuwen, *Inorg. Chem. Commun.*, **3**, 711 (2000).

56. A. W. van der Made, P. W. N. M. van Leeuwen, *J. Chem. Soc., Chem. Commun.*, 1400 (1992).

57. D. de Groot, E. B. Eggeling, J. C. de Wilde, H. Kooijman, R. J. van Haaren, A. W. van der Made, A. L. Spek, D. Vogt, J. N. H. Reek, P. C. J. Kamer, P. W. N. M. van Leeuwen, *Chem. Commun.*, 1623 (1999).

58. S. C. Bourque, H. Alper, L. E. Manzer, P. Arya, *J. Am. Chem. Soc.*, **122**, 956 (2000).

59. M. T. Reetz, G. Lohmer, R. Schwickardi, *Angew. Chem. Int. Ed.*, **36**, 1526 (1997).

60. E. J. H. Put, K. Clays, A. Persoons, H. A. M. Biemans, C. P. Luijkx, E. W. Meijer, *Chem. Phys. Lett.*, **260**, 136 (1996).

61. A. Gong, Q. Fan, Y. Chen, H. Liu, C. Chen, F. Xi, *J. Mol. Catal. Part A: Chem.*, **159**, 225 (2000).

62. P.-A. Jaffrès, R. E. Morris, *J. Chem. Soc., Dalton Trans.*, 2770 (1998).

63. L. Ropartz, D. F. Foster, R. E. Morris, A. M. Z. Slawin, D. J. Cole-Hamilton, *J. Chem. Soc., Dalton Trans.*, 1997 (2002).

64. L. Ropartz, R. E. Morris, D. F. Foster, D. J. Cole-Hamilton, *Chem. Commun.*, 361 (2001).

65. L. Ropartz, R. E. Morris, G. P. Schwarz, D. F. Foster, D. J. Cole-Hamilton, *Inorg. Chem. Commun.*, **3**, 714 (2000).
66. M. C. Simpson, A. W. S. Currie, J. A. Anderson, M. J. Green, D. J. Cole-Hamilton, *J. Chem. Soc., Dalton Trans.*, 1793 (1996).
67. M. Petrucci-Samija, V. Guillemette, M. Dasgupta, A. K. Kakkar, *J. Am. Chem. Soc.*, **121**, 1968 (1999).
68. O. Bourrier, A. K. Kakkar, *Macromol. Symp.*, **210**, 97 (2004).
69. O. Bourrier, J. Butlin, R. Hourani, A. K. Kakkar, *Inorg. Chim. Acta*, **357**, 3836 (2004).
70. A. K. Kakkar, *Macromol. Symp.*, **196**, 145 (2003).
71. I. Angurell, G. Muller, M. Rocamora, O. Rossell, M. Seco, *Dalton Trans.*, 1194 (2003).
72. I. Angurell, G. Muller, M. Rocamora, O. Rossell, M. Seco, *Dalton Trans.*, 2450 (2004).
73. P. N. M. Botman, A. Amore, R. van Heerbeek, J. W. Back, H. Hiemstra, J. N. H. Reek, J. H. van Maarseveen, *Tetrahedron Lett.*, **45**, 5999 (2004).
74. P. N. M. Botman, M. Postma, J. Fraanje, K. Goubitz, H. SChenk, J. H. van Maarseveen, H. Hiemstra, *Eur. J. Org. Chem.*, 1952 (2002).
75. M. van den Berg, A. J. Minnaard, E. P. Schudde, J. van Esch, A. H. M. de Vries, J. G. de Vries, B. L. Feringa, *J. Am. Chem. Soc.*, **122**, 11539 (2000).
76. M. van den Berg, A. J. Minnaard, R. M. Haak, M. Leeman, E. P. Schudde, A. Meetsma, B. L. Feringa, A. H. M. de Vries, C. E. P. Maljaars, C. E. Willans, D. Hyett, J. A. F. Boogers, H. J. W. Henderickx, J. G. de Vries, *Adv. Synth. Catal.*, **345**, 308 (2003).
77. S. C. Bourque, F. Maltais, W.-J. Xiao, O. Tardif, H. Alper, P. Arya, L. E. Manzer, *J. Am. Chem. Soc.*, **121**, 3035 (1999).
78. P. Arya, N. V. Rao, J. Singkhonrat, H. Alper, S. C. Bourque, L. E. Manzer, *J. Org. Chem.*, **65**, 1881 (2000).
79. P. Arya, G. Panda, N. V. Rao, H. Alper, S. C. Bourque, L. E. Manzer, *J. Am. Chem. Soc.*, **123**, 2889 (2001).
80. L. Busseto, M. C. Cassani, V. G. Albano, P. Sabatino, *Organometallics*, **21**, 1849 (2002).
81. L. Busseto, M. C. Cassani, R. Mazzoni, V. G. Albano, P. Sabatino, P. Frediani, E. Rivalta, *Organometallics*, **21**, 4993 (2002).
82. L. Busseto, M. C. Cassani, R. Mazzoni, P. Frediani, E. Rivalta, *J. Mol. Catal. Part A: Chem.*, **206**, 153 (2003).
83. L. Busseto, M. C. Cassani, R. Mazzoni, P. Frediani, E. Rivalta, *J. Organomet. Chem.*, **689**, 2216 (2004).
84. L. Busseto, M. C. Cassani, P. W. N. M. van Leeuwen, R. Mazzoni, *Dalton Trans.*, 2767 (2004).
85. W. Tang, X. Zhang, *Chem. Rev.*, **103**, 3029 (2003).
86. U. Nagel, *Angew. Chem., Int. Ed. Engl.*, **23**, 435 (1984).
87. J. Albrecht, U. Nagel, *Angew. Chem., Int. Ed. Engl.*, **35**, 407 (1996).
88. B. Yi, Q.-H. Fan, G.-J. Deng, Y.-Q. Li, A. S. C. Chan, *Org. Lett.*, **6**, 1361 (2004).
89. V. Cesar, S. Bellemin-Laponnaz, L. H. Gade, *Chem. Soc. Rev.*, **33**, 619 (2004).
90. D. Bourissou, O. Guerret, F. P. Gabbai, G. Bertrand, *Chem. Rev.*, **100**, 39 (2000).
91. T. Fujihara, Y. Obora, M. Tokunaga, S. Hiromichi, Y. Tsuji, *Chem. Commum.*, **36**, 4526 (2005).
92. R. Jira, "Oxidation of Olefins to Carbonyl Compounds (Wacker Process)", in *Applied Homogeneous Catalysis with Ogranometallic Compounds*, B. Cornils, W. A. Herrmann, Eds., pp. 374–393, VCH, Weinheim, Germany, 1996.
93. K. Januszkiewicz, H. Alper, *Tetrahedron Lett.*, **24**, 5159 (1983).
94. H. Alper, K. Januszkiewicz, *Tetrahedron Lett.*, **26**, 2263 (1985).
95. H. A. Zahalka, K. Januszkiewicz, H. Alper, *J. Mol. Catal.*, **35**, 249 (1986).
96. E. Monflier, S. Tilloy, E. Blouet, Y. Barbaux, A. Mortreux, *J. Mol. Catal. Part A: Chem.*, **109**, 27 (1996).
97. S. Antebi, P. Arya, L. E. Manzer, H. Alper, *J. Org. Chem.*, **67**, 6623 (2002).

98. H. Alper, P. Arya, S. C. Bourque, G. R. Jefferson, L. E. Manzer, *Can. J. Chem.*, **78**, 920 (2000).
99. P. P. Zweni, H. Alper, *Adv. Synth. Catal.*, **346**, 849 (2004).
100. N. Brinkmann, D. Giebel, G. Lohmer, M. T. Reetz, U. Kragl, *J. Catal.*, **183**, 163 (1999).
101. D. P. Catsoulacos, B. R. Steele, G. A. Heropoulos, M. Micha-Screttas, C. G. Screttas, *Tetrahedron Lett.*, **44**, 4575 (2003).
102. M. Beller, A. Zapf, T. H. Riermeier, "Palladium-Catalyzed Olefinations of Aryl Halides (Heck Reaction) and Related Transformations," in *Transition Metals for Organic Synthesis*, M. Beller, C. Bolm, Eds., Wiley-VCH, Weinheim, Germany, 2004.
103. I. P. Beletskaya, A. V. Cheprakov, *Chem. Rev.*, **100**, 3009 (2000).
104. D. de Groot, J. N. H. Reek, P. C. J. Kamer, P. W. N. M. van Leeuwen, *Eur. J. Org. Chem.*, 1085 (2002).
105. R. R. Tykwinski, *Angew. Chem. Int. Ed.*, **42**, 1566 (2003).
106. J. Dupont, R. F. de Souza, P. A. Z. Suarez, *Chem. Rev.*, **102**, 3667 (2002).
107. K. Heuzé, D. Méry, D. Gauss, D. Astruc, *Chem. Commun.*, 2274 (2003).
108. K. Heuzé, D. Méry, D. Gauss, J.-C. Blais, D. Astruc, *Chem. Eur. J.*, **10**, 3936 (2004).
109. A. W. Kleij, R. J. M. K. Gebbink, P. A. J. van den Nieuwenhuijzen, H. Kooijman, M. Lutz, A. L. Spek, G. van Koten, *Organometallics*, **20**, 634 (2001).
110. M. Malkoch, K. Hallman, S. Lutsenko, A. Hult, E. Malmström, C. Moberg, *J. Org. Chem.*, **67**, 8197 (2002).
111. G. Rodrígues, M. Lutz, A. L. Spek, G. van Koten, *Chem. Eur. J.*, **8**, 46 (2002).
112. L. Canovese, G. Chessa, C. Santo, F. Visentin, P. Uguagliati, *Organometallics*, **21**, 4342 (2002).
113. I. P. Beletskaya, A. V. Cheprakov, *Chem. Rev.*, **100**, 3009 (2000).
114. A. Dahan, M. Portnoy, *Org. Lett.*, **5**, 1197 (2003).
115. W. Cabri, I. Candiani, *Acc. Chem. Res.*, **28**, 2 (1995).
116. T. Mizugaki, M. Ooe, K. Ebitani, K. Kaneda, *J. Mol. Catal. Part A: Chem.*, **145**, 329 (1999).
117. T. Mizugaki, M. Murata, M. Ooe, K. Ebitani, K. Kaneda, *Chem. Commun.*, 52 (2002).
118. M. Ooe, M. Murata, A. Takahama, T. Mizugaki, T. Mizugaki, K. Ebitani, K. Kaneda, *Chem. Lett.*, **32**, 692 (2003).
119. M. Ooe, M. Murata, T. Mizugaki, K. Ebitani, K. Kaneda, *J. Am. Chem. Soc.*, **126**, 1604 (2004).
120. Y. Ribourdouille, G. D. Engel, M. Richard-Plouet, L. H. Gade, *Chem. Commun.*, 1228 (2003).
121. J. J. Becker, M. R. Gagné, *Organometallics*, **22**, 4984 (2004).
122. S. Hecht, J. M. J. Fréchet, *Angew. Chem. Int. Ed.*, **40**, 75 (2001).
123. D. K. Smith, F. Diederich, *Chem. Eur. J.*, **4**, 1353 (1998).
124. D. K. Smith, F. Diederich, *Topics Curr. Chem.*, **210**, 183 (2000).
125. F. Zeng, S. C. Zimmerman, *Chem. Rev.*, **97**, 1681 (1997).
126. H. Aït-Haddou, S. M. Leder, M. R. Gagné, *Inorg. Chim. Acta*, **357**, 3854 (2004).
127. S. B. Garber, J. S. Kingsbury, B. L. Gray, A. H. Hoveyda, *J. Am. Chem. Soc.*, **122**, 8168 (2000).
128. R. B. Grubbs, S. Chang, *Tetrahedron*, **54**, 4413 (1998).
129. P. Schwab, M. B. France, J. W. Ziller, R. B. Grubbs, *Angew. Chem., Int. Ed. Engl.*, **34**, 2039 (1995).
130. T. M. Trnka, R. B. Grubbs, *Acc. Chem. Res.*, **34**, 18 (2001).
131. P. Wijkens, J. T. B. H. Jastrzebski, P. A. van der Schaaf, R. Kolly, A. Hafner, G. van Koten, *Org. Lett.*, **2**, 1612 (2000).
132. E. Alonso, D. Astruc, *J. Am. Chem. Soc.*, **122**, 3222 (2000).
133. E. M. M. de Brabander-van den Berg, E. W. Meijer, *Angew. Chem., Int. Ed. Engl.*, **32**, 1308 (1993).
134. M. Bardají, M. Kustos, A.-M. Caminade, J.-P. Majoral, B. Chaudret, *Organometallics*, **16**, 403 (1997).
135. J. S. Kingsbury, J. P. Harrity, P. J. Bonitatebus, Jr., A. H. Hoveyda, *J. Am. Chem. Soc.*, **121**, 791 (1999).

136. S. Gatard, S. Nlate, E. Cloutet, G. Bravic, J.-C. Blais, D. Astruc, *Angew. Chem. Int. Ed.*, **42**, 452 (2003).
137. M. Claeys, M. Hearshaw, J. R. Moss, E. van Steen, *Stud. Surf. Sci. Catal.*, **130B**, 1157 (2000).
138. C. K. Rofer-DePoorter, *Chem. Rev.*, **81**, 447 (1981).
139. V. Maraval, R. Laurent, A.-M. Caminade, J.-P. Majoral, *Organometallics*, **19**, 4025 (2000).
140. Y. Ono, *J. Catal.*, **216**, 406 (2003).
141. S. Nishimura, *Handbook of Heterogeneous Catalytic Hydrogenation for Organic Synthesis*, Wiley, Chichester, UK, 2001.
142. G. Zassinovich, G. Mestroni, S. Gladialli, *Chem. Rev.*, **92**, 1051 (1992).
143. T. Naota, H. Takaya, S.-I. Murahashi, *Chem. Rev.*, **98**, 2599 (1998).
144. Y. Chen, T. Wu, J. Deng, H. Liu, X. Cui, J. Zhu, Y. Jiang, M. C. K. Choi, A. S. C. Chen, *J. Org. Chem.*, **67**, 5301 (2002).
145. J. W. Kriesel, T. D. Tilley, *Polym. Prepr.*, **41**, 566 (2000).
146. J. W. Kriesel, T. D. Tilley, *Chem. Mater.*, **12**, 1171 (2000).
147. J. W. Kriesel, T. D. Tilley, *Chem. Mater.*, **11**, 1190 (1999).
148. J. W. Kriesel, T. D. Tilley, *Adv. Mater.*, **13**, 1645 (2001).
149. H. P. Wulff, F. Wattimena, U.S. Patent 4,021,454 (1997).
150. D. Seebach, *Chimia*, **54**, 60 (2000).
151. D. Seebach, A. K. Beck, M. Rueping, J. V. Schreiber, H. Seliner, *Chimia*, **55**, 98 (2001).
152. K. Soai, S. Niwa, *Chem. Rev.*, **92**, 833 (1992).
153. P. B. Rheiner, H. Sellner, D. Seebach, *Helv. Chim. Acta*, **80**, 2027 (1997).
154. H. Sellner, D. Seebach, *Angew. Chem. Int. Ed.*, **38**, 1918 (1999).
155. D. Seebach, R. E. Marti, T. Hintermann, *Helv. Chim. Acta*, **79**, 1710 (1996).
156. H. Sellner, P. B. Rheiner, D. Seebach, *Helv. Chim. Acta*, **85**, 352 (2002).
157. P. B. Rheiner, D. Seebach, *Chem. Eur. J.*, **5**, 3221 (1999).
158. Q.-S. Hu, V. Pugh, M. Sabat, L. Pu, *J. Org. Chem.*, **64**, 7528 (1999).
159. V. J. Pugh, Q.-S. Hu, L. Pu, *Angew. Chem. Int. Ed.*, **39**, 3638 (2000).
160. S. Yamago, M. Furukawa, A. Azuma, J. Yoshida, *Tetrahedron Lett.*, **39**, 3783 (1998).
161. G. E. Keck, K. H. Tarbet, L. S. Geraci, *J. Am. Chem. Soc.*, **115**, 8467 (1993).
162. L. Bourget-Merie, M. F. Lappert, J. R. Severn, *Chem. Rev.*, **102**, 3031 (2002).
163. V. C. Gibson, S. K. Spitzmesser, *Chem. Rev.*, **103**, 283 (2003).
164. R. Andrés, E. de Jesús, F. J. de la Mata, J. C. Flores, R. Gómes, *J. Organomet. Chem.*, **690**, 939 (2005).
165. R. Andrés, E. de Jesús, F. J. de la Mata, J. C. Flores, R. Gómes, *Eur. J. Inorg. Chem.*, 2281 (2002).
166. D. Astruc, K. Heuzé, S. Gatard, D. Méry, S. Nlate, L. Plault, *Adv. Synth. Catal.*, **347**, 329 (2005).
167. D. Astruc, *C. R. Chimie*, **8**, 1101 (2005).
168. S. Rigaut, M.-H. Delville, D. Astruc, *J. Am. Chem. Soc.*, **119**, 11132 (1997).
169. R. Schneider, C. Köllner, I. Weber, A. Togni, *Chem. Commun.*, 2415 (1999).
170. R. Breinbauer, E. N. Jacobsen, *Angew. Chem. Int. Ed.*, **39**, 3604 (2000).
171. E. N. Jacobsen, *Acc. Chem. Res.*, **33**, 421 (2000).
172. N. B. McKeown, *Phthalocyanine Materials: Synthesis, Structure, and Function*, University Press, Cambridge, 1998.
173. M. Kimura, Y. Sugihara, T. Muto, K. Hanabusa, H. Shirai, N. Kobayashi, *Chem. Eur. J.*, **5**, 3495 (1999).
174. G. R. Newkome, C. D. Weis, *Org. Prep. Proc. Int.*, **28**, 242 (1996).
175. N. E. Schore, "The Pauson-Khand Cycloaddition Reaction for Synthesis of Cyclopentenones," in *Organic Reactions*, Vol. 40, pp. 1–90, Wiley, New York, 1991.
176. A. Dahan, M. Portnoy, *Chem. Commun.*, 2700 (2002).

177. P. Bhyrappa, J. K. Young, J. S. Moore, K. S. Suslick, *J. Am. Chem. Soc.*, **118**, 5708 (1996).
178. P. Bhyrappa, J. K. Young, J. S. Moore, K. S. Suslick, *J. Mol. Catal. Part A: Chem.*, **113**, 109 (1996).
179. K. S. Suslick, P. Bhyrappa, *J. Inorg. Biochem.*, **67**, 234 (1997).
180. B. R. Cook, T. J. Renert, K. S. Suslick, *J. Am. Chem. Soc.*, **108**, 7281 (1986).
181. K. S. Suslick, B. R. Cook, *J. Chem. Soc., Chem. Commun.*, 200 (1987).
182. J. Bu, Z. M. A. Judeh, C. B. Ching, S. Kawi, *Catal. Lett.*, **85**, 183 (2003).
183. A. Gilbert, J. Baggot, *Essentials of Molecular Photochemistry*, Blackwell, Oxford, 1991.
184. B. Valeur, *Molecular Fluorescence*, Wiley-VCH, Weinheim, Germany, 2002.
185. R. Ballardini, V. Balzani, A. Credi, M. T. Gandolfi, M. Venturi, *Acc. Chem. Res.*, **34**, 445 (2001).
186. V. Balzani, A. Juris, M. Venturi, S. Campagna, S. Serroni, *Chem. Rev.*, **96**, 759 (1996).
187. V. Balzani, S. Campagna, G. Denti, A. Juris, S. Serroni, M. Venturi, *Acc. Chem. Res.*, **31**, 26 (1998).
188. V. Balzani, A. Credi, F. M. Raymo, J. F. Stoddart, *Angew. Chem. Int. Ed.*, **39**, 3348 (2000).
189. V. Balzani, A. Credi, *Chem. Rec.*, **1**, 422 (2001).
190. V. Balzani, F. Vögtle, *C. R. Chimie*, **6**, 867 (2003).
191. P. Ceroni, G. Bergamini, F. Marchioni, V. Balzani, *Prog. Polym. Sci.*, **30**, 453 (2005).
192. S. Serroni, A. Juris, M. Venturi, S. Campagna, I. R. Resino, G. Denti, A. Credi, V. Balzani, *J. Mater. Chem.*, **7**, 1227 (1997).
193. F. Puntoriero, S. Serroni, A. Licciardello, M. Venturi, A. Juris, V. Ricevuto, S. Campagna, *J. Chem. Soc., Dalton Trans.*, 1035 (2001).
194. M. Sommovigo, G. Denti, S. Serroni, S. Campagna, C. Mingazzini, C. Mariotti, A. Juris, *Inorg. Chem.*, **40**, 3318 (2001).
195. S. Campagna, S. Serroni, S. Bodige, F. M. MacDonnell, *Inorg. Chem.*, **38**, 692 (1999).
196. D. ben-Avraham, L. S. Schulman, S. H. Bossmann, C. Turro, N. J. Turro, *J. Phys. Chem. Part B*, **102**, 5088 (1998).
197. L. S. Schulman, S. H. Bossmann, N. J. Turro, *J. Phys. Chem.*, **99**, 9283 (1995).
198. M. Kawa, J. M. J. Fréchet, *Thin Solid Films*, **331**, 259 (1998).
199. M. Kawa, J. M. J. Fréchet, *Chem. Mater.*, **10**, 286 (1998).
200. T. Seto, M. Kawa, K. Sugiyama, M. Nomura, *J. Synch. Rad.*, **8**, 710 (2001).
201. M. Kawa, K. Motoda, *Kobunshi Ronbunshu*, **57**, 855 (2000).
202. M. Kawa, T. Takahagi, *Chem. Mater.*, **16**, 2282 (2004).
203. T. Förster, *Ann. Phys.*, **2**, 55 (1948).
204. S.-C. Lo, N. A. H. Male, J. P. J. Markham, S. W. Magennis, P. L. Burn, O. V. Salata, I. D. W. Samuel, *Adv. Mater.*, **14**, 975 (2002).
205. T. Gunnlaugsson, J. P. Leonard, *Chem. Commun.*, 2425 (2003).
206. T. Gunnlaugsson, J. P. Leonard, K. Sénéchal, A. J. Harte, *J. Am. Chem. Soc.*, **125**, 12062 (2003).
207. P. Atkinson, Y. Bretonniere, D. Parker, *Chem. Commun.*, 438 (2004).
208. M. H. V. Werts, R. H. Woudenberg, P. G. Emmerink, R. van Gassel, J. W. Hofstraat, J. W. Verhoeven, *Angew. Chem. Int. Ed.*, **39**, 4542 (2000).
209. F. Vögtle, M. Gorka, V. Vicinelli, P. Ceroni, M. Maestri, V. Balzani, *Chem. Phys. Chem.*, **3**, 769 (2002).
210. P. Ceroni, V. Vicinelli, M. Maestri, V. Balzani, S.-K. Lee, J. van Heyst, M. Gorka, F. Vögtle, *J. Organomet. Chem.*, **689**, 4375 (2004).
211. U. Hahn, M. Gorka, F. Vögtle, V. Vicinelli, P. Ceroni, M. Maestri, V. Balzani, *Angew. Chem. Int. Ed.*, **41**, 3595 (2002).
212. V. Vicinelli, P. Ceroni, V. Balzani, M. Gorka, F. Vögtle, *J. Am. Chem. Soc.*, **124**, 6461 (2002).
213. I. Lukes, J. Kotek, P. Vojtísek, P. Hermann, *Coord. Chem. Rev.*, **216–217**, 287 (2001).
214. B. P. Hay, R. D. Hancock, *Coord. Chem. Rev.*, **212**, 61 (2001).

215. C. Saudan, P. Ceroni, V. Vicinelli, M. Maestri, V. Balzani, M. Gorka, S.-K. Lee, J. van Heyst, F. Vögtle, *Dalton Trans.*, 1597 (2004).
216. C. Saudan, V. Balzani, P. Ceroni, M. Gorka, M. Maestri, V. Vicinelli, F. Vögtle, *Tetrahedron*, **59**, 3845 (2003).
217. C. Saudan, V. Balzani, M. Gorka, S.-K. Lee, M. Maestri, V. Vicinelli, F. Vögtle, *J. Am. Chem. Soc.*, **125**, 4424 (2003).
218. C. Saudan, V. Balzani, M. Gorka, S.-K. Lee, J. van Heyst, M. Maestri, P. Ceroni, V. Vicinelli, F. Vögtle, *Chem. Eur. J.*, **10**, 899 (2004).
219. S. H. F. Chong, S. C. F. Lam, V. W. W. Yam, N. Zhu, K.-K. Cheung, S. Fathallah, K. Costuas, J.-F. Halet, *Organometallics*, **23**, 4924 (2004).
220. M. Fischer, F. Vögtle, *Angew. Chem. Int. Ed.*, **38**, 885 (1999).
221. P. Anzenbacher, Jr., D. S. Tyson, K. Jursíková, F. N. Castellano, *J. Am. Chem. Soc.*, **124**, 6232 (2002).
222. B. W. Koo, C. K. Song, C. Kim, *Sens. Actuators, B*, **77**, 432 (2001).
223. J. M. Vanderkooi, G. Maniara, T. J. Green, D. F. Wilson, *J. Biol. Chem.*, **262**, 5476 (1987).
224. W. L. Rumsey, J. M. Vanderkooi, D. F. Wilson, *Science*, **241**, 1649 (1988).
225. S. A. Vinogradov, L.-A. Lo, W. T. Jenkis, S. M. Evans, C. Koch, D. F. Wilson, *Biophys. J.*, **70**, 1609 (1996).
226. R. P. Briñas, T. Troxler, R. M. Hochstrasser, S. Vinogradov, *J. Am. Chem. Soc.*, **127**, 11851 (2005).
227. S. A. Vinogradov, E. Kim, D. F. Wilson, *Proc. SPIE—The International Society for Optical Engineering*, **4626**, 193 (2002).
228. I. B. Rietveld, E. Kim, S. A. Vinogradov, *Tetrahedron*, **59**, 3821 (2003).
229. R. P. Brinas, T. Troxler, R. M. Hochstrasser, S. A. Vinogradov, *J. Am. Chem. Soc.*, **127**, 11851 (2005).
230. M. Albrecht, M. Schlupp, J. Bargon, G. van Koten, *Chem. Commun.*, 1874 (2001).
231. M. Albrecht, R. A. Gossage, M. Lutz, A. L. Spek, G. van Koten, *Chem. Eur. J.*, **6**, 1431 (2000).
232. M. Albrecht, R. A. Gossage, A. L. Spek, G. van Koten, *Chem. Commun.*, 1003 (1998).
233. M. Albrecht, N. J. Hovestad, J. Boersma, G. van Koten, *Chem. Eur. J.*, **7**, 1289 (2001).
234. M. D. Ward, D. A. Buttry, *Science*, **249**, 1007 (1990).
235. U. Schramm, C. E. O. Roesky, S. Winter, T. Rechenbach, P. Boeker, P. S. Lammers, E. Weber, J. Bargon, *Sens. Actuators, B*, **57**, 233 (1999).
236. M.-C. Daniel, J. Ruiz, S. Nlate, J. Palumbo, D. Astruc, J.-C. Blais, *Chem. Commum.*, 2000 (2001).
237. R. Wang, J. Yang, Z. Zheng, M. D. Carducci, J. Jiao, S. Seraphin, *Angew. Chem. Int. Ed.*, **40**, 549 (2001).
238. M.-C. Daniel, J. Ruiz, S. Nlate, J.-C. Blais, D. Astruc, *J. Am. Chem. Soc.*, **125**, 2617 (2003).
239. M.-K. Kim, Y.-M. Jeon, W. S. Jeon, H.-J. Kim, S. G. Hong, C. G. Park, K. Kim, *Chem. Commun.*, 667 (2001).
240. C. Valério, J.-L. Fillaut, J. Ruiz, J. Guittard, J.-C. Blais, D. Astruc, *J. Am. Chem. Soc.*, **119**, 2588 (1997).
241. E. C. Wiener, V. V. Narayanan, "Magnetic Resonance Imaging Contrast Agents: Theory and the Role of Dendrimers," in *Advances in Dendritic Macromolecules*, G. R. Newkome, Ed., pp. 129–247, Elsevier Science, Kidlington, Oxford, UK, 2002.
242. E. C. Wiener, M. W. Brechbiel, H. Brothers, R. L. Magin, O. A. Gansow, D. A. Tomalia, P. C. Lauterbur, *Magn. Reson. Med.*, **31**, 1 (1994).
243. J. P. André, C. F. G. C. Geraldes, J. A. Martins, A. E. Merbach, M. I. M. Prata, A. C. Santos, J. J. P. de Lima, É. Tóth, *Chem. Eur. J.*, **10**, 5804 (2004).
244. S. Laus, A. Sour, R. Ruloff, É. Tóth, A. E. Merbach, *Chem. Eur. J.*, **11**, 3064 (2005).
245. J. B. Livramento, É. Tóth, A. Sour, A. Borel, A. E. Merbach, R. Ruloff, *Angew. Chem. Int. Ed.*, **44**, 1480 (2005).
246. A. Bianchi, L. Calabi, F. Corana, S. Fontana, P. Losi, A. Maiocchi, L. Paleari, B. Valtancoli, *Coord. Chem. Rev.*, **204**, 309 (2000).

247. B. Alonso, D. Astruc, J.-C. Blais, S. Nlate, S. Rigaut, J. Ruiz, V. Sartor, C. Valério, *C. R. Acad. Sci. Paris II,Chemie*, **4**, 173 (2001).

248. S. Nlate, J. Ruiz, V. Sartor, R. Navarro, J.-C. Blais, D. Astruc, *Chem. Eur. J.*, **6**, 2544 (2000).

249. J. Ruiz, C. Pradet, F. Varret, D. Astruc, *Chem. Commun.*, 1108 (2002).

250. V. Balzani, A. Credi, M. Venturi, *Chem. Phys. Chem.*, **3**, 49 (2003).

251. R. L. Carroll, C. B. Gorman, *Angew. Chem. Int. Ed.*, **41**, 4379 (2002).

252. P. J. Low, *Dalton Trans.* **2005**, 2821–2824.

253. G. Maruccio, R. Cingolani, R. Rinaldi, *J. Mater. Chem.*, **14**, 542 (2004).

254. F. M. Raymo, M. Tomasulo, *Chem. Soc. Rev.*, **34**, 327 (2005).

255. A. P. H. J. Schenning, P. Jonkheijm, F. J. M. Hoeben, J. van Herrikhuyzen, S. C. J. Meskers, E. W. Meijer, L. M. Herz, C. Daniel, C. Silva, R. T. Phillips, R. H. Friend, D. Beljonne, A. Miura, S. De Feyter, M. Zdanowska, H. Uji-i, F. C. De Schryver, Z. Chen, F. Würthner, M. Mas-Torrent, D. den Boer, M. Durkut, P. Hadley, Synth. Met., **147**, 43 (2004).

256. J. M. Tour, *Acc. Chem. Res.*, **33**, 791 (2000).

257. S. Nlate, J. Ruiz, J.-C. Blais, D. Astruc, *Chem. Commun.*, 417 (2000).

258. I. J. Macdonald, T. J. Dougherty, *J. Porphyrins Phthalocyanines*, **5**, 105 (2001).

259. J. M. McCaughin, Jr., *Drug Aging*, **15**, 49 (1999).

260. R. B. Veenhuizen, M. C. Ruevekamp, H. Oppelaar, T. J. M. Helmerhorst, P. Kenemans, F. A. Stewart, *Int. J. Cancer*, **73**, 230 (1997).

261. S. J. Madsen, C.-H. Sun, B. J. Tromberg, V. P. Wallance, H. Hischberg, *Photochem. Photobiol.*, **72**, 128 (2000).

262. S. P. Songca, B. Mbatha, *J. Pharm. Pharmacol.*, **52**, 1361 (2000).

263. N. Bousset, V. Vonarx, S. Eleouet, J. Carre, L. Bourre, Y. Lajat, T. Patrice, *Res. Exp. Med.*, **199**, 341 (2000).

264. R. Sadamoto, N. Tomioka, T. Aida, *J. Am. Chem. Soc.* **1996**, *118,* 3978–3979.

265. D.-L. Jiang, T. Aida, *J. Am. Chem. Soc.*, **120**, 10895 (1998).

266. N. Nishiyama, H. R. Stapert, G.-D. Zhang, D. Takasu, D.-L. Jiang, T. Nagano, T. Aida, K. Kataoka, *Bioconj. Chem.*, **14**, 58 (2003).

CHAPTER 11

Metallodendritic Iron Complexes: Design, Catalysis, and Molecular Recognition

DIDIER ASTRUC

Université Bordeaux I, Talence, France

I. INTRODUCTION

Since the concept of hyperbranched polymers, the fathers of dendrimers, by Flory in the 1940s,[1] the polymer chemistry and physics of branched systems has been extensively developed, especially with the advent of well-defined dendrimers in the 1980s.[2,3] Metallodendrimers were introduced by Balzani's group with antenna-type systems[4] and by our group with organometallic redox-stable metallocene-containing dendrimers designed for redox catalysis.[5] Metallodendrimers can indeed be a unique nanodevice for a number of key functions, such as catalysis,[6,7] molecular recognition,[8] sensing,[8] and molecular batteries.[9] In the present chapter, we summarize our strategies toward the design and synthesis of metallodendrimers containing redox organometallic systems and their applications to catalysis and molecular recognition.

II. ORGANO-IRON SYNTHESES OF DENDRITIC CORES AND DENDRONS

In the robust, very easily accessible cationic complexes $[FeCp(arene)][PF_6]$[10] (Cp = h5-cyclopentadienyl), the benzylic protons are more acidic than in the free arene because of the electron-withdrawing character of the 12-electron $CpFe^+$ moiety. For instance, $[FeCp(C_6Me_6)][PF_6]$ is more acidic by 15 pK_a units ($pK_a = 28$ in DMSO) (DMSO = dimethyl sulfoxide) than in the corresponding free arene ($pK_a = 43$ in DMSO). As a result, these complexes are much more easily deprotonated than the free arene.[11] This key proton-reservoir property led us to synthesize stars and dendrimers in an easy way.[12] Indeed, reaction of $[FeCp(C_6Me_6)][PF_6]$, with excess KOH (or *t*-BuOK) in THF (THF = tetrahydrofuran) or DME (DME = 1,2-dimethoxyethane) and excess methyl iodide, alkyl iodide, allyl bromide, or benzylbromide results in the one-pot hexasubstitution (Scheme 11.1a).[5,13,14] With allyl bromide (or iodide) in DME, the hexaallylated

Frontiers in Transition Metal-Containing Polymers, edited by Alaa S. Abd-El-Aziz and Ian Manners.

complex also has been easily isolated and its X-ray crystal structure has been determined. With alkyliodides, the reaction using *t*-BuOK only leads to dehalogenation of the alkyl iodide giving the terminal olefin. Thus, one must use KOH, and the reactions with various alkyl iodides (even long-chain ones) were shown to work very well with this reagent to give the hexaalkylated Fe(II)-centered complexes. The hexaalkylation was also performed with alkyl iodides containing functional groups at the alkyl chain termini. For instance, 1-ferrocenylbutyliodide reacts nicely to give the hexaferrocene star containing the $CpFe^+$ center. The reaction with excess benzylbromide, *p*-alkoxybenzylbromide,[15] or *p*-bromobenzylbromide only gives the hexabenzylated, hexa-*p*-alkoxybenzylated, or hexa-*p*-bromobenzylated complex as the ultimate reaction product. Cleavage of the methyl group in the *p*-methoxybenzyl derivatives synthesized in this way yields the hexaphenolate stars that could be combined with halogen-containing organometallic compounds.[5,16]

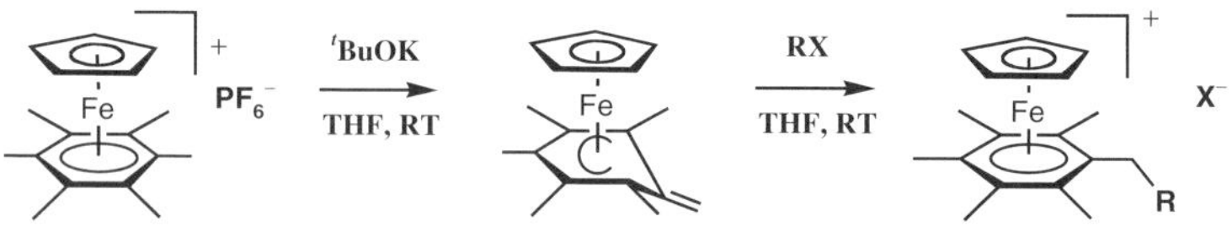

RX = CH_3I, $PhCH_2Br$, $SiMe_3Cl$, PPh_2Cl, $FeCp(CO)_2Br$ (also CO_2 and metal carbonyls)

(a)

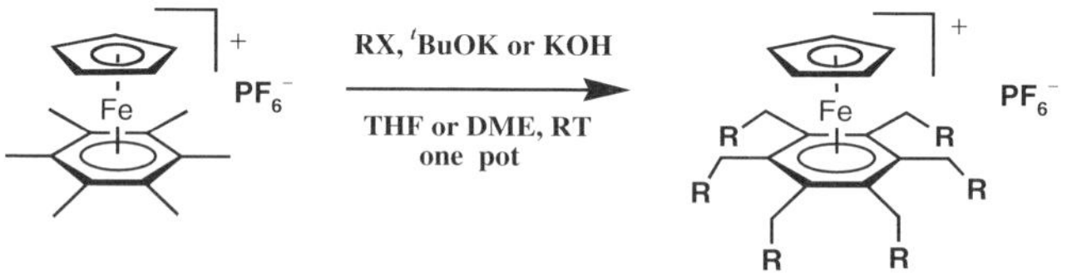

R = CH_3, $(CH_2)_nCH_3$, $(CH_2)_4Fc$ (X = I), CH_2Ph, *p*-CH_2PhOR', $CH_2CH{=}CH_2$, (X = Br)

(b)

Scheme 11.1 Deprotonation of $[FeCp(\eta^6\text{-}C_6Me_6)][PF_6]$ followed by reactions with (a) electrophiles and (b) one-pot hexafunctionalization of this complex under ambient conditions. The reaction in (a) illustrates the mechanism of the one in (b).

It is remarkable that the allyl group (as allyl bromide or iodide) is the only one leading to complete double branching of the C_6Me_6 complex. $CpFe^+$-induced dodecaallylation of C_6Me_6 indeed gives the extremely bulky dodeca-allylation product that can be reached when the reaction is prolonged for two weeks at 40°C. The chains are blocked in a directionality that cannot convert into its enantiomer and makes the metal complex chiral (Scheme 11.2). Both the hexa- and dodeca-allylation reactions are well controlled.[12]

Alkynyl halides cannot be used in the $CpFe^+$-induced hexafunctionalization reaction, but alkynyl substituents can be introduced from the hexa-alkene derivative by bromination followed by dehydrohalogenation of the dodecabromo compound (Scheme 11.3).[17] The hexa-alkene is also an excellent starting point for further syntheses, especially using hydroelementation reactions. Hydrosilylation reactions catalyzed by Speir's reagent, which led to long-chain hexasilanes and hydrometallations, were also achieved using $[ZrCp_2(H)(Cl)]$. The hexazirconium compound obtained is an intermediate for the synthesis of the hexa-iodo derivative. One of the most useful hydroelementation reactions of the hexabutenyl derivatives is the hydroboration leading to the hexaborane. The latter is oxidized to the hexa-ol using H_2O_2 under basic condition. This chemistry can be carried out on the iron complex or alternatively on the free hexa-alkene that may be liberated from the metal by photolysis in CH_2Cl_2 or MeCN using visible

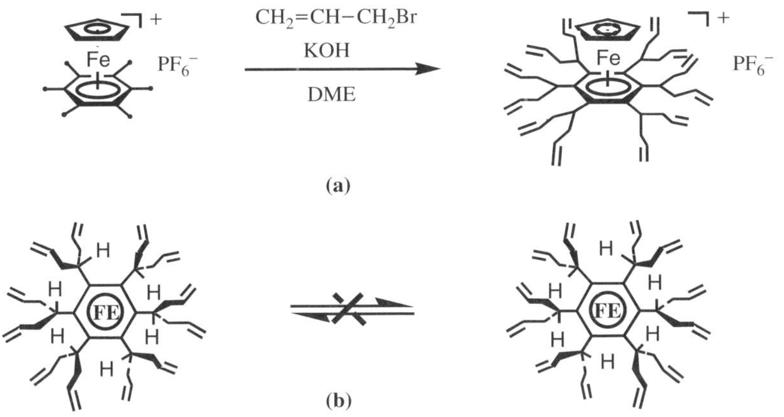

FE =η^{5}-$C_5H_5Fe^+$

Scheme 11.2 One-pot dodecaallylation of hexamethylbenzene (a) induced by the $CpFe^+$ group and (b) hindered rotation of the branches.

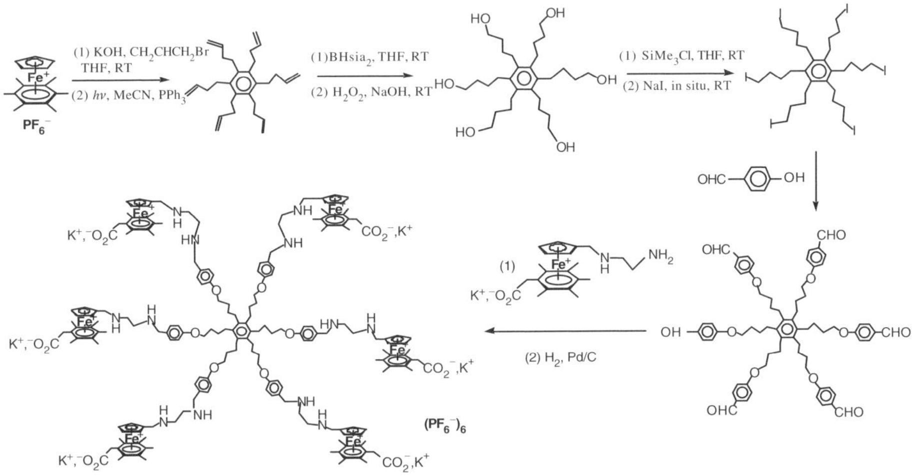

Scheme 11.3 $CpFe^+$-induced hexaallylation of C_6Me_6 and subsequent hexafunctionalization of the aromatic stars with the heterodifunctional, water-soluble organometallic redox catalyst (bottom) for the cathodic reduction of nitrates and nitrites to ammonia in water).[18]

light. The polyol stars and dendrimers can be transformed into mesilates and iodo derivatives that are useful for further functionalization. The hexa-ol is indeed the best source of hexa-iodo derivative either using HI in acetic acid, or even better by trimethylsilylation using $SiMe_3Cl$ followed by iodination using NaI. Williamson coupling reactions between the hexa-ol and 4-bromomethylpyridine or -polypyridine led to hexapyridine and hexapolypyridine, and to their ruthenium complexes.[11] This hexa-iodo star was condensed with *p*-hydroxybenzaldehyde to give a hexabenzaldehyde star, which could further react with substrates bearing a primary amino group. Indeed, this reaction yielded a water-soluble hexametallic redox catalyst that was stable and active in the electroreduction of nitrate and nitrite to ammonia in a basic aqueous solution.[18]

If the hexafunctionalization of hexamethylbenzene leads to stars, the octafunctionalization of durene leads to dendritic cores. The first of these octaalkylation reactions was reported as early as 1982, and led to a primitive dendritic core containing a metal-sandwich unit. Thus, as the hexafunctionalization, this reaction is very specific. Two hydrogen atoms in each methyl group are now replaced by two methyl, allyl, or benzyl groups (Scheme 11.4).[19] Applications to the synthesis of dendrimers containing 8 or 24 redox-active groups have recently been reported.[20]

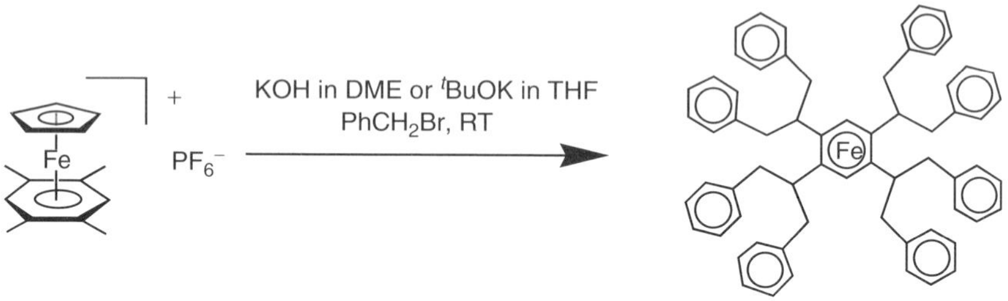

Scheme 11.4 Octabenzylation of durene induced by the $CpFe^+$ group (right: Fe = η^5-C_5H_5, PF_6^-).

Double branching, that is, replacement of two out of three hydrogen atoms by two groups on each methyl substituent of an aromatic ligand coordinated to an activating cationic group CpM^+, in an 18-electron complex is also easily obtained in the pentamethylcyclopentadienyl ligand (in pentamethyl-cobaltocenium and in penta- and decamethylrhodocenium). The interconversion of the two directionalities of decafunctionalized ligands coordinated to $CpCo^+$ or $CpRh^+$ could be observed by ^{1}H NMR (NMR = nuclear magnetic resonance) for the decaisopropyl- and decaisopentyl-cyclopentadienyl cobalt and rhodium complexes (Scheme 11.5).[19]

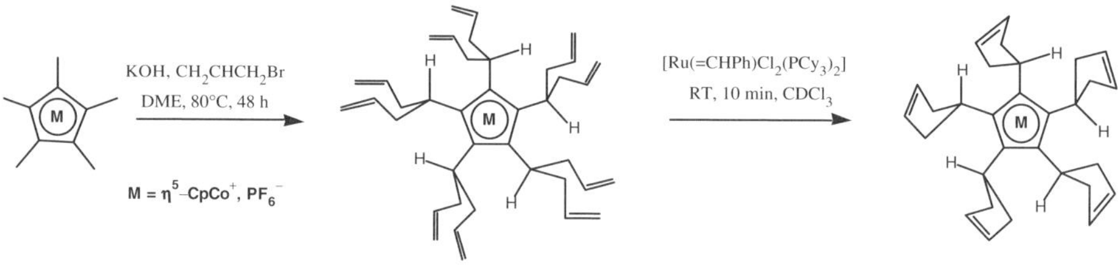

Scheme 11.5 Decaallylation of 1,2,3,4,5-pentamethylcobaltocenium in a one-pot reaction consisting in 10 deprotonation-allylation sequences (steric constraints inhibit further reaction, and the 10 groups introduced are self-organized according to a single directionality) and follow-up RCM (RCM = ring-closing metathesis) of the deca-allylated complex.[21]

In all the preceding examples, the polybranching reaction of arene ligands was limited by the steric bulk. In the toluene and mesitylene ligands, the deprotonation-allylation reactions are no longer restricted by the neighborhood of other alkyl groups. All the benzylic protons, that is, three protons *per* benzylic group, can be replaced by methyl or allyl groups in the one-pot iterative methylation or allylation reactions. Thus, the toluene complex can be triallylated, and the resulting tripod can be disymetrized by stoichiometric or catalytic reaction with transition metals shown in Scheme 11.6. The metathesis reaction, in particular, is complete in 5 minutes at room temperature using the first-generation Grubb's catalyst $[Ru(=CHPh)Cl_2(PCy_2)_2]$ with many polyallylated complexes $[FeCp(arene)]^+$ described earlier in this section as well as the decaallylated cobalt complex. The reaction is very selective, and terminal double bonds remain unreacted using this catalyst at room temperature.[21]

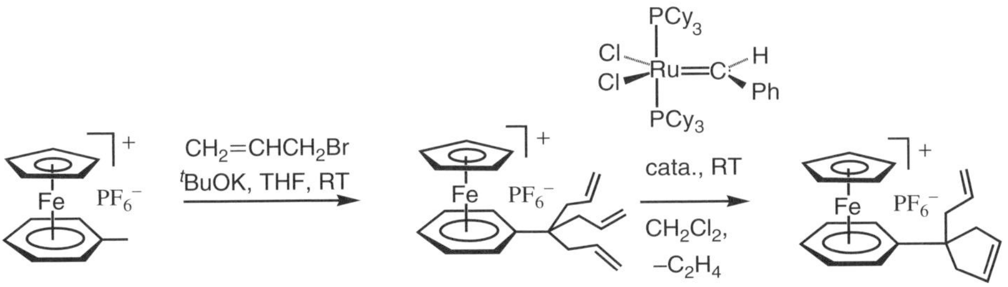

Scheme 11.6 $CpFe^+$-induced triallylation of toluene followed by ring-closing olefin metathesis.

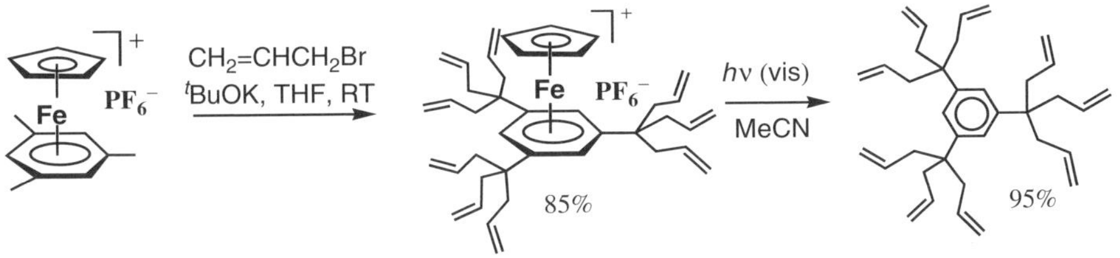

Scheme 11.7 $CpFe^+$-induced nonaallylation of mesitylene followed by photodecomplexation using visible light.

Since these reactions are carried out smoothly at room temperature in the presence of excess KOH and allyl bromide (Scheme 11.7), the mesitylene complex can be nonaallylated. The nonaallyl complex was photolyzed using visible light to remove the metal group $CpFe^+$, then hydroborated using 9-BBN, and the nonaborane was oxidized using H_2O_2/OH^- to the nona-ol.[5] Since the triple branching reaction is very straightforward, we sought a more sophisticated version compatible with a functional group in the para position of the tripod in order to open the access to a functional dendron. Serendipitously, we found that KOH or *t*-BuOK easily cleaved the iron complexes of aromatic ethers under very mild conditions. The activating $CpFe^+$ group again induces this reaction, which is very general for a variety of aromatic ether complexes. Since this cleavage reaction is carried out with the same reagent and solvent as the one used in the trialkylation reaction (ideally *t*-BuOK in THF), we have attempted to perform both reactions in a well-defined order (triallylation before ether cleavage) in a one-pot reaction. Indeed, this works out well, and the CpFe(II) complex of the phenol tripod was made in 50% yield in this way. This complex can be photolyzed in the usual way using visible light, which yields the free phenol tripod. However, we have also further investigated the possibility of obtaining the cleavage of the arene ligand in situ at the end of the phenol tripod construction; *t*-BuOK is a reductant when it cannot perform other reactions. Since the two important reactions are over, there follows the third role of *t*-BuOK follows: single-electron reductant.[22]

Reasoning in this way turned out to be correct: the cleavage of the arene intervenes rapidly at the 19-electron stage, because 19-electron complexes of this kind are not stable with a heteroatom located in the exocyclic position (most probably because the heteroatom coordinates to the metal from the labile 19-electron structure). After optimizing the reaction conditions, a 60%-yield of free phenol dendron from the ethoxytoluene complex could be obtained, and this reaction is now currently used in our laboratory to synthesize this very useful dendron as a starting material (Scheme 11.8).

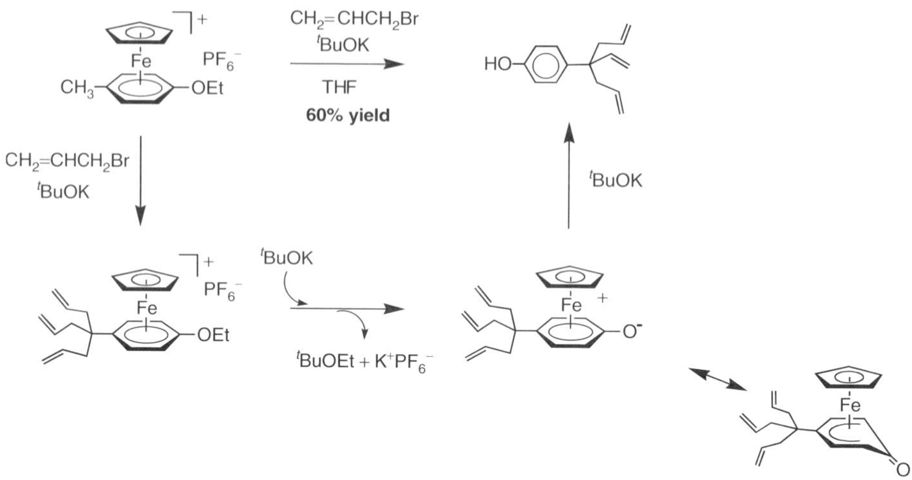

Scheme 11.8 One-pot 60%-yield synthesis of a phenoltriallyl dendron involving 8 steps.

III. CONSTRUCTION OF DENDRIMERS USING THE DENDRITIC CORES AND DENDRONIC BRICKS

This phenoltriallyl dendron has been functionalized at both the phenolic and allylic positions. For instance, the dendron can be bound, after suitable molecular engineering, to the branches of a phenolic-protected dendron (convergent construction), onto stars and dendritic cores (divergent construction), nonaparticles, surfaces, and polymers (Scheme 11.9). An example is

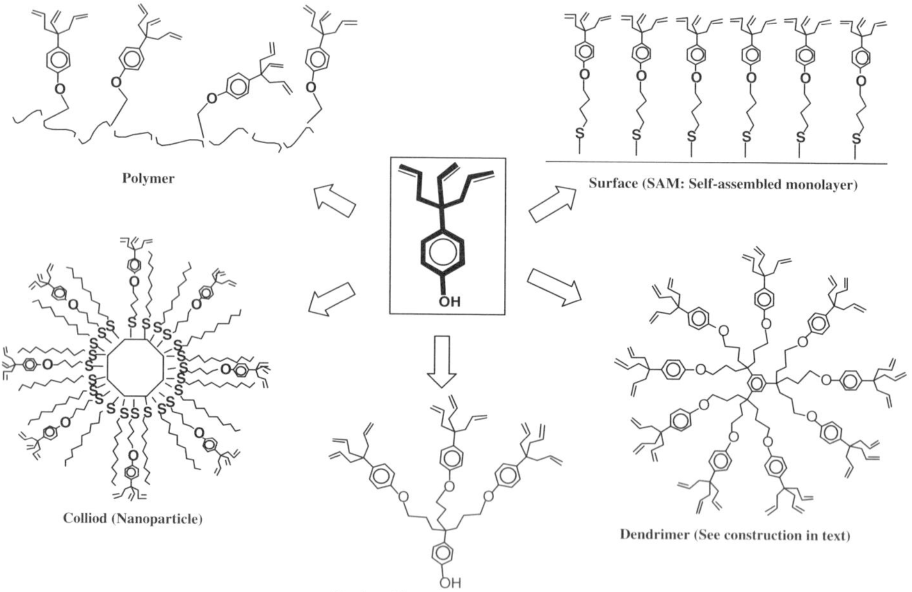

Scheme 11.9 Example of the linkage of the phenoltriallyl dendron to various nanostructures.

provided by the $CpFe^+$-induced hexafunctionalization by a phenol-nonaallyl dendron (prepared according to such a convergent synthesis) that was functionalized in the phenolic position by a tail terminated by a benzylbromide group. This type of strategy allows direct access to large dendrimers by simply using the $CpFe^+$-induced hexafunctionalization reaction that gives hexabranch stars with linear organic halides (Scheme 11.10).[23]

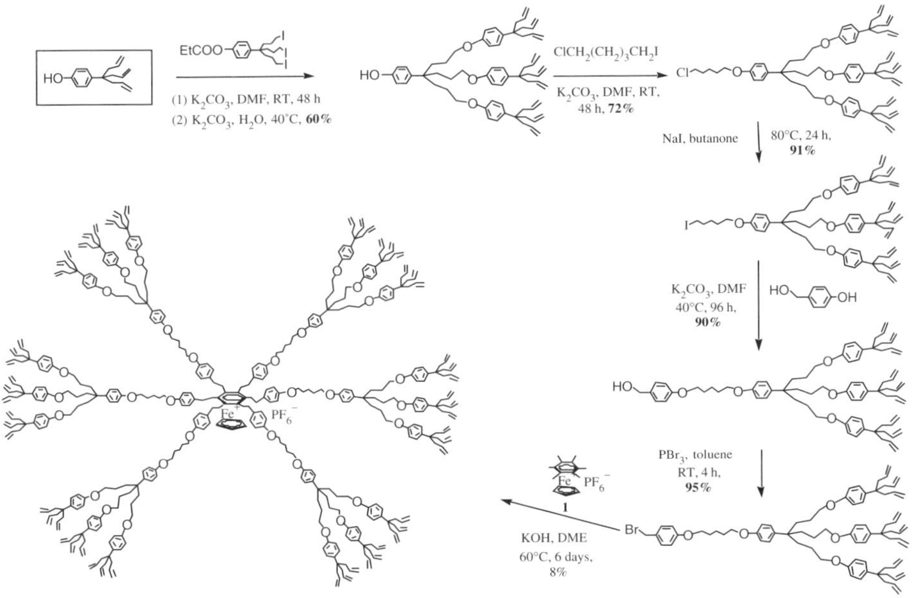

Scheme 11.10 $CpFe^+$-induced hexabenzylation of C_6Me_6 applied to direct convergent dendrimer synthesis of a 54-allyl dendrimer.

The introduction of dendrons onto a core, that is, the convergent dendrimer construction just illustrated rapidly meets limits that are indeed encountered in synthetic Scheme 11.10. A much more powerful construction strategy, however, is the divergent method. In our case, it suffices to use the phenoltriallyl dendronic brick to functionalize a dendrimer at each generation. This strategy has allowed us to synthesize dendrimers of generation 0, 1, 2, and 3 with, respectively, 9 (G_0), 27 (G_1), 81 (G_2), and 243 branches (G_3) (Scheme 11.11 and Figure 11.1). The nonaallyl core (G0) is hydroborated with 9-BBN, and the nonaborane is oxidized to the nonaalcohol that is transformed into the nonamesylate. The latter reacts further with the phenoltriallyl dendron to provide the G1-dendrimer with 27-allyl termini, and so on. This type of construction was persued until G3 (243 branches), although the molecular peak could not be detected for that third generation, presumably due to steric crowding. Matrix-assisted laser desorption ionization–time of flight. The (MALDI–TOF) mass spectrum of the 27-allyl dendrimer shows the molecular peak with only traces of side product. That of the 81-allyl shows a dominant molecular peak, but also important side products resulting from incomplete branching. That of the 243-allyl could not be obtained, possibly signifying that this dendrimer is polydisperse (correct ^{1}H and ^{13}C NMR spectra were obtained, however, indicating that the ultimate reactions had proceeded to completion).

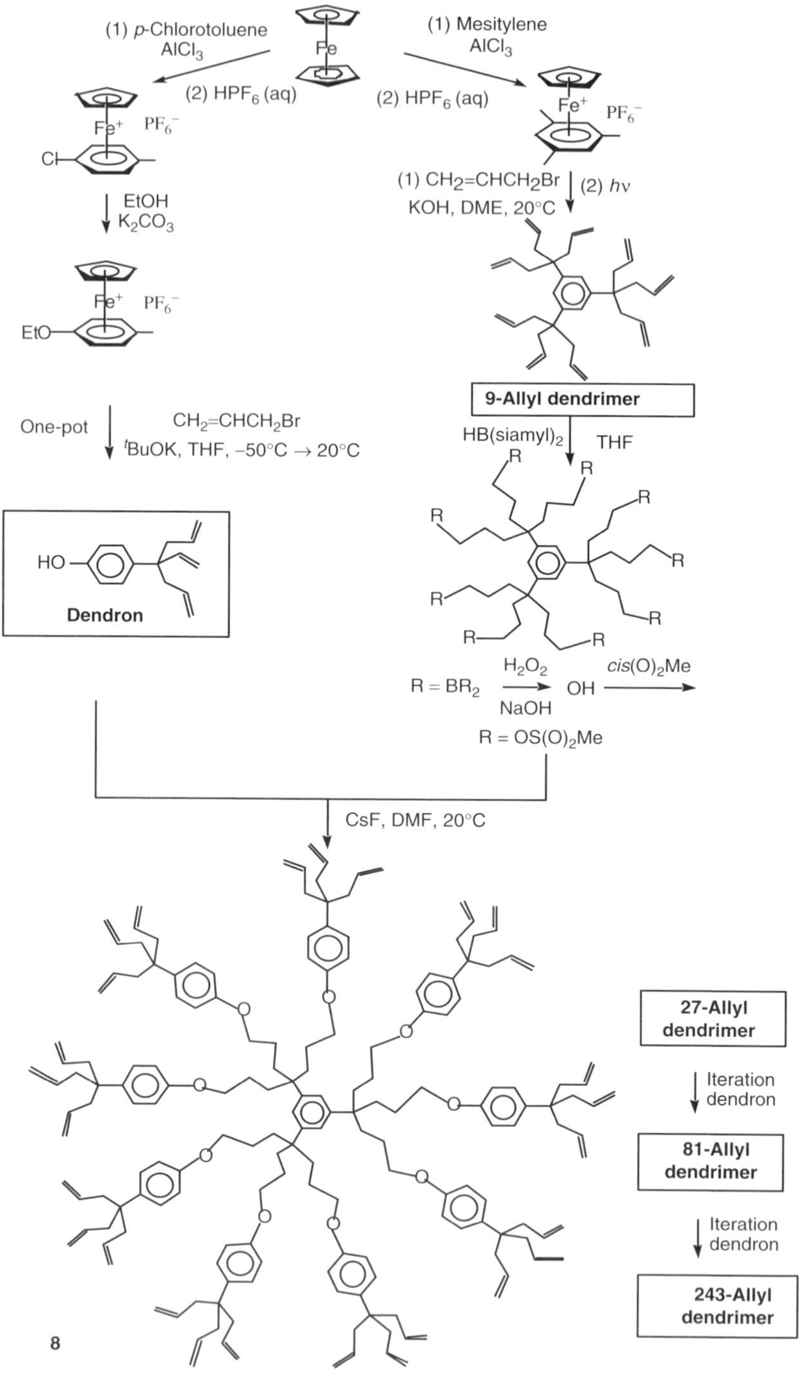

Scheme 11.11 Strategy for the construction of large dendrimers starting from ferrocene.

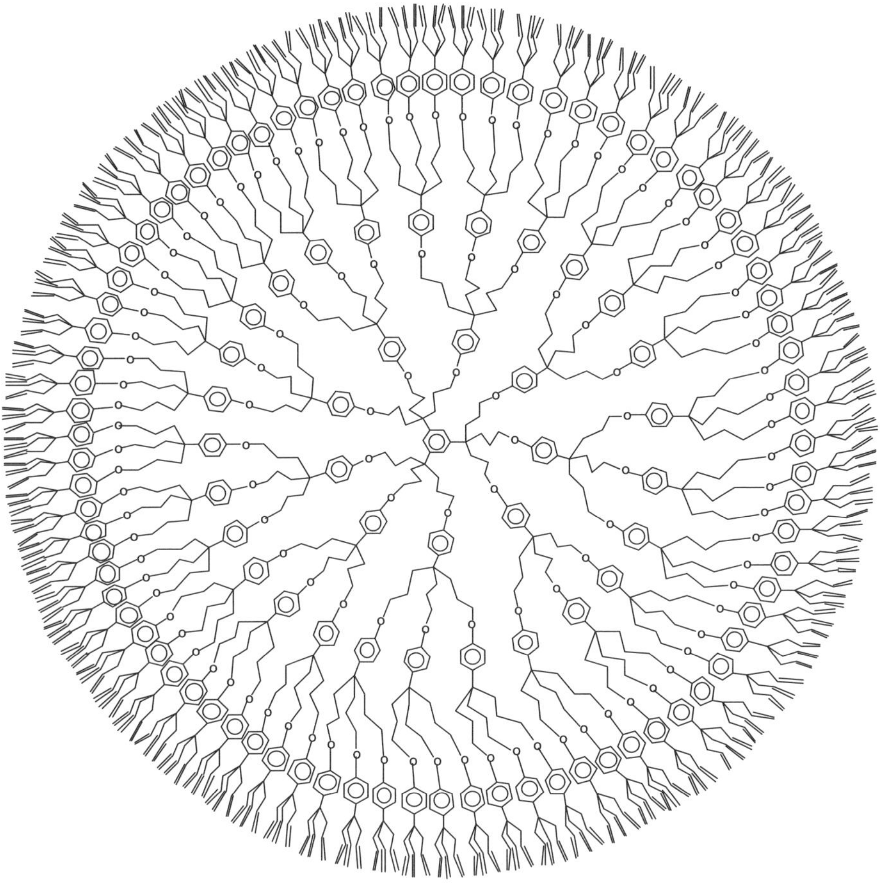

Figure 11.1 243-Allyl dendrimer (third generation, see the construction on Scheme 11.11).

In order to avoid the bulk problem, we decided at this point to switch to the less sterically constraining hydrosilylation series using chloromethyldimethylsilane. This hydrosilylation, carried out at 20–40°C using the Karsted catalyst, turned out to be very regioselective on the terminal carbon atoms, so that the expected isomer was the only one observed by ^{1}H, ^{13}C, and ^{29}Si NMR. It is probable that the bulk provided by the dendrimer is at the origin of this selectivity. The ^{29}Si spectra were particularly simple. The hydrosilylation of the polyallyl dendrimers gave a compound that showed two ^{29}Si NMR signals at 0.73 ppm and 3.8 ppm, and the polyolefin dendrimers showed only one ^{29}Si NMR signal at 3.8 ppm. The dendrimers remained pentane soluble until the ninth generation, and the ^{1}H and ^{29}Si NMR spectra showed the full completion of the reactions at the NMR accuracy (only the two last generations at the periphery are observable by NMR). The construction based on this hydrosilylation is represented on Scheme 11.12. At this time, the MALDI–TOF mass spectra also showed only the molecular peak for G_1-27 allyl (3^3 allyl branches) and a dominant molecular peak for

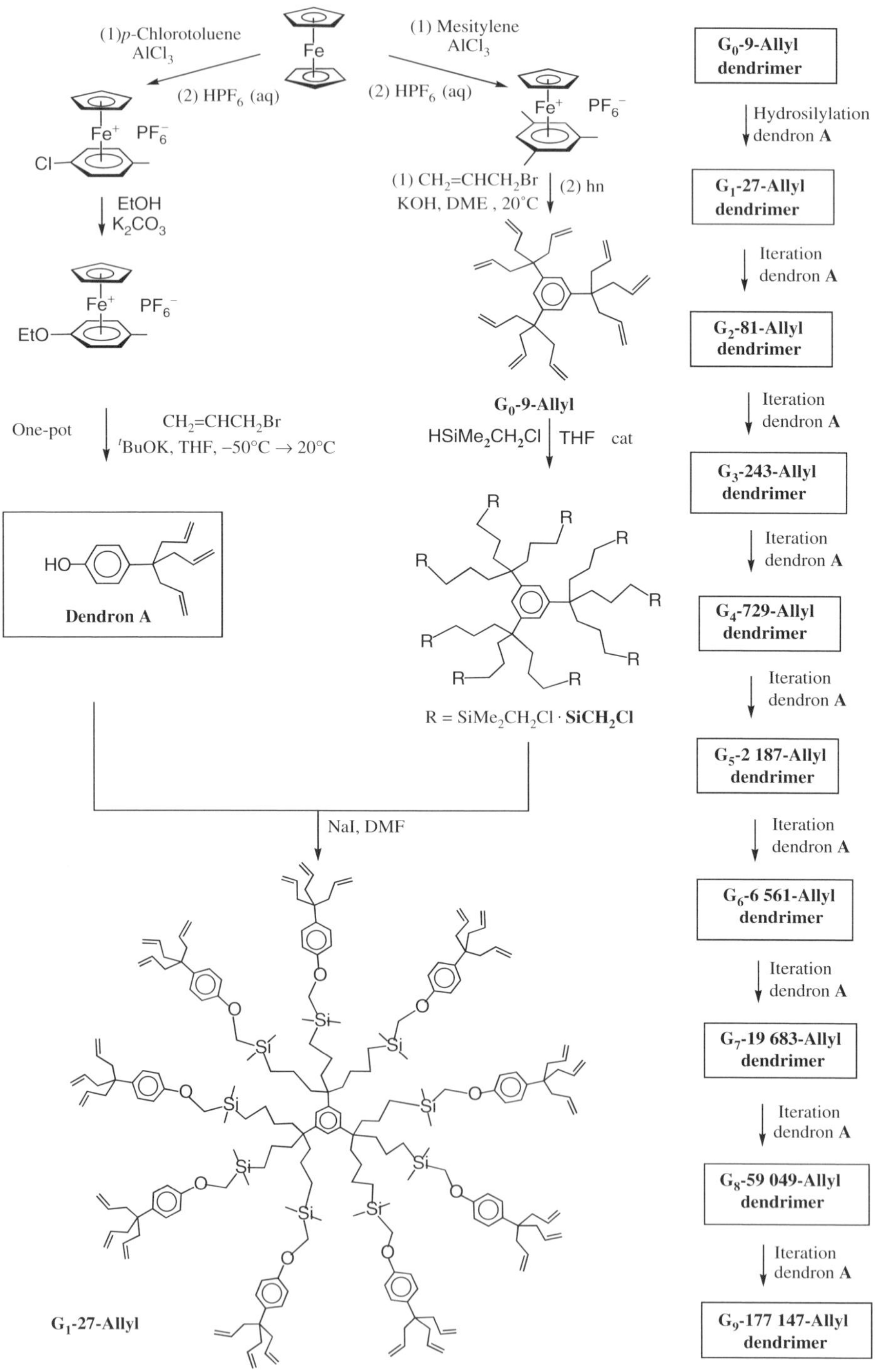

Scheme 11.12 Hydrosilylation strategy for the construction of giant dendrimers.

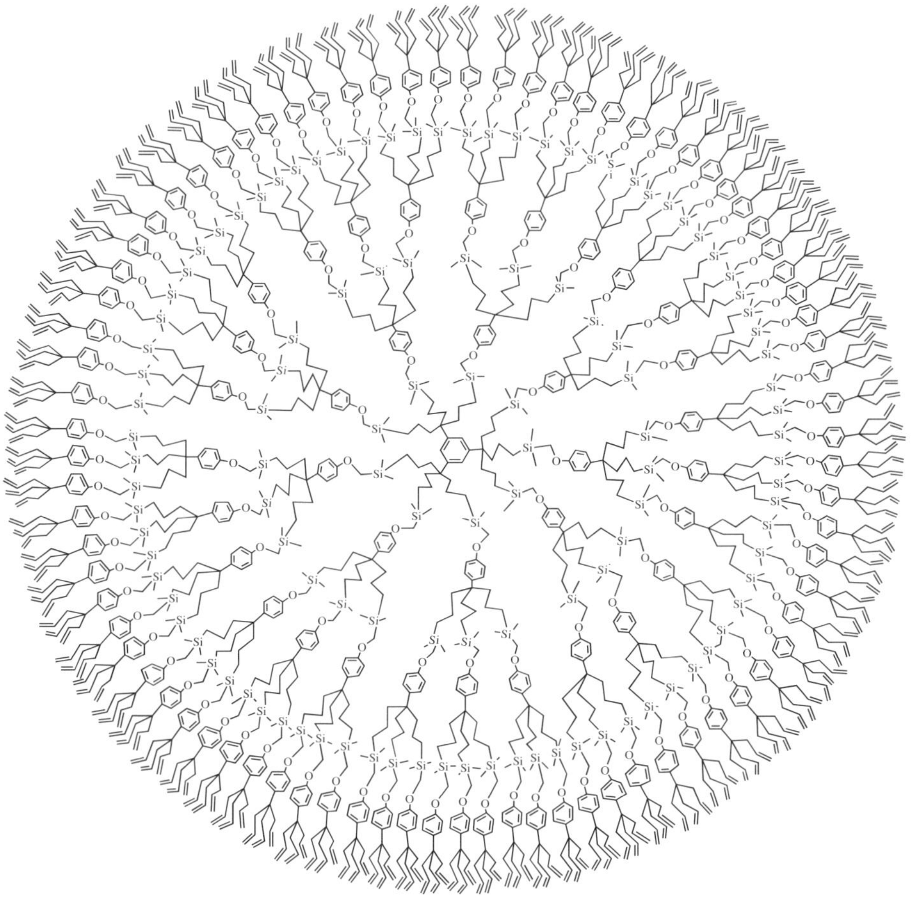

Figure 11.2 Third-generation 243-allyl dendrimer synthesized according to Scheme 11.12.

the G_2-81 allyl dendrimers, but a minor molecular peak was also found for G_3-243 allyl (Figure 11.2).

Moreover, it was possible to continue the construction (G_n has 3^{n+2} allyl branches) and to follow all the reactions by ^{1}H and ^{29}Si NMR until the ninth generation with a theoretical number of branches of $3^{11} = 177,407$. The dendrimers were characterized by a number of techniques besides the multinuclear NMR just indicated. The MALDI–TOF of the G_4-729 allyl shows a massive vanishing near the molecular mass that is not observed as a single peak. The size exclusion chromatograms (SECs) were obtained with low polydispersities (between 1.00 and 1.02) from the G_1-27 allyl to the G_5-2187 allyl. Above this size, the globular shape does not allow overtaking molecular weights larger than 10^5. Some aggregation[20] could be observed by SEC (and other techniques, *vide infra*) when the dendrimer generation increases, and is clear in the SEC of the G_4-729 allyl and the G_5-2187 allyl. Since the dendrimers are organic, they cannot be seen by transmission electron microscopy (TEM), because the samples are lacking heavy atoms. By vaporizing OsO_4 onto the sample in a well-ventilated hood

[*Caution*: OsO_4 is toxic], osmametallocycles are formed with the double bounds at the dendrimer periphery, and the single dendrimers can be observed by TEM. For instance, G_4-729 allyl/OsO_4 was observed in this way, and the diameter matches that obtained using the molecular model. For the largest dendrimer, G_9-177407 allyl, the hydrosilylation could still be carried out and the nucleophilic substitution of chloride by iodide using NaI produced the polyiodo dendrimer G_9-177407 iodo that is visible by TEM due to the heavy iodine atoms. A diameter of 13 nm was found in this way for the G_9-177407 iodo.

One of the most useful microscopy techniques is atomic force microscopy (AFM), which can measure the heights of the dendrimers for all of the generations from the G_1-27-allyl to G_9-177407 allyl.

Monolayers were found for the first four generations, then it appeared that a double layer was observed up to the last one, with a height of 25 nm for the G_9-177407 allyl. The flattening and aggregation of dendrimers recorded by AFM have already been observed in other cases. Until G_5, the heights of the monolayers are slightly lower than those calculated. Finally, we believe that, although the number of defects increases with the generations, the total number of termini in the G_9-177407 allyl is on the order of 10^5.

Although the reaction time increases with generations, it is necessary that the number of back-folding methylene termini increases when the dendrimers become larger than the size corresponding to the "dense-packing limit." We have estimated that in our series this number is around 6000. In the largest dendrimers, the proportion of methylene termini back-folding must be on the order of 90%, although this should be viewed as a dynamic phenomenon. This means that the construction is limited by the volume, not by the surface. The generation limit is thus found much later than the de Gennes "dense-packing limit" if the terminals are small enough and flexible. This finding is indeed true only for dendrimers with small flexible termini.[24] When the size of the termini is larger, the termini do not back fold as was confirmed by calculation,[14] and then the periphery can be severely congested at a much earlier generation.[15]

IV. METALLODENDRIMERS WITH REDOX-STABLE METALLOCENE TERMINALS TOWARD MOLECULAR BATTERIES

The nonaallyl dendritic core could be hydroborated using 9-BBN, then oxidation of the nonaborane in basic medium gave the nona-ol (Scheme 11.13). This compound was the starting point for the introduction of metallocenyl fragments bearing a fluoro substituent on the arene ligand by nucleophilic substitution. With a chlorocarbonyl group on the Cp ring, the metallocene is bound to the dendrimer by amide linkage subsequent to Michael reaction of the nona-ol with acrylonitrile providing the nonaamine. This spectrum of reactions is represented on Scheme 11.13. Note in passing that a Vögtle-type synthesis of amine-terminated dendrimers led to the 72-nitrile and 72-amino dendrimers, providing the possibility to synthesize the polymetallocenyl dendrimers for later generations as well. Indeed, the 18-ferrocenyl and 18-cobaltocenyl dendrimers were also synthesized in this way.[11,25] Likewise, the $CpFe^+$-induced octabenzylation of durene shown in Section II (Scheme 11.4) was the starting point for the synthesis of 24-metallocenyl dendrimers via the 24-amino dendrimers (Scheme 11.14).[20]

The functionalization of the three allyl chains of the phenol dendron could be achieved by hydrosilylation reaction catalyzed by the Karsted catalyst. Indeed, it is very interesting that there is no need to protect the phenol group before performing these reactions. For instance,

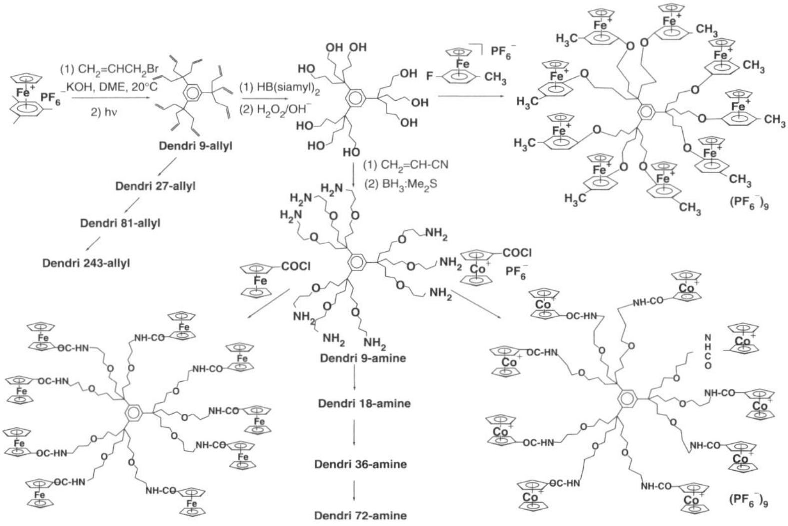

Scheme 11.13 Synthesis of 9-amino- and 9-metallocenyl dendrimers. The 18-amino- and metallocenyl dendrimers are synthesized similarly.

catalyzed hydrosilylation using ferrocenyldimethylsilane gives a high yield of the triferrocenyl dendron $HOp\text{-}C_6H_4C(CH_2CH_2CH_2SiMe_2Fc)_3$ (Scheme 11.15), which is easily purified by column chromatography. Protection of the phenol dendron using propionyliodide gave the phenolate ester, which was hydroborated. Oxidation of the triborane using H_2O_2/OH^- gave the triol, then reaction with $SiMe_3Cl$ gave the tris-silyl derivative. Reaction with NaI yielded the tri-iodo compound, and reaction with the tri-ferrocenyl dendron provided the nonaferrocenyl dendron that was unprotected using K_2CO_3 in DMF. The nonaferrocenyl dendron was allowed to react with hexakis(bromomethyl)benzene, which gave the 54-ferrocenyl dendrimer. This convergent synthesis (Scheme 11.15) is clean, and the 54-ferrocenyl dendrimer gave correct analytical data, although a mass spectrum could not be obtained. This approach is somewhat limited, however, since larger dendrons, which one would like to synthesize in this way, cannot be made because dehydrohalogenation is faster than nucleophilic substitution of the iodo by phenolate when dealing with bulkier, higher generations of dendrons.

Although this problem might be overcome by modifying the iodo branch in such a way that there would be no hydrogens in the β positions, the condensation of higher dendrons onto a core would become tedious or impossible for steric reasons. This well-known inconvenience is inherent in the convergent dendritic synthesis. On the other hand, divergent syntheses are not marred by this problem, since additional generations and terminal groups are added at the periphery of the dendrimer. The limit is that indicated by de Gennes, that is, the steric congestion encountered at a generation where the peripheral branches can no longer by divided. Another obvious limit intervenes if the molecular objects added onto the termini of the branches are large and interfere with one another. The ferrocenylsilylation of all these polyallyl dendrimers was carried out using ferrocenyldimethylsilane in ether or toluene and was catalyzed by the Karsted catalyst at 40°C.

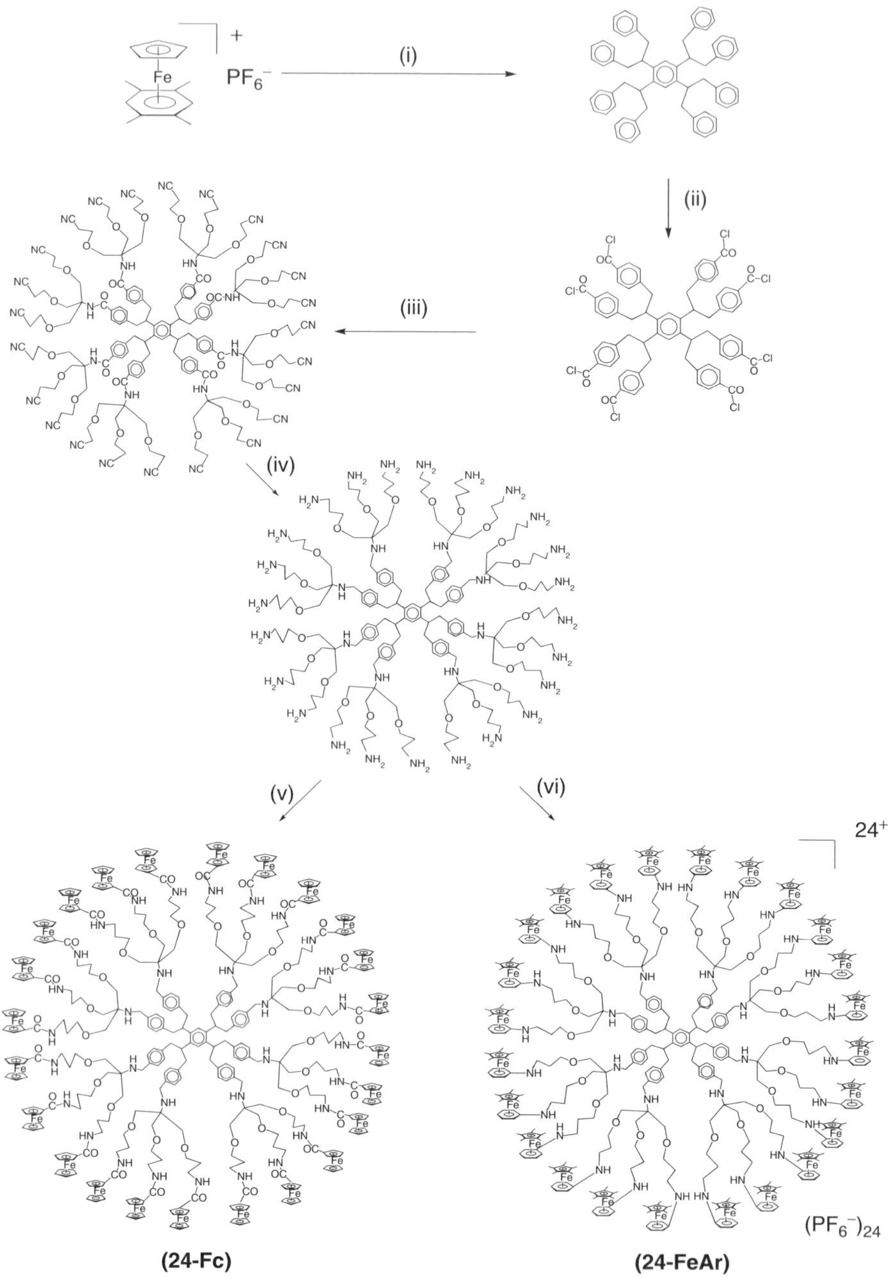

Scheme 11.14 Synthesis of 24-iron-sandwich complexes starting from durene.

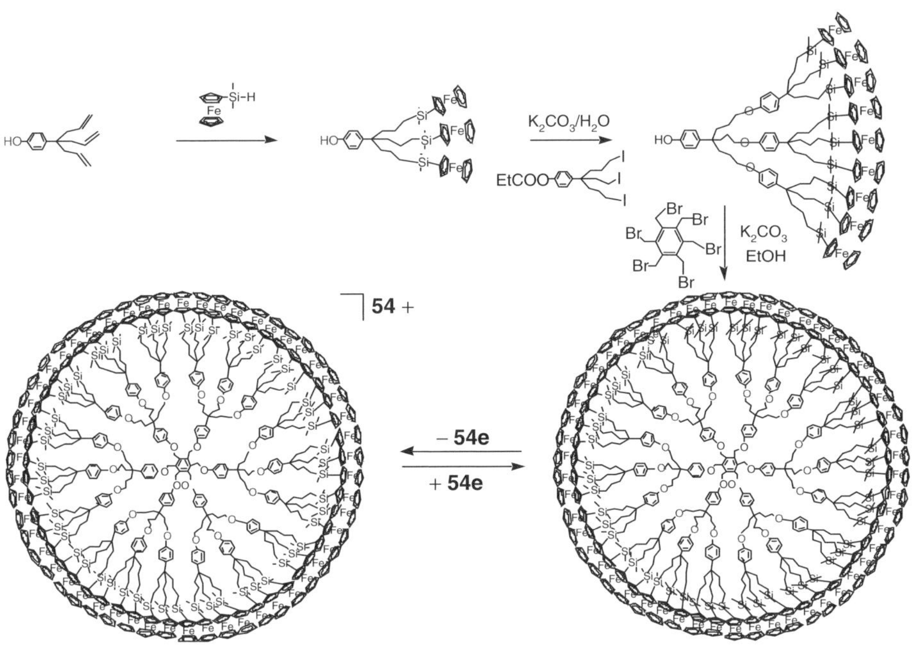

Scheme 11.15 Convergent synthesis of a redox-robust 54-silylferrocenyl dendrimer.

The reactions were complete after two or three days except for the ferrocenylsilylation of 243-allyl, which required a reaction time of one week, indicating some degree of steric congestion (Scheme 11.16).

The ^{1}H and ^{13}C spectra indicated the absence of regioisomer. The solubility in pentane decreased from good for the 9-Fc dendrimer (Fc = ferrocenyl) to low for the 27-Fc dendrimer and nil for the superior dendrimers, but the solubility in ether remained good for all the ferrocenyl dendrimers. Likewise, the retention times on plate or column chromatography increased with generation and no migration was observed for the "243-Fc" dendrimer. The silane use here, $HSi(Fc)Me_2Cl$, reported by Pannel and Sharma,[26] was already used by Jutzi et al.[27] to synthesize the decaferrocenyl dendrimer $[Fe(CCH_2CH_2SiMe_2Fc)_{10}]$ from deca-allylferrocene.

The cyclic voltammetry of all the ferrocenyl dendrimers on the Pt anode shows that all the ferrocenyl centers are equivalent and that only one wave was observed. It was possible to avoid adsorption using even CH_2Cl_2 for the small ferrocenyl dendrimers, but it was required to use MeCN for the medium-size ones (27-Fc, 54-Fc, and 81-Fc). Finally, adsorption was not avoided even with MeCN for the "243-Fc" dendrimer. From the intensity of the wave, the number of ferrocenyl units could be estimated using the Anson–Bard equation,[28] and the number found were within 5% of the branch numbers, except for the "243-Fc"dendrimer, for which the experimental number was too high (250) because of the adsorption.[15]

The first polyferrocenium dendrimers reported by our group in 1994 and characterized *inter alia* by Mössbauer spectroscopy (a "quantitative" technique) were mixed-valence

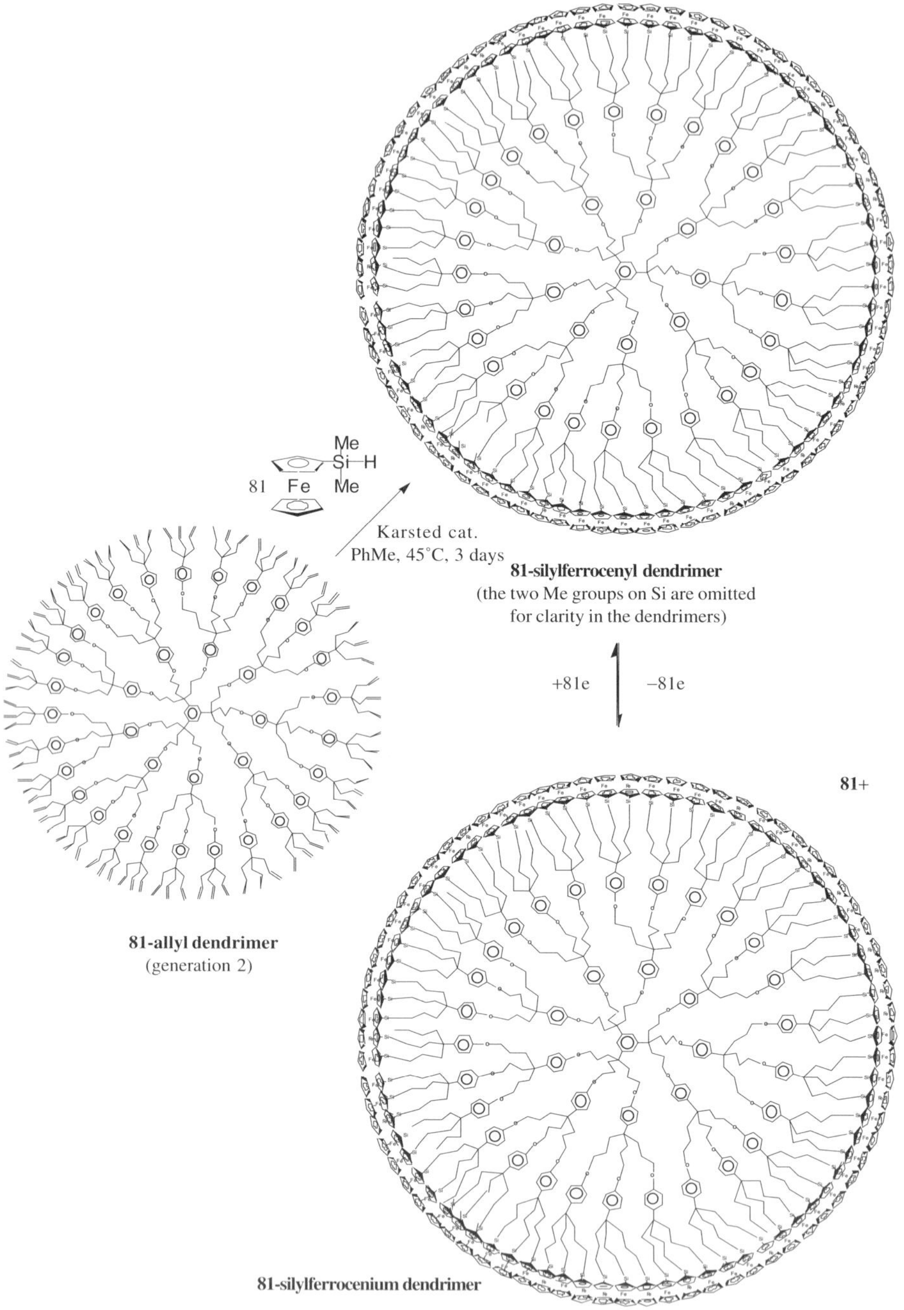

Scheme 11.16 Ferrocenylsilylation of the polyallyl dendrimers synthesized. Example of the second generation 81-allyl dendrimer.

Fe(II)/Fe(III) complexes.[5] Since then, we have been seeking to synthesize larger ferrocenyldendrimers that could also withstand oxidation to their ferrocenium analogs. The syntheses of amidoferrocene dendrimers were reported simultaneously by our group[29,30] and the Madrid group using different cores.[31,32,33] In our reports, we were able to show the use of these metallodendrimers as redox sensors for the recognition of oxoanions, with remarkable positive dendritic effects when the generation increased. The amidoferrocenyl dendrimers are not the best candidates for a stable redox activity on the synthetic scale, however, and thus even less so for molecular batteries. Indeed, although they give fully reversible cyclic voltammetry waves, it is known that ferrocenium derivatives bearing an electron-withdrawing substituent are at least fragile, if stable at all. This inconvenience is probably enhanced in the dendritic structures because of the steric effect that forces ferrocenium groups to encounter one another more easily than as monomers. Thus, we have oxidized our silylferrocenyl dendrimers using [NO][PF_6] in CH_2Cl_2 and obtained stable polyferrocenium dendrimers as dark-blue precipitates, as expected from the known characteristic color of ferrocenium itself.[34] These polyferrocenium dendrimers were reduced back to soluble orange polyferrocenyl dendrimers using decamethylferrocene as the reductant. No decomposition was observed either in the oxidation or in the reduction reactions that were very clean, and this redox cycle could be achieved in quantitative yield even with the "243-ferrocenyl" dendrimer. The zero-field Mössbauer spectrum of the 243-ferrocenium dendrimer (Figure 11.3) showed a single line corresponding to the expected spectrum known for ferrocenium itself, confirming its electronic structure. Thus, these polyferrocenyl dendrimers are molecular batteries that could be used in specific devices. Indeed, as large as they may be, they transfer a very large number of electrons rapidly and "simultaneously" with the electrode. By "simultaneously," we mean that the cyclic voltammogram looks as if it were that of a monoelectronic wave. One must question the notion of the isopotential for the many ferrocenyl units at the periphery of a dendrimer, however. In theory, all the standard potentials of the *n* ferrocenyl units of a single dendrimer are distinct, even if all of them are equivalent and independent. This situation arises because the charge of the overall dendrimer molecule increases by one unit of charge every time one of its ferrocenyl units is oxidized to ferrocenium. The next single-electron oxidation is more difficult than the preceding one due to the dendritic molecule having one more unit of positive charge; it is more difficult to oxidize due to the increased electrostatic factor. Thus, the potentials of the *n* redox units are statistically distributed around an average standard potential centered at the average potential (Gaussian distribution).[28] The differences among all the redox potentials is so tiny, however (probably a fraction of mV), that all the ferrocenyl units seem to be at about the same potential.

Concerning the electrochemical reversibility, one has to take into account the fact that the dendritic molecule, as large as it may be, is rotating much more rapidly than the usual electrochemical timescales.[35,36] Under these conditions, all the redox sites come near the electrode within the electrochemical timescale. Consequently, there is no slowing down of the electron transfer due to the long distance from the electrode, even in large dendrimers. Indeed, the waves of the ferrocenyl dendrimers always appear fully electrochemically reversible, indicating fast electron transfer with all the ferrocenyl sites. When the electrode is derivatized with the ferrocenyl dendrimers, one is then dealing with the solid state (Figure 11.3). The electrochemical reversibility can no longer be explained by the rotation of the dendrimer, because it is no longer rapid. Electron hopping then accounts for the fully reversible heterogeneous electron transfer between the ferrocenyl sites and the electrode.

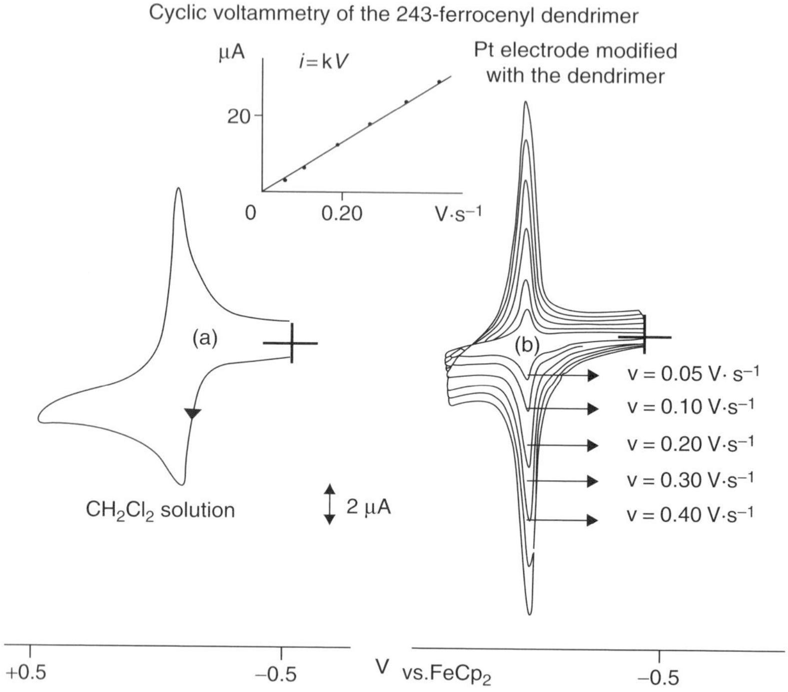

Figure 11.3 Cyclic voltammogram of the 243-ferrocenyl dendrimer ("243-Fc") in CH_2Cl_2 solution containing 0.1 M [n-Bu_4N][PF_6]: (a) in solution (10–4 M) at 100 mV · s^{-1} on the Pt anode; (b) Pt anode modified with 243-Fc at various scan rates, dendrimer-free clear CH_2Cl_2 solution. (Inset: intensity as a function of scan rate: the linearity shows the expected behavior of a modified electrode with a fully adsorbed dendrimer.)

V. DECORATION OF DENDRIMERS WITH RUTHENIUM CLUSTERS CATALYZED BY AN FE(I) ELECTRON-RESERVOIR COMPLEX: TOWARD DENDRITIC CATALYSTS

The clean introduction of clusters onto the termini of polyphosphine dendrimers is a real challenge because of the current interest of dendritic clusters in catalysis and the mixtures usually obtained in thermal reactions of [$Ru_3(CO)_{12}$] with phosphines.[37] The diphosphine $CH_3(CH_2)_2N(CH_2PPh_2)_2$ (abbreviated P-P) was used as a simple, model ligand. The reaction between P-P and [$Ru_3(CO)_{12}$] (molar ratio: 1/1.05) in the presence of 0.1 equiv. [$Fe^{I}Cp(\eta^6$-$C_6Me_6)$] in THF at 20°C led to the complete disappearance of [$Ru_3(CO)_{12}$] in a few minutes and the appearance of a mixture of chelate [P-P. $Ru_3(CO)_{10}$], monodentate [P-P. $Ru_3(CO)_{11}$], and bis-cluster [P-P. {$Ru_3(CO)_{11}$}$_2$]. These reactions were reported by Bruce et al. with simple diphosphines.[38] On the other hand, the reaction of P-P with [$Ru_3(CO)_{12}$] in excess (1/4) and only 0.01 equiv. [$Fe^{I}Cp(\eta^6$-$C_6Me_6)$] in THF at 20°C led, in 20 minutes, to the formation of the air-stable, light-sensitive bis-cluster [P-P. {$Ru_3(CO)_{11}$}$_2$] as the only reaction product. Given the simplicity of this characterization of the reaction product by ^{31}P NMR and the excellent selectivity of this model reaction when excess [$Ru_3(CO)_{12}$] was used, the same reaction between Reetz's dendritic phosphines,[39] derived from DSM's dendritic amines,[40] and

$[Ru_3(CO)_{12}]$ could be more confidently envisaged. This reaction, catalyzed by 1% equiv. $[Fe^ICp(\eta^6\text{-}C_6Me_6)]$ was carried out in THF at 20°C. The dendrimer–cluster assembly was obtained in 50% yield. This shows the selectivity and completion of the coordination of each of the 32 phosphino ligands of P-P to a $Ru_3(CO)_{11}$ cluster fragment (Schemes 11.17 and 11.18).

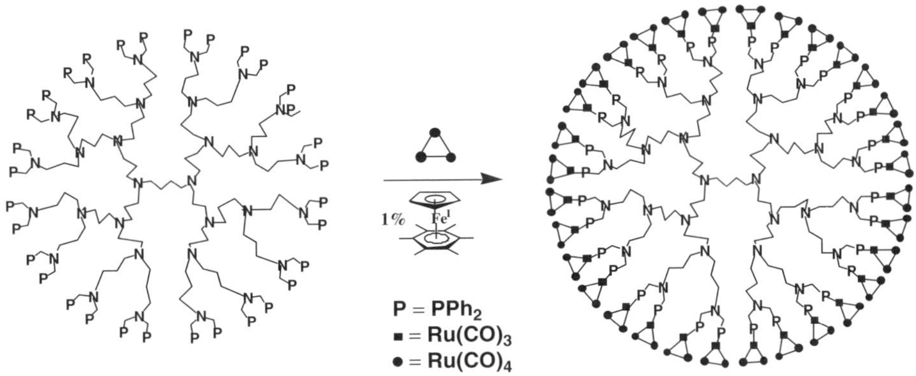

Scheme 11.17 Electron-transfer-chain catalyzed ligand substitution of one Ru-coordinated CO by dendritic phosphine termini in Reetz's 32-phosphine dendrimer under ambiant conditions leading to the 32-$Ru_3(CO)_{11}$ dendrimer-cluster. The ETC mechanism[41,42] proceeds for the introduction of the 32 cluster fragments in the dendrimer for ligation of the first $Ru_3(CO)_{11}$ fragment to the dendritic phosphine. Then, this first complex [dendriphosphine.$Ru_3(CO)_{11}$] would undergo the same ETC cycle as $[Ru_3(CO)_{12}]$ initially does to generate the bis-cluster complex [dendriphosphine.$\{Ru_3(CO)_{11}\}_2$], and so onto Scheme 11.18.

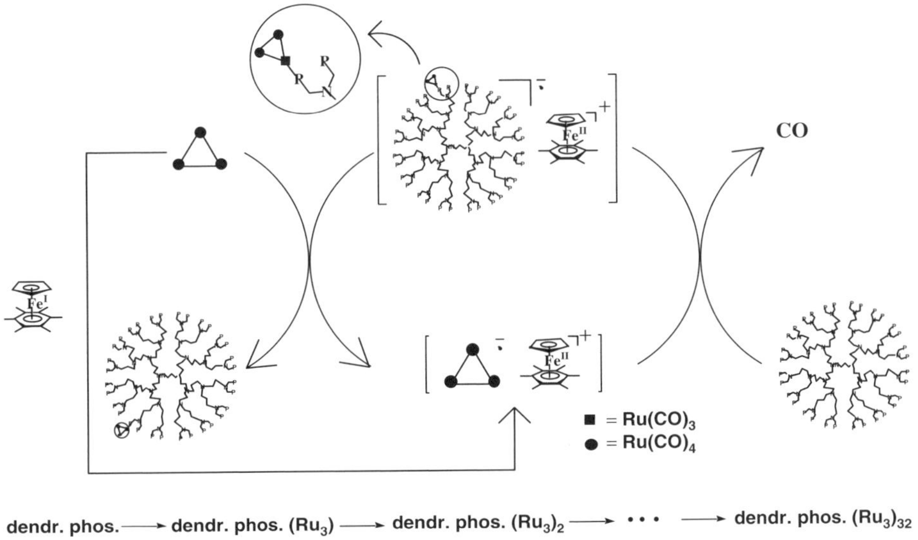

Scheme 11.18 Electron-transfer-chain mechanism for the synthesis of the 96-Ru dendrimer–cluster complex.

Here, we have only emphasized types of dendrimer catalysis involving electron transfer (redox catalysis and electron transfer chain (ETC) catalysis). Other types of dendrimer catalysis are reviewed elsewhere.[6,7]

VI. REDOX RECOGNITION OF ANIONS: DENDRITIC IRON COMPLEXES AS EXO-RECEPTORS

Anions such as adenosine mono-, di-, and triphosphate AMP, ADP, ATP, DNA itself, RNA, etc., are biologically important. Environmental problems also deal with anions (nitrate, phosphate, radioactive pertechnetate), and their recognition by endoreceptors has been extensively studied. The use of metallocenes as electrochemical anion sensors has been in particular developed by Beer through elegant studies with the design of a variety of endoreceptors (chelates, tripods, crowns, porphyrins, calixarenes).[43] The interactions of anions with these metallocene-type endoreceptors are most often based on hydrogen-bonding interactions between the anion and a functional group nearby the redox center, so that the redox potential of this redox center is shifted upon interaction with the anion.[43] Moutet's group has also published several insightful reports along this line, especially with the biologically important ATP anion.[44] Ferrocenyl dendrimers are well suited to function as electrochemical sensors, when all the ferrocenyl termini are equivalent, because they show a single wave in cyclic voltammetry (CV). Dendrimers with receptor sites at the periphery are exoreceptors in contrast with the aforementioned endoreceptors. The specificities of dendritic exoreceptors are the fractality of their surfaces, which recalls that of viruses and cancer cells, and their nanometer dimensions, which allow recovery for polymers and nanosized biomolecules. They can also adsorb much more easily on surfaces than small endoreceptors, which leads to their use for the modification of electrodes and their function as reusable sensors on metal surfaces.

A. Anion Recognition by Covalently Assembled Ferrocenyldendrimers

Titration of 9- and 18-amidoferrocenyldendrimers were carried out using n-Bu_4N^+ salts of $H_2PO_4^-$, HSO_4^-, Cl^-, and NO_3^- and monitored by CV in CH_2Cl_2 (Figure 11.4) and ^{1}H NMR. In CV, two types of behavior were recorded: (1) the appearance of a new CV wave at less positive potentials for $H_2PO_4^-$ while the intensity of the initial CV wave decreases (for one equiv. anion, ΔE was 220 mV with 9-Fc and 315 mV with 18-Fc), and (2) progressive anodic shift of the initial wave for the HSO_4^-, Cl^- and NO_3^- (for one equiv. anion, $\Delta E°$ was, respectively, 65 mV, 20 mV, and negligeable with 9-Fc, and 130 mV, 45 mV, and 30 mV with 18-Fc).

The distinction of CV behavior between strongly and weakly interacting anions has been rationalized in the seminal article by Echegoyen and Kaifer's group with a square scheme:[45] When the strength of the interaction between the anion and the reduced redox form (here ferrocenyl) is significant, a new wave appears, and the variation of ferrocenyl potential between the free and bound forms is related to the ratio of apparent association constants as follows: $E^\circ_{free} - E^\circ_{bound} = \Delta E°(V) = 0.059 \log (K_+/K_0)$ at 25°C. Here E°_{bound} corresponds to the addition of one equiv. anion per ferrocenyl branch, or the stoichiometric amount determined from the breakpoints.

Although only the ratio of apparent association constants is accessible in this way, measurement of K_0 by ^{1}H NMR (using the shifting NH signal) can lead to K_+ as well. For instance, with 9-Fc, $K_+ = (2.2 \pm 0.2) \times 10^5$ in CH_2Cl_2. When the interaction between the ferrocenyl dendrimer and the anion is negligible, only a CV wave shift is observed. The value of K_+, the

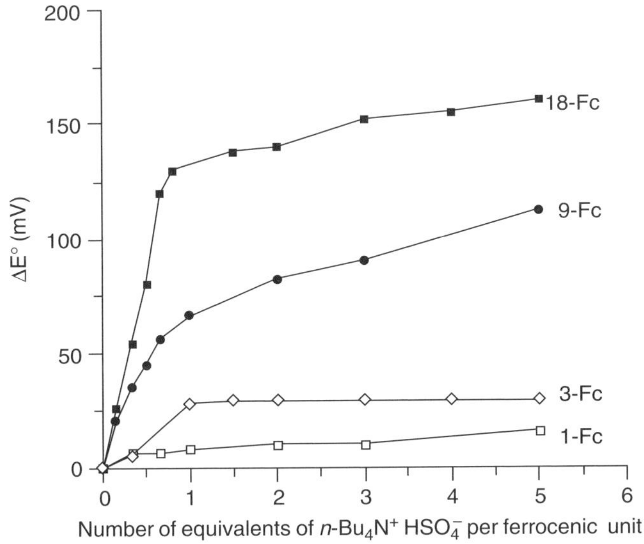

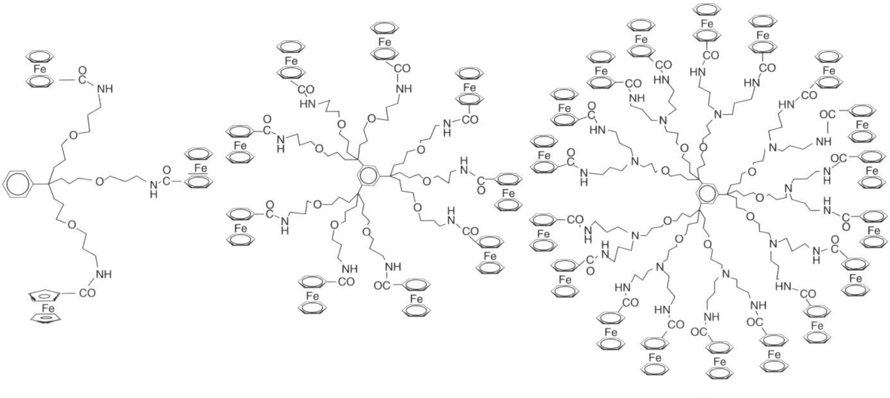

3-Fc 9-Fc 18-Fc

Figure 11.4 Variation (shift, cf. Scheme 11.1, bottom) $\Delta E°$ of the redox potential of the ferrocenyl system recorded by CV along the titration of $[n\text{-}Bu_4N][HSO_4]$ by mono- (1-Fc), tri- (3-Fc), nona- (9-Fc), and octadeca-amidoferrocenyl (18-Fc) compounds showing the marked dendritic effect (1-Fc is the monoamidoferrocenyl derivative $[FeCp(\eta^5\text{-}C_5H_4CONHCH_2CH_2OPh)]$).

apparent association constant between the oxidized (ferrocenium) form of the dendrimer and the anion, is then directly accessible using the value of the concentration c using the equation: $\Delta E°(V) = 0.059 \log cK_+$ at 25°C, which gives $K_+ = 544 \pm 50$, 8500 ± 500, and $61{,}000 \pm 3000$ for, respectively, 1-Fc, 9-Fc, and 18-Fc with HSO_4^- in CH_2Cl_2.

The equivalent point was found by CV to correspond to the interaction of one ferrocenyl branch per equiv. anion in this dendrimer series, but other stoichiometries were observed with other dendrimer series for some anions (*vide infra*).[30]

The 9- and 18-amidoferrocenyl dendrimers (respectively, 9-Fc and 18-Fc) were compared to monomeric (1-Fc) and tripodal trimetallic (3-Fc) amidoferrocenyl derivatives (Figure 11.4), and, as indicated by the numbers just given, a strongly positive dendritic effect (i.e., the strength of the interaction characterized by $\Delta E°$ and K_+ is all the larger as the dendrimer generation is higher) was characterized by the values of $\Delta E°$ and apparent association constants for the anions: $\Delta E°$ (1-Fc) < $\Delta E°$ (3-Fc) < $\Delta E°$ (9-Fc) < $\Delta E°$ (18-Fc). The combined hydrogen bonding and electrostatic factors (attraction between the anion and the ferrocenium cation) alone cannot explain the high $\Delta E°$ values obtained with the 9-Fc and 18-Fc dendrimers, since the values found for the monoamidoferrocenyl derivative 1-Fc are very weak (a few tens of mV). The synergy between these factors and the topology factor in endoreceptors (as pointed out by Beer[43]) or exoreceptors is required to provide large $\Delta E°$ and associated constant values. This feature is to be compared with the positive dendritic effect. Indeed, not only do dendrimers show better recognition of the anions than the monomer and tripod, but this recognition is easily quantified by the increase of the $\Delta E°$ values from 9-Fc to 18-Fc, that is, when the dendrimer generation increases. This can tentatively be taken into account by the narrowing of the channels between the exoreceptor redox sites, which forces a tighter hydrogen-bonding interaction when the dendrimer generation increases (as shown by molecular models). This positive dendritic effect in molecular recognition contrasts with the negative dendritic effect usually observed in catalysis, whereby the reaction kinetics is lowered by the enforced steric constraints around the catalytic metal center inhibiting its approach by substrates when the dendrimer generation increases (Figure 11.5).[4] With the diaminobutane (DAB) dendrimers formulated G_n-DAB-dend-$(NH_2)_n$, amidoferrocenyl and pentamethylamidoferrocenyl[46] dendrimers have been synthesized (Scheme 11.19) for the five generations from G_1 (4 branches) to G_5 (theoretical number; 64 branches).

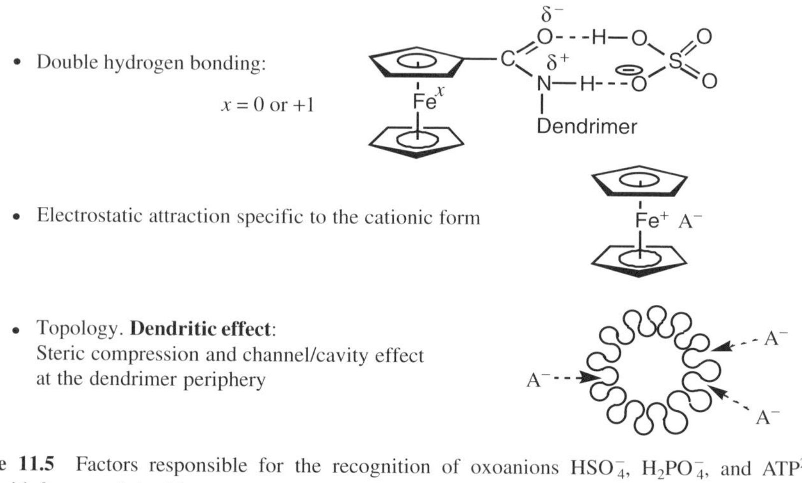

Figure 11.5 Factors responsible for the recognition of oxoanions HSO_4^-, $H_2PO_4^-$, and ATP^{2-} by polyamidoferrocenyl dendrimers.

Recognition of HSO_4^- proceeds best in CH_2Cl_2 with the parent Cp series with, again, a positive dendritic effect (i.e., an increase of $\Delta E°$ when the generation number increases; Figure 11.6).[28,34] In DMF, however, recognition and titration are only possible with the permethylated dendrimer series. It is subjected to another dramatic dendritic effect: the shift of the initial CV wave upon titration (weak interaction) is only observed with G_1 and a new wave

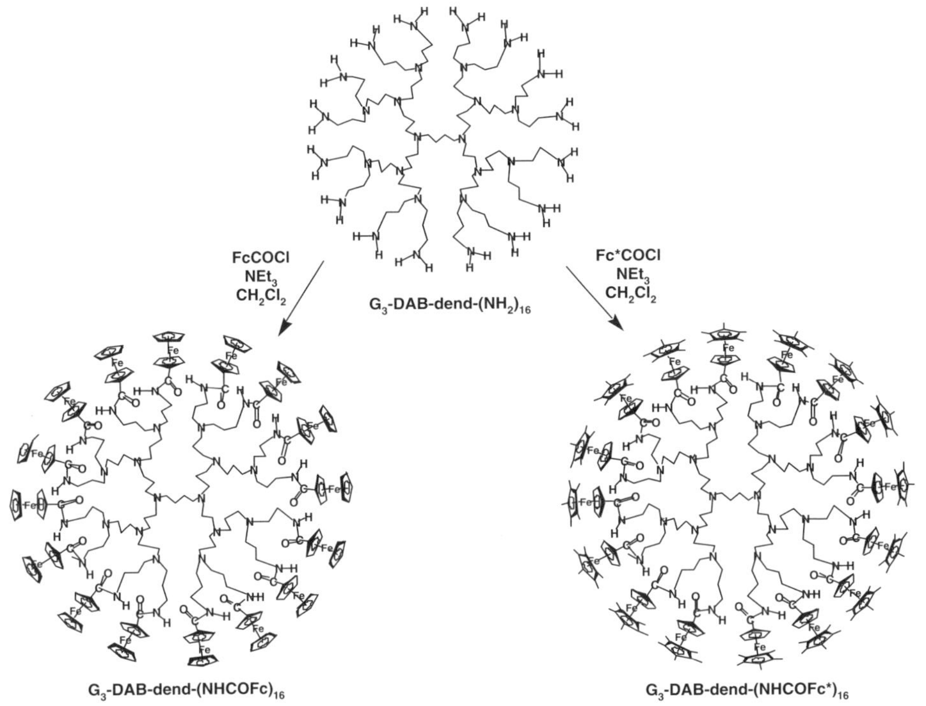

Scheme 11.19 Syntheses of third-generation amidoferrocenyl and pentamethylamidoferrocenyl dendrimers from the corresponding DSM polyamine dendrimer G_3-dend-$(NH_2)_{16}$. For analysis and comparison of their behaviors as hosts of anions, see Reference 46.

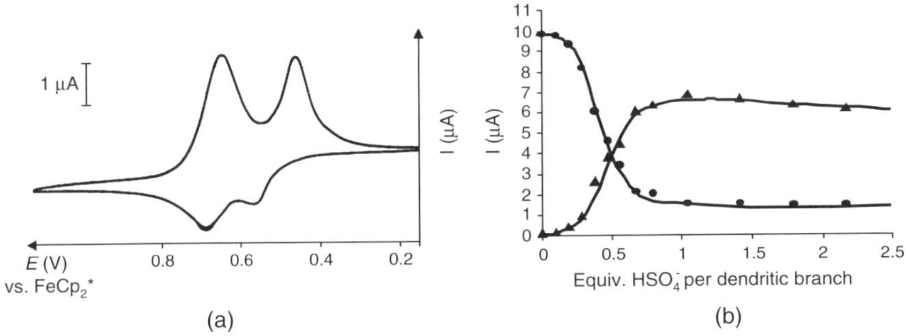

Figure 11.6 Titration of a 3.3×10^{-4} M solution of the G_3-DAB-dend-$(NHCOFc)_{16}$ by a 10^{-3} M solution of $[n\text{-}Bu_4N][HSO_4]$ in CH_2Cl_2 in the presence of 0.1 M $[n\text{-}Bu_4N][PF_6]$, Pt anode, 20°C. (a) Cyclovoltammogram obtained after addition of 0.5 equiv. $[n\text{-}Bu_4N][HSO_4]$ per dendritic branch; (b) variation of the intensities of the initial (●) and new (▲) waves along the titration.

(strong interaction) is observed with G_2 and G_3. With amido ferrocenyl dendrimers, the oxidized ferrocenium form is not very stable due to the electron-withdrawing property of the amido group, but the electron-releasing permethylation in the Cp* ligand fully stabilizes this 17-electron form. As a result of this stabilization and the increased hydrophobicity of the Cp* dendrimers, the recognition of $H_2PO_4^-$ and ATP^{2-} is also much cleaner in the Cp* series than in the parent Cp series even if the electron-releasing character of the permethylated groups causes a slight decrease in the $\Delta E°$ values. With ATP^{2-}, the CVs are also clean in the Cp* dendrimers along the titration, and stoichiometry of 0.5 equiv. ATP^{2-}/ ferrocenyl branch was found, corresponding to the doubly negative phosphate charge (Figure 11.7). It is noteworthy that the dendritic effect on $\Delta E°$ values in this family of dendrimers is considerably less marked and sometimes nil, a feature that was also noted with ferrocenyl-urea dendrimers of this family for the recognition of $H_2PO_4^-$.

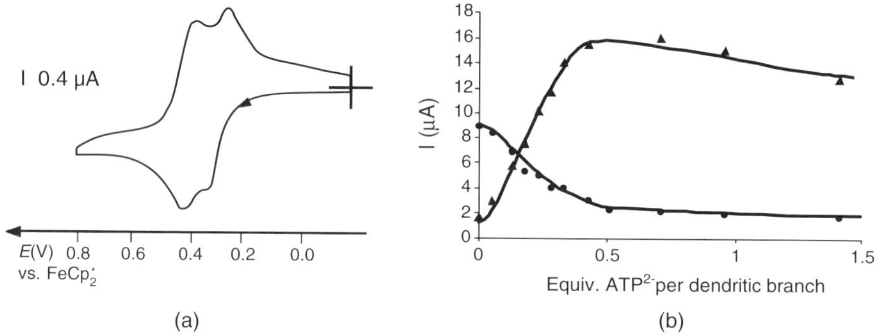

Figure 11.7 Titration of a 1.25×10^{-4} M solution of G_4-DAB-dend-$(NHCOFc^*)_{32}$ (Scheme 11.19) by a 10^{-3} M solution of $[n\text{-}Bu_4N]_2[ATP]$ in CH_2Cl_2 in the presence of 0.1 M $[n\text{-}Bu_4N][PF_6]$, Pt anode, 20°C. (a) CV obtained after addition of 0.25 equiv. $[n\text{-}Bu_4N]_2[ATP]$ per dendritic branch; (b) variation of the intensities of the initial (●) and new (▲) waves along the titration.

B. Anion Recognition by Metallodendrimers Assembled by Hydrogen Bonding Between a Redox-Active Dendronic Phenol and Dendritic Primary Amines

Supramolecular aspects of dendrimer chemistry, especially those involving hydrogen bonding, have been the subject of interest in the last decade.[47–51] We were intrigued by the very simple possibility of hydrogen bonding between primary amines and phenols, and by its potential use in dendrimer chemistry and molecular recognition. The tedious, time-consuming dendritic syntheses often represent an obstacle for research and dendrimers use. Therefore the advantage of forming dendrimers with redox-active termini by simply mixing a commercial polyamine core and a redox-active dendron is obvious. This is especially the case if a specific exoreceptor property with a positive dendritic effect can be obtained as shown previously in the covalently synthesized metallodendrimers. Indeed, there were precedents for hydrogen bonding between primary amines and alcohols with tetrahedral disposition of both O and N valences and 1:1 stoichiometry (i.e., minimal melting point for this stoichiometry), a property that had been used in crystal engineering[52] and chiral recognition.[53]

Thus, mixing a DSM dendritic polyamine and a para-substituted phenol derivative leads to the replacement of the ^1H NMR signals of the OH protons at 5 ppm and NH_2 proton at 1.5 ppm by a common broad, concentration-dependent signal for these three protons located between 2.4 and 4.1 ppm. This means that supramolecular dendrimers involving reversible hydrogen bonding between the DSM polyamine and a phenol dendron form these two components upon mixing (Figure 11.8). The electrochemical timescale of the CV being much larger than that of

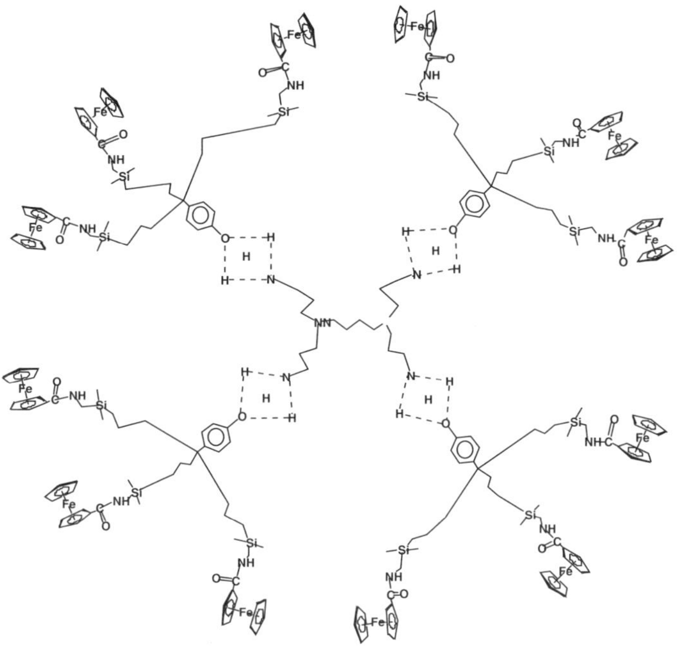

Figure 11.8 Arbitrary representation of the reversible hydrogen bonding between G_1-DAB-dend-$(NH_2)_4$ and a triamidoferrocenyl dendron shown by the concentration-dependent average location of the (broad) NH_2 + OH signal in ^1H NMR between 2.4 and 4.1 ppm vs. TMS (TMS = tetramethylsilane) in $CDCl_3$.[54]

the hydrogen-bond formation and breaking, the CV shows an average situation between the hydrogen-bonded and nonbonded dendrimer branches.[54]

Upon titration of [n-Bu_4N][H_2PO_4] by FcCONHPr, the CV of this ferrocenyl group undergoes the appearance of a new, anodically shifted wave with $\Delta E° = 150$ mV; this value reaches 210 mV with the dendron p-OH-$C_6H_4C\{(CH_2)_3SiMe_2$-$CH_2NHCOFc\}$ alone, and it is not changed in the presence of propylamine. It reaches 250 mV, however, with this dendron + G_1-DAB-dend-$(NH_2)_4$ and 280 mV with this dendron + G_2-DAB-dend-$(NH_2)_8$ or a higher-generation polyamine dendrimer. This increase in the $\Delta E°$ value is the signature of a dendritic effect, as shown previously with covalent redox metallodendrimers. With the monomer FcCONHPr + n-propylamine, it is necessary to add 2.5 equiv. [n-Bu_4N][H_2PO_4] to reach the equivalence point, whereas one equiv. [n-Bu_4N][H_2PO_4] is enough with the covalent dendrimers, because the amine strongly competes with the amidoferrocenyl group in binding $H_2PO_4^-$. With G_1, only 0.5 equiv. [n-Bu_4N][H_2PO_4] is necessary, and a sudden disappearance of the initial CV wave is observed, whereas the intensity of the new CV wave is much reduced (Figure 11.9). This can tentatively be taken into account by the formation of a rather stabilized dendritic supramolecular assembly in which the $H_2PO_4^-$ anion bridges two amidoferrocenyl units (Figure 11.8).

The reduction of the diffusion coefficient responsible for the decrease of the CV wave intensity at the equivalence point can be taken into account by the sudden increase of the mass of this overall supramolecular assembly compared to the much smaller species present in solution before the equivalence point. For higher generations, the number of equiv. [n-Bu_4N][H_2PO_4] necessary to reach the equivalence point increases progressively again (0.8 for G_2

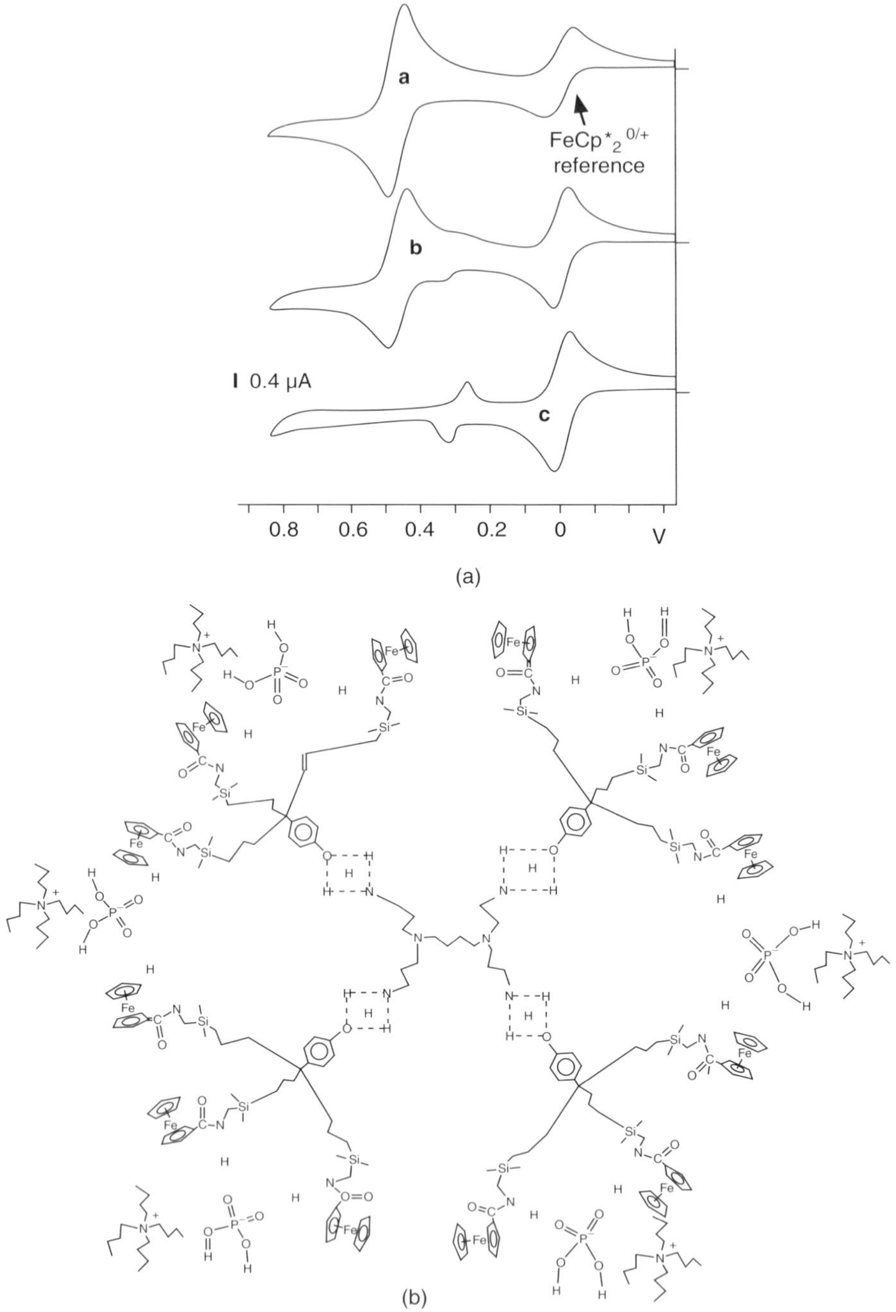

Figure 11.9 (a) Titration of DAB-G_1 in CH_2Cl_2 (Pt, 0.1 M [nBu$_4$N][PF$_6$], 20°C, reference FeCp*$_2$ = decamethylferrocene) by [nBu$_4$N][H$_2$PO$_4$]: (a) before addition; (b) 0.4 equiv.; (c) 0.5 equiv. (b) Proposed supramolecular assembly between the triferrocenyl dendron + DAB-G_1 and 0.5 equiv. [n-Bu$_4$N][H$_2$PO$_4$], taking the sudden drop of CV wave intensity at the equivalent point (i.e., sudden drop of diffusion coefficient) into account.

and 2.0 for G_3 and G_4), probably due to the fact that, for high generations, the steric congestion around the dendrimer partly destabilizes, the hydrogen bonds optimized in G_1 and G_2; this is also consistent with the limit of the $\Delta E°$ increase. Finally, this strategy has also been successfully applied to the recognition of $[n\text{-}Bu_4N]_2[ATP]$ for G_1.[54]

C. Gold-Nanoparticle-Cored Ferrocenyl Dendrimers and Modification of Electrodes for the Fabrication of Useful Sensors

Nanoparticle-cored dendrimers are new materials that were reported for the first time in 2001.[55,56,57] With ferrocenyl termini,[55] they also have a functional use, because they combine the advantages of covalent ferrocenyl dendrimers and amidoferrocenyl gold nanoparticles (AuNPs[58]). Moreover, their large size (in our case) offers the possibility to easily derivatize electrodes with the redox ferrocenyl sensors, which makes them useful for sensing. Thus, ferrocenyl AB_3 and AB_9 dendrons were synthesized with a monothiol for the ligand part A and three or nine amidoferrocenyl or silylferrocenyl groups for B_3 or B_9. The AB_3 dendrons were assembled into AuNP-cored dendrimers either by direct synthesis using the Brust–Schiffrin method[59] using a mixture of dodecanethiol and tripodal thiol ligands (Scheme 11.20) or by the ligand substitution method from dodecanethiol–AuNPs and a given amount of tripodal thiol ligand. Between five and seven AB_3 dendrons were introduced in this way in AuNPs of 2.3-diameter and containing about 150 thiolate ligands overall. For the silylferrocenyl AB_9 dendrons, the ligand-substitution procedure no longer worked, presumably for steric reasons, but the direct synthesis using a mixture of dodecanethiol and AB_9 thiol worked well

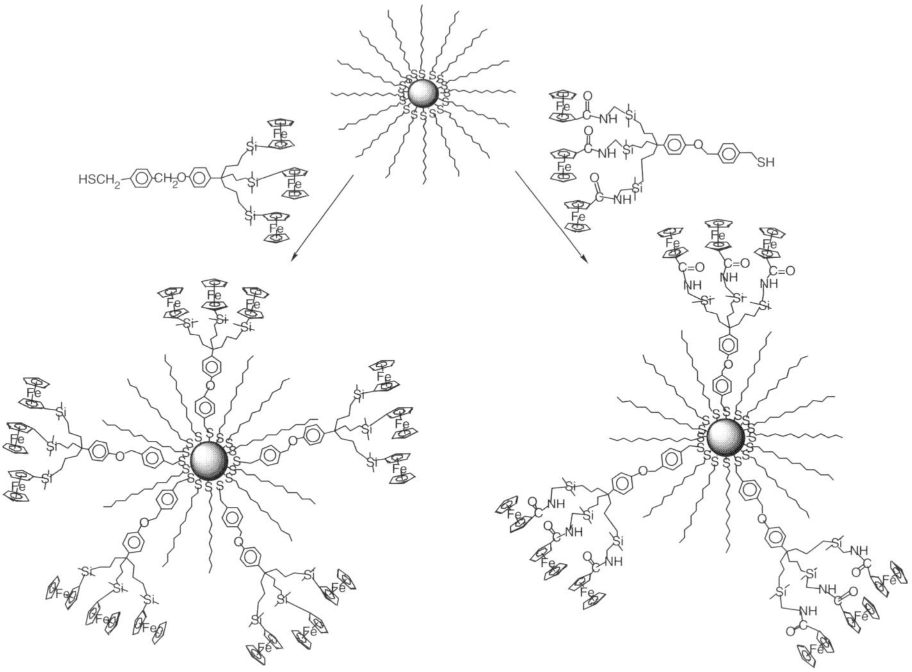

Scheme 11.20 General scheme for the synthesis of triferrocenyldendronized AuNPs (ligand substitution method).

(Scheme 11.21). Two equivalent AB_9 dendrons per core unit were used to assemble about 10%, respectively, 20% AB_9 thiolate dendrons of around 2.9-nm AuNPs, bearing around 200 thiolate ligands overall. This means that these AuNPs contain about 180, respectively, 360 equivalent silylferrocenyl groups at the periphery.[55,60]

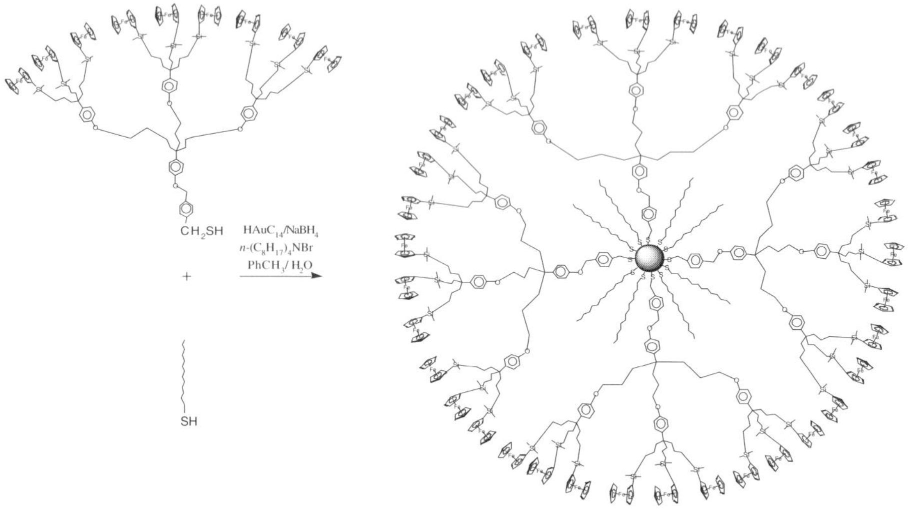

Scheme 11.21 Synthesis of a AuNP-cored silylferrocenyl dendrimer (direct method).

The recognition experiments led to stoichiometries corresponding to one equiv. [n-Bu_4N][H_2PO_4] per ferrocenyl branch with all the AuNP-cored ferrocenyl dendrimers. The $\Delta E°$ values for this anion were 200 mV with the amidoferrocenyl AB_3 dendron-AuNPs, 115 mV for the silylferrocenyl AB_3 dendron-AuNPs and 125 mV for the AB_9 dendron-AuNPs. Recognition studies with [n-Bu_4N]$_2$[ATP] gave results that were similar to those obtained with [n-Bu_4N][H_2PO_4], except that the stoichiometry was reached for only 0.5 equiv. [n-Bu_4N]$_2$[ATP] per ferrocenyl branch because of the double-negative charge of ATP^{2-} (Figure 11.10). The $\Delta E°$ values were usually only slightly lower for [n-Bu_4N]$_2$[ATP] than for [n-Bu_4N][H_2PO_4].

The silylferrocenyl AuNP-cored dendrimers are thus efficient sensors, because the silylferrocenium species is stable, unlike the parent amidoferrocenium species. The silicon atom attached to the ferrocenyl group acts as a Lewis acid that interacts with the $H_2PO_4^-$ anion by hydrogen bonding. With [n-Bu_4N][HSO_4], however, the silylferrocenyl dendrimers do not provide recognition. A $\Delta E°$ value of 42 mV with $K_+ = (18 \pm 4) \times 10^3$ L mol^{-1} was obtained with this anion using the tris-amidoferrocenyl dendron-AuNPs with the same stoichiometry. The weaker interaction of the amidoferrocenyl dendrimers with [n-Bu_4N][HSO_4] than with [n-Bu_4N][H_2PO_4] is due to the fact that NH–oxo anion hydrogen bonding is dominant, and the negative charge on the oxygen atoms is smaller in HSO_4^- than in $H_2PO_4^-$ (sulfur being more electronegative than phosphorus).

Electrodes modified by ferrocenyl polymers have been known for a long time.[54,61,62] More recently, the Madrid group has extensively studied the derivatization of silylferrocenyl dendrimers.[63] Nishihara's group has reported the first example of modified electrodes with AuNPs containing ferrocenyl thiol ligands, in which the stabilization of the modified electrodes is

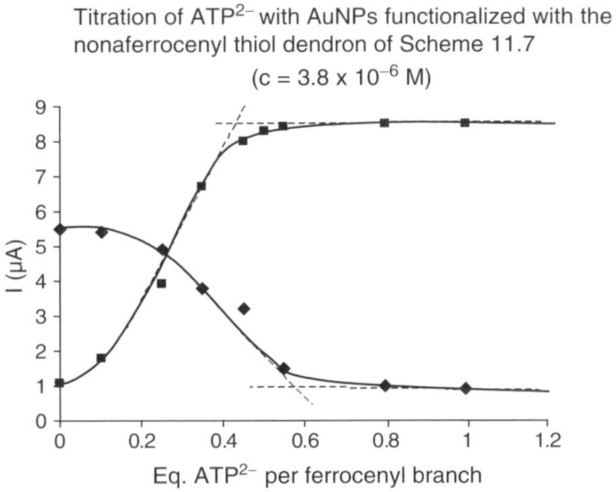

Figure 11.10 Titration of ATP^{2-} with 9-Fc-thiolate dendron-AuNPs shown on Scheme 11.21. Decrease in the intensity of the initial CV wave and increase of the intensity of the new CV wave vs. the number of equiv. $[n\text{-}Bu_4N]_2[ATP]$ added per ferrocenyl branch. Nanoparticles: 3.8×10^{-6} M in CH_2Cl_2.

provided by the biferrocenyl units, whereas monoferrocenyl units do not afford stabilization.[64] We have noticed that the larger the ferrocenyl dendrimers, the more fully they tend to adsorb on Pt electrodes in CH_2Cl_2 solutions, and the easier it is to prepare modified Pt electrodes with ferrocenyl dendrimers by scanning around the region of the ferrocenyl potential. Thus, AuNP-cored ferrocenyl dendrimers discussed earlier are large and readily adsorb on Pt electrodes, forming perfectly stable derivatized electrodes upon scanning about 50 times (for saturation) around the ferrocenyl potential region. These modified Pt electrodes show remarkable changes upon introduction of a solution of $[n\text{-}Bu_4N]_2[ATP]$. The new CV wave observed has undergone a shift of potential; the anodic and cathodic peaks are no longer indentical, showing that a structural rearrangement, due to changes in hydrogen bonding and electrostatic interaction, occurs in the course of the heterogeneous electron transfer. $[n\text{-}Bu_4N][H_2PO_4]$ and $[n\text{-}Bu_4N]_2[ATP]$ are selectively recognized in the presence of other anions such as HSO_4^- and Cl^-; $[n\text{-}Bu_4N][HSO_4]$ alone can also be recognized. The recognized salts can be washed from the electrode using CH_2Cl_2, but the AuN-cored ferrocenyl dendrimers are not removed and can serve again for further similar experiments. Such recycling can be performed several times (Figure 11.11).

VII. CONCLUSION AND OUTLOOK

Metallodendrimers can be assembled in a number of ways (covalent, ionic, hydrogen-bonding, coordination), wherein the metal plays key roles: sensor, catalyst, electronic, magnetic, or optical device. The molecular engineering pertaining to the metallodendritic design is relevant to multiple aspects of nanoscience from fundamental to societal research.[2c] In the present chapter, we have emphasized the problems encountered in dendritic construction (i.e., surface versus volume limit), including synthetic strategies offered by simple organometallic activation and modern methods of characterization. Nanoparticle-cored dendrimers pioneered in 2001 are a particularly attractive category insofar as they facilitate rapid, yet relatively robust constructions.

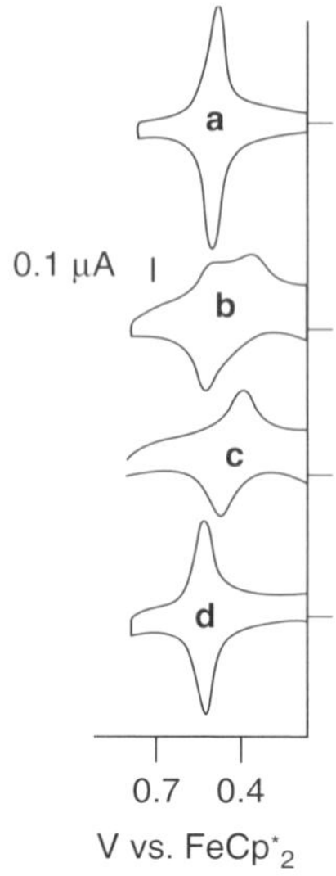

Figure 11.11 Recognition of ATP^{2-} with AuNP functionalized with a nonasilylferrocenyl dendron shown in Scheme 21. CV: (a) modified electrode alone; (b) in the course of titration; (c) with excess of $[n\text{-}Bu_4N]_2[ATP]$; (d) after removal of $[n\text{-}Bu_4N]_2[ATP]$ upon washing the modified electrode with CH_2Cl_2.

Dendritic catalysts offer the possibility of recovery due to their size, but dendritic effects are negative, that is, the kinetics drops when the dendritic generation increases due to the bulk problem (access to the metal center). On the other hand, recognition and sensing with dendritic exoreceptors benefits from positive dendritic effects, because the channels become narrower, and thus better suited for interaction with the substrate when the dendritic generation increases.

It is also interesting to note that metallodendritic assembly by triple H-bonding between phenols and primary amines is sufficient for recognition on the electrochemical timescale (i.e., 0.1 s), which can be a clear advantage over tedious covalent dendrimer construction. One significant advantage of robust dendrimers for sensing, however, is the size providing adsorption on solids. Thus, the larger the metallodendrimer, the better it adsorbs on electrodes for recycling the dendrimer-modified electrode sensor. The bonding interaction leading to sensing must also be strong enough to offer a significant shift in the sensing parameter ($E_{1/2}$), and yet be modest enough to allow easy removal of the sensed substrate for recycling and reuse. This is the case with the hypervalent Si···O bond when the Si atom is directly attached to the silylferocenyl group in ferrocenyl dendrimers.

The design of molecular batteries for molecular electronics will also allow engineering nanoprocesses, and this aspect has a promising future, as do optical and magnetic nanodevices.[2c]

ACKNOWLEDGMENTS

Financial support from the Institut Universitaire de France (IUF), the University Bordeaux I, the Centre National de la recherche Scientifique (CNRS) and the Ministére de la Recherche et de la Technologie (MRT) are gratefully acknowledged. The multiple ideas and hard work of colleagues and students who contributed to the various aspects of the work described in this chapter (see references) is heartily acknowledged.

REFERENCES

1. P. J. Flory, *J. Am. Chem. Soc.,* **63**, 3083, 3091, 3096 (1941); P. J. Flory, *Principles of Polymer Chemistry,* Cornell University, Ithaca, New York, 1953.
2. (a) G. R. Newkome, C. N. Moorefield, F. Vögtle, *Dendrimers and Dendrons: Concepts, Synthesis, Applications,* Wiley-VCH, Weinheim, Germany, 2001; (b) *Dendrimers and Other Dendritic Polymers*, J. M. J. Frechet, D. A. Tomalia (eds.), Wiley, New York, **2002**; (c) "Dendrimers and Nanosciences," *C. R. Chimie,* **6**, 8 (2003).
3. Early comprehensive reviews on dendrimers: D. A. Tomalia, A. N. Naylor, W. A. Godart III, *Angew. Chem. Int. Ed. Engl.*, **29**, 138 (1990); G. R. Newkome, C. N. Moorefield, G. R. Baker, *Aldrichim. Acta*, **25**(2), 21 (1992); N. Ardoin, D. Astruc, *Bull. Soc. Chim. Fr.,* **132**, 875 (1995).
4. V. Balzani, S. Campagna, G. Denti, A. Juris, S. Serroni, M. Venturi, *Acc. Chem. Res.,* **31**, 26 (1998).
5. J.-L. Fillaut, D. Astruc, *Chem. Commun.*, 1320 (1993); F. Moulines, L. Djakovitch, R. Boese, B. Gloaguen, W. Thiel, J.-L. Fillaut, M.-H. Delville, D. Astruc, *Angew. Chem. Int. Ed. Engl.*, **32**, 1075 (1993); J.-L. Fillaut, J. Linares, D. Astruc, *Angew. Chem. Int. Ed. Engl.,* **33**, 2460 (1994).
6. D. Astruc, F. Chardac, *Chem. Rev.*, **101**, 2991 (2001); G. E. Oosterom, J. N. H. Reek, P. C. J. Kamer, P. W. N. M. Van Leeuwen, *Angew. Chem. Int. Ed. Engl.,* **40**, 1828 (2001).
7. M. Dagupta, M. B. Peori, A. K. Kakkar, *Coord. Chem. Rev.,* **233**, 223 (2002); R. Van Heerbeeck, P. C. J. Kamer, P. W. N. M. Van Leeuwen, J. N. H. Reek, *Chem. Rev.,* **102**, 3717 (2002); P. A. Chase, R. J. M. Klein Gebbink, G. Van Koten, *J. Organomet. Chem.,* **689**, 4016 (2004); A. Dahan, M. Portnoy, *J. Polym. Sci., Part A: Polym. Chem.,* **30**, 474 (2005); D. Astruc, K. Heuze, S. Gatard, D. Mery, S. Nlate, L. Plault, *Advan. Syn. Catal.,* **347**, 329 (2005); D. Mery, D. Astruc, *Coord. Chem. Rev.,* 000 (2006).
8. D. Astruc, M.-C. Daniel, J. Ruiz, *Chem. Commun.*, 2637 (2004).
9. J. Ruiz, C. Pradet, F. Varret, D. Astruc, *Chem. Commun.*, 1108 (2002).
10. For a recent review highlighting the large potential of the $[FeCp(\eta^6\text{-arene})]^+$ complexes in organic synthesis and polymer science, see: A. Abd-El-Aziz, *Coord. Chem. Rev.,* **203**, 219 (2000).
11. (a) H. Trujillo, C. Casado, J. Ruiz, D. Astruc, *J. Am. Chem. Soc.,* **121**, 5674 (1999); (b) D. Astruc, *Acc. Chem. Res.,* **33**, 287 (2000).
12. F. Moulines, B. Gloaguen, D. Astruc, *Angew. Chem. Int. Ed. Engl.,* **28**, 458 (1992).
13. D. Astruc, J.-R. Hamon, G. Althoff, E. Roman, P. Batail, P. Michaud, J.-P. Mariot, F. Varret, D. Cozak, *J. Am. Chem. Soc.,* **101**, 5445 (1979); J. R. Hamon, J.-Y. Saillard, A. Le Beuze, M. McGlinchey, D. Astruc, *J. Am. Chem. Soc.*, **104**, 7549 (1982).
14. D. Astruc, *Acc. Chem. Res.,* **19**, 377 (1986).
15. P.-G. De Gennes, H. Hervet, *J. Phys. Lett.,* **44**, L-351 (1983).
16. D. Astruc, *Pure Appl. Chem.,* **75**, 461 (2003).
17. H. W. Marx, F. Moulines, T. Wagner, D. Astruc, *Angew. Chem. Int. Ed. Engl.*, **35**, 1701 (1996).
18. S. Rigaut, M.-H. Delville, D. Astruc, *J. Am. Chem. Soc.*, **119**, 1132 (1997).
19. B. Gloaguen, D. Astruc, *J. Am. Chem. Soc.*, **112**, 4607 (1990).
20. C. Valerio, E. Alonso, J. Ruiz, J.-C. Blais, D. Astruc, *Angew. Chem. Int. Ed. Engl.,* **38**, 1747 (1999).
21. V. Martinez, J.-C. Blais, D. Astruc, *Org. Lett.*, **4**, 651 (2002).
22. V. Sartor, L. Djakovitch, J.-L. Fillaut, F. Neveu, J. Guittard, J.-C. Blais, D. Astruc, *J. Am. Chem. Soc.,* **121**, 2929 (1999).
23. S. Nlate, Y. Neto, J.-C. Blais, J. Ruiz, D. Astruc, *Chem. Eur. J.*, **8**, 171 (2002).
24. J. Ruiz, G. Lafuente, S. Marcen, C. Ornelas, S. Lazare, E. Cloutet, J.-C. Blais, D. Astruc, *J. Am. Chem. Soc.*, **125**, 7250 (2003).

25. S. Nlate, J. Ruiz, V. Sartor, R. Navarro, J.-C. Blais, D. Astruc, *Chem. Eur. J.,* **6**, 2544 (2000).
26. K. H. Pannel, H. Sharma, *Organometallics,* **10**, 954 (1991); S. W. Krsda, D. Seyferth, *J. Am. Chem. Soc.,* **120**, 5323 (1998).
27. P. Jutzi, C. Batz, B. Neumann, H. G. Stammler, *Angew. Chem. Int. Ed. Engl.,* **35**, 2118 (1996).
28. J. B. Flanagan, S. Margel, A. J. Bard, F. C. Anson, *J. Am. Chem. Soc.*, **100**, 4248 (1978).
29. D. Astruc, C. Valerio, J.-L. Fillaut, J.-R. Hamon, F. Varret, in *Magnetism, a Supramolecular Function,* O. Kahn, Ed., p. 1107, NATO ASAI Series, Kluwer, Dordrecht, 1996.
30. C. Valerio, J.-L. Fillaut, J. Ruiz, J.-C. Blais, D. Astruc, *J. Am. Chem. Soc.*, **117**, 2588 (1997).
31. I. Cuadrado, M. Moran, C. M. Casado, B. Alonso, F. Lobete, B. Garcia, J. Losada, *Organometallics,* **15**, 5278 (1996).
32. K. Takada, D. J. Diaz, H. Abruña, I. Cuadrado, C. M. Casado, M. Moran, J. Losada, *J. Am. Chem. Soc.,* **119**, 10763 (1997).
33. C. M. Casado, M. Moran, I. Cuadrado, M. Moran, B. Garcia, B. Gonzales, J. Losada, *Coord. Chem. Rev.,* **185–186**, 53 (1999).
34. R. L. Collins, *J. Chem. Phys.*, **42**, 1072 (1965).
35. S. J. Green, J. J. Pietron, J. J. Stokes, M. J. Hostetler, H. Vu, W. P. Wuelfing, R. W. Murray, *Langmuir*, **14**, 5612 (1998).
36. C. B. Gorman, J. C. Smith, M. W. Hager, B. L. Parhurst, H. Sierzputowska-Grcz, C. A. Haney, *J. Am. Chem. Soc.*, **121**, 9958 (1999).
37. E. Alonso, D. Astruc, *J. Am. Chem. Soc.,* **122**, 3222 (2000).
38. M. I. Bruce, D. C. Kehoe, J. G. Matisons, B. K. Nicholson, P. H. Rieger, M. L. J. Williams, *Chem. Commun.,* 442 (1982); M. I. Bruce, J. G. Matisons, K. B. Nicholson, *J. Organomet. Chem.*, **247**, 321 (1983).
39. M. T. Reetz, G. Lohmer, R. Scwickardi, *Angew. Chem. Int. Ed. Engl.*, **36**, 1526 (1997).
40. E. M. M. de Brabanger-Van Den Berg, E. W. Meijers, *Angew. Chem. Int. Ed. Engl.*, **32**, 1308 (1993).
41. R. Rich, H. Taube, *J. Am. Chem. Soc.,* **76**, 2608 (1954).
42. Reviews on ETC catalysis in transition-metal chemistry: D. Astruc, *Angew. Chem. Int. Ed. Engl.*, **27**, 643 (1988); D. Astruc, *Electron Transfer and Radical Processes in Transition-metal Chemistry*, VCH, New York, 1995, Chapter 6.
43. P. D. Beer, *Angew. Chem. Int. Ed.*, **40**, 486 (2001).
44. O. Reynes, J.-C. Moutet, J. Pecaut, G. Royal, E. Saint-Aman, *New J. Chem.*, **26**, 9 (2002).
45. S. R. Miller, D. A. Gustowski, Z.-H. Chen, G. W. Gokel, L. Echegoyen, A. E. Kaifer, *Anal. Chem.*, **60**, 2021 (1988).
46. J. Ruiz, M. J. Medel, M.-C. Daniel, J.-C. Blais, D. Astruc, *Chem. Commun.*, 464 (2003); M.-C. Daniel, J. Ruiz, J.-C. Blais, N. Daro, D. Astruc, *Chem. Eur. J.*, **9**, 4371 (2003).
47. S. C. Zimmerman, F. Zeng, D. E. C. Reichert, S. V. Kolotuchin, *Science*, **271**, 1095 (1996); F. Zeng, S. C. Zimmerman, *Chem. Rev.*, **97**, 1681 (1997).
48. A. W. Bosman, E. W. Jensen, E. W. Meijer, *Chem. Rev.*, **99**, 1665 (1999).
49. D. K. Smith, F. Diederich, *Topics Curr. Chem.*, **210**, 183 (2000).
50. J. M. J. Frechet, *Proc. Nat. Acad. Sci.*, **99**, 4782 (2002).
51. H. W. Gibson, N. Y. Yamagushi, L. Hamilton, J. W. Jones, *J. Am. Chem. Soc.*, **124**, 4653 (2002).
52. G. R. Desiraju, *Crystal Engineering: The Design of Organic Solids*, Elsevier, New York, 1989.
53. S. Hanessian, M. Simard, S. Roelens, *J. Am. Chem. Soc.*, **117**, 7630 (1995).
54. M.-C. Daniel, J. Ruiz, D. Astruc, *J. Am. Chem. Soc.*, **125**, 1150 (2003); M.-C. Daniel, J. Ruiz, D. Astruc, *Inorg. Chem.*, **43**, 8649 (2004).
55. M.-C. Daniel, J. Ruiz, S. Nlate, J. Palumbo, J.-C. Blais, D. Astruc, *Chem. Commun.*, 2000 (2001).
56. R. Wang, J. Yang, Z. Zheng, M. D. Carducci, *Angew. Chem. Int. Ed. Engl.*, **40**, 293 (2001).

57. M.-K. Kim, Y.-M. Jeon, S. W. Jeon, H.-J. Hong, C. G. Park, K. King, *Chem. Commun.*, **40**, 549 (2001).

58. M.-C. Daniel, D. Astruc, *Chem. Rev.*, **104**, 293 (2004).

59. M. Brust, M. Walker, D. Bethell, D. J. Schiffrin, R. J. Whyman, *Chem. Commun.*, 801 (1994); M. Hasan, D. Bethell, M. Brust, *J. Am. Chem. Soc.*, **125**, 1132 (2003).

60. M.-C. Daniel, J. Ruiz, S. Nlate, J.-C. Blais, D. Astruc, *J. Am. Chem. Soc.*, **125**, 2617 (2003).

61. H. D. Abruña, in *Electroresponsive Molecular and Polymeric Systems*, Vol. 1, T. A. Stotheim, *Ed.*, p. 97, Dekker, New York, 1988.

62. R. W. Murray, in *Molecular Design of Electrode Surfaces*, R. W. Murray, *Ed.*, p. 1, Wiley, New York, 1992.

63. C. M. Casado, I. Cuadrado, M. Moran, B. Alonso, M. Barranco, J. Losada, *Appl. Organomet. Chem.*, **14**, 245 (1999).

64. T. Horikoshi, M. Itoh, M. Kurihara, K. Kubo, H. Nishihara, *J. Electroanal. Chem.*, **473**, 113 (1999); M. Yamada, T. Tadera, K. Kubo, H. Nishihara, *Langmuir*, **17**, 2263 (2001).

CHAPTER 12

Polypeptide-Based Metallobiopolymers

KHALED A. MAHMOUD AND HEINZ-BERNHARD KRAATZ

University of Saskatchewan, Saskatoon, Saskatchewan, Canada

I. INTRODUCTION

The field of bioorganometallic chemistry has rapidly matured over the past decade and has gained significant interest due to potential medical, bioelectronic, and nanomaterial applications.[1–6] Peptide-derived bioconjugates occupy a special position in that they offer the potential for biomolecular recognition and enable probing of biological processes.[7,8] These systems are also of a more fundamental interest for the construction of well-defined scaffolds. The assembly of amino acids and peptides into extended supramolecular three-dimensional structures is controlled by the hydrogen bonding between individual molecules.[9,10] The desire to control the properties of such assemblies has led to a considerable body of work focused on the design of secondary structural elements,[11,12] and on the design of new peptidic materials, ranging from peptide nanotubes[13–17] to peptide hydrogels.[18,19,20]

More recently, bioorganometallic peptide conjugates have allowed additional control over the structural properties by choosing the appropriate organometallic group.[21] Ferrocene (Fc) is a particularly useful in this regard.[21–26] Disubstituted Fc systems, in which both cyclopentadienyl rings are substituted, provide some level of control over the supramolecular assemblies, and offers a useful starting point for the design of electronic biomaterials.[27] In this context, the quest for peptide-like Fc–containing polymers is important.[21,23] The recent preparation of a number of key disubstituted Fc building blocks, such as Fc–amino acid and 1,1′-Fc–diamine, may enable the development of peptide-like ferrocene-based materials with well-defined electronic properties.

Here, we give an overview of recent developments into the preparative and structural aspects of Fc–peptide conjugates. In particular, we highlight the use of disubstituted Fc systems to control the structure and discuss the recent efforts of incorporating Fc into peptide-like oligo- and polyamides and dendrimers.

II. PREPARATION OF FC–PEPTIDE CONJUGATES

Several major synthetic approaches are available for the preparation of Fc–peptide conjugates. The "acid chloride," "active ester," and the oxazolone methods are all viable approaches that will result in the formation of the desired conjugate.[28] The acid chloride method makes use of

Frontiers in Transition Metal-Containing Polymers, edited by Alaa S. Abd-El-Aziz and Ian Manners.

the reaction between Fc–acid chloride, prepared from ferrocenecarboxylic acid and PCl_5 or $SOCl_2$, and amino acids or peptides having an available amino terminus. HCl formed in the reaction is effectively captured by a base. The active ester method works under mild conditions with an often isolable Fc–CO-active ester, in which reactive heterocyclic esters of ferrocene carboxylic acid, such as HOBt esters, HATU esters, and HOSu esters, are formed. The Fc–CO-esters then react with the amino acid or peptide to give the desired Fc–conjugate **I** (see Scheme 12.1). These active esters can be used as a stoichiometric Fc delivery reagent, which makes these reagents amenable to automated solid-phase synthesis of Fc–peptide conjugates. This method has led to the preparation of a number of mono- and disubstituted Fc–peptide derivatives (*vide infra*).

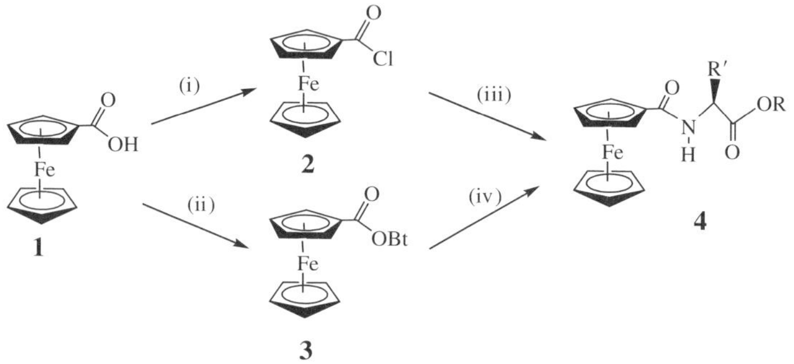

Scheme 12.1 Synthesis of ferrocenoyl amino acid esters (**4**) via the acid chloride and active ester methods: (i) PCl_5 or $SOCl_2$; (ii) carbodiimide such as DCC (N,N′-dicyclohexylcarbodiimide) or EDC (1-ethyl-3-(3-dimethylaminopropyl)carbodiimide), HOBt (1-hydroxybenzotriazole); (iii) and (iv) peptide or amino acid and base.

These methods were employed successfully to prepare a number of mono- and disubstituted Fc–peptide conjugates (Figure 12.1). For example, Herrick and Hirao prepared a series of 1,1′-disubstituted Fc–peptides, such as ferrocenoyl-prolinylmethylester (**5a**), 1,1′-ferrocenoyl-prolinylmethylester (**5b**),[29] and Hirao's ferrocenoyl-alanylprolinylester (**6a**), 1,1′-ferrocenoyl-bis(alanylprolinylesters) (**6b**).[30]

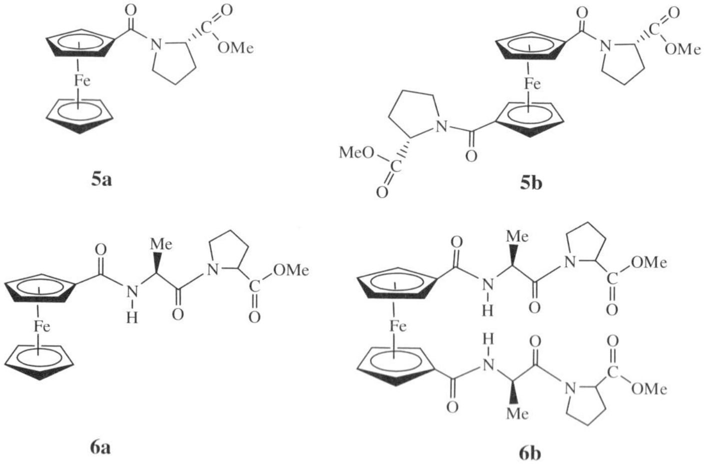

(a)

Figure 12.1 (a) Molecular drawing of compounds Fc-Pro-OMe (**5a**), Fc[Pro-OMe]$_2$ (**5b**), Fc-Ala-Pro-OEt (**6a**) and of Fc[Ala-Pro-OEt]$_2$ (**6b**); (b) Structures of Fc–Ala-Pro-OEt (Ala = alanine, Pro = proline) (**6a**) and of Fc[Ala-Pro-OEt]$_2$ (**6b**), in which the two podant peptides chains are H-bonded and the Fc adopts a 1,2′-conformation.[29,30]

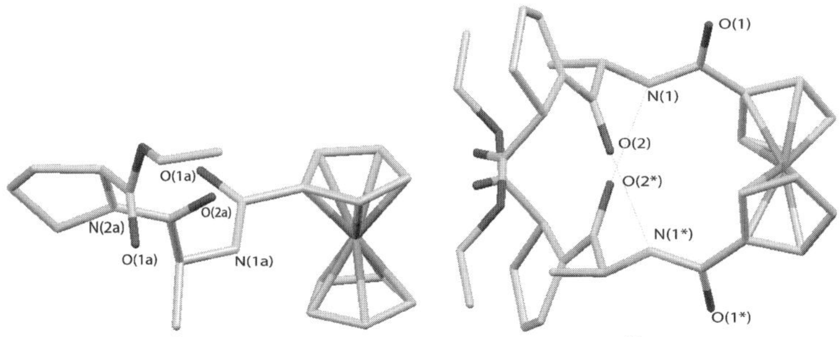

6a

6b

(b)

Figure 12.1 *Continued*

Fc–active esters, initially employed by Degani and Heller for coupling the Fc–group to the ε-amino group of lysines in proteins,[31,32] were successfully employed in the synthesis of symmetrical Fc–peptide–cystamines having two Fc groups linked to the two terminal amino groups.

A. 1,1′-Ferrocenedicarboxylic Acid Conjugates

1,1′-Ferrocenedicarboxylic acid **7**, Fc[COOH]$_2$, serves as the starting material for the synthesis of bisamino acid and bispeptide-substituted Fc–conjugates. The simplest peptide systems based on glycine (Gly) were reported recently.[33] The two compounds Fc[Gly-OEt]$_2$ **8** and Fc[Gly-OH]$_2$ **9** are particularly interesting in that they display two completely different substitutional patterns as well as different H-bonding patterns (Figure 12.2). For Fc[Gly-OEt]$_2$ **8**,

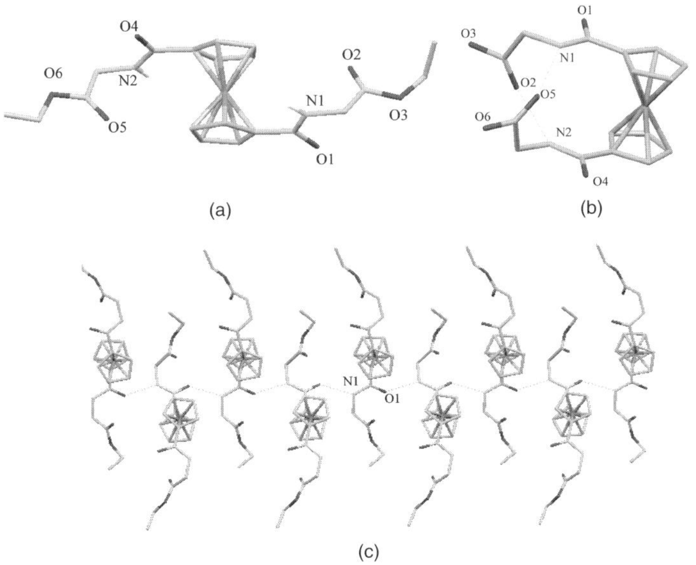

Figure 12.2 (a) Molecular drawing of compound Fc[Gly–OEt]$_2$ **8**, showing the 1,3′-substitution of the ferrocene group; (b) molecular drawing of the centrosymmetric Fc[Gly-OH]$_2$ **9** showing the 1,2′-substitution and the typical H-bonding pattern; (c) formation of a one-dimensional H-bonded polymeric chain involving H-bonding of N(1) and O(1) of adjacent molecules for **9**. (Redrawn from Appoh et al.[33])

a 1,3′-substitution pattern is observed, which results in the formation of intermolecular H-bonding exclusively. The molecules form a one-dimensional chain H-bonded chain, in which O(1) interacts with the N(1) of the Gly group of an adjacent molecule (O(1)···N(1) = 2.839(5)Å; Figure 1b). Interestingly, the amide group on the other Cp ring is not involved in H-bonding and is well separated from other molecules (ca. 4 Å). This H-bonding pattern is reminiscent of that reported by Gallagher et al.,[34,35] and Hirao,[30] with an alternating up–down orientation of the individual H-bonded molecules. In contrast, the free acid Fc[Gly-OH]$_2$ **9** adopts C_2 symmetry with a 1,2′-conformation that allows the formation of strong intramolecular H-bonding between the two podand substitutents (N···O(2A) and O(2)···N(A) = 2.875(3) Å). This pattern provides a rigid framework, in which the amide twist is reduced considerably (2.9°). In addition, intermolecular contacts are established between the Fc–C═O group and the acid O—H of adjacent neighbors, with an H-bonding distance of O(1)···O(3*) = 2.623(3) Å. The related amide Fc[Gly-NH$_2$]$_2$ **10**, described by Mingos[36] engages in intermolecule interactions with adjacent molecules.

A series of related disubstituted Fc–dipeptides was studied by Hirao and co-workers.[30,37,38] It was shown that only intramolecular H-bonding was present in the C_2 symmetrical molecules Fc[Ala-Pro-OR]$_2$ forcing the two Cp rings into a 1,2′-conformation that has been observed frequently in other disubstituted Fc systems. Two strong intramolecular H-bonding interactions were present between the C═O of Ala of the Cp ring and the amide NH of Ala of another strand on the Cp′ ring, giving *d*(O···HN) distances of 1.88 and 2.15 Å (Figure 12.3).

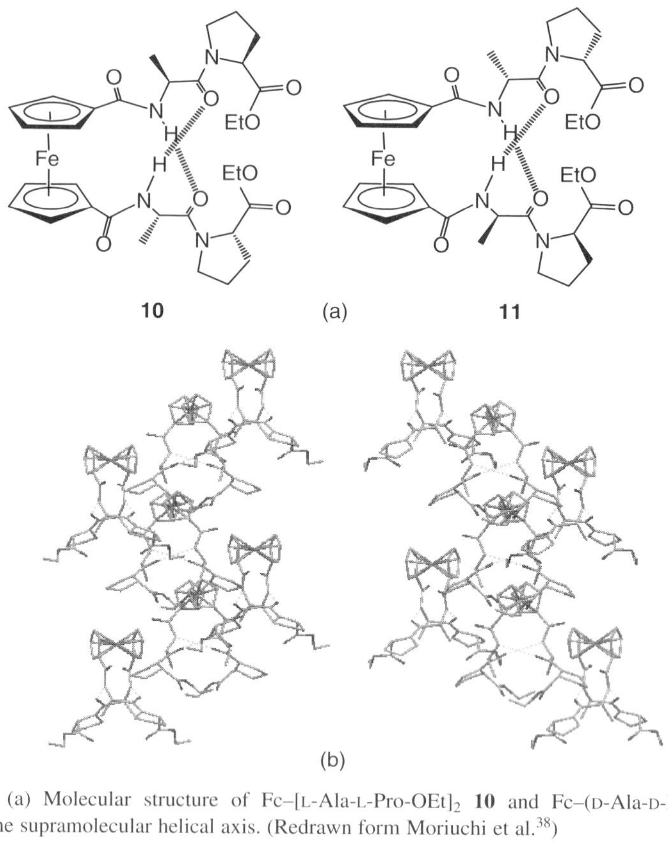

Figure 12.3 (a) Molecular structure of Fc–[L-Ala-L-Pro-OEt]$_2$ **10** and Fc–(D-Ala-D-Pro-OEt)$_2$ **11**. (b) View at the supramolecular helical axis. (Redrawn form Moriuchi et al.[38])

One would expect that the size of the peptide would lead to a significant twist. However, the strong intramolecular H-bonding is the dominant factor, forcing the two amide planes to be coplanar to their respective Cp rings. Changing the ester substituent in this series of compounds has virtually no effect on their molecular conformation. Importantly, the intramolecular H-bonding is preserved in solution, which contrasts the behavior of the monosubstituted Fc–Ala-Pro-OEt **6a**, in which the intermolecular interactions break down in solution. Despite the lack of intermolecular H-bonding, these molecules assemble in a chirality-directed self-assembly into helical structures. The helicity of the superstructure depends on the chirality of the podant peptide. The L-Ala-L-Pro podant peptide chain will result in a helical arrangement, with the podant D-Ala-D-Pro giving the exact mirror image (Figure 12.3). Both helices have a pitch height of >14 Å and a distance between the Fc groups of approximately 4.5 Å. The related GlyPro and ProGly systems form related supramolecular arrangements in which intramolecular H-bonding determines the structure of the assembly.[39] Importantly, the helical chirality of the Fc core itself is preserved in solution.

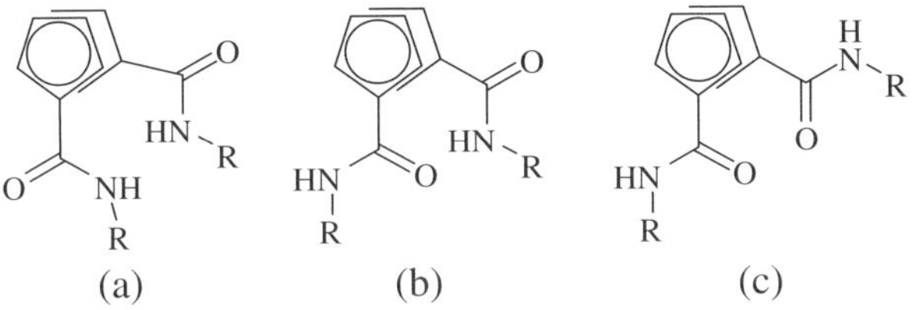

Scheme 12.2 Some of the possible orientations of the amino acid and peptide substituents. Structure (a) is the most commonly found structural motif, having the two C=O antiorientations.

In 1,2′-substituted Fc derivative, the two peptide substituents can adopt three different orientations with respect to each other, as shown in Scheme 12.2. The two Fc–CO groups on the two Cp rings are pointing away from each other (Figure 12.4a), allowing the remaining peptide

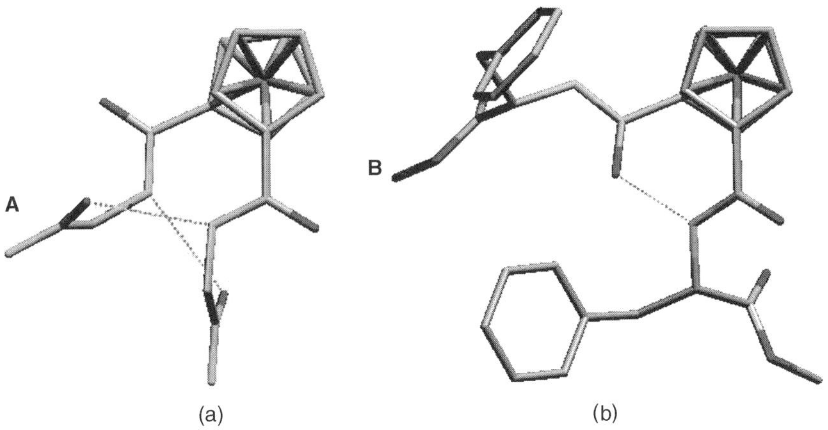

Figure 12.4 1,2′-Substituted ferrocene systems showing the two different conformations of the amino acid substitutents: (a) conformation **A**: drawing of the molecular structure for Fc[Gly-NH_2]$_2$ **10**, showing the two C=O pointing outward, allowing the formation of interstrand H-bonding between the two amide C=O and N—H; (b) conformation **B**: drawing of the molecular structure of Fc[Phe-OMe]$_2$ **13**, showing one of the C=O pointing inward and the other pointing outward, resulting in the formation of only a single H-bond. (Redrawn from Georgopoulou et al.[36] and van Staveren et al.[40])

chain to engage in intramolecular H-bonding interactions between the Fc–amide N–H and the amino acid or peptide carbonyl groups, resulting in the formation of two interstrand H-bonds. This is the typical "Herrick pattern," commonly observed in disubstituted Fc–peptide conjugates. For example, in Mingos' Fc[Gly-NH_2]$_2$ **10**, two H-bonds are formed with N···O distances of 2.88 Å.[36] However, for the sterically more demanding isopropyl substituent in Erker's Fc[Val-OMe]$_2$ **12**, only very long N···O contacts of 3.247 Å are established. Metzler-Nolte[40] described the only case in which isomer **B** is observed. In Fc[Phe-OMe]$_2$ **13**, one of the carbonyl groups points inward toward the amide group of the other amino acid substituent. As a consequence, only a single interstrand C=O···H–N H-bond is formed with a N···O contact of 2.832 Å.

A series of 1,1′-oligoprolinoyl-ferrocenes **14–17** were obtained from 1,1′-ferrocenedicarboxylic acid **7**, HOBt, EDC, and a series of oligoproline esters (Scheme 12.3).[41,42] The proline derivative **5a** was first obtained by Herrick via the acid chloride route.[29] Importantly, this route also offered a way to introduce site differentiation between the two Cp rings and in essence allowed the formation of Fc–peptide conjugates, in which two different peptides could be attached to the two Cp rings. Using only one equivalent of peptideester, ring-differentiated 1-oligoprolinoyl-1′-OBt-ferrocene derivatives **18–20** were obtained.

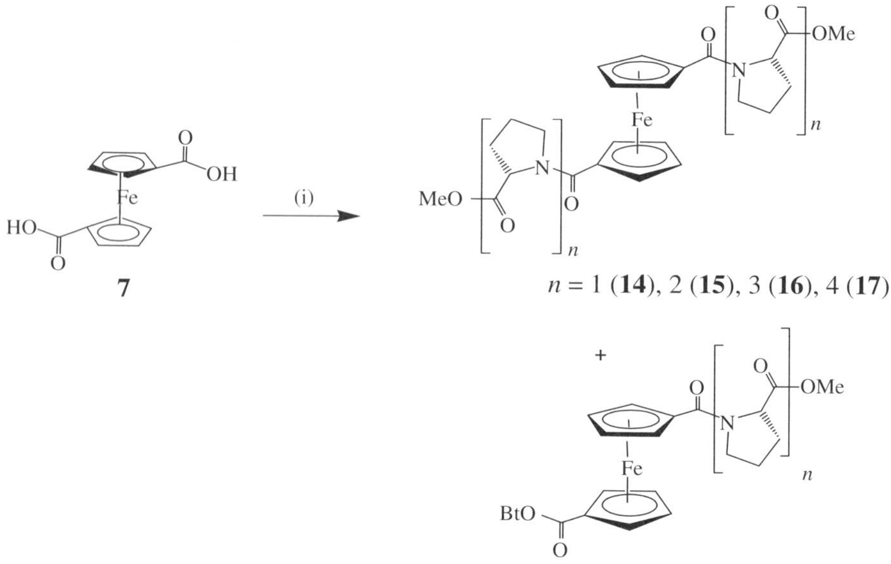

7

n = 1 (**14**), 2 (**15**), 3 (**16**), 4 (**17**)

n = 2 (**18**), 3 (**19**), 4 (**20**)

Scheme 12.3 Synthesis of 1,1′-bispeptide ferrocene derivatives peptides via the HOBt/EDC protocol: (i) HOBt, EDC, CH_2Cl_2, H-Pro$_n$-OMe (n = 1–4).

This class of compounds is a convenient starting material en route to asymmetrically substituted ferrocenoyl derivatives, which may potentially have applications in proteomics and protein sensing. An unusual procedure was reported in 1996 by Erker and co-workers leading to the formation of Fc–Val-OMe (Val = Valine) **22**, Fc[Val-OMe]$_2$ **12** (Scheme 12.4) from two equivalents of the Li-salt of a Cp-Val = OMe conjugate with anhydrous $FeCl_2$ in tetrahydrofuran (THF).[43]

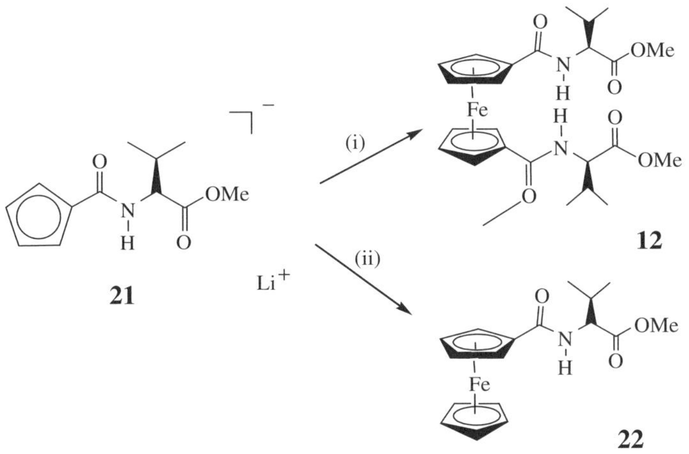

Scheme 12.4 Unusual synthesis of Fc–Val-OMe **22**, Fc[Val-OMe]$_2$ **12** from the Li salt of Cp-Val-OMe **21** formed by the addition of isocyanate: (i) $FeCl_2$, THF, 60°C, 7*d*; (ii) $FeCl_2$/CpLi, 50°C, THF, 7*d*.

Using the same procedure, the Fc–peptide macrocycles **23–27** were obtained in low to moderate yields by the reaction of ferrocenedicarboxylic acid with the appropriate amino acid–cystamine conjugates in the presence of carbodiimide and HOBt under high dilution conditions (Scheme 12.5). At concentrations above 2 mM in ferrocenedicarboxylic acid, the yields are decreased significantly, as expected for the synthesis of macrocycles. Cheng and co-workers reported the synthesis of the related Fc–peptide–cystine macrocycles using the acid chloride method starting from Fc(COCl)$_2$ **27**. In addition to the desired systems, a series of dimeric Fc–macrocycles was obtained (see Section 12.3).[44] Interestingly, the cystine macrocycles were able to coordinate alkali ions very effectively. A number of such Fc–peptide

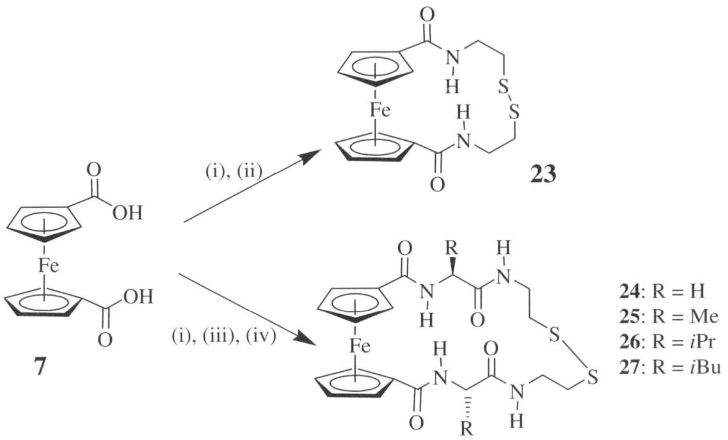

Scheme 12.5 Synthesis of redox active cyclopeptides : (i) EDC, HOBt, CH_2Cl_2; (ii) $(CSA)_2 \cdot 2HCl$, CH_2Cl_2; (iii) (a) Boc(*t*-butyloxycarbonyl)–amino acid CSA (cystamine) dimer, TFA (trifluoroacetic acid); (b) Et_3N, CH_2Cl_2; (iv) dilution.

conjugates were prepared incorporating Gly, Ala, Val, and Leu (Leu-leucine). The structure of the Gly– and Ala–peptide conjugates [Fc–Gly-CSA]$_2$ **24** and [Fc–Ala-CSA]$_2$ **25** is shown in Figure 12.5. The cystamine disulfide was used for the preparation of Fc–peptide films on gold substrates and their subsequent investigations of the electron-transfer properties from the Fc group through the peptide spacer to the gold surface.[45]

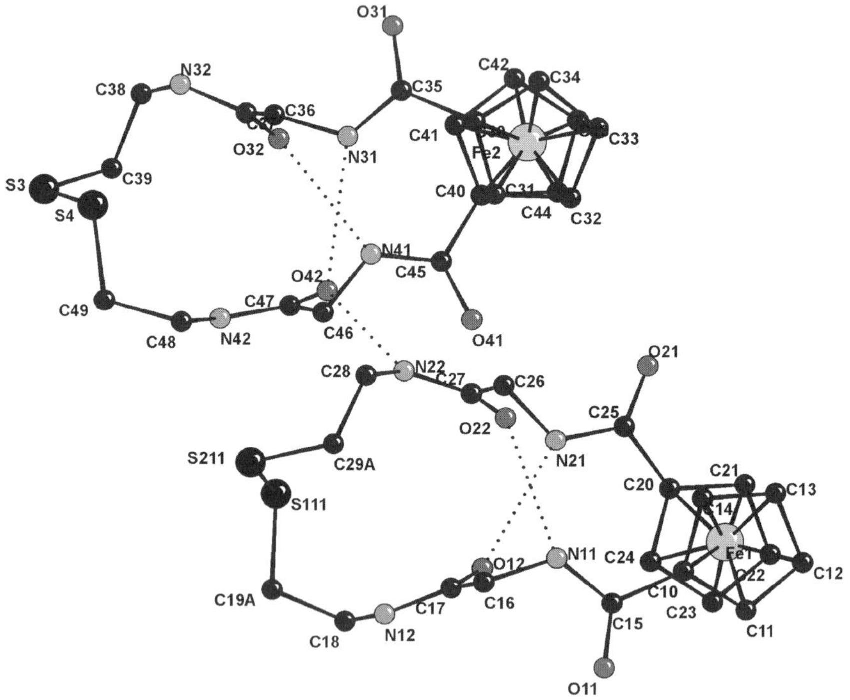

Figure 12.5 Molecular structure of compounds **24**, **25** showing the intra- and intermolecular H-bonding interactions. Note the 1,2′-conformation of the Fc core and its typical Herrick pattern.

Interestingly, expansion of the ring by additional amino acids results in the formation of a fascinating macromolecular assembly, in which the peptide macrocycles associate into "pseudo-β-barrels".[46] Fc[Gly-Val-CSA]$_2$ **28** has Gly as the first amino acid to be attached to the Fc group, giving maximum flexibility to the podant peptide chains, followed by Val, an amino acid with very high propensity to form β-sheets.[47] The nuclear magnetic resonance (NMR) evidence indicates the presence of water in the pseudobarrel. The Fc–system itself adopts a Herrick conformation, and the cystamine acts as a structural restraint, enabling additional H-bonding and establishing the appropriate H-bonding interface on the "exterior" of the macrocycle, which allows β-sheet-type interaction with adjacent molecules (Figure 12.6). The resulting eight β-stranded barrel has the strands arranged in a up–up–down–down–up–up–down–down pattern. However, in contrast to naturally occurring barrel structures, the individual strands are oriented virtually parallel to the axis of the pseudobarrel, while in naturally occurring barrels the strands are tilted with respect to the barrel axis.

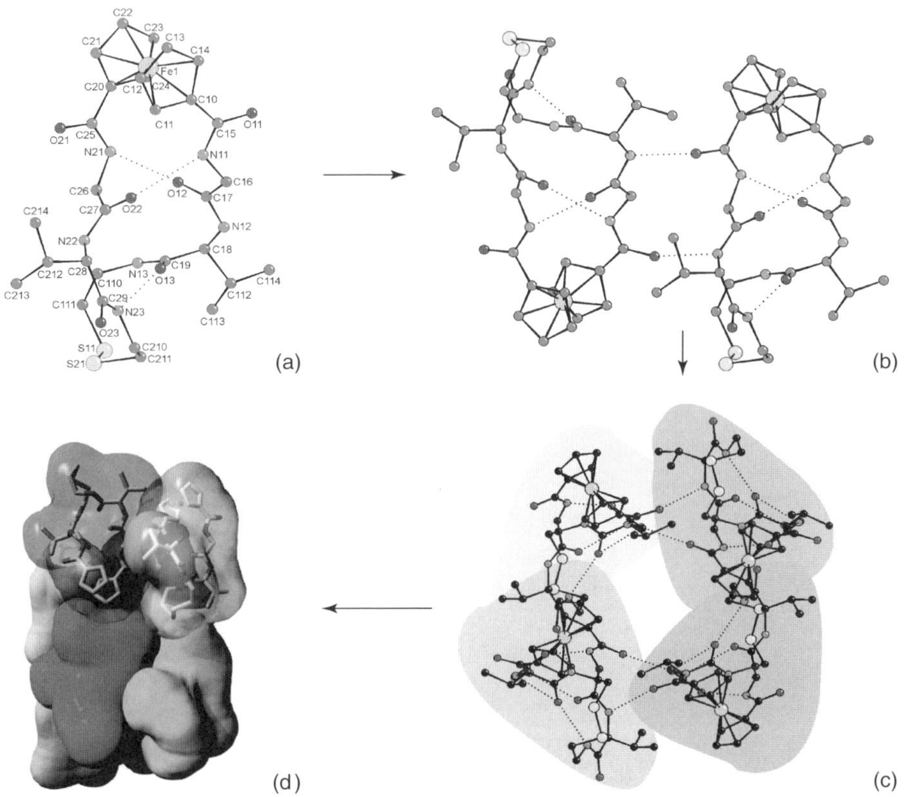

Figure 12.6 The crystal structure of Fc[Gly-Val-CSA]$_2$ **28** and schematic view of the formation of the pseudo β-barrel. (a) Molecular structure of compound **28** showing the *P*-helicity of the ferrocene unit and the intramolecular H-bonding between the two podant peptide strands. The geometry of the Fc group, the amide substitutents and bond lengths and angles are well within established parameters for Fc–peptide conjugates.[48] (b) Formation of β-sheets through intermolecular N(H)···O=C hydrogen bonding. The head-to-tail interaction of the molecules results in an unusual up–up–down–down arrangement of the peptide strands. The arrangement between the two peptide strands on the two Fc conjugates is antiparallel. (c) Four molecules interact via H-bonding to form a β-barrel. A view down the *c* axis is shown. The disordered water molecules inside the cavity of the barrel were omitted for clarity. (d) Molecular surface representation showing a side view of the half-barrel viewed along the *c* axis formed by tiling of cyclic ferrocene peptides. (From Chowdhury et al.[46] Reprinted with permission. Copyright © 2005 Wiley-VCH.)

B. 1′-Aminoferrocene-1-carboxylic acid and 1,1′-Diaminoferrocene Conjugates

Until recently, the ferrocene monomers used to prepare oligomeric and polymeric ferrocene peptide-like systems, like 1′-aminoferrocene-1-carboxylic acid (Fca) **29** and 1,1′-diaminoferrocene **30** were difficult to synthesize and also were not stable.

In several of these synthetic procedures 1-bromo-1′-dilitioferrocene has been used as a key precursor; this compound can be formed by treating of 1,1′-dibromoferrocene **31** with *n*-butyllithium at $-70°C$.[49,50,51] 1′-Aminoferrocene-1-methylcarboxylate **36** was obtained as a major product from the synthetic route described by Butler (Scheme 12.6). This compound can be considered as a promising monomer for the formation of the peptide-like ferrocene backbone polymers. However, the material is readily oxidizable and easily contaminated with butylated by-products.[51]

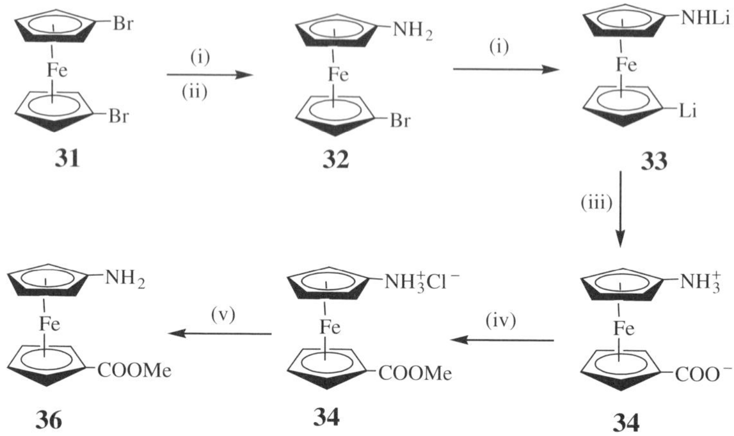

Scheme 12.6 Synthesis of 1′-aminoferrocene-1-methylcarboxylate: (i)BuLi, (ii) $BnONH_2$, (iii) CO_2, (iv) AcCl/MeOH, (v) NaOH.

An alternative synthetic approach to Boc-protected 1′-aminoferrocene-1-carboxylic acid **40** was recently described by Rapic and co-workers, and is shown in Scheme 12.7. Hydrolysis of carbamate-ester by an equimolar amount of sodium hydroxide in aqueous ethanol gave 90% of *tert*-butyl 1′-carboxy-1-ferrocenecarbamate.[52] During the synthesis the symmetrical ferrocenyl-substituted urea, dimethyl 1′,1′-ureylenedi(1-ferrocenecarboxylate) **39** was formed as a by-product (Scheme 12.7).[53] The ferrocenium molecules were conjugated through a urea linker, which upon deprotection may be used in condensation reactions.

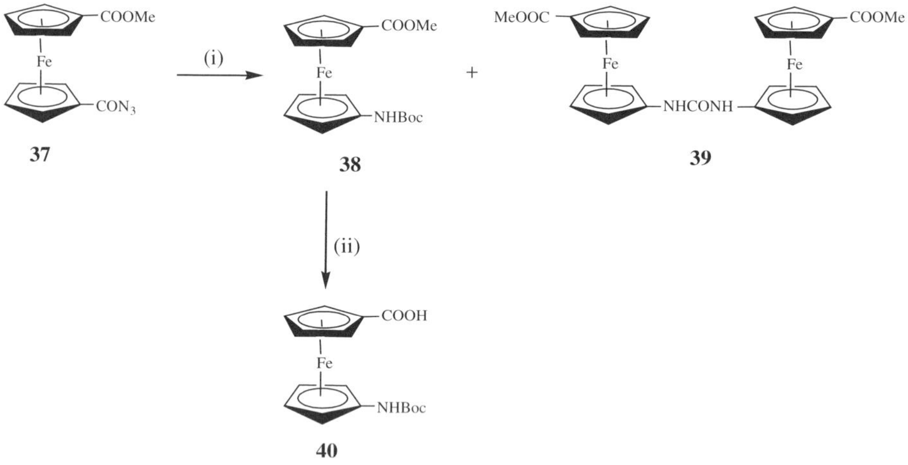

Scheme 12.7 Synthesis of Boc-protected 1′-aminoferrocene-1-carboxylic acid. (i) *t*-BuOH, reflux for 4 hours; (ii) $NaOH/H_2O$, MeOH.

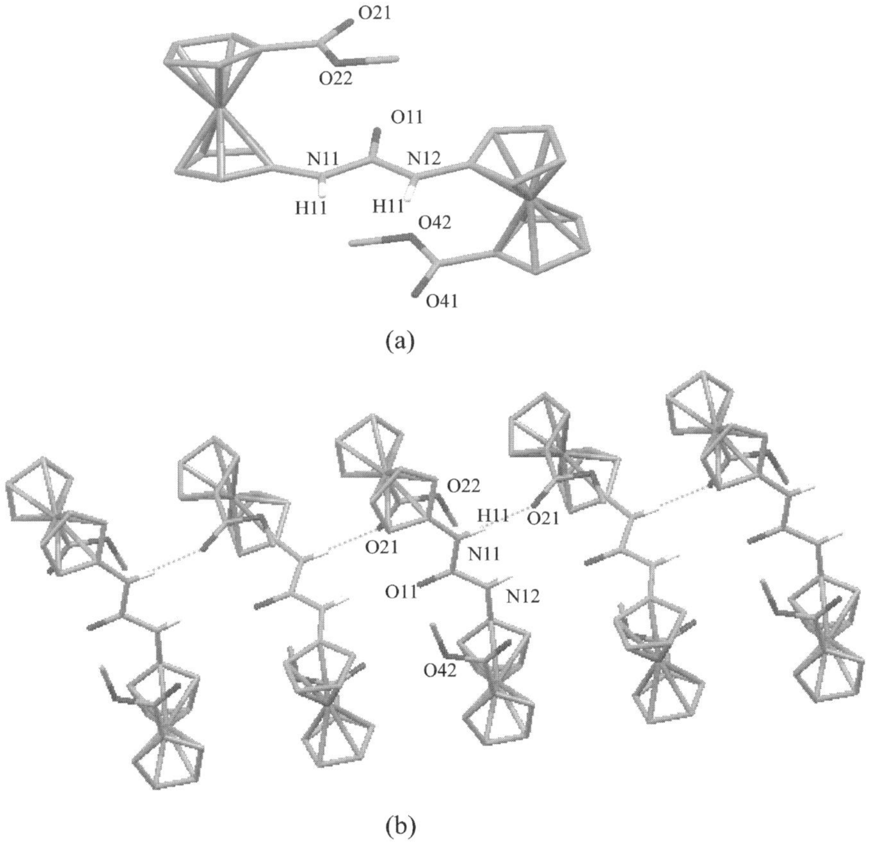

Figure 12.7 (a) A structural drawing of dimethyl-1′,1′-ureylenedi(1-ferrocenecarboxylate) **39**; (b) the crystal packing showing the intermolecular hydrogen bonding pattern. (Redrawn from the crystal structure in Mahmoud and Kraatz.[53])

In comparison to other ureas, the urea O is not involved in intermolecular H-bonding. Instead, intermolecular H-bonding involves the methylester C═O and the urea N—H of an adjacent molecule resulting in a slight asymmetry of the urea group, resulting in the formation of a one-dimensional chain. However, on the molecular level, this introduces a degree of asymmetry in the molecule that manifests itself in two different C—N bond lengths (N(11)—C(15) = 1.367(3) Å, N(12)—C(15) = 1.378(3) Å). Electrochemical studies show that the two Fc groups are electronically coupled with a coupling constant K_c = 207, which classifies this particular system as a Robin–Day class II system.

Heinze proposed a convenient synthetic route starting from ferrocene to give 1-aminoferrocene-1′-carboxylic acid hydrochloride **45**.[54] Scheme 12.8 summarizes the reaction sequence. The key synthon in this synthesis is 1-aminoferrocene provided by the Gabriel synthesis from *N*-ferrocenylphthalimide. However, this route requires the synthesis of 1-haloferrocene or ferrocene boronic acid as precursors of *N*-ferrocenylphthalimide. In addition, the reaction is accompanied by the formation of biferrocenes due to a copper-mediated C—C coupling reaction, even in the presence of excess phthalimide.[55,56,57]

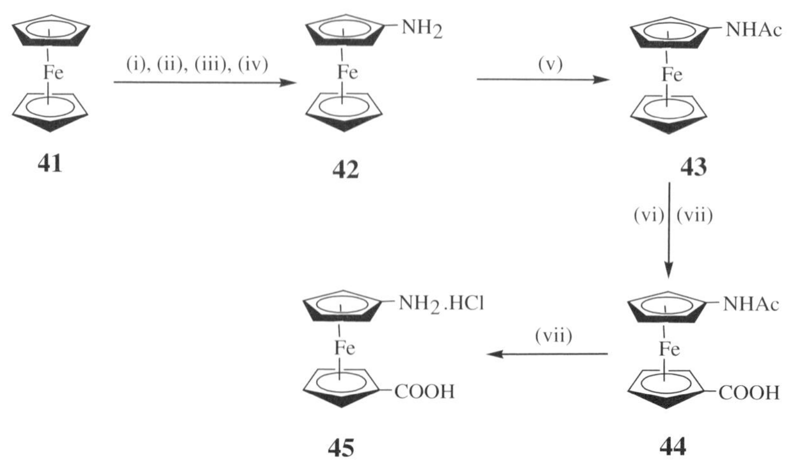

Scheme 12.8 Synthesis of ferrocene amino acid according to Heinze: (i) *t*-BuLi, −70°C; (ii) quenching with iodine; (iii) reaction in pyridine with phthalimide and Cu_2O as catalyst; (iv) hydrazinolysis in ethanol; (v) *N*-acetylation with Ac_2O; (vi) selective Friedel–Crafts acylation with 2,6-dichlorobenzoyl chloride at the unsubstituted Cp ring; (vii) base hydrolysis; (viii) removal of the acetyl protection group with hydrochloric acid.

Rapic and Metzler-Nolte reported the preparation of the first oligopeptide conjugate of Fca as shown in Scheme 12.9.[58] The unnatural organometallic amino acid Fca induces a turn in the structure forming this peptide conjugate with an antiparallel motif. The structure is supported by two intramolecular H-bonds in the solid state, as shown in Figure 12.8. The

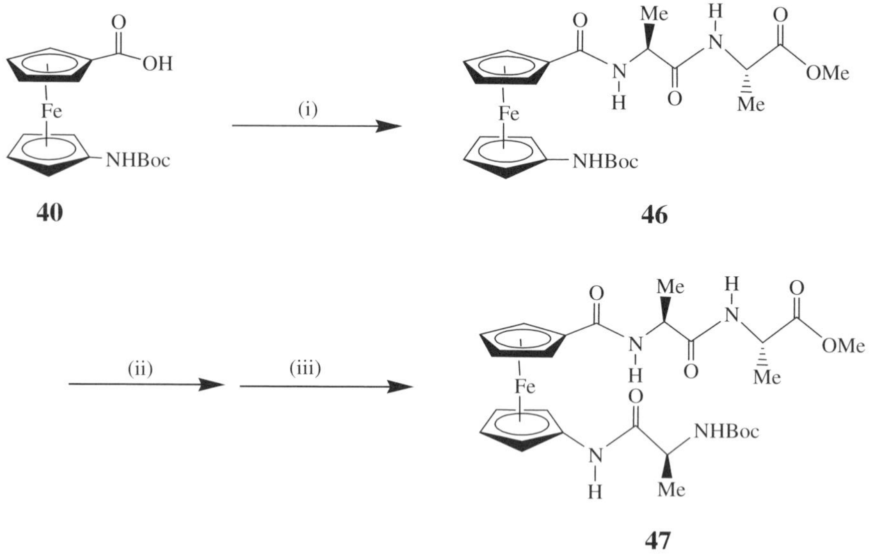

Scheme 12.9 Synthesis of the tetrapeptide Boc-Ala-Fca-Ala-Ala-OMe **47** derived from 1′-aminoferrocene-1-carboxylic (Fca) **40**: (i) +H-Ala-Ala-OMe, EDC/HOBt; (ii) HCl(g)/AcOEt; +Boc-Ala-OH, EDC/HOBt.

N—H—O bond angle deviates only slightly from linearity, and N—O distances around 2.9 Å indicate strong hydrogen bonding. The angle ω between the substituents on the Cp rings is 60.7°, which confirms a typical 1,2′-conformation of the Fc system. The structure adopts a *P*-helical arrangement in the solid state and in solution, which parallels work by Hirao and co-workers on peptide conjugates of ferrocene dicarboxylic acid conjugates.[38] Additional studies show that amino acid and peptide conjugates of Fca in general adopt a β-turn-like structure in which the peptide strands are forced to adopt an antiparallel orientation with respect to each other.[59]

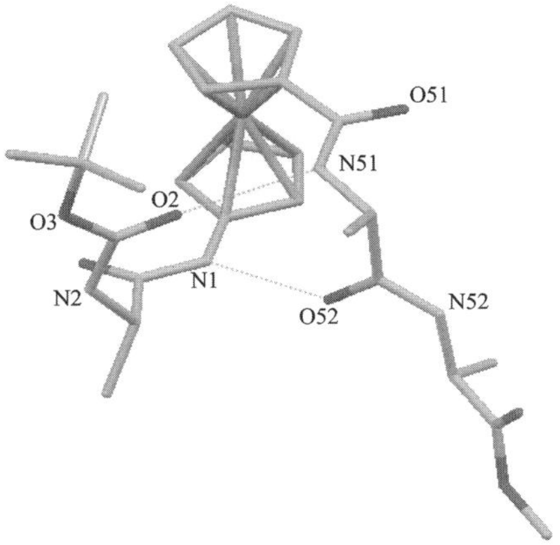

Figure 12.8 X-Ray structure of the first Fca peptide conjugate **47** showing two intramolecular hydrogen bonds. Selected hydrogen bond distances (Å) and angles (°): N1—O52 2.812(3), N51—O2 2.914(3), N1—H1N—O52 169(3), H51—H51N—O2 157(3). (Redrawn from Barisic et al.[58])

Another key compound is 1,1′-diaminoferrocene **30**. Recently, Arnold described an improved high-yield synthesis of diaminoferrocene from diazidoferrocene using a modified Nesmeyanov procedure (Scheme 12.10). Diaminoferrocene (see Figure 12.9) was prepared from diazide by reduction with H_2-Pd/C in MeOH. Removal of the Pd/C catalyst and crystallization at −30°C afforded diaminoferrocene in a good yield. However, extreme care should be exercised when handling solid diazidoferrocene, as it tends to explode if heated rapidly above 56°C.[60]

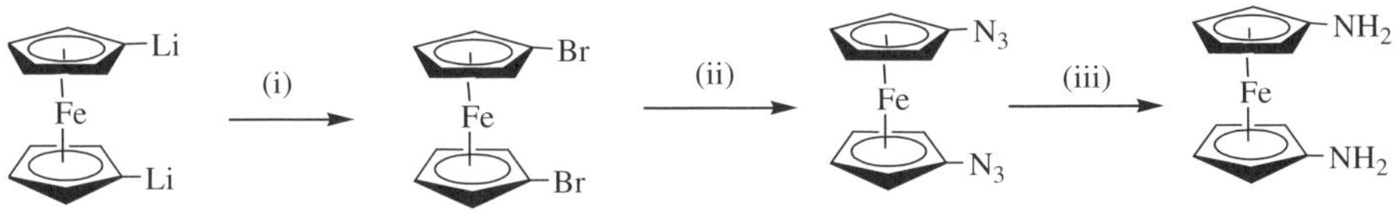

Scheme 12.10 Arnold's synthesis of 1,1′-diaminoferrocene involving the explosive diazidoferrocene: (i) $C_2H_2Br_4$, Et_2O; (ii) NaN_3, CuI, EtOH/H_2O; (iii) H_2, Pd/C, MeOH.

Recently, the synthesis of 1,1′-bis(*tert*-butoxycarbonylamino)ferrocene **52** was reported, a convenient synthon of 1,1′-diaminoferrocene **30**, which circumvents the problems encountered with the explosive diazide. The bis-Boc derivative is compatible with the peptide synthesis protocol, as is demonstrated by the preparation of its first amino acid conjugates. 1,1′-Bis(carbonylazido)ferrocene was synthesized from 1,1′-ferrocenedicarboxylic acid **7** by treatment with ethyl chloroformate, followed by reaction with sodium azide. The resulting carbonylazido 1,1′-Bis(carbonylazido)ferrocene **51** is then reacted with *tert*-butanol to give 1,1′-bis(*tert*-butoxycarbonylamino)ferrocene **52**.[61] Figure 12.10 shows a structural drawing of this material, in which several rotamers coexist within the same material, due to the absence

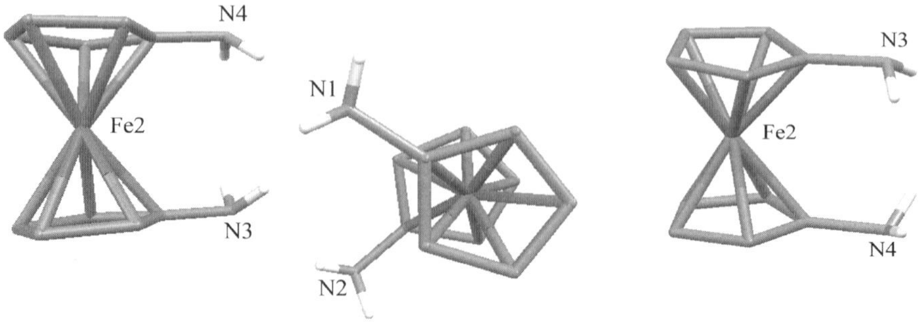

Figure 12.9 Diagram of the two independent $Fc(NH_2)_2$ molecules **30** shown with 50% thermal ellipsoids. (Redrawn from Shafir et al.[60])

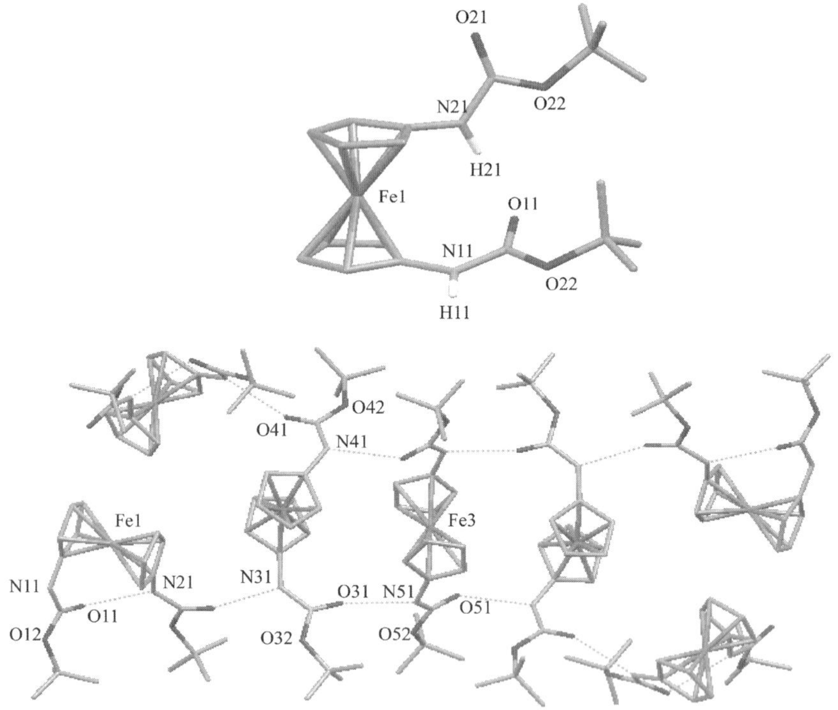

Figure 12.10 Structure of 1,1′-bis(*tert*-butoxycarbonylamino)ferrocene **52** showing three rotamers having the two substituents at different rotational angles with respect to the Cp-Fe-Cp vector with an approximate 1,2′ (70°) and 1,3′ (144°, 180°) conformation. All three rotamers are linked via intermolecular N(H)···O=C bonding. (Redrawn from Chowdhury et al.[61])

of strong intramolecular H-bonding, as encountered in Fc–peptide conjugates that stabilize a particular conformation.

The Cp rings in both ferrocenoyl groups are coplanar (angle between Cp planes: 2.8(2)°). The amide and Cp rings are twisted by 18.9(2)°. Importantly, the molecules maximize their H-bonding through formation of intermolecular and intramolecular N(H)···O═C bonding. Two intramolecular H-bonds (2.805(6) Å) are formed between the amide groups connected to the Cp rings and the adjacent carbonyl of the Boc, causing *M*-helicity of the compound. Importantly, for amino acid conjugates the intramolecular NH···O═C H-bonding results in the formation of a previously unobserved 14-membered H-bonded ring, which is a new structural motif for Fc–peptide conjugates.

III. OLIGOMERIC AND POLYMERIC FERROCENE AMIDES AND DENDRIMERS

Ferrocene-based organometallic polymers have received considerable attention since the first report of poly(vinylferrocene) by Arimoto and Haven at DuPont Inc. in 1955.[22–27,62–72] A large number of attempts to produce ferrocene-containing polymers, including polyalkenes, polyketones, polyesters, polyamines, polyamides, polyurethanes, and polyureas, were reported.[68,73] Polyamidoamine-based conjugates represent a class of polymers that can be readily ferrocenylated through amide-linked Fc in which the metallocene is linked to the side chain.[74] Early synthetic efforts focused on the modification of the polymer side chains with Fc groups. In particular, ferrocenylalkylcarboxylic acids, Fc–$(CH_2)_n$—COOH ($n = 0$–3), were useful, allowing a facile conjugation to the polymer by a variety of methods.[62,66,75–79]

The preparation of polyamides having metallocenes incorporated into the backbone can be achieved by the use of bifunctional ferrocene derivatives, such as diacid chlorides or diamines. In the following, a few examples of such reactions are presented. In the early 1960s Knobloch and Rauscher reported the preparation of polyamides and polyesters by the reaction of 1,1′-ferrocenyldicarbonyl chloride **14** with several diamines and diols by interfacial polycondensation.[80] The synthesis of elastomeric polyamides **54a**, M_n = 10,000–18,000) in high yields was reported by Rausch and co-workers from 1,1′-bis(β-aminoethyl)ferrocene **53** and diacid chlorides (Scheme 12.11). The reaction with bis-isocyanates allows the formation of ferrocene-containing polyureas **54b**.

Related work by Cuadrado and co-workers focused on the interfacial polycondensation protocol to prepare organosilicon polymer **56** in which the amide-linked ferrocenyl moieties are part of the main polymer chain, as shown in Scheme 12.12.[81] The material was prepared by polycondensation of 1,1′-ferrocenyldicarbonyl chloride with the siloxane **55** in the presence of triethylamine as base.

Using 1,1′-ferrocenyldicarbonyl chloride, it should also be possible to prepare polypeptide conjugates in which the Fc group is part of the polymer backbone and linked to the polypeptide through amide linkages. Han and co-workers were able to prepare 1,1′-ferrocenophane derivatives **57** as well as cyclodimers **58** and macrocyclic compounds from **14** and various cysteine derivatives (Scheme 12.13).[82] Presumably, a nondefined polymer is also formed under the reaction conditions. This reaction is akin to the reaction leading to the formation of other Fc–peptide macrocycles discussed earlier (see Scheme 12.5, and Figures 12.5 and 12.6).

The 1,1′-ferrocenylbisalanine system[83–88] is another interesting example for the synthesis of macrocyclic oligopeptides and potentially of ferrocene-containing polypeptides. The fully protected 1,1′-ferrocenylbisalanine derivative **59** contains orthogonal protecting groups, which allow their deprotection by hydrogenolysis, which is followed by cyclization in the presence of the peptide coupling reagent PyAOP.[83,84,88,89] The result is a mixture of the

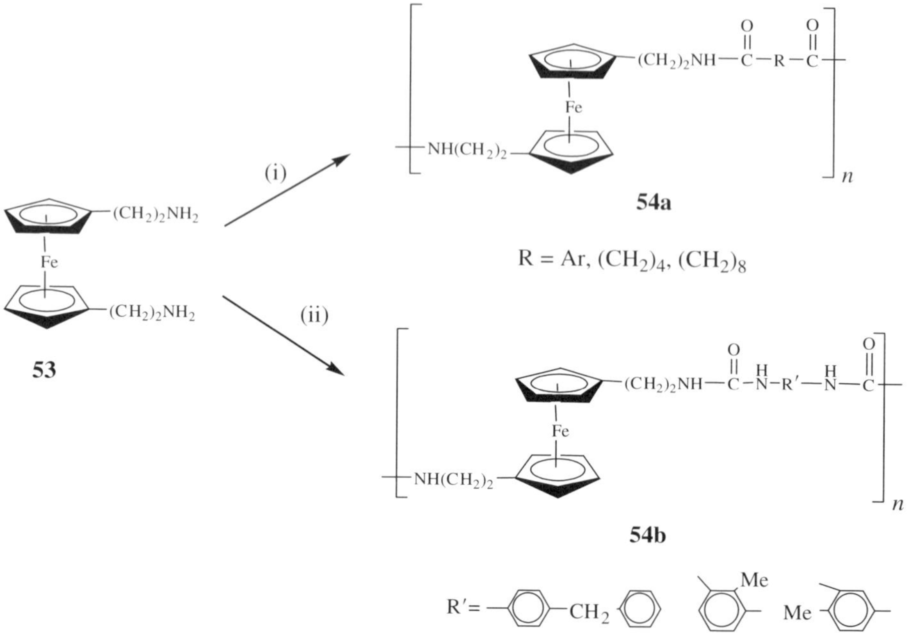

Scheme 12.11 Synthesis of elastomeric polyamides from 1,1′-bis(β-aminoethyl)ferrocene **53** and diacid chlorides. (i) Cl—CO—R—CO—Cl; (ii) OCN—R′—NCO.

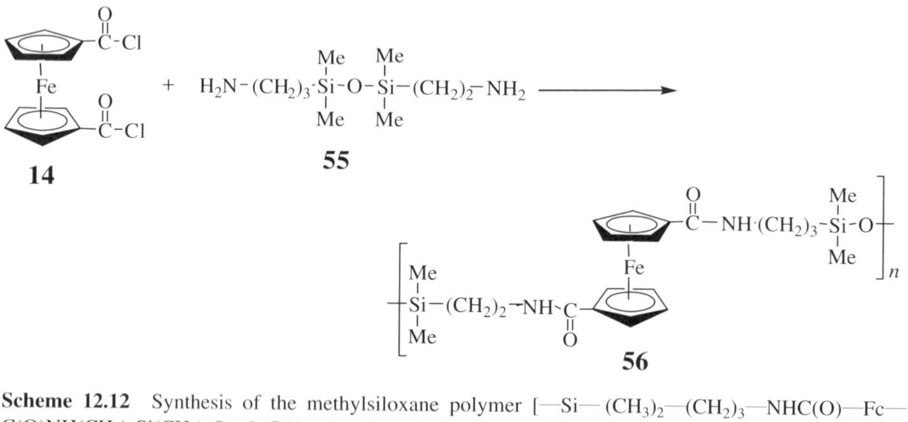

Scheme 12.12 Synthesis of the methylsiloxane polymer [—Si—(CH$_3$)$_2$—(CH$_2$)$_3$—NHC(O)—Fc—C(O)NH(CH$_2$)$_3$Si(CH$_3$)$_2$O—]$_n$ **56** having Fc in the backbone of the polymer chain.

1,1′-ferrocenophane **61** and two macrocyclic peptides, a cyclic "dimer" **62** and a cyclic "trimer" **63** (Scheme 12.14).

An interesting development is the use of Fca for the formation of a ferrocenylene polymer, in which the Fc groups are connected through amide linkages.[51] In essence, Fca polymers could be build up in a stepwise fashion using orthogonalized building blocks. Work by Nakamoto and Heinze are excellent examples for this approach. Starting from a protected Fca derivative, Nakamura was able to obtain its amide dimer.[56] Its structure is related to the symmetrical

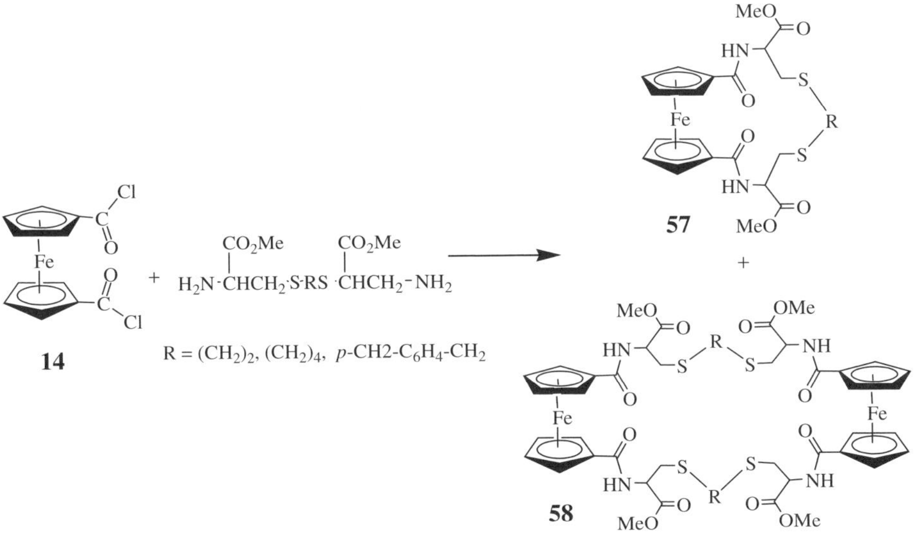

Scheme 12.13 Synthesis of cyclopeptides by condensation of 1,1′-ferrocenyldicarbonyl chloride and bifunctional cysteine derivatives.

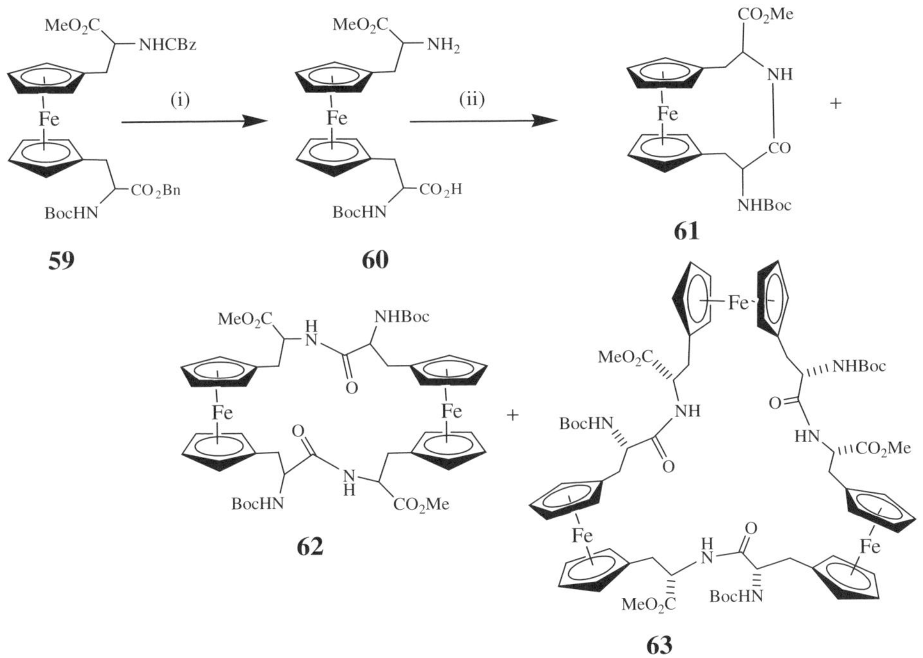

Scheme 12.14 Cyclization of **60** forming macrocyclic oligopeptides **61–63**. (i) H_2, Pd/C(5%), MeOH (ii) 7-azabenzotriazol-1-yloxytris(pyrrolidino)phosphonium hexafluoro phosphate (PyAOP), Dipea, DMF.

ferrocenyl-substituted urea, dimethyl 1′,1′-ureylenedi(1-ferrocenecarboxylate) **39**, which was formed as a by-product in the synthesis of the Boc-protected Fca derivative **38** (see Scheme 12.6).[53] The structure shown in Figure 12.11 clearly shows the amide-linked Fc groups. For the polymeric material, Nakamura proposed a strongly H-bonded structure, shown in Scheme 12.15.

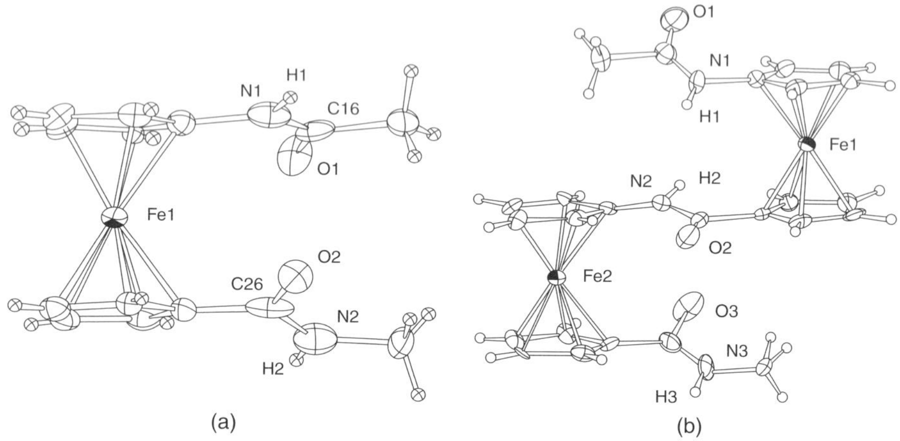

Figure 12.11 (a) ORTEP drawing of the Fca derivative (MeCONH—Cp)Fe(Cp—CONHMe) **64**, (b) ORTEP drawing of the amide dimer (MeCONH—Cp)Fe(Cp—CONH—Cp)Fe(Cp—CONHMe) **65**. (From Okamura et al.[56] Reprinted with a permission. Copyright © 1998 American Chemical Society.)

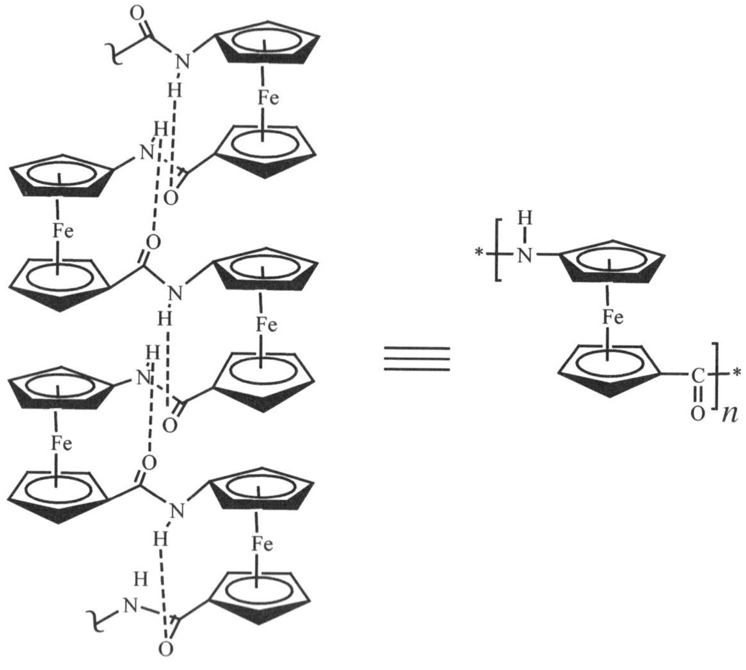

Scheme 12.15 Structure of the organometallic ferrocene-polypeptide proposed by Nakamura.[56]

The ferrocenoyl 1-hydroxybenzotriazole ester **66** reacts with aminoferrocene to give diferrocenyl diamide **68** (Scheme 12.16).[54] Heinze was able to make use of this flexible building block and used it to incorporate it into a peptide backbone by solid-phase synthesis.[90]

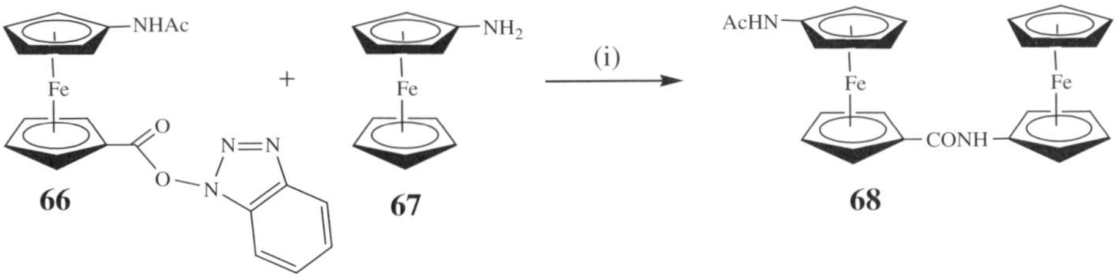

Scheme 12.16 Coupling of ferrocenoyl benzotriazole ester **66** with aminoferrocene **67** to give the diferrocenyl diamide **68**; (i) THF, 12 h.

The crystal structure shows that the diferrocenyl dimer **68** possesses a conformation similar to the Nakamura's dimer. In addition, the individual molecules are connected through intermolecular H-bonding between the NH–acetyl group and the amide C=O (N1···O2 distance: 3.04 Å) and between the NH–amide group and the C=O oxygen atom of the acetyl group (N2···O1 distance: 3.03 Å) resulting in a sheetlike structure (Figure 12.12).

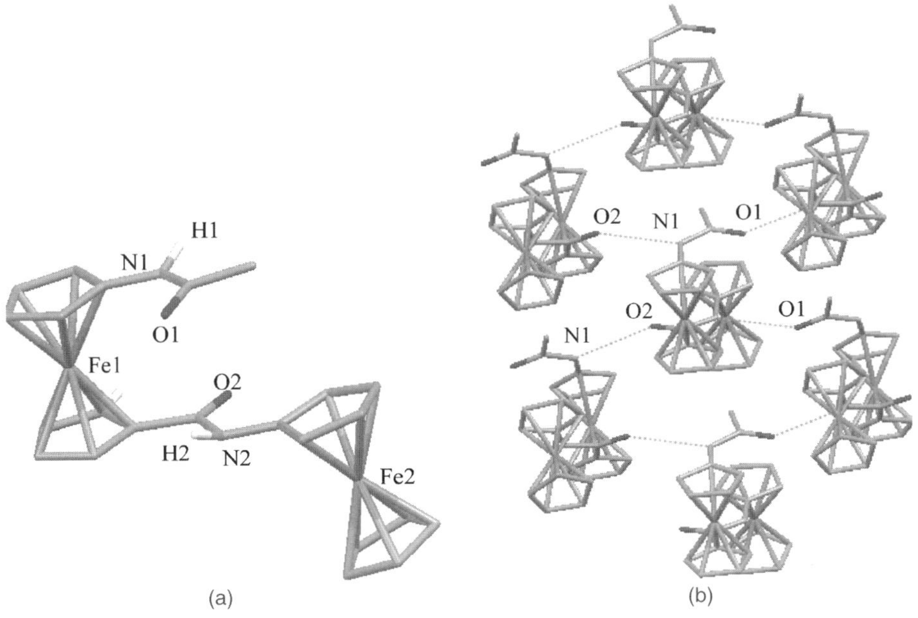

Figure 12.12 (a) Molecular structure of the diferrocenyl peptide **68** in the crystal, (b) hydrogen bonding pattern in the solid state. (Redrawn from Heinze and Schlenker.[54])

In addition to the polyferrocene amide derived from Fca as shown in Scheme 12.15, a second polyferrocene amide **73** should be obtainable by the polycondensation of the recently reported 1,1′-ferrocenediamine **30** and 1,1′-ferrocenyldicarbonyl chloride **14**.

Very recently, this approach was met with success, and a series of novel ferrocene–oligomers **69–72** and polyferrocenyl peptide **73** were reported (Scheme 12.17).[61] Preliminary measurements on the polymeric material indicate a globular structure in aqueous solution and a surprisingly high molecular weight. However, the polymeric material still awaits its full characterization, in particular with respect to its electronic properties.

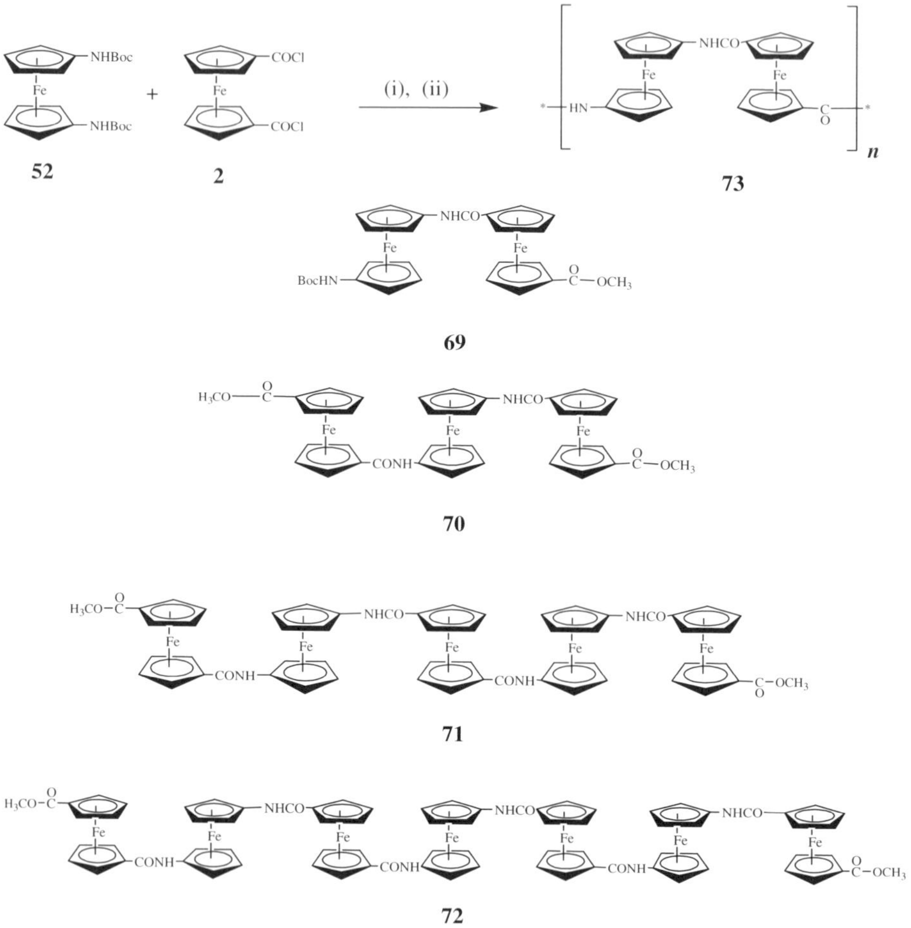

Scheme 12.17 The synthesis of a series of novel ferrocene–oligomers **69–72** and polyferrocenyl peptide **73**, after Boc-deprotection of 1,*n*′-bis(Boc-amino)ferrocene, it was coupled to ferrocene-1,*n*′-dicarbonyl chloride in THF, using the polycondensation protocol.

Conceptually, dendrimers can be viewed as a class of well-defined monodisperse polymers with a globular structure. Dendrimers have been extensively reviewed,[21,62,91,92] so we will take a very narrow view and briefly discuss dendrimers having a ferrocene core. Astruc's ferrocene-terminated dendrimers are reviewed elsewhere and will not be included in this brief discussion.[92]

Although dendrimers are now effectively synthesized by divergent or convergent methods, and their properties are well characterized in many cases, the effect of the dendritic sheath on the electronic properties of a central redox core remains an active area of research. In particular, encapsulation of the central core is one of the major questions, and conflicting information is available. For example, in Kaifer's dendrimers[93,94,95] **74–76**, in which the central Fc group is linked to a dendrimer via an amide linkage, the redox potential exhibits interesting influences as a function of dendrimer size and exterior dendrimer surface. While ester-terminated dendrimers in an organic solvent exhibit a reduction in redox potential of the Fc core, the redox potential for the hydrophilic acid-terminated dendrimers increases with increasing generation in water.

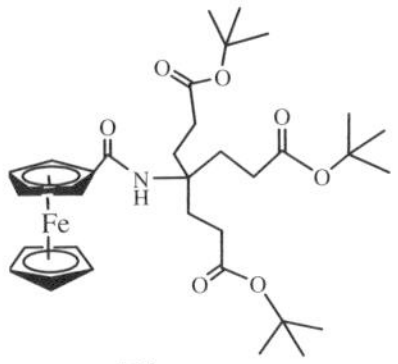

74

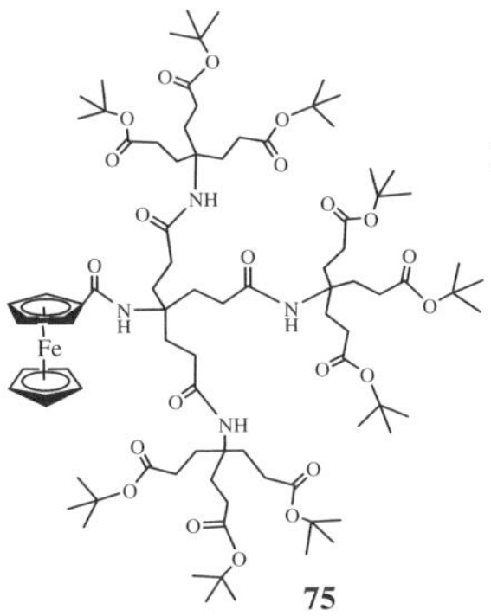

75

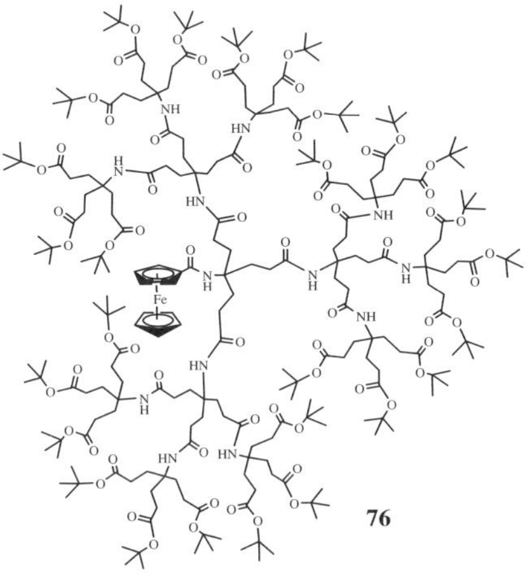

76

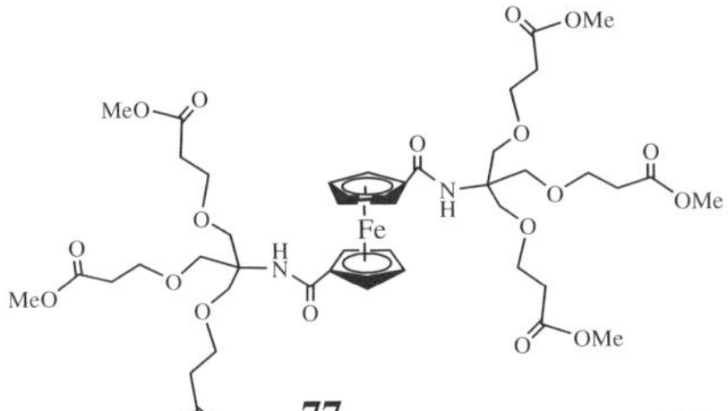

77

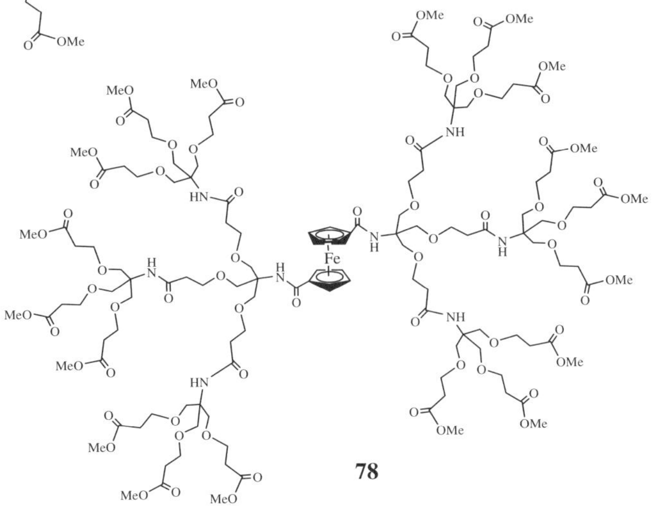

78

Smith reports his Fc–core dendrimers **77**, representing the second generation, and **78** representing the third generation, in which the Fc core is linked to a tris-derived dendrimer.[96] The dendrons act as a nonpolar shield that effectively encapsulates the Fc core and prevents ion penetration to the core. This has the effect of hindering the oxidation of the central Fc core and destabilizing Fc^{+}, causing significant anodic shifts in the Fc redox potential. The synthesis of glutamic acid ester dendrimers having an Fc core by a convergent method was recently reported, leading to dendrimers up to a sixth generation (compounds **78–81** represent the first to fourth generations).[97]

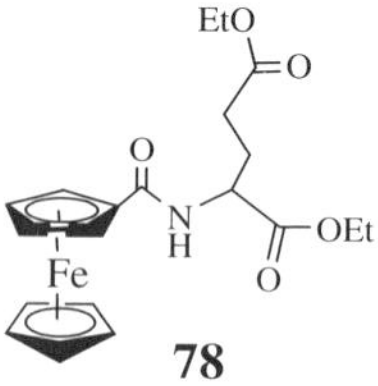

78

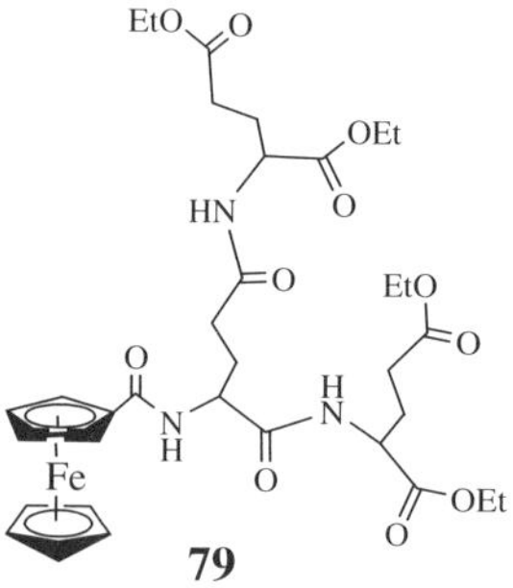

79

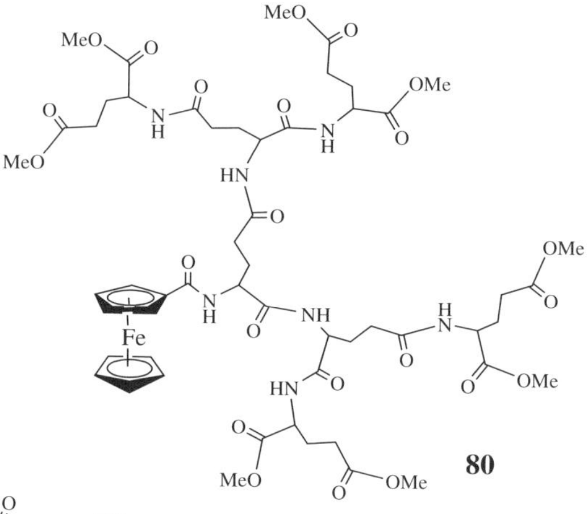

80

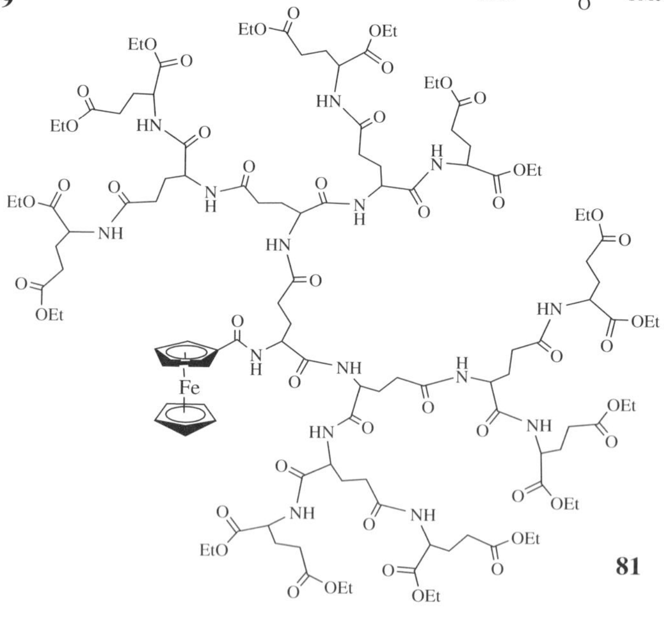

81

It is interesting to note that despite the potential steric problem for the synthesis of higher-generation dendrimers by the convergent method, the Fc group is readily attached to the "buried" amino group on the interior of the dendrimer to give the sixth-generation dendrimer. This is rationalized by a less compact dendrimer structure in organic solvents allowing good solvent penetration. However, in polar solvents, the dendrimer adopts a globular structure, which shields the central core. Upon increasing the dendrimer size, the properties of the central Fc group are attenuated, indicating the progressive encapsulation of the central core. Solution electrochemistry results show that the sixth generation is almost completely encapsulated. Most likely, strong intramolecular H-bonding enables the peptide dendrons to fold upon on themselves, which results in structurally well-defined globular molecules, as is indicated by spectroscopic studies.

IV. CONCLUSIONS AND OUTLOOK

In this chapter, we have reviewed ferrocene peptide conjugates with particular emphasis on derivatives that are amenable to polycondensation reactions and cyclization reactions.

Ferrocene peptide conjugates are readily obtained by a variety of peptide coupling techniques from often commercially available synthons. As would be expected, the peptide backbone plays a major role in that it enables the balancing of inter- and intramolecular hydrogen bonding. It seems that at last, some design principles appear that assist in the rational structural design of such assemblies to emerge slowly. The control of the flexibility of the peptide backbone appears to be a critical aspect. As work on Fc–cyclopeptides suggest, rigidity of the backbone allows the design of an interface that controls the interactions with adjacent molecules. The use of recently reported disubstitued Fc derivatives, which impose specific structural parameters on the peptide backbone, have allowed the design of β-turns and sheetlike structures. But despite these advances, many problems remain unsolved, and in many cases predicting the correct supramolecular assembly remains a difficult task, so that many structural discoveries in this area remain accidental. In the long-term, however, a thorough understanding of the design principles that govern these fascinating redox-bioconjugates may open up the possibility of generating tailor-made redox-active peptidic materials of fixed lengths and electronic makeup with potential application in nanoelectronics. The use of polymeric Fc oligo- and polypeptide systems is of interest in this regard, and may have tremendous potential as all-Fc polymers or Fc–peptide copolymers. Here we find a potentially useful way to generate redox materials by condensation reactions of disubstituted Fc precursors, an area that has been under-explored, largely due to the absence of readily accessible Fc diamino components. Despite a few reports in the literature, Fc–polyamides remain a fairly poorly characterized group of polymers and still face some serious synthetic challenges. Polycondensation results in a series of oligomers and a polymeric material. Their electronic properties, in particular the extent of the electronic delocalization in the polymer, is not known. It would be expected that upon partial oxidation of these materials adjacent sites will interact more strongly.

Peptide–dendrimers, on the other hand, have now become a staple of well-defined macromolecules. Dendrimers having an Fc core are readily synthesized and represent structurally well-defined systems, often stabilized by strong intramolecular H-bonding, in higher generations in particular. As a consequence of the often globular structure, the Fc core is becoming progressively encapsulated with each generation, resulting in an attenuation of the redox properties of the Fc core. Importantly, these dendrimeric systems may offer an environment that is well-suited for a range of guest molecules, potentially making these systems useful receptors for small molecules for the transport and delivery of drugs. Uptake will cause slight changes in the dendrimer environment, which in turn will affect the redox properties, thereby making it possible to electrochemically monitor drug uptake and delivery in vitro and possibly in vivo.

ACKNOWLEDGMENTS

This work was supported by the National Science and Engineering Research Council. One of the authors (H.B.K.) holds the Canada Research Chair in Biomaterials.

REFERENCES

1. G. Jaouen, *J. Organomet. Chem.*, **589**, 1 (1999).
2. G. Jaouen, A. Vessieres, I. S. Butler, *Acc. Chem. Res.*, **26**, 361 (1993).
3. D. R. van Staveren, N. Metzler-Nolte, *Chem. Rev.*, **104**, 5931 (2004).
4. R. H. Fish, G. Jaouen, *Organometallics*, **22**, 2166 (2003).
5. C. Biot, L. Delhaes, H. Abessolo, O. Domarle, L. A. Maciejewski, M. Mortuaire, P. Delcourt, P. Deloron, D. Camus, D. Dive, J. S. Brocard, *J. Organomet. Chem.*, **589**, 59 (1999).
6. C. Biot, G. Glorian, L. A. Maciejewski, J. S. Brocard, *J. Med. Chem.*, **40**, 3715 (1997).
7. F. Noor, A. Wuestholz, R. Kinscherf, N. Metzler-Nolte, *Angew. Chem. Int. Ed.*, **44**, 2429 (2005).
8. J. T. Chantson, M. V. V. Falzacappa, S. Crovella, N. Metzler-Nolte, *J. Organomet. Chem.*, **690**, 4564 (2005).
9. W. A. Petka, J. L. Harden, K. P. McGrath, W. D., D. A. Tirrell, *Science*, **281**, 389 (1998).
10. A. Aggeli, I. A. Nyrkova, M. Bell, R. Harding, L. Carrick, T. C. B. McLeish, A. N. Semenov, N. Boden, *Proc. Natl. Acad. Sci. USA*, **98**, 11857 (2001).
11. J. P. Schneider, J. W. Kelly, *Chem. Rev.* **1995**, **95**, 2169.
12. J. S. Nowick, *Acc. Chem. Res.*, **32**, 287 (1999).
13. (a) M. R. Ghadiri, Granja, J. R., Milligan, R. A., McRee, D. E., Khazanovich, *Nature*, **366**, 324 (1993); (b) J. D. Hartgerink, J. R. Granja, R. A. Milligan, M. R. Ghadiri, *J. Am. Chem. Soc.*, **118**, 43 (1996); (c) T. D. Clark, J. M. Buriak, K. Kobayashi, M. P. Isler, D. E. McRee, M. R. Ghadiri, *J. Am. Chem. Soc.*, **120**, 8949 (1998).
14. (a) D. H. Ranganathan, V. Haridas, C. S. Sundari, D. Balasubramanian, K. P. Madhusudana, R. Roy, I. L. Karle, *J. Org. Chem.*, **64**, 9230 (1999); (b) D. Ranganathan, M. P. Samant, I. L. Karle, *J. Am. Chem. Soc.*, **123**, 5619 (2001).
15. (a) D. P. Pantarotto, C. D. Partidos, R. Graff, J. Hoebeke, J.-P. Briand, M. Prato, A. Bianco, *J. Am. Chem. Soc.*, **125**, 6160 (2003); (b) R. Djalali, J. Samson, H. Matsui, *J. Am. Chem. Soc.*, **126**, 7935 (2004).
16. M. Amorin, L. Castedo, J. R. Granja, *J. Am. Chem. Soc.*, **125**, 2844 (2003).
17. (a) R. C. R. Claussen, B. M. Rabatic, S. I. Stupp, *J. Am. Chem. Soc.*, **125**, 12680 (2003); (b) N. W. Shi Kam, T. C. Jessop, P. A. Wender, H. Dai, *J. Am. Chem. Soc.*, **126**, 6850 (2004).
18. R. P. Lyon, W. M. Atkins, *J. Am. Chem. Soc.*, **123**, 4408 (2001).
19. T. C. Holmes, S. deLasalle, X. Su, G. Liu, A. Rich, A. Zhang, *Proc. Natl. Acad. Sci. USA*, **97**, 6728 (2000).
20. J. H. Collier, B.-H. Hu, J. W. Ruberti, J. Zhang, P. Shum, D. H. Thompson, P. B. Messersmith, *J. Am. Chem. Soc.*, **123**, 2963 (2001).
21. D. R. van Staveren, N. Metzler-Nolte, *Chem. Rev.*, **104**, 5931 (2004).
22. C. U. Pittman, Jr., *J. Inorg. Organomet. Polym. Mater.*, **15**, 33 (2005).
23. E. W. Neuse, *J. Inorg. Organomet. Polym. Mater.*, **15**, 3 (2005).

24. M. Haeussler, Q. Sun, K. Xu, J. W. Y. Lam, H. Dong, B. Z. Tang, *J. Inorg. Organomet. Polym. Mater.*, **15**, 67 (2005).
25. C. E. Carraher, Jr., *J. Inorg. Organomet. Polym. Mater.*, **15**, 121 (2005).
26. A. S. Abd-El-Aziz, C. E. Carraher, Jr., C. U. Pittman, Jr., J. E. Sheats, M. Zeldin, *Macromolecules Containing Metal and Metal-Like Elements*, Vol. 1, Wiley, Hoboken, NJ, 2003.
27. H.-B. Kraatz, *J. Inorg. Organomet. Polym. Mater.*, **15**, 83 (2005).
28. K. Severin, R. Bergs, W. Beck, *Angew. Chem. Int. Ed.*, **37**, 1643 (1998).
29. R. S. Herrick, R. M. Jarret, T. P. Curran, D. R. Dragoli, M. B. Flaherty, S. E. Lindyberg, R. A. Slate, L. C. Thornton, *Tetrahedron Lett.*, **37**, 5289 (1996).
30. A. Nomoto, T. Moriuchi, S. Yamazaki, A. Ogawa, T. Hirao, *Chem. Commun.*, **18**, 1963 (1998).
31. Y. Degani, A. Heller, *J. Am. Chem. Soc.*, **110**, 2615 (1988).
32. Y. Degani, A. Heller, *J. Phys.Chem.*, **91**, 1285 (1987).
33. F. E. Appoh, T. C. Sutherland, H.-B. Kraatz, *J. Organomet. Chem.*, **689**, 4669 (2004).
34. J. F. Gallagher, P. T. M. Kenny, M. J. Sheehy, *Acta Crystallogr.*, **C55**, 1257 (1999).
35. J. F. Gallagher, P. T. M. Kenny, M. Sheehy, *J. Inorg. Chem. Commun.*, **2**, 200 (1999).
36. A. S. Georgopoulou, D. M. P. Mingos, A. J. P. White, D. J. Williams, B. R. Horrocks, A. Houlton, *J. Chem. Soc., Dalton Trans.*, **17**, 2969 (2000).
37. T. Moriuchi, A. Nomoto, K. Yoshida, T. Hirao, *J. Organomet. Chem.*, **589**, 50 (1999).
38. T. Moriuchi, A. Nomoto, K. Yoshida, A. Ogawa, T. Hirao, *J. Am. Chem. Soc.*, **123**, 68 (2001).
39. T. Moriuchi, A. Nomoto, K. Yoshida, T. Hirao, *Organometallics*, **20**, 1008 (2001).
40. D. R. van Staveren, T. Weyhermüller, N. Metzler-Nolte, *J. Chem. Soc., Dalton Trans.*, 210 (2003).
41. Y. Xu, P. Saweczko, H. B. Kraatz, *J. Organomet. Chem.*, **637–639**, 355 (2001).
42. Y. Xu, H. B. Kraatz, *Tetrahedron Lett.*, **42**, 2601 (2001).
43. M. Oberhoff, L. Duda, J. Karl, R. Mohr, G. Erker, R. Fröhlich, M. Grehl, *Organometallics*, **15**, 4005 (1996).
44. H. Huang, L. Mu, J. He, C. J.-P. *J. Organomet. Chem.*, **68**, 7605 (2003).
45. M. M. Galka, H.-B. Kraatz, *ChemPhysChem*, **3**, 356 (2002).
46. S. Chowdhury, D. A. R. Sanders, G. Schatte, H. B. Kraatz, *Angew. Chem. Int. Ed.*, **44**, 1 (2005).
47. C. K. Smith, L. Regan, *Acc. Chem. Res.*, **30**, 153 (1997).
48. L. Lin, A. Berces, H.-B. Kraatz, *J.Organomet. Chem.*, **556**, 11 (1998).
49. L.-L. Lai, T.-Y. Dong, *Chem. Commun.*, 1078 (1994).
50. I. R. Butler, R. L. Davies, *Synthesis*, 1350 (1996).
51. I. R. Butler, S. C. Quayle, *J. Organomet. Chem.*, **552**, 63 (1998).
52. L. Barisic, V. Kovac, V. Rapic, *Croat Chem. Acta*, **75**, 199 (2002).
53. K. A. Mahmoud, H.-B. Kraatz, *J. Organomet. Chem.*, **689**, 2250 (2004).
54. K. Heinze, M. Schlenker, *Eur. J. Inorg. Chem.*, 2974 (2004).
55. D. C. D. Butler, C. J. Richards, *Organometallics*, **21**, 5433 (2002).
56. T.-A. Okamura, K. Sakauye, N. Ueyama, A. Nakamura, *Inorg. Chem.*, **37**, 6731 (1998).
57. A. N. Nesmeyanov, V. A. Sazonova, V. N. Drosd, *Chem. Ber.*, **93**, 2717 (1960).
58. L. Barisic, M. Dropucic, V. Rapic, H. Pritzkow, I. Kirin Srecko, N. Metzler-Nolte, *Chem. Commun.*, 2004 (2004).
59. W. B. Kraatz, V. Rapic, N. Metzler-Nolte, private communication.
60. A. Shafir, M. P. Power, G. D. Whitener, J. Arnold, *Organometallics*, **19**, 3978 (2000).
61. K. A. Mahmoud, H.-B. Kraatz, *J. Inorg. Organomet. Polym.* DOI: 10.1007/s10904-006-9040-0, 2006.
62. A. S. Abd-El-Aziz, I. Manners, *J. Inorg. Organomet. Polym.*, **15**, 157 (2005).
63. K. E. Gonsalves, X. Chen, in *Ferrocenes*, A. Togni, T. Hayashi, Eds., VCH, Weinheim, Germany, 1995, p. 496.

64. I. Manners, *Angew. Chem. Int. Ed. Engl.*, **35**, 1603 (1996).
65. I. Manners, *Can. J. Chem.*, **76**, 371 (1998).
66. E. W. Neuse, *J. Inorg. Organomet. Polym.*, **15**, 3 (2005).
67. E. W. Neuse, L. Bednarik, *Macromolecules*, **12**, 187 (1979).
68. P. Nguyen, P. Elipe-Gomez, I. Manners, *Chem. Rev.*, **99**, 1515 (1999).
69. T. J. Peckham, P. Gomez-Elipe, I. Manners, in *Metallocenes*, A. Togni, R. L. Haltermann Eds., Wiley-VCH, Weinheim, Germany, 1998, p. 723.
70. C. U. Pittman, Jr., *J. Inorg. Organomet. Polym.*, **15**, 33 (2005).
71. H. Nishihara, M. Murata, *J. Inorg. Organomet. Polym.*, **15**, 147 (2005).
72. H. R. Allcock, *J. Inorg. Organomet. Polym.*, **15**, 57 (2005).
73. E. W. Neuse, H. Rosenberg, *J. Macromol. Sci. Rev. Macromol. Chem.*, **C4**, 1 (1970).
74. M. G. Meirim, E. W. Neuse, G. Cadwell, *J. Inorg. Organomet. Polym.*, **8**, 225 (1998).
75. M. A. Buretea, T. D. Tilley, *Organometallics*, **16**, 1507 (1997).
76. C. E. Stanton, T. R. Lee, R. H. Grubbs, N. S. Lewis, J. K. C. Pudelski, M. R., M. S. Erickson, M. L. McLaughlin, *Macromolecules*, **28**, 8713 (1995).
77. A. S. Gamble, J. T. Patton, J. M. Boncella, *Makromol. Chem., Rapid Commun.*, **13**, 109 (1993).
78. R. W. Heo, J.-S. Park, T. R. Lee, *Macromolecules*, **38**, 2546 (2004).
79. K. A. Mahmoud, Y.-T. Long, G. Schatte, H.-B. Kraatz, *Eur. J. Inorg. Chem.*, 173 (2005).
80. F. Knobloch, W. Rauscher, *J. Polym. Sci.*, **54**, 651 (1961).
81. C. M. Casado, M. Moran, J. Losada, I. Cuadrado, *Inorg. Chem.*, **34**, 1668 (1995).
82. Q.-W. Han, X.-Q. Zhu, X.-B. Hu, J.-P. Cheng, *Chem. J. Chin. Univ.*, **23**, 2076 (2002).
83. S. Maricic, A. Ritzen, U. Berg, T. Frejd, *Tetrahedron*, **57**, 5623 (2001).
84. S. Maricic, U. Berg, T. Frejd, *Tetrahedron*, **58**, 3085 (2002).
85. S. Kaluz, S. Toma, *Collect. Czech. Chem. Commun.*, **53**, 638 (1988).
86. R. F. W. Jackson, D. Turner, M. H. Block, *Synlett*, 862 (1996).
87. A.-S. Carlström, T. Frejd, *J. Org. Chem.*, **55**, 4175 (1990).
88. B. Basu, S. K. Chattopadhyay, A. Ritzen, T. Frejd, *Tetrahedron: Asymmetry*, **8**, 1841 (1997).
89. S. Maricic, T. Frejd, *J. Org. Chem.*, **67**, 7600 (2002).
90. K. Heinze, M. Schlenker, *Eur. J. Inorg. Chem.*, 66 (2005).
91. K. Smith, F. Diederich, *Chem. Eur. J.*, **4**, 1353 (1998).
92. D. Astruc, M.-C. Daniel, J. Ruiz, *Chem. Commun.*, 2637 (2004).
93. C. M. Cardona, A. E. Kaifer, *J. Am. Chem. Soc.*, **120**, 4023 (1998).
94. Y. Wang, C. M. Cardona, A. E. Kaifer, *J. Am. Chem. Soc.*, **121**, 9756 (1999).
95. C. M. Cardona, T. D. McCarley, A. E. Kaifer, *J. Org. Chem.*, **65**, 1857 (2000).
96. D. L. Stone, D. K. Smith, P. T. McGrail, *J. Am. Chem. Soc.*, **124**, 856 (2002).
97. F. E. Appoh, D. S. Thomas, H.-B. Kraatz, *Macromolecules*, **38**, 7562 (2005).

CHAPTER 13

Supramolecular Metal Arrays on Artificial Metallo-DNAs and Peptides

MITSUHIKO SHIONOYA

The University of Tokyo, Tokyo, Japan

I. INTRODUCTION

The prime objective of supramolecular construction is to develop a general tool for precise design of molecular building blocks that generate desired self-assembled molecules through noncovalent interactions. Biological systems provide us with a clue as to a programmable arrangement of building blocks without any dispersion in their "number," "composition," "sequence," and "orientation," as seen in biopolymers such as DNA, protein, and polysaccharide. Hence, polymers constructed in the motif of biomacromolecules have great potential as template functions for programmable self-assembly to generate well-defined molecular architectures.

In the field of supramolecular chemistry, metals are known as key components of self-assembled molecular architectures with discrete structures as well as metal-based chemical and physical properties. Metal coordination is directional noncovalent bonding, and hence plays an important role in determining the relative arrangement of ligands. Furthermore, its reversible nature allows us to design molecular switching or motion that shows a dramatic change in structure and hence function by external perturbation.

Recently, we have demonstrated that biopolymers such as DNA and peptides can be chemically reconstructed, and thereby act as template molecules for homogeneous or heterogeneous metal arrays in a programmable manner. This chapter covers our recent approaches to metal arrays on artificial DNAs and cyclic peptides.

II. METAL ARRAYS ON ARTIFICIAL DNAs

Chemical synthesis of DNA strands has been well accomplished, and even unnatural, functionalized building blocks can be incorporated into oligonucleotides at appropriate positions.

Frontiers in Transition Metal-Containing Polymers, edited by Alaa S. Abd-El-Aziz and Ian Manners.

Thus, DNA has a platform for molecular array that could be performed in a programmable way. In view of the most important role of base pairing in the molecular profile of gene expression, among a variety of approaches to DNA-based supramolecular architectures, the replacement of natural base pairs by predesigned artificial base pairs possessing a distinctive shape, size, and function is expected as the most promising method of molecular arrangement within the DNA.[1,2,3]

In the double-helical structure of typical right-handed B-DNA, two kinds of flat hydrogen-bonded base pairs, A–T and G–C, are stacked 3.4 Å apart within the molecule. Inspired by the sophisticated higher-order structure of DNA, we conceived of the substitution of metal-mediated base pairs for hydrogen-bonded base pairs found in nature. In other words, when nucleobases (L) are chemically modified so as to bind to metal ions (M), metal-mediated base pairs (L–M–L) would be aligned along the helix axis. In addition, the use of more than two kinds of metal ligand-type nucleobases would allow heterogeneous metal arrays in a programmable manner.

Since we reported the first Pd^{2+}-mediated base pairing using a metal ligand-type nucleoside having a phenylenediamine base,[4] many types of alternative metal-mediated base pairs have been exploited by our group[5–8] and other groups.[9–13] Some examples are, B^{3+}-induced base paring with catechol,[5] Pd^{2+}-mediated base pairing with 2-aminophenol,[6] Ag^{+}-assisted base pairing with pyridine,[7] Cu^{2+}-mediated base pairing with hydroxypyridone,[8] a base pair with a pyridine-2,6-dicarboxylate nucleobase as a planar tridentate ligand and a pyridine nucleobase,[9] and a Salen base pair.[14] Effects of metal-mediated base pairs on the stabilization of double-stranded[7–11,14] and triple-stranded[7] structures were also investigated.

Hydroxypyridone nucleobase (**H**) quantitatively forms a neutral, square-planar Cu^{2+}-mediated base pair (**H**–Cu^{2+}-**H**) within oligonucleotides.[8] This Cu^{2+}-mediated base pairing was further applied to the alignment of metallo-base pairs along the helix axis of DNA. Recently, we reported the synthesis of a series of oligonucleotides, d(5′-G$\mathbf{H}_n$C-3′) (n = 1–5), carrying one to five hydroxypyridone nucleobases.[15] The photometric titration experiments with Cu^{2+} and circular dichroism study revealed that these oligonucleotides quantitatively produced right-handed double helices of the oligonucleotides, $n\mathrm{Cu}^{2+}\cdot$d(5′-G$\mathbf{H}_n$C-3′)$_2$ (n = 1–5), through Cu^{2+}-mediated base pairing (**H**-Cu^{2+}-**H**) (Figure 13.1). In this metallo-DNA, the Cu^{2+} ions incorporated into each complex are aligned along the helix axes inside the duplexes with the Cu^{2+}-Cu^{2+} distance of approximately 3.7 Å, as determined by the electron spin resonance (ESR) spectrum of $2\mathrm{Cu}^{2+}\cdot$d(5′-G$\mathbf{H}_2$C-3′)$_2$ with a fine structure splitting that is compatible with the dipole–dipole interaction between electron spins over the distance between neighboring base pairs in natural DNA (ca. 3.4 Å). The unpaired d electrons on adjacent Cu^{2+} (d^9) centers couple in a ferromagnetic fashion, with accumulating Cu^{2+} ions attaining the highest spin state, as expected from the number of Cu^{2+} ions incorporated within the oligonucleotides. A proposed right-handed, double-stranded structure is drawn with a pentanuclear Cu^{2+} array inside the DNA (Figure 13.2). Although several excellent examples of metal arrays in the solid states have been reported, metal alignment performed in solution is still rare.

The next possible objective is to incorporate increasing numbers of metal ions into DNA and to arrange metals heterogeneously using DNA templates possessing more than two different kinds of artificial nucleobases aligned in a programmable manner.

III. METAL ARRAYS ON CYCLIC PEPTIDES

Proteins and peptides found in nature consist of twenty kinds of amino acids, and their number and sequence are precisely controlled in biological systems. In terms of the development of

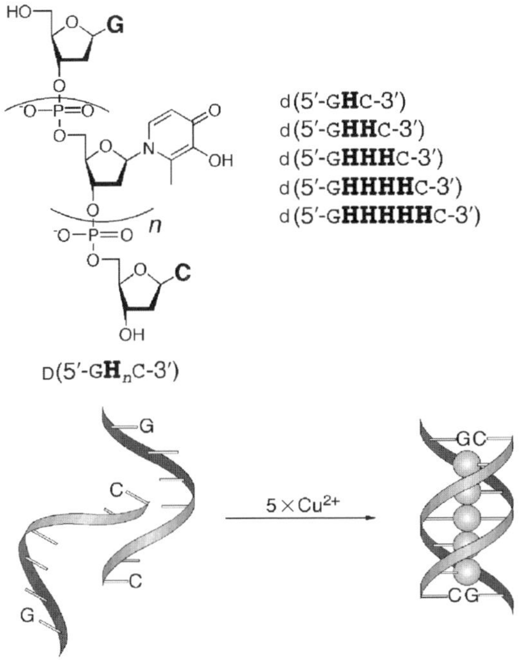

Figure 13.1 Cu^{2+}-mediated DNA duplex formation from two artificial oligonucleotide strands bearing one to five hydroxypyridone nucleobases.

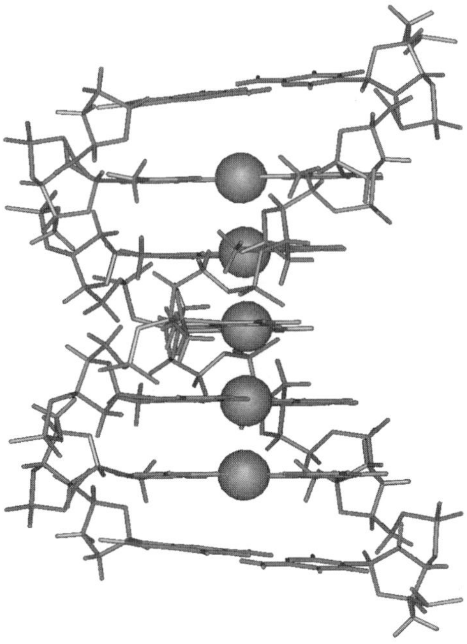

Figure 13.2 A plausible structure of pentanuclear Cu^{2+} complexes within DNA.

functionalized molecules, cyclic peptides with well-defined structures have been extensively studied for ion recognition,[16–19] ion channels,[20,21] metal arrays,[22,23,24] and other applications directed toward metal-dependent functions.[25,26,27] Their functions depend primarily on the number and sequence of their amino acid constituents. Incorporation of metal-binding sites at appropriate side chains and/or utilizing the metal-binding ability of amide functions on the cyclic framework would provide a novel tool for heterogeneous metal arrays in a hierarchical manner.

Cyclic hexapeptides are well known to bind preferentially to alkali and alkaline earth metal ions as tridentate ligands. A cyclic hexapeptide having a repeating L-Ala-L-Met sequence (Ala = alanine; Met = methionine), *cyclo*(L-Ala-L-Met)$_3$ (**L**), was designed so as to have metal binding sites both at the thioether groups (Met) for softer metals (outer circle in pale grey) and the amide carbonyl groups roundly arranged on the cyclic framework for harder metals (inner circle in black) (Figure 13.3). Thus bearing two different types of metal binding sites, the cyclic hexapeptide (**L**) was found to quantitatively and highly selectively form a tetranuclear complex with three Ag^+ and one Ca^{2+} ions in a capsule-like dimeric structure.[28]

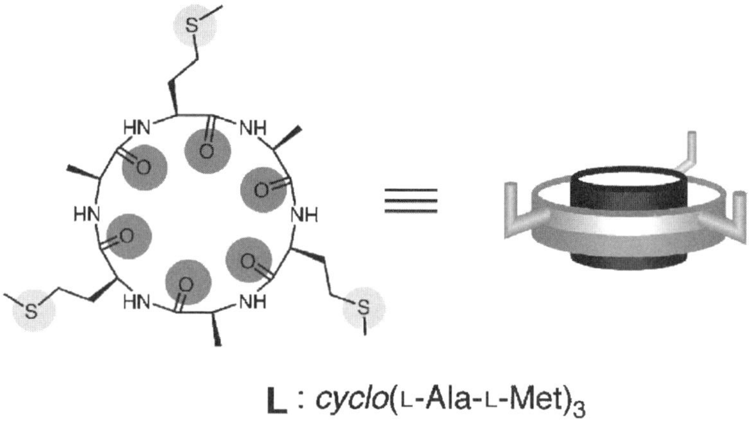

Figure 13.3 Cyclic hexapeptide, *cyclo*(L-Ala-L-Met)$_3$ (**L**).

The cyclic hexapeptide (**L**) with three thioether groups at the side chains and three amide carbonyl groups on the cyclic framework was expected to bind heterogeneously to two different kinds of metal ions in a dimeric structure of the peptides, as shown in Figure 13.4. Interactions between the cyclic hexapeptide (**L**) with metal ions were first examined by ^1H nuclear magnetic resonance (NMR) titration study using Ag^+, which prefers linear coordination, and Ca^{2+} in combination. As a result, when the ratio of **L**/Ag^+/Ca^{2+} reached 2:3:1, the spectrum of **L** completely changed to another set of signals, indicating the binding of the thioether groups to Ag^+ ions and the conformational changes of the cyclic framework upon Ca^{2+} complexation. The co-existence of three Ag^+ and one Ca^{2+} ions was essential for the complexation. The complex formed in solution was finally determined to be $[Ag_3CaL_2]^{5+}$ by the electrospray ionization-time-of-flight (ESI-TOF) mass spectrum of the solution (**L**/Ag^+/Ca^{2+} = 2:3:1). These data clearly indicate that the dimer is quantitatively formed with simultaneous encapsulation of three Ag^+ and one Ca^{2+} ions between the two cyclic peptides (Figure 13.4). Furthermore, the ^1H-^1H correlation spectroscopy (COSY) and nuclear Overhauser enhancement and exchange spectroscopy (NOESY) studies finally determined the head-to-tail stacking of the two cyclic peptides, although there are three possible orientations (head-to-head, tail-to-tail, or head-to-tail) in the stacked structure because of two different ring faces of each cyclic peptide arising from its asymmetric centers.

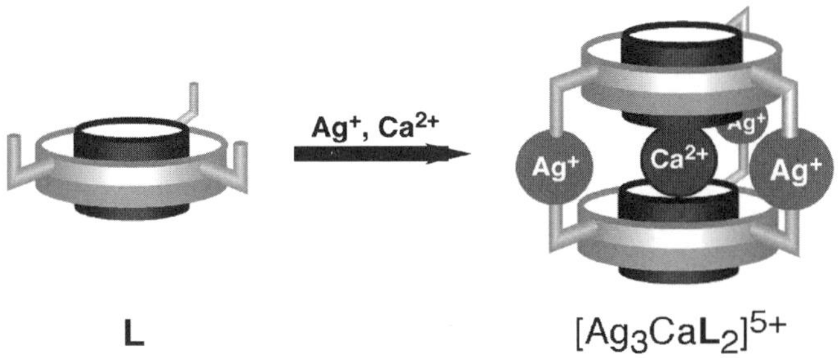

Figure 13.4 Formation of a tetranuclear complex, $[Ag_3CaL_2]^{5+}$, from *cyclo*(L-Ala-L-Met)$_3$ (**L**), three Ag^+ ions, and one Ca^{2+} ion.

The molecular structure was determined by X-ray crystal analysis. We obtained single crystals from a solution of **L**/Ag^+/Ca^{2+} = 2:3:1, which have proven to be a result of heterogeneous crystallization, probably due to the solubility and crystal packing. The resulting complex is a Ca^{2+}-linked dimer of $Ag_3Ca\mathbf{L}_2^{5+}$, $[(Ag_3Ca\mathbf{L}_2)_2Ca]^{12+}$ (Figure 13.5). In the partial structure corresponding to $Ag_3Ca\mathbf{L}_2^{5+}$ in the complex, the central Ca^{2+} ion is bound by the six carbonyl oxygen atoms of the two cyclic peptides in a slightly distorted octahedral geometry, and each Ag^+ ion is coordinated by the two sulfur atoms of the thioether groups of Met in a linear geometry. As suggested by the solution study, the two cyclic peptides face each other in a head-to-tail orientation with the aid of three Ag^+ and one Ca^{2+} ions. Without any support of Ag^+ coordination, the Ca^{2+} binding to the circularly arranged amide carbonyl oxygens on cyclic hexapeptides is known to be relatively weak. So it appears that the Ca^{2+} binding is greatly reinforced in the dimeric structure by the coordination of three Ag^+ ions outside the cyclic framework.

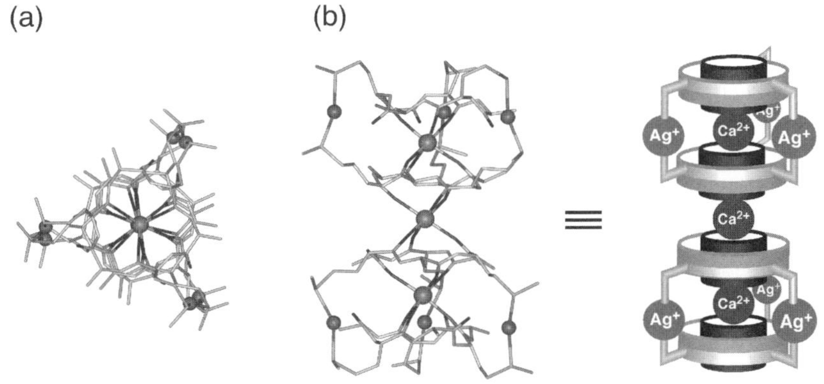

Figure 13.5 The X-ray structure of $[(Ag_3CaL_2)_2Ca]^{12+}$; (a) top view, and (b) side view.

The selectivity of the Ca^{2+} binding to the cavity of the cyclic peptide is quite high, especially when compared with Mg^{2+}. The encapsulation selectivity for the Ag^+-mediated dimer formation was examined by ^{1}H NMR and ESI-TOF mass studies using monovalent alkali, divalent alkaline earth, or trivalent lanthanide ions, instead of Ca^{2+} (Figure 13.6). As a result, divalent Sr^{2+} and Ba^{2+} ions were found to form the dimeric isostructures only when excess

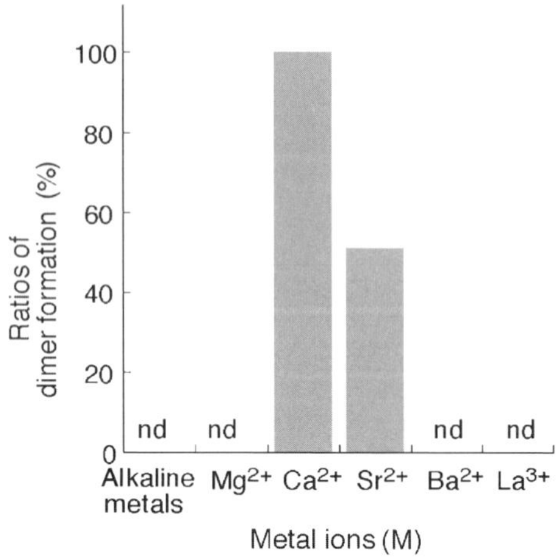

Figure 13.6 Effects of various ions on the dimer complexation. [**L**] = 2 mM, [Ag^+] = 3 mM, [metal ion] = 1 mM in acetone-d_6/CD_3OD (5:1) at 293 K. Alkaline metals include Li^+, Na^+, K^+, Rb^+, and Cs^+. (nd = not detected.)

amounts of these ions were used. However, no complexation was observed under the same conditions with Mg^{2+}, alkali metal ions, or the La^{3+} ion, with an ionic radius similar to Ca^{2+} and Na^+. The degree of affinity of metal ions with the central cavity of the dimer was thus highly dependent on the size, charge, and solvation of metal ions.

In summary, quantitative, selective, and heterogeneous metal assembly was accomplished using a predesigned peptide template. Peptides with the predetermined number and sequence of metal coordination sites would allow template-directed arrays of metal-containing nanosized devices.

IV. CONCLUSION AND PERSPECTIVES

In this chapter, we demonstrated that biopolymers such as DNA and peptides act as excellent molecular templates for homogeneous or heterogeneous assembly of metal ions in a programmable manner by chemically modifying their building blocks or by well-defined ways to arrange building blocks. Such template-directed metal arrays would generate more complicated and more highly functionalized metal clusters with interesting chemical and physical properties.

REFERENCES

1. S. L. Beaucage, D. E. Bergstrom, G. D. Glick, R. A. Jones, *Current Protocols in Nucleic Acid Chemistry*, Wiley, New York, 2001.
2. E. T. Kool, *Acc. Chem. Res.*, **35**, 936 (2002).
3. M. Shionoya, K. Tanaka, *Curr. Opin. Chem. Biol.*, **8**, 592 (2004).
4. K. Tanaka, M. Shionoya, *J. Org. Chem.*, **64**, 5002 (1999).
5. H. Cao, K. Tanaka, M. Shionoya, *Chem. Pharm. Bull.*, **48**, 1745 (2000).
6. M. Tasaka, K. Tanaka, M. Shiro, M. Shionoya, *Supramol. Chem.*, **13**, 671 (2001).

7. K. Tanaka, Y. Yamada, M. Shionoya, *J. Am. Chem. Soc.*, **124**, 8802 (2002).
8. K. Tanaka, A. Tengeiji, T. Kato, N. Toyama, M. Shiro, M. Shionoya, *J. Am. Chem. Soc.*, **124**, 12494 (2002).
9. E. Meggers, P. L. Holland, W. B. Tolman, F. E. Romesberg, P. G. Schultz, *J. Am. Chem. Soc.*, **122**, 10714 (2000).
10. S. Atwell, E. Meggers, G. Spraggon, P. G. Schultz, *J. Am. Chem. Soc.*, **123**, 12364 (2001).
11. N. Zimmermann, E. Meggers, P. G. Schultz, *J. Am. Chem. Soc.*, **124**, 13684 (2002).
12. H. Weizman, Y. Tor, *Chem. Commun.*, 453 (2001).
13. H. Weizman, Y. Tor, *J. Am. Chem. Soc.*, **123**, 3375 (2001).
14. G. H. Clever, K. Polborn, T. Carell, *Angew. Chem. Int. Ed.*, **44**, 7204 (2005).
15. K. Tanaka, A. Tengeiji, T. Kato, N. Toyama, M. Shionoya, *Science*, **299**, 1212 (2003).
16. D. Ranganathan, V. Haridas, I. Karle, *J. Am. Chem. Soc.*, **120**, 2695 (1998).
17. T. Weiß, D. Leipert, M. Kasper, G. Jung, W. Göpel, *Adv. Mater.*, **11**, 331 (1999).
18. D. Yang, J. Qu, W. Li, Y. H. Zhang, Y. Ren, D. P. Wang, Y. D. Wu, *J. Am. Chem. Soc.*, **124**, 12410 (2002).
19. J. Strauss, J. Daub, *Org. Lett.*, **4**, 683 (2002).
20. M. R. Ghadiri, J. R. Granja, L. K. Buehler, *Nature*, **306**, 301 (1994).
21. H. Ishida, Z. Qi, M. Sokabe, K. Donowaki, Y. Inoue, *J. Org. Chem.*, **66**, 2978 (2001).
22. G. Kartha, K. J. Varughese, S. Aimoto, *Proc. Natl. Acad. Sci. USA*, **79**, 4519 (1982).
23. P. Xie, M. Diem, *J. Am. Chem. Soc.*, **117**, 429 (1995).
24. K. Tanaka, K. Shigemori, M. Shionoya, *Chem. Commun.*, 2475 (1999).
25. D. T. Bong, T. D. Clark, J. R. Granja, M. R. Ghadiri, *Angew. Chem. Int. Ed.*, **40**, 988 (2001).
26. S. F. Lopez, H. S. Kim, E. C. Choi, M. Delgado, J. R. Granja, A. Khasanov, K. Kreahenbuehl, G. Long, D. A. Weinberger, K. M. Wilcoxen, M. R. Ghadiri, *Nature*, **412**, 452 (2001).
27. T. Nakanishi, H. Okamoto, Y. Nagai, K. Takeda, I. Obataya, H. Mihara, H. Azehara, Y. Suzuki, W. Mizutani, K. Furukawa, K. Torimitsu, *Phys. Rev. B*, **66**, 165417 (2002).
28. T. Okada, K. Tanaka, M. Shiro, M. Shionoya, *Chem. Commun.*, 1484 (2005).

SUBJECT INDEX

METALS INDEX